Fish Hatchery Management

Second Edition

Financial support for the symposium and
the printing of this book provided by

U. S. Fish and Wildlife Service,
Division of Federal Aid

Fish Hatchery Management

Second Edition

Edited by

Gary A. Wedemeyer

Western Fisheries Research Center
Biological Resources Division, U.S. Geological Survey
6505 NE 65th Street
Seattle, Washington 98115

American Fisheries Society
Bethesda, Maryland

Suggested Citation Formats

Entire book

Wedemeyer, Gary A., editor. 2001. Fish hatchery management, second edition. American Fisheries Society, Bethesda, Maryland.

Chapter within the book

Colt, John E., and Joseph R. Tomasso. 2001. Hatchery water supply and treatment. Pages 91–186 in G. A. Wedemeyer, editor. Fish hatchery management, second edition. American Fisheries Society, Bethesda, Maryland.

Cover illustration provided by Stewart Alcorn

Library of Congress Control Number 2001092152
ISBN 1-888569-26-3

American Fisheries Society
5410 Grosvenor Lane, Suite 110
Bethesda, Maryland 20814-2199

This book is dedicated to Z. P. W. and his millenium generation from whom will come the advances in fisheries and aquatic science needed to write the future editions of Fish Hatchery Management.

Contents

Contributors

Frederick T. Barrows (Chapter 8): Bozeman Fish Technology Center, U.S. Fish and Wildlife Service, 4050 Bridger Canyon Road, Bozeman, Montant 59715, USA.

Julie Bebak-Williams (Chapter 6): Freshwater Institute, Post Office Box 1889, Shepherds-town, West Virginia 25443, USA.

Gary J. Carmichael (Chapter 10): U.S. Fish and Wildlife Service, Mora National Fish Hatchery and Technology Center, Post Office Box 689, Mora, New Mexico 87732, USA.

John E. Colt (Chapter 3): National Marine Fisheries Service, Northwest Fisheries Science Center, 2725 Montlake Boulevard East, Seattle, Washington 98112, USA.

Frank Dalziel (Chapter 4): Department of Fisheries and Aquaculture, Malespina University, 900 5th Street, Nanimo, B.C., V9R 5S5, Canada.

Ronald W. Hardy (Chapter 8): Hagerman Fish Culture Experiment Station, University of Idaho, 3059F Natl. Fish Hatchery Road, Hagerman, Idaho 83332, USA.

Kyle J. Hartman (Chapter 11): Wildlife and Fisheries Program, West Virginia University, College of Agriculture, Forestry, and Consumer Sciences, Division of Forestry, 320 Perceval Hall, Morgantown, West Virginia 26506-6125, USA.

E. D. Lane (Chapter 4): Department of Fisheries and Aquaculture, Malespina University, 900 5th Street, Nanimo, B.C., V9R 5S5, Canada.

Patricia M. Mazik (Chapter 5): U.S. Fish and Wildlife Service, 4401 Fairfax Drive, Mail Stop 840, Arlington, Virginia 22203, USA.

James W. Meade (Chapter 12): U.S. Geological Survey, Northern Appalachian Research Laboratory, RR4, Box 63, Wellsboro, Pennsylvania 16901, USA.

Nick C. Parker (Chapter 5): Texas Cooperative Fisheries and Wildlife Research Unit, Texas Tech University, Department of Range and Wildlife, Lubbock, Texas 79409-2125, USA.

William Pennell (Chapter 4): Department of Fisheries and Aquaculture, Malaspina University, 900 5th Street, Nanaimo, B.C., V9R 5S5, Canada.

Bret Preston (Chapter 11): West Virginia Division of Natural Resources, Wildlife Resources Section 1900 Kanawha Boulevard East, Building 3, Room 814, Charleston, West Virginia 25305, USA.

Thomas E. Schwedler (Chapter 10): Department of Aquaculture, Fisheries, and Wildlife, Clemson University, Clemson, South Carolina 29634, USA.

Robert R. Stickney (Chapter 1): Texas Sea Grant, Texas A&M University, 1716 Briarcrest, Suite 702, Bryan, Texas 77802, USA.

Steven Summerfelt (Chapter 6): Freshwater Institute, Post Office Box 1889, Shepherdstown, West Virginia 25443, USA.

Joseph J. Tomasso, Jr. (Chapter 10): Department of Aquaculture, Fisheries, and Wildlife, Clemson University, Clemson, South Carolina 29634, USA.

Scott Tsukuda (Chapter 6): Freshwater Institute, Post Office Box 1889, Shepherdstown, West Virginia 25443, USA.

Harry Westers (Chapter 2): Aquaculture Bioengineering Corporation, Post Office Box 8, Rives Junction, Michigan 49227, USA.

J. Holt Williamson (Chapter 7): U.S. Fish and Wildlife Service, Dexter National Hatchery and Technology Center, Dexter, New Mexico 88230, USA.

James R. Winton (Chapter 9): U.S. Geological Survey, Western Fisheries Research Center, 6505 NE 65th Street, Seattle, Washington 98115, USA.

List of Species

African catfish	*Clarias gariepinus*
alewife	*Alosa pseudoharengus*
American shad	*Alosa sapidissima*
Arctic char	*Salvelinus alpinus*
Atlantic cod	*Gadus morhua*
Atlantic halibut	*Hippoglossus hippoglossus*
Atlantic herring	*Clupea harengus*
Atlantic salmon	*Salmo salar*
bighead carp	*Hypophthalmichthys nobilis*
black bullhead	*Ameiurus melas*
blue catfish	*Ictalurus furcatus*
bluegill	*Lepomis macrochirus*
blue tilapia	*Oreochromis aureus*
bonytail	*Gila elegans*
brook trout	*Salvelinus fontinalis*
brown bullhead	*Ameiurus nebulosus*
brown trout	*Salmo trutta*
buffalos	*Ictiobus* spp.
bullheads	*Ameiurus* spp.
catfishes	*Ictalurus* spp.
channel catfish	*Ictalurus punctatus*
cherry salmon	*Oncorhynchus masou*
chinook salmon	*Oncorhynchus tshawytscha*
chum salmon	*Oncorhynchus keta*
cods	*Gadus* spp.
coho salmon	*Oncorhynchus kisutch*
Colorado pike minnow	*Ptychocheilus lucius*
common carp	*Cyprinus carpio*
crappies	*Pomoxis* spp.

crucian carp	*Carassius carassius*
cutthroat trout	*Oncorhynchus clarki*
brook trout	*Salvelinus fontinalis*
eels	*Anguilla* spp.
fathead minnows	*Pimephales promelas*
flathead catfish	*Pylodictis olivaris*
Gila trout	*Oncorhynchus gilae*
gizzard shad	*Dorosoma cepedianum*
goldfish	*Carassius auratus*
grass carp	*Ctenopharyngodon idella*
grayling	*Thymallus thymallus*
humpback chub	*Gila cypha*
hybrid striped bass	*Morone saxatilis* × *M. chrysops*
Japanese medaka	*Oryzias latipes*
killifishes	*Fundulus* spp.
kokanee	*Oncorhynchus nerka*
lake sturgeon	*Acipenser fluvescens*
lake trout	*Salvelinus namaycush*
lake whitefish	*Coregonus clupeaformis*
largemouth bass	*Micropterus salmoides*
loaches	family Cobitidae
mackerels	family Scombridae
Macquarie perch	*Macquaria australasica*
minnows	family Cyprinidae
western mosquito fish	*Gambusia affinis*
mountain galaxias	*Galaxias olidus*
Mozambique tilapia	*Oreochromis mossambicus*
muskellunge	*Esox masquinongy*
Nile tilapia	*Oreochromis niloticus*
northern pike	*Esox lucius*
Pacific halibut	*Hippoglossus stenolepis*
Pacific herring	*Clupea pallasi*
Pacific salmon species	*Oncorhynchus* spp.
paddlefish	*Polyodon spathula*
pickerels	*Esox* spp.
pink salmon	*Oncorhynchus gorbuscha*
rainbow trout	*Oncorhynchus mykiss*
razorback sucker	*Xyrauchen texanus*
red drum	*Sciaenops ocellatus*
redear sunfish	*Lepomis microlophus*
Eurasian perch	*Perca fluviatilis*
Rio Grande silvery minnow	*Hybognathus amarus*
Russian sturgeon	*Acipenser gueldenstaedti*

sauger	*Stizostedion canadense*
sea bass	*Dicentrarchus* spp.
sea bream	*Archosargus rhomboidalis*
sheatfish	*Silurus glanis*
silver carp	*Hypophthalmichthys molitrix*
silver perch	*Bairdiella chrysoura*
smallmouth bass	*Micropterus dolomieu*
sockeye salmon	*Oncorhynchus nerka*
southern flounder	*Paralichthys lethostigma*
spotted seatrout	*Cynoscion nebulosus*
steelhead	*Oncorhynchus mykiss*
striped bass	*Morone saxatilis*
striped mullet	*Mugil cephalus*
sturgeons	*Acipenser* spp.
summer flounder	*Paralichthys dentatus*
sunfishes	*Lepomis* spp.
tench	*Tinca tinca*
tiger muskellunge	*Esox lucius* × *E. masquinongy*
turbot	*Scophthalmus maximus*
walking catfish	*Clarias batrachus*
walleye	*Stizostedion vitreum*
white catfish	*Ameiurus catus*
white fish	*Coregonus* spp.
white sturgeon	*Acipenser transmontanus*
yellow bullhead	*Ameiurus natalis*
yellow perch	*Perca flavescens*
yellowtail	*Seriola lalandi*
zebra danio	*Danio rerio*

Preface

The original *Fish Hatchery Management*, written by a team of hatchery biology pioneers lead by Robert G. Piper and published in 1982, has long been considered one of the classic works in fish biology. Over the years, it has served as the basis for successful fish culture operations worldwide and the training of many generations of students. It has been reprinted several times. In 1997, the American Fisheries Society (AFS), under the leadership of Dr. Robert Kendall (now retired), organized a steering committee to begin the work of updating the book. In cooperation with the U.S. Fish and Wildlife Service (FWS), a national survey of hatchery biologists from federal and state agencies and the private sector was conducted to provide guidelines for the content and organization of the needed revision. Based on the survey results, the committee (Drs. Nick Parker, Cooperative Research Unit, Texas Tech University; Patricia Mazik, Cooperative Research Unit, University of West Virginia; Gary Wedemeyer, Western Fisheries Research Center, Biological Resources Division, U.S. Geological Survey; Vincent Mudrak U.S. FWS; Chris Horsch, U.S. FWS; Reginal Harrell, University of Maryland; Thomas Johnson, Private Consultant; Michael Gibson, Arkansas Game and Fish Commission) began the work of developing the outline and selecting the individual chapter authors.

The original book focused on hatchery requirements and operations, broodstock management and spawning, nutrition and feeding, fish health, and fish transportation. Although we attempted to remain true to the spirit of the original book, the reader will notice significant differences in this new edition. Some may comment that the book might have been as aptly titled *Advances in Fish Hatchery Management*. The chapters were written by biologists chosen from academia, government agencies, nonprofit conservation groups, and consultants to the private sector. Thus, a diversity of writing styles has been used to explain the subject matter, reflecting the diversity of the professional experience of the individual authors and their approaches to the biology of present-day fish culture. Some chapters are quite mathematical, while others employ an entirely narrative format. In general, descriptive material on individual fish diseases has been reduced in favor of expanded coverage of the regulatory climate now governing the use of drugs and chemicals for disease control. However, all chapters do include bibliographies meant to either document the information presented or to provide the reader with the opportunity to further explore each topic in greater depth. Finally, appendices such as length/weight tables have not been included in the present edition but have been placed on the Internet at the American Fisheries Society Website (www.fisheries.org) where they can easily be accessed, and updated by the Society as required.

Fish Hatchery Management, Second Edition was written to be useful to those who already have a working knowledge of the field, such as professional fish biologists at public or private conservation hatcheries or aquaculture facilities, as well as to those who are at an earlier stage of their professional development, such as college or university students.

Many people were helpful in the production of this book and deserve special mention. A fitting acknowledgement of the original team of writers (Robert G. Piper, Ivan McElwain, Leo Orme, Joe McCraren, Lori Fowler, and John Leonard) is difficult. Their influence was both fundamental and pervasive. Dr. Robert Kendall (now retired) of the American Fisheries Society developed the initial concept for this revised edition and organized the required funding and the initial steering committee. This dedicated group of people, with the able assistance of Beth Staehle (AFS), did most of the initial work required to get the project underway. Robert Rand and Deborah Lehman of AFS Books provided invaluable editorial support services throughout the duration of the project. The revision was greatly improved by the criticisms and suggestions of the dedicated fishery biologists who served as chapter reviewers. We are indebted to Drs. Jim Meade, Craig Tucker, Ruth Withler, Les Torrans, Michael Masser, Barnaby Watten, Bryan Kenworthy, Ronald Hedrick, Chris Horsch, Dale Bast, Gary Carmichael, Rosalie Schnick, Carole Engle, and Russell Danner. We are especially indebted to Chris Horsch in this regard. Finally, we thank Dr. Frank Shipley, Director of the Western Fisheries Research Center, for providing office space and valuable support services. This book was made possible by a grant from the Federal Aid program of the U. S. Fish and Wildlife Service.

Gary A. Wedemeyer,
Editor

Symbols and Abbreviations

The following symbols and abbreviations may be found in this book without definition. Also undefined are standard mathematical and statistical symbols given in most dictionaries.

A	ampere
AC	alternating current
Bq	becquerel
C	coulomb
8C	degrees Celsius
cal	calorie
cd	candela
cm	centimeter
Co.	Company
Corp.	Corporation
cov	covariance
DC	direct current; District of Columbia
D	dextro (as a prefix)
d	day
d	dextrorotatory
df	degrees of freedom
dL	deciliter
E	east
E	expected value
e	base of natural logarithm (2.71828 ..)
e.g.	(exempli gratia) for example
eq	equivalent
et al.	(et alii) and others
etc.	et cetera
eV	electron volt
F	filial generation; Farad
°F	degrees Fahrenheit
fc	footcandle (0.0929 lx)
ft	foot (30.5 cm)
ft^3/s	cubic feet per second ($0.0283\ m^3/s$)

g	gram
G	giga (10^9, as a prefix)
gal	gallon (3.79 L)
Gy	gray
h	hour
ha	hectare (2.47 acres)
hp	horsepower (746 W)
Hz	hertz
in	inch (2.54 cm)
Inc.	Incorporated
i.e.	(id est) that is
IU	international unit
J	joule
K	Kelvin (degrees above absolute zero)
k	kilo (10^3, as a prefix)
kg	kilogram
km	kilometer
l	levorotatory
L	levo (as a prefix)
L	liter (0.264 gal, 1.06 qt)
lb	pound (0.454 kg, 454g)
lm	lumen
log	logarithm
Ltd.	Limited
M	mega (10^6, as a prefix); molar (as a suffix or by itself)
m	meter (as a suffix or by itself); milli (10^{23}, as a prefix)
mi	mile (1.61 km)
min	minute

mol	mole
N	normal (for chemistry); north (for geography); newton
N	sample size
NS	not significant
n	ploidy; nanno (10^{29}, as a prefix)
o	ortho (as a chemical prefix)
oz	ounce (28.4 g)
P	probability
p	para (as a chemical prefix)
p	pico (10^{212}, as a prefix)
Pa	pascal
pH	negative log of hydrogen ion activity
ppm	parts per million
qt	quart (0.946 L)
R	multiple correlation or regression coefficient
r	simple correlation or regression coefficient
rad	radian
S	siemens (for electrical conductance); south (for geography)
SD	standard deviation
SE	standard error
s	second
T	tesla
tris	tris(hydroxymethyl)-aminomethane (a buffer)
UK	United Kingdom
U.S.	United States (adjective)
USA	United States of America (noun)
V	volt
V, Var	variance (population)
var	variance (sample)
W	watt (for power); west (for geography)
Wb	weber
yd	yard (0.914 m, 91.4 cm)
a	probability of type I error (false rejection of null hypothesis)
b	probability of type II error (false acceptance of null hypothesis)
Ω	ohm
μ	micro (10^{26}, as a prefix)
′	minute (angular)
″	second (angular)
°	degree (temperature as a prefix, angular as a suffix)
%	per cent (per hundred)
‰	per mille (per thousand)

1

History and Purpose of Fish Culture

Robert R. Stickney

Introduction

The first publication on the subject of fish culture was by Fan Li who wrote during the fifth century B.C. about carp production in China. Avault (1996) suggests that the Chinese and, perhaps, even the Egyptians may have been involved in fish culture as long ago as 4,000 B.C. The practice was undoubtedly well established before Fan Li's summary was produced. Hieroglyphs associated with the tombs of the Pharaohs of ancient Egypt show fish that are thought to depict the culture of tilapia.

Early Europeans also practiced aquaculture. There was oyster culture in ancient Rome and Gaul (Bardach et al. 1972). Fish culture, first with common carp and later brown trout, began in Europe during the Middle Ages. A comprehensive summary of European marine fish culture is included in a book on aquaculture history by Kirk (1987).

The first comprehensive summary of U.S. fish culture methods was published by the U.S. Fish and Fisheries Commission (1897). This volume can be considered the great grandfather of the book you now hold. An excellent summary of a century of federal government fish culture activity in the United States was written by Bowen (1970).

This brief history focuses on fish culture in the United States. A few of the important people and developments that created modern fish culture in the United States are described, with a bit more detail provided for the formative years than with respect to more recent events. The fish culturists mentioned by name were selected from among hundreds who could have been included. The majority were pioneer culturists of the late nineteenth century, though a few names of more modern culturists are included. Much more detail on people and events can be found in Stickney (1996a). A short chronology is presented in Table 1.

The enormous contributions of many skilled state and federal hatchery biologists and, in recent times, private producers are acknowledged, but not detailed in this short history. It is to those individuals, the backbone of the discipline, that this chapter is dedicated.

The Birth of Fish Culture in the United States

The first published report of fish culture activity in the United States resulted from a paper presented by the Reverend John Bachmann of Charleston, South Carolina, who discussed hatching and pond rearing of brook trout at a meeting of the South Carolina State Agricultural Society in 1804. Several years later, in 1853, Theodatus Garlick, M.D., and Professor H. A. Ackley put together a display for the Ohio State Fair that contained brook trout the two men had produced from eggs fertilized and hatched on Ackley's farm.

Brook trout captured the majority of fish culture attention for several years. The legendary Seth Green (Figure 1) began his fish culture activities with brook trout in 1864. Another true fish culture pioneer was the Reverend Livingston Stone (Figure 2), who established his first hatchery in 1866. He produced edible fish but also sold Atlantic salmon eggs and bass fry.

In 1866, the Canadian culturist Samuel Wilmot, became the first culturist to spawn Atlantic salmon, and, by 1870, he was selling salmon eggs to various states in the United States. In 1866, W. W. Fletcher began hatching salmon eggs procured from Canada. The fry produced were used in conjunction with enhancement efforts on the Merrimac River. Livingston Stone became involved in salmon hatching in 1868, and Charles G. Atkins first spawned Atlantic salmon taken from U.S. waters in 1871.

Beginnings of Aquaculture as a Science

On 20 December 1870 the American Fish Culturists' Association was formed (Stickney 1996a) to provide fish culturists with an opportunity to present information in a public forum. Its first president was William Clift. Summaries of the business meetings and the text of presentations were published. The association changed its name to the American Fish

TABLE 1. Selected historical events in North American fish culture.

Year	Event or activity
1864	Seth Green begins rearing brook trout
1866	Livingston Stone establishes his private hatchery; Atlantic salmon spawned by Samual Wilmot in Canada
1868	Charles G. Atkins spawns Atlantic salmon in the United States
1870	American Fish Culturists' Association formed
1871	U.S. Fish and Fisheries Commission established under Spencer F. Baird
1872	McCloud River salmon hatchery established by Livingston Stone
1878	American Fish Culturists' Association becomes American Fish Cultural Association
1884	American Fish Cultural Association becomes American Fisheries Society
1885	U.S. Fish and Fisheries Commission is stocking 17 species of fish around the nation
1904	U.S. Fish and Fisheries Commission becomes Bureau of Fisheries
1928	First commercial trout hatchery established in Hagerman Valley, Idaho
1938	Mitigation hatchery program established for salmon recovery
1950	Bureau of Fisheries split into Bureau of Commercial Fisheries and the Bureau of Sport Fisheries and Wildlife
1958	H. S. Swingle demonstrates economic feasibility of rearing catfish commercially
1960s	Technology for net pen salmon culture developed; commercial aquaculture receives the attention of the scientific community, leading to development of new industries
1969	World Mariculture Society formed
1971	Bureau of Commercial Fisheries becomes the National Marine Fisheries Service
1973	Fish Culture Section of American Fisheries Society formed
1970s	Commercial rainbow trout and catfish industries develop rapidly
1980s	Salmon successfully introduced into Great Lakes and multi-billion dollar recreational fishery established
1986	World Mariculture Society becomes World Aquaculture Society
1990s	Interest in offshore aquaculture develops; net-pen culture expands from salmon to other species

FIGURE 1. The pioneer fish culturist, Seth Green.

Cultural Association in 1878. In 1884, the name was changed, once more, to the American Fisheries Society.

Membership dues were initially $3.00 a year. Dues were increased to $5.00 in 1872, but reduced to $3.00 once again in 1875, probably in response to cries of protest from the membership. Life membership was $15.00, a princely sum in the last half of the nineteenth century. Mem-

FIGURE 2. The "Indiana Jones" of fish culture, Livingston Stone.

bers of the society in the early years included such recognizable names as Fred Mather, Livingston Stone, and Seth Green.

Government Involvement

The first state fish commission was created in Massachusetts in 1856, 15 years before the federal government became involved in fish culture. By the end of 1870, fish commissions had also been created in Connecticut, New Hampshire, New Jersey, New York, Pennsylvania, Rhode Island, and Vermont. Perhaps typical of the state fish commissions was the New Hampshire Fish Commission, formed with the purpose of restoring migratory fishes to the Merrimac and Connecticut rivers (Stickney 1996b). Among the important activities of the commissions were the spawning, hatching, and releasing of various species of fishes into state waters to enhance local fisheries. Commercial fish culturists, such as those mentioned above, were typically retained by the commissions to supply those fish.

Declines in U.S. capture fisheries attributed to overfishing were being reported as early as 1762. Striped bass and sturgeon had been eliminated from the Exeter River in New Hampshire by that time. Dam building, currently receiving much of the blame for salmon declines in the Columbia River Basin of Washington, Oregon, and Idaho, had prevented alewife from spawning in the Exeter River in New Hampshire by 1790.

Nearly one hundred years passed before state fish commissions began to form and respond to the declines that had been first reported in 1762. In January 1871, Spencer F. Baird, a naturalist and staff member at the Smithsonian Institution in Washington, D.C., was interested in the nation's continuing fisheries decline and initiated a plan he had developed to get the federal government involved in fish restoration.

In 1870, Baird approached a House member, Representative Henry L. Dawes of Massachusetts, seeking funding for a Smithsonian study of the problem. Dawes said the Smithsonian was not mandated to undertake such studies, so Baird became more ambitious; he went back to Dawes on 3 January 1871 with the idea of creating a new government agency. A resolution was put before Congress by Dawes, and, on 22 January of the same year, President Ulysses S. Grant signed into law the bill creating the U.S. Fish and Fisheries Commission.

In 1904, the commission became the U.S. Bureau of Fisheries, housed in the Department of Commerce. In 1950, the Bureau of Fisheries was divided into the Bureau of Commercial Fisheries (BCF) and the Bureau of Sport Fisheries and Wildlife, which later became the U.S. Fish and Wildlife Service (USFWS) in the Department of the Interior. In 1971, BCF was renamed the National Marine Fisheries Service (NMFS), a division within the National Oceanic and Atmospheric Administration and was transferred into the Department of Commerce.

Spencer F. Baird was the first commissioner of the U.S. Fish and Fisheries Commission and served from the time the commission was created until his death in 1887. During much of that time, Baird also served as secretary of the Smithsonian Institution. Baird authored over 1,500 publications. He employed many of the practicing fish culturists of the day to establish government hatcheries aimed at producing both marine and freshwater species for enhancement stocking. Within a few years both Atlantic and Pacific salmon were being cultured, as were bass, carp, Atlantic cod, American shad, striped bass, various species of trout, along with rainbow trout, walleye, lake whitefish, American lobster, American oyster, and others. Baird also believed in the desirability of carp as a food fish. He imported them from Europe, established hatcheries, and saw to it that the fish were stocked throughout the nation.

Recognizing that basic biological and environmental information was critical to any successful hatchery and stocking effort, Baird developed a research program. The keystone facility for that part of the effort was the Woods Hole Laboratory (now the Woods Hole Oceanographic Institution), a brainchild of Baird's. Various additional federal fish hatcheries and laboratories were established during and after Baird's reign and were not restricted to the seacoast or saltwater locations. Many were located on the Great Lakes.

Within a few years, billions of larvae, fry, older fish, and shellfish were being transported from hatcheries, often by railroad (Figures 3 and 4), for indiscriminate stocking of the nation's waters. Over several decades, the total numbers of aquatic animals released was in the trillions!

The inability of early fish culturists to feed newly hatched larvae of fishes with very small eggs meant that the majority of the animals were stocked as larvae or fry. Ultimate survival appears to have been virtually zero for many species. There were some successes, however, though success often meant establishing populations of fish outside of their native ranges, a questionable practice by today's standards. That was true for American shad along the Pacific coast, striped bass in California, largemouth bass in the western states, and rainbow trout east of the Cascades. Brown trout and, as previously mentioned, carp were introduced to the United States from Europe. Many additional introductions were made subsequent to Baird's tenure as commissioner, with movement of fish from place to place being largely unrestricted for many decades.

International traffic goes both ways, and by the early 1880s, Baird was responsible for shipping brook trout, whitefish, Atlantic salmon,

FIGURE 3. During the late nineteenth and early twentieth centuries, trains criss-crossed the nation carrying fish and shellfish for stocking.

and largemouth bass to Germany; brook trout, lake trout, and Atlantic salmon to France; brook trout and Atlantic salmon to England; and brook trout to South America. In 1877, chinook salmon were being shipped to

FIGURE 4. A truck picks up cans of fish for distribution by the Oregon Fish Commission in the early part of the twentieth century.

New Zealand where they flourished and are still known in that country by their Native American name, quinnat salmon. Chinook salmon were soon followed by rainbow trout, which are prized sport fish in New Zealand today. Most international shipments were in the form of eggs.

Dr. G. Brown Goode, an assistant secretary at the Smithsonian, was appointed to assume the position of commissioner (by President Grover Cleveland) following Baird's death. The appointment lasted only six months, after which time the commission's authorizing legislation was changed:

- the commissioner was to be appointed by the president with advice and consent of the Senate,
- the commissioner was to have scientific and practical acquaintance with fish and fisheries, and
- the commissioner was to be paid $5,000 per year and could hold no other office or employment with the government.

Goode left the position, and on 20 January 1888 Marshall McDonald was appointed commissioner.

McDonald is reported to have conducted experiments that led to the development of an automatic fish hatching jar, which was patented by McDonald and Stephen C. Brown in 1882 (Stickney 1996a). Development of the McDonald jar is one of the most important advancements in fish culture credited to Marshall McDonald. Yet, the design seems to have been purloined. In 1879, Oren M. Chase, who ran the state of Michigan's Detroit Whitefish Hatchery, invented the Chase hatching jar. The jar was displayed at an international exhibition in Berlin in 1880, and Chase was awarded a golden medal of honor (Westers and Stauffer 1973).[1] The Chase and McDonald jars are remarkably similar.

The "Indiana Jones" of Fish Culture

The army rejected the Reverend Livingston Stone from serving as a chaplain during the Civil War for health reasons. Seeking outdoor work, he became a fish culturist. It is not clear from the literature what his health problem was, but it apparently was not too debilitating for over the years Stone faced severe hardships that included physical stress and a few confrontations with disgruntled Native Americans.

Spencer F. Baird searched out the leading fish culturists in the United States and enticed them to go to work for the new commission. Seth Green

[1] This information was not in my possession at the time the 1996 book was written (Stickney 1996a). I gratefully acknowledge having had it brought to my attention by Harry Westers, a fish cultural pioneer of the twentieth century.

and W. Clift were put to the task of spawning shad in 1872. Whitefish spawning was assigned to Nelson W. Clark of Clarkston, Michigan. Charles Atkins was asked to expand his Atlantic salmon culture facilities on the Penobscot River in Maine and to begin producing salmon. Between 1874 and 1880, landlocked Atlantic salmon[2] were stocked in the states of California, Connecticut, Illinois, Iowa, Kansas, Kentucky, Maine, Maryland, Massachusetts, Michigan, Minnesota, New Hampshire, New Jersey, New York, North Carolina, Ohio, Pennsylvania, Rhode Island, South Carolina, Vermont, Virginia, West Virginia, and Wisconsin. None of those stocking programs resulted in establishment of the species.

To augment the depleted salmon stocks of the northeast, Baird tapped Livingston Stone and asked the former preacher to establish a salmon spawning station in California. On Stone's first trip in 1872, he established a primitive facility, collected eggs from chinook salmon, and made the first shipments of eggs to the east coast.

The facility Stone established (Figure 5) was on the McCloud River, a tributary of the Sacramento. It was located some 200 km (120 mi) north-northwest of San Francisco, near a stagecoach road. The facility was at least 15 km (9 mi) from the nearest town or village and 30 km (18 mi) from a sawmill (Stickney 1996c). Originally called the McCloud Station, it was later named Baird Station in honor of the first commissioner. A second station, upriver from the salmon hatchery, was established to spawn rainbow trout.

In 1873, Baird charged Stone with transporting a variety of fishes and invertebrates to the West Coast in what was called an aquarium train. The baggage car of the train had been fitted with tanks in which fishes and invertebrates, including oysters and lobsters, were to be carried to the West Coast for stocking. A trestle collapsed under the train when it was crossing the Elkhorn River in Nebraska (Stickney 1996a). Stone and his colleagues in the aquarium car were battered by various kinds of flying and floating objects, including tanks of fish and oysters, but managed to escape and climb onto a corner of the car that protruded above the water. At least two men died in the mishap, though none of the commission employees was seriously injured.

All the aquatic animals in the aquarium car either escaped or were released by the fish culturists. Stone indicated that some of the species may not have survived. He contacted Baird by telegraph to report the incident and was told to report back to Washington, D.C., pick up another load of animals, and repeat the process. Stone reportedly threw a

---

[2] Anadromous Atlantic salmon and those that were unable to migrate to the sea were considered as separate species for many years. At present, there is no distinction between Atlantic salmon that have access to the sea and those that do not.

FIGURE 5. A large water wheel was designed into the Baird Station on the McCloud River in California by Livingston Stone to lift water from the river into the hatching house (U.S. Fish and Fisheries Commission 1897).

fish from the train each time he crossed the Mississippi River on his subsequent annual trips to the West Coast.

Once Stone had established salmon and rainbow trout hatcheries on the McCloud, he was dispatched to find another hatchery site to the north. After a long search, he established the Clackamas Hatchery in Oregon.

While the Elkhorn River incident was traumatic, it was not the only tribulation experienced by Stone and his men[3] during their summers in California. They faced hostile Indians, poisonous snakes, and extremely hot weather while they worked in cold water, ponds that filled with mud, and floods that sometimes destroyed major portions of their facilities.

Chinook salmon and rainbow trout were not the only species produced by the Baird Station. By the 1890s, coho were also being spawned.

[3] The fish culturists involved in the California activities were, without exception, men. Women were sometimes involved as egg pickers in the Atlantic salmon hatcheries on the east coast, but otherwise were not participants in the fish culture process.

Between 1891 and 1899, 1.6 million fry and 560,000 yearling coho salmon were distributed.

Moving Fish Around

Contemporary fish geneticists often decry the movement of fish from one region to another because of concerns over loss of genetic diversity. Yet, beginning in the last quarter of the nineteenth century, billions of fish were being moved around the nation. By 1885, the commission was involved with the distribution of carp, channel catfish, cod, grayling, brook trout, brown trout, herring, lake trout, rainbow trout, Atlantic salmon, chinook salmon, shad, smelt, tench, tomcod, white perch, and whitefish.

Salmon eggs from the McCloud (Baird) Station in 1874 (a typical year) were sent to Colorado, Connecticut, Illinois, Indiana, Iowa, Louisiana, Maine, Missouri, Massachusetts, Michigan, Minnesota, New Jersey, New York, Ohio, Pennsylvania, Rhode Island, Texas, Utah, and Virginia. The total number of eggs shipped that year was in excess of 2,200,000, with numbers ranging from about 7,000–304,000 per state (Stickney 1996c). From 1891 to 1899, over 100 million eggs, 150 million fry, and 2 million yearling chinook salmon were distributed.

Chinook salmon did not become established outside of its native range except in New Zealand, but many of the fish distributed from the Baird Station were shipped to locations where the species was native or could possibly be expected to become established. In 1891, for example, chinook salmon were shipped from the Baird Station to the following West Coast locations or entities: California Fish Commission (2,852,000 eggs), McCloud River (25,500 fry), Clackamas River, Oregon (1,332,400 fry), Trinity River and tributaries, California (140,000 fry), Redmond Creek, California (150,000 fry), and Supply Creek, California (25,000 yearlings).

By the early twentieth century, production by the commission had expanded to include sockeye and Atlantic salmon. Atlantic salmon were routinely stocked in the Pacific Northwest for many years, though, as was true with respect to stocking Pacific salmon in the east, Atlantic salmon did not become established on the Pacific coast.

Numbers of chinook and sockeye salmon produced in some years between 1901 and 1940 exceeded 100 million (Stickney 1997b). The numbers of salmon being released became much more prodigious after World War II when both state and federal mitigation hatcheries proliferated in the Columbia River Basin and, more recently, when the private, not-for-profit salmon hatchery program became established in Alaska.

Brook trout, native to New England, Pennsylvania, and the Appalachians to northeast Georgia, were shipped by the millions, at various life stages, beginning in 1872 and extending through the 1920s to Alaska, Arkansas, California, Colorado, Montana, New Mexico, Oregon, South Dakota, Utah, Washington, and Wyoming (Stickney 1996d). From 1901 through 1940, brook trout distributions from federal hatcheries ranged from nearly 4.5 million to over 35 million a year, with a total of approximately 575 million.

The brown trout is a species that, unlike various other nonnatives, has been widely accepted and even demanded by anglers. Brown trout were widely distributed by the U.S. Fish and Fisheries Commission. In 1896, brown trout were sent to Arkansas, Colorado, Connecticut, Idaho, Illinois, Iowa, Maine, Maryland, Massachusetts, Michigan, Minnesota, Montana, New Hampshire, New Jersey, North Dakota, Ohio, Oregon, Rhode Island, South Dakota, Utah, Vermont, Washington, Wisconsin, and Wyoming.

The Baird trout station that Stone established upstream from the salmon station was designed to accommodate up to 6 million rainbow trout eggs. The facility was put into operation in 1880 when the egg take was 338,000 (Stickney 1996e). A portion of the eggs (70,000) were hatched on-site and subsequently stocked in the McCloud River. Most of the remaining eggs were shipped to a hatchery in Baltimore, Ohio, where they were hatched and then further distributed to Washington, D.C. (1 location), Iowa (1 location), Maryland (212 locations) Michigan (6 locations), Minnesota (1 location), New Hampshire (2 locations) North Carolina (3 locations), Pennsylvania (1 location), South Carolina (2 locations), West Virginia (1 location), and Wisconsin (1 location). Flooding of the Baird trout station in 1881 prevented egg take, but the facility was put back into operation in 1882.

Production never did reach Stone's estimate of its full capacity. In 1883, Stone relinquished management of the facility to Loren W. Green. Activities at the station were ended in 1888 as rainbow trout hatcheries had been established in the east by that time. For example, Frank Clark was rearing rainbow trout in his Northville, Michigan station by 1883. Clark had by then developed broodfish from among the young fish he had received from California. Some of Clark's eggs were shipped to Fred Mather in New York for transshipment to Europe—Germany, France, and England were recipients in 1883, while Scotland, along with the other three nations, received rainbow trout eggs in 1885.

The Wytheville, Virginia station also began shipping rainbow trout eggs in 1883. Fish were sent to Tennessee, Alabama, and Georgia. In 1886, the Washington, D.C. Central Station of the U.S. Fish and Fisheries Commission began distributing rainbow trout fry and larger fish. From 1889 through 1900 the commission reportedly distributed nearly 2.5 million rainbow trout eggs, 800,000 fry, and 1.4 million fingerlings or larger fish. Trout were being produced in seven commission hatcheries from Maine to California by 1894. From 1901 through 1940, a total of over 335 million rainbow trout were distributed from federal hatcheries (Stickney 1997c). By 1901 the anadromous steelhead variety of the rainbow trout was being reported separately. Between 1901 and 1940 a total of some 135 million steelhead were distributed by the federal government.

There is a some question as to the true identity of the trout that Stone shipped from the Baird Station. He called the fish black-spotted trout, a name that was applied at the time to the cutthroat trout. Stone recognized two varieties; fine-scaled and coarse-scaled. One thought is that one variety was steelhead and the other a resident rainbow strain. Needham and Behnke (1962) theorized that the fine-scaled variety was a hybrid between rainbow and cutthroat trout. In either case, since Stone did not attempt to keep the varieties separate during his spawning activities, the trout that he distributed may have been some type of hybrid.

Cutthroat trout made the government fish distribution list in 1892 (Stickney 1996e). Native to the western United States, the fish was soon being reared at hatcheries in Colorado and Virginia. As the nineteenth century closed, cutthroat trout could be found in several states east of the Rocky Mountains, including on the eastern seaboard in such places as Maine, Washington, D.C., and Virginia. Cutthroat trout distributions from federal hatcheries between 1901 and 1940 ranged from slightly less than two million to over 51 million fish annually with a total of nearly 580 million over the four decades (Stickney 1998).

Largemouth bass are native to the Mississippi Valley and Great Lakes region, as well as along the Atlantic coast from Florida to North Carolina. Smallmouth bass were originally found in the Great Lakes region southward to north Georgia and Alabama and eastward to the Appalachians and westward to Oklahoma.

The first successful bass stocking program (if we exclude the Elkhorn River incident) was in 1887 (Stickney 1997a). That year, 550 smallmouth bass were collected and redistributed in Virginia and Washington, D.C. By 1892, smallmouth were not only being caught and restocked, but there were also attempts to spawn them. In 1894, the Quincy, Illinois station distributed over 13,000 bass to 16 states and the District of Columbia. Numbers climbed rapidly thereafter reaching over 225,000 by 1901 and well in excess of 10 million by 1940 (Stickney 1998). By 1915, bass had been distributed within at least 46 states.

Yellow perch distribution began in 1894 when fry, fingerlings, yearlings, and adults were shipped around the nation. Similarly, northern pike were distributed, beginning in 1892, to California, Idaho, Illinois, Indiana, Iowa, Kansas, Missouri, and Ohio. Esocids seem to have been of less interest to the federal hatchery program than many other species. Numbers of northern pike and pickerel[4] produced in the period from 1901 through 1940 ranged from zero to slightly over 9 million (Stickney 1998).

Yellow perch were not greatly emphasized in federal hatcheries until 1903 when nearly 29.5 million eggs, fry, and older fish were distributed as compared with 1,700 in 1902 and only 621 in 1901. By 1905, distribution had climbed to over 140 million and surpassed 260 million two years later. Thereafter, through 1940, numbers produced ranged from slightly over 90 million to as high as nearly 330 million annually (Stickney 1998).

Walleye, called pike–perch in many of the early U.S. Fish and Fisheries Commission reports, were produced in extraordinary numbers. Most were undoubtedly released as eggs or newly hatched fry, and it is probable that few actually recruited into the fisheries. By 1901, the number of walleye distributed from federal hatcheries exceeded 240 million. During the period from 1901 to 1940, in excess of 13 billion walleye were released (Stickney 1998).

Many other examples of fish distribution can be found in the commission reports and subsequent publications by the Bureau of Fisheries, and the U.S. Fish and Wildlife Service, but the examples mentioned here underscore the point. Such unrestricted movement of aquatic animals would not be undertaken today, but when the majority of the unrestricted movement of aquatic species, both within and outside of their native ranges, was occurring, the reasons for the stocking programs were thought to be in the national interest.

Another interesting activity that involved the movement of hundreds of millions of fish was the stranded fish rescue program, initiated in 1888 (Stickney 1998). The program was conducted in areas where annual flooding of the Mississippi River caused the stranding of fish in transient ponds and in ponds that froze to the bottom in the winter. Fish were commonly restocked into the portion of the river adjacent to where they were captured, but significant numbers were also used to stock other waters in response to applications from various states for fish. Among the fishes rescued were catfish, largemouth and smallmouth bass, crappie, sunfish, walleye, yellow perch, and northern pike. In 1914, about 2.5 million fish were rescued. The number had climbed to nearly 26 million

---

[4] Listings in government publications combined northern pike and pickerel. Northern pike are also called pickerel in some locations, but the references to pickerel in the fish distribution literature (Stickney 1998) could also, perhaps, refer to the chain pickerel *Esox niger*.

in 1918, and peaked at 182.5 million in 1931. Low numbers, such as the 5.2 million fish collected in 1929, were associated with years in which summer rainfall was sufficient to allow the flood waters to recede slowly, thereby providing fish with the means of returning to the main stem. The program was phasing out by 1939 because of channelization and dam-building activities that were bringing the Mississippi River under control. Activities were curtailed after 1940.

Increased State Involvement in Stocking Fish

Following a hiatus in fish distribution during World War II, federal hatcheries once again moved into full production. They had been joined for many years, even several decades in many cases, by state hatchery programs, ultimately established in every state. Some state commissions were in the fish culture business, in fact, before the commission was formed.

Reservoir construction waned and the states assumed increased responsibility for stocking their own waters; many federal hatcheries were closed and programs abandoned during the last quarter of the twentieth century. Currently, the primary activities of federal fish hatcheries are in conjunction with stocking waters on federal lands and Indian reservations, salmon production in the Pacific Northwest, a few restoration projects, and recovery of various threatened and endangered species (Stickney 1999). There are also various research scale projects with striped bass, such flatfish as southern flounder, Pacific halibut, and other species.

Hatchery construction at the state level has waned considerably in recent years, though it has not totally ceased. Many states have constructed hatcheries within the past several years, conducted significant renovations, considered construction of new facilities, or have had federal hatcheries turned over to them. Increasingly, stocking of private waters (e.g., farm ponds and private lakes) is devolving to the private sector. Fish for stocking farm ponds, once readily available free for the asking, are now often not available from state fisheries agencies but must be purchased from private producers.

Commercial Aquaculture

Shortly after World War II, the first rainbow trout farm was established in the Hagerman Valley near Twin Falls, Idaho. The Thousand Springs area of the Snake River Canyon (Hagerman Valley) features an abundant supply of groundwater that erupts from the north side of the canyon wall. High in quality and of nearly perfect temperature for some species of trout, the water is now diverted through raceways heavily stocked with trout and then discharged into the river. Several farms sprang up in the Hagerman Valley and by the 1980s they supplied the majority of the trout sold in the United States, though trout farming has developed in many other states. While modern trout farms employ large numbers of concrete raceways (Figure 6), they basically do not differ from raceways used in the hatcheries of the late nineteenth century (Figure 7).

Minnows and goldfish were being produced by commercial farmers before the 1960s, a decade during which commercial aquaculture began to develop rapidly. Baitfish continue to be produced in many states, but the industry is centered in Arkansas and other southern states (Stone et al. 1997). The fledgling baitfish industry grew from only about 80 ha (200 acres) in 1950 to over 24,000 ha (60,000 acres) today. Arkansas alone has nearly 12,000 ha (30,000 acres) devoted to baitfish culture.

Perhaps the first warmwater food fishes to gain the attention of people interested in commercial fish farming were buffalo. Several farmers constructed ponds or converted rice ponds into fish ponds and were producing modest numbers of buffalo by the 1950s (Meyer et al. 1968).

Homer S. Swingle, an entomologist with a keen interest in sportfishing and a man who had developed an aquaculture research program at Auburn University, Alabama, aimed at learning more about production of recreational fishes, turned his attention toward food fish. He ex-

FIGURE 6. Concrete raceways at a modern commercial trout farm in the Hagerman Valley, Idaho.

FIGURE 7. Hatchery raceways from the nineteenth century are readily recognizable even though concrete and fiberglass have replaced wood.

amined buffalo production (Swingle 1957b), but also demonstrated that profit could be made from the rearing of channel catfish (Swingle 1957a). Swingle projected profitability if the fish could be sold in the round for $0.50/lb ($1.10/kg).[5] Whether the 1957 publication on the potential profitability of channel catfish farming triggered the explosion of a new industry or just happened to coincide with rapid growth in interest and the targeting of a new species is debatable. In any event, buffalo fell by the wayside and fish farmers, both the established ones and those who were just entering the business, turned their attention toward catfish. There was some interest and culture activity associated with blue catfish and white catfish, but the industry ultimately became almost totally devoted to channel catfish production. Federal laboratories created to conduct research, primarily on channel catfish culture, were established at Stuttgart, Arkansas, and Marion, Alabama, by the U.S. Fish and Wildlife Service in the 1960s.[6]

Other federal laboratories had been previously created to conduct research, most of which supported development of technology required

[5] Pondbank prices did not reach $0.80/lb ($1.76/kg) until the mid-1990s.

[6] The Marion laboratory was closed and turned over to the state of Alabama and the Stuttgart laboratory was transferred to the U.S. Department of Agriculture during the 1990s.

for rearing sport and commercial fish for stocking. Those laboratories often had specific subject areas. For example there were the Eastern Fish Disease Laboratory in Leetown, West Virginia (1931), Western Fish Disease Laboratory in Seattle, Washington (1935), and the Tunison Fish Nutrition Laboratory in Cortland, New York (1932), to name just three.

Development of catfish farming in the 1960s undoubtedly provided some of the spark that ignited attempts to commercially culture other species. Information was available from state and federal research laboratories, but only a few universities offered curricula that included commercial fish culture. Universities with departments having curricula in fisheries might have had a hatchery course or two involving operation of public hatcheries (for example, beginning in the 1930s, courses in salmonid production at the University of Washington in Seattle conducted by Lauren Donaldson, or spawning and rearing bass and bluegills for stocking ponds and reservoirs by Swingle at Auburn University), but courses in commercial aquaculture were not common until the 1970s.

With the blossoming interest in food fish aquaculture came very rapid development of research activities in universities. During the 1960s and 1970s, specialized curricula in aquaculture were developed at a number of universities and groups of faculty were formed to support the new teaching and research programs. A few examples from various locations with well-established programs, in addition to those previously mentioned, are Clemson University, Mississippi State University, Louisiana State University, Texas A&M University, Southern Illinois University, and the University of California at Davis. Many others have at least some classes or even degree options in aquaculture or have aquaculture research and classes as a part of graduate curricula. Included are programs in community colleges, trade schools, and introductory courses in many high schools.

By the late 1960s, marine shrimp culture was attracting venture capital and farms were being established, though the technology for producing shrimp was virtually nonexistent. Shrimp farms sprang up in Florida, literally by the dozens, but few lasted more than a year or two.

Breakthroughs in shrimp culture technology pioneered by the Japanese and successfully modified in the United States, particularly by biologists working at the Galveston, Texas NMFS laboratory, ultimately helped to establish shrimp farming. While it was domestic interest in shrimp farming that led to the research, the industry developed primarily in Asia and Latin America, beginning in the 1960s but exploding in the 1980s, when many technological advances were introduced. The growing season in the tropics meant much higher rates of production than were possible on the U.S. mainland. Also, the native species of shrimp from the Gulf of Mexico did not adapt well to culture conditions. Interest was aroused once again in the United States when various nonnative species were introduced. Shrimp farms were established in Hawaii, Puerto Rico, South Carolina, and Texas. Expensive coastal land, strict environmental laws, the availability of suitable sites, and various other factors have severely limited the number of shrimp farms that have been developed.

It was not only penaeid shrimp that demanded attention. Research was conducted and farms were established that focused on such species as the American lobster, Florida spiny lobster, freshwater shrimp, tilapia (blue, Mozambique, and Nile), striped bass and hybrid striped bass[7], summer flounder, southern flounder[8], yellow perch, sturgeon, paddlefish, various species of clams, blue mussels, and a variety of other species.

With the proliferation of research activity in aquaculture over the past 40 years, what was once an art has developed into a recognized science. Great strides have been made in water quality management, feeds have been developed on the basis of the defined nutritional requirements of various species, an understanding has been gained on the types of diseases aquatic animals contract and how to treat most of them, selective breeding techniques and biotechnological approaches to sex control have been developed, and new types of culture systems have been developed. Earthen ponds are still the culture unit of choice for many species, but wooden raceways have been replaced with linear or circular concrete, aluminum, and fiberglass raceways and tanks. Closed recirculating water system technology, while still on the margins of economic feasibility for many species and in many locations, has improved significantly over the primitive systems of the 1970s.

Floating net-pen culture is a relatively new development that has broad applications in the marine environment. Much of the net-pen technology currently employed by the salmon farming industry around the world is based on work conducted in Washington State beginning in the 1960s (Novotny 1975). That technology was adopted and improved upon by Norway, which became the world's leading net-pen salmon produc-

---

[7] A great deal of research had been conducted on striped bass, so the basic techniques were well established by the time commercial aquaculture began.

[8] Research on flounder was discontinued in the early 1970s, largely because the fish brought a low price in the marketplace. The economics in the 1990s have made those fishes very attractive and commercial ventures have been established, as have new research programs.

ing nation during the 1980s. Japan independently developed floating net-pens that can be seen in bays throughout that nation.

During the 1990s, aquaculturists began exploring the potential for moving into exposed waters off the coast of the United States where hostile conditions present enormous engineering challenges. Work to date has involved system development and the establishment of small demonstration farms. Flounder are being cultured off New England, while red drum, red snapper, and cobia are among the species of interest in the Gulf of Mexico.

Marine fish enhancement is also gaining favor in some circles. Today we know how to rear to fingerling size many of the same species that the U.S. Fish and Fisheries Commission could stock only as eggs or newly hatched larvae. Thus, chances for survival are greatly improved. Some successes have already been achieved with striped bass and red drum on the U.S. mainland and with striped mullet in Hawaii. The possibility of establishing enhancement hatcheries that work in collaboration with the commercial (or recreational) fishing industry is actively being discussed. A template for such collaboration currently exists in the form of the not-for-profit salmon ranching operations in Alaska.

The ornamental fish business depends on imports of fish collected from nature, mostly in Latin America, Asia, and Oceania. A significant ornamental fish culture industry exists in the United States. Most of the species produced are freshwater, but there is increasing interest and activity associated with marine ornamental fish culture. Florida leads the nation in ornamental fish production. The climate allows many species to be produced in ponds, whereas other states usually have to depend upon indoor facilities to avoid winterkill. Goldfish are an exception.

Backlash to Aquaculture

In the 1980s, aquaculturists began facing opposition from environmental groups, commercial fishermen, and individual landowners (Stickney 1988, 1990). Branded by various individuals and groups as polluters, as despoilers of natural beauty as in the case of salmon net-pens, as potential disrupters of nature's balance when nonnative species escaped, introducers of diseases into wild fish, or as just creating a nuisance, commercial aquaculturists were caught by surprise. While they certainly had an economic incentive for producing their products, food fish culturists were also dedicated to providing wholesome food for the tables of consumers. If asked, most aquaculturists would undoubtedly call themselves conservationists, so the criticisms were taken rather hard, and, at first, were often rejected out of hand.

Much of the criticism was aimed at those practicing aquaculture in public waters, though land-based systems were not immune and were coming under increasing scrutiny. After a period of denial, the aquaculture community began to take the criticisms seriously and to address each of them. Some response was mandated under permitting requirements by state and federal agencies, but action was also promulgated in response to research results and the desire of the producers to be responsible neighbors. The criticism has not ended, nor have all the problems been resolved. The aquaculture community continues to address the issues.

Purposes of Fish Culture

Commercial fish culture preceded government involvement, but was quickly overshadowed when the state fish commissions and U.S. Fish and Fisheries Commission were formed. The situation changed beginning in the 1960s, and private enterprise continues to expand today. The private sector produces not only food fish, but also such additional items as ornamental fishes and invertebrates (including live rock), baitfish, aquatic macrophytes, and algae.

Display Fish

Putting fish on display may be a reason that fish culture was originally developed. Koi carp (a highly selected breed of common carp) have been around for millennia. Originally, Chinese emperors had sole rights of possession. Today, Koi are popular with fish hobbyists, the more serious among whom may pay hundreds of thousands of dollars for a single fish (though Koi can be obtained for as little as a few dollars each). Coloration and other characteristics dictate market prices for individual fish. Many species of fish that are being displayed today do not carry high price tags, but the facilities in which they are displayed may be elaborate and very expensive. Among the places where fish can be viewed on display are public aquaria; zoos and exhibits; and in home, office, and commercial settings.

Glimpses of the underwater world and its inhabitants are available to many people through visits to public aquaria. Such facilities range from very modest to extravagant. Tens to hundreds of millions of dollars have been spent on some modern aquaria that may feature tanks that are 7–10 m (20–30 ft) or more deep and contain large sharks, adult sea turtles, marine mammals, and a wide variety of other large aquatic animals. Aquaria open to the public may be operated by private entities (including foundations) or local, state, or federal government agencies or universities (Figure 8). Most have an education theme.

FIGURE 8. A touch tank in the public aquarium area of a Texas Parks and Wildlife Department red drum hatchery.

Many public aquaria operate to a large extent, or solely, using recirculating water systems. The use of recirculating technology is a necessity for public aquaria located away from the coast if marine displays are included. Artificial seawater is usually used unless the facility is close enough to the coast to make pumping or trucking seawater economically feasible. The biomass of animals in a public aquarium tank is typically quite low so water quality is less difficult to maintain than in commercial recirculating systems.

Most zoos feature primarily terrestrial animals, including a wide variety of species that are not native to the part of the world in which the zoo is located. Aquarium displays may be incorporated into a zoo as a relatively small part of the overall array of exhibits. An option is to incorporate fish into small streams and ponds associated with terrestrial animal displays.

The fish bowl or small aquarium is a nearly ubiquitous feature in American homes. Many offices now feature aquaria. Waiting rooms in physician's offices and hospitals, bank lobbies, executive suites, and various other types of offices feature tanks of colorful fishes. In many cases such display aquaria are set up, stocked, and maintained by a firm specializing in that activity.

Commercial Species

The U.S. commercial aquatic food industry is led by channel catfish, now an industry with farmgate value in excess of $350 million annually (FAO 1997). Other species with significant levels of aquaculture production include rainbow trout, Atlantic salmon, tilapia, Pacific oysters, American oysters, blue mussels, and crawfish. There is some sturgeon, striped bass, hybrid striped bass, red drum, Chinese carp, abalone, clam, and penaeid shrimp culture as well. Summer flounder culture is rapidly developing and a number of other species are either being produced in small amounts or are being actively researched.

Goldfish are produced not only for the home aquarium and backyard fish pond, but also as bait and as live food for various predatory aquarium species. Some farmers in Florida produce over 100 species of ornamental fishes and that industry now exceeds $100 million a year in Florida alone (Anonymous 1998). From an economic standpoint, the value of ornamental fishes on a per unit area or weight basis is much higher than for any other food or bait species currently being produced.

As fish and wildlife agencies in many states have moved away from providing free fish for the stocking of private waters, commercial fish farmers have filled the void. Even in cases where the state will provide fish to stock a new private pond, supplemental stocking may have to come from commercial producers. The most readily available fishes from commercial growers for stocking private ponds are largemouth bass, sunfish, and channel catfish. Grass carp as diploids or triploids, are commercially available for use in aquatic vegetation control, though their use is not permitted in some states.

Fee fishing has existed in the United States for at least a few decades. Private ponds or lakes are stocked with catchable fish—often largemouth bass, trout, or catfish—and the public pays for the opportunity to catch them. In most cases anglers are asked to keep every fish caught and fees are assessed by either the number or weight of fish in the creel. The fee-fishing operator usually provides a number of other services, including sale or rental of tackle, bait sales, fish cleaning, snacks, and beverages.

Operators of fee-fishing lakes may develop their own broodstock, hatcheries, and rearing facilities or they may purchase fish from a culturist who grows fish to catchable size. Ponds are kept well stocked and the fish are fed only a maintenance ration, if at all, so they will be hungry and thus, more likely to be caught.

Commercial Fisheries

Attempts to increase or at least sustain the stocks of several commercial fish species through enhancement programs began, as we have seen, in the nineteenth century through the activities of state fish commissions and the U.S. Fish and Fisheries Commission. With the exception of salmon production to augment both the commercial and recreational fisheries, enhancement of commercial fish stocks was curtailed early in the present century. The concept is, however, currently being reconsidered as fish culturists now have the ability to rear many species of commercial importance to sizes at which the animals have a realistic chance for survival upon release. Successful stock restoration efforts, augmented by enhancement stocking, include those associated with red drum in Texas (which provides good sportfishing; commercial fishing is banned) and striped bass along the Atlantic coast. The potential for enhancement has been discussed for flounder, cod, haddock, and various other commercially valuable species. It is likely that more emphasis will be placed on the use of enhancement programs to restore marine fish populations in the future. Such efforts should go hand in hand with research to determine proper stocking rates under existing environmental conditions, compatibility of wild and cultured fishes, and maintenance of genetic diversity.

Recreational Fisheries

Various approaches are used in conjunction with supplying fish for stocking public waters. One is called put-and-take. This involves stocking catchable fish that immediately become available to anglers. It is common for anglers to follow hatchery trucks to stocking locations on trout streams, for example, where fish may be caught within minutes of being stocked.

Put-and-take stocking sometimes involves providing an alternative fishing experience during a portion of the year when environmental conditions are suitable. An example is the spring trout season offered in several states. Trout are stocked in selected lakes during the spring when water temperatures will support survival of the fish. Anglers are expected to catch most of the trout before the water temperature increases to the point that it becomes lethal. A trout stamp may be required, the proceeds from which are used to support the program.

Another option is known as put-grow-and-take, which involves stocking fingerlings and keeping the water body closed to fishing or imposing a minimum length limit until the fish grow to catchable size. Stocking fingerling northern pike in Midwestern lakes outside their native range is an example. Others are the coho salmon and steelhead fisheries that were developed in the Great Lakes in the 1980s.

Both of the options mentioned are used to provide angling opportunities in waters that do not support sustainable populations of the species being stocked for the put-grow-and-take fisheries. Small impoundments, along with lakes and segments of streams that receive high fishing pressure, are candidates for put-and-take and put-grow-and-take fisheries. Large lakes may also be candidates when appropriate spawning habitat or environmental conditions are not available to sustain natural populations.

Augmentation stocking may be employed in water bodies that can support sustainable fisheries but where selective fishing pressure has led to unbalanced fish populations. An example might be a lake that contains largemouth bass as the primary species of angling interest, but that also contains a good forage base. If angling pressure reduces the bass population to the point that reproduction does not provide sufficient recruits to sustain the fishery, augmentation stocking with bass may be appropriate.

Mitigation

Human activities sometimes result in the diminution or destruction of fish habitat. One of the most notable examples was the loss of upstream access by anadromous fishes to much of the upper Columbia River Basin as a result of high dam construction beginning early in the twentieth century. Additional impacts on portions of the river that remained accessible to salmon included conversion of riverine habitat to reservoirs, increased turbidity due to runoff from agricultural lands, and the establishment of heavy industry. In 1938, Congress passed legislation that mandated mitigation for losses of renewable aquatic resources and associated recreational and economic values associated with the removal of upstream access to migratory salmonids. Mitigation in that case involved the development of a hatchery program (Figure 9). This is an example of in-kind mitigation; that is, the same species that had been impacted were cultured and used for restocking.

Mitigation may also be imposed when human activities take a water body completely out of production. Filling a lake to make way for a construction project is an example. Mitigation may involve construction of a water body elsewhere to replace the one lost or it might involve development of a stocking program to improve fish production in an existing location. The same species of fish that inhabited the destroyed water body might be stocked (again, in-kind mitigation), or a different

FIGURE 9. An agency hatchery in Washington that produces salmonids for mitigation stocking.

species mix may be used if that is more appropriate to existing conditions (out-of-kind mitigation).

Recovery of Threatened and Endangered Species

A number of fishes are listed as threatened or endangered under the Endangered Species Act (ESA). Various actions can be taken under ESA to enhance the opportunities for listed species to recover to sustainable levels. Actions may include banning the taking of a listed species, rehabilitating critical or essential habitat, and captive propagation. The latter option, in the case of fishes, often means hatchery rearing. One or more generations of a listed species may be maintained in a hatchery to establish a sufficient population for stocking into native habitats. Repeated stocking over several years may be required to re-establish populations, even in the absence of fishing pressure.

When sufficient numbers of fish are available, attempts should be made to develop a broodstock that contains as much genetic diversity as possible. This cannot always be done, however. For example, sockeye salmon returns in the early 1990s to Redfish Lake, Idaho, were as low as one to two individuals in some years. In one instance, a single male returned to Redfish Lake. Milt from that fish, dubbed Lonesome Larry, was cryopreserved and saved for use during the next spawning season. A population certainly must be considered endangered when it is represented by only a handful of individuals. Even the most heroic efforts may not bring such populations back from the brink of extinction.

Recovery efforts are being conducted by state and federal agencies to establish fish populations that can provide some level of regulated harvest. Hatcheries involved in the recovery of threatened and endangered species take every step possible to keep mortality to an absolute minimum. This is necessary since the populations or species being cultured often have few representatives. Once a species has been restored to the point that captive breeding is no longer required, the hatchery operations can be curtailed.

Population Assessment

Cultured fish can be mass marked or tagged for use in population assessment studies. It is often assumed that the cultured fish will, upon release into the natural environment, behave the same as their wild counterparts, which is often not the case. Therefore, differential mortality or differences in susceptibility to capture can affect conclusions drawn from studies in which cultured fish are used to infer how wild populations behave. Tens of millions of hatchery-reared Pacific salmon have been implanted with coded wire nose tags over periods of several years before release into the waters of the Pacific Northwest. Tags recovered from returning fish can be used to determine the percentage of hatchery fish in a given spawning run, the amount of straying that occurs, and other types of information.

There are various other ways to tag or mark fish, including external tags that may be metal or plastic disks or streamers (spaghetti tags) attached to the outside of the fish, tags inserted into the eye tissue, (passive integrated transponder) PIT tags, fin clips, otolith marking through immersion of fry in tetracycline baths or exposing them to rapid temperature changes, and branding. By releasing a known number of tagged fish into a lake and then comparing the numbers of tagged relative to untagged fish captured, one can calculate the number of fish in the untagged population with a relatively high degree of accuracy so long as certain assumptions are met.

With PIT tags and other types of tags that allow fisheries scientists to identify individual fish, it is possible to obtain detailed information on growth and migration, and to gather other types of data relating to population dynamics. Wild fish can be captured, tagged, and released, but having a captive population of many hundreds, thousands, or even millions of fish available for tagging can be a distinct advantage.

Research

In addition to the thousands of studies conducted by fish culturists to learn more about how to properly produce and maintain the species in which they are interested, scientists are increasingly using fish as laboratory animals for various types of research projects unrelated to aquaculture. With increasing pressure by animal rights groups to curtail experimentation using mammalian models (e.g., mice and primates), fish have become popular for a variety of biomedical studies. Fish are also being used in genetics research and are widely employed in bioassays of biocides, contaminants, and so forth. One of the most popular species currently is the Japanese medaka. Zebra danio, killifish, goldfish, and various other species have also been used as research models. Tilapia is one among many groups of fishes that have been used in basic physiological studies. Cell lines have been developed from various fish species and used in the study of diseases.

Small fish such as medaka and zebra danio are desirable for use in research that requires rapid generation times, relatively large numbers of animals, or many replications of experimental treatments in confined volumes of water. Those and other fishes used in research tend to be hardy and can be easily reproduced in captivity.

Clearance of drugs and chemicals for use on food fish must be obtained in most cases on a species by species basis. In other words, if an antibiotic is approved for channel catfish, it is not automatically approved for any other species, although it may be approved for other ictalurids that live within the same thermal regime. Approval for use of the same chemical on, for example, trout would require independent efficacy and safety trials. Thus, at least in the case of drug clearances, a general fish model will not lead to general clearance for use, at least not in food fish.

References

Anonymous. 1998. Florida aquaculture sales total $102 million in 1997. Aquaculture News (June):1, 12. Vidalia, Louisiana.

Avault, J. W., Jr. 1996. Fundamentals of aquaculture. AVA Publishing, Baton Rouge, Louisiana.

Bardach, J. E., J. H. Ryther, and W. O. McLarney. 1972. Aquaculture: the farming and husbandry of freshwater and marine organisms. Wiley, New York.

Bowen, J. T. 1970. A history of fish culture as related to the development of fishery programs. Pages 71–93 *in* N. G. Benson, editor. A century of fisheries in North America. American Fisheries Society, Special Publication 7, Bethesda, Maryland.

FAO (Food and Agriculture Organization of the United Nations). 1997. Aquaculture production statistics 1986–1995. FAO Circular 815, Rev. 9, Rome.

Kirk, R. 1987. A history of marine fish culture in Europe and North America. Fishing News Books, Surrey, England.

Meyer, F. P. P., D. L. Gray, W. P. Mathis, J. M. Martin, and B. R. Wells. 1968. Production and returns from the commercial production of fish in Arkansas during 1966. Proceedings of the Annual Conference Southeastern Association of Game and Fish Commissioners 21(1967):525–531.

Needham, L. P. R., and R. J. Behnke. 1962. The origin of hatchery rainbow trout. Progressive Fish-Culturist 24:156–158.

Novotny, A. J. 1975. Net-pen culture of Pacific salmon in marine waters. Marine Fisheries Reviews 37:36–47.

Stickney, R. R. 1988. Aquaculture on trial. World Aquaculture 19:16–18.

Stickney, R. R. 1990. Controversies in salmon aquaculture and projections for the future of the aquaculture industry. Pages 455–461 *in* Proceedings of the fourth Pacific congress on marine science and technology. Tokyo.

Stickney, R. R. 1996a. Aquaculture in the United States: a historical survey. Wiley, New York.

Stickney, R. R. 1996b. Fish production and distribution in the United States, part 1. Government initiatives. World Aquaculture 27(1):31–41.

Stickney, R. R. 1996c. Fish production and distribution in the United States, part 2. Pacific salmon distribution prior to 1900—McCloud River salmon activities. World Aquaculture 27(2):32–38.

Stickney, R. R. 1996d. Fish production and distribution in the United States, part 3. Trout distribution begins. World Aquaculture 27(3):33–40.

Stickney, R. R. 1996e. Fish production and distribution in the United States, part 4. Rainbow and cutthroat trout to 1900. World Aquaculture 27(4):62–66.

Stickney, R. R. 1997a. Fish production and distribution in the United States, part 5. Early activities with nonsalmonids. World Aquaculture 28(1):53–58.

Stickney, R. R. 1997b. Fish production and distribution in the United States, part 6. Salmon distributions from 1901 to 1940. World Aquaculture 28(2):55–58.

Stickney, R. R. 1997c. Fish production and distribution in the United States, part 7. Trout distributions from 1901 to 1940. World Aquaculture 28(4):55–58.

Stickney, R. R. 1998. Fish production and distribution in the United States nonsalmonid fish distribution between 1901 and 1940. World Aquaculture 29(4):10ff.

Stickney, R. R. 1999. Fish production and distribution in the United States, part 9. Activity in the post World War II period. World Aquaculture 30(1):14ff.

Stone, N., E. Park, L. Dorman, and H. Thomforde. 1997. Baitfish culture in Arkansas: golden shiners, goldfish and fathead minnows. Cooperative Extension Program, University of Arkansas at Pine Bluff.

Swingle, H. S. 1957a. Preliminary results on the commercial production of channel catfish in ponds. Proceedings of the Annual Conference Southeastern Association of Game and Fish Commissioners 10(1956):160–162.

Swingle, H. S. 1957b. Revised procedures for commercial production of bigmouth buffalo fish in ponds in the southeast. Proceedings of the Annual Conference Southeastern Association of Game and Fish Commissioners 10(1956):162–165.

U.S. Fish and Fisheries Commission. 1897. Methods of the United States Commission of Fish, and Fisheries with chapters on the cultivation of oysters and frogs. Pages 1–340 *in* U.S. Fish Commission Report for 1897. U.S. Fish and Fisheries Commission, Washington, D.C.

Westers, H., and T. M. Stauffer. 1973. A history of fish culture in Michigan. Pages 107–125 *in* Michigan fisheries centennial report 1873–1973. Michigan Department of Natural Resources, Lansing.

2

Production

Harry Westers

Production Goals

Purpose of the Production Program and Methods

The two main purposes of fish production are for private (for profit) programs and for public programs (e.g., stocking of natural waters for recreation, or enhancement and restoration purposes). Private, for profit programs include the production of fish for the aquarium trade, baitfish, food fish, and fish for stocking private waters. The production of fish involves many different programs, production strategies, and species; new species are continuously being added to the list of aquaculturally produced fish.

There is an enormous variety of production methods in public and private programs, each requiring its own specific type of hatchery facility. The term hatchery facility is very broad and includes facilities that neither incubate nor hatch eggs. One broad classification for methods of fish production is extensive versus intensive. However, there is no clear demarcation line from one method to another, rather it can be viewed as a continuum of culture systems with unfertilized, natural ponds requiring little management at one extreme, and closed, recirculating water systems requiring continuous monitoring at the other extreme. Another broad classification of production methods is static versus flowing water culture, each with its own level of intensity.

Whatever method is used, successful fish production requires management of water quality compatible with the species—the most important factor in successfully producing a hatchery crop of fish.

Good water quality must be maintained throughout the production cycle. As the biomass increases, the challenge and ability to maintain acceptable water quality also increases and is dependent on the type of culture system used.

Flowing water systems afford the greatest control because they allow for a quantitative approach to determine production capability based on selected water quality measurements and the amount of waste generated. Earthen pond systems are the most difficult to manage. They are dynamic ecosystems that respond to many intrinsic and extrinsic factors; one major factor is weather, which is both unpredictable and uncontrollable. There are innumerable types of hatchery facilities and their corresponding production and management strategies. This chapter will cover the major production methods—intensive, semi-intensive, and extensive systems.

Target Species, Number, and Size

To a large extent, the type, design, and size of a production system is determined by the species selected, their numbers, and final desired size. For instance, a public program to produce Atlantic salmon smolt for release in natural streams requires a somewhat different facility than a program to rear rainbow trout as a food fish. Both are salmonids, which are coldwater fish. The difference in production systems relates, to a large degree, to production programs, species, and available resources. Fish for release into natural waters may be of a different quality, genetically and physiologically, than fish reared for immediate human consumption. Whenever possible, private hatcheries use sequential, multiple cohort production programs in order to realize the greatest output, versus the batch, single cohort program used by most public hatcheries.

The selection of species by the private, for profit industry is mostly driven by market demand and value of the product. High value species, such as striped bass hybrids and yellow perch are good candidates for recirculation aquaculture systems (RAS), which are more expensive to construct and operate.

Production Flexibility

Fish production for stocking into natural waters often must meet rather precise size requirements within narrow time periods. Increasing or lowering the water temperature during incubation offers a practical and economic way to manipulate the time required to reach the target size for stocking. Public hatcheries should be equipped with water recirculating systems for egg incubation, hatching, and, possibly, for early rearing whenever the ambient temperature regimes do not match program objectives. Private, for profit programs often can greatly increase production by using sequential rearing programs, where a new batch of eggs or fish is brought into the facility at optimum time periods, depending on growth rate and harvest size requirements. Many species used in aquaculture spawn only once a year, however, proper use of photoperiods, water temperatures, and hormones, either singly or in combination, allow out-of-season spawning.

Facilities can be designed for the greatest degree of production flexibility. However, the cost may be prohibitive. The challenge is to strike the best balance between capital costs, labor and energy requirements, and production flexibility, as well as manageability of the facility.

Feeding, Energy, and Labor Efficiency

In fish culture operations, feed, energy, and labor are the most expensive components. They often add up to 60–80% of the total production cost. The relative cost ratios are intimately associated with the type of production method used. Because of the high cost of feed, it is important to accomplish a high degree of feed efficiency—not an easy task. Knowledge of the amount of food, the frequency, and the method of feed application under a variety of production methods is essential. Feed loss must be kept to a minimum, but in the aquatic environment some loss is unavoidable. Reported feed losses have been as high as 20% in cage culture of rainbow trout in Scotland. Feed cost varies from 10 to 50% of the total production cost. The low value reflects high energy or labor costs with a concomitant high production cost. High percent feed costs are indicative of relatively low production costs. For example, a feed cost of $0.55/kg ($0.25/lb) and a feed conversion of 1.2 results in a feed-related production cost of $0.66/kg fish ($0.30/lb). When feed costs represent 15% of the production cost, the cost of fish produced is $4.40/kg fish ($2.00/lb), but at 50%, the cost is only $1.32/kg ($0.60/lb).

The goal of most fish farmers is to grow the fish as quickly and as cheaply as possible and to produce a quality product. For the best growth and health, a nutritionally complete diet is required (see Chapter 9, page 483). For economic reasons, the cost should be as low as possible. These two requirements—high growth rate and low feed costs—can be mutually exclusive. Furthermore, the desire for maximum growth can tempt farmers to overfeed, thus contributing to feed loss and more waste. Improving feeding efficiency through less costly diets is an ongoing goal of research. Of special interest is the development of cheaper substitutes for fish meal, the most expensive component in today's fish diet.

Energy costs are closely related to the type of production system being used and available water resources—whether delivered by gravity or pumped. The energy fraction of production cost varies from a low of 2% to more than 20%. Often there is a direct relationship between energy and labor costs. As the intensity of a production system increases, the demand for energy also increases, while labor costs may decrease. Large systems are more energy and labor efficient than small ones due to economy of scale.

Unless the water is heated or chilled, the greatest demand for energy is for moving water, whether for mechanical aeration or delivering flows to rearing units. In certain situations, increased production can be achieved more inexpensively using pure oxygen rather than by pumping additional water. The use of variable rate pumps or a series of smaller pumps may also reduce energy costs. Energy efficiency must be carefully considered in facility design in relation to labor requirements.

In most cases, at least in developed countries, labor costs exceed energy costs, thus, where energy can replace labor, it probably should be applied. Labor costs range from 10 to 60% of the production cost. A higher labor cost per unit production typically is encountered with public fish production programs. There are two reasons for this. First, wages and benefits for public hatchery workers are usually higher than in the private industry. Second, due to their mission, public hatcheries are used inefficiently. Their program is one of single batches; more often than not, public hatcheries produce a smaller product compared to most commercial programs, which, when possible, use a multiple batch, sequential rearing strategy. Annual production can be twice or more the maximum carrying capacity, yet labor and energy costs may not differ much from a single batch rearing program.

Production Methods

Fish Physiology

In aquaculture, it is the responsibility of the culturist to keep the fish healthy. This means he or she must manage the hatchery environment—water quality—to make it compatible with the physiological requirements of the fish. Water quality is the most important factor in successfully growing and maintaining a healthy crop of fish. To manage water quality, the culturist requires a knowledge about the fish's needs, the fish's impact on its environment through its metabolic processes, and potential other impacts on water quality. In intensive, flow-through aquaculture systems, the rearing environment only serves to provide physical living space. Water quality changes relate to the feed input, which can be quantified and managed to a large extent. In extensive culture, many physical, chemical, and biological factors interact making it much more difficult to manage water quality.

Oxygen

All fish require oxygen (O_2), which, in flow-through systems, is delivered with the incoming flow of water. This flow may have to be oxygenated or oxygen may be supplemented with "in tank" aeration. Only a portion of the dissolved oxygen (DO) is used by the fish—available oxygen (AO). Available oxygen is the difference between what is delivered (DO_{in}) less the selected minimum concentration (DO_{out}). How much oxygen the fish need depends on their metabolic rate, which is primarily a function of water temperature and fish size. These are the same factors that determine feed requirements, hence the amount of oxygen required per unit of feed is relatively constant and is independent of water temperature and fish size. For salmonids, the oxygen required per kilogram of feed falls in the range of 200–250 g/kg (90–114 g/lb). The highest rates of metabolism are usually observed several hours after cessation of feeding (postprandial increase) except in the case of a feeding frenzy, which has an impact of short duration. To extract oxygen, water must flow over the gills in one direction. This is accomplished by the action of muscular "pumps" that create the breathing movements observable in fish. During inspiration (inhaling), water is taken in through the mouth, enters the buccal cavity, and passes over the gills to the opercular cavities, which are closed. During expiration (exhaling), the mouth is closed, the opercular covers are opened, and water leaves through the opercular openings.

Breathing in water has a substantial energy cost associated with it. Estimates for routine ventilation of around 10% of calories expended have become generally accepted, but this may increase to as high as 70% with ventilation volumes three times routine. By way of comparison, in resting humans, the energy cost of breathing is estimated to be around 2% of the resting metabolic rate. The higher costs for aquatic vertebrates is related to the higher viscosity (840 × air) and density (60 × air) of the medium. Furthermore, O_2 saturated water has only about $^1/_{30}$ the O_2 content of air.

In actively swimming fish or fish that maintain their position in a fast current, the problem of increasing ventilation cost is circumvented by transferring the work of ventilation from respiratory muscles to the muscles of locomotion (ram ventilation), which all but eliminates the cost of ventilation. Salmonids can use ram ventilation when swimming at velocities over 1 body length per second (BL/s). Unexercised salmonids can swim continuously at 1–2 BL/s without reducing intestinal blood flow and, therefore, nutrient absorption and growth. There is an increasing body of evidence suggesting that growth rates of salmonids increase when fish are forced to swim at moderate speeds for prolonged periods of time. Exposure of salmonids to water currents also leads to behavioral changes, including a reduction in the frequency of aggressive encounters—further benefiting growth.

Because gills are also used for ion and water exchange, the effective surface of fish gills may be physiologically restricted in order to reduce this exchange. Freshwater fish must balance the advantage of maximizing O_2 diffusion at the gills with the osmoregulatory disadvantage resulting from the accompanying increases in ionic loss and water influx. Studies with rainbow trout have shown a clear increase in sodium ion (Na^+) loss through diffusion with increasing oxygen uptake. Osmotic pressures are continuously acting to change the chemical composition of blood and tissue. Freshwater continuously diffuses inward through the gills diluting the blood. Consequently, large amounts of very diluted urine are produced, but there still is a loss of blood electrolytes with the urine; these ions must continuously be replaced, either through diet or from the water by means of an active transport through the gills. In saltwater fish, these actions are reversed. Gills also exchange gases other than O_2. Both ammonia (NH_3) and carbon dioxide (CO_2) are major byproducts of metabolism excreted by the gills. Both of these can also have a substantial impact on the quality of the rearing environment.

It is the culturist's responsibility to make certain that the fish are adequately supplied with O_2. But how much is enough? When rainbow trout were reared under different conditions of DO at 15°C (59°F), food intake was reduced when the oxygen saturation of the water fell below 60% (90 mm Hg partial oxygen pressure), however, growth and feed conversion efficiency appeared to be affected when saturation fell below 70%. Rates of food intake and growth of channel catfish held at 26–27°C (78.8–80.6°F) were lower for groups of fish reared in water at 60% saturation (4.9 mg/L) than those held in 100% O_2 saturated water. Among factors that give the better response to environmental pollutants, oxygen is the most important. Raising DO concentrations to saturation can increase the toxic tolerance level of un-ionized ammonia by a factor of 2–3. On the other hand, low DO lowers the tolerance level for NH_3, as well as that of CO_2. Managing DO concentrations is of great importance to accomplish the maximum production potential of any aquaculture system. Most often it will be the first factor limiting the carrying capacity. A desirable goal for the fish culturist is to expose the fish, at all times, to DO concentrations at or near saturation.

Carbon Dioxide

Carbon dioxide (CO_2) is generated by fish as a byproduct of metabolism. Each unit of O_2 consumed generates about 1.40 units of CO_2. As O_2 consumption can be related directly to the feed consumed, so can the production of CO_2. Salmonids require from 200 to 250 g O_2 per kg food, thus 1 kg food generates 280–350 g of CO_2 (127–159 g/lb).

The dynamics of CO_2 in its reaction with water are complex and primarily controlled by pH and alkalinity. It is in an acid-base equilibrium with the total carbonate system ($CaCO_3$). As CO_2 dissolves, it forms a mixture of free carbon dioxide gas (CO_2), which is the toxic portion, carbonic acid (H_2CO_3), and the nontoxic ions bicarbonate (HCO_3^-) and carbonate (CO_3^{-2}).

$$CO_2 + H_2O \rightarrow H_2CO_3 \leftrightarrow H^+ + HCO_3^- \leftrightarrow H^+ + CO_3^{-2}$$

As hydrogen ions are released, the pH of the water is reduced and this, in turn, determines the relative proportions of each compound present and, thus, the degree of toxicity. If the environmental concentration of free CO_2 increases to a level where it interferes with the ability of the fish to excrete CO_2, then CO_2 builds up in the blood (hypercapnia); at the same time, the blood pH is lowered (acidosis). Acidosis can be life-threatening because it affects the capacity of the hemoglobin to bind with oxygen. Conditions of low DO concentrations may aggravate this situation; however, there seems to be no full agreement on this.

A study showed that a DO of 6.0 mg/L can support a moderate swimming activity for salmonids if little or no CO_2 is present; however, in the presence of 30 mg/L CO_2, the DO must be in excess of 11.0 mg/L. This again points out the importance of maintaining DO concentrations near saturation.

In another study, using Coho salmon, virtually no effects were observed at free CO_2 concentrations as high as 40 mg/L and minimum DO concentrations. These fish were kept at a water temperature of 21°C (70°F). Fish may be able to acclimate to CO_2 levels as high as 60 mg/L with little adverse effect. The toxic effect of CO_2 on fish is difficult to predict. Recommendations for maximum levels range from a low of 10 mg/L to a high of 60 mg/L.

In the hatchery environment, the production of CO_2 reduces the pH and this, in turn, reduces the proportion of toxic un-ionized ammonia. This can be especially important at the gill boundary layer, protecting the gill by substantially lowering the pH (up to 1.5 units) in this microenvironment. However, as the ambient CO_2 concentration increases, as is often the case in intensive culture, the proportional increase in CO_2 at the surface of the gill is small and, consequently, that of the pH as well. This can be the reason why it is believed that ammonia is more toxic when ambient CO_2 concentrations are high. Excretion of CO_2 through the gills is in the form of the bicarbonate ion (HCO_3^-) in exchange with the chloride ion (Cl^-).

In a supersaturated DO environment, the fish can obtain adequate oxygen with reduced ventilation rates, but this reduced rate can result in reduced CO_2 excretion causing hypercapnia. Whenever the cumulative oxygen consumption exceeds 20 mg/L, the culturist should consider the need to manage CO_2 by means of vigorous aeration or packed columns (see Chapter 6).

Ammonia

Ammonia (NH_3) is the major end product of nitrogen (protein) metabolism, often representing over 80% of the waste nitrogen. It is excreted through the gill epithelium, but the mechanism of brachial ammonia excretion remains highly controversial. The main mechanisms are thought to be by passive outward diffusion of both NH_3 and ammonium (NH_4^+) down their respective diffusion gradients and excretion via a carrier-linked ionic exchange involving Na^+ with NH_4^+. As soon as NH_3 enters the water, it reacts with the hydrogen ions and changes into an equilibrium mixture of NH_4^+ and the un-ionized NH_3, which is believed to be the toxic form.

$$NH_3 + H^+ + OH^- \leftrightarrow NH_4^+ + OH^-$$

The lower the pH (more hydrogen ions), the lower the proportion (percent) of the toxic NH_3. Lower temperatures also favor the nontoxic ionized form, resulting in less un-ionized ammonia.

Chemical analysis is used to measure the total ammonia nitrogen (TAN), which is the sum of both forms of ammonia present ($NH_3 + NH_4^+$). (Refer to online spreadsheets at *http://www.fisheries.org/publications/epubs/fhm2e/appendices.shtml* for an ammonia calculator.)

The problem for the fish culturist is the extreme variability in the values for safe concentrations that have been obtained in different studies. The variability of ammonia toxicity is the result of interactions with other factors that either increase or decrease the tolerance of fish toward ammonia. These are water quality measurements that are most likely to vary in hatchery operations. Already mentioned is that high DO levels may decrease ammonia toxicity and low concentrations may increase toxicity.

Whether elevated CO_2 levels increase or decrease ammonia toxicity depends on other water quality factors. However, most often, high ambient CO_2 concentrations aggravate the problem because it reduces the gradient at the gill boundary layer.

Various management activities, such as handling and other types of disturbances, (e.g., predators, visitors) can be stressful. The physiological responses involve increased levels of adrenaline in the blood that trigger disturbances, such as depletion of the blood Na^+ and Cl^- ions. These are simple stress responses, the kind most likely encountered in routine aquaculture practices. They result in net ion losses at the gills and, at least temporarily, increase the toxic effects of ammonia. Ammonia toxicity is also related directly to basic, source water quality of which hardness, alkalinity, and pH are the most important factors. These are critical water chemistry variables for stress management in aquaculture. Natural soft water tends to be acidic and low in Ca^{2+}, Mg^{2+}, Na^+, and Cl^- ions. Such conditions are suboptimal for the regulation of ionic balance and osmoregulation. Low pH and low external sodium chloride (NaCl) limit Na^+ and Cl^- uptake, and low Ca^{2+} increases the permeability of the gills. But, in addition to their negative effects on stress, Na^+ also appears to function in the ionic exchange with NH_4^+. The issue of ammonia toxicity must be considered in light of all of these factors, many of which are under the control of the culturist.

As much as possible, the culturist should use the quantitative approach in predicting safe water quality values and synchronize them with maximum production potential and visa versa. This is possible with intensive, flow-through production systems where the rearing environment is impacted by the food introduced to the system. These systems, whether single pass, multiple pass, or circulation, can be quantitatively rated as to how much feed they can process before exceeding predetermined water quality values. This will be the approach to establish carrying capacities and basic facility design concepts for intensive, flow-through systems.

Extensive systems, on the other hand, are more complex, and it is more difficult to manage the water quality. These systems are not only impacted by the feed introduced, but, more, by the complex interactions of biological, chemical, and physical factors, many of which are difficult to control.

Extensive and Intensive Production

Production methods can be classified broadly as extensive and intensive, but it is difficult to determine where extensive methods end and intensive methods begin. Figure 1 illustrates a range from extensive to very intensive production levels.

Extensive Methods of Production

Production in large, natural (earthen) ponds is mostly associated with the extensive method, but such ponds can have their own level of intensity.

Figure 1 shows intensification as fish culture moves from unfertilized ponds to fertilized ponds, ponds with supplemental feeding, ponds with full feeding and aeration (e.g., catfish), and limited water exchange to fast water turnover rates. Carrying capacities increase from a low of 100–200 kg/ha to over 1 million kg/ha. Densities increase from 0.01 to

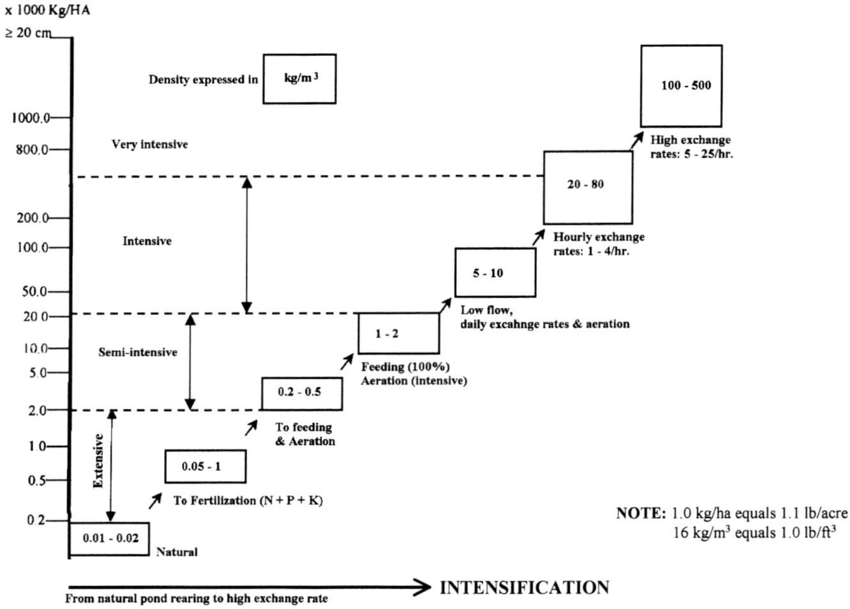

FIGURE 1. Schematic relation between the level of intensification and fish production in kg/ha and density in kg/m³.

0.02 kg/m³ (0.000625–0.00125 lb/ft³), to well over 100 kg/m³ (6.25 lb/ft³). Production systems can also be viewed as "internally" versus "externally" driven. In internally driven systems, the processes taking place within the culture system are the primary determinants of water quality. In contrast, a flow-through system would be considered externally driven because water quality factors are primarily determined by external factors that can be managed by adjustments of water flow, aeration, and feed input. Systems that are primarily internally driven are extensive; externally driven systems are intensive.

Extensive systems are dynamic ecosystems that respond to and interact with physical, chemical, and biological factors. Although weather is a major external factor with its variations in temperature, sunlight, wind, and rainfall, it is beyond the control of the fish culturist. Any and all of these factors trigger physical, chemical, and biological responses—complex internal responses to external influences, including actions taken by the culturist such as aeration, fertilization, and water flows.

Pond water quality studies have increasingly led researchers to conclude that phytoplankton-related processes are the main determinants of water quality. To enhance and direct primary (phytoplankton) production, organic and inorganic fertilizers are added, which can result in a 5- to 10-fold increase in production of fish. However, optimal pond fertilization regimes can be extremely complex due to the dynamics of the intrinsic and extrinsic factors. This will be discussed in greater detail under pond management.

Predicting carrying capacity limits of extensive, internally driven systems is extremely difficult and production often varies from year to year and from pond to pond. Successful high density pond culture can only occur when the culturist can make predictions concerning changes in pond water variables (e.g., pH, DO) in response to minor changes in pond inputs, such as fertilization and feeding rates.

Intensive Methods of Production

Intensive methods of production are primarily externally driven because most water quality factors are influenced by external factors under the control of the culturist. These are flowing water systems that offer much control over water quality through a quantitative approach.

The effects of fish metabolism on water quality can be predicted when the rate of O_2 consumption and NH_3 and CO_2 production are known. To predict this with a high degree of exactness is limited due to the variability of life itself. Also, good data are lacking for new species being cultured. Nevertheless, simple models can be used to determine space and flow requirements, carrying capacities, and facility designs based on the available data. As more precise data become available, the new values can be used in these models.

Successful and efficient intensive terrestrial animal production systems have been developed. These systems depend, in large part, on a precise understanding of the environmental requirements of the species being reared and the ability to maintain the environment through input of clean, well oxygenated air and output of the waste products (CO_2, NH_3, liquids, and solids). Disease organisms are controlled through vaccination, therapeutants, sanitation, and biosecurity measures. To accomplish the same level of success with aquatic species is more complicated. The contrast between the aquatic environment and the terrestrial environment is considerable. However, the worldwide trend in production methods is toward intensification, and increased levels of manageability, predictability, and environmental sustainability are, therefore, required.

Life History Production Phases

Two types of production strategies are commonly used—single batch rearing and multiple batch rearing. Single batch rearing strategies are used primarily by public hatcheries, which rear fish for stocking into natural waters. These hatcheries mimic the natural, once-a-year, spawning cycle of wild fish. In those cases when fish are grown beyond one year, there will be an overlap of two cohorts. Single cohort rearing results in inefficient use of a production facility. Private, for profit production programs aim at the most efficient use of a facility, i.e., the maximum possible output. Multiple batch rearing strategies can double, even quadruple, annual output compared to the single batch rearing approach. In either case, hatcheries are often designed to support different life history phases, each phase having its own rearing facility. Examples of phases are: incubation, hatching, and yolk absorption; feed training and early rearing; fingerling rearing; and final rearing or outgrowing. Providing separate facilities for many phases is expensive, although easier to manage than a mixed cohort rearing program with rearing phases mixed together in a single rearing unit. Examples of the following strategies will be presented: single batch, multiple batch, and mixed multiple batch rearing.

Single Batch Rearing Strategy

Fall Chinook salmon spawn in the fall, smolt at a size of 8–9 cm (3.1–3.5 in) the following spring, and migrate to the ocean or Great Lakes. The hatchery life cycle is only 6 months and consists of two phases: incubation, hatching, and yolk absorption; and early rearing to ±5.0 g, ±8.0 cm. Hatchery facilities for this species are used only for 6 months out of a year.

Coho salmon also spawn in the fall. In most cases, they smolt after about 18 months. Six months prior to smoltification, a new cohort enters the hatchery. This program consists of three phases: incubation, hatching, and yolk absorption; early rearing; and fingerling (smolt) rearing. Early rearing is completed after the previous cohort has been stocked. The hatchery is used more efficiently than is the case for the fall chinook. Rearing multiple species can further improve facility efficiency.

Multiple Batch Rearing Strategy

In a multiple batch rearing strategy where each cohort retains its identity, separate facilities are required for each phase. Ideally, cohorts overlap in harvesting resulting in a more or less continuous output of fish.

Commercial growers of food fish often start their production with fingerlings purchased from fingerling producers, who themselves may start their program with eyed eggs or fry purchased from egg or fry producers. Specialization simplifies facility and equipment requirements, reduces complexity of tasks, enhances specialized professional skills, and results in increased production efficiency.

Mixed Multiple Batch Rearing Strategy

In this rearing strategy, rearing phases are mixed together. Fewer rearing units are required, but frequent sorting of harvest size fish is needed.

Because all cohorts are mixed together within a single rearing unit or pond, new cohorts can be introduced more frequently than with multiple batch rearing strategy where they retain their identity in separate rearing units. It is not practical, economically, to construct facilities that must support many phases.

Mixed cohort rearing can continuously maintain a biomass near a rearing unit's maximum capacity with frequent harvesting and new introductions. This approach is practiced with "cull" harvesting of catfish, allowing an annual production of 11,200 kg/ha (10,000 lb/acre), three to four times the maximum biomass the pond can support. Life history phase rearing will be revisited in the discussion on carrying capacity and facility requirements.

Growth Rates

Fish growth is a most important factor in aquaculture for commercial, as well as public programs. Most of the time it is the goal of the culturist to grow the fish to its maximum growth potential.

Length–Weight Relationships

An increase in fish length provides an easily measured index of growth. Length data are needed for several aspects of hatchery work; for example, production commitments are often specified by length. On the other hand, much hatchery work, such as feed projections, is based on fish weight and its changes. It is very useful to be able to convert back and forth between length and weight without having to make measurements each time. Standardized length-weight conversion tables were developed for this purpose and have been available for many years. These tables are based on the condition factor that is the ratio of fish weight to its length cubed:

$$K = (W_g)/L^3_{\mathrm{cm}} \text{ or } C = (W_{\mathrm{lb}})/L^3_{\mathrm{in}}.$$

The value for a condition factor varies according to how length is measured and, more importantly, according to the units of measurement—metric or English. For the purpose of this chapter, lengths are total lengths (TL), measured from the tip of the snout (or lower jaw, whichever projects farther forward) to the tip of the tail when the tail is spread normally. When measurements are made in metric units (centimeters, grams), the symbol used is K. For English measurements (inches, pounds), the symbol used is C. The two types of condition factors can be converted by the formula $C = 0.03613K$. (Piper et al. 1982 uses length measured in millimeters and reports the conversion factor as 36.13). In either case, the values are quite small. For example, for one sample of channel catfish, condition factors were $K = 80.76 \times 10^{-4}$ (0.008076) and $C = 2918 \times 10^{-7}$ (0.0002918).

The reader should be aware that some authors use length measured in millimeters instead of centimeters. A well fed fish will have a higher ratio than a poorly fed one of the same length; it will be in better condition, hence, the term condition factor.

Each fish species has a characteristic range of condition factors. This range will be small if fish do not change their bodily proportions as they grow (some species do change these proportions, but not the commonly cultured ones). Relatively slim fish, such as trout, have smaller typical condition factors than do stouter fish such as sunfish.

Whenever fish are sampled, their size is most often recorded as weight, not as length. Length in centimeters can be calculated from the following formula (based on the condition factor equation above),

$$L = (W/K)^{1/3}. \quad (1)$$

Once the condition factor is known, the length can be determined from the weight. (Refer to online spreadsheets at *http://www.fisheries.org/publications/epubs/fhm2e/appendices.shtml* for a length-weight calculator.)

Growth Rate

Since fish are cold-blooded animals (poikilotherms) and take on the temperature of their environment, they perform best at their species-specific standard environmental temperature (SET). At this temperature, the best growth rates and feed utilizations can be realized. Growth rates can be expressed as weight gain and length gain. Weight gain is exponential, whereas length gain is linear and relatively constant for a particular temperature.

The exponential growth for weight and linear growth for length occurs during the early life of the fish; as it approaches maturation, more and more of its energy is channeled into reproductive activities. Figure 2 shows how the exponential curve (A) becomes linear (B) and, eventually even asymptotic (C). Most of the time, the hatchery life of the fish covers the time period of exponential growth. The growth of trout, when measured in length units, shows a linear relationship with temperature. This is true for salmonids, in general, for a temperature range between 5 and 15°C. It also is true for other species of cultured fish.

Knowing the potential growth rates of the fish will help in determining rearing space needs, water-flow projections, and production goals. The ability to project the size of the fish in advance is necessary for determining feed orders, egg requirements, and stocking dates. A key principle underlying size projections is that well-fed and healthy fish grow at predictable rates determined by water temperature. At a constant temperature, the daily, weekly, or monthly increment of length is nearly constant for some species of fish during the first few years of

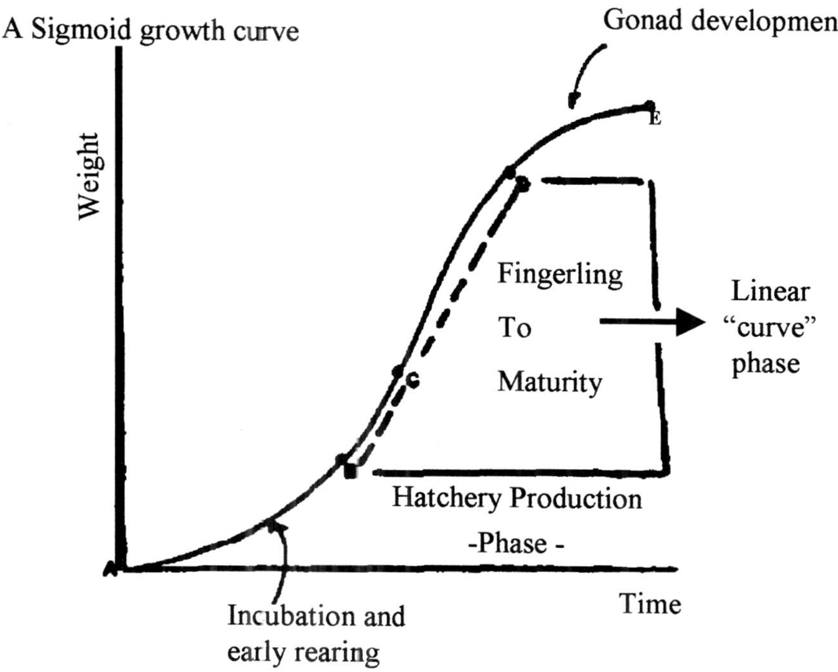

FIGURE 2. Typical growth curve.

life. Carefully maintained production records will reveal this growth rate for a particular species and hatchery.

Example: On 1 November, a sample of 240 fish weighs 5.465 kg. The water temperature is a constant 10.0°C. From past hatchery records, it is known that the fish have a condition factor of 0.0110 and that their average daily growth is 0.056 cm. Will it be possible to produce 20 cm fish by next 1 April?

1. The average weight of fish is 5.465 kg/240 fish = 0.02277 kg per fish = 22.77 g per fish.
2. The average calculated length of the fish on 1 November is

$$L = (W/K)^{1/3};$$

$$(22.77\ \text{g}/0.0110\ \text{cm/g})^{1/3} = 12.71\ \text{cm}\ .$$

3. From 1 November through 31 March there are 151 days
4. The average increase in fish length from 1 November through 31 March is 151 d × 0.056 cm/d = 8.456 cm
5. Average length on 1 April will be 12.71 cm + 8.456 cm = 21.166 cm .

   Yes, 20-cm fish can be produced by 1 April.

Variable Water Temperatures

In the previous example, a growth of 0.056 cm/d was used. If all factors remain constant at the hatchery, growth can be expected to remain at 0.056 cm/d and can readily be projected for any given period of time. However, not all hatcheries have a water supply that maintains a constant temperature from 1 month to the next. Unless water temperature can be controlled, a different method for projecting growth must be used.

The daily length increase (DL) is temperature dependent. Fish grow most efficiently within rather narrow ranges of temperature. Within this range, their growth is quite constant and can be expressed per unit (degree) of temperature. This is the temperature unit growth rate (TUG) expressed as length increase (L_i) in cm/°C. To determine a TUG value the following equation is used:

$$\text{TUG} = L_i/(\text{t} \times {}^\circ\text{C}) \text{ or TUG} = [(W_t/K)^{1/3} - (W_0/K)^{1/3}]/(\text{t} \times {}^\circ\text{C}) , \qquad (2)$$

where L_i is the change in length for time period t, at an average temperature °C, for the period; W_t is the ending weight; and W_0 is the starting weight.

Growth can be projected if water temperature and length (or weight) gain are known for several months.

Example: On 15 April, a sample of channel catfish yields an individual average fish weight of 20 g. The fish are sampled again on 15 May when the average individual fish weight is 50 g. The average daily water temperature has been 22.2°C for this 30-d period. From previous hatchery records it is found that the condition factor (K) is 85.0×10^{-4} (0.0085). Using the above equation for weight, the TUG for this 30-d period can be calculated.

1. $\text{TUG} = [(50 \text{ g}/0.0085)^{1/3} - (20 \text{ g}/0.0085)^{1/3}]/(30 \text{ d} \times 22.2{}^\circ\text{C/d})$.
2. $\text{TUG} = [18.05 \text{ cm} - 13.30 \text{ cm}]/666{}^\circ\text{C}$.
3. $\text{TUG} = 4.75 \text{ cm}/666{}^\circ\text{C} = 0.0071 \text{ cm}/{}^\circ\text{C}$.

The fish are sampled a third time on 15 June and the average weight is 112.7 g. The average daily water temperature has been 25.5°C for this 31-d period.

1. $\text{TUG} = [(112.7 \text{ g}/0.0085)^{1/3} - 18.05]/(31 \text{ d} \times 25.5{}^\circ\text{C/d})$.
2. $\text{TUG} = [23.66 \text{ cm} - 18.05 \text{ cm}]/790.5{}^\circ\text{C}$.
3. $\text{TUG} = 5.61 \text{ cm}/790.5{}^\circ\text{C} = 0.0071 \text{ cm}/{}^\circ\text{C}$.

So these catfish can be expected to grow 0.0071 cm/°C per day.

In August, the average daily water temperature is 29.4°C. If the fish are 26.0 cm on 1 August, their length on 30 August can be calculated:

1. $\text{TUG} = 0.0071$; $t = 30$ d; $^\circ\text{C/d} = 29.4$.

2. $L_i = \text{TUG}\ (t \times °\text{C}) = 0.0071\ \text{cm} \times 30\ \text{d} \times 29.4\ °\text{C/d} = 6.3\ \text{cm}$.
3. 26.0 cm + 6.3 cm = 32.3 cm on 30 August.

The following growth models are offered:

$$W_t = K\,[(W_i\,/K)^{1/3} + (\text{TUG} \times C \times t)]^3\,; \tag{3}$$

$$L_i = \text{TUG} \times t \times °\text{C}\ . \tag{4}$$

To determine the required average temperature to derive a specific weight (or length increase) within a predetermined time period we can use

$$°\text{C} = [(W_t\,/K)^{1/3} - (W_1\,/K)^{1/3}]/(\text{TUG} \times t) \text{ or } °\text{C} = L_i/(\text{TUG} \times t)\ .$$

If you are using English units (°F, pounds, inches) the monthly temperature units (MTU) required per inch of growth must first be determined. Monthly temperature units are the average water temperature for a 1 month period, minus 32°F (the freezing point of water). Thus, a hatchery with a monthly average water temperature of 50°F would have 18 MTU (50°F–32°F) available for growth, To determine the number of MTU required for 1 inch of growth, the MTU for the month is divided by the monthly gain in inches (available from past records).

Consider a hatchery with a water temperature that fluctuates from a low of 41°F in November to a high of 56°F during June. June would have 24 MTU (56°F–32°F), but November would have only 9 MTU (41°F–32°F). Let us assume from past records that the fish in November grew 0.33 in and in June grew 0.89 in. How many MTU are required to produce 1 inch of growth?

1. In November, 9 MTU/0.33 in = 27 MTU/in of growth.
2. In June, 24 MTU/0.89 in = 27 MTU/in of growth.

Once the number of MTU required for 1 inch of growth is determined, the expected growth for any month can be calculated using the equation

$$\frac{\text{MTU for the month}}{\text{MTU required per inch growth}} = \text{monthly growth in inches}$$

Example: From past hatchery records it is determined that 27 MTU are required per inch of growth, and the average water temperature for the month of October is expected to be 48°F. What length increase can be expected for the month of October?

1. The MTU available during the month of October will be 16 (48–32°F).
2. Since 27 MTU are required for 1 inch of growth, the projected increase for October is 0.59 in (16/27):

$$16\ \text{MTU}/27\ \text{MTU/in} = 0.59\ \text{in}\ .$$

If fish at this hatchery were 3.41 in on 1 October, the size can be projected for the end of October. The fish will be 4 in long (3.41 in + 0.59 in).

Monthly variations occur in the number of MTU required per inch of growth; an average value can be determined from past records.

Another common way to measure growth performance is by the specific growth rate (SGR) in percent per day:

$$SGR = 100\ [(\mathrm{Ln}\ W_t - \mathrm{Ln}\ W_1)/t]\ . \qquad (5)$$

Where Ln W_t is the natural logarithm of the weight at time t, and Ln W_1 is the natural logarithm of the initial weight. This reflects the exponential growth in weight typical of young fish.

Example: Using the previous length increase of channel catfish from 26.0 cm on 1 August to 32.3 cm on 30 August, what is its SGR in percent per day?

1. Convert length to weight, using the K-values of 0.0085:

$$W_1 = 26.0^3 \times 0.0085 = 149.4\ \mathrm{g}\ ;$$

$$W_t = 32.3^3 \times 0.0085 = 286.4\ \mathrm{g}\ .$$

2. $SGR = 100\ [(\mathrm{Ln}\ 286.4 - \mathrm{Ln}\ 149.4)/29] = 100\ [(5.657 - 5.006)/29] = 2.24\%/\mathrm{d}$.

There are many factors influencing growth rates; various techniques have been used to increase growth rates, such as manipulating photoperiod regimes, altering salinities, diet improvements, and behavioral modifications. With respect to behavioral changes, salmonids, when exposed to water velocities, orient themselves against the current and form schools. This activity reduces the level of aggression and makes more of the energy input (food) available for growth.

Prolonged exercise induces marked increases in rates of protein synthesis, turnover, and deposition in muscle tissue. It can lead to a dramatic increase in muscle fiber size. Salmonids can also switch to ram ventilation when exposed to velocities of 1 or more BL/s, thus sparing the energy required for "pumping" water across the gills. This, again, constitutes a net energy gain since pumping water across the gills has such a substantial energy cost. For instance, brook trout exhibit better growth rates and feed conversions when reared in water velocities from 1.5 to 2.0 BL/s. These facts have very important implications for design and operational modes of salmonid rearing units. These same principles may apply to many other species as well.

Other factors impacting growth rates are DO, un-ionized ammonia, CO_2, alkalinity, pH, and dissolved and suspended solids. These water quality factors must be managed properly. Of special interest for this chapter are the effects of rearing density on fish growth and health. This will also be discussed in the section covering carrying capacity, because rearing density is also the most important economic factor for an aquaculture facility—the required rearing space.

Fish Feeding and Feed Conversions

As stated earlier, fish grow most efficiently within rather narrow ranges of temperature. Fish perform best at their species-specific SET. At this temperature, the best feed utilization can be realized.

Feed utilization is measured by feed conversion ratios (FC) or feed efficiencies (FE). The first represents food fed divided by weight gain, the latter is its reciprocal, weight gain over feed fed and is expressed as a percent. For example, a feed conversion of 1.4 is the same as a feed efficiency of 71%.

Temperature and fish size determine metabolic rate and these factors are, from a practical fish cultural point of view, the main ones responsible for feed intake (appetite). Depending on temperature and species, daily length increases may be as little as 0.01 cm (0.004 in) or as high as 0.10 cm (0.04 in). Table 1 shows percent weight gains for 1.0 cm fish ($L_1 = 1.0$) and daily length increases (ΔL) from 0.01 cm to 0.10 cm at 0.01 cm intervals ($\Delta L = 0.01$ cm – 0.10 cm).

The percent gain is the new weight (W_t) minus the starting weight (W_1) divided by the starting weight times 100:

$$\%\ \text{gain} = [(W_t - W_1)/W_1] \times 100\ . \tag{6}$$

The percent gain multiplied by FC represents the feed fed as a percent of the original weight (W_1). Feeding levels are commonly expressed in percent body weight or biomass (%BW). The percent gain values of Table 1 also represent feeding levels in percent body weight for feed conversions of 1. For feed conversions below 1, feeding levels are less, but they are greater for feed conversions over 1 for the corresponding length and weight gains of Table 1. The values are based on a starting length (L_1) of 1.0 cm (0.4 in).

To determine percent gain values for other lengths, the values must be divided by those respective lengths. For each 0.01-cm length increase, the percent gain values increase by a value of 3.0. In other words, the percent gain is 3 times the daily length increase times 100 for 1.0-cm fish:

$$\%\ \text{gain/d} = 3.0 \times\ \Delta L \times 100\ . \tag{7}$$

For other lengths (L),

$$\%\ \text{gain/d} = (3.0 \times \Delta L \times 100)/L\ , \tag{8}$$

where ΔL is TUG × °C .

TABLE 1. Percent gain for 1.0-cm fish with daily length increases from 0.01 to 0.10 cm, for two condition factors: 0.010 (W_0 = 0.010) and 0.015 (W_0 = 0.015).

		W_0 = 0.010		W_0 = 0.015	
ΔL (cm)	L_1 (cm)	W_1 (g)	%Gain	W_1 (g)	%Gain
0.01	1.01	0.0103	3.0	0.01545	3.0
0.02	1.02	0.0106	6.0	0.01592	6.0
0.03	1.03	0.0109	9.0	0.01639	9.0
0.04	1.04	0.0112	12.0	0.01680	12.0
0.05	1.05	0.0115	15.0	0.01725	15.0
0.06	1.06	0.0118	18.0	0.01770	18.0
0.07	1.07	0.0121	21.0	0.01815	21.0
0.08	1.08	0.0124	24.0	0.01860	24.0
0.09	1.09	0.0127	27.0	0.01905	27.0
0.10	1.10	0.0130	30.0	0.01950	30.0

Because %BW equals % gain times the feed conversion, the equation for %BW is

$$\%BW = (3.0 \times \Delta L \times 100 \times FC)/L\,. \tag{9}$$

As stated before, there is evidence that the TUG is greatest at SET. Under these conditions, the best feed utilization is realized—the lowest feed conversion and highest feed efficiency.

The product TUG x FC is relatively constant and independent of temperature. Values range from 0.004 to 0.008, depending on species, quality of the rearing environment, and feed quality, as well as other factors impacting appetite and fish health. However, overall, a value of 0.0066 appears to fit most species and diets. This gives a constant of 1.98 or rounded to 2.0 for ease.

The feeding equation is now simplified to

$$\%BW = (2.0 \times {}^{\circ}C)/L\,. \tag{10}$$

There is a need to develop sound, scientific feeding programs because nutritionally enriched and expensive feeds will continue to be partially wasted. Feeding levels should be based on the fish's energy requirements and the best protein to energy ratio of the diet (see also Chapter 8, page 493). Based on these concepts, a feeding standard has been computed for rainbow trout grown at 5, 10, and 15°C (41, 50, and 59°F). These feeding levels are 20–40% less than most feeding tables provided by feed manufacturers. Table 2 compares some of the computed values with the feeding equation proposed for salmonids. From equation (1), $L = (W/K)^{1/3}$; therefore,

$$\%BW = (2.0 \times {}^{\circ}C)/(W/K)^{1/3}\,. \tag{11}$$

TABLE 2. Comparing Cho's (1992) recommended feeding levels for salmonids with Westers's values using %BW = (2.0 × °C) / 100 $W^{1/3}$. The value of 2.0 represents a TUG × FC of 0.00666 × 300 = 2.0.

W gram	5.0°C		10°C		15°C	
	Cho	Westers	Cho	Westers	Cho	Westers
2.5	1.51	1.58	3.22	3.16	5.17	4.74
5.2	1.21	1.25	2.53	2.50	3.96	3.75
9.2	1.02	1.03	2.10	2.06	3.23	3.10
14.9	0.88	0.88	1.80	1.75	2.75	2.64
32.6	0.70	0.67	1.42	1.35	2.14	2.01
40.6	0.66	0.63	1.33	1.25	1.90	1.89
60.5	–	–	1.18	1.10	1.77	1.65
100.9	–	–	1.02	0.93	1.51	1.40
190.4	–	–	0.86	0.75	1.26	1.15
$\overline{X}$	1.00	1.01	1.72	1.65	2.63	2.48
Difference:	>1.0%		<4.0%		<5.7%	

Cho's values are for rainbow trout.
Westers uses a *K* value of 0.01 ($100W^{1/3}$); or, $L = (100W)^{1/3}$

In another case listed in Table 3, feeding levels for catfish, a warm-water species, are compared with equation (11). This table also shows the estimated TUGs for each 15-d period.

Considerable variation will be noticed. To determine the TUG values, equation (2) was used:

$$\text{TUG} = [(W_t/K)^{1/3} - (W_1/K)^{1/3}]/(t \times °\text{C}) .$$

where W_t is the weight in grams at the end of the period, W_1 is the weight at the beginning of the period, t is the number of days, and °C is the average temperature for the period. For catfish, a *K* value of 0.0085 is used.

The TUG value for the entire rearing period of 180 d and an average temperature of 26.7 °C is

$$\text{TUG} = (38.8\text{ g} - 13.3\text{ g})/(180\text{ d} \times 26.7\text{ °C/d}) = 0.0053\text{ cm}^2/\text{d} . \quad (12)$$

If an FC of 1.3 is assumed, the feeding equation based on this TUG value is identical to the one used for salmonids. Conversions in catfish ponds are usually much higher, probably reflecting lost, uneaten feed through overfeeding. During most of the production cycle, the feeding levels reported are greater than those based on equation (11); however, during the early part of the cycle, the levels reported are substantially less!

TABLE 3. Comparing Westers's daily feed requirements in %BW (Eq. 11) with values reported by Lovell for catfish. Also shown are the estimated TUG values for each 15-d period. The growing period is April 15 through October 15 ($K = 0.0088$).

		%BW			
°C	W_g	Lovell	Westers	% ±	TUG
20.0	20	2.0	3.0	+50	0.0061
22.2	30	2.5	2.9	+16	0.0033
25.5	50	2.8	2.8	0	0.0055
26.7	70	3.0	2.6	−13	0.0062
28.3	100	3.0	2.5	−17	0.0048
28.9	130	3.0	2.3	−27	0.0041
29.4	160	2.8	2.2	−21	0.0035
29.4	190	2.5	2.1	−16	0.0077
30.0	270	2.2	1.9	− 4	0.0056
30.0	340	1.8	1.8	0	0.0045
28.30	400	1.6	1.6	0	0.0043
26.1	460	1.4	1.4	0	0.0031
22.8	500	1.1	1.2	+ 9	$\overline{X}$ ± SE

All fish have a maximum growth potential that is observable within a population ("runts" versus "hogs"). Through selective breeding and other manipulations, growth rates of certain species, such as channel catfish, rainbow trout, and Atlantic salmon, have been improved and accelerated. Fish such as these require high feeding levels, therefore, the product of TUG × FC will be greater.

Another feeding program method similar to the above, using English units of length in inches and weight in pounds, is the hatchery constant (HC) method. When water temperature, diet, and species remain constant, all factors in the numerator of equation (9), ($[3.0 \times \Delta L \times 100 \times \text{FC}]/L$) remain constant. Multiplying the numerator factors establishes the HC:

$$\text{HC} = 3.0 \times \text{FC} \times \Delta L \times 100\,. \qquad (13)$$

The daily percent of body weight (%BW) to feed for any length of fish can then be obtained by substituting HC into equation (9):

$$\%\text{BW} = \text{HC}/L\,. \qquad (14)$$

Example: Feed requirements for 10 April need to be calculated. There are 20,000 fish, weighing 2,000 lb, and 6.3-in long on 1 April. The anticipated growth for April is 0.60 in and the FC is set at 1.2.

$$\Delta L = 0.02 \text{ in}$$

$$\text{HC} = 3 \times 1.2 \times 0.020 \times 100 = 7.2\,,$$

where 1.2 is FC and 0.020 is ΔL.

$$\%BW = HC/L = 7.2/6.50 = 1.1\% \ (0.011)$$

$$2{,}200 \text{ lb fish} \times 0.011 = 24.2 \text{ lb of feed required on 10 April for 20,000 fish}$$

The "hatchery constant" method only works for constant water temperatures, while the earlier discussed method can be used for hatcheries with variable temperatures.

New species used in aquaculture, such as yellow perch, are still far removed from domestication; these are still "wild" fish. As relative newcomers to aquaculture, little data on growth rates and feed conversions are available.

Another problem with yellow perch, in addition to its wild characteristic, is that males grow considerably slower then females. Add to this the lack of knowledge of nutritional requirements (optimal diets) and it becomes obvious that much research is still required before this fish species can be as successfully reared as are channel catfish, salmon, and trout.

The same challenge pertains to all newcomers and it is, therefore, very important to establish sound databases, while simultaneously pursuing research that can lead to increased growth rates and domestication of species grown for human consumption.

Table 4 shows recommended feeding levels for yellow perch reared at 22°C, three TUG values, and FC of 1.4. The constants (300 × TUG × FC) are 1.05, 1.47, and 1.89.

Carrying Capacity

Flow-Through Rearing Units, Flow, and Space Requirements

There are several ways to express and determine carrying capacity of intensive, flow-through aquaculture systems. Capacity can be expressed as the maximum allowable weight of fish per unit of flow, loading (Ld); per unit of space, density (D); or as maximum production per year, annual production (AP).

Relationship between flow, space, and exchange rates. The relationship between flow and space requirements is the relationship between loading (Ld) and density (D) with the water exchange rate (R) as the variable.

Loading is expressed in kg/Lpm fish (lb/gpm), density in kg/m^3 fish (lb/ft^3), and exchange rate is in number of water turnovers per hour:

$$\text{Ld} = (D \times 0.06)/R; \quad D = (\text{Ld} \times R)/0.06; \quad R = (D \times 0.06)/\text{Ld}. \quad (15)$$

The constant 0.06 represents m^3 calculated from 1.0 Lpm × 60 min = 60 L × 0.001 m^3/L = 0.06 m^3 and 8.0 represents the English equivalent in ft^3 calculated from 1.0 gpm × 60 min = 60 gal × 0.1337 ft^3/gal = 8.02 ft^3. This

TABLE 4. Recommended feeding levels for yellow perch, reared at 22°C. A condition factor of 0.012 is assumed and a feed conversion of 1.4. Three TUGs are used: 0.0025, 0.0035, and 0.0045, the resulting hatchery constants (HC) are 1.05, 1.47, and 1.89.

		%BW		
		HC 1.05	HC 1.47	HC 1.89
W_g	L_{cm}	TUG 0.0025	TUG 0.0035	TUG 0.0045
10	9.4	2.4	3.3	4.4
20	11.8	1.9	2.6	3.5
30	13.6	1.7	2.3	3.0
40	14.9	1.5	2.1	2.7
50	16.1	1.4	1.9	2.6
60	17.1	1.3	1.8	2.4
70	18.0	1.3	1.7	2.3
80	18.8	1.2	1.7	2.2
90	19.6	1.2	1.6	2.1
100	20.3	1.1	1.5	2.0
110	20.9	1.1	1.5	2.0
120	21.5	1.1	1.5	1.9
130	22.1	1.0	1.4	1.9
140	22.7	1.0	1.4	1.8

relationship equation is extremely useful to establish rational design and operational values for intensive, flow-through, fish production systems.

Loading (flow requirements). The maximum allowable loading depends on the fish's tolerance to the rearing water quality changes caused by metabolic activity. Of particular importance are impacts of DO reductions and NH_3, CO_2, and suspended solids increases. Flow-through systems are externally driven, and water quality factors are determined by external forces, such as feed input, flow rates, and aeration, which are under the control of the culturist. The culturist must make certain that the water quality is compatible with the requirements of the fish throughout the rearing period. This requires a knowledge of these requirements and the impacts on quality by external forces under the culturist's control. Feed is responsible for the negative changes in water quality. How much feed can be added (fed) per unit of flow depends on how much oxygen is required and how much NH_3, CO_2, and solid waste is generated and at what point these components render the water quality unacceptable.

Dissolved oxygen. The oxygen required per kilogram of feed ranges from 200 to 250 g/kg (90–114 g/lb) for salmonids. This is also the range for many other species. Depletion of DO usually is the first limiting water quality factor. The concept of available oxygen (AO) as the difference between inflow DO in mg/L (DO_{in}) and effluent concentrations (DO_{out}) was presented earlier.

Equation (16) can be used to determine the amount of feed that can be fed per Lpm inflow based on available oxygen:

$$Ld_F = AO/O_F, \tag{16}$$

where Ld_F is kg/Lpm feed and O_F is g O_2 required per kg feed per day.

For instance, 1 Lpm at 1.0 mg/L DO delivers 1.44 g of O_2 in a 24-h day (1.0 Lpm × 1.0 mg/L × 60 min/h × 24 h/d = 1,440 mg/d = 1.44 g/d). If 1 kg feed fed per day requires, for example, 250 g of oxygen, 0.00575 kg feed can be fed per Lpm (1.44g/d × 1kg/250 g) per AO.

However, the greatest metabolic activity takes place during a "feeding day" of approximately 17 h, which represents 70% of the 24-h day. In a period of 17 h, 1.0 Lpm at 1.0 mg/L DO delivers 1 g O_2. The use of this measurement results in the more realistic, conservative equation presented above (eq. 16).

For English equivalents:

$$Ld_{Fe} = 3.8\ AO/O_{Fe}, \tag{17}$$

where 3.8 represents 3.8 g of O_2 from 1.0 gpm at 1.0 ppm DO for a 17-h "feeding day" (1.0 gpm × 1.0 ppm × 3.785 g × 60 min/h × 17 h/d = 3860 mg/d = 3.8 g/d).

The feed loading equation can be changed to the fish loading equation (Ld):

$$Ld = (AO \times 100)/(O_F \times \%BW). \tag{18}$$

For example, for an AO value of unity, an O_F value of 250 (O_{Fe} = 114), and a feeding level of 1% body weight, the Ld is 0.4 kg/Lpm (Ld_e = 3.3 lb/gpm). For an inflow DO of 10.0, an effluent DO of 6.0, AO is 4.0, and the maximum allowable loading is 1.6 kg/Lpm (13.2 lb/gpm). The question is how much additional oxygen, if any, can be made available before toxic un-ionized ammonia accumulation renders the water unacceptable. It is assumed that after oxygen, ammonia is the next limiting factor.

Ammonia. To determine ammonia content, we must know how much total ammonia nitrogen (TAN) is generated per kg feed (TAN_F), the percent un-ionized ammonia (%UA), and the allowable maximum concentration of un-ionized ammonia (AUA). The amount of TAN_F depends on protein content of the diet, its composition (protein-energy ratio), and the species reared. Generally it ranges from 20 to 30 g/kg (9.1–13.6 g/lb). In the examples to follow, 30 g will be assumed (TAN_F = 30; TAN_{Fe} = 13.6).

The feed loading equation (eq. 16), $Ld_F = AO/O_F$, indicates that 0.004

kg feed can be fed per Lpm per 1 AO for an O_F value of 250 g/Lpm (0.033 lb/gpm). For 30 g TAN_F, 0.12 g TAN is generated. The concentration of TAN in mg/L (TAN_C), based on the 17-h feeding day is:

$$TAN_C = TAN_F/O_F. \tag{19}$$

This is per 1 AO. The equation must be multiplied by other AO values:

$$TAN_C = AO\ (TAN_F/O_F)\ . \tag{20}$$

The concentration of un-ionized ammonia (UA_C) is TAN_C times %UA (see Table 1) Ammonia Ionization Table values divided by 100:

$$UA_C = (TAN_C \times\ \%UA)/100\ . \tag{21}$$

Finally, a decision must be made regarding the maximum allowable concentration of un-ionized ammonia (AUA). As mentioned already, the problem is the extreme variation in the values for safe concentrations.

There are several water quality factors believed to lessen the toxicity of un-ionized ammonia. To some extent these are under the control of the culturist. Already mentioned is the fact that maintaining DO concentrations near saturation can increase the toxic tolerance level substantially. There also is considerable evidence that an influx of Na^+ can result in an efflux of NH_4^+ through an ion exchange mechanism. This may enable the fish to excrete ammonia against an ambient gradient. Soft water, low in NaCl, can be treated with NaCl. This can be practical with semi-recirculation systems (SRAS), which will be discussed later. Controlling CO_2 levels can also ameliorate the effects of NH_3 through acidification of the gill boundary layer as discussed earlier.

Maximum safe concentrations are, therefore, difficult to define for fish under intensive culture conditions. Maximum concentrations of 0.02–0.03 mg/L are recommended for salmonids reared in flow-through culture systems. Once the values for TAN_F, TAN_C (UA_C), and AUA have been established, it is possible to determine how much oxygen can be made available, the maximum available oxygen (MAO):

$$MAO = AUA/UA_C. \tag{22}$$

Applying equations (18 through 22) with the following assigned values (O_F = 250, TAN_F = 30, %BW = 1.0, AUA = 0.025, %UA = 1.0 and AO = 1.0) produces these results: The feed loading (Ld_F) is 0.004 kg/Lpm; the loading value (Ld) is 0.4 kg/Lpm; the TAN concentration (TAN_C) is 0.12; the un-ionized ammonia concentrations (UA_C) is 0.0012 and the maximum available oxygen (MAO) is 20.8 mg/L.

Using equation (18), the maximum allowable loading for MAO of 20.8 and 1.0% BW is

$$Ld = (20.8\ mg/L \times 100)/(250\ g/Lpm \times 1) = 8.32\ kg/Lpm\ . \tag{23}$$

Recall the relationship of loading with density and exchange rate (eq. 15). How can we best balance these values? For this, we need to select the maximum allowable rearing density (D) and an optimum exchange rate (R) for the rearing unit.

Rearing density (space requirements). In intensive, flow-through aquaculture systems, high water exchange rates permit high rearing densities because of the excellent flushing action. This is analogous to high air turnover rates in the high density production of broilers (chicken), or the opposite, as most of us have experienced at one time or other, the uncomfortable conditions in a room full of people without sufficient air exchange.

Nevertheless, it is difficult to determine safe, optimum rearing densities in a rational, quantitative and qualitative, way, as is possible for loading. There still is a great lack of understanding about optimum rearing densities for a particular species and size. It is of utmost importance that appropriate water quality is maintained, so loadings do not violate the water quality limits.

Loading density relationships. It is important to strike a proper balance between loading (flow), density (space), and the water exchange rate. In the example for loading, a maximum of 8.3 kg/Lpm (69 lb/ft^3) was possible. If it is assumed that these fish can be reared at a maximum density of 80 kg/m^3 (5 lb/ft^3) then, according to equation (15), the exchange rate is a low 0.576 (every 104 min). On the other hand, if the goal is to operate at an exchange rate of 4 per hour ($R = 4$), the resulting density is an unrealistic 555 kg/m^3 (35 lb/ft^3).

If these two criteria, exchange rate ($R = 4$) and density ($D = 80$ kg/m^3) represent the best choices, than the loading is 1.2 kg/Lpm:

$$\text{Ld} = (80 \text{ kg/m}^3 \times 0.06)/4 = 1.2 \text{ kg/Lpm} .$$

A seven-pass serial reuse design would meet the maximum allowable loading of 8.3 (8.3/1.2 = 7). Each pass requires 3.0 mg/L available oxygen (20.8/7 = 3.0), the incoming DO needs to be 9.0 mg/L to meet a minimum effluent DO of 6.0 mg/L.

A seven-pass reuse is equivalent to a 700% use of the water; each 1 Lpm is used seven times over. This can also be stated differently, namely, that each pass requires only 14.3% of the total combined flow rate. Instead of a serial reuse design, it is possible to design an SRAS, which requires no treatment for NH_3. The 14.3% represents the new water inflow, replacing a 14.3% water outflow and 100% of water is continuously recirculated. For example, each of seven rearing units has a rearing volume (RV) of 6.0 m^3 (212 ft^3). To operate at an exchange rate of 4 ($R = 4$), the flow rate (Q_{RV}) must be 400 Lpm (106 gpm). Seven rearing units require a total flow rate of 2,800 Lpm (Q = 2,800 Lpm; 740 gpm). The new water inflow is the 400 Lpm (Q_{new} = 400 Lpm; 106 gpm), and the outflow is the same. The recirculating flow rate is 2,800 Lpm. The

400 Lpm must remove the daily contribution of NH_3, which is 1,008 g/d, representing 2.5 mg/L TAN on average per 17-h day or 1.75 mg/L TAN average during a 24-h period. The concentration of un-ionized ammonia is 0.025 mg/L for 1.0% UA, and meets the predetermined maximum allowable concentration.

The new water inflow of 400 Lpm equates to 576,000 L per day (400 Lpm × 60 min/h × 24 h/d) and the volume of 579 m^3 is equal to 13.7 rearing volumes of 42 m^3. This rates the efficiency of the SRAS as 1,370% (%RV = 1,370)! The percent volume replacement per day (%RV) can be determined with equation (24):

$$\%RV = (Q_{new} \times 144)/RV\ , \qquad (24)$$

where Q represents the inflow of new water in Lpm. For English equivalents,

$$\%\ RV_e = (Q_e \times 1440)/7.48\ gal/ft^3 \times RV\ , \qquad (25)$$

where Q_e represents the inflow of new water in gpm and 7.48 equals gal/ft^3.

In contrast, conventional RAS accomplish daily rearing volume replacements from 5% to 100%. In these systems, ammonia nitrogen (NH_3) is converted, via nitrite (NO_2) to the less toxic nitrate (NO_3) by nitrifying bacteria (biofilters).

In a RAS, as TAN is converted to NO_3, nitrates continue to accumulate in the rearing environment, eventually reaching a maximum allowable nitrate concentration (ANO_3). Once this has occurred, any daily addition must be removed to maintain steady state. This is accomplished by removing the proper volume of water, which must then be replaced with new make-up water. Often, the daily removal of solids represents the required amount of water that must be removed.

For example, at a loading of 1.0 kg/Lpm and a feeding level of 1.0% BW, 0.01 kg feed is introduced per day, generating 0.3 g/d TAN.

Stoichiometrically, 1 g TAN equals 4.2 g NO_3, therefore, 0.3 g TAN generates 1.26 g NO_3.

If the maximum allowable concentration of NO_3 can be 200 mg/L (ANO_3 = 200 mg/L), then, once this concentration is reached at steady state, any daily input of NO_3 must be removed. For the above example, this is 1.26 g. One liter of 200 mg/L NO_3 represents 0.2 g NO_3. If 1.26 g must be removed, then 6.3 L water at 200 mg/L NO_3 must be removed. This represents a certain percentage of the rearing volume. Per 1 exchange per hour ($R = 1$), 1 Lpm exchanges 60 L/h, a rearing volume of 60 L (RV = 60 or 0.06 m^3). Sixty liters at 200 mg/L NO_3 represents 12.0 g of NO_3. To this is added 1.26 g NO_3 per day, which equals 10.5% of the 12.0 g (10.5% of the RV). To maintain steady state at 200 mg/L NO_3, 10.5% of the rearing volume must be removed daily and be replaced with make-up water. This makes the recycle system 89.5% efficient.

The daily percent rearing volume (% RV) replacement for a RAS can be determined with equation (26):

$$\%RV = (Ld \times \%BW \times TAN_F \times 4.2 \times R \times 1{,}000)/(60 \times ANO_3)\,. \quad (26)$$

For English equivalents,

$$\%RV = (Ld_e \times \%BW \times TAN_{Fe} \times 4.2 \times R \times 1{,}000)/(227 \times ANO_3)\,. \quad (27)$$

In this latter equation, 227 is derived from 60 × 3.785 (1 gal equals 3.785 L). For example, for a maximum loading of 1.2 kg/Lpm, a feeding level of 1.0%, TAN_F is 30, *R* is 4.0 and ANO_3 is 500, the percent daily replacement is 20%, and the density is 80 kg/m_3.

Finally, what is the criteria to determine maximum allowable densities? The equation expressing the relationships of loading, density, and exchange rate offers the best approach. Because maximum loading can be determined rationally on the basis of water quality, it leaves density and exchange rate as the "unknowns." Rearing units in intensive culture systems should be operated at exchange rates from 1 to greater than 10 per hour. Circular units should be operated at *R* values of 1–3, and plug-flow units at *R* values from 2–14 (see Table 5). The high exchange rates may result in exceeding the allowable density. In that case, rearing units are either operated in series or as SRAS. Once optimum loading has been determined and the exchange rate decided upon, the density is fixed.

For circular rearing units, loadings can be greater because supersaturated oxygen rapidly blends in with the circulating flow pattern and is quickly mixed into the lower DO concentration rearing water, preventing hyperoxic conditions.

Bioprogramming: carrying capacity, design, and operation. Carrying capacity, design, and operational mode must synchronize, according to the relationship expressed by equation (15).

The equations previously developed can be used to derive rational carrying capacities, designs, and operations for flow-through production systems. Analysis performed for a point in time can be done with a calculator, but when there is the need for many projections over a time period, frequent feed adjustments, for example, calculations are best performed with computer spreadsheet programs.

The sample data presented in Tables 6 A, B, and C involve single batch rearing strategy programs for two species, fall chinook salmon and coho salmon, reared simultaneously. Production, flow, and space values are derived using variables based on previous records, rational assumptions, factual information, and data. The production goal is for 1

TABLE 5. Upper and lower dimensions for raceways, tanks, and troughs and their recommended maximum flow rates in Lpm (gpm) per cross-sectioned area (Lpm/m² and gpm/ft²). Also shown are the corresponding velocities in cm/s (ft/s) and exchange rates (R values).

Rearing units	*L* m (ft)	*W* m (ft)	*D* m (ft)	RV m^3 (ft^3)	*A* m^2 (ft^2)	*v* cm/s (ft/s)	*R*	*Q* Lpm (gpm)	*Q/A* Lpm/m^2 (gpm/ft^2)
Raceways									
Upper	50	5.0	0.8	200	4.0	3.5	2.5	8333	2083
	(164)	(16.4)	(2.6)	(7062)	(43)	(0.11)	2.5	(2201)	(51)
Lower	18	1.8	0.8	26	1.44	3.5	7.0	3033	2312
	(59)	(5.9)	(2.6)	(905)	(14.6)	(0.11)	7.0	(801)	(55)
Tanks									
Upper	12	1.2	0.6	8.64	0.72	2.5	7.5	1080	1500
	(39.3)	(3.93)	(2.0)	(308)	(7.86)	(0.08)	7.5	(285)	(36)
Lower	8	0.8	0.6	3.84	0.48	2.5	11.25	720	1500
	(26.2)	(2.62)	(2.0)	(137)	(5.24)	(0.08)	11.25	(190)	(36)
Troughs									
Upper	6	0.6	0.4	1.44	0.24	1.5	9.0	216	900
	(19.7)	(1.97)	(1.3)	(50.5)	(2.56)	(0.05)	9.0	(57)	(22)
Lower	4	0.4	0.4	0.64	0.16	1.5	13.5	144	900
	(13)	(1.3)	(1.3)	(22.4)	(1.7)	(0.05)	13.5	(38)	(22)

Velocity can be determined with $v = (16.666 \times Q)/A$
Where *v* is cm/s, 16.66 is Lpm which, in one hour, equates to 1.0 m^3
Q is Lpm flow rate and *A* is cross-sectional area in cm^2 (1.0 m^2 = 10,000 cm^2).

million chinook smolts and 700,000 coho smolts. The chinook are grown to 5.12 g (8.0 cm for K = 0.010), they are released in the spring following the fall egg take. Cohos are reared to 35 g (15.2 cm for K = 0.010). They remain in the hatchery 1 year longer than the chinook.

Phase I is incubation, hatching, and yolk absorption. The eggs of both species are collected in the fall and incubated simultaneously. Mortalities during this phase are 40% for chinook and 30% for coho. Heath-tray (16-stack) incubators are used. Phase II is the indoor rearing phase; for chinook this is also the final rearing phase to 5.12 g. The cohos are reared indoors to 7.8 g (9.2 cm). Chinook losses are 20%, coho losses 12%. Phase III is the final rearing phase for the coho. They are reared in outdoor rearing ponds; losses are 3%.

Mortalities are based on previous records. To determine the number of eggs/fish at the start of each phase, the following procedure can be used. Start with the final numbers required for stocking and, for ease, use "per 1,000 fish." Losses for chinook during its final rearing phase

TABLES 6, A, B, AND C. Production, space, flow, and design values for a single batch rearing strategy in three phases for chinook and coho salmon grown to smolt size. Total production goal is for 1 million chinook and 0.7 million coho at 5.12 g (5,120 kg = 11,288 lbs) and 35 g (24,500 kg = 54.013 lbs) final weight, respectively.

TABLE 6A. Phase I: Incubation, hatching and yolk absorption to feeding fry (10/15-12/31)

Value	Chinook	Coho	Source
Number of eggs × 1,000	2,083 –1,250	1,167–817	Mortality records (40% and 30%)
Temp. °C (°F)	8.0 (46.4)	8.0	Records
Time in days (*t*)	75	75	Records
Number of incubation trays	208	117	10,000 eggs / tray
Number of incubation stacks	13	7.3	16-tray stock
Lpm (gpm)	208 (55)	117 (31)	16 Lpm / stock

are 20%. Per 1,000 fish output, the phase must start with 1,250 fish (1,000/0.80) and phase I with 2,083 eggs (1,250/0.60). Thus, for a goal of 1 million smolts, the program must start with 2,083 million green eggs. Coho must start with 817,000 green eggs, based on the mortalities during the three phases. Critical values for water quality, production capability, flow, and space requirements and design are presented in Table 6 along with appropriate equations.

It is important to realize that many of the numbers and values used are "more-or-less" a characteristic of life with its variables. They are, however, rational numbers based primarily on data from salmonids, which have been reared intensively for over 100 years. Unfortunately, there are still gaps in the data. Examples presented have safe concentration values for un-ionized ammonia and carbon dioxide. As discussed earlier, these values are difficult to determine because other water quality values can have a negative or positive influence. Another controversial value is the one for maximum density. What is important is that the process is valid for different numbers and different species reared intensively. The main value of this exercise is to present the culturist with the means to rationally establish critical production, design, and operational values.

After the chinook have been stocked, indoor rearing with coho salmon can continue, as they can also occupy the vacated 10 rearing units, but only 5 units are required to complete the indoor rearing cycle. Final rearing occurs in outdoor ponds, which receive variable temperature water. Winter temperatures go down to a low of 4°C (39.2 °F).

TABLE 6B. Phase II: indoor rearing for chinook and coho. Final rearing for chinook (1/1–5/18).

Value	Chinook	Coho	Source
Number of fish × 1000 start to finish	1250-1000	817-752	Mortality records (20% and 8%)
Temp°C (°F)	8.0 (46.4)	8.0	Records
DO saturation. (mg/L)	11.3	11.3	Temp. and elev. records
pH	7.7	7.7	water quality records
%UA	0.791	0.791	records
t	138	138	1/1–5/18 records
W_1 (g)	0.156	0.156	records
L_1 (cm) (in)	(2.5) (1.0)	(2.5) (1.0)	K = 0.0100 records
TUG (cm)	0.0050	0.0035	records
L_t (cm) (in)	(8.02) (3.15)	(6.36) (2.5)	(TUG × °C × t) + L_1
W_e (g)	5.12	2.57	K = records
FC	1.0	1.0	records
%BW	1.5	1.3	(300 × TUG × °C × FC) + L_t
O_F (g)	220	220	assumed
TAN_F (g)	25	25	assumed
TAN_C (mg/L)	0.114	0.114	TAN_F / O_F
UA_C (mg/L)	0.00089	0.00089	(TAN_C × %UA) / 100
AUA (mg/L)	0.020	0.025	assumed safe
MAO (mg/L)	22	28	AUA /UA_C
MLd kg/Lpm (lbs/gpm)	6.7 (55)	9.8 (82)	(100 MAO) / (O_F × %BW)
DO_{in}–DO_{out}	11.3–6.0	11.3–6.0	selected
AO	5.3	5.3	11.3–6.0 = 5.3
Ld	1.60	1.85	(100 AO) / O_F × %BW)
Number of passes	4	5	MLd/Ld
% Efficiency	412	530	(MLd/Ld) × 100
MBM kg (lbs)	5,720 (11,288)	1933 (4261)	(number of fish × W_t) / 1000
Q_{new} Lpm (gpm)	776 (205)	197 (52)	MBM / MLd
Q_{circ} Lpm (gpm)	3,104 (820)	985 (260)	Q_{nes} × number of passes
D kg/m^3 (lb/ft^3)	40 (2.5)	50 (3.1)	selected
R	1.5	1.6	R = (D × 0.06) / Ld
MRV m^3 (ft^3)	128 (4518)	39 (1377)	MBM / D
RV_{RU} m^3 (ft^3)	12.5 (441)	12.5	selected
Number of RU	10	3	MRV / RV_{RU}

MLd is maximum allowable loading based on maximum available oxygen.
MRV is minimum rearing volume required based on maximum allowable rearing density.
RV_{RU} is rearing volume per rearing unit.
RU is rearing unit.

TABLE 6C. Phase II: continued indoor rearing of coho (5/1-8/30). Phase III: final outdoor rearing of coho (9/1–5/15).

Value	Phase II	Phase III	Source
Number of fish ×1,000	761–45	745–700	Mortality records (4% and 3%)
Temp. °C (°F)	8.0 (46.4)	4–10 (33.8–50)	records
DO_{sat}	11.3	final 10.7	records
pH	7.7	7.9	records
%UA	0.791	1.46	records
t	102	255	records
W_1	2.57	7.8	records
L_1	6.36 (2.5)	9.2 (3.6)	$K = 0.0100$
TUG	0.0035	0.0035	records
L_t	9.2 (3.6)	15.2	(TUG × °C × t) + L_1
W_t	7.8	35	$K = 0.0100$
FC	1.2	1.3	records
%BW	1.0	1.0	records
O_F	220	220	selected (assumed)
TAN_F	25	25	selected (assumed)
TAN_C	0.114	0.114	TAN_F / O_F
UA_C	0.00089	0.0017	(TAN_C × %UA) / 100
AUA	0.025	0.030	assumed safe
MAO	28	17.6	AUA / UA_C
MLd	105	8.0	(100MAO) / (O_F ×%BW)
DO_{in}–DO_{out}	11.3–6.0	10.7–6.0	selected DO_{out}
AO	5.3	4.7	DO_{in}–DO_{out}
Ld	2.4	2.1	(100 × AO) / (O_F ×%BW)
Number of passes	5.0	3.8	MLd / Ld
% Efficiency	529	380	Number of passes × 100
MBM	5,811	24,500 (54,013)	(Number of fish × W_t) / 1,000
Q new	208 (55)	3,062 (809)	MBM / MLd
Q circ.	1,038 (274)	12,248 (3236)	Q new × number of passes
D	60	80	selected
R	1.5	2.3	(D × 0.06) / Ld
MRV	97 (342)	306 (10,811)	MBM / D
RV_{RU}	12.5 (441)	30 (1059)	selected
Number of RU	8	10	MRV / RV_{RU}

The average temperature during outdoor rearing is 8.3 from: °C = [L_t–L_1] / (TUG × t)

Many of the values are dependent on good record keeping and most values will vary somewhat from year to year. With public programs there can be a tendency to hedge with respect to mortality rates and start a program with more eggs than actually required. This can cause a problem with rearing space, because weight increase is exponential and administrators are faced with disposal of surplus fish. On the other hand, shortages may reflect poorly on the ability of the hatchery staff to execute a successful production program, which makes it tempting for a hatchery manager to overestimate the egg requirement.

Several times during a production program, fish culturists take stock of the performance of the fish in order to determine whether or not the program is on target. This requires handling the fish to determine growth performance. At times, total inventories are taken. Handling fish is stressful to them, as well as the culturist. Proper procedures must be followed.

Inventory, handling, and harvest. Managing a fish hatchery properly and efficiently depends on an accurately maintained inventory. Weight data are not only important for the proper management of the hatchery, but are also used by the administrators. Hatchery procedures based on weight data include feed calculations, loadings and densities, and loadings for transportation and for drug applications.

Administrative functions based upon the weight of fish include preparation of annual reports, budgeting, estimating production capability of rearing facilities, monthly production records, feed contracting, and planning for distribution (stocking and processing).

Past record data can be used to project growth for several months with a reasonable degree of accuracy. An inventory is essential after production fish have been thinned and graded, when moved to new rearing units (see phase rearing), or when it is important to provide accurate data.

Fish can be weighed by either the wet or dry method. The wet method involves weighing the fish in a container of water that has been preweighted. Care must be exercised that water is not added to, nor splashed from the container during weighing. This method is used with small fish. Dry weighing is used for larger fish. A dip net is hung from a hook at the bottom of a suspended scale. The scale should be tared to compensate for the weight of the net. Dry weighing reduces fish handling; with a little practice, its accuracy is equal to that of wet weighing. The most common ways to determine inventory weights are the sample count, total weight, and pilot tank weight. In the sample count method, the total number of fish is obtained initially by counting and weighing the entire lot. In subsequent inventories, a sample weight per fish is determined. To calculate the weight of fish in grams, the weight of the sample in kilograms is multiplied by 1,000 (or by 454 for pounds) and divided by the number of fish in the sample. The total weight of fish in the lot is then estimated by multiplying the total number of fish (adjusted for recorded mortality) by the weight per fish in grams divided by 1,000 to convert to kilograms (divide by 454 for pounds).

This method can be inaccurate, but may be the only way to estimate the weight of a group of fish. To assure the best possible accuracy the following steps are suggested:

1. The fish should be crowded and sampled while actively moving and mixing throughout.
2. Once a sample of fish is taken in the dip net, the entire sample should be weighed. This is particularly true if the fish vary in size. The practice of weighing an entire net full of fish will obtain more representative data than that of weighing preset amounts. Relatively light net loads should be taken to prevent injury to the fish.
3. When a fish is removed from the water it retains a surface film of water. For small fish, the weight of water film makes up a larger part of the recorded weight than it does for larger fish. The net full of fish should be carefully drained and the net bottom wiped several times before the fish are weighed.
4. At least five samples should be taken. If there is much discrepancy between samples, more samples should be taken, until there is some consistency in the calculations. The sample values are then averaged and applied to the total lot. Larger samples are required for large fish. Even with care, the sample count method can have an inaccuracy as high as 15–20%.

In the total weight method, as the name implies, all of the fish in a lot are weighed, avoiding sampling error. However, initial sample counting must be conducted to determine the number of fish in the lot. This is done when the fish are small and more uniform in size. It is the most accurate method of inventorying fish, but is not practical with fish grown in ponds. Pond fish may still be sampled, but this is more to determine their condition and health than for inventorying purposes. Samples can also help to adjust the feeding rates. To sample fish in a pond, attract them with bait and then capture them with a lift net, for instance. Usually the sample will be biased because only the more active, aggressive fish will be collected.

More representative samples may be collected with a seine. Data from such samples can be extrapolated to the whole pond if (1) a known area is covered by the seine sweep, (2) if few fish escape the net, and (3) if the population is uniformly distributed throughout the pond. Also, large fish often outrun the pulled seine. It is best to seine several areas of the pond to overcome some of the inherent bias. Generally, pond fish should be sampled when information is needed to check performance and health for management purposes. Handling fish should be kept to a minimum to avoid injury and stress, which can lead to disease or death. Mortalities may well occur after fish are stocked and can go unnoticed.

Water quality should be optimum when handling fish for sampling, grading, or harvesting. Dissolved oxygen concentrations should be kept at or near saturation, because the metabolic rate of the fish may double or triple when they are handled. The use of a 1–3% salt solution during handling and moving fish can compensate for loss of blood electrolytes. Containers should be full of water so they cannot slosh and throw fish against the sides of the container.

Grading to size is one activity that requires considerable handling of fish. The fish must be in good health and condition prior to grading. Grading is a common practice in commercial aquaculture for economic reasons because many fish are sold by size or grade. Grading fish to increase hatchery production by allowing smaller fish to increase their growth rate is questionable. In any fish population there are fish that are small because of their genetic background.

Harvesting involves grading as well because the faster growing fish reach harvest size before the slower growing ones. This is mentioned under phase rearing, where multiple cohort, sequential rearing is practiced and each faster growing fish of one cohort may catch up with the slower growing fish of the previous cohort.

In commercial catfish production, the practice of cull harvest is often applied. Multiple cohorts are reared together in the pond and at 3-week intervals, for instance, large, market-size fish are seined, removed, and replaced with a new batch of small fish. More or less continuous harvesting is practiced, preserving water and obtaining annual productions of up to 11,200 kg/ha (10,000 lb/acre), well in excess of the maximum standing crop or biomass. Harvesting a crop of fish by draining a pond has the advantage of removing the entire crop in a relatively short time. However, there is a substantial loss of water and, as the pond volume becomes smaller, the water becomes shallower, the fish become more concentrated and more vulnerable to predators, and there is the possibility of oxygen depletion.

Some ponds cannot be drained. For instance, the State of Michigan uses over 100 walleye culture ponds statewide, many undrainable. Walleye are reared from 4 to 8 weeks before the zooplankton crashes, and they are stocked as 3–4 d old fry. When conditions permit, seining is used to harvest the fingerlings, but for most undrainable ponds trapnetting is the only practical method for harvest. Fyke or V-traps are set along the shore and copper sulfate crystals, to a concentration of 0.15–0.30 mg/L, are, at times, applied to irritate the fish and increase their movement. Success is variable and frequently fish are not moving enough to allow the traps to intercept a high percentage of them.

Pond Space and Water Requirements

Earthen ponds come in many sizes and shapes and are used for many different species. There is a important relationship between level of management inputs and production capacity, as illustrated below. Five categories are considered.

The first two categories of management are considered extensive pond rearing; the third one could be considered semi-intensive pond rearing—ponds that are not aerated to improve water quality and where oxygen depletion is the limiting factor. The last two categories are considered intensive pond culture—aeration and new water inflow counter the problem of oxygen depletion and shift the limiting water quality factor to the accumulation of ammonia and or nitrite.

In these pond systems, the water quality is dramatically influenced by management techniques, but there is still a substantial effect on water quality by factors beyond the control of the culturist. These are the internal "ecosystem" and weather-related factors. Critical water quality conditions can develop quickly and the response must be fast in order to prevent losses. Additional aeration, water circulation, increased water influx, cessation of feeding, and adding NaCl are some potential countermeasures. Because of the difficulty of predicting the safe carrying capacity of a natural pond, it is important to conduct frequent water quality analyses, especially when fish densities approach critical levels or when natural food supplies start to diminish.

This problem can be especially acute when rearing fry to the fingerling stage with species such as walleye and yellow perch. Because of their extremely small size, they are frequently stocked into fertilized ponds in order to start feeding on microscopic organisms, such as protozoans and rotifers, soon switching to larger zooplankton. Although consumption (numbers) of zooplankton may be low initially, it increases exponentially over time. At the same time, it is important to maintain a sufficient food base of phytoplankton for the zooplankton to consume. During this early, critical feeding phase, a period of 2–3 weeks of cold, cloudy weather can severely limit the required zooplankton bloom and massive mortalities will result due to a lack of food. Thus, the technical aspects of predicting where carrying capacity limits lie are extremely complex.

A natural pond can function as a recirculation system because of the considerable biological activity that occurs at the water-soil interface, including nitrification of ammonia. The major difference between the pond and a RAS, with respect to carrying capacity, relates to available space and to the rearing density.

For instance, a 0.1 ha pond (0.247 acres) with an average depth of 1.0 m (3.25 ft), has a rearing volume of 1,000 m^3 (35,300 ft^3). A 15% volume (150 m^3) replacement per day requires a flow rate of 104.2 Lpm over a

24-h period. If the pond can support 800 kg (8,000 kg/ha), the loading (Ld) would be 7.7 kg/Lpm, but the density (*D*) would be a low 0.8 kg/m^3, and the exchange rate (*R*) only 0.006 (equation 15).

Rearing Unit Management

Water Flow Rates

There is considerable variation in the size and shape of fish rearing units. However, there are only two basic flow patterns: plug-flow and circulating flow. The raceway is a typical representative of a plug-flow rearing unit, so are troughs and linear tanks. These are shaped like channels, often with a length-to-width ratio of about 10:1. Water enters at one end and discharges at the opposite end. The higher the R value, the greater the velocity. Velocities of plug-flow rearing units can be determined with

$$v = (R \times L_m)/36 \,, \tag{28}$$

where v is the velocity in cm/s, L_m, the length in meters, and time, 36 from 3600 s/hr. Because velocity is reported as cm/s and the length is reported in meters, 3600 is divided by 100 (100 cm/m).

For English equivalents,

$$v_e = (R \times L_{\text{ft}})/3{,}600 \,, \tag{29}$$

where v_e is water velocity in ft/s and L_{ft} is length in feet. Raceway velocities are below cleaning velocities and, therefore, they function as settling basins. This is problematic. Fish activity often stirs up accumulated solid waste, causing the fecal material to break apart into smaller fractions that take longer to settle. It makes the interception and removal of solids from the effluent more difficult.

For a raceway to be self-cleaning, it needs a water velocity of 10 cm/s (0.33 ft/s) or more. There are two problems with this. First, a velocity this high requires a very high exchange rate and, as a result, rearing densities are either too great or the raceway is used inefficiently because there is no proper balance between loading (Ld), density (*D*), and exchange rate (*R*) as discussed previously. For instance, a 30 m (100 ft) raceway with a width of 3.0 m (10 ft) and a depth of 0.8 m (2.6 ft) has a rearing volume (RV) of 72 m^3 (2,542 ft^3). A velocity of 10 cm/s requires an exchange rate of 12 according to equation (30):

$$R = (v \times 36)/L_m \,. \tag{30}$$

For a loading of 1.0 kg/Lpm, the rearing density is 200 kg/m^3 (equation 16).

The second problem is a physical one. Foot screens are used to keep fish in the raceway. The screen area is the cross-sectional area of the raceway, which, in this example is 2.4 m². The flow rate (Q) for an exchange rate of 12 is 14,400 Lpm (3,804 gpm):

$$Q = (R \times \mathrm{RV})/0.06\,. \tag{31}$$

The screen area must be able to handle 4,800 Lpm/m² (147 gpm/ft²), well beyond the recommended flows of 2,000 Lpm/m² (50 gpm/ft²) for debris-free water and 1,000 Lpm/m² (25 gpm/ft²) for open water sources that carry debris loads.

Other variables are the types of screen and their percent open area. The most popular screens for raceways are slotted aluminum plates with about 40% open area. The size of the slots is also an important factor in their capacity to "process" flows. Recommended slot sizes for aluminum plates are presented in the following table:

Slot size		Fish size	
cm	inch	g	Number/lb
0.16–0.32	1/16–1/8	< 0.5	to 1,000
0.32–0.64	1/8–1/4	0.5–2.8	1,000 to 200
0.64–1.28	1/4–1/2	2.8–18	200 to 30
1.28–1.92	1/2–3/4	>18	>30

Large screen openings can handle greater flow per area than small ones. In plug-flow rearing units, screen openings usually increase with increasing size of the rearing units in combination with increased fish sizes, from troughs to tanks to raceways.

Table 5 is based on the assumption that raceways can be operated at flow rates of about 2,000 Lpm/m² cross-sectional area (±50 gpm/ft²), tanks at 1,500 Lpm/m² (±37 gpm/ft²), and troughs up to 900 Lpm/m² (±22 gpm/ft²). These flow rates translate into velocities of 3.5, 2.5, and 1.5 cm/s, respectively, and, depending on the size of the various units, into exchange rates of 2.5–7.0 for raceways, 7.5–11.25 for tanks, and 9.0–13.5 for troughs. All three types of rearing units have an assumed length-to-width ratio of 10–1 with water depths of 0.8 m (2.6 ft) for raceways, 0.6 m (2.0 ft) for tanks, and 0.4 m (1.3 ft) for troughs. Although the designation "raceway," "tank," and "trough" and their size ranges and dimensions are somewhat arbitrarily chosen, they do represent examples of desirable upper and lower ranges of these plug-flow rearing units. When operated at the suggested velocities, the hydraulic characteristics will be conducive to the creation of a healthy rearing environment. As discussed earlier, high maximum AO values result in high loading values. In order to accomplish the required balance between Ld, D, and R (equation 15), raceways may have to be arranged in serial reuse fashion.

Round tanks are the most common representatives of circulating-flow rearing units. Circular tanks can be self-cleaning at proper rotational velocities; the overall rate of flow exiting the central drain in the bottom is aided by the swimming motion of the fish. The cleaning flow can be of relatively low volume, which has resulted in designing round tanks with dual drains, with the bottom drain discharging 5–20% or more of the total flow, carrying with it the solids, while the remainder of the flow is discharged at an elevated drain. This solid-free water can then be recirculated and mixed with 5–20% or more new water. This is the principle of the SRAS described earlier. Plug-flow rearing units are better suited for serial reuse; round tanks, because of their self-cleaning ability with minimal flow rates, can be incorporated into an SRAS.

Finally, round tanks can be operated at selected water velocities, largely independent of the rate of water inflow (R value) analogous to the circulation of air in crowded, closed spaces. Water velocities of 0.5–2.0 times BL/s can benefit fish health and growth. The flow through a round tank does not have to be greater than what is required to support a selected biomass provided the water inlet and outlet structures are designed properly.

The advantages of round tanks over plug-flow rearing units:

1. They can be self-cleaning, with a small fraction of the total flow removing solids.
2. Because they can be self-cleaning, the solid-free water can be recirculated, the percentage depending on water quality criteria discussed earlier.
3. Loadings can be high because oxygen supersaturated water can be introduced as it mixes instantly with the rearing water. As a matter of fact, it is not difficult to maintain DO concentrations near 100% at all times throughout the rearing unit, a very important feature. This is not possible with plug-flow rearing units.
4. High loadings can still result in relatively high rearing densities (economic use of rearing space) despite relatively low R values, yet velocities can be much greater than those attainable with plug-flow rearing units.
5. Feed distribution is easier, round tanks require fewer feeding stations.

The major disadvantage of round rearing units relates to fish handling. Fish cannot be "cornered." Special crowding screens can be designed, however. Also, floor space use is not as economical as with plug-flow units. The exception is the square, Swedish circulating rearing tank with rounded corners on the inside.

Sanitation

Opportunistic, as well as obligate pathogens are naturally present in the environment. Many of the opportunistic pathogens that can cause problems in intensive fish culture occur naturally in soil. Examples are columnaris, bacterial gill disease, and coldwater disease organisms. Soils can be readily transferred by footwear. Foot baths placed at strategic locations can be effective in preventing the transfer of these and similar potential pathogens into the fish-rearing area. Airborne (aerosol) transmission can also occur. Washing hands and arms with a disinfectant soap can reduce the chance of bringing potential pathogens into contact with the rearing environment. These methods, along with meticulous husbandry practices can substantially reduce the potential for diseases caused by opportunistic pathogens. The most effective way to prevent diseases caused by obligate pathogens is by preventing these organisms from invading the hatchery. These pathogens depend on fish to survive. A fish-free, or specific-pathogen free, water source is ideal. If not available, the water source can be treated with either ozone or ultraviolet radiation, or both. In addition to the above mentioned recommended measures, it is of great importance to prevent introduction of obligate pathogens. Only "disease-free" certified eggs should be used. Do not allow any potentially contaminated material to enter the facility. Even certified eggs should be disinfected with an iodophor solution, and all utensils must be disinfected routinely in a chlorine solution. Culture activities should be strategically scheduled, minimizing the number of personnel working with a particular group of fish.

Despite all reasonable efforts, serious disease outbreaks can occur. These may require decontamination of all equipment and rearing units. The most effective way is a solution of 200 mg/L active chlorine. Chlorine gas is an irritant and is poisonous, therefore, disinfection must take place in a well-ventilated area. High test hypochlorite (HTH) is a powder that contains either 15, 50, or 65% available chlorine. To produce a solution that has 200 mg/L available chlorine:

For 15%, add 1.4 g/L (2 oz per 10.5 gal) ;

For 50%, add 0.4 g/L (1 oz per 18 gal) ;

For 65%, add 0.32 g/L (1 oz per 23 gal) .

Add 10 mL of commercial household bleach per liter of water (35 mL/gal) to obtain 200 mg/L available chlorine. Chlorine can be neutralized with sodium thiosulfate. Seven mg of sodium thiosulfate neutralizes 1 mg of chlorine. One liter of 200 mg/L available chlorine is neutralized by 1.5 g of sodium thiosulfate. Chlorine levels are reduced by organic matter. For full effectiveness, it is necessary to thoroughly clean equipment before it is exposed to the disinfecting solution. An exposure time of 30–60 min is required for full effectiveness.

Quaternary ammonia compounds (Roccal; Hyamine 1622; Hyamine 3500) are disinfectants that have also been used as antiseptics. Prolonged immersion of nets and other utensils in a solution of 5 mL/L Roccal (19 mL/g) is quite effective. However, chlorination is a better overall disinfectant than the quaternary ammonium compounds, but chlorine will destroy netting, rapidly corrode metal, and damage many plastics. Eyed eggs can be disinfected with an iodophor solution. Only pathogens on the surface of the eggs will be killed. A concentration of 100 mg/L of active iodine is recommended. In poorly buffered water (<50 mg/L total alkalinity), add 1 g/L sodium bicarbonate (3.8 g/gal) because the iodine will lower the pH and kill the eggs. Fresh solution is brown to amber in color, but after batches of eggs have been treated, the solution fades to yellow, indicating that it is no longer a potent disinfectant and should be discarded.

Unless husbandry is practiced meticulously with regard to cleanliness at all times, all efforts to rid the system of obligate pathogens may well be in vain.

Pond Management

Pond management pertains to the management of earthen ponds, ponds that serve the dual role of providing natural feed and rearing space for the fish. Because of the complex interactions in the pond's ecosystem, it is much more difficult to apply a quantitative approach as described for intensive, flow-through fish production systems. In a pond, the impact of fish metabolism on water quality is minor compared with that of phytoplankton, which serves as the primary food base. The pond must be prepared properly and free of competing fish. To make it a good feed (fish) producer, the right kinds and amounts of nutrients must be available.

Preparation

Proper management of earthen ponds begins before water is introduced. It is advisable to dry and disk ponds to promote aerobic breakdown of the nutrient-rich bottom sediments. Although some nutrients are desirable for fingerling culture, to promote algal growth that zooplankton graze on, an overabundance of nutrients tends to produce more undesirable blue-green and filamentous algae. Relatively new ponds with little buildup of organic material, or those with sandy, permeable bottoms, which allow nutrients to escape to the groundwater, are less likely than older or more impermeable ponds to require drying and disking. They may actually leak if the bottom is disturbed, and it may be necessary to compact the bottom soil rather than disk them.

If a pond is to remain dry for several months, it should be seeded around the edges with rye grass at 20–25 kg/ha (8–10 lb/acre). This cover prevents erosion of pond dikes; it can be flooded in the spring to serve as a source of organic fertilizer. The grass should be cut and partially dried before the pond is reflooded or its rapid decay in water may deplete the dissolved oxygen.

Application of 1,000 kg/ha (890 lb/acre) of agricultural lime during the fallowing period, followed by disking, may improve the buffering capacity of a soft-water pond. Fertilizers are often spread on the pond bottom prior to filling; nuisance vegetation may also be sprayed at this time.

Wild Fish Control

Wild fish must be kept from ponds when they are filled, as they compete with cultured species for feed, complicate sorting during harvest, may introduce diseases, or confound hybridization studies. Proper construction of the water system and filtration of inlet water can prevent the entrance of wild fish and their eggs. This may require filtering the incoming water. It is recommended passing water through a 0.25 mm (0.01 in) mesh microstrainer or sand filter to remove eggs of fish such as gizzard shad or common carp. Ponds can also be filled first and then treated with chemicals to kill wild fish. Rotenone is relatively inexpensive and is registered and labeled for this purpose. It should be applied to give a concentration of 0.5–2.0 ppm throughout the pond. Rotenone does not always control some fishes, such as bullheads and mosquitofish, and it requires up to 2 weeks to lose its toxicity in warm water and even longer in cold water. However, 2–2.5 ppm potassium permanganate ($KMnO_4$) can be added to detoxify rotenone.

Antimycin A is a selective poison that eliminates scaled fishes in the presence of catfish. It does not kill bullheads, however, which are undesirable in most ponds and especially in channel catfish ponds. The chemical varies in activity in relation to water chemistry and temperature; the instructions on the label must be closely followed. Expert advice should be sought in special cases.

Chlorine in the form of HTH, used at concentrations of 5 ppm for as little as 1 hour, will kill most wild species of fish that might enter the pond. Chlorine deteriorates rapidly and usually loses its toxicity after 1 day at this concentration. Chlorine is a nonspecific poison and will kill not only the fish, but most of the other organisms in the pond as well.

Fertilization

The production of phytoplankton, as the primary producer (i.e., the base of the food pyramid), is the key to successful fish production in ponds. A large number of nutrients are required to stimulate phytoplankton growth. Most of the micronutrients (e.g., iron (Fe), sulfur (S), magnesium (Mg), copper (Cu), aluminum (Al), selenium (Se), zinc (Zn), etc.) are naturally present, but macronutrients such as nitrogen (N), phosphorus (P), potassium (K), and carbon (C) may need to be supplied on a regular basis. In most waters, carbon, in the form of CO_2, is adequate. Potassium requirements are either nonexistent or very limited. The soil (mud), when acidic, may need to be limed by adding calcium compounds, to provide for the proper pH and the required alkalinity level. Commonly used compounds are quicklime (CaO), hydrated lime ($Ca(OH)_2$), agricultural lime ($CaCO_3$), or dolomite ($CaMg(CO_3)_2$).

Most fertilization regimes involve N and P. Under, theoretically, optimal conditions a N:P ratio of 10:1 is recommended, but, depending on local situations, this ratio is extremely variable. Natural cycles of N and P differ even from pond to pond. Optimum pond fertilization is complex because of the dynamics of the pond's ecosystem. The average nutrient composition of phytoplankton biomass is approximately 45–50% C, 8–10% N, and 1% P, giving the typical C:N:P ratio of about 50:10:1. Two types of fertilizers are available for pond fertilization—organic and inorganic.

Organic fertilizers. Organic materials such as composted plant residues, manure, stable drainage, slaughterhouse waste, and municipal sewage are very good sources of nitrogen. They also contain a large percentage of organic carbon, as well as other minerals in small amounts. Typical analyses are shown in Table 7. Values may vary slightly depending on the conditions under which the crops were grown or the products were processed.

Organic fertilizers are recommended for new or sterile ponds. Their use is limited by cost and labor requirements for application. The advantages of organic fertilizers: (1) a shorter cycle for plankton production than inorganic fertilizers, (2) decomposition to liberate CO_2, which is used by plants for growth, (3) aid in clearing silt-laden waters, and (4) use as a supplemental feed. Their disadvantages: (1) more expensive than inorganic fertilizers, (2) may deplete the oxygen supply, (3) may stimulate filamentous algae growth, and (4) require more labor to apply than inorganic fertilizers.

Inorganic fertilizers. Inorganic fertilizers are relatively inexpensive sources of N, P, and K (which stimulate algal growth); and Ca (which helps to control water hardness and pH). In nitrogen-free water, 0.3–1.3 mg/L N must be added to stimulate phytoplankton growth. To sustain this growth, about 1 mg/L must be applied at weekly intervals. In a normal hatchery pond, this comes to about 9 kg/ha N (8 lb/acre). Because nitrogen can enter the pond system from the atmosphere,

Table 7. Composition of several organic fertilizer materials. (Source: Snow 1957)

Fertilizer	Nitrogen	Phosphorus	Potassium	Protein	Carbohydrate
Alfalfa hay	2.37	0.24	2.05	14.8	33.5
Grass hay	1.12	0.21	1.20	–	–
Peanut vine hay	1.62	0.13	1.25	10.1	38.5
Cottonseed meal (36%)	5.54	0.83	1.22	34.6	24.5
Cottonseed meal (43%)	7.02	1.12	1.45	43.5	15.8
Fish meal	10.22	2.67	0.40	63.9	2.1
Peanut meal	6.96	0.54	1.15	43.5	31.3
Meat scrap	8.21	5.15	–	51.0	3.5
Soybean oil meal	7.07	0.59	1.90	44.2	29.0
Horse manure	0.49	0.26	0.48	–	–
Cow manure	0.43	0.59	0.44	–	–
Chicken manure	1.31	0.40	0.54	–	–
Sheep manure	0.77	0.39	0.59	–	–
Cladophera sp.	2.90	0.32	–	–	–
Potamogeton sp.	1.30	0.13	2.08	–	–
Najas flexilis	1.90	0.30	2.19	–	–
Chara spp.	0.70	0.27	0.58	–	–
Wood yeast (Torula)	6.9[a]–8.6	0.82–1.96	–	43.54	37.4–43.9
Green Italian rye grass	0.50	0.09	0.40	3.1	11.5
Green rye	0.42	0.10	–	2.6	12.9
Green oats	0.42	0.09	0.50	2.6	13.5
Green vetch	0.67	0.07	0.41	4.2	8.1
Green, white clover	0.82	0.09	0.38	5.1	6.6

[a]Calculated by dividing protein content by 6.25.

watershed, and decomposing organic matter, it is not always necessary to add more.

For the operation of warmwater hatchery ponds, it is recommended that N be included in the fertilizer applications during the late spring and summer months for all ponds except those that have been weed-free for at least 3 years. If development of phytoplankton is delayed longer than 4 weeks, N should be added. Forms of N available for pond fertilization are listed in Table 8.

Phosphorus is an active chemical and cannot exist alone except under very specialized conditions. It is generally considered to be the most essential single element in pond fertilization and the first nutrient to become a limiting factor for plant growth. Plankton require from 0.018 to 0.09 mg/L as a minimum for growth. Several workers have recommended applications of about 1.0 mg/L phosphorus pentoxide (P_2O_5) periodically during the production season.

Phosphorus will not exist for long in pond water solution. Although

TABLE 8. Nitrogen fertilizers for pond enrichment.

Source material	Chemical formula	Percent nitrogen	pH of aqueous solution
Ammonium metaphosphate	$(NH_4)_3PO_3$	17[a]	
Ammonium nitrate	NH_4NO_3	33.5	4.0
Ammonium phosphate	$(NH_4)_3PO_4$	11[b]	4.0
Ammonium sulfate	$(NH_4)_2SO_4$	20	5.0
Anhydrous ammonia	NH_3H_2O	82	
Aqua-ammonia	NH_3H_2O	40–50	
Calcium cyanamide	$CaCN_2$	22	
Diammonium phosphate	$(NH_4)_2HPO_3$	21[c]	8.0
Urea	H_2HCONH_2	46	7.2
Sodium nitrate	$NaNO_3$	16	7.0

[a] Also contains 73% P_2O_5
[b] Also Contains 48% P_2O_5
[c] Also contains 48-52% P_2O_5

both plants and animals remove appreciable amounts of the added phosphate, the majority of applied phosphorus eventually collects in the bottom mud. Here, phosphorus may be bound in insoluble compounds that are permanently unavailable to plants. Some 90–95% of the phosphorus applied to field crops in fertilizers becomes bound to the soil and the same may hold true in ponds.

Several phosphate fertilizers are available for use in ponds. Ordinary superphosphate is readily available commercially and is satisfactory for pond use. More concentrated forms may save labor in application, however. Sources of P_2O_5 in commercial phosphate fertilizers are listed in Table 9. When N also is desired, ammoniated phosphates are recommended as they are completely water-soluble and generally should give a more rapid response. An application rate of 9 kg/ha P_2O_5 (8 lb/acre) is normal in pond fertilization. This amount supplies about 1 mg/L in a pond averaging about 1.0 m deep. In the United States, the usual practice is to supply the needed P periodically throughout the growing season.

In some cases, fish production can be increased 25–100% by adding lime at the rate of 2–3 metric tons per ha (2–3 tons/acre). Calcium is available in three principal forms. It is 71% of calcium oxide (CaO) or quicklime, 54% of calcium hydroxide ($Ca(OH)_2$) or hydrated lime, and up to 40% of calcium carbonate ($CaCO_3$) or ground limestone.

Waters softer than 10 mg/L total hardness generally require applications of lime, whereas waters harder than 20 mg/L seldom respond to liming. The need for lime may be indicated when inorganic fertilization fails to produce a substantial plankton bloom. However, analysis of the water or, preferably, of the bottom mud, should be made for total hard-

TABLE 9. Sources of P_2O_5 in commercial phosphate fertilizers.

Source material	Chemical formula	% P_2O_5	Availability
Ammonium metaphosphate	$(NH_4)_3PO_3$	73	Variable solubility; has 17% nitrogen
Basic slag	$(CaO)_5P_2O_5SiO_2$	9	Poor in calcium-rich waters
Bone meal		15	Not readily available
Calcium metaphosphate	$Ca(PO_3)_2$	60–65	Equal to superphosphate in acid and neutral soil
Defluorinated rock	$Ca(PO_4)_2$	41.3	Used primarily in livestock feeds; insoluble in water
Diammonium phosphate	$(NH_4)_2HPO_3$	53	Completely water- soluble has 21% nitrogen
Enriched superphosphate	$Ca(H_2PO_4)_2$	32	About the same as ordinary superphosphate
Monoammonium phosphate	$NH_4H_2PO_4$	48	Completely water-soluble in form of amonophos-phate; has 11% nitro-gen
Ordinary superphosphate	$Ca(H_2PO_4)_2$	18–20	Not completely water-soluble
Phosphoric acid	H_3PO_4	72.5	Water-soluble and acid in reaction
Potassium metaphosphate	KPO_3	55–58	Equal to or superior to ordinary superphos phate; has 35–38% K_2O
Rock phosphate	$(Ca_3(PO_4)_2)_3CaF_2$	32	Least soluble of calcium salts; availability varies from 0–15%
Triple phosphate	$(Ca_3(PO_4)_2)_3$	44–51	A major portion is water soluble

ness and alkalinity before lime is applied. A state agricultural experiment station or extension service can assist with these (and other) analyses.

Liming can be done with the pond either dry or filled with water. Suitable mechanical equipment is needed to assure uniform dispersion. If the pond contains water, additional amounts of lime may be added to satisfy the needs of the water, as well as of the bottom mud. It may take 3–6 months before the pond responds. In some situations, limed ponds revert to an acid condition within 2 years after the initial application.

Combining fertilizers. In making a decision on whether to use organic or inorganic fertilizers, the advantages and disadvantages should be carefully considered. Comparative tests have been attempted, but conclusive answers as to which material is best often will depend upon the individual situation or production cycle involved.

Combining organic and inorganic fertilizers is a common practice. Many culturists have found that a combination of an organic meal and superphosphate, in a ratio of 3:1, gives higher fish production than the organic material alone. In hatchery rearing ponds where draining is frequent and time for development of a suitable food supply often is limited, combining organic and inorganic fertilizers appears to be advantageous. Although the cost of such a procedure is greater than with inorganic fertilization, the high value of the fish crop involved normally justifies the added cost, particularly in the case of bass and catfish rearing.

The ratio of 4:4:1, N_2:P_2O_5:K_2O, is needed to produce favorable plankton growth for fish production ponds. The fertilizer grade most commonly used is 20–20–5, which gives the 4:4:1 ratio with relatively little filler.

The type of fertilizer program chosen will be determined by such factors as species of fish reared, time of year, cost, availability of product, and past experience. If the species to be reared is a predator such as largemouth bass, striped bass, or walleye, a typical program might be as follows.

In spring, while the pond is still dry, disk the pond bottom. Apply lime, if needed, to bring the pH into a favorable range. The fertilizer can be spread on the dry pond bottom and the pond then filled, or the pond is filled and then the fertilizer spread; the following example assumes it is on the pond bottom.

Spread 500 kg/ha (446 lb/acre) chopped alfalfa hay, 200 kg/ha meat scraps, 200 kg/ha ground dehydrated alfalfa hay, 50 kg/ha superphosphate, 10 kg/ha potash, and 1,000 kg/ha chicken manure. Fill the pond and wait 3–5 d before stocking fish.

This fertilizer program for sandy loam soils and slightly acid waters will produce an abundance of zooplankton needed for rearing the predator species. Usually this amount is added only one time and will sustain the pond for 30–40 d. If the fish crop is not of harvest size by that time, a second application of all or part of the components may be needed.

If the species to be reared is a forage species such as bluegill, redear sunfish, goldfish, or tilapia, the following program might be used: 100 kg/ha ammonium nitrate, 200 kg/ha superphosphate, 50 kg/ha potash, 100 kg/ha chopped alfalfa, and 300 kg/ ha chicken manure. This fertilizer program will produce more phytoplankton than the one outlined for predators. As with the one above, this program will have to be repeated about every 30–45 d. The type of fertilizer program that works best at any particular station will have to be developed at that station. The program that works best at one station will not necessarily work well at another. The examples given above are strictly guidelines.

A successful fertilization regime was used for the production of 3–4 cm walleye fingerlings in an Ohio hatchery. Liquid inorganic fertilizers were added once a week to achieve an inorganic N:P ratio of 20:1 by weight, but with no more than 30 mg/L phosphate to discourage the growth of large filamentous green algae and blue-green algae. Instead, it is the goal to produce single cell algae such as diatoms, coccoid greens, and flagellates to serve as food for the right kind of zooplankton, cladocerans, and copepods, which serve as food for walleye.

Because consumption of zooplankton by walleye increases exponentially with time, frequent monitoring is mandatory. Weekly algal counts, secchi disk readings, fish growth performance, N and P measurements, along with regular checks on pH and DO, all can contribute to a successful production program. For the Ohio hatchery, a 10-fold increase in fingerling production was realized and variation in survival and growth was reduced, thus production became much more predictable.

Fertilization protocols for ponds must be tuned to specific conditions of climate, local water and soil characteristics, and species production goals. It must not be a hit and miss approach, but rather a carefully applied quantitative and qualitative effort.

Vegetation Control

Aquatic plants must have sunlight, food, and carbon dioxide in order to thrive. Elimination of any one of these requirements inhibits growth and eventually brings about the death of the plant. The majority of the common water weeds start growth on the bottom of the pond. Providing adequate depth to ponds and thus excluding sunlight essential to plant growth may prevent weeds from becoming established. Water plants are most easily controlled in the early stages of development. Control methods applied when stems and leaves are tender are more effective than those applied after the plant has matured. In most cases, seeds or other reproductive bodies are absent in early development and control at this time minimizes the possibility of reestablishment.

The first step in controlling aquatic vegetation is to identify the plant. After the weed has been identified, a method of control can be selected. Control methods may be mechanical, biological, or chemical, depending upon the situation.

Biological weed control is based on natural processes. Exclusion of light from the pond bottom by adequate water depth and turbidity resulting from phytoplankton is one method. Production of filamentous algae that smother submerged, rooted types of weeds is another.

The most inexpensive form of weed control for many ponds is prevention through the use of fertilizers. When an 8-8-2 grade fertilizer is applied at a rate of 100 kg/ha (89 lb/acre), every 2–4 weeks during the warm months of the year, microscopic plants are produced that shade the bottom and prevent the establishment of weeds. Although up to 14 applications may be needed each season, fish production is increased along with the weed control achieved. Generally, most aquatic weeds may be controlled by fertilization in properly constructed ponds. However, such a program of fertilization will be effective in controlling rooted weeds only if the secchi disk reading already is 45 cm (18 in) or less.

Winter fertilization is a specialized form of biological control effective on submerged rooted vegetation if the ponds cannot be drained. An 8-8-2 grade fertilizer or equivalent is applied at a rate of 100 kg/ha (89 lb/acre) every 2 weeks until a dense growth of filamentous algae covers the submerged weed beds. Once the algae appears, an application of fertilizer is made at 3–4 week intervals until masses of algae and rooted weeds begin to break lose and float. All fertilization is then stopped until the plants have broken free and decomposed. This will start in the late spring and generally takes from 4 to 6 weeks. Phytoplankton normally replaces the filamentous algae and rooted weeds and should be maintained by inorganic fertilization with 100 kg/ha (89 lb/acre) of 8-8-2 grade fertilizer applied every 3–4 weeks.

Lowering the water level of the ponds in the late fall has been helpful in achieving temporary control of weed watershield. This practice also aids in the chemical control of alligator weed, water primrose, southern water grass, needlerush, knotgrass, and other resistant weeds that grow partially submerged and have an extensive root system.

Plant-eating fish that convert vegetation to protein have been considered in biological control. Among these are grass carp, Israeli carp, and tilapia. Experiments have indicated that the numbers of Israeli carp and tilapia required to control plants effectively are so large that these fish would compete for space and interfere with the production of other, more desirable species.

Extensive development of herbicides makes chemical control of weeds quite promising in may instances. When properly applied, herbicides are effective, fast, relatively inexpensive, and require less labor than some of the other control methods. Chemical control, however, is not a simple matter. Often the difference in toxicity to weeds and to fish in the pond is not great. Some chemicals are poisonous to humans or to livestock, and they may have an adverse effect on essential food organisms. Decay of large amounts of dead plants can exhaust the oxygen supply in the water, causing the death of fish and other aquatic animals. It is essential that discretion regarding treatment be followed if satisfactory results are to be obtained.

An important aspect of vegetation control is the rate of dilution of applied herbicides and the effect of substances present that may neutralize the toxicity of the chemical used. The rate of water exchange by seepage or outflow and the chemical characteristics of the water and pond bottom also affect the success of chemical control measures. Often the herbicide must reach a high percentage of the plant surface before a kill is obtained; the chemical must be applied carefully if good results are to be achieved.

The herbicide is applied directly on emergent or floating weeds and to the water where submerged weeds are growing. The first type of treatment is called a local treatment, the second is termed a solution treatment applied, either to a plot or to the entire pond.

Conventional sprayers are used to apply the local treatment and in some instances may be suitable for solution treatments. Chemicals for solution treatment are sometimes diluted with water and poured into the wake of an outboard motor, sprinkled over the surface of the pond, or run by gravity into the water containing the weed beds. Crystalline salts may be placed in a fine woven cotton bag and towed by boat, allowing the herbicide to dissolve and mix with the pond water. Some herbicides are prepared in granular form for scattering or broadcasting over the areas to be treated. Generally, the more rapidly the chemical loses its toxicity, the more uniformly it must be distributed over the area involved for effective results. Also, if the chemical is at all toxic to fish, it must be uniformly distributed. Emergent or floating vegetation receiving local treatments applied with spray equipment should be uniformly covered with a drenching spray applied as a fine mist.

Several precautions should always be taken when herbicides are used. An applicator's permit is required. All instructions on the label must be followed and chemicals can only be stored in the original, labeled container. Avoid inhalation of herbicides and prevent their repeated or prolonged contact with the skin. Wash thoroughly after handling herbicides and always remove contaminated clothing as soon as possible. Prevent livestock from drinking the water during the posttreatment period specified on the label. Do not release treated water into locations that may be damaged by activity of the chemical. Avoid overdoses and spillages. Avoid use near sensitive crops and reduce drift hazards as much as possible; do not apply herbicides on windy days. Clean all application equipment in areas where the rinsing solution will not contaminate other areas or streams.

Fish culturists must also be aware of the current registration status of herbicides. Continuing changes in the regulation of pesticide and drug use in the United States has created confusion concerning what chemicals may be used in fisheries work.

Pathogens

Microbes such as viruses, bacteria, fungi, and protozoa, are an ever-present segment of the environment. These microbes play a vital role in ecosystems and most of them are not only harmless, but extremely beneficial, without them life, as we know it, would not be possible. However, some can become pathogenic; these are the opportunistic pathogens, others are obligate pathogens that cannot survive for long without a suitable host.

Opportunistic pathogens, as the name implies, take advantage of a situation that has placed the potential host, in this case the fish, at a disadvantage or weakened position. Therefore, to protect the fish against these potential pathogens, the fish must be maintained in a healthy rearing environment. Water quality must be kept within required standards. Good sanitation must be practiced, and the fish must receive a nutritionally complete, well-balanced diet. Management practices must be such that fish are not placed under conditions of severe or chronic stress. Obligate pathogens are best controlled by excluding them altogether from the rearing environment. This requires a specific pathogen-free source of water along with rigorously practiced sanitary and biosecurity measures. In addition, management practices should include those recommended for the control of opportunistic pathogens. These are the basic preventative measures.

Should pathogens become established and cause disease, these organisms may have to be attacked directly with chemicals, drugs, antibiotics, or other therapeutants. This is a last resort approach, but that does not mean that treatment should be delayed until losses have taken on epizootic proportions or after a long period of chronic losses. It is best to treat early. This requires early identification of the pathogen. Whenever applicable, treatment should include improvements in water quality and management practices. Few therapeutants are available to the fish culturist. Consult Chapter 9 (page 559) for diagnostic procedures, pathogen identification, and treatment protocols.

Record Keeping

Records

Record keeping, in any business or organization, is an integral part of the system. It is the means by which we measure and balance input and output, evaluate efficiency, and plan future operations. Listed below are factors that should be considered in effective record keeping. These factors are particularly applicable to trout and salmon operations, but many of them pertain to warmwater hatcheries as well.

Water

1. Volume in m^3 (ft^3) for each rearing unit and for the entire hatchery.
2. Liter per minute (gpm) flow into each unit and for the entire hatchery.
3. Rate of change for each unit and for the total hatchery.
4. Water temperature.
5. Water quality: DO, TAN, CO_2, alkalinity, and pH, are of primary importance.

Fish

1. Weight and number of fish and eggs on hand at the beginning and end of accounting period.
2. Fish and eggs shipped or received.
3. Gain in weight in grams, kilogram, or in percentage.
4. Date eggs were taken, number per liter, and source.
5. Date of first feeding of fry.
6. Number per kilogram of all lots of fish or weight per fish in grams.
7. Data on broodstock.

Mortality

1. Fish or eggs actually collected and counted (daily pick-off).
2. Unaccountable losses (predation, cannibalism) determined by comparison of periodic inventories.

Food and diet

1. Composition.
2. Cost per kilogram of feed and cost per kilogram of fish gained.
3. Amount of food fed as percentage of fish body weight.
4. Kilogram of food per kilogram of fish produced (feed conversion ratio).

Disease

1. Occurrence, kind, and possible contributing factors.
2. Type of control and results.

Costs (other than fish food)

1. Maintenance and operation.
2. Interest and depreciation on investment.

3. Analysis of all cost and production records.

Production Summary

Some additional records that should be considered for extensive (pond) culture follow.

Water

1. Area in hectares of each pond.
2. Volume in cubic meter.
3. Average depth.
4. Inflow required to maintain pond level.
5. Water temperatures.
6. Source and quality.
7. Weed control (dates, kind, amount, cost, results).
8. Fertilization (dates, kind, amount, cost, results).
9. Algae and zooplankton blooms (dates, secchi disk readings, kinds of plankton).

Fish

1. Broodstock
   a. Species, numbers.
   b. Stocked for spawning (species, numbers, dates).
   c. Replacements (species, numbers, weight).
   d. Feeding and care (kind, cost, and amounts of food, including data on forage fish production).
   e. Diseases and parasites (treatments, dates, results).
   f. Fry produced per hectare and per female.
2. Fingerlings
   a. Species (numbers stocked, size, weight, date).
   b. Number removed, date, total weight, weight per thousand, number per kilogram, individual weight in grams.
   c. Supplemental feeding (kind, amount, cost).
   d. Disease and predation, including predacious insects during the fry stage.
3. Production per hectare by species (numbers and kilogram).
4. Days in production.
5. Weight gain per hectare per day.
6. Cost per kilogram of fish produced at hatchery.
7. Cost per kilogram including distribution costs.

Forms can best be designed by the managers to fit the data to be collected. Today's computer technology offers total flexibility, as well as computations of the data.

Computerization

Computer technology and development of new software programs are growing at an ever rapid pace. The business and science of fish production can no longer ignore this technology. In many cases, managers and hatchery personnel can select appropriate software programs for their type of operation, but in certain situations it is advisable to make use of computer consultant services to have programs designed that specifically address the uncommon features of their operation.

Several commercial software programs are available for fish production, but many of the private and public hatcheries have written their own specific programs, designed to meet the needs for their operation.

In addition to fish culture (biology) programs, most facilities can also benefit greatly from preventative maintenance and production programs. Computers are now everyday tools in businesses large and small, including the business of fish production; public and private, large and small.

References

Avault, J. W. 1996. Fundamentals of aquaculture. AVA Publishing Company, Inc., Baton Rouge, Louisiana.

Avella, M., and M. Bornancin. 1989. A new analysis of ammonia and sodium transport through the gills of the freshwater rainbow trout (*Salmo gairdneri*). Journal of Experimental Biology 142:155–175.

Barton, B. A. 1996. General biology of salmonids. Pages 29–86 *in* W. Pennell and B. A. Barton, editors. Principles of salmonid culture. Elsevier, Amsterdam, The Netherlands.

Bebak, J. 1998. The importance of biosecurity in intensive culture. Pages 245–253 *in* G. S. Libey and M. B. Timmons, editors. Proceedings of the second international conference on recirculating aquaculture. Virginia Polytechnic Institute and State University, Blacksburg.

Boyd, C. E. 1990. Water quality in ponds for aquaculture. Alabama Agricultural Experiment Station, Auburn University, Auburn.

Brune, D. E., and J. R. Tomasso, editors. 1991. Aquaculture and water quality. World Aquaculture Society, Baton Rouge, Louisiana.

Cho, C. Y. 1992. Feeding systems for rainbow trout and other salmonids with reference to current estimates of energy and protein requirements. Aquaculture 100:107–123.

Clark, W. C., and J. E. Shelbourn. 1989. Temperature requirements for the production of zero-age coho salmon (*Oncorhynchus kisutch*) smolts. Pages 813–819 *in* N. DePauw, E. Jaspers, H. Ackefors, and N. Wilkins, editors. Aquaculture, a biotechnology in progress. European Aquaculture Society, Bredene, Belgium.

Culver, D. A. 1996. Fertilization procedures for pond culture of walleye and saugeye. Pages 115–122 *in* R. C. Summerfelt, editor. Walleye culture manual. North Central Regional Aquaculture Center Culture Series 101, Iowa State University, Ames.

Diana, J. S. 1997. Feeding strategies. Pages 245–260 *in* H. S. Egna and C. E. Boyd, editors. Dynamics of pond aquaculture. CRC Press, Boca Raton, Florida.

Downey, P. C., and G. W. Klontz. 1981. Aquaculture techniques: oxygen (O_2) requirements for trout quality. Idaho Water and Energy Resources Research Institute, University of Idaho, Moscow.

Egna, H. S., and C. E. Boyd, editors. 1997. Dynamics of pond aquaculture. CRC Press, Boca Raton, Florida.

Goddard, S. 1996. Feed management in intensive aquaculture. Chapman and Hall, New York.

Gustafson, G. 1996. Pond culture of walleye in Michigan: fry to advanced fingerlings. Pages 53–55 *in* R. C. Summerfelt, editor. Walleye culture manual. North Central Regional Aquaculture Center Culture Series 101, Iowa State University, Ames.

Haskell, D. C. 1959. Trout growth in hatcheries. New York Fish and Game Journal 6:204–237.

Heinen, J. M., J. A. Hankins, A. L. Weber, and B. J. Watten. 1996. A semi-closed recirculating water system for high density culture of rainbow trout. Progressive Fish-Culturist 58:11–22.

Holeton, G. F., and D. J. Randall. 1967. The effect of hypoxia upon the partial pressure of gases in the blood and water afferent and efferent to the gills of rainbow trout. Journal of Experimental Biology 64:317–327.

Iwama, G. K. 1996. Growth of salmonids. Pages 467–505 *in* W. Pennell and B. A. Barton, editors. Principles of salmonid culture. Elsevier, Amsterdam, The Netherlands.

Jackson, A. 1992. Growth, nutrition and feeding. Pages 202–216 *in* L. M. Laird and T. Needham, editors. Salmon and trout farming. Ellis Horwood Limited, Chinchester, United Kingdom.

Jensen, G. L. 1990. Sorting and grading warmwater fish. Southern Regional Aquaculture Center Publication 391, Louisiana State University Agriculture Center, Baton Rouge.

Jobling, M. 1994. Fish bioenergetics. Chapman and Hall, New York.

Johansen, K. 1982. Respiratory gas exchange of vertebrate gills. Pages 99–127 *in* D. F. Houlihan, J. C. Rankin, and T. J. Shuttleworth, editors. Gills, volume 16. Society for Experimental Biology Seminar Series, Cambridge University Press, Cambridge, Massachusetts.

Johnson, F., and A. Wandsvik. 1991. The impact of fish energy diets on pollution control in the fish farming industry. Pages 51–63 *in* C. B. Cowey and C. Y. Cho, editors. Nutritional strategies and aquaculture waste. University of Guelph, Ontario, Canada.

Jones, D. R., and D. J. Randall. 1978. The respiratory and circulating systems during exercise in fish. Pages 425–501 *in* W. S. Hoar and D. J. Randall, editors. Fish physiology, volume VII. Academic Press, New York.

Kinnunen, R. E. 1996. Walleye fingerling culture in undrainable ponds. Pages 135–145 *in* R. C. Summerfelt, editor. Walleye culture manual. North Central Regional Aquaculture Center Culture Series 101, Iowa State University, Ames.

Landau, M. 1991. Introduction to aquaculture. Wiley, New York.

Leon, K. A. 1986. Effect of exercise on feed consumption, growth, food conversion and stamina of brook trout. Progressive Fish-Culturist 48:43–46.

Lin, C. K., D. R. Teichert-Coddington, B. W. Green, and K. L. Veverica. 1997. Fertilization regimes. Pages 73–103 *in* H. S. Egna and C. E. Boyd, editors. Dynamics of pond aquaculture. CRC Press, Boca Raton, Florida.

Lin, H., and D. Randall. 1991. Evidence for the presence of an electrogenic proton pump on the trout gill epithelium. Journal of Experimental Biology 161:119–134.

Lovell, T. 1988. Nutrition and feeding of fish. Van Nostrand Reinhold, New York.

Martinez-Cordora, L. R., M. A. Porchas-Cornejo, H. Villarreall-Colmenares, and J. A. Calderon-Perez. 1998. Effect of aeration on chlorophyll a, zooplankton, and benthos in yellow leg shrimp, *Penaeus californiensis*, ponds. Journal of Applied Aquaculture 8:17–23.

McDonald, G., and L. Milligan. 1997. Ionic, osmotic and acid-base regulation in stress. Pages 119–144 *in* G. K. Iwama, A. D. Pickering, J. P. Sumpter, and C. B. Schreck, editors. Fish stress and health in aquaculture. Cambridge University Press, Cambridge, Massachusetts.

Meade, J. W. 1985. Allowable ammonia for fish culture. Progressive Fish-Culturist 47:135–145.

Meade, J. W. 1989. Aquaculture management. Van Nostrand Reinhold, New York.

Nath, S. S., and J. P. Boolte. 1998. A water budget model for pond aquaculture. Aquaculture Engineering 18:175–188.

Payan, P. 1978. Study of the Na+/NH4+ exchange across the gill of the perfused head of the trout (*Salmo gairdneri*). Journal of Comparative Physiology (B) 124:181–188.

Pennell, W., and W. E. McLean. 1996. Early rearing. Pages 365–466 *in* W. Pennell and B. A. Barton, editors. Principles of salmonid culture. Elsevier, Amsterdam, The Netherlands.

Perry, S. F., and G. McDonald. 1993. Gas exchange. Pages 251–278 *in* D. H. Evans, editor. The physiology of fishes. CRC Press, Boca Raton, Florida.

Petit, J. 1989. Water supply, treatment and recycling in aquaculture. Pages 87–92 *in* G. Barnabe, editor. Aquaculture, volume I. Ellis Horwood, New York.

Piedrahita, R. H. 1991. Modeling water quality in aquaculture ecosystems. Pages 322–362 *in* Brune and Tomasso (1991).

Pillay, T. V. R. 1990. Aquaculture, principles and practices. Fishing News Books, University Press, Cambridge, UK.

Piper, R. G., and five coauthors. 1982. Fish hatchery management. U.S. Fish and Wildlife Service, Washington, D.C.

Plumb, J. A. 1994. Health maintenance of cultured fishes: principal microbial diseases. CRC Press, Boca Raton, Florida.

Poston, H. A. 1983. Effect of population density of lake trout in cylindrical jars on growth and oxygen consumption. Progressive Fish-Culturist 45:8–13.

Randall, D. 1991. The impact of variations in water pH on fish. Pages 90–104 *in* Brune and Tomasso (1991).

Russo, R. C., and R. V. Thurston. 1991. Toxicity of ammonia, nitrite, and nitrate to fishes. Pages 58–89 *in* Brune and Tomasso (1991).

Saunders, R. L. 1962. The irrigation of the gills of fishes II. Efficiency of oxygen uptake in relation to respiratory flow and carbon dioxide. Canadian Journal of Zoology 40:817–862.

Saunders, R. L., and P. R. Harmon. 1990. Influence of photoperiod on growth of juvenile Atlantic salmon and development of salinity tolerance during winter-spring. Transactions of the American Fisheries Society 119:689–697.

Schreck, C. B., and H. W. Li. 1991. Performance capacity of fish: Stress and water quality. Pages 21–29 *in* Brune and Tomasso (1991).

Schumann, D., and J. Piiper. 1966. Der Sauerstoff-bedarf der atmung bei fischen nach messungen an der narkotisierten Schlere (*Tinca tinca*), Archieven Gesamte Physiologie. 288:14–26.

Severson, R. F., J. L. Stark, and L. M. Poole. 1987. Use of oxygen to commercially reared coho salmon. Pages 25–34 *in* G. Bouck, editor. Papers on the use of supplemental oxygen to increase hatchery rearing capacity in the Pacific Northwest. Bonneville Power Administration, Portland, Oregon.

Shelton, G. 1970. The regulation of breathing. Pages 293–359 *in* W. S. Hoar and D. J. Randall, editors. Fish physiology, volume IV. Academic Press, New York.

Shilo, M., and S. Sarig. 1989. Fish culture in warmwater systems: problems and trends. CRC Press, Boca Raton, Florida.

Smith, S. A. 1998. Biosecurity and fish health monitoring for aquaculture facilities. Pages 22–24 *in* G. S. Libey and M. B. Timmons, editors. Proceedings of the second international conference on recirculating aquaculture. Virginia Polytechnic Institute and State University, Blacksburg.

Snow, J. R. 1957. Algae control in warmwater hatchery ponds. Proceedings of the Annual Conference Southeastern Association of Game and Fish Commissioners 10(1956):80–85.

Soderberg, R. W. 1994. Flowing water fish culture. Lewis Publishers, Boca Raton, Florida.

Soderberg, R. W. 1997. Factors affecting fish growth and production. Pages 199–211 *in* H. S. Egna and C. E. Boyd, editors. Dynamics of pond aquaculture. CRC Press, Boca Raton, Florida.

Solomon, D. J. 1992. Producing fish for release. Pages 217–230 *in* L. M. Laird and T. Needham, editors. Salmon and trout farming. Ellis Horwood Limited, Chichester, United Kingdom.

Stevens, E. D., and D. J. Randall. 1967. Changes in gas concentrations in blood and water during moderate swimming activity in rainbow trout. Journal of Experimental Biology 46:329–337.

Stickney, R. R. 1991. Effects of salinity on aquaculture production. Pages 105–132 *in* Brune and Tomasso (1991).

Stickney, R. R. 1993. Introduction. Pages 1–31 *in* R. R. Stickney, editor. Culture of nonsalmonid freshwater fishes. Advances in fisheries science. CRC Press, Boca Raton, Florida.

Summerfelt, S. T., and E. M. Wade. 1997. Recent advances in water treatment processes to intensify fish production in large recirculating systems. Pages 350–367 *in* M. B. Timmons and T. Losordo, editors. Advances in aquacultural engineering. Northeast Regional Agricultural Engineering Service, Ithaca, New York.

Summerfelt, S. T., M. B. Timmons, and B. J. Watten. 1998. Culture tank design to increase profitability. Pages 263–269 *in* G. S. Libey and M. B. Timmons, editors.

Proceedings of the second international conference on recirculation aquaculture. Virginia Polytechnic Institute and State University, Blacksburg.

Timmons, M. B., and S. T. Summerfelt. 1997. Advances in circular culture tank engineering to enhance hydraulics, solids removal, and fish management. Pages 66–84 *in* M. B. Timmons and T. Losordo, editors. Advances in aquacultural engineering. Northeast Regional Agricultural Engineering Service, Ithaca, New York.

Tomasso, J. R., and D. E. Brune. 1991. Aquacultural water quality: the emergence of an applied discipline. Pages 11–20 *in* Brune and Tomasso (1991).

Tvinnereim, K., and S. Skybakmoen. 1989. Water exchange and self-cleaning in fish rearing tanks. Pages 1041–1047 *in* N. DePauw, E. Jaspers, H. Ackefors, and N. Wilkins, editors. Aquaculture, a biotechnology in progress. European Aquaculture Society, Bredene, Belgium.

Wedemeyer, G. A. 1996. Physiology of fish in intensive culture systems. Chapman and Hall, New York.

Wedemeyer, G. A. 1997. Effects of rearing conditions on the health and physiological quality of fish in intensive culture. Pages 35–72 *in* G. K. Iwana, A. D. Pickering, J. P. Sumpter, and C. B. Schreck, editors. Fish stress and health in aquaculture. Cambridge University Press, Cambridge, Massachusetts.

Westers, H. 1979. Controlled fry and fingerling production in hatcheries. Pages 32–52 *in* E. A. Huisman and H. Hogendoorn, editors. European Inland Fisheries Advisory Committee workshop on mass rearing of fry and fingerlings of freshwater fishes/papers. Ministry of Agriculture and Fisheries Department of Sport and Professional Fisheries, The Hague, The Netherlands.

Westers, H. 1987. Feeding levels for fish fed formulated diets. Progressive Fish-Culturist 49:87–92.

Westers, H. 1991. Modes of operation and design relative to carrying capacities of flow-through systems. Pages 151–159 *in* M. B. Timmons, editor. Engineering aspects of intensive aquaculture. Cornell University, Ithaca, New York.

Westers, H. 1991. Operational waste management in aquaculture effluents. Pages 231–238 *in* C. B. Cowey and C. Y. Cho, editors. Nutritional strategies and aquaculture waste. University of Guelph, Guelph, Ontario, Canada.

Westers, H., and K. M. Pratt. 1977. Rational design of hatcheries for intensive salmonid culture, based on metabolic characteristics. Progressive Fish-Culturist 39:157–165.

Wilson, R. W., and E. W. Taylor. 1992. Transbranchial ammonia gradients and acid-base responses to high external ammonia concentration in rainbow trout (*Oncorhynchus mykiss*) acclimated to different salinities. Journal of Experimental Biology 166:95–112.

Wilson, R. W., P. M. Wright, S. Munger, and C. M. Wood. 1994. Ammonia excretion in freshwater rainbow trout (*Oncorhynchus mykiss*) and the importance of gill boundary layer acidification: lack of evidence for Na+/NH4+ exchange. Journal of Experimental Biology 191:37–58.

Wood, C. M. 1996. Ammonia and urea metabolism and excretion. Pages 379–426 *in* D. H. Evans, editor. The physiology of fishes. CRC Press, Boca Raton, Florida.

Wright, P. A., and C. M. Wood. 1985. An analysis of brachial ammonia excretion in the freshwater rainbow trout: effects of environmental pH change and sodium uptake blockade. Journal of Experimental Biology 114:329–353.

3

Hatchery Water Supply and Treatment

John E. Colt and Joseph R. Tomasso

Introduction

Fish, like other organisms, exist in intimate contact with their environment. The primary components of an aquatic or aquaculture environment may be classified into three general groups: the water quality components (mainly consisting of temperature and substances dissolved in the water), physical components (excluding physical water quality components), and population and community components. In order to maximize the productivity of an aquaculture operation (measured as survival, growth, feed conversion efficiency, or reproductive success), culture environments must be optimized. Nonoptimal or stressful environments will reduce productivity.

The relationship between environment and productivity of fishes is well described by Schreck (1981) who argues that fishes have some maximum, genetically determined, ability to perform. In aquaculture, performance may be viewed in the context of the integrated responses, such as growth and reproduction. The genetically based performance capacity (potential performance capacity) is reduced by nonoptimal environments and stress. The concept of performance capacity can act as a guide in aquaculture operations as we try to provide environments that allow animals to minimize energy and nutrient expenditures for maintenance activities and maximize energy and nutrient use for growth or reproduction.

This chapter will focus on water quality and quantity as they relate to culture environments. Three general topics will be addressed. First, what are the water quality components of importance to fish culture? Second, what are acceptable sources of water for fish culture operations and how is the water delivered to the culture operation? Third, how can the water be treated for reuse in culture operations or release into public surface water?

Water Quality Components of Importance to Fish

General Comments

The water quality components of an aquaculture environment may be divided into those that directly affect the fish being cultured (primary components) and those that indirectly affect the fish (secondary components). Water quality interactions in a warmwater fish pond offer good examples (Boyd 1979a). The dissolved oxygen (DO) concentration of the pond water has a direct effect on the well-being of the fish. Phytoplankton populations have no direct effect on the fish (unless the culturist is relying on natural food to support aquaculture production or a toxic algal species is present), but phytoplankton directly affect DO concentrations by net production of oxygen during the day. Similarly, extreme pH levels can directly affect culture animals (and the toxicity of ammonia), but alkalinity affects animals because it influences the daily pH fluctuations in a pond. Culturists should manage for acceptable or optimal (depending on the economics of specific situations) levels of primary components. In pond aquaculture, this often means managing a secondary component that controls a primary component of interest. For example, extreme pH fluctuations in ponds can be managed by increasing alkalinity through liming (Boyd 1979a). Raceway culturists can usually manage the primary components directly—often by modifying the water flow or oxygenating the system. Some of the primary water quality components for selected species of interest to culturists are discussed below. The relationship between water quality and fish culture is discussed in a general sense with some examples. Recently, several books have been published detailing the environmental requirements of individual species or genera (Table 1).

Temperature

Temperature is the predominant physiochemical characteristic in aquaculture operations. It affects the amount of oxygen a given amount of

TABLE 1. References that detail water quality requirements for the culture of various fishes.

Genus or species	Reference
Salmonids	Barton (1996)
Temperate basses	Tomasso (1997)
Channel catfish	Boyd and Tucker (1998) Stickney (1993)
Black basses	Williamson et al. (1993)
Sunfishes	Williamson et al. (1993)
Walleye	Nickum and Stickney (1993)
Pikes	Westers and Stickney (1993)
Tilapias	Costa-Pierce and Rakocy (1997)

water will hold, the rate of photosynthesis and decomposition, the ionization of ammonia, and the metabolic rate of fish. Seasonally changing temperatures act as cues for reproduction in many fishes. Immunocompetency of fishes and virulence of pathogens is also related to temperature. Table 2 provides examples of temperature preferences for selected species.

Fry (1947, 1971) characterized temperature as an environmental controller of metabolism. Modulation of the metabolic rate affects the amount of oxygen fish consume per unit of time and body mass, the rate at which animals feed and should be fed, the rate of growth, and feed conversion ratios.

Metabolic rate, as approximated with oxygen consumption rates, can be used as a tool to determine optimal temperatures for the culture of a species. As temperature increases, the standard metabolic rate (the metabolic rate of resting, fasted fish) increases exponentially while the metabolic rate of actively swimming fish increases sigmoidally (Neill and Bryan 1991). The difference between active and standard metabolic rates at a given temperature is the metabolic scope (Figure 1). Culturing a species at the temperature of maximum metabolic scope allows the greatest amount of energy to be partitioned to growth.

Due to the high specific heat of water, a great deal of energy is required to raise or lower its temperature. Therefore, modifying water temperatures at aquaculture facilities is rarely economical, and the importance of siting aquaculture facilities becomes clear. Raceway operations using surface water should be sited where expected high and low temperatures of the water source are well within the survival and growth ranges of the species to be cultured. Pond operations should be sited in areas where prevailing climates are compatible with the species to be cultured. Particular interest should be paid to expected low tempera-

TABLE 2. Examples of temperatures important for the culture of selected species.

Activity	Temperature (°C)	Reference
	Striped bass	
Recommended for hatching	17.8–20.0	Rees and Harrell (1990)
Optimal fingerling growth	26.9–30.3	Kellogg and Gift (1983)
Maximum growth of juveniles	24.0	Cox and Coutant (1981)
	Channel catfish	
Optimal for hatching	25.0–28.0	Busch (1985)
Best growth	26.0–30.0	Kilambi et al. (1971); Andrews and Stickney (1972); Andrews et al. (1972)
	Largemouth bass	
Over 75% survival of eggs	17.0–21.0	Kelly (1968)
Maximum growth of fry	27.0	Coutant and DeAngelis (1983)
Mortality of Florida subspecies	2.0–3.0	Guest (1985); Carmichael et al. (1988)
	Rainbow trout	
Optimum for growth	16.5–17.2	Jobling (1994)
Spawning temperature	10.0–13.0	Piper et al. (1982)
	Walleye	
Best survival through yolk sac absorption	8.9–12.0	Koenst and Smith (1976)
Optimal for fingerling growth	20.0–25.0	Nickum and Stickney (1993)

tures in ponds. Early attempts to overwinter red drum in ponds in Texas sometimes resulted in mortality due to low temperatures even though wild red drum were able to overwinter in bay systems just a few miles from the culture ponds. Apparently, the wild fish were able to seek thermal refugia, such as deep water, although the pond fish could not avoid the low temperatures.

Recirculating culture systems are often cited as a means of growing a species in an area that would be unsuitable for outdoor culture. However, high-density recirculating systems commonly require 10–100% water turnover per day to make up for evaporation and for backwashing filters. When siting a recirculating system, the volume and temperature of incoming water, the need for heat exchangers, and the cost of heating and cooling the water should be considered.

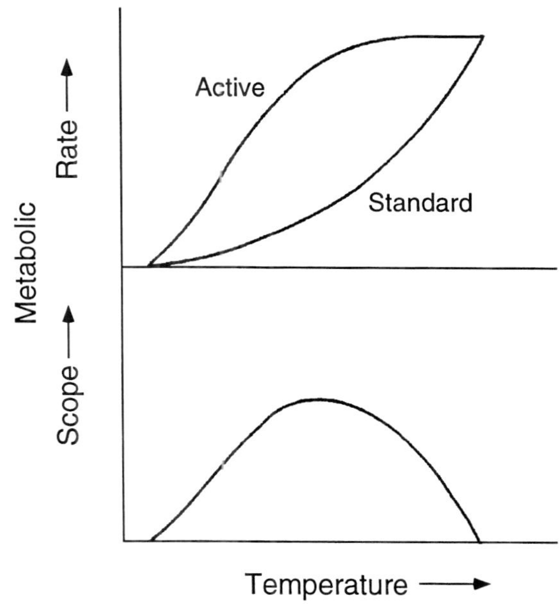

FIGURE 1. The effect of temperature on metabolic rates and metabolic scope (from Neill and Bryan 1991).

Dissolved Oxygen

Fish, like other animals, require oxygen as a terminal electron acceptor in cellular metabolism. Without oxygen, the ability to efficiently convert energy into a usable form is curtailed, usually to the point that life cannot be sustained. Oxygen enters the fish by diffusion across the gills, so adequate DO concentrations must be maintained in the culture water to provide an adequate pressure gradient to drive the diffusive process. Less-than-optimal DO concentrations will result in reduced growth and increased susceptibility to infectious disease. Acceptable DO values vary from species to species and within each species (Table 3). However, some basics apply to all species when planning and operating oxygenation systems:

1. Fish of a given species and size require more oxygen in warmer water than in cooler water (Figure 1).
2. Fish require more oxygen after feeding than before.
3. Oxygen consumption is proportional to the size (Figure 2, upper panel) and number of fish in a system.
4. Smaller fish consume more oxygen per unit weight than larger fish (Figure 2, lower panel).

Table 3. Examples of dissolved oxygen concentrations important for the culture of selected species.

Activity	Dissolved oxygen (mg/L)	Reference
	Striped bass	
Best egg hatching percentage	Saturation	Turner and Farley (1971)
Recommended for phase I pond culture (1–2 inches)	>6.0	Brewer and Rees (1990)
Preferred for phase II pond culture (6–10 inches)	>5.0	Kerby et al. (1983)
	Channel catfish	
Begin supplemental aeration in production ponds	2–3	Tucker and Boyd (1985)
Minimum in hatchery	4–5	Tucker (1991)
	Largemouth bass	
No effect on egg survival	>2.8	Dudley and Eipper (1975)
Minimum for survival at 30°C	1.2	Moss and Scott (1961)
Gradient avoidance threshold	3.0	Whitmore et al. (1960)
	Rainbow trout	
Minimum level at outflow of production raceway	6.0	Hinshaw (1990)
	Walleye	
Minimum in transport containers	4.0	Nickum and Stickney (1993)

5. Fish require more oxygen if they have impaired gill function (either damaged by exposure to irritating chemicals or suspended solids), impaired oxygen transport capacity (as in anemia or nitrite toxicosis), or are exposed to other stressors (Tucker 1996; Boyd and Tucker 1998).

A reciprocal relationship between the relatively low solubility of oxygen in water at higher temperatures (Figure 3) and higher oxygen consumption rates in fish at higher temperatures (Figure 1) often presents problems for aquaculturists, especially in warmwater pond culture. This problem is even more acute in warmwater and tropical mariculture operations where salinity reduces the solubility of oxygen (Figure 4).

Salinity and Salt

The term salinity is often incorrectly used interchangeably with the term salt. Salinity refers to the solid material dissolved in seawater (Sverdrup et al. 1942 as described in Stickney 1991). Salinity is expressed as g/kg

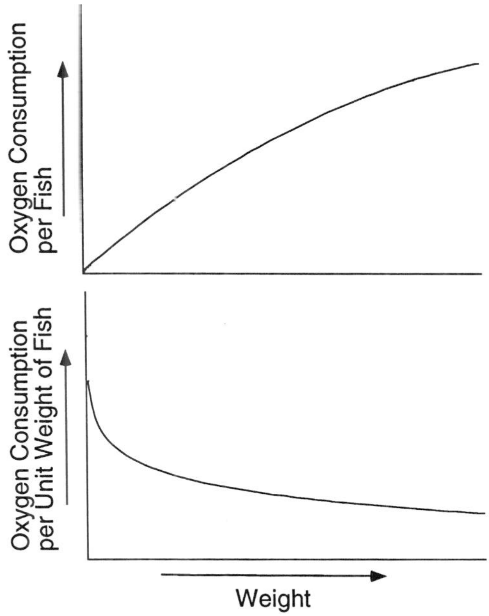

FIGURE 2. Oxygen consumption rates of fish as a function of weight. The upper panel is based on weight of the individual. The lower panel is oxygen consumption per unit mass. Upper panel calculated as $M = aW^b$ where M = total oxygen consumption of fish, W = weight of fish, a and b are species-specific constants; lower panel calculated as $M/W = aW^{(b-1)}$ (Hill and Wyse 1989). The exponent b was assumed to be 0.8, which is common, but not exclusive, for fishes (Fry 1971). One was substituted for a in all calculations.

or ppt (abbreviated as either ppt or ‰). Bowen (1966) lists 66 elements found in "sea salts." Salt, on the other hand, usually refers to simple salts such as sodium chloride (NaCl) or calcium chloride.

Full strength seawater has a salinity of about 35 ppt. Most estuaries have salinities less than 35 ppt ranging downward to 0 ppt. However, parts of some estuaries may actually become hypersaline (salinities >35 ppt). A good example is Laguna Madre on the south Texas coast. During the summer, freshwater inflow is reduced due to the low rainfall characteristic of the area. Meanwhile, the mixing of Laguna Madre with ocean water is constrained by small tidal amplitudes and an almost uninterrupted line of barrier islands. Evaporation concentrates the water in Laguna Madre; salinities of 50 ppt or higher may occur by late summer.

Many fishes traditionally of interest to fishery managers are euryhaline or anadromous (e.g., salmonids and striped bass) capable of living in a wide range of salinities during all or specific stages of their lives. Recently, purely euryhaline species that grow well over a range of sa-

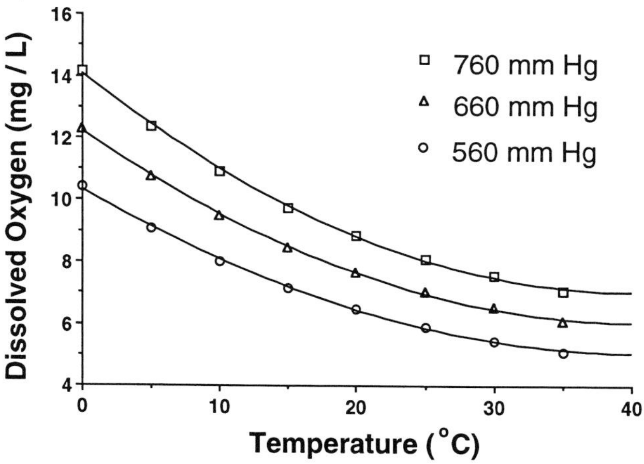

FIGURE 3. The solubility of oxygen in pure water under three pressures of air. The 760, 660, and 560 mm Hg curves approximate the solubility of oxygen in pure water at sea level, 1,200 m and 2,400 m, respectively (calculated as $DO_c = DO_t\,[P_o - P_w]/[760 - P_w]$ where DO_c = corrected solubility of oxygen, DO_t = solubility of oxygen at 760 mm Hg, P_o = observed atmospheric pressure in mm Hg, P_w = vapor pressure of water in mm Hg). The 760 mm Hg curve was adapted from tabular values in Boyd (1979a), the formula for determining solubilities in pure water at different atmospheric pressures was taken from Boyd (1979a).

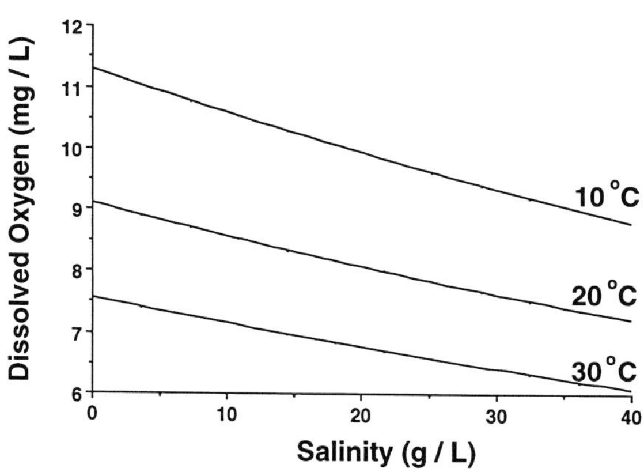

FIGURE 4. The solubility of oxygen in three salinities under 760 mm Hg air pressure. Curves adapted from tabular values of Murray and Riley (1969) as presented in Spotte (1979).

linities have been cultured in hatcheries (e.g., red drum and spotted seatrout). If euryhaline fishes are to be produced in a hatchery, it is important for the hatchery manager to have an understanding of the salinity requirements of the species being cultured. This knowledge may be applied two ways. First, providing the proper salinity for a species increases the likelihood that the culture cycle will be concluded successfully. Second, a detailed knowledge of salinity requirements may offer the culturist a means of improving production or reducing losses. For example, red drum grow well over a range of salinities. However, above a salinity of about 5 ppt the pathogenic dinoflagellate *Amyloodinium* causes substantial losses on fish farms and hatcheries. At salinities below 5 ppt, *Amyloodinium* is not active (Cheung 1993). By culturing red drum in salinities below 5 ppt, the problem can be avoided.

Manipulating salinity (and salt) concentrations is also an effective way of reducing losses during handling and transport of euryhaline (and even stenohaline) fishes. As Figure 5 indicates, the osmolality (an indication of total dissolved particles per unit volume) increases as salinity increases. Most teleost fishes have a plasma osmolality of about 320

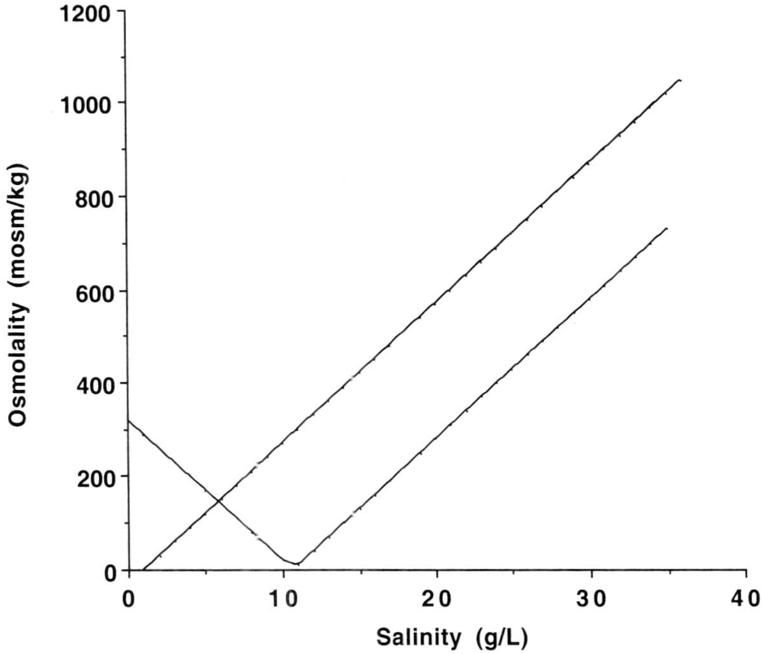

FIGURE 5. The effect of increasing salinity on osmolality of water (straight line), and the difference between plasma osmolality and environmental osmolality (angled line, assuming a plasma osmolality of 320 mosm/kg).

mosm/kg. By adjusting the salinity or salt concentration to equal the osmolality of the plasma, both osmotic and ionic gradients are reduced to the point where fish do not experience the severe ion and water imbalances characteristic of fish stressed by transport in freshwater or seawater (see Chapter 11). Figure 5 also clearly shows that seawater is a more ionically and osmmotically "hostile" environment than freshwater (the difference between plasma and environmental osmolality is greater in seawater).

Limiting or eliminating the osmotic and ionic gradients during growout periods has also been suggested as a means to reduce the amount of energy needed to iono- and osmoregulate, allowing more energy to be partitioned into growth. In general, this has not been demonstrated to occur experimentally in several species tested (reviewed in Brett 1979; Jobling 1994).

Water Hardness and Calcium

Water hardness is a term that refers to the divalent cations (primarily calcium and magnesium) present in the water (Boyd 1979a). Hardness is chemically expressed as equivalent calcium carbonate ($CaCO_3$) in mg/L. Calcium is probably the most important component of hardness with respect to fish culture due to its ability to reduce water flux across the gills. A reduced water flux lowers the amount of energy required to maintain proper water and ion balance during stress (see discussions in Weirich et al. 1992; Grizzle et al. 1993; Seals et al. 1994). Environmental calcium can also be pumped into fishes by active uptake sites on the gills, perhaps offering an alternative to diet as a calcium source (reviewed in Seals et al. 1994).

Calcium concentrations required for successful culture vary from species to species and, generally, are not well defined. Channel catfish appear to do well with very little calcium; however, many hatcheries increase the concentration in water supplies to increase early survival (Tucker and Steeby 1993). The recommended hardness for striped bass prolarvae, larvae, and juveniles is greater than 150 mg/L as calcium carbonate (Hall 1991). Calcium levels can be increased in culture water by the addition of calcium sulfate (gypsum) or calcium chloride. Calcium chloride has the further advantage of adding chloride, another physiologically active compound (see discussion of nitrite toxicity and chapter on fish transport).

Alkalinity and pH

Alkalinity is a measure of the buffering capacity of water (Boyd 1979a). It affects both the initial pH of water and the change in pH due to activities such as photosynthesis in ponds or carbon dioxide release by fish in raceways (Figure 6). The effect of alkalinity on pond pH levels and liming as a means of controlling pH are discussed extensively in Boyd and Tucker (1998).

The sensitivity of fishes to extremes of pH varies; however, a range of 6.5–9.0 is recommended (Piper et al. 1982) for most species. Cool water fish appear to be more sensitive to pH values above 9 than warmwater fish (Piper et al. 1982). Environmental pH affects the toxicity of ammonia (see Ammonia section), hydrogen sulfide (see Hydrogen Sulfide section), and metals (see Heavy Metals section). Fromm (1980) reviewed the literature on the effects of acidity on fishes and suggests, as other researchers have, that a pH of less than 6.5 will affect reproduction in many species. Generally, eggs and larvae are more sensitive to acid stress than juveniles and adults.

Ammonia

Ammonia is the primary nitrogenous waste product excreted by most fishes. It exists in an ionized form (ammonium) and un-ionized form (ammonia). Wet chemical tests measure the sum of the two forms. The

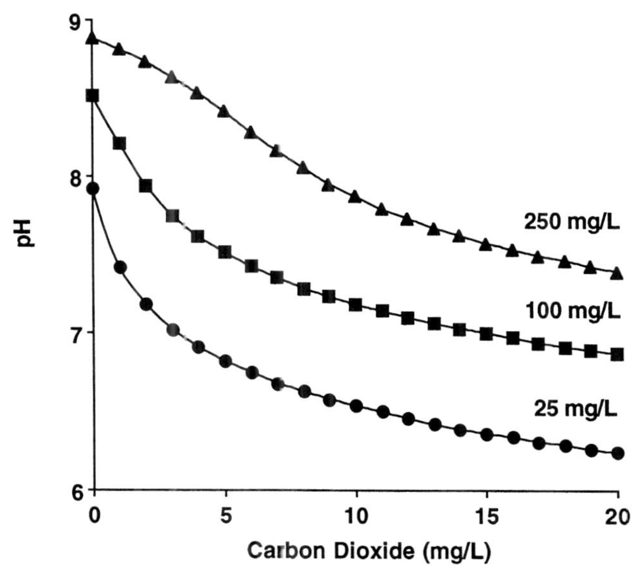

FIGURE 6. Impact of alkalinity on the change in pH due to the addition of carbon dioxide. Assumes water was initially in equilibrium with atmosphere and no loss of carbon dioxide (based on Colt and Orwicz 1991a).

total ammonia concentration is commonly expressed on a nitrogen basis and is called the total ammonia nitrogen (TAN). In an aqueous environment, the two forms exist in an equilibrium that is determined by pH, temperature, and salinity (Figure 7). The pH of the medium has a strong influence on the equilibrium; temperature exerts a moderate influence (Figure 8). Salinities of 18–22 ppt have little effect on ammonia ionization; however, full-strength seawater will contain up to 20% less un-ionized ammonia than freshwater of the same pH and temperature (Bower and Bidwell 1978).

The degree to which the total ammonia in fish culture water is ionized or un-ionized is important to culturists because the un-ionized form is toxic to fish. Un-ionized ammonia is freely diffusable across gill membrane although the ionized form is not. Consequently, when the concentration of un-ionized ammonia is high in the environment, it rapidly diffuses across the gill membrane into the blood. Ammonia is excreted from fish by the same mechanism, so even slightly elevated concentrations of un-ionized ammonia in the culture water will decrease the diffusion gradient, impairing ammonia excretion (discussed in Tomasso 1994).

At high environmental concentrations, ammonia will kill fish (Table 4). Sublethal concentrations will impair growth and may stimulate infectious disease outbreaks (Tomasso 1994). Safe concentrations of ammonia are not well established (Meade 1985). Wedemeyer (1996) recommends that un-ionized ammonia concentrations not exceed 0.02–0.03 mg/L in salmonid raceways. The establishment of safe levels for pond culture is even more problematic due to the confounding effect of daily pH cycles that cause the concentration of the un-ionized ammonia to cycle as well.

Managing ammonia in fish culture systems usually involves managing ammonia input (by adjusting feeding rates) or managing ammonia removal. Inputs from feeds may be adjusted by changing feeding

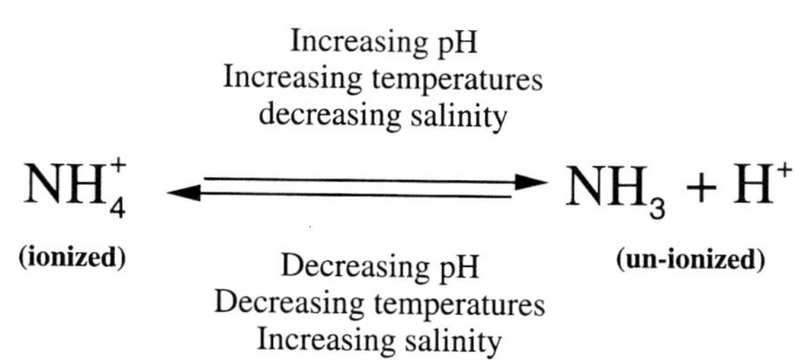

FIGURE 7. The effect of pH, temperature, and salinity on the equilibrium between ionized ammonia (ammonium) and un-ionized ammonia.

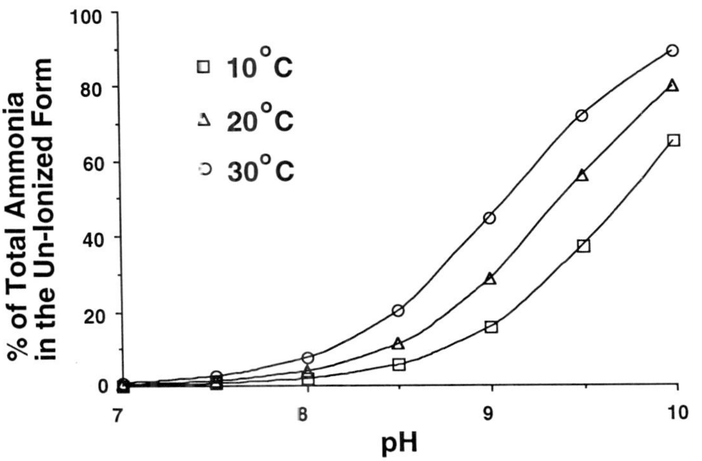

FIGURE 8. The percent of total ammonia in the un-ionized form as a function of pH at three temperatures. Curves and equations are based on Emerson et al. (1975).

rates or the protein levels in the feeds. The concentration of ammonia in raceways may be managed by adjusting either flow rates or nitrogen inputs in feeds. Often, ammonia problems in raceways are self-correcting as pH is decreased due to carbon dioxide excreted by the fish. Since potential ammonia problems are related to nitrogen inputs during feeding and feeding rates are related to standing crops, stocking rates have a large influence on the probability of developing high ammonia concentrations in ponds during a culture cycle and should be carefully considered.

Nitrite

Nitrite (NO_2^-) is produced by the bacterial-mediated oxidation of ammonia in aquatic systems. Under normal circumstances, nitrite is further oxidized by these bacteria to nitrate. In fish culture systems, however, nitrite concentrations may become elevated if oxidation rates of ammonia exceed oxidation rates of nitrite.

Increases in concentrations of nitrite in culture water are of concern to culturists because nitrite can reach toxic concentrations to many species of fishes. Nitrite, like ammonia, exists in a pH-dependent equilibrium with nitrous acid (HNO_2). Nitrous acid is freely diffusable across the gill membrane and is, thus, considered the more toxic form. However, very little nitrous acid is present in aqueous solutions at pH values considered acceptable for fish culture (Figure 9). Above a pH of 6, almost all of the total nitrite present is in the ionized form (nitrite). Nitrite is not freely diffusable across gill membranes, but it can enter many fresh-

TABLE 4. Examples of lethal levels of un-ionized ammonia (ammonia-N). Values are median-lethal concentrations (the concentration of un-ionized ammonia that will kill 50% of the exposed animals in a given period of time).

Activity	Ammonia-N (mg/L)	Reference
	Striped bass	
96 h, juveniles (20–22°C)	1.01	Oppenborn and Goudie (1993)
	Channel catfish	
96 h, fingerlings (26°C)	2.90	Colt and Tchobanoglous (1976)
	Largemouth bass	
72 h, fingerling (25°C)	0.82	Carmichael and Tomasso (unpublished)
	Rainbow trout	
96 h, fingerlings (12–14°C)	0.60	Thurston and Russo (1983)
	Walleye	
96 h, fingerlings	1.06	Mayes et al. (1986)

water fish species by being actively pumped into the fish by the mechanism that ordinarily transports chloride ions into the fish. Apparently, nitrite and chloride ions have enough characteristics in common (they are both monovalent anions and about the same size) that the chloride uptake mechanism cannot discriminate. The active uptake process al-

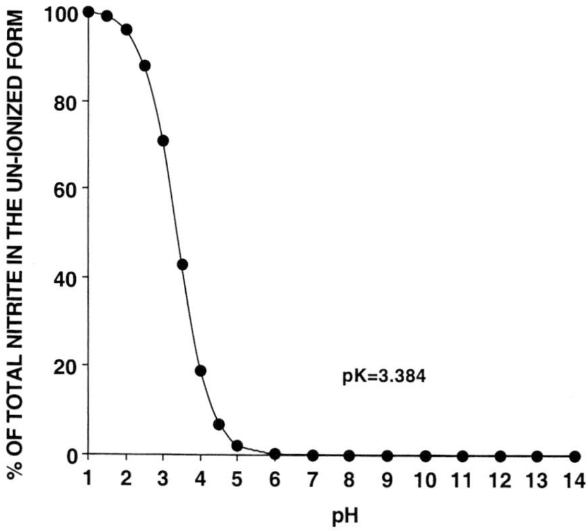

FIGURE 9. The percent of total nitrite in the un-ionized form (nitrous acid) as a function of pH. Calculations assume a pK of 3.384 at 20°C (Colt and Tchobanoglous 1976).

lows nitrite to be concentrated in plasma with respect to the environment and, in effect, magnifies the toxic effect of nitrite. The toxicity of nitrite to fishes varies widely (Table 5), and care should be taken to learn the toxic levels of nitrite to all species being cultured on a facility.

Once inside the fish, nitrite apparently has several mechanisms of toxicity. The most obvious is the oxidation of hemoglobin to methemoglobin. Fish with high levels of methemoglobin have brown blood and suffer from a functional hypoxia because methemoglobin is not capable of binding and transporting oxygen.

Because nitrite is pumped into many freshwater species by the chloride uptake mechanism, addition of chloride to the culture water will competitively exclude nitrite from entering the fish. In some fishes (salmonids and hybrid striped bass, for example), the addition of calcium chloride to the water is more effective than the addition of sodium chloride (Tomasso 1994). Channel catfish respond equally well to both calcium chloride and sodium chloride.

Nitrate

Nitrate is produced by bacteria-mediated oxidation of nitrite in aqueous systems. It is not considered toxic to fishes at levels that can reasonably be expected in aquaculture systems.

Carbon Dioxide

Carbon dioxide concentrations in aquaculture systems are generally low because the carbon dioxide in the water will normally near equilibrium

TABLE 5. Examples of lethal levels of nitrite (mg/L nitrite-N). Values are 96 h median-lethal concentrations (LC-50, the concentration of nitrite that will kill 50% of the exposed animals in a given period of time). All species were tested under the same water quality conditions (23°C, pH 7.8–8.2, hardness 200 mg/L as calcium carbonate, chloride 22 mg/L). All information from Tomasso (1986).

Species	LC-50 (mg/L nitrite-N)
Fathead minnow	50
Goldfish	58
Channel catfish	8
Largemouth bass	154
Smallmouth bass	179
Green sunfish	179
Bluegill	89
Blue tilapia	18

with atmospheric carbon dioxide, which is low. However, in some instances the carbon dioxide concentrations in culture systems can rise to levels that detrimentally affect culture animals. This may occur in intensive systems where the culture animals and decomposers are producing carbon dioxide faster than it can move into the atmosphere. Carbon dioxide levels in transport tanks or bags may also increase because of a buildup of the gas in the atmosphere above the water in sealed transport systems.

High environmental carbon dioxide concentrations will detrimentally affect fish because it will cause carbon dioxide concentrations in the blood to increase and react with water in the plasma to produce carbonic acid, which reduces blood pH. Low blood pH will reduce the capacity of hemoglobin to transport oxygen (Wedemeyer 1996). The resistance of fishes to carbon dioxide varies greatly; however, chronic concentrations of less than 10 mg/L are generally considered acceptable, provided DO concentrations are near saturation (Tucker and Boyd 1985).

Gas Supersaturation

Dissolved gases in water are normally in near-equilibrium with the atmosphere above the water. Gases diffuse either into or out of the water depending on the relative pressure of gases (often referred to as (ΔP)). Under these normal conditions, total gas pressure (the sum of the partial pressures of nitrogen, oxygen, carbon dioxide, and other trace gases) near the surface of a body of water is nearly equal to the gas pressure of the atmosphere. Total gas pressure of water may increase to above the atmospheric pressure under certain conditions. In hatcheries, conditions that commonly lead to supersaturated conditions are introducing air to water under pressure and heating water. Introducing air under pressure may occur when charging empty pipes with water without properly venting or when air leaks into an operating pump head. Heating water leads to supersaturation because warm water will dissolve less gas than cold water (Figure 3).

Gas supersaturation is of concern to culturists because the gases diffuse across the gills into the blood of the fish. Emboli (bubbles) may form in the blood as the excess gases come out of solution causing severe physiological complications and, commonly, death (Bouck 1980). Total gas concentrations above 100% of saturation should be avoided since some species are sensitive to total gas concentrations of below 102% (Colt 2000).

Clinical signs of gas bubble disease include protrusion of the eyes and gas bubbles in the gill filaments of adults. Eggs exposed to supersaturated water may float due to attached air bubbles; exposed larval fishes may develop hyperinflated swim bladders and exhibit bubbles in the yolk sac (Boyd and Tucker 1998).

Hydrogen Sulfide

Hydrogen sulfide (H_2S) is a poisonous gas produced by anaerobic activity of bacteria. It may be present in well water, in water pumped from the bottom of lakes and reservoirs, and in anaerobic sediments of production ponds (Piper et al. 1982; Tucker 1993; Wedemeyer 1996). Like ammonia, hydrogen sulfide exists in a pH and temperature-dependent equilibrium with its ionization products, HS_2 and H^+. At 20ºC and pH 5, 99% of the total is in the un-ionized form, although only 1% of the total is in the un-ionized form at 20ºC and pH 9 (Wedemeyer 1996). Only the un-ionized form is freely diffusable across the gills, so the toxicity of hydrogen sulfide increases with decreasing pH.

Piper et al. (1982) lists several ways hydrogen sulfide problems can be corrected (and prevented): remove excess organic matter from ponds, raise the pH of the culture water, oxygenate the culture water, and add an oxidizing agent such as potassium permanganate. Providing water circulation in culture ponds to prevent the establishment of anoxic zones or anoxic sediments will also limit the production of hydrogen sulfide.

Toxicity of hydrogen sulfide varies among species; recommendations to culturists also vary. Generally, concentrations above 2 µg/L are considered detrimental and concentrations below 0.1 µg/L are considered safe for most fishes (Piper et al. 1982; Tucker 1993; Wedemeyer 1996). Fish exposed to hydrogen sulfide increase their ventilatory rate and may become hyperactive. During production of fingerlings in small (<0.25 ha) ponds, hydrogen sulfide may be a concern during harvesting. Harvesting is usually accomplished by draining the pond water and fish into a catch basin. Properly constructed catch basins allow the water to drain while retaining the fish in a pool of water of acceptable quality. The fish are then dipped from the basin. This approach is efficient and probably less stressful to the fish than multiple seinings of the pond. However, the draining process disrupts the sediments and may allow the release of hydrogen sulfide sequestered in the sediments. The concentrations of hydrogen sulfide in the pond and basin water may become lethal. These events emphasize the need to circulate water near the bottom of ponds to minimize the formation of anoxic areas.

Suspended Solids and Turbidity

Turbidity refers to the degree of light penetration through water with turbidity inversely proportional to light penetration. In fish culture operations, suspended solids such as soil particles in the incoming water, uneaten fish feed, and fecal material contribute to turbidity. Algae are also a major contributor to turbidity in culture ponds. When organic solids settle to a pond bottom, they contribute to the biological oxygen demand of the sediments that may lead to anoxic areas and hydrogen sulfide production. In hatcheries, solids may settle on eggs, causing heavy mortality. Solids in the water column may interfere with fish finding food or may damage the gills of fishes (Piper et al. 1982; Wedemeyer 1996).

Wedemeyer (1996) suggests that 80–100 mg/L total suspended solids is a reasonable chronic exposure level for salmonids and other sensitive species, but he also points out that guidelines have not been developed for solids with sharp and irregular edges that might damage gills. Based on a review of the literature, Hall (1991) recommended that striped bass larvae be held in water with less than 1,000 mg/L suspended solids and juveniles be held in less than or equal to 10 mg/L clay or silt and less than or equal to 2,000 mg/L fine sand. Turbidity may be beneficial in some cases. For instance, walleye fry survived and grew better in turbid water than in clear water (Bristow et al. 1996). Turbidity apparently made feed more visible in the water column and decreased reflected light from the sides of the tanks (reducing the number of fish that oriented with the tank sides rather than the surface of the water).

Chlorine

Chlorine, in various forms, is used as a disinfectant in fish hatcheries and is a common additive to municipal water supplies. Chlorine reacts with water to form hypochlorous acid (HOCl), which forms an equilibrium with OCl^-. This equilibrium is pH dependent. Both forms cause gill damage and anemia in fishes (reviewed in Boyd and Tucker 1998). When exposed to acutely toxic levels of chlorine, fish may increase ventilatory rates and die with opercula flared. Most analytical techniques measure the sum of HOCl and OCl^- (referred to as free chlorine, free residual chlorine, or total chlorine). Recommended chronic exposure levels for salmonids vary from below 1 μg/L (USEPA 1976) to below 3–5 μg/L (Wedemeyer 1996). For other aquatic organisms, USEPA (1976) recommends below 10 μg/L. Chlorine is relatively unstable and can be removed by aeration, sunlight, or activated carbon. Hypochlorous acid and OCl^- will react with ammonia or other nitrogenous compounds to form chloramines, which are generally more toxic and more difficult to remove than HOCl and OCl^-.

Heavy Metals

The term heavy metals generally refers to the transition metals and metalloids in the periodic table (e.g., aluminum, antimony, arsenic, barium, beryllium, cadmium, chromium, copper, manganese, mercury, nickel, selenium, silver, zinc). Most of these elements are toxic to fish. Many are found as contaminants or natural constituents of surface and ground waters. However, with the exception of copper, metal toxicity does not appear to be a major problem in aquaculture (Boyd and Tucker 1998). Heavy metal toxicity and allowable levels are discussed in Wedemeyer (1996), and Boyd and Tucker (1998).

Copper is a common contaminant in water and is highly toxic to fish with 96 h median-lethal concentrations in the μg/L to mg/L range. Copper binds to the gills of fishes and brings about ionoregulatory disturbances that result in death (Laurén and McDonald 1986). Toxicity is highly dependent on water hardness, alkalinity, pH, and dissolved organic carbon, with increases in these values resulting in increased tolerance by fishes (Sorensen 1991; Straus and Tucker 1993; Welsh et al. 1993; Erickson et al. 1996). Two common sources of copper in aquaculture systems are copper pipes and copper water additives used to control algae and disease.

Iron is a common constituent of well water. Aquaculturists have historically been advised to avoid iron in culture water; however, Boyd and Tucker (1998) could find no clear evidence that iron concentrations up to 1 mg/L are toxic to fish. Iron is quickly oxidized and precipitates in well aerated environments.

Pesticides

Pesticide is a term that describes chemicals used to control unwanted animals, plants, and fungi (e.g., insecticides, herbicides, fungicides). They may be inadvertently introduced to aquaculture systems in the incoming water, or by overspraying ponds or exposed tanks. Some herbicides are intentionally introduced to aquaculture ponds to control growth of aquatic plants.

Many insecticides are toxic to fish in the (mg/L range. Organochlorine (e.g., DDT, endrin, toxaphene) and pyrethum (permethrin) insecticides are the most acutely toxic followed by the organophosphate insecticides (e.g., diazinon, malathion, guthion). Carbamate insecticides (e.g., sevin) are generally the least toxic with lethal levels in the mg/L range. Hatchery managers should avoid the use of pesticides as much as possible. Topical application with hand sprayers is the best way to limit contamination of water supplies when pesticides must be used. Hatcheries near agricultural fields may be oversprayed as neighboring farms are treated. Hatchery managers should be aware of what kind of pesticides are used and when spraying occurs in the area. Managers of hatcheries that use surface water should be aware of pesticide use practices in the watershed of their water source.

Generally, herbicides are not as toxic to fishes as insecticides, and herbicides are not as persistent as insecticides. Lethal levels for 2,4-D, 2,4,5-T, simazine, paraquat, and diquat are all in the mg/L range. Several common agricultural herbicides have been adapted and labeled for use in ponds. Formulations of pesticides and regulations concerning their use change often. The local university cooperative extension service can provide up-to-date information on choice and use of pesticides and the possible consequences of pesticide use by neighboring farms.

Water Sources

General Comments

An adequate supply of high quality water is critical for hatchery operations. Sources of water for hatcheries are groundwater, surface waters, and domestic water supplies. The characteristics of the different types of water sources are presented in Table 6 (see pages 112–113). Many hatcheries have two or more water supplies, often with different water temperatures and quality. Hatcheries use large amounts of water. Hatchery location should be based on a thorough knowledge of local and regional hydrology, hydrogeology, and climatology. Whenever possible, a hatchery should control a substantial amount of land around its water supply. There are salmon hatcheries that have operated at the same site for more than 100 years. The water supplies for many hatcheries are threatened with encroachment by residential, agricultural, or industrial development. A hatchery that was far out in the country when initially developed may now be surrounded on all sides by housing and warehouses.

Groundwater

Groundwater is usually the best water source for hatcheries. A spring occurs when the water flows out of ground without pumping. Some springs reach the surface from a single point; others cover a wide area. When springs are used as hatchery water supplies, they should be enclosed in a concrete box or building to prevent airborne contamination and the growth of aquatic plants, and all wild fish and aquatic animals should be eliminated. Springs often have relatively constant temperature and flow. However, they may be supersaturated with nitrogen and carbon dioxide gases. The DO concentrations of springs are variable, depending on the hydrogeology. Most spring waters can be considered pathogen-free except in areas with solution cavities that allow rapid movement of water. The development of residential, agricultural, and industrial wells in an aquifer may substantially reduce the flow of springs over the long term.

Wells and springs from the same aquifer are of comparable quality, but wells are less desirable because of the need to pump. When a well is drilled into a confined aquifer, the pressure in the aquifer may be high enough so that pumps are not needed. Sites that can be developed with wells are much more common than sites with springs. When wells are used for intensive culture, standby generators or direct-drive engines must be provided. For critical applications, a comprehensive maintenance program is needed for the wells, controls, and standby units. The yield of a well depends on the permeability and thickness of the aquifer and the depth and diameter of the well. The yield of a well can be evaluated by monitoring the change in water level with pumping time. The operating level in the well may be 3–31 m (10–100 ft) below the static level (no pumping). For a given well and pump, the flow depends on the static lift (difference between the ground level and the operating water level in the well) and pressure losses in the distribution system. With time, water flow may decrease due to clogging of well screens from corrosion or growth of iron bacteria. The screens can be cleaned with a varying degree of success.

Surface Water

Surface water supplies include flowing sources (rivers and streams) and static sources (lakes, ponds, and reservoirs). The quality and quantity of surface waters are much more variable than groundwater.

Rivers and streams are exposed to the prevailing climatic conditions; flows change seasonally. Because of turbulence, the DO concentration is usually high. Streams are generally free of high levels of carbon dioxide or hydrogen sulfide, but may be supersaturated with atmospheric gases due to entrainment of air below dams or photosynthesis. At high elevations and latitudes, icing in the winter can be a serious operational problem. Rivers and streams with resident or migratory fish populations cannot be considered free of fish pathogens. Streams and rivers can provide large qualities of water that may not require pumping if adequate fall is available. The minimum streamflow may be important for many applications. Historic streamflow data are published by the U.S. Geological Survey, state agencies, and water districts. Although hatchery water use is nonconsumptive (all the water is returned), the amount of water that

TABLE 6. Characteristics of common water supplies.

Supply source	Quality available	Pathogen content	Temperature	Suspended solids
Spring	may vary with season and year	generally none if covered and fish excluded	may be stable but can change seasonally	low
Well	generally stable but may very with year		variation depends on depth of well	low
Rivers and streams	generally seasonal	variable	seasonal changes	high during peak runoff
Lakes, ponds, and reservoirs	large	variable	seasonal changes	turbidity can be high during some periods
Domestic water supply	adequate for incubation or experimental rearing but expensive		depends on source of water and distribution system	low

can be used from smaller streams is typically limited to less than 60%, due to regulatory limitations.

The temperature variation in lakes and ponds is more variable than in rivers and streams. Within a season, temperatures also may vary with depth. During the summer, limited vertical mixing may allow bacterial respiration to reduce the DO concentration in the bottom waters.

Domestic Water Supply Systems

Domestic water supplies may a convenient source of water for small-scale facilities where no other source is practical. Domestic water is treated to protect human health and, therefore, may contain lethal concentrations of chlorine or chloramines. Gas supersaturation is common in domestic water supplies. Contamination with copper, zinc, or lead may result from leaching of pipes and fittings. Potential toxicity problems from chlorine, metals, or gas supersaturation depend strongly on operation of the system. Because of the extreme variability in the potential

TABLE 6. (CONTINUED).

Dissolved oxygen	Carbon dioxide	Total gas pressure	Metals	Chlorine
variable	variable depending on recharge conditions and subsurface geology	variable	iron and manganese may be problem in water with low DOs	
variable	same as above	variable	same as above	
near saturation or above	low	high during spill periods in dammed rivers	depends on industrial and domestic discharges	
can be low during summer periods	low, but may depend on impact of photosynthesis and therefore change on a daily basis	can be super-saturated near the thermocline in the spring or on sunny afternoon near the surface	iron and manganese common in summer when using low dissolved oxygen bottom waters	
near saturation or above	low	variable, but more of a problem in the summer	copper and zinc may leach from pipes and fittings	commonly added to control bacteria in water and on pipes

toxicity problems, it may be difficult to justify the needed treatment systems.

Selection of Water Supplies

Selection of water supplies is generally based on the following three factors: quantity, pathogen content, and temperature. Additional information on bioengineering reconnaissance for new sites and water quality sampling is presented by Shepherd (1991) for salmonid hatcheries. For hatcheries using springs, wells, and small streams, the long-term water yields are generally not well known prior to construction, and low water flows may be substantially less than what was anticipated. Even when the flow remains close to the estimated design flow, changes or increases in rearing programs will commonly result in water shortages, especially during periods of abnormal climatic conditions. There are few hatcheries with too much water.

Temperature has a critical impact on developmental rates of eggs and growth rates of aquatic animals. Heating and cooling of water is very expensive and substantial changes in water temperature are only possible for incubation and early rearing where small flows are needed.

Most water supplies show a seasonal change in water temperature; surface waters can show both seasonal and diel (daily) temperature changes. The annual temperature variation in groundwater and springs depends on the depth of the well or subsurface geology. The average annual temperature of groundwater, at 10–20 m (33–66 ft) below the ground surface, is 1.1–1.7°C greater than the mean annual air temperature (Meisner et al. 1988). This relationship does not apply well to the northeastern United States, Canada, and Alaska, due to the insulating effects of snow cover.

In surface waters, the annual temperature variations can be modeled by a harmonic function (Collings 1973):

$$T = M + A(\sin[0.0172d + C]) \,, \tag{1}$$

where

T = Stream temperature (8°°C) on day d,
M = Mean annual temperature (°C),
A = Annual amplitude of the stream-temperature curve (°C),
d = Day of year (varies from 1 to 366), and
C = Phase coefficient of the cycle (radians).

Separate values of M, A, and C can be computed for the maximum daily, mean daily, and minimum daily temperature. This equation is useful for computing the daily or weekly temperatures for use in growth modeling.

The values of the M and A can be computed directly from the temperature data. The value of the phase coefficient should be computed by minimizing the sum of the squares between the observed and predicted temperature, but can be estimated from one the following relationships:

$$C = 0.0172\, d_{max} - 2.081 \,; \tag{2}$$
$$C = 0.0172\, d_{min} - 4.712 \,, \tag{3}$$

where

d_{max} = Date of maximum temperature;
d_{min} = Date of minimum temperature.

For convenience, 2π (6.283) should be added to all negative phase coefficients.

Small streams may have substantial diel temperature changes in the range of 5–10°C/d or greater. Activities such as logging and riparian vegetation removal can increase both the daily and yearly temperature variation. Extreme temperatures may have a substantial impact on fish health and growth, even though the daily average temperature is acceptable. It has been shown that in the wild salmon will avoid high temperatures by seeking out cool-water refuges (Berman and Quinn 1991), an option usually not available in a hatchery environment.

Water Transmission

General Comments

Water must be conveyed from its source to the hatchery and then distributed to individual rearing units. Water may be conveyed by open channels or pipes.

Open Channel Flow

Open channels may be cheaper to construct for large flows, but they require that the point of use be lower than the elevation of the water source, and the slope of the channel must always be downhill. Many open channels have high maintenance costs due to the growth of aquatic or emergent vegetation, deposition of sand and gravel, and bank scouring. Hatcheries with open channels may have more problems with diseases due to the introduction of pathogens by birds or greater difficulty in the eradication of resident fish.

Most open channels are either trapezoidal or rectangular in cross-section and are lined with concrete. A pipe partially filled with water (and open to the atmosphere at some point) functions as an open channel system. The capacity of an open channel is proportional to the water depth and slope and inversely proportional to the roughness of the channel lining (Linsley and Franzini 1972). Head losses (see Pumps and Pumping section for definition) in open channels are generally low, so in open channels used for water distribution (raceways or incubation systems), the difference in the water surface elevation is negligible along its length. This is important because it allows the water distribution channel to function as a head tank and facilitates the equal distribution of water between rearing units (see Water Flow Control section).

Pipe Flow

Piped water systems have two basic advantages over open channel systems. First, for a gravity system (source higher than the point of use), it is possible to convey water uphill as long as maximum elevation in the distribution system is less than the height of the source. Compared to an open channel system, a piped system may require substantially less earthwork because a continuous downhill slope is not needed. Second, if a pump is used, the elevation of the hatchery can be placed above the elevation of water supply. Although the use of pumps increases the operational costs and complexity of a hatchery, it may also allow for flexibility in the location of the hatchery with regard to the distance from the water supply and its elevation. A hatchery using an open channel system may be more subject to flooding due to the design constraints for the channel that require the hatchery be located downgrade from the intake. In a pumped system, it is possible to locate the hatchery at a much higher elevation and further away from the intake.

Piped water supply systems are typically designed using a maximum water velocity of 1.5 m/s (5 ft/s). This velocity criterion is based on water hammer and pressure loss considerations (see Water Flow Control section for discussion of water hammer). The minimum required pipe diameter is presented in Table 7 for a wide variety of flows. Pipes are only made in fixed sizes and not all manufactured sizes will be available in a given area. Therefore, it may be necessary to increase the size of a pipe selected from Table 7 to the next available physical pipe size. Hatchery piping may be constructed from concrete, steel, cast iron, HDPE, or PVC. The choice of pipe material depends on subsurface conditions, pressure requirements, and local availability of materials and contractors.

Pumps and Pumping

Most pumps used in hatcheries are electric-driven centrifugal pumps. The shape of the impeller depends on the required lift. In the design of hydraulic systems, it is common to use the term "head" to characterize the performance of pumps. The head is the elevation of a free water surface above or below a horizontal reference datum (Metcalf and Eddy, Inc. 1981). For a system under pressure, the head can be measured by connecting a small clear tube vertically; the head is the distance from the centerline of the pipe to the free water surface in the tube (Metcalf and Eddy, Inc. 1981). Head and pressure are related by the following relationship:

$$P = \gamma Z \,, \tag{4}$$

where

P = Pressure (Pa, lb/ft^2),
γ = Specific weight of water (kN/m^3, lb/ft^3), and
Z = Head (m, ft).

The specific weight of water depends slightly on temperature (Colt 1984) and at 15°C is equal to 9.799 kN/m^3 or 62.4 lb/ft^3 (or 9.799 kPa/m or 0.434 psi/ft).

TABLE 7. Required pipe diameter for a maximum water velocity of 1.5 m/s. Lpm = liters per minute; m^3/s = cubic meters per second; gpm = gallons per minute; cfs = cubic feet per second.

Lpm	Pipe diameter (cm)	Pipe diameter inches (in)	m^3/s	gpm	cfs
1,000	12	5	0.02	265	0.6
2,000	17	7	0.03	529	1.2
3,000	21	8	0.05	794	1.8
4,000	24	9	0.07	1,058	2.4
5,000	27	10	0.08	1,323	2.9
6,000	29	11	0.10	1,587	3.5
7,000	31	12	0.12	1,852	4.1
8,000	34	13	0.13	2,116	4.7
9,000	36	14	0.15	2,381	5.3
10,000	38	15	0.17	2,646	5.9
15,000	46	18	0.25	3,968	8.8
20,000	53	21	0.33	5,291	11.8
25,000	59	23	0.42	6,614	14.7
30,000	65	26	0.50	7,937	17.7
35,000	70	28	0.58	9,259	20.6
40,000	75	30	0.67	10,582	23.6
45,000	80	31	0.75	11,905	26.5
50,000	84	33	0.83	13,228	29.5
60,000	92	36	1.00	15,873	35.4
70,000	100	39	1.17	18,519	41.3
80,000	106	42	1.33	21,164	47.2
90,000	113	44	1.50	23,810	53.1
100,000	119	47	1.67	26,455	58.9

The amount of power needed for pumping is equal to

$$\text{Power} = \left[\frac{\gamma Q H_t}{K E_p}\right] \tag{5}$$

where

Power = kW (hp),
γ = Specific weight of water (kN/m^3, lb/ft^3),
Q = Water flow (m^3/s, ft^3/s),
H_t = Total dynamic head (m, ft),
K = Constant (1 for kW, 550 for hp), and
E_p = Pump efficiency (dimensionless).

Pump efficiencies usually range from 60 to 85%. For example, to pump 1 m^3/s with 10 m total dynamic head and a pump efficiency of 60% would require 163 kW (220 hp) of power. The total dynamic head is equal to static lift of the pump (elevation of discharge – elevation of suction), plus friction losses due to pipes and fittings. The static lift of the pump cannot be changed, although the friction losses due to pipes and fittings can be modified by changes in pipe size and fitting types. Information on the design of pumps and hydraulic systems can be found in standard references (Metcalf and Eddy, Inc. 1981).

The suction lift (measured from the centerline of the impeller to the level of suction water surface) of centrifugal pumps is typically limited to 5–7 m (16–23 ft); a limitation that depends on the specific pump used and design conditions.

Pumps at hatcheries are generally powered by three-phase electrical motors. When the primary electrical power is lost, standby pumping capacity can be provided by mechanical coupled engines or engine-generator sets. For pumps used in wells, it is possible to mount the motor above ground and use a vertical shaft to turn the impellers or use a submergible motor/pump combination.

Water Flow Control

In a pressurized water system, the flow to any rearing unit can be controlled by valves (Figure 10a). Ball valves are the most common types of valves used in hatcheries. The flow goes from full off to full on in a ¼ turn, and as a result, fine adjustment is difficult. Gate valves are a better choice for an application requiring accurate flow adjustment, but are more expensive than ball valves. In systems with high pressures, a ball valve can be turned off rapidly enough to result in water hammer problems. When a flow liquid in a pipe is abruptly stopped by the closing of a value, dynamic energy is converted to elastic energy and a series of positive and negative pressure waves travel back and forth in the pipe until they are damped out by friction (Linsley and Franzini 1972). These pressure waves sound like a series of hammer blowers and result in pressure transients high enough to break pipes or loosen tees and elbows.

A more serious problem with the use of valves to control flow in a pressurized distribution system is the interrelationship of flows between the different units. Figure 10a shows a pressurized water system for a series of vertical stack incubators. If there are substantial changes in pressure between points 1 and 6, then for equal valve settings, incubator 1 will get more water than incubator 6. In addition, changing a single valve setting will change the flow to all the incubation units. Accurate adjustment of flows to all units can be a very time-consuming and aggravating process, especially if the system pressure also changes with time.

A constant head system offers superior regulation of flows to a series of units. In Figure 10b, an open-channel header supplies water to a

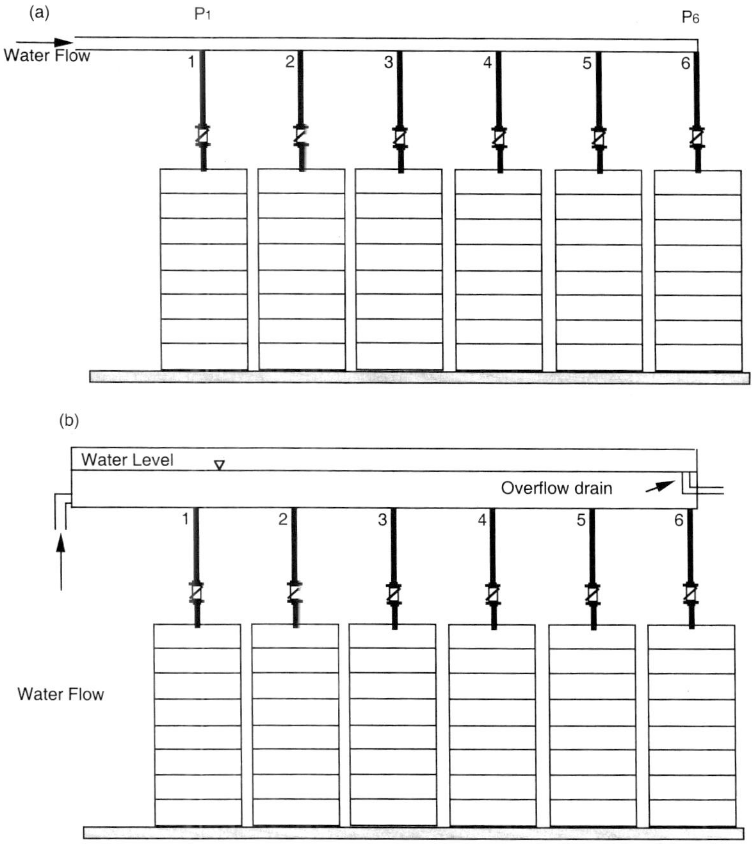

FIGURE 10. Comparison of (a) pressurized versus (b) constant headwater supply system for a series of vertical stack incubators. In the constant head system the water level in the distribution head is maintained a constant elevation (head) by the standpipe overflow drain. The symbol ∇ is used to indicate the water level. In the pressurized system, P_1 and P_6 are the pressures in the distribution pipe at points 1 and 6.

series of vertical stack incubators. The water velocity in the header is low enough to produce negligible head losses along the header. Since all incubators are exposed to the same pressure, all incubators will receive the same flow if valve settings are similar. To maintain the constant head in the header, enough water must be supplied to maintain a small flow over the standpipe drain.

The best way to control flow is use a constant head supply and orifice controls of water flow. In Figure 10b, this would require insertion of an orifice plate between the downstream side of the valve body and the valve union. The orifice can be constructed from 3 to 5 mm sheet PVC or other materials. The valve is not used for adjustment, only to turn the flow on and off. To change the flow, the union is removed and a new orifice plate is exchanged for the existing one. Orifice plates can be color coded for simplicity. The flow through an orifice plate is equal to

$$Q = KC_vA_o\sqrt{2gH}, \qquad (6)$$

where

Q = Flow (L/min, gal/min),
K = Conversion constant (60,000, 448),
C_v = Orifice coefficient (0.6–0.7 for both sets of units),
A_o = Area of orifice (m^2, ft^2),
g = Acceleration due to gravity (9.8 m/s^2, 32.2 ft/s^2), and
H = Difference in head between head tank and discharge (m, ft).

The constant-head reservoir does not have to be adjacent to the units if size of the distribution piping is large enough to reduce the pressure losses to negligible values.

The flow to a series of raceways is generally controlled by removable dam boards set in vertical slots in the concrete walls. Equal distribution of flow in this type of system requires that the elevations of the tops of the upper dam boards in each raceway are identical. Increasing the flow to a single raceway requires removal of one or more of the dam boards. Accurate adjustment of flow in these systems is difficult.

Water Flow Measurement

Although flow information would seem critical to the operation of hatcheries, flowmeters in many large hatcheries are not operated much past the end of their warranty period. Unless the flow information is actually useful to hatchery staff, the meters will not be used.

Open-Channel Flow

The discharge of water through a hatchery channel can be measured easily with a Cippoletti, rectangular, broad-crested, or triangular weir built into the channel (Figure 11). It is also possible to install weirs in dam board slots of a raceway discharge. The only measurement needed is that of the water elevation (head) with respect to the crest of the weir. Because of rapid changes in the water elevation near the weir, the use of the actual height of water above the weir will substantially underestimate water flow. Reference of the measured head to a calibration chart (Table 8) gives the corresponding discharge in liters per minute or gallons per minute.

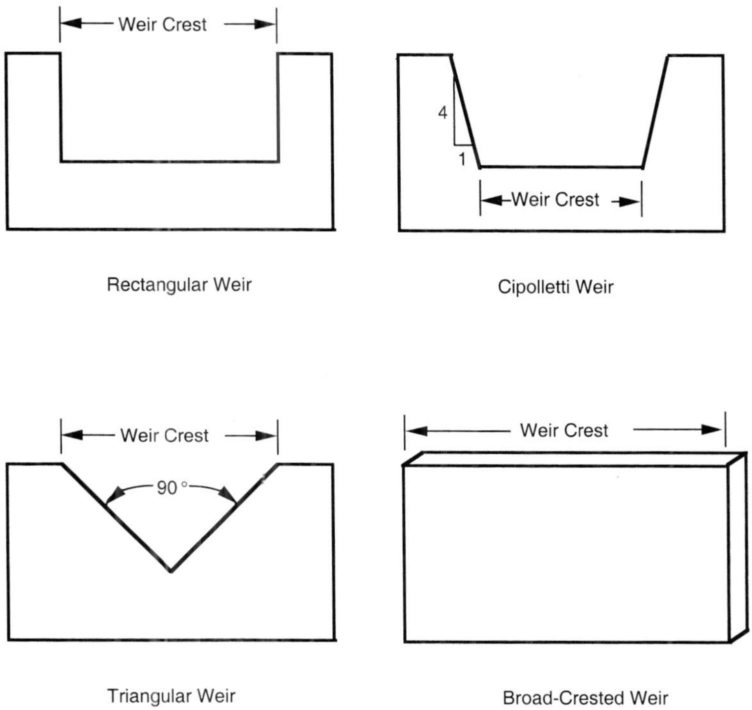

FIGURE 11. Diagram of rectangular, Cipolletti, 90° triangular, and broad-crested weir plates.

The water flow through a weir depends on the shape of weir (Table 8). The triangular weir can be used to measure smaller flows than the either the rectangular weir or Cippoletti weir. Weirs are accurate over a range of flows; the accuracy of a measurement will be substantially reduced when the required conditions are not met. The accuracy of the broad-crested weir is probably less than the other three types of weirs.

Water-flow determinations will be inaccurate if the head is measured at the wrong point or if the weir has not been constructed carefully (Figure 12). The following considerations must be met if weir operation is to be successful:

- The head must be measured at a point sufficiently far behind the weir. Near the weir, the water level drops as water begins its fall over the weir crest. The head should never be measured closer to the weir than 2.5 times the depth of water flowing over the crest. For example, if 5 cm (2 in) of water are flowing over the weir crest, the head should be measured 12.5 cm (5 in) or more behind the weir. A

TABLE 8. Water flow as a function of head and weir types for a 0.5 m crest length (see Figure 11 for a diagram of the different types of weirs). For the rectangular weir, it is assumed that the weir does not extend completely across the channel (i.e., it is a contracted weir). Computations are based on Metcalf and Eddy Inc. (1981).(Gallons per minute = liters per minute/3.78; cubic meters/second = liters per minute/60,000.)

	Water flow (L/min)			
Head (cm)	Rectangular	Cipolletti	90° Triangular	Broad-crested
0.50	19	20	0.15	16
1.00	55	56	0.83	44
1.50	101	102	2.28	81
2.00	156	158	4.68	125
2.50	218	220	8.18	175
3.00	286	290	13.00	230
3.50	361	365	19.00	290
4.00	441	446	27.00	354
4.50	526	532	36.00	422
5.00	615	623	46.00	494
5.50	710	719	59.00	570
6.00	809	820	73.00	650
6.50	912	924	89.00	733
7.00	1,019	1,033	107.00	819
7.50	1,129	1,145	128.00	908
8.00	1,244	1,262	150.00	1,001
8.50	1,362	1,382	174.00	1,096
9.00	1,484	1,506	201.00	1,194
9.50	1,609	1,633	230.00	1,295
10.00	1,737	1,763	262.00	1,398
11.00	2,003	2,035	332.00	1,613
12.00	2,282	2,318	413.00	1,838
13.00	2,572	2,614	505.00	2,073
14.00	2,873	2,921	607.00	2,316
15.00	3,185	3,240	722.00	2,569
16.00	3,507	3,569	848.00	2,830
17.00	3,840	3,909	987.00	3,100
18.00	4,182	4,259	1,138.00	3,377
19.00	4,533	4,619	1,303.00	3,662
20.00	4,894	4,988	1,481.00	3,955
21.00	5,263	5,367	1,674.00	4,256
22.00	5,641	5,754	1,880.00	4,563
23.00	6,028	6,151	2,101.00	4,878
24.00	6,422	6,557	2,337.00	5,199
25.00	6,825	6,971	2,588.00	5,528
26.00	7,236	7,393	2,855.00	5,863
27.00	7,654	7,824	3,137.00	6,204
28.00	8,080	8,262	3,436.00	6,552
29.00	8,513	8,709	3,751.00	6,906
30.00	8,954	9,163	4,082.00	7,266
35.00	11,260	11,547	6,002.00	9,157
40.00	13,730	14,108	8,380.00	11,187
50.00	19,110	19,716	14,639.00	15,635

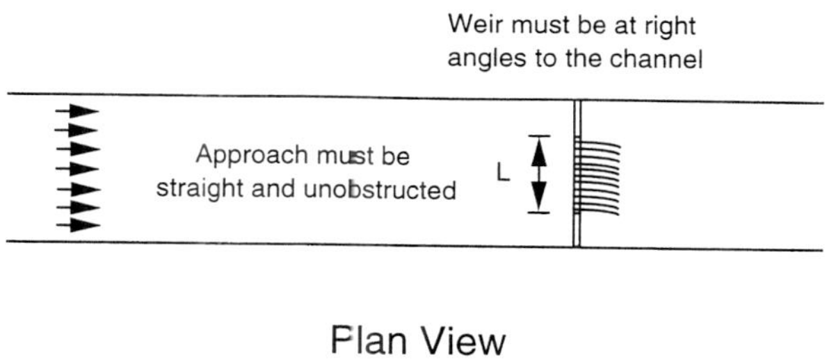

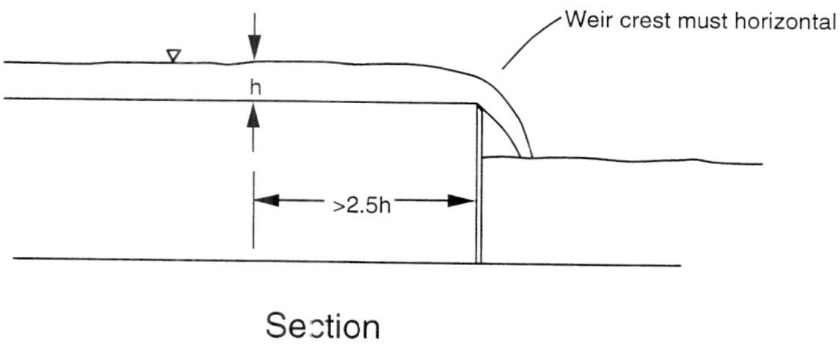

FIGURE 12. Plan and section of weir installation. The symbol ∇ is used to indicate the water level. L is the length of the weir crest and h is the head on the weir crest.

practical measuring technique is to drive a stake into the channel bottom so that its top is exactly level with the weir crest. Then, the head can be measured with a thin ruler as the depth of water over the stake. A ruler also can be mounted permanently on the side of a vertical channel wall behind the weir, if such a wall is available. For constructed weir stations, it is common to build a vertical stilling basin to increase the accuracy of the head reading. It is also possible to automatically measure and record water height information from a weir.

- The weir crest must be exactly level and weir faces exactly vertical, or the standard head-to-discharge calibration will not apply.
- The weir crest, formed from a metal plate, must be leakproof, sharp or square-edged, and no thicker than 3 mm (1/8 in). The distance of the weir crest above the bottom of the channel should be at least 2.5 times the water head of the weir to minimize approach water velocity.
- Air must have access to the underside of falling water as it flows over the weir crest. Otherwise, air pressure may force water against the downstream face of the weir, increasing the flow rate above the standard table values.

- The channel above the weir must be straight, level, and clean to ensure smooth water flow. Sediment and debris should not be allowed to collect on or behind the weir.

Piped Systems

There are a wide variety of flowmeters that can be used to measure flow in pipes. If necessary, it is possible to transmit the flow information to a central site, at substantially increased costs. One of the most commonly used flowmeters is a simple, inexpensive paddlewheel flowmeter (Figure 13). One type of this meter is self-powered. Most flowmeters assume that the flow is fully turbulent; the meter must be mounted in a straight section of pipe (10 pipe diameters in the upstream direction and 5 pipe diameters in the downstream direction). In many hatchery applications, it is impossible to satisfy these conditions and the accuracy of the measured flow is seriously degraded. There are many other types of flowmeters that are more accurate than the paddlewheel, but at substantially higher prices and complexity. Highly accurate flow information is probably not critical for most applications and a simple flowmeter may be preferred over an expensive, highly accurate meter.

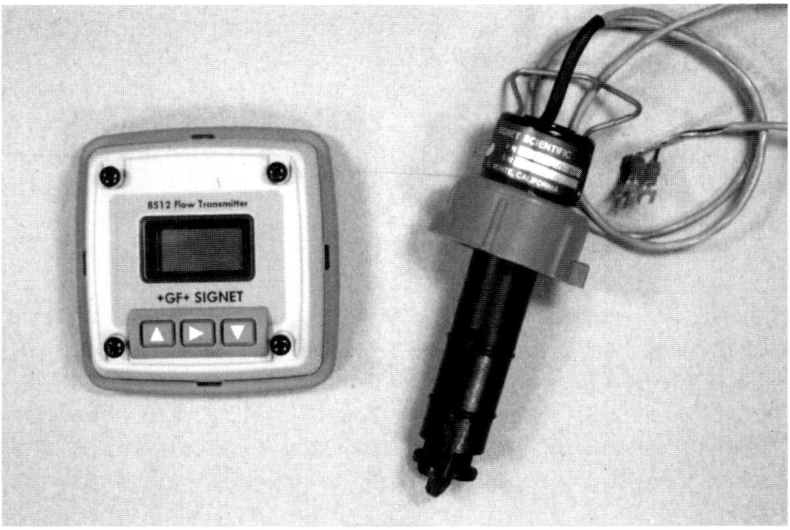

FIGURE 13. Photograph of paddlewheel flow sensor and readout/transmitter (Signet Instrument Company).

Water Treatment Systems

General Comments

An adequate supply of high quality water is critical for hatchery operations. Aquatic rearing systems may be classified as static, flow-through, reuse, or closed depending on the characteristics of water usage (Figure 14). A warmwater pond is an example of a static system (Figure 14a). Although there may be a discharge from a pond during rainy periods (or periodic draining), water is generally added only to make up for evaporation and percolation. In a flow-through system, a continuous flow of water is used to provide DO and remove metabolic products (Figure 14b, 14c). In a reuse system, water quality is maintained by treatment and reuse of process water (Figure 14d). The water requirements of a reuse system can be reduced to 10–20% of an equivalent flow-through system. A closed system is a reuse system that only provides new water for evaporation, spillage, and solids losses (Figure 14e). Large inland marine aquariums and small research aquariums are very close to true closed systems. Closed systems are not generally cost effective for production-oriented systems.

Even though the water supply may be abundant and of high quality, most hatcheries require some type of water treatment. The location of water treatment units can be divided into pretreatment, internal processing, and posttreatment. Pretreatment of the water supply may be needed to remove specific compounds (iron, manganese, nitrogen gas) or organisms prior to its use in the hatchery, or to add specific substances (oxygen, calcium). Posttreatment is commonly required because of discharge limitations or regulatory permits. Internal processing occurs in the rearing units (for example, aeration or sedimentation of fecal solids) or prior to reuse of the water (carbon dioxide and ammonia removal).

Because of the complexity of the different types of systems and water supplies, it is impossible to develop a general process flow sheet for even a specific type of hatchery. Some of the common processes that are needed in the three locations are presented in Table 9. These are the most common locations of hatchery treatment processes, but not all treatment processes will be needed at a specific site or application.

Information will be provided in the following sections on solids removal, heating and chilling, aeration, degassing, oxygen supplementation, disinfection, dechlorination, iron and manganese removal, ammonia removal, and alarms and control systems. The information is presented on a unit-process basis where the operation of a specific process is considered independently of other potential treatment processes that may be included. The design of the overall treatment system must

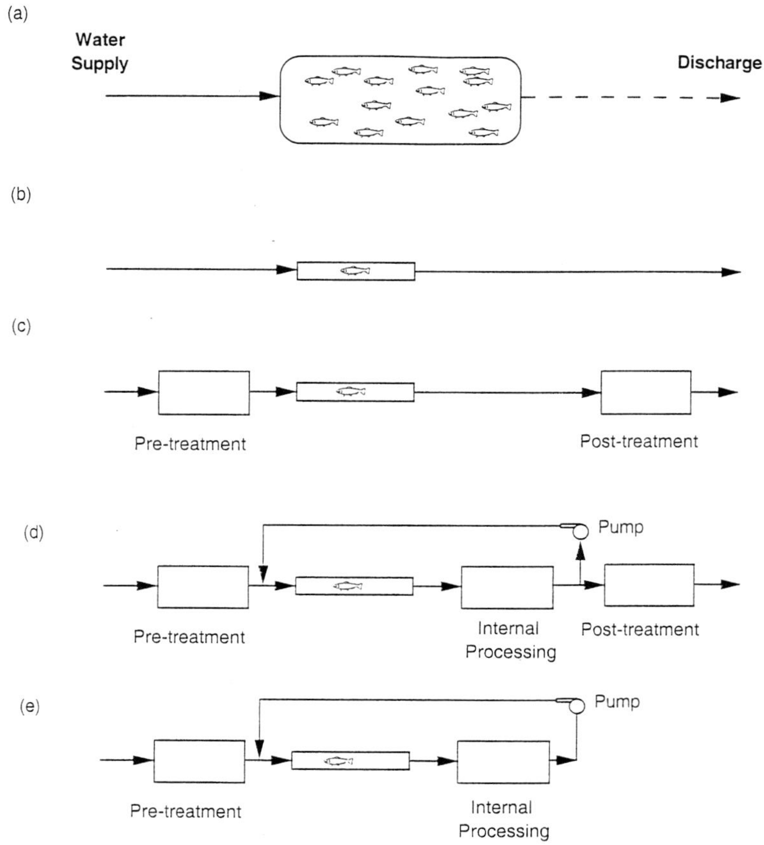

FIGURE 14. Classification of culture system by water usage characteristics. (a) static pond system, (b) simple flow-through system, (c) complex flow-through system, (d) reuse system, and (e) closed system.

consider the interaction of all the treatment processes and operation. For example, improved removal of uneaten feed can reduce ammonia production and the size of the ammonia removal unit.

Many of the treatment processes used in hatchery applications have been developed for wastewater treatment or industrial applications. Because of substantial differences in concentrations and operational requirements for hatchery applications, the operational performance of these treatment systems may be quite different for hatchery applications. Because of these potential problems, information on operations under similar hatchery applications is highly desirable. A short telephone call to a hatchery manager who is using a similar unit may be more valuable than a five-page color brochure from a sales representative.

TABLE 9. Overview of potential hatchery treatment processes.

Pretreatment	Internal processing	Posttreatment
Aeration and degassing	Aeration	Disinfection
Sedimentation	Sedimentation	Sedimentation
Filtration	Ammonia removal	
Disinfection	Carbon dioxide removal	
Dechlorination	Organic removal	
Iron and manganese removal	Heating and cooling	
Heating and chilling	Chemical addition	
Chemical addition		

Solids Removal Systems

Solids removal will commonly be required in aquaculture systems. Sources of solids include the water supply, algae, uneaten feed, and fecal solids. Suspended solids are measured by filtering a known volume of water through a standard filter (generally, 0.45 m pore size) and weighing the dried material retained. The two most important physical characteristics of solids from the standpoint of water treatment are specific gravity and size. The biological effects of suspended solids are best defined for inorganic solids (silt and sand). Generally, smaller particles are more difficult to remove than larger particles. Turbidity refers to reduction in light transmission due to very fine particles or color compounds. Suspended solids and turbidity measure two fundamentally different physical characteristics of a water, although the two variables are commonly highly correlated.

The selection of the specific removal processes for solids will depend on the quality of water needed, size and density of particles present, and costs. The general characteristics of different solids removal processes are presented in Figure 15. Detailed information is presented below for the commonly used solids removal processes.

Sand Filtration

These filters use a granular medium of sand, garnet, or anthracite coal. Straining is the major mechanism of removal. The most common sand filters used in hatchery applications are pressure sand filters (Figure 16a). A single media or dual media (e.g., anthracite and sand, or sand and garnet). There is no one mixed media design that is optimum for all applications. For critical applications, pilot-scale testing of different media may be necessary.

Sand filtration involves two phases: filtration and backwashing. The water is applied to the filter until the solids in the effluent reach an unac-

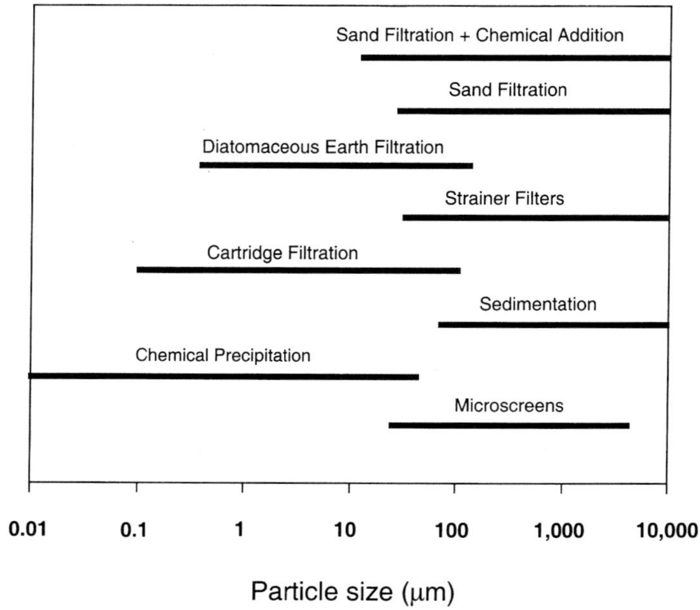

FIGURE 15. Removal characteristics of common solid removal processes as a function of particle size.

ceptable concentration or the head losses across the filter reach a limiting value. At that point, the filter is cleaned or backwashed by reversing the water flow through the filter. The water flow is increased until the filter medium is expanded and the accumulated material in the filter is removed. A clear section of pipe may be provided so the quality of the backwash water can be seen.

Filters are commonly designed for loadings of 300–400 Lpm/m^2 (7–10 gpm/ft^2) for filtration and 600–800 Lpm/m^2 (15–20 gpm/ft^2) for backwashing. The horizontal orientation of the tanks may offer a smaller footprint (floor area/gpm) than vertical tanks. Filters with a capacity of 200–2,000 Lpm (50–500 gpm) are commercially available. For marine applications, fiberglass filters offer longer lifetime due to absence of corrosion. It is generally necessary to provide three or more filters so the output of two filters can be used to backwash the third filter. If the captured solids are not adequately removed during the backwash cycle, "mudballs" can form in the filters. This results in channeling of water flow through the filter and reduced solids removal. In extreme cases, it is necessary to physically remove the media from the filter and replace it with new media. The pressure losses for a clean filter are in the range of 20–35 kPa

FIGURE 16. Photograph of (a) pressurized sand filter (Astral Products Limited) and (b) microscreen (Engineered Products, used with permission).

(3–5 psi) and increase to 150–250 kPa (20–35 psi) prior to backwashing. It is commonly necessary to provide a booster pump for sand filters.

Microscreens

This method uses a horizontally oriented cylindrical screen that rotates at 1–3 rpm (Figure 16b). The solids are removed by straining and collect on the inside of the cylinder. A high pressure spray system removes the accumulated solids from the screen material and deposits them in an effluent channel. The spray system can be controlled by an increase in the water level in the microscreen vault or by a timer. The water flow in the effluent channel ranges from 5 to 20% of the total flow. The solids in this flow may be further concentrated by sedimentation or other processes.

The capacity of a microscreen depends on the solid concentration, cross-sectional area of the screens, and screen size. Recommended capacities for typical salmonid applications are presented in Table 10. Microscreens have been widely used to remove uneaten feed and fecal matter for salmon and trout facilities. The head losses through microscreens are in the range of 5–10 cm (2–5 in) so it may be possible to operate the system without additional pumping.

Gravity Sedimentation (Clarifiers)

The settling velocity of a particle depends on its diameter and density. The larger the particle size and density, the faster the particle will settle. Very small particles such as clay may not settle at all, as the random movement of water molecules is greater than the settling velocity.

The design of gravity sedimentation can be based on batch settling testing (Schroeder 1977) or pilot-scale experiments. Information on the design procedures can be found in standard references (Metcalf and Eddy, Inc. 1979). Sedimentation basin designs are based on the overflow rate or surface loading rate in $m^3/m^2{\cdot}d$ (gpm/ft^2). The operation of the ideal sedimentation basin is independent of depth. Therefore, shallow basins

TABLE 10. Loading rate of typical microscreen as a function of mesh size and solids type (EMA-Engineered Produces Division, used with permission).

Mesh size (μm)	Loading rate in Lpm/m² (gpm/ft²) of submerged screen area: Clean water 10 mg/L TSS	Effluent water 25 mg/L TSS
18	1,800 (45)	980 (24)
21	2,100 (51)	1,100 (28)
27	2,811 (69)	1,500 (38)
30	2,811 (69)	1,500 (38)
37	3,500 (86)	1,900 (47)
60	4,500 (110)	2,400 (60)
89	5,500 (134)	3,000 (73)
120	6,300 (154)	3,400 (84)
150	6,400 (158)	3,500 (86)
285	7,700 (188)	4,200 (103)

should work as well as deeper basins. Although practical considerations limit the use of shallow basins, tube settlers have been developed to allow higher overflow rates (Hansen and Culp 1967).

The design of sedimentation basins for aquatic culture systems is quite different from those used in wastewater treatment. Hatchery wastes are denser, settle faster, and tend to form a heavy, viscous sludge (Mudrak 1981). The solids loadings to the sedimentation basin are also lower than wastewater applications so that continuous sludge removal is not generally required. A cross-section of a typical rectangular clarifier is presented in Figure 17.

Typical overflow rates range from 40 to 60 $m^3/m^2{\cdot}d$ (1,000–1,500 gpd/ft^2) for coldwater fish and 160–175 $m^3/m^2{\cdot}d$ (4,000–4,500 gpd/ft^2) for warmwater fish. The overflow rate should be based on peak hourly flow. The basin depth typically ranges from 1.2 to 1.8 m (4–6 ft), but may depend on the type of sludge removal used. Mudrak (1981) has recommended that a earthen stabilization pond follow the sedimentation basin. The pond should be designed with a 4 h retention time and an average depth of 1.2–1.5 m (4–5 ft). This pond will serve as a buffer zone should any operational problems occur with the sedimentation basin.

The amount of sludge recovered in the sedimentation basins ranges from 3.2 to 25.8% of the dry weight of the feed (Mudrak 1981). Each 1,000 kg (2,205 lb) of dry feed will produce 2.5 m^3 (88 ft^3) of waste sludge at approximately 8% solids. Direct land application is the recommended procedure for sludge disposal. On-site storage of sludge may be required during the winter when rain or snow make land application impossible. Sludge can be applied to agricultural land at an annual rate of or 6–13 metric ton dry sludge per hectare (3–6 ton/acre) or 80–160 m^3 wet sludge per hectare (9,000–17,000 gal/acre) depending on crop type (Mudrak 1981).

In concrete raceways, in-raceway settling can be provided by installing screens 3–5 m (10–15 ft) in front of the overflow weir (JRB 1984). The screens exclude fish from the settling area and provide a quiet area for

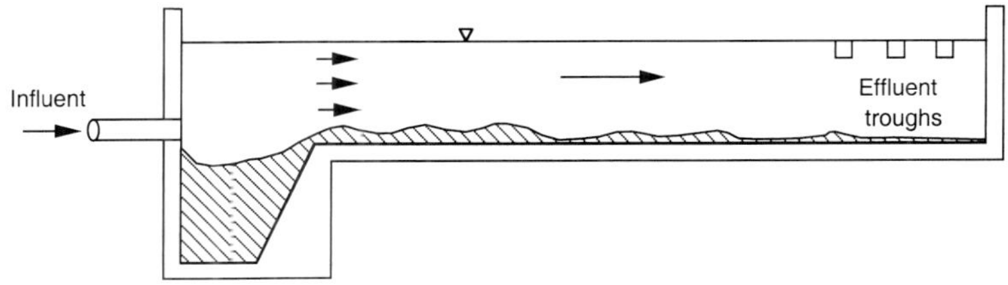

FIGURE 17. Cross-section of typical rectangular clarifier (Mudrak 1981). The symbol ▽ is used to indicate the water level.

uneaten feed and fecal matter to settle out. The screens can be constructed from PVC pipe, aluminum pipe, or plastic mesh. Solids are removed by siphoning or pumping once- or twice-weekly. The installation of screens and their proper operation can reduce the solids discharged by up to 75% (JRB 1984). Because of the low concentrations of solids present in hatchery effluent and high sampling variability, the performance of these types of systems can be highly variable.

Other Types of Solids Removal Systems

There are a variety of other types of solids removal systems that may have application in some hatcheries. Some of the key characteristics are presented in the following table:

Diatomaceous earth filter	Can remove some bacteria and parasites. Can remove particles as small as 2 to 20 μm; expensive to operate. Used in aquarium applications where clarity is critical.
Cartridge filter	Limited capacity and very expensive; can remove 0.1 to 40 μm particles. Can be used to polish effluent from sand or diatomaceous filter.
Strainer filter	Simple to clean and cheap. Can remove particles in the 0.5 to 5mm range. Large mesh strainers can be used to remove fish, crustaceans, and debris from surface waters or discharge.
Chemical precipitation	Only method for removal of clay turbidity. May require lime addition to prevent depression of pH and akalinity.

Additional information on these methods for removing solids can be found in Boyd (1979b), Huguenin and Colt (1989), and Spotte (1979).

Heating and Chilling

Heating and chilling are two of the most common processes used in water treatment. Because most cultured species are temperature conformers, temperature can be used to control developmental and growth rates.

Energy Requirements

Energy is commonly measured in joules (J), British thermal units (BTU), calories, kilowatt-hour (kW-h), and horsepower-hour (hp-h). Power, the rate of energy use, is expressed in watts (W), kilowatts (kW), BTU/s,

BTU/h, and horsepower (hp). Various types of heating and cooling equipment are rated in specific units by practice and convention.

The energy required to heat or cool water can be computed from the following equation:

$$q = mC_p\,(DT), \qquad (7)$$

where

q = Energy required (watts, BTU/s),
m = Mass flowrate (g/s, lb_m/s),
C_p = Heat capacity of water at constant pressure (Joule/[g·°C], BTU/[lb_m·°F]), and
ΔT = Change in temperature (°C, °F).

The heat capacity of water is equal to 4.1855 J/(g·°C) or 1 BTU/(lb_m·°F). Heating and cooling of process water requires a huge amount of energy. For example, heating of 1,000 liters (264 gal) per minute of water by 10°C, requires 698,000 W (698 kW or 935 hp). Heating and cooling of water is commonly restricted to static systems or incubation and early rearing systems where the flows are small.

Heating

In aquatic systems, heating is commonly provided by electrical resistive heaters or boilers. Electrical heating depends on the conversion of electrical energy to heat. The heating elements are high resistance wires that may be directly immersed into the water or covered with an outer covering to prevent corrosion. Materials used to cover the heating elements are stainless steel, quartz, titanium, plastics, or Teflon. The choice of material depends primarily on chemical resistance, resistance to mechanical damage, and resistance to burnout if not immersed in water. Electrical heaters are available in the range of 20–36,000 W. Electrical heaters are almost 100% efficient in the conversion of electrical energy to heat. Generally, electrical heaters are limited to very small applications due to costs.

Boilers may be fueled by natural gas propane, butane, or oil and can be obtained in the size range of 15–500 kW and greater. The efficiency of boilers ranges from 60 to 70%. They can be purchased as a unit, including controls. The higher volumetric density of propane, butane, and heating oil allows substantial storage of fuel on site. On-site storage can be used to prevent disruption of the supply during bad weather, but involves some additional risk. The operation of boilers requires trained personnel and, in large installations, may require licensed operating engineers. The operation of boiler systems is much more complex than electrical heating systems.

Cooling

Cooling or chilling requires the use of a compressor-based refrigeration system (Figure 18). Chillers are commonly rated in tons of capacity. A ton of capacity is equal to 12,000 BTU/h. The energy requirements of chillers ranges from 0.70 to 1.0 tons/kW. The performance of a chiller depends on the ambient temperature.

Energy Recovery

It is possible to recover up to approximately 50% of the added energy using a heat exchanger. In Figure 19a, the chilled incubation water is pumped through a heat exchanger where the incoming process water is chilled. Because of the relative small difference in temperature between the two waters, a plate and frame heat exchanger is needed. There is no mixing between the two water streams.

Another way to reduce energy consumption is to use a second water source of different temperature to heat or chill the process water (Figure 19b). Examples of this approach might include the use of cold stream water to chill warmer groundwater or the use of warm seawater to heat cold stream water.

FIGURE 18. Photograph of 100-ton chiller at Manchester Field Station, Manchester, Washington.

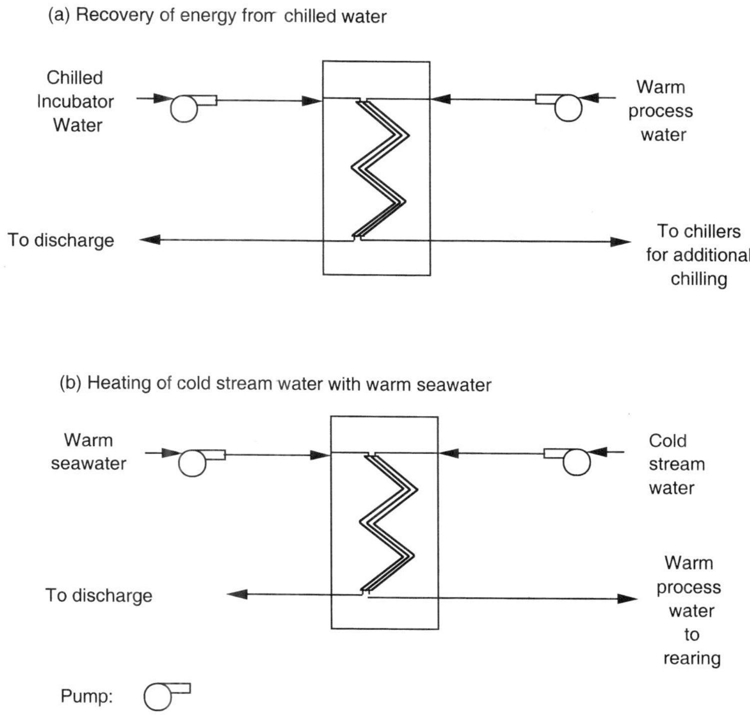

FIGURE 19. Two potential heat recovery options. (a) recovery of energy from chilled water and (b) heating of cold water with warm seawater.

Aeration

Aeration or the addition of DO is one of the most common processes used in aquaculture. The "best" aeration system for a given application depends on site conditions, production schedules, layout of the rearing units, and operational procedures. The design of an aeration system must consider the potential impacts on all dissolved gases in solution. The depletion of oxygen may not always be the most limiting measurement and, when ammonia or carbon dioxide is more limiting, aeration will have little effect on carrying capacity (Colt and Orwicz 1991a).

Types of Aerators

Surface aerators spray or splash water into the air and disrupt the surface film of the water. The paddlewheel aerator is commonly used to aerate commercial ponds (Figure 20a). Subsurface aerators mix water and air together in an aeration basin and transfer oxygen from air bubbles into the water. Diffused aerators (Figure 20b) and airlift pumps (Figure

FIGURE 20. Common types of aerators used in hatcheries. (a) paddlewheel aerator (Craig Tucker; used with permission), (b) diffused aerator (Four Points Systems Inc., used with permission).

20c) are used in transport and static culture systems. Gravity aerators are a special type of surface aerator that uses gravity rather than mechanical power to transfer oxygen. This type of aerator is commonly used in flow-through systems where adequate head is available (Figure 20d).

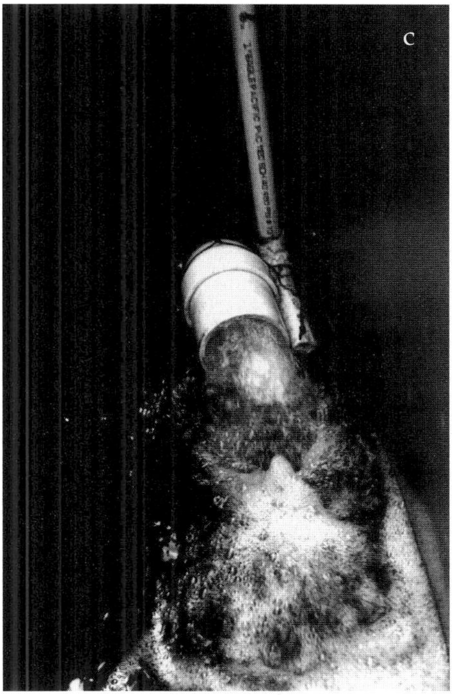

FIGURE 20 (CONTIUNED). Common types of aerators used in hatcheries. (c) airlift pump, and (d) gravity aerator.

Aerator Configuration and Location

Depending on the system configuration or operational limitations, aerators may be placed at different locations (Figure 21). The aerators may

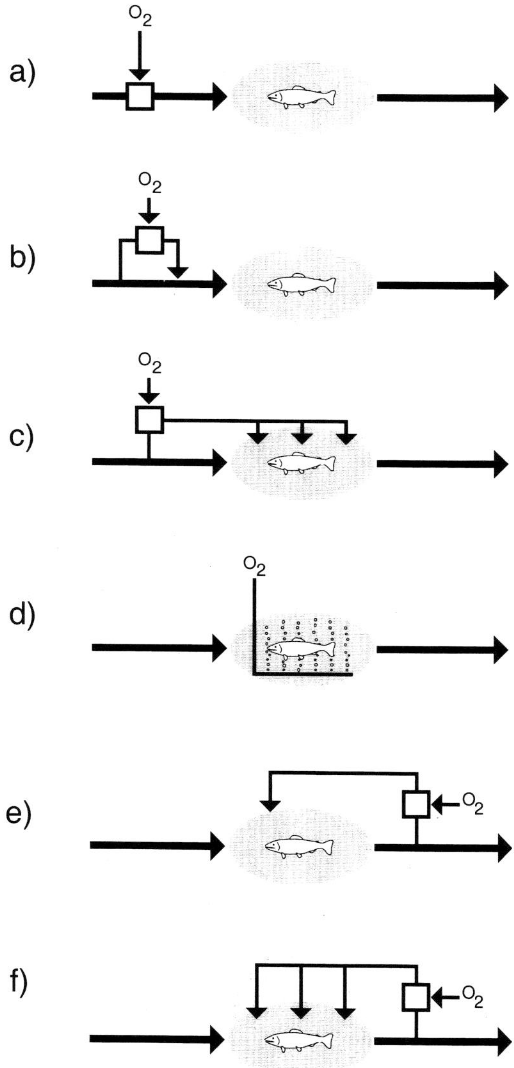

FIGURE 21. Location of aerators in hatchery applications. (a) influent aeration—full-flow, (b) influent aeration—side-stream, (c) influent aerator with multipoint introduction of high DO water into rearing unit, (d) diffused aeration in rearing unit, (e) side-stream aeration of used process water, and (f) side-stream aeration of used process water with multipoint introduction of high DO water into rearing unit.

be located in the influent stream (Figures 21a, 21b, and 21c), recycle stream (Figures 21e and 21f), or rearing unit (Figure 21d). In the recycle configuration (Figures 21e and 21f), the recycle flow may be several times larger than the influent flow. In a side-stream system (Figures 21b and 21c), only a part of the water flow passes through the aeration system.

Gas Transfer and Aeration Rating

The rate at which oxygen is transferred into water is proportional to the area of the gas-liquid interface and the gradient between saturation and the existing concentration of gas in the water. A positive gradient transfers oxygen into the liquid phase, and, conversely, a negative gradient transfers oxygen into the gas phase.

Aerators are rated in terms of standardized aeration efficiency (SAE) using standardized testing procedures (Colt and Orwicz 1991b). The SAE measures the amount of oxygen transferred per energy input (kg O_2/kW·h, lb O_2/hp·h) at a DO concentration of 0 mg/L, 20°C, and 760 mm Hg (atmospheric pressure). The standardized procedure for the determination of SAE reduces the uncertainty in aerator rating and allows more meaningful comparisons between units. Values of SAE are presented in Table 11 for aerators typically used in aquaculture.

Actual performance under field conditions depends primarily on the required DO concentration (C) and to a lesser extent on temperature and pressure. The rating of aerators under field conditions requires the computation of the field aeration efficiency (FAE). The computation of these variables assumes that the field installation is identical to the unit tested under standard conditions. FAE is the oxygen transfer per unit power input under field conditions and is derived as follows:

$$\mathrm{FAE} = \mathrm{SAE}\left[\frac{1.024^{(t-20)}(C^* - C)}{9.092}\right] \tag{8}$$

where

FAE = Aeration efficiency under field conditions (kg O_2/kW·h, lb O_2/hp·h),
SAE = Aeration efficiency under standard conditions of 0 DO, temperature = 20°C, and clear water (kg O_2/kW·h, lb O_2/hp·h),
t = Water temperature (°C),
C^* = Saturation concentration at local barometric pressure and temperature (mg/L), and
C = Concentration of DO in water (mg/L).

The FAE values for aquaculture systems will be substantially less than the listed SAE values due primarily to the necessity of maintaining

TABLE 11. Typical standardized aerator efficiency (SAE) for aerators used in aquaculture (Adapted from Colt and Orwicz 1991b). (lb O_2/hp·hr = kg O_2/kWhr x 1.6440)

Type	SAE (kg O_2/kW·hr)
Surface aerators	
Low-speed surface	1.2–2.4
Low-speed surface with draft tube	1.2–2.4
High-speed surface	1.2–2.4
Paddlewheel	
triangular blades	2.7–2.9
PVC pipe blades	1.2–1.9
tractor powered	1.3–2.0
Gravity aerators	
Cascade weir (45°)	1.5–1.8
Corrugated inclined plane (20°)	1.0–1.9
Horizontal screens	1.2–2.6
Lattice aerator	1.8–2.6
Packed column	
zero head	1.2–2.4
0.5–1.0 m head	10–80[a]
Aeration cone	2.5
Submerged aerators	
Air-lift pump	2.0–2.1
Diffused-air	
fine-bubble	1.2–2.0
medium-bubble1.0–1.6	
coarse-bubble	0.6–1.2
Nozzle aerator	1.3–2.6
Propeller-aspirator-pump	1.7–1.9
Static tube	1.8–2.4
U-tube	
zero head	0.72–2.3
0.5–1.0 m head	10–40[a]
Venturi aerator	2.0–3.3

[a] Does not include pumping power.

a DO concentration of 5–7 mg/L. For example, at 30°C, and C = 5 mg/L, FAE is equal to only 36% of the SAE value.

Process Selection and Design

A wide range of aeration devices is available for aquaculture. The actual type selected will depend on a variety of factors related to the characteristics of the aerator, the culture system, site conditions, and system operations.

Aerator efficiency and effluent DO levels are important for many systems. Systems with high values of FAE will reduce electrical costs. Another consideration is that lethal dissolved gas pressures may be produced by some types of submerged aerators (Colt and Westers 1982).

In a many systems, oxygen transfer rate is more important than efficiency. Good examples are tractor-powered paddlewheel aerators, which have been widely used in catfish ponds for emergency aeration and can easily be moved from pond to pond when needed. These aerators are used when conditions are critical, so the ability to transfer large amounts of oxygen is more important than efficiency. Diffused aeration with pure oxygen is widely used in transportation systems and emergency systems for high-intensity systems due to its ability to transfer large amounts of oxygen without any power input.

The selection of aerators will also be based on the physical characteristics of (a) the site; (b) the number, size, and configuration of rearing units; (c) the hydraulics of rearing units; and (d) the mode of operation. In many cases, the construction of the system may be completed before the need for an aeration system is realized. Therefore, it is commonly necessary to design or retrofit an aeration system around a given system, rather than design a complete culture system from scratch. Some of the most common site considerations are presented in Table 12.

The number, size, and configuration of rearing units is important in the selection process. In large tanks or ponds, individual mechanical or floating aerators may be used. In the aquarium trade, where a large number of small tanks must be aerated, diffused aeration is commonly used. Although the amount of air available is relatively fixed, the air can be inexpensively distributed to a large number of individual units, and flow to an individual unit may be varied easily.

Aerators that interfere with the normal operations of a culture system or require extensive maintenance will probably not be used too long. Operational personnel may lack the time, knowledge, or tools to operate and repair some types of aerators. Aerators such as gravity aerators or paddlewheels that can be constructed and repaired on site may be better choices for some operations even if their overall efficiency is lower than other types of aerators.

Degassing

Supersaturated dissolved gases are common in water from wells, springs, streams, and lakes and may vary considerably between seasons and operations. Chronic exposure of hatchery fish to supersaturated gases may result in developmental problems, increased incidence of disease, and mortality from gas bubble disease.

TABLE 12. Common design considerations for aeration systems.

Item	Considerations
	Site
Head	If enough head is available, operation of a gravity aerator may be possible without the use of external power.
Power	If electrical power is unavailable or unreliable, it may be necessary to use a motor generator system or engine-powered aerators.
Location	In remote locations, the lack of spare parts and trained personnel may favor simple systems.
Subsurface	In areas with rocky or unstable soils, excavation may be expensive.
Layout	Retrofitting existing hatcheries should involve careful consideration of problems associated with installing additional electrical lines, piping, and pumps between or around existing structures and utilities. A side-stream pure oxygen system may be easier to retrofit at some sites.
	Operational
Fouling of diffusers	Diffusers (airstones) may foul due to the growth of algae or bacteria. Fine-bubble diffusers may require special air filters and nonmetal air lines to prevent clogging due to rust and scale. Diffusers may foul rapidly if not operated continuously.
Icing	Surface aerators and some types of gravity aerators produce enough spray to develop ice on walkways and roads. This may present safety hazard to personnel.
Safety	Electrical lines, fuel tanks, or rotating shafts may present a safety hazard to personnel. Diffused aeration systems may be safer than other systems because the electrical lines are only required for the central blower unit. Electrical safety is a major concern in marine systems due to the high conductivity of seawater.
Harvesting/feeding	In ponds, static tube or surface aerators may need to be removed prior to harvesting. Aerators should not interfere with the daily operations of the facility.
Repair	The ease of repair may be an important consideration in remote locations. This includes both the skills and tools required and local availability of spare parts.
Reliability	A simple and highly reliable aerator is desirable. When adequate head is available, gravity aerators will operate during power failures.

Types of Aerators Used for Degassing

The most commonly used aerator for degassing is the packed column aerator (PCA). The PCA consists of a column filled with high surface-area packing (Figure 22). Water flows down over the media in a thin film. Because of the large gas-liquid surface area, this system is highly efficient for transfer of oxygen and nitrogen, but will not produce dissolved gas supersaturation.

Other types of surface aerators (cascade, inclined plane, tray, screen, and lattice) may be useful in some applications, but are not as effective as the PCA and require substantially more space and head. Submerged aerators can actually produce gas supersaturation and should be avoided for most applications. The use of vacuum for degassing and oxygen addition will be discussed in the following section on pure oxygen systems.

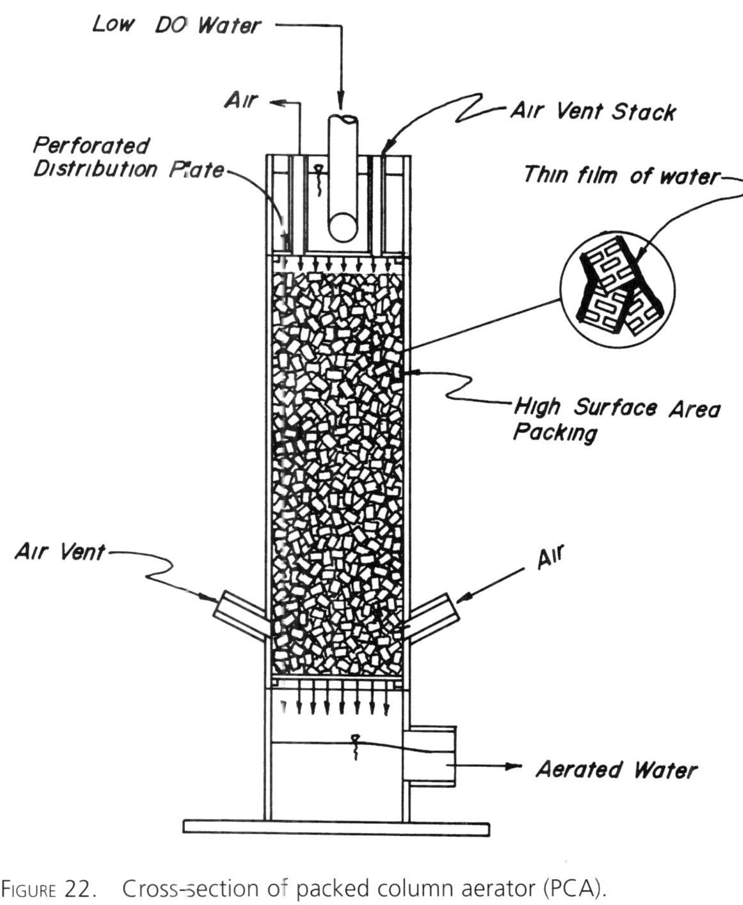

FIGURE 22. Cross-section of packed column aerator (PCA).

Design of Packed Columns for Degassing

The performance of packed columns can be computed as a function of temperature, barometric pressure, DO, and ΔP. The ΔP is the difference between the total gas pressure and local barometric pressure (Colt 1984), commonly expressed as millimeters of mercury (mmHg). The reporting of gas supersaturation levels in concentration units (mg/L) is not valid. For most degassing applications, the effluent must meet both a DO and ΔP criteria.

Column height. Column height has a substantial impact on the effluent ΔP (Table 13). For reasonable column heights, packed column aerators can reduce the ΔP to 10–20 mm Hg. To achieve ΔPs below this range, a very high column or a pure oxygen column is needed. A pure oxygen PCA operated at a vacuum is the only type of system that can reduce the dissolved nitrogen concentration below saturation.

Hydraulic loading. For a given flow, the cross-section area of the column can be computed from the water flow and maximum loading rates listed below:

Media type/ size (cm)	Maximum loading m^3/m^2 (h)	Minimum column diameter (m)
Pall ring/2.54	150	0.2
Pall ring/3.81	300	0.3
Pall ring/5.03	>340	0.4
Pall ring/8.89	>340	0.7

$$\text{gpm}/\text{ft}^2 = \text{m}^3/\text{m}^2 \bullet \text{h} \times 0.4090$$

Media type. The required column height depends on the media selected and effluent criteria for both gas supersaturation and dissolved oxygen (Colt and Bouck 1984). Although in general, the small-sized media are more efficient at transferring gas, column diameter for these media must be increased due to its reduced hydraulic capacity.

Degassing of Other Gases

Other gases may need to be stripped out of aquaculture waters. These include methane and hydrogen, carbon dioxide, hydrogen sulfide, and radon.

Methane and hydrogen. Methane and hydrogen are two insoluble gases that may be found in groundwater and bottom water from lakes and reservoirs. Conventional packed column aerators are effective in removing these gases. In applications where these gases may be present, the columns should be vented to the outside.

TABLE 13. Effluent ΔP as a function of temperature and packing height for highly supersaturated nitrogen and low oxygen (based on 3.81 cm pall rings, influent ΔP = 250 mm Hg, and atmospheric partial pressure ratio of nitrogen/oxygen = –2.87). To account for inflow and discharge piping, another 0.5 to 0.7 m should be added to the column. Above the stair-stepped line, the dissolved oxygen is less than 90% of saturation. Slightly less packing height is needed when both oxygen and nitrogen are supersaturated, so this information will result in conservative degassing for that application.

Packing height (cm)	Effluent ΔP (mm Hg)						
	5°C	10°C	15°C	20°C	25°C	30°C	35°C
0	250	250	250	250	250	250	250
10	184	177	169	161	152	143	133
20	167	159	150	141	131	120	110
30	152	143	133	123	112	102	90
40	139	129	118	108	97	86	75
50	126	116	105	94	83	72	61
60	115	104	93	82	71	61	51
70	105	94	83	72	61	51	42
80	95	84	73	63	52	43	34
90	87	76	65	55	45	36	28
100	79	68	58	48	39	30	23
110	72	61	51	42	33	25	19
120	65	55	45	36	28	21	16
130	59	49	40	32	24	18	13
140	54	44	35	28	21	15	10
150	49	40	31	24	18	13	9
160	44	36	28	21	15	11	7
170	40	32	25	18	13	9	6
180	37	29	22	16	11	7	5
190	33	26	19	14	9	6	4
200	30	23	17	12	8	5	3
210	27	21	15	10	7	4	3
220	25	19	13	9	6	4	2
230	23	17	12	8	5	3	2
240	21	15	10	7	4	3	1
250	19	13	9	6	4	2	1
260	17	12	8	5	3	2	1
270	15	11	7	5	3	1	1
280	14	10	6	4	2	1	1
290	13	9	6	3	2	1	1
300	11	8	5	3	2	1	0
400	4	3	1	1	0	0	0
500	2	1	0	0	0	0	0

Carbon dioxide. The carbon dioxide gas concentration in groundwater is highly variable and can typically range from below saturation to 30–40 mg/L. Concentration depends on the depth and characteristics of the soil, subsurface geology, and chemical reactions of dissolved carbon dioxide gas. Carbon dioxide gas is difficult to remove because of its high solubility. Gas-to-liquid ratios in the range of 10–20 are needed to remove a substantial amount of carbon dioxide (Piedrahita and Grace 1994). Because of reactions of carbon dioxide with water, it is impossible to remove more than about 50% of the dissolved carbon dioxide gas with a single pass through a packed column.

Hydrogen sulfide. Hydrogen sulfide is similar to carbon dioxide, with high solubility and liquid-phase reactions with water. A forced-draft packed column is needed to strip substantial amounts of hydrogen sulfide (Howe and Lawler 1989); these types of columns should be vented to the outside. Because of the liquid-phase reactions, the amount of hydrogen sulfide depends very strongly on pH.

Radon. High levels of radon gas have been found in hatchery buildings resulting from degassing of groundwater prior to use (Dwyer and Orr 1992). The off-gas from stripping applications should be discharged to the outside.

Oxygen Supplementation

The use of pure oxygen can increase the carrying capacity of an aquatic culture system when DO is the most limiting factor. Pure oxygen systems can economically saturate or supersaturate water with DO. Unlike air contact systems, pure oxygen systems have the capability of reducing dissolved nitrogen to or below saturation for the purposes of controlling gas bubble disease.

Sources of Pure Oxygen

Enriched oxygen may be obtained from three sources: high-pressure oxygen gas, liquid oxygen (LOX), and on-site oxygen generators. In many facilities, at least two (in some cases all three) sources of oxygen are used.

High-pressure oxygen gas (98–99%) can be obtained in cylinders containing 3–7 m^3 (100–250 ft^3) at 17.6 MPa (2,550 psi). Several cylinders can be connected together using standard manifold assemblies to increase the total capacity.

Liquid oxygen (98–99%) is produced on a large-scale basis by distilling liquefied air. Liquid oxygen (LOX) is generally trucked to the use site in bulk, then pumped into the storage tank. A liquid oxygen supply system will consist of a storage tank, evaporator, filters, and pressure regulator (Figure 23a). The maximum gas pressure in these containers is generally in the range of 1,000–1,400 kPa (150–200 psi). If the pressure exceeds these values, gas is vented to the atmosphere through a safety valve. Approximately 0.25% of the liquid will be lost per day, even if no oxygen is used.

Manufacturers have adapted industrial pressure-swing adsorption (PSA) systems for aquaculture applications (Figure 23b). Single units are available to produce 0.50–10.0 m^3 oxygen/h (15–400 ft^3 gas/h) and may be connected in parallel to produce larger quantities. These units require a source of 600–1,000 kPa (90–150 psi) filtered air and generate enriched oxygen gas of 85–95% oxygen. Oxygen generators work on a demand basis, so oxygen is produced only when needed. When used as a primary source of oxygen, a standby electrical generator is generally needed.

Types of Aerators Used for Pure Oxygen Systems

The most common types of pure oxygen systems used in aquaculture are "Michigan" columns, low-head oxygenators, and airstones. Additional information on other types of pure oxygen systems is presented by Colt and Watten (1988).

The "Michigan" column is similar to a packed column except it does not contain media and the size of the discharge pipe is reduced (Figure 24a). This type of column may be required in reuse systems due to problems with media fouling. The reduced discharge produces a vacuum of 50–120 mm Hg (Colt et al. 1993) and results in excellent degassing of nitrogen.

The low head oxygenator (LHO) is a proprietary oxygen transfer unit that can be readily retrofitted into serial raceway systems. The unit consists of a distribution plate positioned over seven rectangular chambers (Figure 24b). Water flows over the dam boards at the end of a raceway, through the distribution plate, and then falls through the rectangular chambers. The LHO is most efficient when the required change in DO is in the range of 3–5 mg/L (Wagner et al. 1995).

Pure oxygen diffused aeration generally has a low absorption efficiency and has been limited to emergency use in production systems or for transport systems. Some recently developed fine-bubble diffusers have absorption efficiencies in the range of 40–60% when submerged at 1–2 m (3.3–6.6 ft) or more.

Performance of Pure Oxygen Systems

Important measures of performance for a pure oxygen system include

ΔDO Change in DO through the contact system,
ΔP_{out} Effluent ΔP from the contact system,
DC_{out} Effluent dissolved carbon dioxide concentration from the contact system,

FIGURE 23. Typical pure oxygen supply systems. (a) liquid oxygen system (LOX) and (b) pressure-swing absorption system (PSA) (SeQual, used with permission).

FIGURE 24. Common types of pure oxygen systems. (a) "Michigan" spray column. (b) Low head oxygenator (LHO) system. Water falls vertically through the perforated distribution plate into the oxygen-enriched gas zone and then flows out the bottom and down the raceway. The pure oxygen gas is introduced on the left side and exits on the right side (top view). The unit is divided into seven compartments, so there is minimal backmixing of the gas as it moves from the first compartment to last compartment. Within each compartment, the gas phase is well-mixed. (Water Management Technology Inc., used with permission).

COC Cumulative oxygen consumption, and
AE Absorption efficiency (% of oxygen transferred into water).

It is important to note that substantial trade-offs may be required between certain limits. For example, it may not be possible to simultaneously produce a high ΔDO, a high absorption efficiency, and a low variable cost.

ΔDO (change in dissolved oxygen in mg/L). One of the major advantages of pure oxygen systems is that the effluent DO concentration can be varied by changing the gas-to-liquid (G/L) ratio (Figure 25). In contrast, the effluent DO concentration from typical surface or gravity aerators in contact with the atmosphere cannot exceed the air saturation concentration. Adjustment of the gas flow rate (and G/L ratio) can be used to vary the capacity as a function of the (1) time of day, (2) production cycle, or (3) operational requirements. Although it is possible to adjust the ΔDO value, increasing the G/L ratio does reduce the absorption efficiency and increase oxygen gas costs.

ΔP_{out} *(effluent* ΔP*).* The allowable ΔDO from the absorber is also a function of the criteria for ΔP and the influent ΔP (Watten 1990; Watten et al. 1991). This is because the addition of oxygen increases the total gas pressure of the water. The increase in total gas pressure is offset by the removal of nitrogen gas. In general, higher ΔP can be used with circular tanks than raceways. In a well-mixed circular tank, the high ΔP influent water is diluted with the lower ΔP water in the tank.

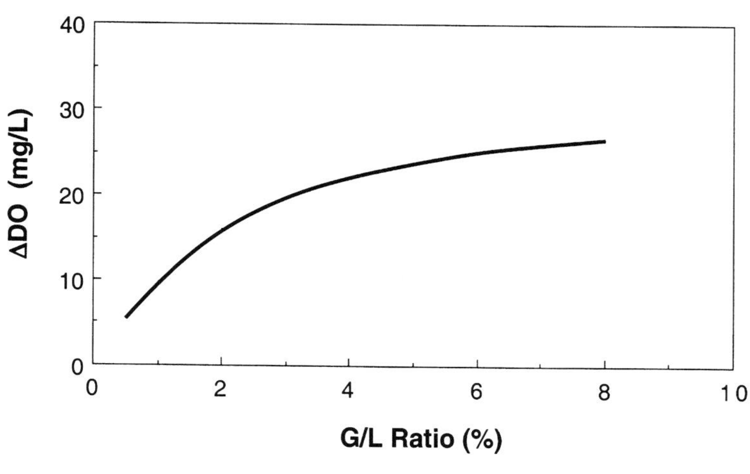

FIGURE 25. Impact of gas-to-liquid ratio (G/L ratio) on change in DO through pure oxygen system (Colt and Watten 1988).

DC_{out} (effluent dissolved carbon dioxide concentration). Due to the high solubility of carbon dioxide, little dissolved carbon dioxide gas will be removed in the off-gas from the absorber unit; other types of aeration equipment will be needed to remove carbon dioxide gas.

COC (cumulative oxygen consumption). The metabolic use of oxygen by aquatic animals produces carbon dioxide and ammonia. The total amount of oxygen used in a rearing system is referred to as cumulative oxygen consumption and is a measure of the intensity of the operation. The depletion of oxygen may not always be the most limiting factor, and when ammonia or carbon dioxide is more limiting, aeration will have little effect on the carrying capacity (Colt and Orwicz 1991a). Maximum cumulative oxygen consumption (DO_{in} – DO_{out}) based on pH, carbon dioxide, and un-ionized ammonia limitations are presented in Figure 26 for water quality criteria typical of salmon and trout culture.

AE (absorption efficiency). The absorption efficiency is an important operating limit and has a substantial impact on operating costs. The highest absorption efficiencies occur at low values of the G/L ratio. If a small amount of oxygen is slowly introduced into water, a high proportion of it would be absorbed, but relatively little oxygen would be added. Therefore, the system can produce either high ΔDOs at low absorption efficiencies or low ΔDOs at high absorption efficiencies (Watten et al. 1991). For nitrogen stripping applications, the absorption efficiency must be limited to allow removal of nitrogen gases. Injection of oxygen gas into a pressurized pipeline can result in very high transfer efficiency and potentially lethal total gas pressures.

Selection of Pure Oxygen Systems

Due to differences in site conditions, operational modes, and specific absorber units, the modeling and design of a pure oxygen system is complex; additional information on design is presented by Colt and Watten (1988), Watten et al. (1991), and Watten (1990).

Disinfection Systems

Disinfection is the process by which the concentration or number of bacteria, viruses, cysts, or spores is reduced. Sterilization is the complete removal of all microorganisms and may be possible only under special conditions. The use of surface waters for culture may require disinfection to prevent transmission of disease from wild populations to the culture animals. Disinfection may also be required for eggs and larvae, especially for small marine fish and crustaceans. Disinfection of hatchery effluent may also be needed for the holding of out-of-basin stocks, nonnative species, or genetically modified animals to prevent transmis-

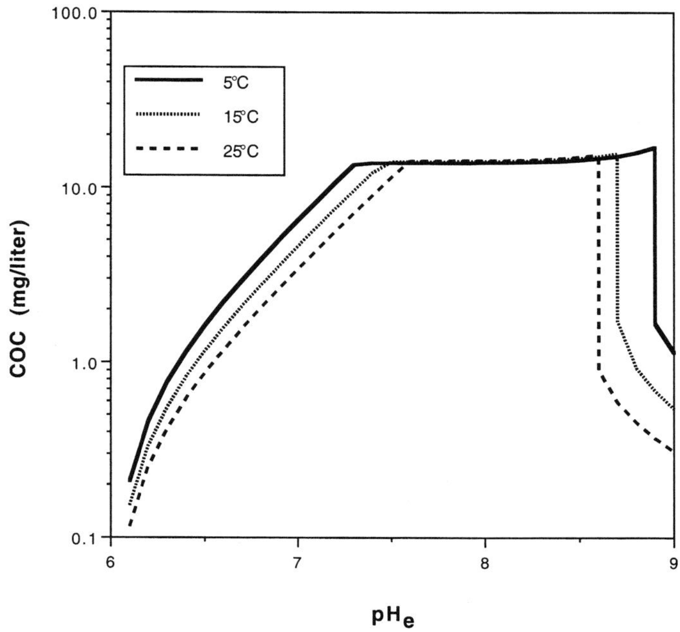

FIGURE 26. Maximum cumulative oxygen consumption (mg/L) as a function of equilibrium pH_e system using pure oxygen aeration. The pH_e is pH of a solution in equilibrium with the atmosphere. For example, the pH_e of rainwater is approximately 5.5 in the absence of pollution. A pH_e of 8.5 corresponds to an alkalinity of about 100 mg/L as $CaCO_3$. At low pH_es COC is limited by low pH, at intermediate pH_e by high carbon dioxide, and at high pH_e by the un-ionized ammonia concentration (Colt and Orwicz 1991a).

sion of pathogens to wild fish or escape of culture animals. Three processes are commonly used for disinfection in aquatic systems: ultraviolet light (UV), ozone, and chlorine. Each of these processes is highly effective and the selection of the best disinfection process will depend on the culture system, mode of operation, and culture species.

The rate of disinfection often is first order with respect to the concentration of organisms (Schroeder 1977):

$$\frac{dN}{dt} = -kN \tag{9}$$

where

$\frac{dN}{dt}$ = The rate of organism die-off,

N = The concentration of organism, and

k = Rate constant that depends on pH, temperature, and disinfectant concentration.

Integration of this equation between time = 0 and time = t, yields

$$\ln\left[\frac{N}{N_0}\right] = -kt \quad \text{or} \quad N = N_0 e^{-kt}, \tag{10}$$

where

N_0 = Concentration of organism at time = 0.

The number of organisms is zero only at time = infinity. Therefore, in real systems, the concentration of organisms can be reduced, but never completely eliminated. The concentration of organisms required to result in an infection is called the infective concentration. If N_0, k, and infective concentration are known, then the required t can be computed. In practice, information on the infective concentration is generally lacking and N_0 and k can vary widely. Therefore, the typical design is based on pilot-scale tests using growth and mortality of the culture animal as the criteria. As a result, design criteria for disinfection can vary widely.

The ideal disinfectant would be highly toxic to microorganisms, have low toxicity to the culture animals, and maintain a residual concentration in the water so microorganisms would be killed in all parts of the system. Chlorine can be used to maintain a relatively stable residual, but this residual is highly toxic to most aquatic animals. Ozone is very reactive, and it is difficult to maintain a residual. Ultraviolet light has no substantial residual activity.

The use of disinfection in flow-through systems requires extreme reliability as failure of the system for a short time may result in exposure of the animals to an effective concentration of pathogens and development of disease problems. In reuse systems, the concentration of pathogenic organisms depends on the growth rate of the organism within the system and the flow rate through the disinfection system (Spotte and Adams 1981). In recycle systems for marine mammals or adult amphibians without gills, chlorine can be used without substantial toxicity to the culture animals

Ultraviolet Light

Exposure to light with a wavelength near 253.7 nm is an effective disinfectant. The UV dose is defined as intensity × time and is expressed in units of microwatts per square centimeter per second (mW/cm^2 per s). The required dose is related to the size and transparency of the organism to UV radiation. As little as 3,600 mW/cm^2 per s is required to kill *Aeromonas salmonicida,* which is 0.5–1.0 6 (m wide, but 1,720,000 mW/cm^2 per s were required to kill adult *Ichthyophthirus multifilis,* which are approximately 800 mm wide (Hoffman 1974).

Effective use of UV will require removal of particles larger than 70 μm by chemical addition and coagulation or filtration. These large particles may harbor pathogens. Filtration may have little change on UV adsorption as most of adsorption is due to dissolved compounds such as humic substances (Qualls et al. 1983). The use of UV in highly colored waters may not be effective.

The lamps used in UV systems age and must be changed yearly. Some systems use an ultraviolet sensor to monitor the UV output of the system. Lamp output depends on temperature, so the lamps are commonly placed inside a quartz sleeve. In some systems, a wiper is used to clean the sleeve. In systems with sleeves, the lamps can be replaced without shutting down the system. The lamps can either be placed inside a pressurized chamber (Figure 27) or placed in an open channel. The open channel type of UV system has very little head loss and can be installed in an existing system without repumping.

Ozone

General information on the uses of ozone in water treatment can be found in Langlais et al. (1991); the comprehensive bibliography prepared by Rosenthal and Wilson (1987) is also very useful. The ozone molecule contains three atoms of oxygen (O_3), is colorless, and has a distinctive pungent odor. Ozone gas is unstable and must be generated on site.

Ozone can react with ammonia, nitrite, reduced iron and manganese, bromide, dissolved organic compounds, and microorganisms. The ammonia oxidation rate is pH dependent and is slow for pH less than 9 (Colberg and Lingg 1978). Under ambient pH and operating conditions,

FIGURE 27. Pressurized ultraviolet radiation systems.

only 5–15% of the ammonia will be oxidized to nitrate by ozone. Nitrite is quickly oxidized by ozone to nitrate. Reduced iron and manganese (Fe^{2+} and Mn^{2+}) are common in groundwater and at the bottom of surface water reservoirs. Oxidization of these compounds to their insoluble forms (Fe^{3+} and Mn^{4+}) allows their removal by sedimentation or filtration. In seawater, ozone can react with the bromide ion and ammonia (Grguric et al. 1994) to form the hypobromite ion, bromate ion, hypobromous acid, and bromamine. Bromate ions and bromamines are stable compounds. In seawater systems using marine water sources, residual bromide compounds may become critical concerns when attempting to culture sensitive species. The reactions of ozone with cell constituents of microorganisms are not well understood, even under laboratory conditions. It is thought the disinfection reaction of ozone involves proteins contained in the cell membrane.

When ozone is added to water there is commonly an immediate ozone demand (reduction in ozone concentration). This chemical demand must be satisfied before the disinfecting chemical oxidation reactions can proceed. This demand is related to the chemical composition of the water, solids content, and water temperature and must be determined by actual testing.

A general schematic of an ozone disinfection system is presented in Figure 28a. Key components are (a) an air preparation or oxygen supply system, (b) two or more ozone generators, (c) an ozone contact system, (d) an ozone destruction system for off-gas, and (e) an ozone destruction system for removing ozone from the process water. Not all of these components may be present in a given system.

The corona discharge method is the most widely used ozone generation system in aquaculture applications (Figure 28b). Ozone is formed in a corona discharge by the interaction between an electrical discharge and oxygen molecules. Typically, the energy requirements of an ozone generator are approximately 3 kW·h/kg (6–7 kW·h/lb) of ozone.

Ozone is difficult to measure accurately, especially in liquid phase. In most facilities, it is necessary to measure ozone concentrations in both the gas and liquid phases. Additional information on the measurement of ozone is presented by Colt and Cryer (2000).

Lethal levels of ozone can be as low as 9.3 μg/L, and sublethal effects have been observed below 5 μg/L. Under production conditions, ozone concentrations in the range of 10–20 μg/L (measured at the discharge from the contactor) have been found to have no impact on salmonid eggs and fry (Tipping 1988) because the chemical ozone demand from uneaten feed and waste products will quickly reduce the ozone residual to inconsequential levels.

The most common ozone contactors are the deep-tank contact basin

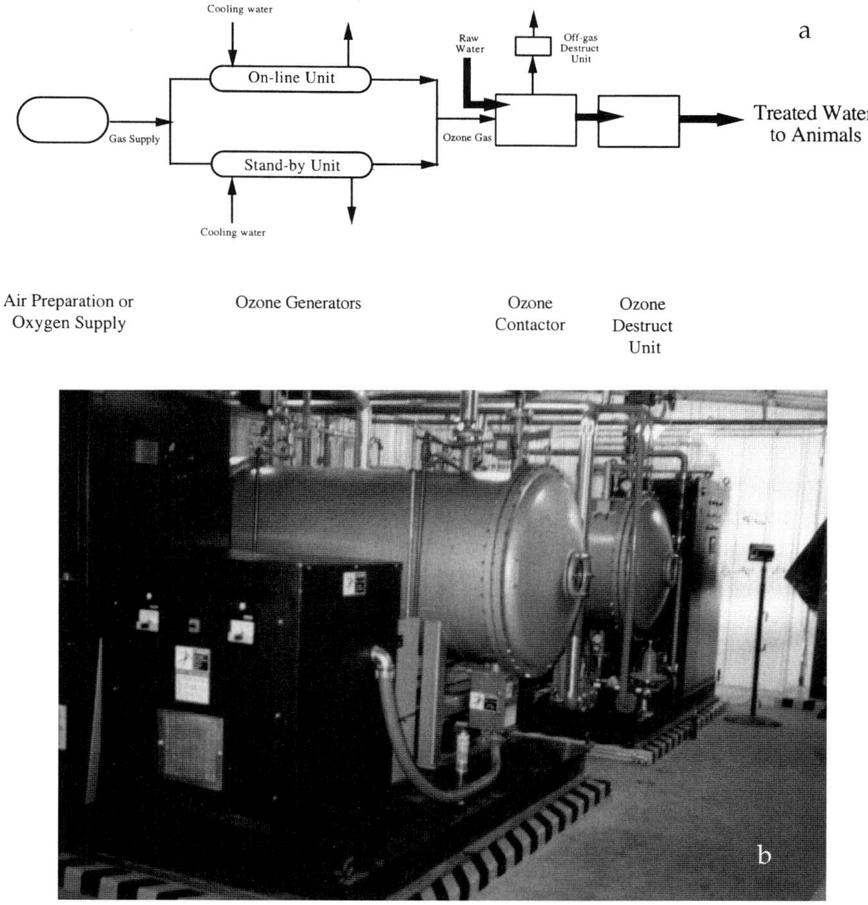

FIGURE 28. Ozone Treatment System (a) Typical process schematic diagram for an ozone system and (b) Two 250 lb/d ozone generators at Cowlitz Fish Hatchery (Ed Cryer, used with permission).

and the injector system. Ozone removal and destruction is needed for both undissolved ozone gas from the contact system and for the residual ozone in the process water before use. Destruction of excess ozone gas is generally accomplished by passing the ozone-gas mixture through either a thermal ozone destruction unit or a catalytical destruction unit. Ozone in the process water can be removed by detention time (decay), air stripping, granular activated carbon, UV radiation, and chemical addition. The extreme oxidation potential of ozone requires careful selection of construction materials to prevent corrosion and deterioration.

Generally, the most serious problems occur with a mixture of ozone gas and water vapor.

Ozone has been shown to be effective at inactivating bacterial, viral, and protozoan agents at low dosages (0.5–1.5 mg/L applied dose) in comparison to other chemicals disinfectants. The C_t value is the actual ozone residual (mg/L) multiplied by the detention time of the contact unit. This measurement is commonly used in the design and rating of ozone disinfection systems. The design value depends strongly on the pathogen of interest; several large-scale hatchery disinfection systems have been constructed in Pacific Northwest using C_t that ranged from 1 to 2 mg(min/L).

Ozone has attracted much interest recently in enhancing water quality in reuse systems for shrimp and fish. These systems have generally not included effective ozone removal, so the applied ozone doses have been limited to prevent ozone toxicity problems. These systems have not been designed to provide a given level of disinfection, although some disinfection will occur. Improvements in water quality and solids removal have resulted from applied doses in the range of 0.025–0.045 kg ozone/kg of feed (Summerfelt and Hochheimer 1997); but the difference between the required dose for water quality improvements and toxicity problems is very narrow.

Other Chemicals

Chemical disinfectants are commonly used for cleaning tanks, nets, and other culture tools. Commonly used compounds include calcium hypochlorite ($Ca(OCl)_2$), sodium hypochlorite ($Na(OCl)_2$), quaternary ammonia (Roccal or Hyamine), and iodophors (Providone iodine, Betadine, etc). The following active concentration and exposure times can be used:

Compound	Concentration of active ingredient (mg/L)	Contact time (minutes)
Iodophor	200	30–60
Hypochlorite	200	30–60
Quaternary ammonium	600	30–60

When chlorine compounds are added to water hypochlorus acid (HOCl) and hypochlorite ions (OCl^-) are formed. The relative amount of HOCl and OCl^- is very important because HOCl has 40–80 times the disinfectant power of OCl^- (Metcalf and Eddy, Inc. 1979). The concentration of HOCl + OCl^- is called the free residual chlorine. For a given amount of added chlorine, a pH above 7.00 substantially reduces its effectiveness. In very alkaline water, it may be necessary add an acid to reduce the pH before adding the chlorine.

Another use of chlorine compounds, is the disinfection of effluent water, especially from laboratories or quarantine facilities. Typical concentrations and contact times are 10–20 mg/L and 30–60 min. Removal of chlorine (dechlorination) may be necessary in some cases to prevent toxicity problems in the receiving water. Hypochlorus acid and hyprochlorite are relatively unstable compounds. In municipal water, it is becoming more common to use chloramines rather than chlorine compounds. Chloramines are formed by reacting chlorine with ammonia. This compound is much more stable and, although somewhat less toxic than chlorine, it still may be present in lethal concentrations in some municipal water supplies.

Dechlorination

Four main methods are used to remove chlorine from water: (1) natural decay, (2) granular activated carbon, (3) chemical addition, and (4) granular activated carbon/chemical addition.

Natural Decay

In nonflowing systems, natural decay of chlorine may be adequate if animals are not added for 2 or 3 d after filling the system. Aeration and sunlight will increase the rate of decay. When chlorine is added as chloramines (see previous section on disinfection), up to 20–30 d may be needed for the residual chlorine concentration to decay to nontoxic levels.

Granular Activated Carbon

Granular activated carbon (GAC) can be used to remove both free and combined residual chlorine. Generally, the GAC is contained in a pressure filter; typical loading rates and detention times are 400 Lpm/m^2 (10 gpm/ft^2) and 5 min based on empty bed volume, respectively. The carbon should be replaced on a yearly basis.

Activated carbon will efficiently and consistently remove 90–95% of the chlorine residual. The residual chlorine concentrations in the effluent from commercial dechlorination units vary from 5 to 70 mg/L (Seegert and Brooks 1978). These concentrations are high enough to cause serious damage to most aquatic animals, especially for eggs or small fish where organic levels are low. However, these levels are probably not acutely lethal and under production conditions, adverse impacts may not be evident.

Chemical Addition

Sodium thiosulfate ($Na_2S_2O_3$) and sodium sulfite (Na_2SO_3) have been used to remove residual chlorine. These compounds instantaneously react with both free and combined chlorine. Sodium thiosulfate and sodium sulfite are not toxic to humans, are stable in dry form, and can be delivered to dechlorination systems using inexpensive metering pumps. Due to the reaction of these chemicals with oxygen, solutions should be prepared on a daily basis.

Sodium sulfite is probably the safest and most economic chemical for dechlorination. Approximately 2 mg of sodium sulfite are required for each mg of residual chlorine removed (Metcalf and Eddy, Inc. 1979). The use of sodium sulfite requires an accurate knowledge of the residual chlorine concentration, which may vary widely depending on time of day and chlorination operations. Improper dosing of sodium sulfite can result in residual chlorine or sodium sulfite in the effluent. Excess sodium thiosulfate or sodium sulfite is not too toxic to aquatic animals, but reacts with oxygen to reduce the DO.

Granular Activated Carbon/Chemical Addition

Seegert and Brooks (1978) recommend a combined system using activated carbon plus sulfite addition (Figure 29). The activated carbon column will remove 90–95% of the chlorine residual. Then sodium sulfite is added on a 6:1 weight basis. The important feature of this system is that it is nearly foolproof. In the event of a power failure or metering pump, the carbon column will continue to remove 90–95% of the residual chlorine. This will prevent serious damage to the culture animals (on a short-term basis) until the dosing can be resumed. Due to the small amount of chlorine residual that must be removed after the carbon column, excess sodium sulfite can be added to assure complete removal of residual chlorine without problems with sulfite toxicity or DO depletion.

Iron and Manganese Removal

Groundwater or anaerobic waters commonly contain reduced iron and manganese. Upon aeration, these elements react with oxygen:

$$Fe^{+2} + 1/4O_2 + 5/2H_2O \rightarrow Fe(OH)_3 + 2H^+$$

$$2Mn^{+2} + O_2 + 2H_2O \rightarrow 2MnO_2$$

These reactions will occur in the culture system after aeration. Although $Fe(OH)_3$ and MnO_2 are relatively nontoxic, they form solids that tend to clog valves, filters, and egg rearing systems and may cause gill irritation.

To prevent problems with formation of solids, it is necessary to oxidize these compounds and remove the solids before the water is used in the culture system. Within the range of typical pH, the rate of oxidation of iron and manganese is slow. The rate of oxidation can be increased by

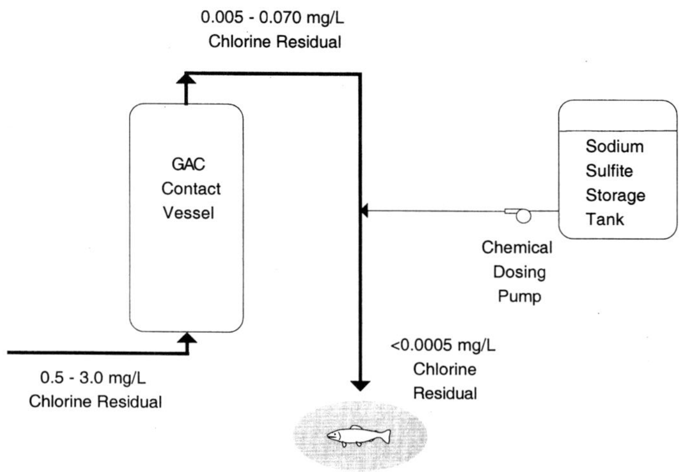

FIGURE 29. Process flow sheet for granular activated carbon/chemical addition system for dechlorination.

addition of ozone, hydrogen peroxide, chlorine, or potassium permanganate. Potassium permanganate may be the treatment of choice, due to its low cost, ease of mixing and application, noncorrosive nature, and lack of harmful residue if used properly. The resulting solids can be removed by sedimentation or sand filtration. The design of systems for the removal of iron and manganese is often quite site-specific and pilot-scale testing may be required. Many package treatment systems are available, but their application may be difficult.

Ammonia Removal

After DO, ammonia is commonly the next most limiting water quality variable. Ammonia is the primary nitrogenous byproduct of protein metabolism in many aquatic animals and its un-ionized form (NH_3) can build up to toxic concentrations. Removal of ammonia will allow greater production from a given water supply. Ammonia removal is a substantial process in recycle systems and ponds. In flow-through systems, limited ammonia removal occurs. For a discussion of the chemistry and toxicity of the different forms of ammonia, see Ammonia section.

Biological Nitrification

The literature on biological nitrification is extensive; the review articles by Wheaton et al. (1994a, 1994b) are highly recommended. Ammonia is oxidized to nitrate in a two-step process by two groups of bacteria:

Bacterial group	Reaction
Nitrosomonas	$NH_4^+ + 1.50_2 \rightarrow NO_2^- + 2H^+ + H_2O$
Nitrobacter	$NO_2^- + 0.50^2 \rightarrow NO_3^-$
Overall reaction	$NH_4^+ + 2O_2 \rightarrow NO_3^- + 2H^+ + H_2O$

The oxidation of ammonia can be modeled by Monod kinetics:

$$\frac{dC_s}{dt} = -K_{max}\left[\frac{C_s}{K_s+C_s}\right]C_{bacteria} \tag{11}$$

where

C_s = Concentration of ammonia (mg/L),
t = Time (seconds),
K_{max} = Maximum specific oxidation rate of ammonia (1/d),
K_s = Half velocity constant for ammonia oxidation (mg/L), and
$C_{bacteria}$ = Concentration of *Nitrosomonas* (mg/L).

The values of K_{max} and K_s depend on temperature and salinity. A similar equation can be written for the oxidation of nitrite to nitrate, although the specific values of the constant are different.

Many of the important characteristics of bacterial nitrification systems can demonstrated from the above equations.

(1) Nitrification consumes DO and reduces the alkalinity due to H^+ addition. In warmwater applications, pure oxygen aeration of filter influent may be necessary to supply enough oxygen in the filter. In systems with little makeup water, it may be necessary to add alkalinity to prevent pH depression.
(2) The values of K_{max} and K_s depend strongly on temperature; the optimum temperatures for ammonia and nitrite oxidation are above 28°C. Below 5°C, nitrification is very slow. For the same capacity, much smaller nitrification systems are needed for coldwater temperatures.
(3) Since the oxidation rate of both compounds are a direct function of $C_{bacteria}$, systems with large concentrations of nitrifying bacteria will have larger oxidation rates.
(4) The maximum ammonia nitrogen and nitrite-nitrogen concentrations in aquaculture systems are typically in the range of 0.1–1.0 mg/L. In this range, the oxidation rate of both compounds are almost first order with respect to substrate (ammonia or nitrite). Therefore, the rate of oxidation will depend strongly on the concentration of ammonia and nitrite.

The growth rate of nitrifying bacteria is slow. The time required to reach steady-state can be reduced by seeding the filter with filter media from an established filter. Marine filters should be seeded with material from a marine filter rather than a freshwater filter. During the startup of the filter, typically, an ammonia peak is observed followed by a nitrite peak. These peaks have frequently resulted in heavy mortality of aquatic species (Liao and Mayo 1974; Mevel and Chamroux 1981). When the filter is started, the oxidation rate of ammonia is zero and the concentration of ammonia will start to increase. The biomass of *Nitrosomonas* increases in response to the increase in ammonia concentration and after a period of time will reduce the ammonia concentration. The increased rate of oxidation of ammonia will increase the concentration of nitrite in the system. In response to this increase, the biomass of *Nitrobacter* will increase and reduce the concentration of nitrite. At steady-state, the rate of ammonia oxidation is equal to the rate of nitrite oxidation and the concentration of nitrite is low.

The rate of nitrification depends on several environmental limits and the concentration of the two compounds. The overall temperature optimum for the nitrification process is in the range of 28–36°C (Sharma and Ahlert 1977). At temperatures below 5–6°C, the rate of ammonia oxidation is too low to maintain *Nitrosomonas* and nitrification is unstable. The optimum pH for pure cultures of *Nitrosomonas* and *Nitrobacter* are 8.3–8.6 and 7.2–8.2, respectively (Sharma and Ahlert 1977). At pH values in the range of 5.4–6.3, nitrification is completely inhibited. The nitrification process requires 4.57 mg of oxygen per mg of TAN oxidized. Dissolved oxygen concentrations below 0.6 mg/L result in the inhibition of nitrification (Wild et al. 1971; Forster 1974). Rapid increases in salinity reduces the K_{max} value, but do not change K_s (Akai et al. 1983). At least 10 d are required to adjust to a new salinity (Rosenthal and Otte 1979). Nitrite oxidation is more sensitive to changes in salinity. Many of the chemicals used for treatment of diseases are toxic to the nitrifying bacteria. After use of a long-term nitrification inhibitor, 5–30 d may be required before filter operation returns to normal.

Due to the slow growth of nitrifying bacteria, most nitrifying systems used in aquaculture are based on bacteria attached to physical surfaces, where large populations can be maintained. Commonly used media include rock, oyster shell, random-dump plastic media, sand, and small plastic beads (Figure 30). Reducing the size of media increases the unit surface of the media (m^2/m^3) and its capacity to remove ammonia.

A commonly used nitrifying filter is the trickling filter. The water is distributed over the top surface of the filter and trickles down over the media and out the bottom. In the submerged trickling filter, the filter is filled with water (Figure 31). Water flow through the submerged trickling filter may be upward or downward. Expanded bed or fluidized bed filters use sand or positively buoyant plastic media. Water velocity is used to fluidize the bed and prevents clogging. Rotating biological contactors are a type of nitrifying filter where a series of large plastic disks are attached to a rotating

FIGURE 30. Photograph of common media (a) oyster shell, (b) random dump plastic media, (c) gravel, and (d) bead media.

FIGURE 31. Photograph of outdoor trickling filter at Dworshak National Fish Hatchery. This system consists of 6 filters 25 x 75 ft with 6 ft of 3 1/2 pall rings (D. Owsley; used with permission).

shaft. A recent development is the bead filter. This filter is filled with small positively buoyant plastic beads. The water flows upward through the beads. When the head loss starts to build up, the flow is stopped and the beads are mixed using a rotating impeller or compressed air (Figure 32). The solids are allowed to settle to the bottom and are removed by opening a drain valve and the flow is resumed. Because of the excellent solids removal characteristics of bead filters, these units have been used in conjunction with fluidized bed filters.

Clogging is more of a problem with smaller-sized plastic media (diameter < 2.5 cm) and the rock or slag media. The larger-sized plastic media is more resistant to clogging and solids removal is generally not required. Clogging is not a problem for the expanded bed filters, bead filters, or the rotating biological contactors (RBCs).

The performance of a nitrifying filter will depend on the surface area of the filter, depth of filter, hydraulic loading, temperature, and concentration of influent ammonia. The detailed design of nitrifying filters

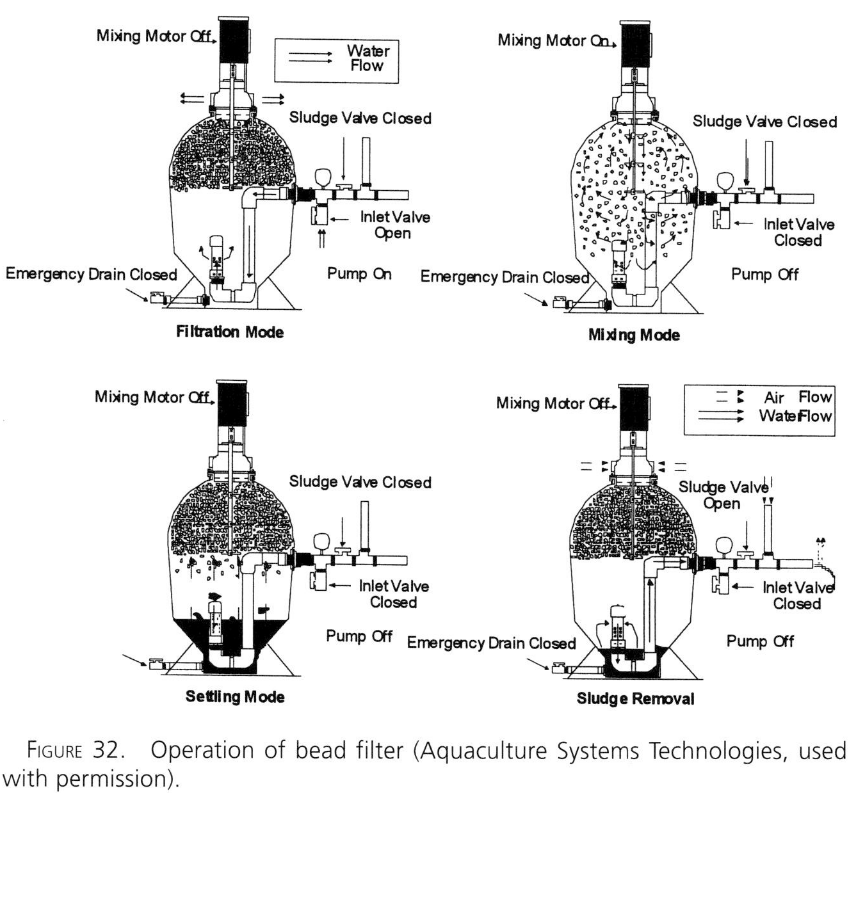

FIGURE 32. Operation of bead filter (Aquaculture Systems Technologies, used with permission).

is beyond the scope of this book (see Wheaton et al. 1994a, 1994b), but Table 14 can be used in the preliminary sizing of different types of media.

Other Methods

The ion exchange method offers a purely chemical means of ammonia removal. The most commonly used ion exchange media for ammonia is

TABLE 14. Physical characteristics of different types of media and unit volumetric nitrification rates at 5–15°C (cold water) and 20–30°C (warm water). Based on assumed unit nitrification rate of 100 mg TAN/m²·d for cold water and 270 mg TAN/m²·d for warm water.

				Ammonia removal in g/TAN·m³·d (g/TAN·ft³·d)	
Type of media	Size cm (inch)	Weight kg/m³ (lb/ft³)	Unit surface area m²/m³ (ft²/ft³)	5–15°C (41–59°F)	20–30°C (68–86°F)
Pall ring	8.89 (3.5)	44 (2.75)	102 (31)	10 (0.9)	28 (2.6)
Pall ring	3.81 (1.5)	68 (4.25)	115 (35)	12 (1.1)	31 (2.9)
Pall ring	5.08 (2)	76 (4.75)	128 (39)	13 (1.2)	35 (3.2)
Pall ring	2.54 (1)	88 (5.5)	207 (63)	21 (1.9)	56 (5.2)
Pall ring	1.59 (0.6)	112 (7)	341 (104)	34 (3.2)	92 (8.6)
Bio-saddles	7.62 (3)	56 (3.5)	89 (27)	8.9 (0.8)	24 (2.2)
Bio-saddles	5.08 (2)	60 (3.75)	108 (33)	11 (1.0)	29 (2.7)
Bio-saddles	2.54 (1)	96 (6)	207 (63)	21 (1.9)	56 (5.2)
LANPAC	5.84 (2.3)	99 (6.2)	223 (68)	22 (2.1)	60 (5.6)
IMPAC	8.38 (3.3)	83 (5.2)	213 (65)	21 (2.0)	58 (5.3)
TriPack	8.89 (3.5)	53 (3.3)	125 (38)	13 (1.2)	34 (3.1)
TriPack	5.08 (2)	63 (3.9)	138 (42)	14 (1.3)	37 (3.5)
Oyster shell (whole)	5–10 (2–4)	320 (20)	100 (20)	10 (0.9)	27 (2.5)
Rock	2.5–6.5 (1–3)	1,250–1,450 (78–91)	181–230 (55–77)	18–23 (1.7–2.1)	49–62 (4.6–5.8)
Sand (60 mesh silica sand)	0.025 (0.010)	2,000 (125)	9,000[a] (3,000)[a]	900 (84)	2,400 (230)
Sand (100mesh silica sand)	0.015 (0.006)	2,000 (125)	15,000[a] (4,500)[a]	1,500 (150)	4,100 (380)
Polyethylene beads	0.32 (0.125)	530 (33)	1,300 (400)	130 (12)	350 (33)
KMT biofilm carrier elements	0.72 (0.28)	180 (11)	500 (152)	50 (4.6)	140 (13)

[a] at 100% bed expansion

clinoptilolite, a naturally occurring zeolite. Clinoptilolite is highly selective for the ammonium ion (NH_4^+). The capacity of this mineral to adsorb ammonium in ranges from 1.6 to 2.0 meq/g or 22.4–28.0 mg TAN/g (Klieve and Semmens 1980). The capacity or exchange capacity is the maximum amount of ammonium ion that the mineral can adsorb at high ammonia concentration. In practice, the effective field capacity is approximately 1.1 mg TAN/g (Bruin et al. 1981).

The maximum ammonia concentration in the effluent is commonly called the breakthrough concentration. When the effluent concentration is equal to the breakthrough concentration, the capacity of the column is exhausted and the ion-exchange material must be regenerated.

Although clinoptilolite is highly selective for ammonia, the material can be regenerated by a strong salt (NaCl) solution. Typically, a solution of 10% NaCl is used, although seawater may be used in coastal areas (Bruin et al. 1981). The brine solution is circulated through the column for 1–3 h. Before the column can be used, it must be rinsed until the pH and turbidity return to normal. The rinse requirement is in the range of 2–4 bed volumes. The rinse water is generally discharged. The brine solution and rinse water must be treated before discharge. Several processes using nitrifying bacterial for regeneration have been proposed (Semmens and Porter 1979). In this process, the absorbed ammonia is converted to nitrate by attached bacteria.

Bacterial fouling of clinoptilolite may be important in aquatic systems (Johnson and Sieburth 1974; Watten and English 1985). Fouling can reduce both the capacity of media and the kinetics of ammonia uptake. Regeneration with 10% NaCl + 1% NaOH (sodium hydroxide) or 10% NaCl + 0.5% Cl_2 (chlorine) can be used to remove fouling. In addition, treatment with NaOH appears to increase the capacity and selectivity of clinoptilolite (Klieve and Semmens 1980; Watten and English 1985).

Temperature has little effect on the ion exchange process (Marking and Bills 1982). A hardness of 300 mg/L as $CaCO_3$ reduces the exchange capacity by 50%. The exchange capacity of clinoptilolite is optimum near a pH of 7.0 and decreases for pH values higher and lower. Due to the high concentration of Na^+ in seawater, clinoptilolite is ineffective for ammonia removal.

Alarms and Control Systems

The majority of control systems previously installed in hatcheries were expensive proprietary systems that could not be modified except at great expense. In addition, spare parts and support service for many of these systems are no longer available. Because of the inflexibility of these systems, few systems were operated long after their warranty period expired. Advances in computers, input/output boards, sensors, and software have made complex alarm and control systems cost-effective for hatcheries. A complete microcomputer-based supervisory control and data acquisition system (Figure 33) consists of (1) sensors, (2) input/output multiplexer, (3) computer, (4) control software, and (5) various alarm and control outputs. Most systems used in hatcheries are based on open architecture, components from several manufacturers can be mixed.

Sensors, Meters, and Outputs

Sensors convert a measurement into an output that can be displayed locally or transmitted to a remote site. Common outputs are on/off, 0–100 mv, and 0–5 mA. Although on/off sensors present less information than the variable outputs, their operation is much easier to test. Some

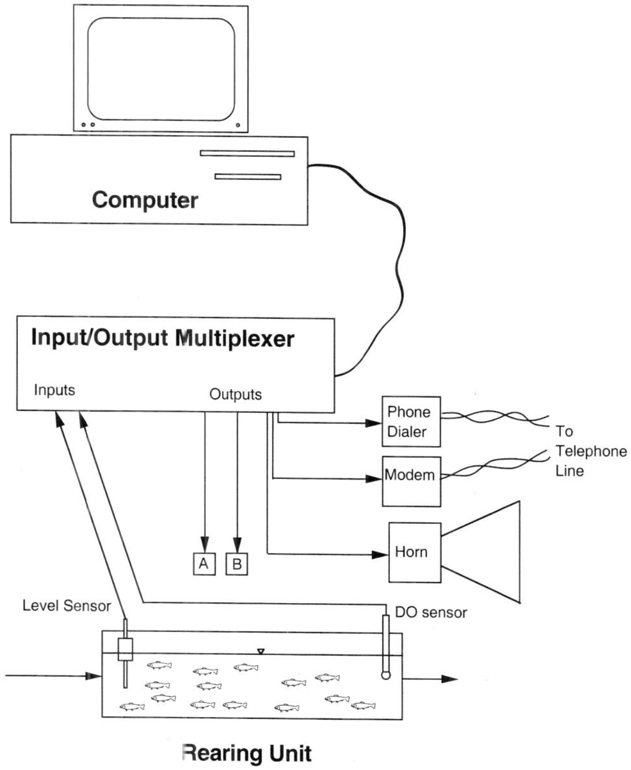

FIGURE 33. Major components of a microcomputer-based supervisory control and data acquisition system. Outputs A and B could be used to turn on an emergency aeration and stop automatic feeders.

sensors also contain local control outputs that can be used to turn on pumps or open valves. Common sensors used in hatcheries include: flow, level, security, temperature, DO, total gas pressure, and light levels. It is possible to obtain stand-alone security systems, but it is better to integrate the security system into the overall hatchery monitoring and control system.

Input/Output Multiplexer

All of the inputs and outputs are routed through an input/output (I/O) multiplexer. This unit is under control of control software on the microcomputer. The I/O multiplexer is used to scan through all of the inputs on an user-defined interval, measure each variable, and control the outputs. Both digital and analog inputs and outputs are provided. A digital input or output is either on or off. An analog input or output can take a range of values (e.g., 0–100 mv). Additional serial outputs (RS-232, IEEE-488) may be provided to allow digital communication between the unit and other computers. The digital outputs from the I/O multiplexer may not be powerful enough to switch large electrical loads. In those cases, an ancillary relay board is connected to the I/O multiplexer. The I/O multiplexer outputs control a high-capacity relay that can start and stop large electrical loads.

Computer/Software

The computer power needed for monitoring and control applications in aquaculture is modest. Software is available for either IBM-compatible or MacIntosh computers. Most software is menu-driven and complex programming is not necessary. Using a point and click approach, complex monitoring and control schemes can be developed.

For a single rearing unit equipped with a low-water and DO sensor, the following monitoring and control scheme could be used:

If (level = low.or.DO <5 mg/L), then phone dialer =
  on.and.horn = on.
If (DO < 4 mg/L,) then A = on.
If (DO < 3 mg/L), then B = on.

In the first line, the phone dialer and local alarm horn would be turned on if either the water level drops below the setpoint or the effluent DO drops below 5 mg/L. The second and third lines turn on controls A and B when the DO drops below 4 and 3 mg/L, respectively. These two outputs could control an emergency aeration system or secondary water supply.

The control sequence is all in software and can be easily modified. If a different sensor is needed at an existing location, only signal information associated with that physical line must be changed. When a modem connection is provided, it is possible to log on to the system from a remote site and review the system status and change control settings. This can save many unnecessary trips to the hatchery in the middle of the night.

The software also can be used to archive important system information for review at a later time. The location and output of the various sensors can be shown on a graphical representation of the process flow sheet. For the weak of heart, there are a growing number of system integrators that can assist with system conceptualization, goals, design, installation, documentation, and operation (Lee 1995).

Hatchery Effluents

Hatchery effluent may pollute receiving streams and lakes. As with any other source of wastewater, hatcheries are subject to federal, state, and local regulations regarding pollution. Hatchery operators are responsible for knowing the regulations that apply to their facilities and filing the required monitoring information. Some smaller hatcheries may be exempt from state and federal regulations. Some treatment of hatchery effluent is required of almost every hatchery. This is true even for reuse systems; their advantage lies in the greatly reduced volume of effluent to be treated compared with single-pass hatcheries.

Overview of Discharge Regulations and Permitting

The discharge of waters from aquaculture projects is regulated by the U.S. Environmental Protection Agency (USEPA) through the National Pollutant Discharge Elimination System (NPDES). These regulations apply to the discharges from aquaculture facilities maintaining or growing freshwater, estuarine, or marine plants or animals. The following types of aquaculture facilities require a NPDES discharge permit (40CFR122):

1. A facility that discharges to surface waters of the United States during at least thirty (30) days per year *and*
   a. A coldwater aquaculture facility which has a production of 20,000 lb or more per year or which feeds 5,000 lb of food or more in any one calendar month.

or

   b. A warm water aquaculture facility that has a production of 100,000 lb per year.
2. Any exempted facility may be required to obtain a discharge permit if it is determined that the facility is a significant contributor of pollution. In making this designation, the following factors are considered:

a. The location and quality of the receiving waters,
b. The capacities of the facility,
c. The quantity and nature of the pollutants from the facility, and
d. Other factors relating to the significance of the pollution problems in the area.

For those states that have demonstrated legal and administrative ability, the USEPA can delegate the administrative responsibility to the states. Most states have assumed the regulatory responsibility for permitting of hatchery effluents. The NPDES regulations apply to discharges to the waters of the United States. A facility discharging to a sanitary sewer would not require a NPDES permit. Additional treatment may be required for disease or containment concerns, even if the discharges are to a sanitary sewer. An application for a discharge permit must be filed before the pollutants are discharged. There are substantial differences in the permits between different USEPA regions and states.

In aquaculture facilities, simple settling of solids may be the only type of treatment required to achieve discharge standards. It may also be necessary in the permit to develop a Best Management Practices (BMP) to be followed in operating the facilities, harvesting the fish, and removing and disposing of the solids from the system. Information on the types of drugs and chemicals used in the facility may also be requested. The use and discharge of chemicals from aquaculture facilities will probably be regulated more intensely in the future.

Although the federal requirements for NPDES permits are straightforward, state regulations may be more stringent and actual permit conditions are up to each state agencies. The permits for a given variable may be commonly written in terms of

Daily minimum (mg/L),
Daily maximum (mg/L, mL/L, kg/d, lb/d),
Instantaneous maximum (mg/L, mL/L),
Monthly average (mg/L, kg/month, lb/month), and
Yearly maximum (kg/year, lb/year).

Depending on the state, the permit limit may be written on an absolute basis (effluent) or net addition from the hatchery basis (effluent-influent). If the hatchery has an absolute daily suspended solids limit of 6 mg/L, it will be in violation of its permit when the influent suspended solids concentration is 35 mg/L and the effluent concentration is 10 mg/L. Some hatcheries using a stream water source may not be able to meet suspended solids standards during periods of high runoff. The permit will specify frequency of analysis, sample type, analytical methods, and reporting intervals. The permit may also contain a narrative standard (for example, "the receiving water shall contain no unnatural turbidity, color, oil film, floating solids, foams, settleable solids, or deposits as a result of this discharge") and other various permit requirements. Three examples of NPDES permits are presented in Tables 15a, 15b, and 15c to show the range of potential regulations and the variation between different states and different facilities.

Effluent Management

Although a variety of water quality measures could be regulated, the most commonly regulated measurement is suspended solids. Typical concentrations of pollutants in aquaculture facilities are very low, although the mass of pollutants (kg/d) may be large because of the large flows involved. Suspended solids can readily accumulate in a relatively concentrated area and, as a result, can cause a substantial impact on environmental quality. Control of phosphorus discharges will be covered separately.

Uneaten feed and fecal solids are regulated under the NPDES permitting process (see previous section on regulations). Historically, solids management in raceways was driven by fish health concerns. The build up of uneaten feed, fecal matter, and dead fish on the bottom of rearing units was unsanitary and could result in increased disease problems. The most common approach was to drop the water level and manually sweep the solids to the discharge with a broom. This approach is labor-intensive and can expose the fish to high levels of solids during the cleaning process. The hatchery solids were discharged to the receiving water without treatment. Below the discharges of some hatcheries, accumulations of uneaten fed and fecal matter covered the stream bot-

TABLE 15a. Discharge limitation for Platte River Hatchery, Michigan (Michigan DNR, Production Salmon Hatchery). TSS = total suspended solids; SS = settleable solids; TP = total phosphorus.

Parameter	Monthly average	Daily maximum	Yearly maximum	Monthly average	Daily maximum	Frequency of analysis	Type of sample
Flow	44 x 10^6 g/d	report				weekly	total flow
TSS	1,470 lb/d	2,200 b/d		4 mg/L	6 mg/L	weekly	24 h composite
Outfall observation						daily	visual
TP (as P)			440 lb/y net 121 lb per 3-month period (net)	~0.008 mg/L net		weekly	24 h composite

TABLE 15b. Discharge limitation for Snake River Hatchery, Idaho (Clear Springs, Production Rainbow Trout Hatchery). TSS = total suspended solids; SS = settleable solids; and TP = total phosphorus.

Parameter	Monthly average	Daily maximum	Yearly maximum	Monthly average	Daily maximum	Frequency of analysis	Type of sample
Flow		report				weekly	grab
TSS				5 mg/L net	10 mg/L net	monthly	4-sample composite
TSS (harvest)				67 mg/L net	100 mg/L net	monthly	4-sample composite
SS (harvest)				0.7 mg/L net	1.0 mg/L net	monthly	4-sample composite
TP (as P)	46.54 lb/d net	53.86 lb/d net	17,000 lb/y net	0.10 mg/L net	0.16 mg/Lnet	weekly	4-sample composite
Temperature		report				monthly	grab

tom to a depth of 10–20 cm (4–8 in). Although most large cool and coldwater hatcheries have installed some type of effluent treatment, there are still hatcheries that discharge without any type of treatment.

TABLE 15c. Discharge limitation for Upland Finfish Hatcheries—State of Washington general permit. TSS = total suspended solids; SS = settleable solids; TP = total phosphorus.

Parameter	Monthly average	Daily maximum	Yearly maximum	Monthly average	Daily maximum	Frequency of analysis	Type of sample
Flow		report				2–4/ month	grab
SS-rearing				0.1 mL/L	net	weekly	grab
TSS-rearing				5.0 mg/L net	15.0 mg/L net	month	composite (flow-proportional)
SS-offline settling basin					1.0 mL/L	weekly	grab
TSS-offline settling basin					100 mg/L	quarterly	grab
SS-release					1 mg/L	1/release	grab
TSS-release					100 mg/L	1/release	grab

Raceways. Three approaches have been used for treatment of solids from cool- and coldwater hatcheries. The first approach treats all the discharges from the hatchery in a settling pond with a 1–2 h detention time (Figure 34a). For a large flow, this approach requires a substantial amount of land for the settling pond. The settling pond is generally lined with asphalt or concrete to allow yearly removal of accumulated solids with a front-end loader or tractor. An alternative is to pass all of the hatchery effluent through a microscreen. The solids from the microscreen can be treated in a clarifier (Figure 34b).

The second approach uses a dual drainage system. During normal operation, the water is discharged without treatment. During cleaning, the water level in a single raceway is lowered and this water is diverted to a settling basin or treatment system. The raceway is then cleared manually

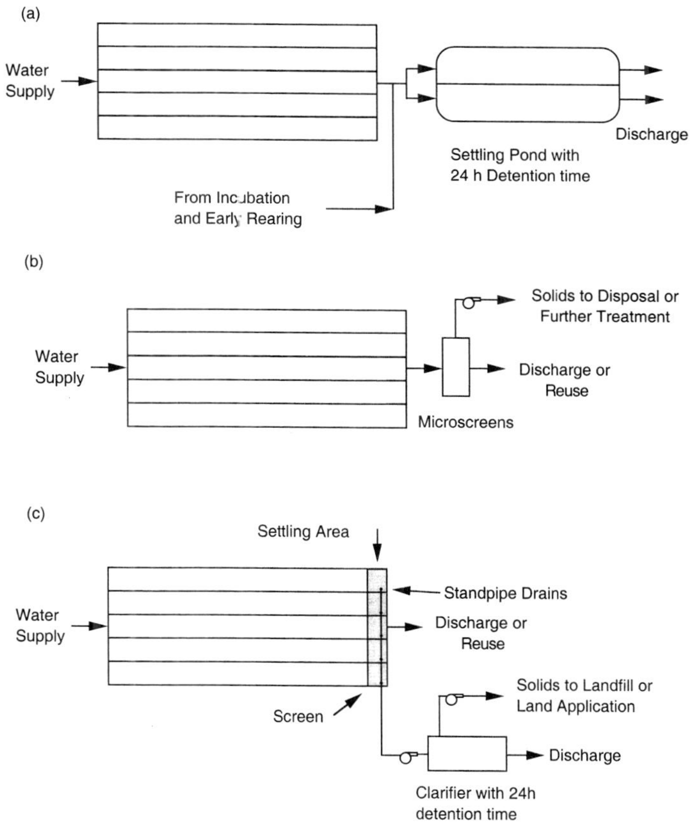

FIGURE 34. Process flow sheet for solids collection systems (a) total hatchery flow system —settling pond, (b) total hatchery flow system—microscreens, and (c) cleaning waste system—clarifier.

by sweeping the solids toward the discharge end of the raceway. The cleaning flow is relative small as only one raceway is cleaned at a time.

The third approach settles the solids in the end of each raceway. A screen is installed 3–5 m (10–16 ft) in front of the overflow weir (Figure 34c). The screen excludes fish and provides a quiescent area for the solids to accumulate. One to two times a week, the solids are manually vacuumed into the solids drainage system and conveyed to a clarifier. The detention time of the clarifier is 24 h.

In terms of the percent of solids removed, the in-unit system is the most desirable as the solids have moved the minimum distance and have not subjected to excessive velocities or turbulence. There are many systems that serially reuse the water 2–5 times. In these systems, many of the solid particles from the upper raceways will be broken down into smaller pieces by the time they arrive at the end of the last raceway.

There has been great amount of effort directed toward the development of self-cleaning rearing units. One is the baffled raceway (Boersen and Westers 1986). The distance between the baffles is approximately equal to the width of the raceway (Figure 35a, 35b). The clearance between the bottom of the baffle and the raceway is adjusted to provide a velocity ranging from 0.1 to 0.4 m/s, depending on fish size and species. Using simple continuity relationships (see Figure 35c), the mean velocity under the baffles is equal to

$$V_{\text{baffle}} = \left[\frac{Q}{W \bullet D_1}\right]\frac{D_1}{D_2} \qquad (12)$$

where

V_{baffle} = Mean cross-sectional velocity (m/s, ft/s),
Q = Water flow (m^3/s, ft^3/s),
W = Width of the raceway (m, ft),
D_1 = Water depth in raceway (m, ft), and
D_2 = Water depth between the bottom of the baffle and the raceway (m, ft).

The term within the bracket is the mean upstream velocity in the raceway.

Further work with a variety of species has shown that the baffles are effective in moving the uneaten feed and fecal matter to the end of the raceway. They appear to work better with domestic rainbow trout and should be used with care when raising wild trout and salmon (Kindschi et al. 1991; Barnes et al. 1996; Tipping 1998). From an operational standpoint, baffles restrict movement of fish culturists up and down the raceway and may increase problems with enumeration due to nonuniform distribution of fish within the raceway.

Circular and semicircular tanks. Circular tanks can be self-cleaning over

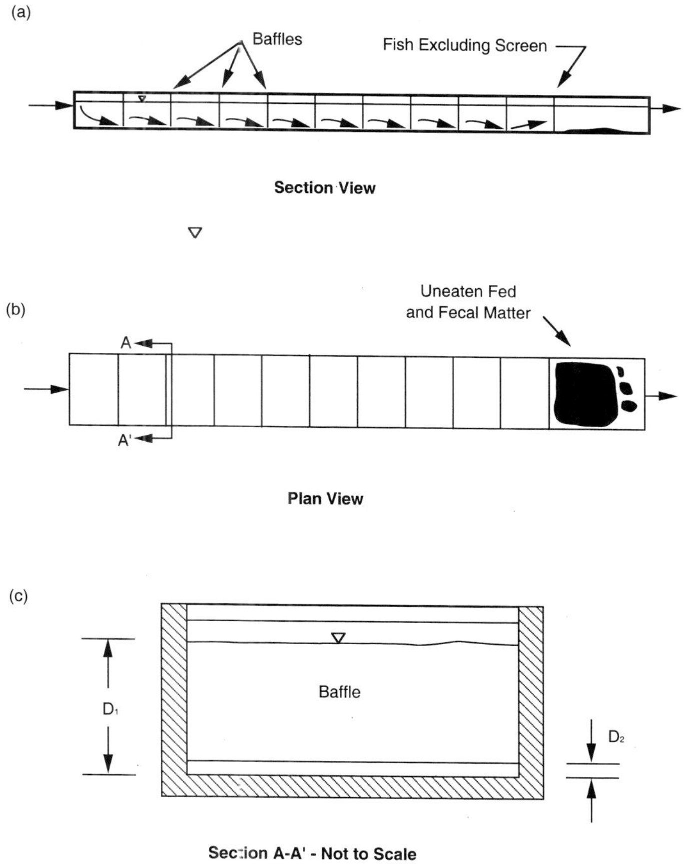

FIGURE 35. Baffled raceway (a) section view, (b) plan view, and (c) cross-section. D_1 is the water depth and D_2 is the distance between the bottom of the baffle and the bottom of the raceway. The symbol [∇] is used to indicate the water level.

a limited flow range. Solids may accumulate at lower and higher flows. If the circular tank is operated in a self-cleaning mode, the full hatchery flow must be treated by use of microscreens or settling ponds because solids are continuously discharged from the rearing unit. It is also possible to design a dual discharge system to separate cleaning wastes from normal discharge (Figure 36a). During normal operation, water flows through the center screen and over the external standpipe drain. During cleaning, the cleaning water standpipe is removed, which greatly increases the discharge. Increased water flow, coupled with manual sweeping of the tank can rapidly remove any solids from the tank. An alternative is to install a dual level discharge structure. The lower discharge pulls water off the bottom of the tank and contains higher levels of solids than the upper discharge (Figure 36b).

Ponds. Most commercial ponds only discharge during pond draining or extreme rainfall events. In most states, discharges from ponds are not regulated. Because of the infrequent nature of pond discharges, water management techniques offer the greatest potential to reduce pollution from ponds. In addition, natural in-pond processes greatly reduce the amount of pollution discharge. Three common water management techniques are (a) addition of water only to make up for evaporation and seepage losses, (b) use of ponds for several years without draining, and (c) operation of the pond with a minimum of 7.5 cm (3 in) of water storage potential during the rainy period (Tucker et al. 1996). Managing pond water levels to maintain a minimum 7.5 cm (3 in) water storage potential reduced the average discharge of nitrogen, phosphorus, and

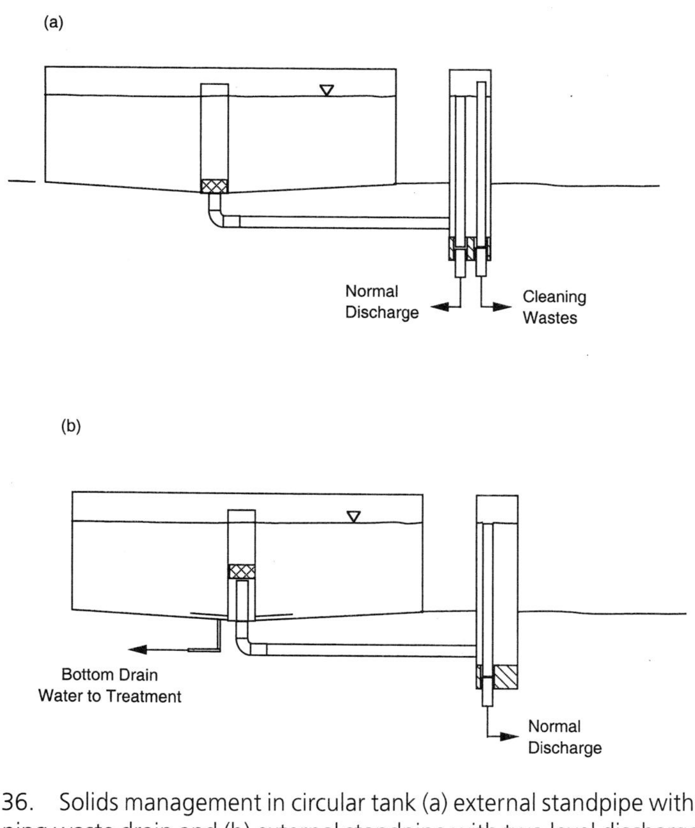

FIGURE 36. Solids management in circular tank (a) external standpipe with separate cleaning waste drain and (b) external standpipe with two level discharge from rearing unit. The symbol ∇ is used to indicate the water level.

organic matter in pond discharges by about 70% compared with ponds not managed to maintain storage potential (Tucker et al. 1996). Harvesting fish without draining ponds between fish crops could reduce the average annual nutrient and organic matter discharge by over 60% relative to annually drained ponds. These techniques appear to be simple and effective methods of reducing waste discharge from aquaculture ponds with the added benefits of substantially reducing pumping costs. Using these three techniques, less than 6% of nitrogen and 2% of the phosphorus in the feed are discharged. When ponds are discharged, much of the nitrogen, phosphorus, and solids are contained in the last 5–20% of the effluent (Schwartz and Boyd 1995).

Net-pens and cages. In typical net-pens, uneaten feed and fecal solids will pass through the bottom of the rearing unit and collect beneath the rearing system. The impact of the solids will depend on the amount of uneaten fed, amount of fecal solids produced, local currents, and the number of net-pens in the complex. Based on environmental concerns, there have been several commercial attempts to collect and treat solids from net-pens. The technical and economic feasibility of these systems remains to be demonstrated. Control of the environmental impacts of net-pens at this time requires careful feeding to reduce overfeeding and, if necessary, reduction in the overall facility production.

Effluent Treatment

With the exception of solids removal, treatment of hatchery effluent typically has not been required. This is because the concentration of most pollutants in hatchery effluents is very low and the cost of potential treatment processes is high. Constructed wetland (Schwartz and Boyd 1995) or grass filter strips (Ghate et al. 1997) may be the most economic type of treatment for temperate areas where a large part of the *N* and *P* is soluble rather than particulate. If needed, these treatment options may be economical if used in conjunction with an effective water management program to reduce the amount of flow needing treatment.

Special Treatment Issues

Under certain conditions, special treatment of some types of hatchery effluent may be needed. Because of the capital and operating costs involved with these types of treatment, other nontreatment-based options should be considered.

Viruses, bacteria, and parasites. The transmission of pathogenic organisms from hatchery to wild animals is not well understood, mainly because of experimental problems in trying to document the process. Disinfection of hatchery effluent is uncommon. The fish health policies of some states suggest effluent disinfection when holding out-of-basin animals due to the potential disease transmission concerns. Chlorine or ozone disinfection has been used. Although it is possible to produce a substantial kill, these processes do not result in sterilization and some number of pathogenic organisms will be released. Therefore, it is better to try to avoid out-of-basin holding in sensitive areas.

Discharge from laboratories working with pathogenic organisms requires special treatment. These flows are generally in the range of 20–400 Lpm (5–100 gpm). Two approaches have involved (a) two-stage treatment of the waste stream with very high levels of ozone followed with filtration through a granular activated carbon and deep-bed sand filter or (b) superchlorination (10 mg/L chlorine with 10 min detention time) followed by discharge into the sanitary sewer.

Animal containment or confinement. Two levels of animal containment or confinement may be needed in hatcheries:

Application	Concern	Approach
Rearing of out-of-basin animals	Genetic impacts of large releases; potential disease transmission to wild fish	Double screening of drains to prevent escape of animals
Exotic animals and plants	Ecological impacts and interactions with wild animals and plants	Discharge to sanitary sewer, septic tank, or land application No discharge of water outside of facility; or Location of facility in area with interior drainage

The U.S. Fish and Wildlife Service's fishery laboratory at Gainesville, Florida, was specially designed for the study of nonnative animals. The facility is surrounded by an earthen berm that prevents any culture water from leaving the site except under very exceptional rainfall events. The voluntary guidelines for holding transgenic aquatic animals (ABRAC 1995) are an excellent source of information on containment and confinement techniques.

Chemical and drugs. The number of chemicals and drugs that can be used for food animals is very limited and, as a result, treatment is not generally needed prior to discharge. In pond culture, natural biological processes may deactivate many chemicals. If needed, granular activated carbon (Marking et al. 1990) can be used to remove many drugs from hatchery water prior to discharge.

Phosphorus. When hatchery effluent is discharged in lakes (or streams that run into lakes), very low phosphorus concentrations may increase the growth of photoplankton and filamentous algae. The impact of a given amount of phosphorus depends on existing nitrogen and phosphorus concentrations, depth of the lake, and detention time. In one case in Michigan, the effective average annual phosphorus criteria was set at 0.008 mg/L as *P* (8 µg/L). Conventional wastewater treatment systems can attain an effluent in the range of 0.5–1.0 mg/L. Control of phosphorus in hatchery effluent will require (a) use of low phosphorus diets, (b) careful feeding to avoid excessive feed wastes, (c) effective removal of uneaten fed and fecal solids to minimize solubilization of phosphorus, and (d) if necessary, reduction in production.

References

ABRAC (Agricultural Biotechnology Research Advisory Committee–U.S. Department of Agriculture). 1995. Performance standards for safely conducting research with genetically modified fish and shellfish. Documents 95-04 and 95-05. (Available from Biotechnology Information Center, National Agriculture Library, 10301 Baltimore Avenue, Beltsville, Maryland 20705, USA; http://www.nbiap.vt.edu/perfstands.)

Akai, D., O. Miki, and S. Ohgaki. 1983. Nitrification model with an inhibitory effect of sea water. Ecological Modeling 19:189–198.

Andrews, J. W., and R. R. Stickney. 1972. Interactions of feeding rates and environmental temperature on growth, food conversion, and body composition of channel catfish. Transactions of the American Fisheries Society 101:94–99.

Andrews, J. W., L. H. Knight, and T. Murai. 1972. Temperature requirements for high density rearing of channel catfish from fingerling to market size. Progressive Fish-Culturist 34:240–246.

Barnes, M. E., W. A. Sayler, and R. J. Cordes. 1996. Baffle usage in covered raceways. Progressive Fish-Culturist 58:286–288.

Barton, B. A. 1996. General biology of salmonids. Pages 29-95 *in* W. Pennell and B. A. Barton, editors. Principles of salmonid culture. Elsevier, Amsterdam.

Berman, C. H., and T. P. Quinn. 1991. Behavioural thermoregulation and homing by spring chinook salmon, *Oncorhynchus tshawytscha* (Walbaum) in the Yakima River. Journal of Fish Biology 39:301–312.

Boersen, G., and H. Westers. 1986. Waste solids control in hatchery raceways. Progressive Fish-Culturist 48:151–154.

Bouck, G. R. 1980. Etiology of gas bubble disease. Transactions of the American Fisheries Society 109:703–707.

Bowen, H. J. M. 1966. Trace elements in biochemistry. Academic Press, New York.

Bower, C. E., and J. P. Bidwell. 1978. Ionization of ammonia in seawater: effects of temperature, pH and salinity. Journal of the Fisheries Research Board of Canada 35:1012–1016.

Boyd, C. E. 1979a. Water quality in warmwater fish ponds. Alabama Agricultural Experiment Station, Auburn.

Boyd, C. E. 1979b. Aluminum sulfate (alum) for precipitating clay turbidity from fish ponds. Transactions of the American Fisheries Society 108:307–313.

Boyd, C. E., and C. S. Tucker. 1998. Pond aquaculture water quality management. Kluwer Academic Publishers, Boston.

Brett, J. R. 1979. Environmental factors and growth. Pages 599–675 in W. S. Hoar, D. J. Randall, and J. R. Brett, editors. Fish physiology, volume VIII. Academic Press, New York.

Brewer, D. L., and R. A. Rees. 1990. Pond culture of phase l striped bass fingerlings. Pages 99–120 *in* R. M. Harrell, J. H. Kerby, and R. V. Minton, editors. Culture and propagation of striped bass and its hybrids. American Fisheries Society, Southern Division, Striped Bass Committee, Bethesda, Maryland.

Bristow, B. T., R. C. Summerfelt, and R. D. Clayton. 1996. Comparative performance of intensively cultured larval walleye in clear, turbid and colored water. Progressive Fish-Culturist 58:1–10.

Bruin, W., J. W. Nightingale, and L. Mumaw. 1981. Preliminary results and design criteria from an on-line zeolite ammonia removal filter in a semi-closed recirculating system. Pages 92–96 *in* L. J. Allen and E. C. Kinney, editors. Proceedings of the bio-engineering symposium for fish culture. American Fisheries Society, Fish Culture Section, Bethesda, Maryland.

Busch, R. L. 1985. Channel catfish culture in ponds. Pages 13–84 *in* C. S. Tucker, editor. Channel catfish culture. Elsevier, New York.

Carmichael, G. J., J. H. Williamson, C. A. Caldwell-Woodward, and J. R. Tomasso. 1988. Responses of northern, Florida and hybrid largemouth bass to low temperature and low dissolved oxygen. Progressive Fish-Culturist 50:225–231.

Cheung, P. 1993. Parasitic diseases of marine tropical fishes. Pages 646–658 *in* M. K. Stoskopf, editor. Fish medicine. Saunders, Philadelphia.

Colberg, P. J., and A. J. Lingg. 1978. Effect of ozonation on microbial fish pathogens, ammonia, nitrate, nitrite, and BOD in simulated reuse hatchery water. Journal of the Fisheries Research Board of Canada 35:1290–1296.

Collings, M. R. 1973. Generalization of stream-temperature data in Washington. U.S. Geological Survey, Water-Supply Paper 2029-B, Washington, D.C.

Colt, J. 2000. Gas bubble disease. Pages 403–408 *in* R. R. Stickney, editor. The encyclopedia of aquaculture. Wiley, New York.

Colt, J. 1984. Computation of dissolved gas concentrations in water as functions of temperature, salinity, and pressure. American Fisheries Society, Special Publication 14, Bethesda, Maryland.

Colt, J., and G. R. Bouck. 1984. Design of packed columns for degassing. Aquacultural Engineering 3:251–273.

Colt, J., and E. Cryer. 2000. Ozone. Pages 622–628 *in* R. R. Stickney, editor. The encyclopedia of aquaculture, Wiley, New York.

Colt, J., and K. Orwicz. 1991a. Modeling production capacity of aquatic culture systems under freshwater conditions. Aquacultural Engineering 10:1–29.

Colt, J., and K. Orwicz. 1991b. Aeration in intensive culture. Pages 198–271 *in* D. E. Brune and J. R. Tomasso, editors. Aquaculture and water quality. World Aquaculture Society, Baton Rouge, Louisiana.

Colt, J., J. Sheahan, and G. R. Bouck. 1993. Evaluation of the "Michigan" type pure oxygen for oxygen addition and nitrogen removal. Aquacultural Engineering 12:141–154.

Colt, J., and G. Tchobanoglous. 1976. Evaluation of short-term toxicity of nitrogenous compounds to channel catfish, *Ictalurus punctatus*. Aquaculture 8:209–224.

Colt, J., and B. J. Watten. 1988. Applications of pure oxygen in fish culture. Aquacultural Engineering 7:397–404.

Colt, J., and H. Westers. 1982. Production of gas supersaturation by aeration. Transactions of the American Fisheries Society 111:342–360.

Costa-Pierce, B. A., and J. E. Rakocy, editors. 1997. Tilapia aquaculture in the Americas. World Aquaculture Society, Baton Rouge, Louisiana.

Coutant, C. C., and D. L. DeAngelis. 1983. Comparative temperature-dependent growth rates of largemouth and smallmouth bass. Transactions of the American Fisheries Society 112:416–423.

Cox, D. K., and C. C. Coutant. 1981. Growth dynamics of juvenile striped bass as functions of temperature and ration. Transactions of the American Fisheries Society 110:226–238.

Dudley, R. G., and A. W. Eipper. 1975. Survival of largemouth bass embryos at low dissolved oxygen concentrations. Transactions of the American Fisheries Society 104:122–128.

Dwyer, W. P., and W. H. Orr. 1992. Removal of radon gas liberated by aeration columns in fish hatcheries. Progressive Fish-Culturist 54:57–58.

Emerson, K., R. C. Russo, R. E. Lund, and R. V. Thurston. 1975. Aqueous ammonia equilibrium calculations: effect of pH and temperature. Journal of the Fisheries Research Board of Canada 32:2379–2383.

Erickson, R. J., D. A. Benoit, V. R. Mattson, H. P. Nelson, and E. N. Leonard. 1996. The effects of water chemistry on the toxicity of copper to fathead minnows. Environmental Toxicology and Chemistry 15:181–193.

Forster, J. R. M. 1974. Studies on nitrification in marine biological filters. Aquaculture 4:387–397.

Fromm, P. O. 1980. A review of some physiological and toxicological responses of freshwater fish to acid stress. Environmental Biology of Fishes 5:79–93.

Fry, F. E. J. 1947. Effects of environment on animal activity. University of Toronto Studies Biological Series 55:1–62, Toronto.

Fry, F. E. J. 1971. The effects of physiological factors on the physiology of fish. Pages 1–98 *in* W. S. Hoar and D. J. Randall, editors. Fish physiology, volume VI. Academic Press, New York.

Ghate, S. R., G. J. Burtle, G. Vellidas, and G. L. Newton. 1997. Effectiveness of grass strips to filter catfish *Ictalurus punctatus* pond effluent. Aquacultural Engineering 16:149–159.

Grguric, G., J. H. Trefry, and J. J. Keaffaber. 1994. Ozonation products of bromine and chlorine in seawater aquaria. Water Research 28:1087–1094.

Grizzle, J. M., K. A. Cummins, and C. J. Ashfield. 1993. Effects of environmental concentrations of calcium and sodium on the calcium flux in stressed 34-day-old striped bass. Canadian Journal of Zoology 71:1379–1384.

Guest, W. C. 1985. Survival of adult, Florida and northern largemouth bass subjected to cold temperatures. Proceedings of the Annual Conference Southeastern Association of Fish and Wildlife Agencies 36(1982):332–339.

Hall, L. W., Jr. 1991. A synthesis of water quality and contaminants data on early life stages of striped bass, *Morone saxatilis*. Reviews in Aquatic Sciences 4:261–268.

Hansen, S. P., and G. L. Culp. 1967. Applying shallow depth sedimentation theory. Journal of the American Water Works Association 59:1134–1148.

Hill, R. W., and G. A. Wyse. 1989. Animal physiology, 2nd edition. Harper Collins, New York.

Hinshaw, J. M. 1990. Trout farming: a guide to production and inventory management. Southern Regional Aquaculture Center Publication 222, Stoneville, Mississippi.

Hoffman, G. L. 1974. Disinfection of contaminated water by ultraviolet irradiation, with emphasis on whirling disease (*Myxosoma cerebralis*) and its effect on fish. Transactions of the American Fisheries Society 103:541–550.

Howe, K. J., and D. F. Lawler. 1989. Acid–base reactions in gas transfer: a mathematical approach. Journal of the American Water Works Association 81:61–66.

Huguenin, J. E., and J. Colt. 1989. Design and operating guide for aquaculture seawater systems. Elsevier Publishers, Amsterdam.

Jobling, M. 1994. Fish bioenergetics. Chapman and Hall, London.

Johnson, P. W., and J. M. Sieburth. 1974. Ammonia removal by ion exchange, a backup system for microbiological filters in closed-system aquaculture. Aquaculture 4:61–68.

JRB Associates. 1984. Development of effluent limitations for Idaho fish hatcheries. Submitted to the U.S. Environmental Protection Agency, Office of Water Enforcement (EPA Contract 68-01-6514), Washington, D.C.

Kellogg, R. L., and J. J. Gift. 1983. Relationship between optimal temperatures for growth and preferred temperatures for the young of four fish species. Transactions of the American Fisheries Society 112:424–430.

Kelly, J. W. 1968. Effects of incubation temperature on survival of largemouth bass eggs. Progressive Fish-Culturist 30:159–163.

Kerby, J. H., L. C. Woods, and M. T. Huish. 1983. Pond culture of striped bass x white bass hybrids. Journal of the World Aquaculture Society 14:613–623.

Kilambi, R. V., J. Noble, and C. E. Hoffman. 1971. Influence of temperature and photoperiod on growth, food consumption, and body composition of channel catfish. Proceedings of the Annual Conference Southeastern Association of Game and Fish Commissioners 24(1970):519–531.

Kindschi, G. A., R. G. Thompson, and A. P. Mendoza. 1991. Use of raceway baffles in rainbow trout culture. Progressive Fish-Culturist 53:97–101.

Klieve, J. R., and M. J. Semmens. 1980. An evaluation of pretreated natural zeolites for ammonium removal. Water Research 14:161–168.

Koenst, W. M., and L. L. Smith. 1976. Thermal requirements of the early life history stages of walleye and sauger. Journal of the Fisheries Research Board of Canada 33:1130–1138.

Langlais, B., D. A. Reckhow, and D. R. Brink, editors. 1991. Ozone in water treat-

ment - application and engineering. Lewis Publishers, Chelsea, Michigan.
Laurén, D. J., and D. G. McDonald. 1986. Influence of water hardness, pH, and alkalinity on the mechanisms of copper toxicity in juvenile rainbow trout, *Salmo gairdneri*. Canadian Journal of Fisheries and Aquatic Sciences 43:1488–1496.
Lee, P. G. 1995. A review of automated control systems for aquaculture and design criteria for their implementation. Aquacultural Engineering 14:205–227.
Liao, P. B., and R. D. Mayo. 1974. Intensified fish culture combing water reconditioning with pollution abatement. Aquaculture 3:61–85.
Linsley, R. K., and J. B. Franzini. 1972. Water-resource engineering, 2nd edition. McGraw-Hill, New York.
Marking, L. L., and T. D. Bills. 1982. Factors affecting the efficiency of clinoplitolite for removing ammonia from water. Progressive Fish-Culturist 44:187–189.
Marking, L. L., D. Leith, and J. Davis. 1990. Development of a carbon filter system for removing malachite green from hatchery effluents. Progressive Fish-Culturist 52:92–99.
Mayes. M. A., H. C. Alexander, D. L. Hopkins, and P. B. Latvaitis. 1986. Acute and chronic toxicity of ammonia to freshwater fish: a site-specific study. Environmental Toxicology and Chemistry 5:437–442.
Meade, J. W. 1985. Allowable ammonia for fish culture. Progressive Fish-Culturist 47:135–145.
Meisner, J. D., J. S. Rosenfeld, and H. A. Regier. 1988. The role of groundwater in the impact of climate warming on stream salmonines. Fisheries 13(3):2–8.
Metcalf and Eddy, Inc. 1979. Wastewater engineering: treatment, disposal, reuse, 2nd edition, McGraw-Hill, New York.
Metcalf and Eddy, Inc. 1981. Wastewater engineering: collection and pumping of wastewater. McGraw-Hill, New York.
Mevel, G., and S. Chamroux. 1981. A study on nitrification in the presence of prawns (*Penaeus japonicus*) in marine closed systems. Aquaculture 23:29–43.
Moss, D. D., and D. C. Scott. 1961. Dissolved oxygen requirements of three species of fish. Transactions of the American Fisheries Society 90:377–393.
Mudrak, V. A. 1981. Guidelines for economical commercial fish hatchery wastewater treatment systems. Pages 174–182 *in* L. J. Allen and E. C. Kinney, editors. Proceedings of the bio-engineering symposium for fish culture. American Fisheries Society, Fish Culture Section, Bethesda, Maryland.
Murray, C. N., and J. P. Riley. 1969. The solubility of gases in distilled water and in seawater. II Oxygen. Deep-Sea Research 16:311–320.
Neill, W. H., and J. D. Bryan. 1991. Responses of fish to temperature and oxygen, and response integration through metabolic scope. Pages 30–57 *in* D. E. Brune and J. R. Tomasso, editors. Aquaculture and water quality. World Aquaculture Society, Baton Rouge, Louisiana.
Nickum, J. G., and R. R. Stickney. 1993. Walleye. Pages 231–250 *in* Stickney (1993).
Oppenborn, J. B., and C. A. Goudie. 1993. Acute and sublethal effects of ammonia on striped bass and hybrid striped bass. Journal of the World Aquaculture Society 24:90–101.
Piedrahita, R. H., and G. R. Grace. 1994. Carbon dioxide control. Pages 209–233 *in* M. B. Timmons and T. M. Losordo, editors. Aquacultural reuse systems: engineering design and management. Elsevier, New York.

Piper, R. G., and five coauthors. 1982. Fish hatchery management. U.S. Fish and Wildlife Service, Washington, D.C.

Qualls, R. G., M. P. Flynn, and J. D. Johnson. 1983. The role of suspended particles in ultraviolet disinfection. Journal of the Water Pollution Control Federation 55:1280–1285.

Rees, R. A., and R. M. Harrell. 1990. Artificial spawning and fry production of striped bass and hybrids. Pages 43–72 *in* R. M. Harrell. J. H. Kerby, and J. V. Minton, editors. Culture and propagation of striped bass and its hybrids. American Fisheries Society, Southern Division, Striped Bass Committee, Bethesda, Maryland.

Rosenthal, H., and G. Otte. 1979. Adaptation of a recirculating fish culture system to salinity change. Wissenschaftlichen Kommission fuer Meeresforschung 27:203–206.

Rosenthal, H., and J. S. Wilson. 1987. An updated bibliography (1945-1986) on ozone, its biological effects and technical applications. Canadian Technical Report, Fisheries and Aquatic Sciences 1542.

Schreck, C. B. 1981. Stress and compensation in teleostean fishes: responses to social and physical factors. Pages 295–321 *in* A. D. Pickering, editor. Stress and fish. Academic Press, New York.

Schroeder, E. D. 1977. Water and wastewater treatment. McGraw-Hill, New York.

Schwartz, M. F., and C. E. Boyd. 1995. Constructed wetlands for treatment of channel catfish pond effluents. Progressive Fish-Culturist 57:255–266.

Seals, C., C. Kempton, J. Tomasso, and T. Smith. 1994. Environmental calcium does not affect production or selected blood characteristics of sunshine bass reared under normal culture conditions. Progressive Fish-Culturist 56:269–272.

Seegert, G. L., and A. Brooks. 1978. Dechlorination of water for fish culture: comparison of the activated carbon, sulfite reduction, and photochemical methods. Journal of the Fisheries Research Board of Canada 35:88–92.

Semmens, M. J., and P. S. Porter. 1979. Ammonium removal by ion exchange: using biologically restored regenerant. Journal of the Water Pollution Control Federation 51:2928–2940.

Sharma, B., and R. C. Ahlert. 1977. Nitrification and nitrogen removal. Water Research 11:897–925.

Shepherd, B. G. 1991. On choosing well: bioengineering reconnaissance of new hatchery sites. Pages 354–364 *in* J. Colt and R. J. White, editors. Fisheries bioengineering symposium. American Fisheries Society, Symposium 10, Bethesda, Maryland.

Sorensen, E. M. 1991. Metal poisoning in fish. CRC Press, Boca Raton, Florida.

Spotte, S. 1979. Fish and invertebrate culture. Wiley-Interscience, New York.

Spotte, S., and G. Adams. 1981. Pathogen reduction in closed aquaculture systems by UV radiation: fact or artifact? Marine Ecology Progress Series 6:295–298.

Stickney, R. R. 1991. Effects of salinity on aquaculture production. Pages 105–132 *in* D. E. Brune and J. R. Tomasso, editors. Aquaculture and water quality. World Aquaculture Society, Baton Rouge, Louisiana.

Stickney, R. R. 1993. Channel catfish. Pages 33–80 *in* R. R. Stickney, editor. Culture of nonsalmonid freshwater fishes, 2nd edition. CRC Press, Boca Raton, Florida.

Straus, D. L., and C. S. Tucker. 1993. Acute toxicity of copper sulfate and chelated copper to channel catfish *Ictalurus punctatus*. Journal of the World Aquaculture Society 24:390–395.

Summerfelt, S. T., and J. N. Hochheimer. 1997. Review of ozone processes and applications as an oxidizing agent in aquaculture. Progressive Fish-Culturist 59:94–105.

Sverdrup, H. U., M. W. Johnson, and R. H. Fleming. 1942. The oceans: their physics, chemistry and general biology. Prentice-Hall, Englewood Cliffs, New Jersey.

Thurston, R. V., and R. C. Russo. 1983. Acute toxicity of ammonia to rainbow trout. Transactions of the American Fisheries Society 112:696–704.

Tipping, J. M. 1988. Ozone control of *Ceratomysosis*: survival and growth benefits to steelhead and cutthroat trout. Progressive Fish-Culturist 50:202–210.

Tipping, J. M. 1998. Return rate of hatchery-produced sea-run cutthroat trout reared in a pond versus a standard or baffled raceway. Progressive Fish-Culturist 60:109–113.

Tomasso, J. R. 1986. Comparative toxicity of nitrite to warmwater fishes. Aquatic Toxicology 8:129–137.

Tomasso, J. R. 1994. Toxicity of nitrogenous wastes to aquaculture animals. Reviews in Fisheries Science 2:291–314.

Tomasso, J. R. 1997. Environmental requirements and noninfectious diseases. Pages 253–270 *in* R. M. Harrell, editor. Striped bass and other *Morone* culture. Elsevier, New York.

Tucker, C. S. 1991. Water quantity and quality requirements for channel catfish hatcheries. Southern Regional Aquaculture Center Publication 461, Stoneville, Mississippi.

Tucker, C. S. 1993. Water analysis. Pages 166–197 *in* M. K. Stoskopf, editor. Fish medicine. Saunders, Philadelphia.

Tucker, C. S. 1996. The ecology of channel catfish ponds in northwest Mississippi. Reviews in Fisheries Science 4:1–55.

Tucker, C. S., compiler. 1998. Characterization of finfish and shellfish aquaculture effluents. Southern Regional Aquaculture Center (SRAC 600), Stoneville, Mississippi.

Tucker, C. S., and C. E. Boyd. 1985. Water quality. Pages 135–227 *in* C. S. Tucker, editor. Channel catfish culture. Elsevier, New York.

Tucker, C. S., S. K. Kingsbury, J. W. Pote, and C. L. Wax. 1996. Effects of water management practices on discharge of nutrients and organic matter from channel catfish (*Ictalurus punctatus*) ponds. Aquaculture 147:57–69.

Tucker, C. S., and J. A. Steeby. 1993. A practical calcium hardness criterion for channel catfish hatchery water supplies. Journal of the World Aquaculture Society 24:396–401.

Turner, J. L., and T. C. Farley. 1971. Effects of temperature, salinity and dissolved oxygen on the survival of striped bass eggs and larvae. California Fish and Game 57:268–273.

USEPA (U.S. Environmental Protection Agency). 1976. Quality criteria for water. USEPA, EPA-440/9-76-023, Washington, D.C.

Wagner, E. J., T. Bosakowski, and S. A. Miller. 1995. Evaluation of the absorption efficiency of the low head oxygenation system. Aquacultural Engineering 14:49–57.

Watten, B. J. 1990. Design of packed columns for commercial oxygen addition and dissolved nitrogen removal based on effluent criteria. Aquacultural Engineering 9:305–328.

Watten, B., J. Colt, and C. Boyd. 1991. Impact of dissolved nitrogen and carbon dioxide on the operation of pure oxygen systems. Pages 474–481 *in* J. Colt and R. J. White, editors. Fisheries bioengineering symposium. American Fisheries Society, Symposium 10, Bethesda, Maryland.

Watten, B. J., and M. J. English. 1985. Effects of organic matter in aquacultural waste on the ammonium exchange capacity of clinoptilolite. Aquaculture 46:221–235.

Wedemeyer, G. A. 1996. Physiology of fish in intensive culture systems. Chapman and Hall, New York.

Weirich, C. R., J. R. Tomasso, and T. I. J. Smith. 1992. Confinement and transport-induced stress in white bass *Morone chrysops* X striped bass *M. saxatilis*: effect of calcium and salinity. Journal of the World Aquaculture Society 23:49–57.

Welsh, P. G., and six coauthors. 1993. Effect of pH and dissolved organic carbon on the toxicity of copper to larval fathead minnow (*Pimephales promelas*) in natural lake waters of low alkalinity. Canadian Journal of Fisheries and Aquatic Sciences 50:1356–1362.

Westers, H., and R. R. Stickney. 1993. Northern pike and muskellunge. Pages 99–214 *in* Stickney (1993).

Wheaton, F. W., and five coauthors. 1994a. Nitrification filter principals. Page 101–126 *in* M. B. Timmons and T. M. Losordo, editors. Aquaculture water reuse systems: engineering and management. Elsevier, Amsterdam.

Wheaton, F. W., and six coauthors. 1994b. Nitrification filter design methods. Page 127–172 *in* M. B. Timmons and T. M. Losordo, editors. Aquaculture water reuse systems: engineering and management. Elsevier, Amsterdam.

Whitmore, C. M., C. E. Warren, and P. Doudoroff. 1960. Avoidance reactions of salmonid and centrarchid fishes to low oxygen concentrations. Transactions of the American Fisheries Society 89:17–26.

Wild, H. E., Jr., C. N. Sawyer, and T. C. McMahon. 1971. Factors affecting nitrification kinetics. Journal of the Water Pollution Control Federation 43:1845–1854.

Williamson, J. H., G. J. Carmichael, K. G. Graves, B. A. Simco, and J. R. Tomasso. 1993. Centrarchids. Pages 145–198 *in* Stickney (1993).

4

Open Systems: The Culture of Fish for Release into Natural Systems

William Pennell, E. D. Lane, and Frank Dalziel

Introduction

Many species of fish are now intensively cultured throughout the world. For example, in Japan over 100 species of marine fish have had their life cycles closed through culture and nearly 50 species are cultured commercially. In the United States and Canada (see Chapter 1), new species are added to the culture list annually for commercial aquaculture and a variety of conservation and enhancement purposes. There is no clear boundary between fisheries management, ocean ranching, and commercial aquaculture in Japan; there are many situations where culture is the key factor in the production of seed for a variety of wild fisheries. In North America, these categories are much more distinct with commercial aquaculture of fish for food or for ornamental purposes operating within their own regulatory frameworks entirely within the private sector. The production of fish for release into the natural environment is overseen by various governmental bodies. It is the latter category that will be examined in this chapter.

Until recently, salmonids have formed much of the foundation of most European and all North American fish culture. The many nonsalmonid species have, with a few exceptions such as catfish, emerged as subjects for culture only within the last two decades. Salmonids, as noted in Chapter 1, have been cultured for at least 200 years and have proven relatively amenable to culture technology. Almost all of the early salmonid culture had augmentation of natural stocks and introduction of new species as its main goal, and, over the last century, the annual output of hatcheries has risen to several billion fry and smolts released in total from Japan and Russia, North America, and Europe. Rainbow trout now live in every country in the world where water temperatures below 15°C exist, including tropical countries with cool montane waters. Many strains of salmonid species are maintained solely by hatcheries, and there are major commercial fisheries entirely created and maintained by hatchery output. Despite this success, there are now many critics of salmonid hatcheries and much literature discussing the failures of certain hatchery approaches including ecological and genetic damage to wild stocks. Hatcheries may contribute to management complexity and actually cause extinction of local strains and even species through overfishing of wild stocks when they are intermixed with the more productive hatchery strains.

Thus, even as fish culture has expanded beyond any possible early imaginings and has had enormous successes, it is now being reevaluated for potential negative effects. Hatcheries and seminatural culture (e.g., spawning channels, enrichment) of salmonids and many other species are now permanent fixtures in fisheries management, but we have learned many hard lessons along the way. It is the purpose of this chapter to explore some of these lessons, to discuss the ways in which hatcheries are most effectively used for conservation and enhancement, and to point out that what is true for salmonids is likely to be true, in varying degrees, for all other fish species. In the course of this discussion the differences between food production hatcheries (e.g., private salmon smolt or trout fingerling facilities) and release hatcheries will be noted.

The fish culture activities discussed here are known under a variety of titles. Ocean, or sea, ranching refers to the release of anadromous salmonids (Pacific and Atlantic salmon and some species of trout and char) as fry or smolts into marine waters. The fish then behave as wild fish, travel on extensive feeding migrations, and return with precision to the release point—the hatchery of origin. In general, the idea is to take advantage of free ocean production by sending fish out to forage as free-range cattle do. Ocean ranching in North America is nearly always undertaken by governmental organizations (Canada, United States) or by nongovernmental organizations under the supervision of government (Alaskan cooperatives). In a few states, private ocean ranching is permitted, but it has not gained much popularity. Ranching strategies are also applied to lakes where the target users are usually sport fishers. In these cases, fish may or may not return to a point of origin. Various other terms are general substitutions for ocean ranching. These include enhancement, augmentation, and mitigation, the latter referring to a hatchery program designed to replace damaged natural production potential. The term enhancement usually implies increasing the population beyond its current size, but such goals are viewed with doubt today.

The success of an ocean ranching program is usually measured by the number of adult fish that result from the juveniles released by the hatchery. For anadromous salmonid hatcheries, this is called the ocean survival rate and is usually calculated as the proportion of the hatchery release (fry or smolts) returning as adults from the high seas migration to the coastal fishery and spawning areas, including the hatchery of origin. The proportion of adults arriving in the home streams or rivers (or hatchery) is called the escapement. The interplay between ocean survival, hatchery culture techniques, and hatchery selection regimes (including broodstock selection and egg-to-fry survival) remains poorly understood. Ocean survival rates range from less than 1% to as high as 10–15%. This rate determines the cost effectiveness of the hatchery. Benefit-cost ratios are complicated however, since they may take into account social benefits provided by the returning fish, or because the hatchery may be seen as preserving a genetic heritage with value beyond the level of fish production available for harvest.

In the Pacific Northwest and western Canadian waters, high ocean survival was common two decades ago, but much lower survival appears to be the rule in recent years. The possible reasons for this decline have been debated intensely and range from changes in ocean regimes (e.g., temperature, primary and secondary productivity) to hatchery techniques and conditions (e.g., crowding, disease buildup). It is possible that wild fish also had higher than normal ocean survival at this time.

An ocean ranching hatchery using a closed broodstock with no input from wild spawning stocks effectively alters or eliminates the natural selection regime acting upon freshwater migration and reproduction of adults and the rearing period of juveniles. Since these two life history stages are characterized by high mortality and, hence, potentially strong natural selection, it is reasonable to expect genetic changes in the returning fish. On the other hand, the relatively low ocean survival rates, as defined above, also indicate the potential for strong natural selection to take place during the marine life history phase. Ocean ranching facilities usually use local wild stocks as the basis of initial production, thereby, working with fish adapted to local freshwater conditions and an appropriate, potentially genetically based, marine migration pattern. To the extent that the hatchery uses broodfish originating from previous hatchery outputs, the fish may become partially domesticated in the freshwater phase. There is evidence that artificial selection regimes (such as smolt size and time of release) can increase ocean survival, although long-term benefits of hatchery selection regimes are unclear. Selection regimes could also be aimed at developing desirable fish characteristics for the market. This approach is accepted and effective in captive fish-farming, but less well understood in ranching programs. Consequently, fish in some ocean ranching hatcheries may become genetically differentiated from their wild stock ancestors. In such cases, hatchery management goals will depart from those of wild stock management concerns and conservation for greater fish production.

A few statistics will serve to emphasize the size of artificial production of wild stocks (see also Chapter 1). Japan now produces about 2 billion chum salmon fry each year and appears to have reached an ocean production limit for these stocks of fish, seen as greater average age at return and lower average adult weight. Japanese chum stocks are now produced almost totally in hatcheries and, due to many transfers of eggs among hatcheries, may now consist of one or just a few genetic strains. This program has been in operation since the late nineteenth century when Charles Atkins, of Maine, introduced European and North American hatchery methods to the Japanese. Very large Japanese style hatcheries are now being brought on-line in Russia, and both Japan and Russia release other species of salmon and trout, the Russian emphasis being on pink salmon, as well as chum.

In Alaska, fisheries cooperatives operate some of the world's largest and most mechanized hatcheries, some producing 100 million released fry per year. The total release of Alaskan hatcheries is about 1.2 billion pink and chum fry and several million smolts of coho salmon, sockeye salmon, and chinook salmon. The Alaskan hatcheries are operated by harvesters' cooperatives, and fish are caught by the cooperative members in marine waters near the hatcheries. In British Columbia, Canada, there are many hatcheries, ranging from large facilities operated by the federal government to many small community-operated hatcheries. There are also provincial hatchery programs for steelhead and cutthroat trout. About 20% of the total wild salmonid harvest originates from these hatcheries, but the sport fishery is increasingly the beneficiary of hatchery production. In British Columbia, a reevaluation of federal hatcheries has led to the closure of some facilities producing fish with low ocean survival rates.

The very large Columbia River watershed has always dominated salmonid production in the Pacific Northwest, but the long history of extensive dam building in this watershed (136 major dams and many hundreds of smaller impoundments) has eliminated or severely reduced the abundance of the majority of wild salmon runs. Hatchery production now accounts for much of the fish production in the Columbia watershed in a large-scale attempt to maintain the original runs and strains, but this has not been a notably successful enterprise. It is estimated that the total biomass of adult salmon produced annually is only a fraction of that seen in earlier times, in fact, equaling about the weight of the hatchery smolts and fry released each year.

Another type of fish culture facility is the conservation hatchery designed to save endangered strains of fish. These can include anadromous fish or completely freshwater strains. Some of the Columbia Basin hatcheries can be seen in this light. To be made the target of a hatchery rescue effort, a strain of fish is usually identified as being at risk due to habitat loss, pollution, or some other impact. A hatchery program is then designed to prevent extinction or decline to dangerously low population size. In theory, such hatcheries are temporary, intended to safeguard the strain until the causes of population decline (e.g., overfishing, habitat damage) have been remedied. Often, however, the hatchery becomes a permanent feature of the population and its management, whereupon it becomes a mitigation hatchery.

Conservation hatcheries are increasingly used for both salmonids and other types of fish. In Europe, nearly all major Atlantic salmon strains are supported to some extent by hatcheries, and the Baltic strains of Atlantic salmon are almost entirely maintained by hatchery output in what has now become a permanent fish culture mitigation program. This is because hydroelectric dams have destroyed much of the natural spawning and rearing habitat. On the east coast of the United States, there are large programs designed to rescue Atlantic salmon from extinction or local extirpation. In Maine, in a new program, wild smolts are captured and reared to maturity in a rapid amplification effort to increase certain strains. Very similar work is underway in British Columbia for endangered steelhead strains. The Dexter Fish Hatchery in New Mexico works with a wide variety of nonsalmonid endangered fish (see below) and there are similar programs for species of sturgeon in both the United States and Canada and for paddlefish in the United States.

A final type of hatchery is mainly of historical importance. This is the colonization hatchery designed to introduce new species or strains of fish into targeted regions. The combination of such hatcheries and fish transportation technology led to the introduction of eastern brook trout (also known as brook char) into western states and of the rainbow trout throughout most of the world. About 95% of western mountain lakes in Canada and the United States have been stocked with salmonids, often of eastern origin; these are now known to have had significant ecological effects in these lakes. Two of the most famously successful nonsalmonid fish introductions were those of the American shad and striped bass from eastern to western United States coasts around the turn of the century.

It is now generally considered a bad idea to intentionally introduce nonnative species into new habitats. When fish transfers occur today, sterilization methods are used to prevent colonization or fish are stocked into isolated water bodies. Although the vast majority of introduction attempts have failed, the successes have sometimes been spectacular (e.g., the American shad and striped bass introductions). The negative aspects of introductions are ecological (competition with native species and strains), genetic (introgression of nonnative genes into the local conspecific gene pool), and pathological (introduction of novel diseases into naïve populations; see Chapter 9).

If the historically important introduction programs are set aside, the major concerns over the effects of hatcheries today are ecological, genetic, and managerial. Overstocking of one species or strain into an area may negatively affect local strains through competition for food or predation. Production of conspecifics may result in many genetic changes in the wild population, ranging from loss of genetic diversity in the population to alteration of gene frequencies due to hatchery selection. Because hatcheries enjoy high egg-to-fry survivals, relatively few broodstock animals are required to maintain production. This means that hatchery fish can be harvested at the highest rates (90–95%). If these hatchery fish are mixed with wild fish during the harvest, the resulting fishing pressure may reduce or eliminate wild strains, which has often happened. This is known as the mixed-stock fishing problem and takes place even without hatcheries, but hatcheries exacerbate the problem dramatically.

Life Histories of Salmonids and Implications for Hatcheries

Overview of the Salmonid Life Cycle

The well known and phenomenal homing behavior of salmonids (trout, salmon, char) provides fisheries and hatchery managers with opportunities and problems. Virtually all salmonids return from feeding migrations or feeding areas to their regions of origin to spawn, usually with great geographical and temporal precision. For many species, this entails a return to the very spawning gravels where they began life as eggs. It, thus, has been attractive to use hatchery technology to intervene in the life cycles in several ways. Hatchery programs may attempt to start new populations of fish from hatchery plantings (e.g., eggs or fry) with the expectation that adults will later return to the site of release to spawn. New runs can be created that return directly to the hatchery eliminating the need for collecting wild broodstock. Existing strains can be augmented by hatchery production, and hatchery managers know when and where to expect broodstock to be used as egg sources for various purposes. As discussed below, homing behavior, over as little as 10,000 years, has created the evolution of many genetically distinct populations within each species (referred to here as strains), and these populations must be preserved if genetic diversity and evolutionary resilience are to be management goals for wild strains. Such preservation is challenging, however, since nearly all human activities within watersheds have negative effects on salmonid populations, as does commercial fishing and sportfishing. Fish culture activities themselves have made many changes in natural populations, including elimination or alteration of genetically distinct salmonid strains.

The formation, through homing behavior, of many separate breeding populations has often resulted in precise adaptations of populations to their environments so that generalizations about a species, let alone the whole group, are difficult. For example, the size of conspecific adults may vary by 100-fold, and the same species may include some strains that are anadromous and others that are entirely resident in freshwater. Migratory behavior may include oceanic feeding migrations many thousands of kilometers in length in some populations or species, although in others lifelong movement may be only a few meters up and down a small mountain stream. The following description presents a general view of salmonid life history and is followed by two more detailed examples. Comments on the implications of life history for hatcheries are offered throughout, and it is proposed that nonsalmonid fishes will present similar problems and opportunities to fish culture. Although most salmonids fit into the following descriptions, variations are many and wide.

This overview of salmonid life histories and the more detailed accounts of two salmonid species, the Atlantic salmon and the coho salmon, are intended to convey the complexity of salmonid life cycles within the group, as well as within species. This is done with reference to fish culture science and technology to demonstrate the influence of fish biology on fish culture practices and vice versa. Throughout the following accounts of wild salmon biology, aspects of hatchery relevant procedures are noted, some general and some quite specific. The intent is to show that correct hatchery strategy and technique must be developed according to the dictates of the cultured species' life history and that to ignore these dictates may have grave consequences for the populations being cultured. This is especially true for a conservation hatchery.

Neither of the following two specific life history accounts should be taken as complete. Many books have been written and countless papers published about these two species, but the summaries that follow should give the reader further appreciation of challenges offered to the hatchery manager by these and other salmonids. By logical extension, the same should be true of any fish species brought into hatchery culture as shown in the discussion of nonsalmonid fish culture later in the chapter. Table 1 presents some life history characteristics of the major salmonid species in North America. In some cases, strains that some authors accord species status have been considered as strains or subspecies.

The basic life cycle of salmonids provides the theme around which the pronounced variation can be viewed. All salmonids have large, yolk-filled eggs that usually incubate in gravel beds under running freshwater.

TABLE 1. Some key characteristics of important North American salmonids.

Oncorhynchus–A. Pacific salmon, semelparous, fall spawning. Native range, North Pacific Ocean and surrounding landmasses.

Chinook salmon *O. tshawytscha.* Rivals sockeye in complexity and variation. Homing precise and many genetically distinct strains within large watersheds. Many of these at dangerously low population levels. Hatcheries very successful with the simpler coastal strains ("fall chinook") that have a three month freshwater phase. Many failures with the interior "spring" chinook strains that have a one or two year freshwater rearing period. Other distinct life history differences found in different strains. Wide size range to 45 kg (100 lb). Precocious male parr found in some interior strains. Return time to home watershed highly variable—sometimes months before spawning. Capture and handling of broodstock may be challenging.

The intense sportfishing pressure on this species has led to hatchery production. Conservation hatcheries also employed to rescue endangered strains.

Coho salmon *O. kisutch.* (See text). Less variable than chinook or Atlantic salmon, but with distinct strains in different geographical areas. Life history shows high level of ecological opportunism. Genetic influences on differences in life history poorly known, but current work suggests genetic factors are important. Hatcheries have been very successful in producing fish; a large fraction of Pacific NW coho are maintained by hatcheries.

Sockeye salmon *O. nerka.* The most variable of the Pacific salmon. Precise homing has led to formation of countless genetic strains. Large runs support important fisheries, but many smaller populations exist with distinct characteristics. Fry rear in pelagic zone of lakes, one to three years, but a few strains head directly to sea. Adult size usually between 2.5–3.0 kg. Early hatchery programs in Pacific NW were aimed at increasing sockeye population, but these failed due to lack of life history knowledge, disease, and lack of assessment technology. Relatively little hatchery effort today except in Alaska due to problems with IHN virus and alternative enhancement strategy (spawning channels, lake enrichment).

Kokanee salmon *O. nerka.* The nonanadromous form of sockeye, analogous to landlocked Atlantic salmon. Essentially noninterbreeding with anadromous forms, many strains, probably of independent origin; taxonomic status is debated. Usu-

TABLE 1. (Continued).

ally very small, miniature versions of sockeye; adult size (100–500 grams at age 3–5 years) limited by food. Breeding in streams, feeding, and growth usually take place in pelagic zone of lakes. Many transplants made to nonnative lakes, some naturalization. Some hatchery work for stocking.

Chum salmon *O. keta*. Simplified freshwater life history; fry go directly or very soon to sea. Hatcheries accelerate incubation with temperature and fry to eight times normal size before release. Major ocean ranching species for Japan. Adults 2, 3, or 4 sea winters. Adult freshwater migration usually short, fish spawning in lower reaches of streams (some dramatic exceptions). Wild fish may spawn quickly making broodstock capture difficult. Fish are usually dark and ripe when they enter freshwater. Some strains very large (over 20 kg) but most adults about 2–6 kg.

Pink salmon *O. gorbuscha*. A small fish (adults 1.5–2.0 kg). Very simple life history—nearly all fish are two years old when spawning (one and three year old fish only rarely have been seen). Two separate cycles in the same stream (odd and even year fish) with little to no gene flow between them. One usually dominant (one year class always much greater in number than the other). Fry go directly to sea. Most populations spawn in lower reaches of streams. Very large hatchery programs in Alaska and Russia. A large proportion of North Pacific Ocean salmonid biomass.

***Oncorhynchus, B.* Pacific trout.** Iteroparous, spring spawning. Native range, North Pacific Ocean and surrounding landmasses. (Note: These species were for many decades classified as genus *Salmo*.)

Rainbow trout and steelhead trout *O. mykiss*. The most complicated of all salmonid species. Native to both shores of the North Pacific Ocean. Both anadromous (steelhead) and freshwater strains, the latter being either stream fish or lake fish. Broad taxonomic division (including both steelhead and freshwater forms) into colourful redband trout (interior) and coastal trout. Many distinct strains, some given species status by some authors (e.g., golden, Apache, Gila). Size ranges from a few ounces to 50 pounds. This is the most widely cultured fish worldwide and has been naturalized in all parts of the world where water temperatures are less than about 15°C. Domesticated hatchery strains used for aquaculture and for stocking.

Cutthroat trout *O. clarki*. Nearly as complicated and diverse as the rainbow trout. Native to both shores of the North Pacific Ocean. Anadromous forms have limited saltwater migrations, generally about 6 months long in coastal zones near home streams. The more numerous freshwater forms are bewilderingly variable, many having been given species status by some authors. Coastal and interior taxonomic division noted. Large piscivorous lake fish of 15 kg may inhabit the same watersheds as tiny stream residents of 150 g or less. Interbreeding with rainbow trout often noted and hatchery introductions of both rainbow and cutthroat have played havoc with native strains. Some strains cultured for stocking and some conservation hatcheries used to preserve endangered strains.

Salmo Atlantic trout. Iteroparous, fall spawning. Native range, North Atlantic Ocean and surrounding landmasses.

TABLE 1. (Continued).

Atlantic salmon *S. salar* (see text). Native to both eastern and western shores of North Atlantic Ocean. Much within-strain variation, genetically based. Lengthy freshwater phase (1–8 years) and long, wide ranging ocean migration. A large fish, 5–30 kg. In the native range, most stocks reduced and many extinct or endangered strains. Hatcheries commonly employed to maintain or salvage strains. Parr-smolt transformation is complex. Nonanadromous strains common in lakes ("landlocked"). This species has not responded well to the many attempts to introduce it outside its range, especially anadromous strains. The most important of species for sea farming. Many hatchery techniques have been developed to rear an essentially difficult hatchery fish.

Brown trout *S. trutta*. A well-known sport fish, native to western north Atlantic shores and Eurasia. There are several life history forms including anadromous (relatively local ocean migrations usually with wintering in freshwater), lake and river dwellers. Size varies from less than a kilogram to 12 or more kilograms. Many well-defined strains, some once given species status. May hybridize with Atlantic salmon. Has been naturalized in many parts of the world, second to rainbow trout in introductions.

***Salvelinus*, the Chars.** Iteroparous, fall spawners. Holoarctic distributions with relict populations further south.

Brook trout *S. fontinalis*. Relatively small stream dwelling fish native to eastern North America. Much variation among strains with notable color differences. Some strains in lakes and some with very limited saltwater migrations. Size from 200 g to one kg, usually 0.5 kg. But larger (up to 5–6 kg) piscivorous strains exist in lakes. A popular sport fish, it has been transplanted to many parts of western North America where it has naturalized and caused problems with native species. This species has also been introduced into the southern hemisphere, into New Zealand Waters.

Arctic char *S. alpinus*. Highly variable circumpolar fish resident in many lakes and streams. Summer feeding migrations to the ocean in many strains. May attain considerable age (over 30 years, often over 20), and there may be more than one strain in a lake. Dwarf strains are noted. Size at first breeding varies from less than a kilogram to over 20 kilograms. Recently has become a popular aquaculture fish. Can be cultured at very high rearing densities (e.g., 100 kg/m^3) and does best at high densities. Little enhancement work. Hatchery knowledge is recent and continues to develop.

Lake trout *S. namaycush*. Resident in Arctic lakes, circumpolar, and many relict populations in deeper lakes to the south. Highly diverse with many different strains. Unique in salmonids for group spawning over gravel beds with no redds or egg burying. The only salmonid with no strains showing any tolerance for sea water. May attain very large size, and be up to 40 years old and over 40 kilograms. Subject to sport fishing. Hatcheries used for stocking and stock rebuilding. The splake, a hatchery created hybrid between lake trout and brook trout, is stocked in many lakes. Many introductions of lake trout outside of original range.

TABLE 1. (Continued).

Dolly Varden *S. malma*. Lake and stream residents with some coastal anadromous forms (limited salt water feeding migration). Native to both shores of the North Pacific. May become large (up to 10–12 kg) but most strains are small and insectivorous. Relatively little interest in culturing this fish. The recent taxonomic split of this species into bull trout (see below) and Dolly Varden has rendered some generalizations difficult.

Bull trout *S. confluentus*. Recently split as new species from the Dolly Varden, hence much literature may have confused the two. May become very large, 15 kg, and inhabits lakes and large rivers, generally in the interior of Canada and Alaska. Listed as an endangered species in some regions, culture for restoration is now being done. Many strains exist.

Spawning adults engage in courtship activities over gravel beds in streams and rivers. These behaviors may be elaborate and species specific and result in complex mating patterns not easily duplicated in the hatchery. Breeding fish display various degrees of tertiary sexual characteristics such as hooked snouts, bright skin colors, enlarged jaws, and protruding teeth. Courtship may last several days and includes territorial displays, combat, and other premating behavior, including the digging of depressions, called nests, in the gravel (Figure 1). The spawning act itself is very brief, primarily between one female and one main male, although there are usually accessory males involved. Sperm and eggs are released together into the nest with nearly 100% fertilization rates, after which the female covers her eggs with several layers of gravel. The process is repeated until several nests, each one upstream from the last, have been created and the female has shed nearly all of her eggs. This series of nests is collectively termed the redd. After spawning, salmonids of some species may recover from the very heavy energy expenditures and survive to spawn again 1 or 2

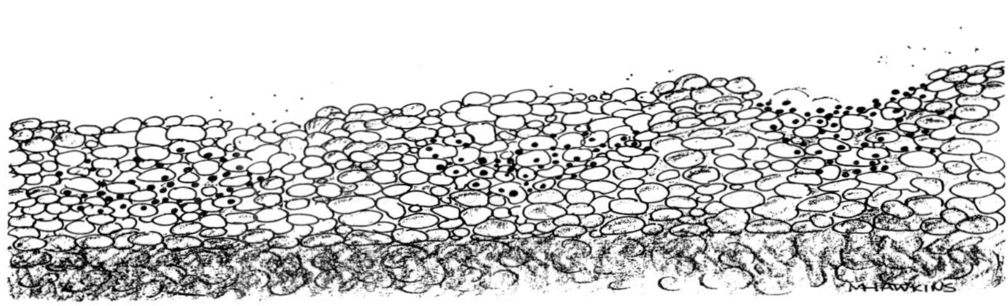

FIGURE 1. Diagram of spawning redd (drawing by Melannie Hawkins; black dots indicate eggs).

years hence (iteroparity) or may die (semelparity). Breeding occurs at different times of the year in different species, but autumn and spring are, by far, the most common breeding seasons.

The large salmonid egg, 4–9 mm (0.2–0.4 in) in diameter (Figure 2), is surrounded by a tough, nonsticky, keratinous chorion that develops considerable turgor due to osmotic gradients in the perivitelline space between the chorion and the egg membrane. The chorion, permeable to gases and ammonium, is resistant to bacterial and fungal attack, but the fertilized eggs go through a period of increasing sensitivity to mechanical shock that creates dangers in the hatchery if eggs are handled at this time. Only one sperm enters the egg through the micropyle in the chorion; this must occur within a few seconds of the egg being extruded into freshwater. Although polyspermy is not a problem, the timing of fertilization and egg activation in freshwater requires attention in the hatchery (Chapter 7). Once the eggs have completed epiboly (denoted in practice by the formation of visible dark eyes), they become very resistant to mechanical shock. They can then be safely handled for sorting and disinfecting. Eyed eggs have been shipped all over the world in moist air containers. Egg transport is very important to commercial aquaculture in which the sale of genetically selected eggs is a part of their business.

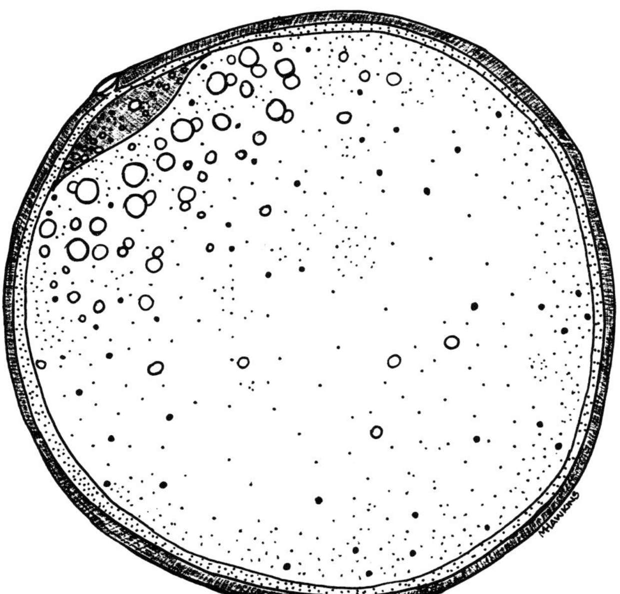

FIGURE 2. Schematic drawing of an unfertilized salmon egg (drawing by Melanie Hawkins).

Because of the large egg size, specific fecundity of salmonids is quite low compared to many other fish. Within a species, larger salmonids have somewhat larger eggs, as well as more eggs than smaller fish, that is, the specific fecundity is somewhat reduced. Larger eggs produce larger fry, presumably with higher survival rates. Because of surface to volume relationships, the larger salmonid egg requires higher dissolved oxygen (DO) levels in the surrounding water. In nature, this translates into a need for high flow rates through interstitial gravel spaces, a situation most common in large-sized, well-sorted gravel. Theory suggests that the larger fish can spawn in larger gravel and, therefore, will have larger eggs. Some studies have suggested that hatchery fish have larger eggs than their wild counterparts, although this is debated. Certainly, any tendency of hatcheries to choose large broodstock (as discussed below), could result in larger hatchery eggs, although not greater specific fecundity. High DO levels maintained in incubators would make development of large eggs possible and without disadvantage. There are other possibilities of hatchery selection related to egg size. Large eggs produce large fry, and these could have a selected advantage in highly competitive hatchery tanks. In British Columbia, it has been noted that size variation in chinook salmon eggs is often pronounced and that this creates a wide range of best ponding times (see below). In this case, incorrect ponding times (later for more slowly developing large eggs) could result in poor success for either very large or very small eggs depending upon the hatchery's method for assessing yolk utilization and readiness of the alevin (Figure 3, and see below for description) to feed.

Natural egg development in gravel beds proceeds at a rate deter-

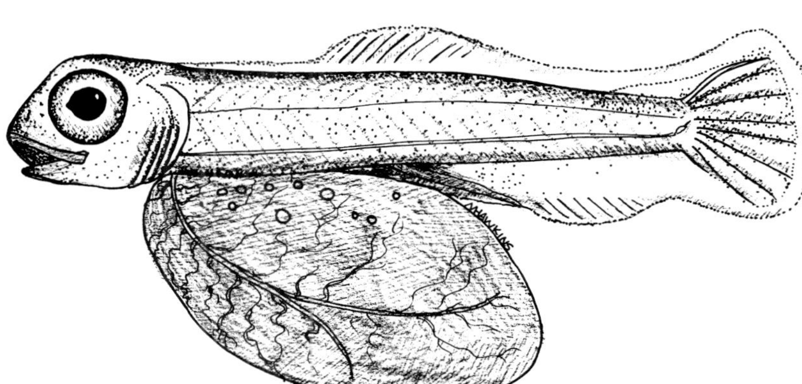

FIGURE 3. Coho salmon alevin (drawing by Melanie Hawkins).

mined chiefly by temperature and influenced by egg size, water quality, and species characteristics. The schedule of development is usually expressed as degree days or accumulated thermal units (ATUs) and can be used to predict key embryological events such as hatch and emergence, both in the hatchery and in nature (see Chapter 7). As noted earlier, the large egg requires high DO concentrations, generally low temperatures, and a steady flow of intragravel water around the chorion. These requirements are translated into all artificial incubator designs. It is well known that salmonid eggs require very high DO concentration just before hatching; asphyxiation of eggs in the hatchery can easily occur at this stage if incubation flows are too low. Upon hatch, a larval fish, called the alevin (Figure 3), emerges with fully developed eyes, gills, considerable musculature, and a large yolk sac. It has an amazing ability to move through very small spaces in gravel to seek better water quality conditions. This behavior has been the undoing of some incubator designs. The alevin has several other behaviors that must be understood if the artificial incubation system is to produce large, high quality fry. This includes the larva's need to remain upright, undisturbed, and in the dark throughout development, which takes several weeks.

The alevin gut is present, but not functioning; all nutrition comes from the large ventral yolk sac. When this yolk is nearly, but not completely, used, the alevin's behavior changes from negative geotropism and avoidance of light to that of swimming cautiously upward through the gravel and out into the water column above the gravel bed, usually at night. This is a critical time of complete behavioral change during which the small fish, 25–30 mm (1.0–1.2 in) in length, must fill its swim bladder by swallowing air (salmonids are physostomes), learn to feed, select its preferred habitat, avoid predators, and deal with conspecific rivals. When this process is complete, the small fish is called a fry (Figure 4). First food consists of small invertebrates such as planktonic crustaceans and, more commonly, insect larvae. The determination of the timing of first feeding involves a complex of three major factors—temperature, time of volitional emergence, and amount of yolk remaining in the yolk sac—so that determining the time of first feeding in hatcheries is a critical and complex fish culture judgment. For example, the amount of yolk remaining at the optimal time of first feeding changes with the average incubation temperature. A mistake in the timing of ponding (moving the new fry from incubators into rearing tanks) may result in heavy mortality, gill disease, and pronounced variance in growth rates (see earlier discussion on egg size). The large salmonid egg results in a large fry, allowing use of artificial feeds in the hatchery and eliminating the need for live feeds such as rotifers and brine shrimp (see Chapter 8). This is a major convenience for salmonid fish culturists.

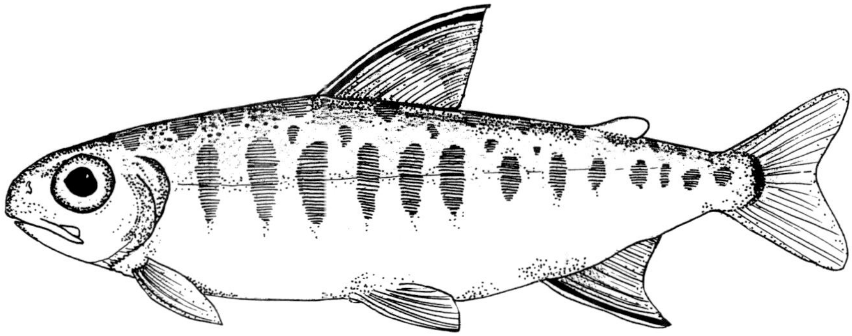

FIGURE 4. Coho salmon 0+ fry (drawing by Melannie Hawkins).

Total time of egg and alevin development in the gravel beds ranges from several weeks to several months, mainly depending on temperature. Emergence is almost always in the spring to early summer, and the timing of emergence has evolved to coincide with the availability of food. Emergence then is a temporal anchor point in the life cycle, and spawning time, which is extremely variable, pivots to allow adequate time for development depending on mean incubation water temperature. This relationship helps explain a major variable among conspecific salmonid breeding populations spawning time, which often acts as a reproductive isolating mechanism between populations. It also may explain why so many attempts at strain introduction have failed; the donor strain must have the appropriate genetically determined spawning time to succeed in the recipient stream environment.

The gravel bed and the water moving through it, whether spring fed in a lake or under a stream, constitutes the first major distinct habitat required by a life stage of salmonids, the developing egg and alevin. Hatchery regimes may be chosen to imitate natural conditions or may be adjusted to manipulate events, usually by changes in water temperature. Thus, eggs can be made to hatch and alevins to emerge many weeks earlier than in nature. This allows prerelease feeding in order to create larger fry, with high survival potential, to be released at the normal time when ocean conditions are best. Alternately, development can be delayed by many months through the use of refrigerated water. Warming or cooling of incubation water, especially in combination with altered photoperiod regimes later in the rearing cycle, can create offset breeding groups of fish that can be artificially spawned in different seasons of the year. Temperature and photoperiod control can also be used to produce winter smolts (juvenile fish ready to go to sea) useful in farming operations.

The water flowing above the gravel beds during courtship and spawning constitutes the last habitat required by most salmonids in their life cycle. A few salmonids, notably some sockeye salmon strains and lake trout, spawn in lakes where there is little current. Key environmental factors of stream spawning salmonids are current speed, water depth, and temperature-all aspects of rivers and streams easily altered by human activities in watersheds. For lake spawners, groundwater flow, shoreline currents, water depth, and substrate are the key variables. The most obvious fish culture application for stream spawners is the construction of artificial spawning channels in attempts to optimize conditions for spawning and egg survival, while maintaining as much of the natural life history as possible (Figure 5). The holding of broodstock, either through the capture of wild fish or creation of domestic stocks in the hatchery, is a key component of salmonid hatcheries. Failure to meet the environmental and behavioral requirements of prespawning fish may create serious problems, including broodstock mortality, reduction in egg fertility, and reduced egg quality (Chapter 7).

Between the time of fry emergence and the return of spawners to the natal stream or lake, salmonids may require several additional habitats. Wild salmonid juveniles go through early rearing in diverse habitats including fluvial, lacustrine, and estuarine. The early fry are sometimes called parr and show distinctly different behaviors from species to spe-

FIGURE 5. Weaver Creek spawning channel, British Columbia (photograph, W. Pennell).

cies. Behavior can be strongly territorial. Nipping behavior is typical in stream environments where food drifts with the current to stationary fish, whereas schooling behavior is common in lakes and estuaries where fish move through water seeking food. Many salmonid hatchery practices, especially feeding protocols, have been designed to cope with territorial behavior that may result in eye and fin damage and high variance in growth. Essentially, this is to create a schooling type of behavior in otherwise territorial fish. Fry that would, in nature, show schooling behavior (e.g., lake rearing sockeye salmon fry) are to some extent easier to rear in high density. Proper adjustment of feeding regimes, light levels, and current speeds and patterns makes it possible to rear all species at high density while minimizing agonistic behavior.

The fry stage may merge without notable transition to the juvenile and subadult stages. In such cases, habitat may change continuously, young fish moving into deeper or faster water as they become larger. In other cases, there may be abrupt changes in habitat, behavior, and even physical form. The most dramatic example is the smolt formation seen in many anadromous species. During the parr-smolt transformation, fish exhibit many changes in body form and color (for example scales and teeth protrude further, the body becomes silvery blue or green, and the condition factor is reduced). Behavior changes drastically as well. Smolts are more active, show downstream migratory behavior, seek salt water, and usually swim in schools. Physiological changes are many, including significant increases in metabolic rate, increased susceptibility to stress, and reduced resistance to disease. The smolt is considered to be a most delicate stage in the hatchery and special protocols are usually put into effect at the time of this transformation.

A major physiological change in smolts is the switching over from the hyperosmoregulation of freshwater to hypoosmoregulation in salt water. In freshwater, the fry or parr must contend with a steady influx of water that enters via the gills and must be removed by the excretion of copious amounts of dilute urine. In salt water, the situation is reversed, and the blood (usually about 10 ppt or one-third as salty as ocean water) contains less salts than the external environment. Now the fish will tend to lose water via the gills and must drink seawater to replace it. This requires the excretion of large amounts of salts, chiefly by special "chloride" cells in the gills that are enhanced during the parr-smolt transformation. There are numerous tests used in hatcheries to determine smolt status in hatchery fish, the best among them focusing on blood salt regulatory ability when fish are placed in salt water.

The entire smolting process is endocrine based and usually orchestrated by changes in photoperiod. The visible changes take place in just a few weeks, but the physiological process is usually several months in duration, and events 6 months or more earlier may determine whether a parr will become a smolt the following spring. The many physiological changes noted must occur in the correct sequence or smolts may not be fully able to osmoregulate in salt water despite outward appearances. Many fish culture practices may interrupt smolting, including some medications, acute stress (e.g., handling), and chronic stress (e.g., crowding or poor water quality); these must be carefully avoided. Even artificial lights in hatcheries may interfere with photoperiod control of the process.

Another notable feature of smolting species is that olfactory imprinting appears to take place at the time of the transformation so that smolts may be used to begin new runs of fish by acclimation in nonnative water bodies during smolting. This is likely just one of several sequential imprinting events taking place during development that help guide salmonids back to their places of origin. The smolting process can be manipulated, usually through use of artificial photoperiod regimes combined with feeding and temperature adjustments. Smolting can be accelerated, saving hatcheries an entire year of rearing, or off-season smolts can be produced, mainly for farming purposes.

Most smolts migrate downriver to the sea, although analogous migrations also occur in freshwater in certain species of trout. In Pacific salmon that have been transplanted into large lakes (e.g., the Great Lakes), the parr-smolt transformation still takes place even though there is no need for the fish to hypoosmoregulate. Some species of Pacific salmon (most notably pink and chum salmon) migrate directly to the ocean after fry emergence. They are not considered to have a true smolting transformation. Other species have smolting processes less strictly controlled by photoperiod; these may be dealt with differently in the hatchery. Habitat requirements of out-migrating (wild) smolts include adequate flows of cool water. Reservoir impoundments in the run of major rivers create serious challenges to smolt survival, as is well demonstrated in the Columbia watershed. The timing of hatchery smolt releases is also critical to smolt survival as discussed below.

The juvenile stages of salmonids (somewhat arbitrarily taken to be fish past the parr or fingerling stages) may be spent in a wide variety of habitats both in fresh and marine waters. In freshwater, some species and strains of species live in lakes where they may be pelagic, near-bottom, or littoral. They may also reside wholly in streams and rivers or go through transitions from one habitat to the other, either by seasons or with increase in age and size. Habitat definition may be quite precise and differ among species. There may be requirements for a certain depth and velocity of water or stream refugia in the form of back eddies created by large boulders or large woody debris. In marine waters, the feeding migrations may be very large, spanning an entire ocean over several years, or may be local and of short duration. In some cases, fish travel no further than the bay receiving the river of spawning. Successful management of all wild salmonids in a watershed essentially requires preservation of nearly all notable aquatic habitats, since damage to any one habitat is likely to create a bottleneck in the life history of at least one salmonid species or strain living in that watershed.

The long ocean feeding migrations of anadromous salmon begin with simple directional (compass) navigation seaward and quite precise bicoordinate navigation in the open sea and during the homeward journey. Possible mechanisms used to determine position include internal clocks and different forms of light and magnetic fields. Return migrations are determined by the onset of maturation, and, in spite of enormous distances traveled by some species, arrival on the spawning grounds takes place within a narrow time frame each year. There appears to be a strong heritable component to all migratory behavior, but learning, often in the form of imprinting, influences migration, at least in freshwater. Nonanadromous strains and species also undertake migrations within freshwater. Since homeward migration is initiated by the onset of maturation, sterile fish could be expected to remain in the feeding habitats, perhaps exposed to fishing. This has proven to be the case in experiments using sterilized coho salmon, but the technique has not been widely used.

Maturation in salmonids may occur as early as 6 months after fry emergence or up to 5 to 7 years later. Late maturing fish are usually large, but size at time of spawning varies greatly and is influenced by food resources, water temperature, and genetic factors. Thus, anadromous fish are usually much larger than their freshwater counterparts because of the rich feeding resources in the ocean. The largest salmonid spawners (exclusive of the European genus *Hucho*) are chinook salmon, some strains of which contain fish exceeding 45.4 kg (100 lb). The tiny precocious male parr of chinook and Atlantic salmon may be the smallest reproductively mature salmonids. Resident trout of several species, especially those in small mountain streams, may also be very small, both males and females maturing at weights of a few ounces, and kokanee of some strains are equally small. Within a species, piscivory may occur in some strains and not in others, the former often becoming very large. Salmonids are exclusively carnivorous, but their diet varies greatly from small zooplankton and insect larvae to large fish.

Atlantic Salmon

Introduction, year classes, and marine life. The Atlantic salmon, native to countries surrounding the North Atlantic Ocean, is one of the most studied of all fish and enjoys one of the highest profiles of all salmonids. It has been the subject of both commercial and sport fisheries for centuries in Europe and was one of the very first species of fish to be cultured, as early as the eighteenth century and possibly before. The Atlantic salmon is also one of the most important fish in commercial aquaculture, being reared in North and South America, Europe, Australia, and New Zealand. It has one of the most complicated life histories of all salmonids and, although much is known about it, much also remains to be discovered. Fish culture practice and research have been major pathways to knowledge of this species' biology. The current status of wild stocks of Atlantic salmon is precarious in most parts of its range and fish culture technology is being employed as a means of preserving and, in some cases, rebuilding many strains of this species. Unlike the rainbow trout, which has been introduced into cool waters worldwide, Atlantic salmon have proven resistant to adapting to new places. Despite many hundreds of introduction attempts, no anadromous strains have been fully naturalized outside the native range without continual hatchery support. Only a relatively few landlocked strains of Atlantic salmon have been successfully introduced outside their native location. Domesticated anadromous Atlantic salmon are reared in many sea cage farms and, next to rainbow trout, this is the favored salmonid for aquaculture.

Atlantic salmon can be categorized in several ways. For example, European strains are differentiated from North American strains by chromosome number (56 versus 58 pairs, respectively). The European strains are divided into, at least, two major groupings thought to have been separated during the last glaciation. As with other salmonids, many separate breeding populations of Atlantic salmon exist throughout the range, each with its own distinct characteristics. For example, about 1,000 such populations exist in North America, excluding those that have become extinct. Within a region, sometimes a large watershed, the many separate strains may be closely related (may exchange individuals), forming a "metapopulation." Many of the strains are very small, a few dozen breeding pairs. This raises concerns about the maintenance of genetic variation and evolutionary viability of these populations.

In most regions of the native range, there are two types of Atlantic salmon populations. Landlocked populations reside in lakes and never go to sea, whereas anadromous populations undertake significant ocean feeding migrations of 1 to 3 or more years. Some of the anadromous strains have very long ocean migrations completely crossing the North Atlantic Ocean; others have more restricted migrations (although all travel from many hundreds to thousands of kilometers). While in the ocean, Atlantic salmon prefer cool waters, usually less than 10°C, which has implications for salmon farm siting for this species. All strains of this species display strong and accurate homing behavior.

Spawning. Spawning takes place in the fall and occurs earlier in higher latitudes than in more southerly parts of the range. In anadromous strains, adults enter freshwater from the ocean over a wide span of time from early spring to just before spawning. The early run fish migrate various distances upriver and hold for up to 6 months (in some cases much longer) in deep pools before making the final ascent to the spawning grounds. During this time, the adults do not feed and rely on stored energy reserves both to live and to develop ovaries and testes to full reproductive maturity. It is not known why some Atlantic salmon adults reenter freshwater so much in advance of spawning since there is no further growth potential in freshwater. Possibly this behavior evolved at times when stream ascent was possible only at certain seasons of the year with specific flow regimes. Broodfish may lose as much as a third of their body weight between freshwater entry and spawning, after which they are called kelts. Kelts have the potential to migrate back to the sea, recover, and return to freshwater to spawn 1 or 2 years hence (iteroparity). Some kelts make only a short ocean migration, returning the next year to spawn; others (the majority) spend two winters at sea before returning to spawn.

The age of spawners is quite variable with fish having spent only 1 year at sea (grilse, usually males) returning with older fish having spent two or three winters at sea. A few fish stay in the ocean even longer than 3 years (up to 6 years), but these are rare, possibly due to historical fishing pressure. Returning with first time spawners are repeat spawners, fish that have already made a previous return to freshwater. These are rare in most rivers, possibly due to fishing pressure and human induced challenges in the freshwater environment where these fish will reside for several months. In some rivers, however, large numbers of both grilse and repeat spawners are normal. In most salmon rivers, the number of spawning age groups present in any one breeding season may be quite large, as will be discussed further below.

Although maturing fish enter freshwater rivers sea-bright, they gradually darken and take on the coppery-brown coloration and thickened skin of the spawner. Males show an extended and upwardly hooked lower jaw, the kype, and elongated, exposed teeth used in fighting with other males; the female jaw changes only slightly. The female chooses a redd site, takes possession and guards it against other females. Although her combat is less noticeable than that of the males, she will vigorously repulse other females from the redd area. Males devote so much energy to courtship and spawning that they jeopardize their survival and are much underrepresented in the repeat spawning population.

The youngest of all spawners are the precocious male parr, small stream fish that have never gone through the parr-smolt transformation nor been to sea. These tiny versions of the anadromous adults do produce viable sperm (more spermatozoa per unit body weight than full-sized male spawners do), and there is evidence that they participate successfully in reproduction on the spawning beds. Although at 10–15 cm (4–6 in) in length, these tiny males cannot defend a female nor stimulate her to spawn, they can dash in during the several seconds of the spawning act and fertilize some eggs. Precocious male parr may spawn more than once as parr and may also become smolts and return to spawn yet again as adults some years later. It is reasonable to assume that precocious males constitute a risk-spreading adaptation in wild populations and that their absence from hatchery broodstock programs represents a departure from normal conditions. Certain hatchery regimes (e.g., high temperature and feeding rates, as well as some photoperiod regimes) may induce many precocious male parr in fingerling rearing tanks. This is a problem for the hatchery's production plan and avoided if possible.

Spawning behavior is similar to the generic account given earlier. The female digs a nest, and the male and female release gametes simultaneously into the depression of the nest that is located in fast flowing shallow water (approximately 50 cm/s [19.7 in/s] and 30–40 cm [12–16 in] deep). Several hundred fertilized eggs are deposited in each nest and then buried to a depth of at least 12–20 cm (5–8 in), but sometimes much deeper, as the female moves upstream to begin the next nest in the series.

The eggs are 5–7 mm (.2–.3 in) in diameter and will incubate over the winter. Water temperatures at spawning vary from about 5 to 10°C, but in most parts of the range soon cool, often to temperatures just above freezing, so that development is very slow for both egg and alevin.

As noted earlier, the large eggs of salmonids reduce fecundity compared to such fish as centrarchids or gadoids. However, egg to fry survival in the redd is relatively high, generally 5–10% and the resulting large fry have a significant headstart in life. Larger Atlantic salmon females have larger eggs so that total fecundity increases although specific fecundity declines with size. Specific fecundity ranges from about 1,000–2,000 eggs per kg (per 2.2 lb) of body weight; there are genetic differences in fecundity among strains as well. Spawner size varies greatly with sea age and ranges from about 2 kg (4.4 lb) for grilse up to 30 kg (66.1 lb) for older fish. Repeat spawners do not generally grow as much per additional year at sea as fish that have not spawned; much of their energy intake is used for recovery and development of the next batch of gametes. Some very large Atlantic salmon have been recorded, but most are between 4 and 10 kg (8.8 and 22.1 lb). For example, in Maine, three sea-winter spawners are about 7 kg (15.4 lb) and two sea-winter spawners about 4 kg (8.8 lb). These are the two most common spawner age classes for these populations. If hatcheries tend to choose large broodstock, they will, in effect, be selecting only one of the many life history patterns present in the population.

Freshwater rearing. Fry emerge in the spring, swallow air to fill the swim bladder, and begin feeding on small stream insects. Atlantic salmon fry (1st year or 0+ juveniles) and parr (later years) are territorial, but tend to reside motionless in fast water close to the bottom, usually in coarse rubble substrate. Fry tend to be found in stream depths of 10–15 cm (4–6 in) and current speeds of about 0.5 m/s (1.64 ft/s). Older parr move to slightly deeper (ca. 20–40 cm [8–16 in]) and faster parts of the stream and require larger substrate, cobble to small boulder size stones. Overall, parr move relatively little in streams and use their very large pectoral fins to maintain stationary positions near the bottom where they wait for drift insects to come to them. The near bottom holding behavior combined with innate aggressive tendencies led fish culturists to rear Atlantic parr in shallow containers for many years. In this two-dimensional fish culture, rearing density was expressed as fish per unit area. Fish culturists have since learned to modify fish behavior and now rear Atlantic parr in three dimensions and at quite high densities.

Growth of fry and parr is usually slow due to the oligotrophic status of most Atlantic salmon streams. They usually require several years to attain smolt threshold size in the fall before final spring smolting. Smolt size varies from about 14 cm (6 in) to over 20 cm (8 in) depending on the age of the smolt, temperature regime, and productivity of the natal river. Atlantic salmon parr can withstand quite high temperatures, up to 18°C, without stress. In the south of the range, especially in Europe, many parr may become smolts after only 1 year of freshwater rearing. More often 2 to 3 years are required, and, in the higher latitudes, 4–5 year old smolts are common. Ungava Bay has Atlantic salmon populations requiring up to 8 years of rearing before becoming smolts. Research on European Atlantic salmon smolts has shown that there is a genetic basis for the number of years required for smolt formation, but that this trait is also strongly influenced by the environment. Some fish are more aggressive and, if given enough food, will attain the necessary size in time to become S1 smolts (that is, 1+ or 1-year smolts), but all fish can become S1 smolts if given the opportunity (abundant food and adequate temperature). Many hatcheries can now produce all S1 smolts by providing these conditions, but if S1 smolts are produced by grading, by discarding smaller fish, then genetic changes will be imposed on the population. In commercial aquaculture this change might be desirable, but it would not be desirable in a conservation hatchery.

The number of smolts produced per 100 m^2 (1,076 ft^2) of stream area is low, about 2–4 smolts per year for most populations. For stream stocking, it is vital to know the carrying capacity of streams and different habitat types needed by the stages of parr. Because of the Atlantic salmon parr's tendency to hide among larger substrate and be stationary, stocking must be done along the whole of the stream or river rather than at point sources.

Smolting occurs in those fish that attain a body length of about 12–13 cm (4.7–5.1 in) by the end of the summer growing season. Thus, the physiological "decision" to become a smolt is made many months before the visible changes begin to appear in the spring. Much of the knowledge of smolt physiology has come from fish culture research, and it is now possible to produce hatchery smolts with great assurance. Failure to understand the smolting process, however, can lead, and has led, to the production of many pseudosmolts, fish unlikely to survive after release.

To conclude this brief discussion of a complicated species, it is interesting to note that a synoptic look at a single breeding population reveals many possible age classes, fish born in different years, spawning in any 1 year. For example, a spawning group might include precocious parr males of either 2 or 3 years of age; male and female grilse (one sea-winter age fish); two, three, four and even five sea-winter males and females; and repeat spawners (usually, but not always, females). Some spawners might repeat spawn up to three times, although this is rare. Thus, fish of many age classes and life history patterns appear on the spawning beds each year, and several life history patterns will arise from the eggs of each female. If there is a catastrophe in the production of any age-class (or phase of age-class) there will be several other age-classes available as backup. In other words, it would take many bad years to destroy such a population or to eliminate the contribution of one female to the population. In terms of genetics, even a very small population includes the genes from several spawner year classes and, presumably, genetic information for very different life histories. Thus, scientists have suggested that the effective population size for Atlantic salmon is much greater than it would appear from a simple spawner enumeration. This may be an adaptation to small population size dictated by habitat limitations in highly oligotrophic stream systems. The goal of hatcheries is to maintain or replenish salmon strains without reducing genetic variation in the strain. However, any hatchery selection regime that reduces the number of age-classes in the population will have the opposite effect. Hatchery breeding protocols also fail to replicate assortative breeding patterns (mate selection) of wild salmon spawners. This failure also likely affects the genetic structure of the wild populations being augmented or conserved by hatcheries.

Coho Salmon

Introduction, year classes, and marine life. The coho salmon is distributed along the coasts of the North Pacific Ocean from California to northern Japan and, like the Atlantic salmon, tends to inhabit many streams in an area rather than forming a few large spawning populations. Also, like the Atlantic salmon, the coho has a significant freshwater rearing period during juvenile stages, which means that it is vulnerable to habitat damage during this phase of its life history. It has been introduced to the Great Lakes and a few other places, but, despite many attempts, anadromous runs have not been created outside the native range. Coho are semelparous and die after spawning unlike Atlantic salmon, and, in this regard, they have a less complicated life history. They are not known to form precocious male parr, and they almost always remain in salt water for either 6 months ("jacks") or for 1.5 years. They are generally small, about 1.5–3.0 kg (3.3–6.6 lb), but some strains are larger with adults up to 14 kg (30.9 lb). Ocean migrations are variable with some strains remaining within confined areas of coastal waters, others wandering along the coastal shelf, and some, especially the northern stocks, moving far off shore. There are reports of some coho in captivity becoming 2+ ocean fish, but this has not been verified in the literature. Coho home to their streams of origin, but there are reports of higher straying rates than seen in other salmon species. This has led to a regional management approach in which coho spawning in several nearby streams are lumped together as one functional population. This is an area requiring more investigation.

The coho was one of the first Pacific salmon species to be reared successfully, en mass, in hatcheries. Large hatchery release programs began many decades ago along the Columbia River system and in California to mitigate damage done by large river-run dams. Coho are relatively easy to rear using intensive fish culture techniques, and because they are a prized saltwater sport fish, as well as an important troll caught commercial fish, they are now reared in large numbers in many hatcheries from Oregon through British Columbia and Alaska. Only in recent years has the variation in life history among strains, and within strains, become appreciated so that some traditional hatchery practices may be open to review. In recent years, coho have declined to dangerously low levels all along the west coasts of Canada and the United States. Coho are also reared commercially in open sea-pens, but to a lesser extent than Atlantic salmon.

Spawning. Coho are fall spawners, usually spawning earlier with higher latitude and later in the more southern parts of the range. There are many exceptions to this, however. For example, some stocks in southern British Columbia spawn in late summer. Adults within a breeding population (strain) may enter freshwater over a time period of several weeks before spawning (but not the several months seen in some strains of Atlantic salmon). Most coho spawn in the lower reaches of their streams, although there are important strains that migrate some hundreds of kilometers upriver to spawning grounds. When entering freshwater, coho are usually silver-bright and, thus, are important to in-river sports and native fisheries. Within a few weeks they darken, resorb scales, and develop a thick leathery skin, dark brown-red pigmentation, and a pronounced hook on the upper jaw, especially noticeable in males that also develop large fang-like teeth. Breeding color and pigmentation pattern vary among strains.

Coho usually spawn in very small channels, in small clusters, spread out over the lower reaches of a watershed. Spawning is similar to that of the Atlantic salmon. A female selects, guards, and prepares a redd site in a gravel bed and spawns with a dominant male of her choosing or one that has fought off other male contenders. There may be an accessory male attempting to spawn on the female's side away from the dominant male. At the same time, small jacks may dash in to participate, just as precocious male parr do in Atlantic salmon. This is a description of the most typical of coho spawning, which takes place in small streams and side channels using pockets of gravel with room for only a few fish. In these cases, the males are often smaller than the female and there are many refugia for accessory males and jacks promoting surreptitious spawning.

Precocious female spawners are extremely rare. Jacks are now considered to be an important part of the coho life history strategy. Jacks, about 200–400 g (7–14 oz) in weight, do not display the full coho breeding color, but rather appear more like juvenile, nonreproductive fish. This may be in keeping with their "sneak" spawning strategy in which it does not pay to attract the attention of large, formidable males. Jacks probably result from both genetic and environmental influences. Heavy feeding regimes and large hatchery smolt size may result in very high proportions of jacks in the returning adult population. Until very recently, jack sperm was never used in hatcheries, but recent research indicates that fish with a genetic tendency to become jacks also have the highest growth rates. Avoiding the use of jacks could, therefore, not only alter the life history options of the population, but actually select for a lower average population growth rate.

Some strains of coho use large open gravel beds for spawning rather than small channels characterized by habitat complexity and shelter. In these locations, large aggregations of spawning fish form and, at least in some populations, males are of comparable size with females. In this situation, theory suggests that there is less opportunity for smaller males and jacks to participate in sneak spawning and the advantage goes to the larger male. Presumably, the differences in jack numbers and male sizes are genetically determined and strain specific. Differences in body size and proportions have been described between lower and upper watershed coho strains, the latter being more fusiform and possessing larger caudal peduncles.

The protracted time of freshwater entry of adults (10–20 weeks) seen in most coho strains presents a challenge to hatcheries. It is more expensive to include fish from the entire run, especially as the fish entering last do so during heavy winter rains and high freshets making them difficult to capture. Early-run fish are easier to capture, but must be kept alive longer as they complete their gonadal maturation. This is a common problem to the hatchery production of many salmonid species (see discussion below) and there are many examples of hatchery-supported runs becoming truncated, usually toward the early- to middle-run fish. This clearly represents a loss of genetic information in the population and could well compromise the strain's ability to adapt in the face of changing environmental conditions.

Freshwater rearing. Eggs and alevins develop in the gravel redd over the winter in typical salmonid fashion. The eggs are large, 6–8 mm (0.24–0.31 in) diameter and egg-to-fry survival in stable streams is generally high, but much reduced in areas where development destabilizes stream banks and where resulting washouts or siltation occur.

Alevins emerge in the spring to make the transition to fry. Coho fry feed largely on insect drift and become territorial, but unlike Atlantic salmon, they prefer the quiet waters of pools below riffles, small side channels, beaver ponds, and even lake shores. The fry, especially in moving water, maintain small territories, force excess fry to move out, and colonize all available habitat. In fact, virtually any body of still or slowly flowing water will be used by coho. Excess fry in streams have been observed to move downstream as "nomadic" fry and eventually are pushed into the estuary. It was long thought that these fish perished, but recent research indicates that they may survive and that such fry migration may be a genetically programmed behavior in some strains. Rearing in freshwater usually requires 1 or 2 years depending on water temperature and food resources. In a few northern steams, 3 years may be required. Coho can tolerate quite high water temperatures (ca. 18–20°C), not uncommon in many summer rearing habitats. Hatchery rearing temperature regimes are usually a few degrees lower to reduce stress and disease incidence.

The freshwater rearing period can be less than a year in the case of estuarine fish, but usually is either 1 or 2 full years in duration. The shorter, 1-year period is most common in the south of the range, but stream temperature and food resources are the important factors. Fish requiring 3 years of freshwater rearing and, rarely 4 years, are found in northern waters. This is similar to the S1-S2 or S3-S4 smolting patterns seen in Atlantic salmon and may be an evolutionary means of risk spreading. In the hatchery, 1 year of rearing always suffices, and by using a hatchery protocol of altered photoperiod and increased temperature, coho can be induced to become smolts in the first spring (after about 3–4 months of rearing). These are called 0+ smolts and have been used chiefly for salmon farming. They are attractive to hatchery managers because of the great savings in rearing costs and tank space, but they have not been used significantly for ocean ranching.

Smolting takes place in the spring and the silver-green smolts head out to sea for the ocean feeding migration. The physiological decision to become a smolt takes place the previous late summer to early fall and is dependent on attaining a critical size and possibly a certain level of energy reserves. Smolts are usually greater than 10 g (0.35 oz) in weight. The downstream migration takes place both by passive drifting and intentional downstream swimming. Peaks in out-migration have been linked to specific hormonal surges and often coincide with the new moon.

The ability of coho to use a variety of habitats, both for spawning and freshwater rearing, is at once the strength and the Achilles' heel of this species. Much important coho rearing habitat does not, to most people, resemble fish habitat in the slightest and is easily eliminated by developments such as logging, farming, and urban expansion. Coho habitat tends to disappear by many small increments, no single one of which is critical, but in summation is devastating to the strains. This combined with the highest of exploitation rates and periodic shifts in oceanic environmental conditions, has progressively weakened coho populations along the eastern Pacific shores. Hatcheries have played a part in over-exploitation of coho. Because hatchery stocks may be harvested at a rate of over 90% of the returning adults, the average exploitation rate tends to be adjusted to that of the hatchery stocks, leaving the intermixed wild stocks overharvested. There is currently much reexamination of coho management in Canada and the United States.

The coho life cycle, despite an earlier perception of simplicity, is quite complicated. Coho salmon are composed of many genetically distinct strains, displaying different life history strategies, but also with considerable variation within strains. This includes estuarine and lake rearing fry; upstream and downstream spawners; S1, S2, and S3 smolts; 0+ and 1+ ocean fish; and fish with either local or well offshore ocean migrations. The ease with which coho can be reared in intensive culture has led to a proliferation of both large and small hatcheries. Most of these, whatever their initial goals, have become permanent producers of fish. There is much concern over the long-term effects of hatcheries on wild coho strains. In the upper Columbia watershed, for example, many wild stocks are extinct due to habitat losses, dams, overfishing, and use of hatcheries. In the lower Columbia watershed, remaining wild stocks have been mixed with hatchery stocks to the point of genetic homogeneity. Some extinctions of strains may have occurred. There have been many decades of fish transfers among hatcheries and a general selection of early run fish that has eliminated most of the spectrum of the once wild lower Columbia coho populations. Although it is not possible to apportion causality to any one influence, hatcheries appear to have done their share of damage to Columbia coho despite the intention of mitigation and enhancement.

The above accounts of salmonid life histories are intended to highlight the complexity of these fish and the implications for hatcheries of life history characteristics. The homing behavior, formation of many distinct strains (breeding populations), and resulting life history complexity of salmonids make them especially diverse genetically. This diversity presents challenges to hatchery programs for which the goal is to enhance or preserve wild strains of fish. Each breeding population is unique, and there are many ways to unintentionally distort the gene pool. On the other hand, knowledge of salmonid life history creates many opportunities to successfully rear these fish. The very homing behavior that creates so much variation within the species allows biologists to accurately work with single strains within the species. Although the following discussion focuses on salmonids, most species of fish have complex life histories and analogous population genetics, and they can be affected by fish culture in similar ways.

Genetic Aspects of Hatchery Management

Conservation hatcheries aim to protect specific salmon strains that may be endangered for various reasons. The main goal is to increase numbers of fish through hatchery technology while maintaining genetic diversity and creating the least possible changes in gene frequencies in the population. There are alternative approaches such as producing more fish without regard to genetic variation within and among strains and even among regional stocks, but such an approach may have negative effects in the long term. There is also an issue of bioethics. It may not be defensible to deliberately reduce biological variation that has built up over tens of thousands of years.

As noted above, each subpopulation of a salmonid species is, to some extent, genetically isolated from the others because nearly all salmonids return to their place of origin. Species with precise homing to spawning beds will show a finer scale of geographical strain formation than fish returning to a watershed, but then choosing from several possible spawning locations. The relatively small amount of straying observed among strains takes place most often between nearby populations and constitutes an important source of genetic variation counterbalancing the selective forces to closely adapt to the local environment and reducing the loss of genetic variation due to genetic drift. Occasionally, local catastrophes or unusual oceanographic temperature patterns may promote large scale straying and new colonization, but these are rare events of uncertain importance. Freshwater run timing to ensure optimal emergence timing, body size, dietary preferences, and innate migratory behavior are but a few of the genetically determined adaptations possessed by most strains within a species. To properly manage a salmonid species, all of its many strains must be conserved since they, together, constitute the species and give the species the best chance of survival over the long term as environmental change, both anthropogenic and natural, inevitably takes place. Again, it is important to see the many strains as interconnected by straying so that each provides long-term genetic variation to others. This is especially relevant in the face of concern that small populations may lose their ability to adapt. Estimates of critical population size vary from about 100 pairs to 500 pairs of reproducing adults, but as noted with Atlantic salmon, a very small breeding population in any 1 year may be supported by many different year classes within the strain, as well as by genetic input by low level straying. Since some species include hundreds to thousands of discrete breeding strains, management of a species must include all strains and regional metapopulations. Hatcheries are increasingly used in salmonid management, but they can be either helpful or damaging depending upon hatchery goals and practices.

Although every conceivable biological characteristic of a salmonid may vary from one breeding subpopulation to another, several are of key importance to hatchery strategists. Run timing has been mentioned and is often the first apparent change to be induced by a hatchery. Choice of broodstock from one portion of the run (e.g., early, middle, or late) may have a profound effect on the population. The tendency, as noted above, is to choose fish from the early part of the run. It is convenient for the hatchery to capture all fish at once and to have all fish mature in a narrow time frame. This reduces staff requirements and increases efficiency. If all fish eggs are taken within a few days, incubation systems and, later, rearing containers are efficiently used. Less water is required. In contrast, taking throughout the period of the run requires many trips to the capture site and many smaller incubation and rearing units. This is especially true if the run is prolonged, lasting several weeks or months. However, the maintenance genetic variation causing an extended run or return period is presumably an adaptation to environmental variability;

a narrow time of return would increase population vulnerability to environmental variation.

The selection of broodstock is a key issue in all hatcheries. For production of marketable flesh in a commercial aquaculture facility, for example, a salmon farm, managers use standard selection systems to produce highly domesticated fish with low stress responses, high growth and conversion rates, and other enhanced qualities desired for the market. Conservation or augmentation of wild fish stocks is much more complicated. The hatchery manager wishes to avoid (or minimize) new selective pressures while using the hatchery to give the population a boost in recruitment. In early hatcheries, staff almost universally selected the largest broodstock assuming that these were the best fish. Growth rates and fish size are not simply related, however, being confounded by age at maturation, ecological traits (e.g., breeding strategies), and feeding opportunities. For example, fish of the same age attaining different sizes may reflect differences in risk taking, or differences in ecological niche choice within a strain. The reproductive value of jacks and precocious parr has been noted. Any variation in size may reflect genetic differences that are of value to the population. It is, of course, rarely possible to distinguish the relative importance of environmental and genetic variation observed in salmonid populations, but it has been shown that there is usually a heritable component in phenotypic traits. In terms of hatchery management, experience has shown the need for a formal plan at the beginning of broodstock collection, otherwise human tendencies will result in the bias in broodstock selection as noted above.

Linked to the selection of broodstock is the need to take an adequate number of fish to ensure a full sampling of genetic variability from the parent population. To avoid inbreeding and to capture as much of the genetic diversity of the population as possible, a large number of broodstock should be taken each year. The usual recommendation is 50–100 breeding pairs. If a hatchery is to persist for many years, a still larger number is preferable. These estimates assume that equal numbers of males and females are used and that there is no use of pooled sperm. If the sex ratio is different than one:one, a much larger number of animals is needed to attain the same effective population number in genetic terms. The collection of broodstock is the first, the most difficult, and perhaps the most important step in ensuring the best genetic representation of the target population.

There are situations in which the number of breeding pairs is so reduced that effective hatchery intervention may require using a large proportion of the total breeding stock. This may be considered too invasive and risky, or the population may be considered so close to extinction that the intervention is justified. In such cases, one solution may be to take a proportion of the wild juvenile population, rear these for several years to maturation and then create a larger hatchery population of juveniles to be released as smolts. This reduces the immediate impact of the hatchery on the wild population, but the population of juveniles may still not be representative of the population's genetic variation.

Another difficulty is the unknown degree of assortative breeding taking place on the spawning grounds. Do large females only breed with large males? How much jack or grilse sperm actually contributes to the next generation? What is the nature of sperm competition in natural spawning? With few answers to such questions, the best the hatchery manager can do is to include all reproductive forms in a best guess approximation of what may be going on in nature. If fertilization regimes are designed to avoid sperm competition (Chapter 7), and the number of animals of each sex is set to ensure an adequate, effective population size, then the hatchery will not create additional genetic drift or inbreeding.

Hatchery rearing conditions will alter the normal phenotype in various ways from behavior to meristics, and physical body measurements and proportions. The high energy hatchery diets increase growth rates beyond what is usually seen in the wild. Adipose tissue is likely to increase, and a proportion of hatchery-produced fish will mature earlier than they would in the wild. The hatchery strives to produce large numbers of healthy, high quality, juveniles for release and in so doing alters the normal behavior, which is usually aggressive and territorial (with the possible exception of some naturally schooling species). There are a few studies that indicate hatchery-induced behaviors may have detrimental effects on the released fry or smolts. For example, association of overhead shadows with feeding may predispose fry or smolts to bird predation. However, in general, it appears that fry and smolts released from modern hatcheries survive about as well as wild fry and smolts.

Little is known about selection regimes during the hatchery rearing period. Modern practices emphasize feeding protocols that reduce the ability of fish to outcompete other fish for food. However, there is significant cumulative mortality from first feeding to release that indicates the potential for selection. It is not possible to suggest that an aggressive, risk-taking strategy would be an advantage or a disadvantage in a hatchery rearing tank.

That hatchery intervention will cause change is inevitable. However, it is important to note that all human activity, including commercial and sport fishing and watershed alteration through land use, also creates genetic and ecological effects on populations. It is the ability of strains to adapt to these changes that is important. The environment presents a moving target to evolving populations, and the hatchery manager must strive to avoid removing the genetic flexibility of populations to adapt to the changing environment. The fish culturist should not strive to maintain a static gene pool in a rigid environment, for neither exists.

Most of the above concerns are related to the use of hatcheries as conservation tools either attempting to preserve endangered species or to augment wild strains without unduly changing them. Hatchery induced genetic change can be reduced but probably not eliminated. Domestication, for example, may occur regardless. In the case where pure ocean ranching is to take place with no reference to a conserved wild stock, suitable species and strains of fish are selected, reared in large numbers and released. All survivors return either to the hatchery terminus or are caught on their return migration. Ideally, few stray into rivers to join wild spawning populations. In this type of operation, the genetic concerns are to maintain a large, effective population number and to inaugurate a breeding program similar to those used in commercial aquaculture. Maximizing ocean survival is of paramount economic importance and ongoing databases are kept to determine the major factors. These include size and time of release, the effects of crowding and other stresses on the smolting process, and the health status of the released smolts or fry. Although larger fish usually survive better than smaller fish, this is not always the case. Time and size release studies have produced complex and sometimes counterintuitive results. The difference between a 4% and 6% return rate has an enormous economic implication and in these hatcheries there should be more emphasis placed on the quality and nature of the smolt or fry output than on numbers of fish released.

The mixed-stock fishing problem and its exacerbation by hatcheries has been mentioned. Salmon ranching facilities should be designed so that incoming fish can be selectively harvested at locations distant from the many other wild stocks entering coastal waters in the summer and fall. Alternatively, they can be harvested entirely at the hatchery, but this requires use of species that return sea-bright to freshwater if market value is to be maximized.

Description of Ocean Ranching and Conservation Facilities

Hatchery design and operation is described in Chapter 7. A brief outline of ocean ranching facilities, as distinct from conservation hatcheries, will be useful here to make clear some of the options available to this type of fish culture. Large ocean ranching facilities may deal with only one or with several species. They use large quantities of water and the outflow usually connects to a fish ladder or fishway bringing the returning adults directly back into the hatchery. There is no labor expended in capturing broodstock on spawning beds or in river channels. Fish are sorted, usually with the help of hydraulic lifters and fish pipes (Figure 6), and sent to holding raceways separated by species and sex. Excess fish are sold and the remaining broodstock are held through the last maturation stages (final vitellogenesis, ovulation, and spermiation). This may take only a few days for some species and several weeks for others. In the latter cases, great care must be taken to maintain fish in low stress environments. Fish density, light, water temperature, disturbance, and handling are key variables. Holding raceways may have covers to reduce light and disturbance or water sprays to diffract surface images.

Because preservation of wild strain characteristics or gene frequencies is not the primary issue, it is beneficial to have fish mature in the shortest possible time span. When fish are mature, eggs are taken en mass. Large incubators, some of them holding 500,000 eggs each are filled to capacity, best flows set, and fungus treatments carried out at frequent intervals. At the eyed stage, eggs are sorted mechanically to remove those that were not fertilized. Fertilized eggs are then placed in secondary incubators where they will soon hatch and where alevins will develop. There are many incubator designs. Large facilities are automated, and there are many laborsaving systems.

When the fry are ready to feed, they are piped into early feeding tanks, either raceways or circular tanks, and fed special rations of the correct size and nutritional content (Chapter 7). Small fish are fed fre-

FIGURE 6. Fish sorting system at a large salmonid hatchery (photograph, W. Pennell).

quently, usually by programmable automatic feeders. The use of heated water (naturally warmer groundwater or water heated by boiler and heat exchanger) makes acceleration of growth possible. Chum fry, for example, can be reared for up to 2 months and still be released at the same time that naturally produced wild stocks leave their natal rivers. Fry can be induced to smolt during the summer of ponding or within 1 year instead of 2. The size of the smolts at release, as well as the time of releases, can be controlled within quite wide limits.

It is important to leave the reader with a sense of the size of many ocean ranching hatcheries. Water is generally used just once (see Chapters 3 and 6) and the effluent is often the size of a small river. Although impacts on water quality in effluent-receiving waters are slight, large hatcheries require major engineered structures to bring water to the facility. The outflow, in effect, becomes the natal river to which hatchery fish return; the return of many thousands of broodfish to a point source is a spectacular sight. Water sources for large facilities (see Chapters 2 and 3) are usually rivers, but may be pumped groundwater from aquifers.

Such large facilities do not usually stand alone, but are parts of production programs, so that many are located within a region (e.g., Hokkaido, Japan; Prince William Sound, Alaska). The ecological impacts of many millions of fry and smolts released into natural waters may be significant and cause competition and food limitation, especially in semienclosed areas during the out-migration. Assessment of the hatchery's success relies on identification of the hatchery fish, as discussed below, and is a vital aspect of large facility operations.

Ocean ranching facilities come in all sizes, and there are many small facilities operated by community groups, often consisting of volunteers. The goal of small community hatcheries is usually to boost production of one or two species; assessment may or may not be done. These facilities may be temporary or may persist for decades. They generally use less expensive materials and are less automated than the large permanent facilities described above. Such hatcheries do produce high quality fish and, in the aggregate, many of them also engender local and political support for salmonid management concerns. However, it is clear that genetic interactions between community hatchery fish and wild populations are inevitable and undesirable. There are thousands of such small- and medium-sized facilities along the west coasts of the United States and Canada.

Modern conservation hatcheries are designed to be different from either of the above types of facilities. They are capable of holding many strains of fish separately from broodstock to fry and smolts, and they may hold valuable and rare fish contained in quarantine systems. They often use pumped groundwater and entail staffing and equipment costs far in excess of the efficient production facilities previously described. Although the physical structures of these conservation hatcheries may be large and complex, the number of fish produced is quite small. Typically, broodstock collection takes place in distant field locations and either gametes are collected in the field or the broodfish are brought to the hatchery for maturation in special holding containers. Great care is taken to minimize genetic effects of the hatchery on the wild fish populations. The Provincial Hatchery system of British Columbia is based on these principles and provides a good model for such hatcheries.

Assessment Technology

As noted in Chapter 1, there have been several early waves of hatchery production in North America, but these programs were eliminated due to a combination of economic difficulties and complete lack of assessment. There was no way to determine if the hatchery fish were actually surviving and contributing to the wild population or to the target fisheries. For example, although sockeye salmon hatcheries were constructed throughout the western states and Canada at the turn of the century and produced millions of sockeye fry, it is possible that very few fry survived to adulthood. When the utility of these hatcheries was challenged, the operators and supporters had only faith as evidence of their worth. Most of these hatcheries were closed down in the 1930s. The early Canadian walleye production program met a similar fate.

Not only is assessment vital to determination of benefit-cost ratios of hatcheries, but assessment techniques are also needed to estimate impacts of hatchery production on wild salmon populations and to detect trends in ocean survival, which may be due to climate regime changes or other factors having a bearing on fisheries management. Assessment technology also allows hatchery managers to perform experiments to determine the effects of hatchery practices on ocean survival. There is a variety of assessment methods and new ones are being developed, but all rely on fish identification. The most common identification techniques can be summarized as external marks such as fin clips or tags, internal tags with an external indicator, and biochemical identifiers.

The earliest tag was the fin clip (although Izaak Walton described a yarn caudal peduncle tag in 1654 that, although fraudulent, shows that the concept is not new). The fin clip entails various combinations of removal of the bilateral fins in combination with the median adipose fin common to all salmonids. This yields 12 combinations of removal (e.g., right pectoral or left pectoral, right or left pelvic, diagonal left/right pelvic-pectoral and the above with or without the adipose). Considering the importance of the bilateral fins for swimming maneuverability, it is surprising that most of these combinations worked at all, and there is evidence that mortality due to the operation and to impaired swimming occurred, skewing results. There are various other external tags that can be applied to fish. These range from numbered ribbons or filaments to plastic disks and metal opercular clips. They are all useful, but seldom last long enough to span the growth cycle of fish from fry to adult and so are usually used for wild spawner studies. Aluminum external tags were used as early as the 1870s to study the life history of Atlantic salmon in New England, and the Carlin tag is still used to mark Atlantic salmon smolts. The Carlin tag has a subdermal anchor and a piece of plastic leader to allow for fish growth between the anchor and the metal tag. Such tags may be lost or may affect marine survival; they are only useful on fish with a large smolt size.

Other external marks include tattoos and cold brands. Ink tattoos are effective on some species, but have a tendency to fade over time as fish grow. Cold brands are applied with a piece of metal supercooled in liquid nitrogen. They leave a very distinct mark (e.g., several bands), but these also may fade over time and are most useful on broodstock or larger fish. A recent innovation is the fluorescent visible tag injected into adipose tissue of the posterior portion of the eyelid. These can take the form of a thin thread (visual implant filament) or a liquid elastomer that hardens after injection. The tags apparently have no effects on the fish and are very bright and can be seen with or without ultraviolet light. Recent innovations of color banding and use of different body regions may increase the information that can be conveyed. At a minimum, within a spawning population, they allow simple identification of hatchery and wild fish and may have many other management uses.

The most successful modern tag system is the binary-coded wire tag (CWT), which has been used on the Pacific coast for 3 decades. A small (1.1 mm × 0.25 mm [0.04 in × 0.01 in]) stainless steel tag marked with distinctive binary bands is injected into the nose cartilage of juvenile fish. The adipose fin was traditionally clipped as an external marker for later tag recovery, although the development of auto-sorting systems with electronic tag detectors may make this an option. There is enough information on the standard tag to identify over 60 agencies and over 4,000 codes per year. Modern tagging machines inject the tags from continuous spools of wire at a very rapid rate. About 5% of most large hatchery releases are so marked (higher proportions from smaller releases), and people in the capture or sport industries are encouraged to return the heads of marked fish to a central processing laboratory. The result is one of the world's largest fisheries databases (over 1 billion fish tagged) used to assess hatchery effectiveness, hatchery-wild stock interactions, migratory routes, exploitation rates, and changes in ocean conditions.

Smaller versions of the CWT have been made for injecting into very small fish (e.g., chum fry of about 1–2 g [0.04–0.08 oz] weight), and special codes are available to allow separate marks for batches of fish used for hatchery experiments. No tagging system is perfect and there are indications of some pathology related to the tag injection. Tagging mortality and tag loss is assessed in the hatchery and, in the short term, is slight. A recent innovation is a coded wire tag with a laser etched number on each tag allowing easier and more efficient reading.

A more recent internal tag is the passive induced transponder tag (PIT). About the size of a grain of cooked rice, this tag emits a radio signal when stimulated by a transducer. The signal is capable of conveying a number to an LED panel on the transducer "wand" thus identifying an individual fish. The tags are injected into the coelom of fish as small as about 4 g (0.14 oz). These tags are expensive, but ideal for precise work when the identity of an individual fish is more important than identity of a group of fish. They are used often in breeding programs.

Radio tags have been used for some years to track individual fish. They emit a battery powered, continuous signal that can be followed by an above-water tracking device and are mainly used to determine freshwater spawning migration paths and swimming behavior. Some radio tags also transmit information such as fish heartbeat rate and body core temperature. In recent years, such tags have become ever smaller and more effective, but they are very expensive and used for special purposes. Some recently developed radio tags are small enough to be placed in the guts of salmon smolts.

Another class of tags uses biochemical or anatomical features of the fish themselves. Otolith marking is one example. Application of increased water temperature during critical stages of development creates a distinctive band on the otolith that can be used to distinguish hatchery fish from wild populations. Certain minerals or isotopes may serve as marks for groups of fish, or these may be created by dietary formulation. Nuclear magnetic resonance technology is used to identify the isotopes in scales or otoliths.

Biochemical tags consist of identifiable protein sequences detected with electrophoretic gels or segments of DNA that vary among individuals, strains, and species of fish. The technology of DNA fingerprinting has developed over the past decade and can now be done rapidly. It is used to differentiate wild strains from each other as runs converge in coastal waters. These techniques can also be used to assess genetic variation and can provide many shortcuts in selection programs. Special DNA probes and primers have been developed to determine fish sex before the onset of sexual maturation. In the future, biochemical tags (which do not have to be applied and are carried by all fish) may supersede mechanical tags as the tools of choice for massive tagging programs.

The development of tagging methodologies is very active; fish culturists can expect many new options in the years to come.

Nonsalmonid Culture

Nonsalmonid fish species have been cultured for enhancement, food, and ornamental use for as long as fish culture has existed, mainly in Europe and Asia. Recently, however, many species have been reared in North America and elsewhere for both aquaculture and release into wild populations. This appears to be a trend on the increase as populations face increasing exploitation pressure and environmental degradation. It is critical to assess whether the lessons learned from a century of salmonid culture are being applied to the current diversification of culture and release activities or whether history simply will be repeated.

In this short section of the chapter, only two areas of nonsalmonid culture will be described, namely, culture of some nonsalmonid fish for conservation purposes (Dexter National Fish Hatchery) and marine culture of three fish species for enhancement purposes (Texas and California). Dexter was chosen for inclusion because of its uniqueness in culturing rare and endangered species of what historically were classed as coarse fish. The culture of marine sciaenids was chosen to include marine species in a book that otherwise primarily discusses freshwater and anadromous species. Marine species are increasingly being cultured worldwide and the inclusion of three species that are being cultured in the United States is intended to allow the reader an introduction to these activities. Described are some of the culture methods used, successes achieved, and difficulties encountered in the culture of marine species released to the wild. Comprehensive treatment of the methods of marine fish culture would require a whole book, however, it is hoped that this small section will capture the imagination of aquaculture students.

Red Drum Culture on the Texas Coast

Life History

Red drum attain a size of 95 to 150 cm (37 to 59 in; the latter size is rare) and spawn in the Gulf of Mexico in offshore waters from August to December. Eggs, larvae, and early juveniles are carried along shores into estuaries and bays from Tuxpan, Mexico, in the Gulf of Mexico, to Massachusetts, in the Atlantic Ocean, by wind-driven and tidal currents. The juveniles live in estuaries, bays, and inshore waters until the age of 3–6 years when they move offshore to spawn. Year-class strength of natural populations appears to be dependent upon larval recruitment into estuaries and bays and upon the environmental conditions in these shallow waters. The movement of larvae into the estuaries and bays is cited as a production bottleneck and provides the rationale for the enhancement of red drum.

Fishery

The fishery and enhancement program described here takes place along the coast of Texas. The enhancement program is managed by the Texas Parks and Wildlife Department (TPWD) with cooperation of the Coastal Conservation Association, Central Power and Light Company, and Dow Chemical Company.

Historically, red drum supported both a recreational and commercial fishery along the Texas and other coasts within its range. In 1977, because data indicated a significant decrease in population size, the Texas State legislature set catch limits for both commercial and sportfishing and imposed mandatory reporting of sales. In 1981, the legislature designated red drum and spotted seatrout as game fish and prohibited their sale, ending the commercial fishery. The population continued to decline, and, in 1983, the first hatchery dedicated to marine stocking opened near Corpus Christi. Today there is a large sport fishery centered in the Texas bays, which is regulated by both daily limits and size restrictions.

Culture

Red drum from the Atlantic coast exhibit some genetic differences from those in the Gulf of Mexico. Within the gulf, regional genetic differences may occur, but the hatcheries treat them as a single genetic stock. It is thought that because the adults migrate off shore to spawn, there is likely considerable gene flow around the Gulf of Mexico. There were concerns that stocking large numbers of juveniles in Texas bays could reduce heterozygosity in natural populations. To alleviate this concern, wild broodstock for the hatchery are captured, more or less randomly, along the coast and held in spawning tanks with 5–6 broodfish per tank. Each year captive broodfish are randomly mixed into new groups; each year at least 25% of the broodfish are rotated out of production and replaced with newly captured wild broodfish. No broodfish are kept in the program for more than 4 years. Meristic and genetic information from the offspring is recorded to enable comparison between stocked fish and the wild population.

Hatchery methods are well established and enable production all year, if required. Normally hatchery production of red drum occurs in late March to October, producing out-of-cycle juveniles. Hatchery spawning tanks are equipped with photoperiod and temperature control. Drum in the tanks spawn naturally at temperatures from 25 to 27°C and salinity from 28 to 32 ppt. The eggs are positively buoyant and are skimmed from the surface by an egg collector and moved to small incubation tanks (379–1,514 L; 100–400 gal). Hatching occurs in 24 h at 25–27°C, and the larvae are transferred to large outdoor ponds (temperature, 25–27°C and salinity 28–32 ppt) in 1.5 d. They remain in the ponds for about 30 d until the juveniles are about 30–35 mm (1.2–1.4 in). Ponds are lined with heavy black rubber and are fertilized to produce polychaete larvae, and calanoid and cyclopoid copepods, the main initial food for the larvae and fry (Figure 7). Near the end of pond rearing, the red drum juveniles become large enough to eat the water boatman (Corixidae) living in the ponds. There is a potential for a high level of cannibalism if pond food is low. Ponds produce 45–68 kg (100–150 lb) of 25–30 mm (1.0–1.2 in) juveniles per 30-d cycle.

Estimation of Success

The Texas Parks and Wildlife Department releases 20–30 million red drum juveniles each year into coastal bays. Success of this large program can be measured in different ways, the most obvious being any increase in

FIGURE 7. Postlarval and juvenile ponds for red drum and spotted seatrout at Lake Jackson Center, Texas. Note pond liners and hatchery building in the background (photograph by E. D. Lane).

the population and catch of red drum along the Texas coast. Since 1990, it is estimated that the abundance of red drum 1–5 years of age in the bays is double that before the 1980s. Recreational catches have increased by about 18% and the number of large red drum, larger than 711 mm (28 in), total length in the catch has increased by 750%. Although these figures are impressive, the ending of the commercial fishery in 1981 could be partly responsible for population increases. The enhancement success is supported by experimental catches done to measure the effect of out-of-phase stocking. Hatchery red drum are raised to fingerling size earlier in the season than wild juveniles by the manipulation of temperature and photoperiod in the hatchery. The release of hatchery fish results in a bimodal length frequency in the 1st year in bays that are stocked, whereas fish in unstocked bays show unimodal length frequency distributions. Comparisons made between Cedar Lakes (unstocked) and Matagorda Bay, Aransas Bay, and others (stocked) show this difference in length frequency distributions indicating survival, at least for the 1st year, of stocked drum. Experiments are also underway using oxytetracycline (OTC) marking to identify stocked hatchery fish, apparently with preliminary success. A comparison of the populations in stocked and unstocked bays indicated that the subadult populations in the stocked bays was, on average, 21.5% higher than it would have been had no stocking occurred.

Other dimensions of success are benefits to society and achievements in organizational cooperation, for example, the partnership commitment of non-TPWD organizations. The red drum enhancement program has involved external partners from the start of the program, namely the Gulf Coast Conservation Association (presently the Coastal Conservation Association), Central Power and Light Company, and Dow Chemical Company. There is also the perceived effect of this enhancement program on the "user group," in this case, the sport fishers. The hatchery program enjoys almost complete support of both fishers and conservationists along the Texas coast. The hatcheries operate with a professional staff, but also use a large cadre of volunteers. This allows resource users direct involvement in the management of their resource and a link between the interested public and the professional managers.

Spotted Seatrout

Life History

Spotted seatrout attain a size of approximately 90–125 cm (35–49 in), the latter size being rare. Adults spawn from April to October in the bays along the Texas coast. The post larvae and young of the year often are found on or near grass flats in the bays. Adults (1–3 years of age) are usually found in deeper waters in the bays, occasionally moving through the passes to the immediate onshore area of the Gulf of Mexico. Their range encompasses bay and inshore areas from Cape Cod, Massachusetts, to the lower reaches of the Bay of Campeche. Spotted seatrout differ from red drum in spending their entire life in bays. Year-class strength is determined by conditions in the bays, most likely by winter conditions. In winter, strong cold winds from the north, "northers," sweep across the shallow waters occasionally causing rapid drops in temperature. Mortalities are also caused by periodic blooms of the red tide dinoflagellate *Ptychodiscus brevis*. Spotted seatrout subadults and adults are piscivorous, mullet being a common food item.

Fishery

Like red drum, the spotted seatrout historically supported both a commercial and sports fishery, and also like the red drum, the commercial fishery was closed in 1981 when the legislature declared both seatrout and red drum to be game fish, which stopped all sales of these species. However, despite the cessation of the commercial fishery, recreational landings of seatrout declined throughout the 1980s. The culture of spotted seatrout by TPWD has paralleled the culture of red drum. Both species are raised in the same facilities and in cooperation with the Coastal Conservation Association, Central Power and Light Company, and Dow Chemical Company. The catch of the present sport fishery, regulated by both bag and size limits, is considered to have increased from the low levels seen in the mid-1980s. Although the impact of stocking has not been fully quantified, some success is claimed.

Culture

Spotted seatrout was one of the first saltwater species to be cultured in the United States. Early efforts in pond culture in South Carolina (1947–1951) were unsuccessful. The first efforts (before 1976) to culture hatchery spawned seatrout produced variable returns. Between 1983 and 1988, TPWD operated an ongoing program to develop hatchery and pond culture for spotted seatrout at the Perry R. Bass Marine Fisheries Research Station, Palacios, Texas. In 1999, all three Texas marine hatcheries produced about 5 million spotted seatrout annually to stock in Texas bays.

Culture methods for seatrout are the same as those for red drum. In summary: 5–6 broodfish held under controlled photoperiod and temperature, natural spawning in tanks, positively buoyant eggs skimmed and hatched in 379–1,514 L (100–400 gal) hatching tanks, hatching in 18–24 h, held for 1.5 d, then moved to ponds for about 30 d, and stocked at 25–30 mm (1.0–1.2 in) length. The broodfish are handled in the same way as red drum. The principal differences between spotted seatrout and red drum involves the separation of broodfish stocks. Spotted seatrout exhibit clinal genetic variation along the coastal bays of Texas, and this genetic population structure is taken into consideration. Broodfish and juveniles from particular bay systems are kept separate throughout the culture period until and including stocking. For example, the broodfish for Lake Jackson Hatchery are taken from Galveston Bay and the offspring from these broodfish stocked only into the Galveston Bay. Galveston Bay broodstock are replenished only from wild Galveston Bay stocks.

Photoperiod and temperature are used to control growth rates to allow off-cycle stocking similar to that of the red drum stocking program. Spotted seatrout spawn in hatcheries in April-October and juveniles are stocked 1 month after spawning.

Estimation of Success

Assessment of the spotted seatrout hatchery program has been attempted using off-cycle stocking and subsequent capture length-time data, OTC marking data, and genetically marked releases of seatrout. Only 5 million spotted seatrout are released (compared to 20–30 million red drum), and the off-cycle results are only valid for a short period. Early results from OTC marking trials show promise, but work is preliminary. Spotted seatrout possess a rare marker allele allowing a genetically identifiable hatchery stock to be developed and a preliminary estimate of a stocking program's impact.

The population of spotted seatrout was considered to be low in the early 1980s due to overfishing and poor recruitment, the latter probably due to severe cold winter weather and blooms of red tide. The seatrout population and its fishery are considered to have recovered from the low levels of 1980, however, the contribution of the stocking program on this recovery is unknown. The results of the genetic marking have indicated that stocked seatrout do survive in nature, based on experimental catch and creel census recovery. The data from returns also suggest that stocking will have a limited and localized impact on the population and fishery. This conclusion is based on allele and genotype frequencies in the catch from the stocking area (Port Mansfield) and the remainder of lower Laguna Madre, Texas. The genetic baseline for the populations in this area were established in 1994, stocking occurred in 1995, and monitoring continued to 1997, when the observed genotype frequencies returned to approximately those of the 1994 baseline. Stocking spotted seatrout was shown to have an effect on the local population and genetically marked fish did enter the recreational fishery.

Like the red drum program, the seatrout enhancement involves several partners (as noted earlier) and the stocking program enjoys a very high profile and strong support from the recreational fishing community and conservation groups.

White Seabass Culture in Southern California

Life History

The white seabass is one of the larger members of the Sciaenid family, with reports of it reaching a size of 152 cm (59.8 in) and 41 kg (90.4 lb). Its range is from Juneau, Alaska, in the north to Baja, California, in the south and extending into the Gulf of California. It is rare north of San Francisco. The species is a shorefish, most commonly found around kelp beds or over rocky shore from the surf zone to about 125 m (410 ft). White seabass spawn from March to August, the first spawning occurring when they reach between 60 and 75 cm (24 and 30 in), reportedly near kelp beds. The young inhabit bays and the shoreline along sand beaches.

The fish is a large terminal predator feeding on other fish (anchovies, pilchards, herring, and smelt) and squid.

Fishery

White seabass have been fished both commercially and by sport fishers in California since records have been kept. It was reported that there was a fourfold decrease in the harvest between 1918 and 1928 with a high of 1,361,000 kg (3,000,000 lb) taken in 1922. Regulations to control the catch of white seabass have been in place since 1931, but the decline in catches has continued. Between the 1950s and 1980s, the California catch declined from around 55,000 fish (approximately 120,000 kg or 264,560 lb) to less than 3,500 fish. In 1983, the Ocean Resources Enhancement and Hatchery Program (OREHP) Advisory Panel identified white seabass as an appropriate species for a stocking program. The panel membership includes representation from the sports and commercial fishers' associations, academic institutions, aquaculture associations, and federal and state government departments and commissions (principle among which are the California Department of Fish and Game and private industrial contributors).

Culture

California white seabass culture was planned for hatchery production to start with a pilot phase of approximately 50,000 fish annually (to 1993) then to increase to an annual hatchery production of 400,000–500,000 fish. White seabass is a sciaenid and hatchery methods used for this species are similar to those used for other sciaenids (i.e., red drum, see above). However, there are specific differences: incubating temperature is 18°C, with the salinity kept between 32 and 34 ppt. The eggs hatch in 2 days and are moved to early rearing tanks and raceways for 18 d (Figure 8). During this time, they are fed enriched brine shrimp nauplii. There are some disease losses in the hatchery, including occasionally severe mortality in rearing tanks due to *Rickettsia* spp.

After hatchery rearing, the fry are reared in net cages either moored in open water or alongside a dock. Submerged fiberglass raceway cages attached to a pier or floating dock are also used (Figures 9 and 10). Seabass are reared in the sea-pens for 6 months to a year until they reach a size of 160–280 mm (6–11 in). During the time in sea-pens, they are fed manufactured pellets formulated for mahimahi. Before stocking, all fish are injected with a coded wire tag and 10% of fish are also given an external tag (Figure 11).

Genetic concerns have been raised and are considered in the hatchery program. Broodstock are placed in mass spawning tanks with 25 males and 25 females. After each annual cycle, five of each sex are removed to the wild and five of each sex are added from the wild. It is considered that this schedule will maintain the genetic variation in fish produced by the hatchery.

Estimation of Success

The goals of this hatchery-based enhancement program are to release juvenile seabass to the wild in such numbers and with sufficient survival that increases in catch to both the sport and commercial fisheries will result. Another goal is to maintain the genetic heterogeneity of both the wild and cultured stocks of white seabass. Compared to the pro-

FIGURE 8. Juvenile white seabass in culture, 19 d old at the California Fish and Game Hatchery in Carlsbad, California (photo courtesy of Hubbs–Sea World Research Institute).

FIGURE 9. Cage culture of seabass. Volunteer group helps state biologists tag white seabass juveniles with coded wire tags. Santa Barbara, California (photo courtesy of Hubbs–Sea World Research Institute).

FIGURE 10. Volunteer grow-out facility cages in Dana Point Harbor, California (photo courtesy of Hubbs-Sea World Research Institute).

FIGURE 11. Transfer of juvenile seabass to release sites in bays along the Texas coast. Similar transfers occur for both red drum, and spotted seatrout (photograph compliments of Texas Parks and Wildlife Department).

grams for sciaenid enhancement in Texas, California white seabass enhancement is relatively recent and the number of fish released is comparatively small. It is not surprising that the final results of this enhancement operation are still not known. However, preliminary results have indicated that stocked fish do survive in the wild. During test fishing in 1993, about 10% of the fish collected from the release site (bay) were hatchery fish. Three fish recovered from local kelp beds were near the sport's legal size (711 mm or 28 in) and had survived up to 4 years. Anecdotal reports in 1999 have indicated that a small number of recoveries of subadult fish continues to occur, but no ripe or spawning adults of hatchery origin have been captured, and there has not been a notable increase in the fishing success to date.

At present, the OREHP for white seabass cannot be fully assessed in terms of its stated goals, but there is considerable success in the involvement of user groups in the fishery and in the potential for commercial aquaculture.

Conservation Fish Culture

In North America, one of the most notable and, perhaps, laudable restoration hatcheries is Dexter National Fish Hatchery and Technology Center located in the Pecos River Valley in Dexter, New Mexico. Staff of this hatchery concentrate their work on fish species that are threatened or endangered, primarily from the U.S. Southwest and northern Mexico. During the late 1800s and throughout the twentieth century, the unique fish fauna of the U.S. Southwest was dramatically changed due to human activity—notably, dams, groundwater and surface water removal, overgrazing, and introductions of nonnative fish species. These historic and presently occurring human disturbances imperiled many species, threatening some with extinction.

Dexter National Fish Hatchery began operations in 1932 to produce warmwater sport fish. However, in 1974, following the passage of the Endangered Species Act (1973), the hatchery began working with endangered and threatened species. Dexter staff then developed hatchery techniques for captive rearing and redistribution of endangered species to the wild.

Culture facilities at Dexter include a laboratory, hatchery, tank house, and 46 lined earthen ponds, all of which are adjustable to suit the requirements of many different fish species and combinations of species in the hatchery.

This hatchery complex has worked on species from the families Cyprinidae, Catostomidae, Ictaluridae, Cyprinodontidae, Poecilidae, and Percidae, and at any one time contains as many as 20 species under culture. In January 1999, there were 15 species being cultured (Table 2). Rearing many species at Dexter simultaneously needs special operating procedures to maintain the genetic integrity of closely related species to prevent hybridization and to prevent cross-contamination by pathogens and parasites. Species integrity and health is especially important in a hatchery dedicated to endangered species.

Out-planting is conducted to enhance or reestablish populations for species for which a sufficient number of individuals have been reared and for which suitable habitat is available within the species range. The goal is to allow the removal ("de-listing") of a species from the endangered species list (the Red Book). As Dexter, by its mandate, must work with species in which population numbers have become dangerously low, there is always a concern of starting a project with a small gene pool. Often there is no opportunity to increase the gene pool size and enhancement must be done with a founder population of relatively few individuals, a problem associated with all captive rearing programs. Dexter now specializes in developing management strategies to conserve genetic variation and heterogeneity to produce fish that are behaviorally and physiologically adaptive following stocking. Although Dexter has provided fish at various life stages for stocking for many years, there is a conspicuous lack of data regarding the success of most introductions. Results from an ongoing reintroduction of Colorado pike minnow

TABLE 2.

Species held at Dexter Fish Hatchery and Technology Center on 01/01/1999.

Catostomidae	
Razorback sucker	*Xyrauchen texanus*
Yaqui sucker	*Catostomus bernardini*
Warner sucker	*Catostomus warnerensis*
Cyprinidae	
Colorado pikeminnow	*Ptychocheilus lucius*
Bonytail chub	*Gila elegans*
Chihuahua chub	*Gila nigrescens*
Pahranagat river chub	*Gila robusta jordani*
Virgin river chub	*Gila robusta seminuda*
Woundfish	*Plagopterus argentissimus*
Guzman beautiful shiner	*Cyprinella formosa formosa*
Ictaluridae	
Yaqui catfish	*Ictalurus pricei*
Cyprinodontidae	
Leon Springs pupfish	*Cyprinodon bovinus*
Desert pupfish	*Cyprinodon macularius*
Poecilidae	
Gila topminnow	*Poeciliopsis occidentalis occidentalis*
Big Bend gambusia	*Gambusia gaigei*

to the San Juan River indicate survival exceeding expectations of management biologists. Present follow-up protocols, hopefully, will produce more data on the success of future introductions.

Conclusion

We have attempted to review the scope of hatchery production for salmonids to demonstrate how pervasive hatcheries have become in the management of salmonids over the last century. Salmonids were chosen as examples because of the long history of their culture and because research on both hatchery technology and on the potential dangers of hatchery programs is most advanced for this group of fish. There is no question that hatcheries can be effective and that they can produce viable populations of anadromous and nonanadromous salmonid fish for a variety of purposes and that this production has continued for many decades in some hatchery systems. However, depending on management goals, hatcheries clearly have the potential to do harm. In some cases, hatchery production makes management of wild stocks more difficult. Mitigation hatcheries may not achieve their goals. Alteration of genetic structure of wild fish populations may occur, possibly with negative results. Some negative outcomes of salmonid hatchery programs have been well documented and others are theoretically likely, but are less well empirically established.

Once hatchery programs are established, it is difficult to eliminate them even if they are found to be doing more harm than good. Public perceptions lend strong political support to the production of fish. Benefit-cost analysis of hatcheries may be done on a narrow and short-term basis, but some of the possible negative effects may be long term and difficult to detect. Despite these dangers, hatchery production can be very useful to preserve endangered strains or species and to produce fish for recreational or commercial capture in some carefully considered situations.

Although scientific understanding of interactions between hatchery produced and wild fish populations is still in its formative stages, much has been learned and most hatchery program proposals are scrutinized carefully before being put into operation. The question, posed earlier, remains to be answered: Is the knowledge we have gained, by experience and research, of cultured and wild salmonid population interactions now being applied to nonsalmonid fish?

The above discussions of recent marine fish hatchery programs in Texas indicates an awareness and concern for possible genetic effects of cultured fish on wild populations and for program assessment. Facilities such as the Dexter conservation hatchery in New Mexico clearly demonstrate a sophisticated approach to the problem of endangered fish species and their preservation through hatchery intervention. There is, however, a vast collage of freshwater fish culture programs showing a full spectrum of precautionary strategies ranging from none at all to very well developed approaches. For example, in some parts of the United States sport fish such as centrarchid species are stocked over large regions for fishing tournaments with little regard to resident, wild populations. Both ecological and genetic effects of these stocked fish must be significant. In contrast, there are well designed programs dealing with sturgeon, paddlefish, and other endangered or threatened fish species in which there are well developed research programs on the target species' biology, ecology, and genetics with respect to impacts of culture.

The answer to the question is "sometimes, in some places."

References

Bartley, D. M., D. B. Kent, and M. A. Drawbridge. 1995. Conservation of genetic diversity in a white seabass hatchery enhancement program in southern California. Pages 249–258 *in* H. L. Schramm Jr., and R. G. Piper, editors. Uses and effects of cultured fishes in aquatic ecosystems. American Fisheries Society, Symposium 15, Bethesda, Maryland.

Barton, B. A. 1996. General biology of salmonids. Pages 29–96 *in* W. Pennell and B. A. Barton. Principles of salmonid culture. Elsevier Science, Amsterdam.

Baum, E. 1997. Maine Atlantic salmon, a national treasure. Atlantic Salmon Unlimited, Herman, Maine.

Behnke, R. J. 1992. Native trout of western North America. American Fisheries Society, Monograph 6, Bethesda, Maryland.

Billard, R., and J. Jensen. 1996. Gamete removal, fertilization, and incubation. Pages 291–394 *in* W. Pennell and B. A. Barton. Principles of salmonid culture. Elsevier Science, Amsterdam.

Chamberlain G. W., R. J. Miget, and M. G. Haby, compilers. 1990. Red drum aquaculture. Texas A&M University, Sea Grant College Program, Galveston.

Colura, R. L., T. L. King, J. D. Gray, and B. W. Bumguardner. 1992. Analyses of six years of spotted seatrout (*Cynoscion nebulosus*) pond culture trials. Aquaculture 107:313–332.

Drawbridge, M., C. Ulrich, and M. Chen, editors. 1996. Procedures manual for the growout of juvenile white seabass (*Atractoscion nobilis*). Hubbs-Sea World Research Institute.

Groot, C. 1996. Salmonid life histories. Pages 97–230 *in* Pennell and Barton (1996).

Groot, C., L. Margolis, and W. C. Clarke, editors. 1995. Physiological ecology of Pacific salmon. UBC Press, Vancouver.

Heard, W. R. 1996. Ocean ranching: an assessment. Pages 833–870 *in* W. Pennell and B. A. Barton. Principles of salmonid culture. Elsevier Science, Amsterdam.

Holtby, B., C. Groot, and L. Margolis, editors. 1991. Pacific salmon life histories. UBC Press, Vancouver.

Holtby, L. B., and G. F. Hartman. 1982. The population dynamics of coho salmon (*Oncorhynchus kisutch*) in a west coast rain forest stream subject to logging. Pages 308–344 in G. F. Hartman, editor. Proceedings of the Carnation Creek workshop: a ten-year review.

Jorstad, K. E., and G. Naevdal. 1996. Breeding and genetics. Pages 655–726 *in* W. Pennell and B. A. Barton. Principles of salmonid culture. Elsevier Science, Amsterdam.

Karel, B., I. Blandon, R. Ward, and T. King. 1999. An evaluation of stocking spotted seatrout in the lower Laguna Madre, Texas. In manuscript Perry R. Bass Marine Fisheries Research Station.

Kent, D. B., and M. A. Drawbridge. 1995. Accomplishments and roadblocks of a marine stock enhancement program for white seabass in California. Pages 492–498 *in* H. L. Schramm Jr., and R. G. Piper, editors. Uses and effects of cultured fishes in aquatic ecosystems. American Fisheries Society, Symposium 15, Bethesda, Maryland.

Matlock, G. C. 1990. Preliminary results of red drum stocking in Texas. NOAA (National Oceanic and Atmospheric Administration) Technical Report NMFS (National Marine Fisheries Service) 85:11–15.

McEachron L. W., C. E. McCarty, and R. R. Vega. 1995. Beneficial uses of marine hatcheries: enhancement of red drum in Texas coastal waters. Pages 161–166 *in* H. L. Schramm Jr., and R. G. Piper, editors. Uses and effects of cultured fishes in aquatic ecosystems. American Fisheries Society, Symposium 15, Bethesda, Maryland.

McEachron L. W., R. L. Colura, B. W. Bumguardner, and R. Ward. 1998. Survival of stocked red drum in Texas. Bulletin of Marine Science.

Meerburg, D. J. 1986. Salmonid age at maturity. Canadian Special Publication of Fisheries and Aquatic Sciences 89.

National Research Council. 1996. Upstream, salmon and society in the Pacific Northwest. National Academy Press, Washington, D.C.

Pennell, W., and B. A. Barton, editors. 1996. Principles of salmonid culture. Elsevier Science, Amsterdam.

Pennell, W., and W. E. Mclean. 1996. Early rearing. Pages 365–466 *in* W. Pennell and B. A. Barton. Principles of salmonid culture. Elsevier Science, Amsterdam.

Acknowledgments

The authors thank the people listed below, but accept all responsibility for any errors or omissions in this chapter.

Connie J. Stolte, Conservation Scientist V and Assistant Hatchery Manager, Texas Sea Center, Texas Parks and Wildlife Department, Lake Jackson, Texas.

Dale D. Lyon, Conservation Scientist III, Texas Sea Center, Texas Parks and Wildlife Department, Lake Jackson, Texas.

Holt Williamson, Dexter National Fish Hatchery, Dexter, New Mexico.

Rockey Ward, Genetics Laboratory, Perry R. Bass Marine Research Station, Texas Parks and Wildlife Department, Palacios, Texas

Mark A. Drawbridge, Hubbs–Sea World Research Institute, San Diego, California.

Martin F. Chen, California Department of Fish and Game, Carlsbad, California.

Ruth Withler, Department of Fisheries and Oceans, Canada Pacific Biological Station, Nanaimo, British Columbia.

5

Semicontrolled Systems

Patricia M. Mazik and Nick C. Parker

Introduction

The efficient operation of a hatchery system depends on a number of chemical, physical, and biological variables. Rearing units for fish can be of various types including ponds, raceways, silos, cages, and net-pens. The selection of a rearing unit depends on the fish species, water quality characteristics (e.g., water availability and flow), oxygen demand, and soil characteristics. This chapter will discuss types, operation, and production of semicontrolled systems, which include ponds, raceways, cages, and net-pens.

Ponds

Earthen ponds have traditionally been used for fish culture. In the southeastern United States, 95% of the catfish produced are grown in ponds. Ponds can be found in many shapes and sizes, but the most common are levee-type and watershed ponds. Levee ponds have four raised levees (sides), are usually rectangular, have a flat bottom, and can range in size up to 8 ha (20 acres) or more (Figure 1). Their maximum depth is usually 1.2–2.4 m (4–8 ft). A watershed pond is a hillside pond with a main levee that dams a valley, an irregular shape, and is, generally, deeper than an earthen pond.

For pond culture to be profitable, suitable resources must be available. Fish culture in ponds is water intensive and requires large volumes of good quality water. The water must be free of contaminants and wild fish. Water must be pumped into levee ponds, either from a well or from

FIGURE 1. Fish ponds in Marion, Alabama.

a surface water source. While watershed ponds may also have a pumped water source, much of their water is collected through runoff. Since water temperature is not controlled in ponds, the growing season must be long enough to produce a market size fish. For catfish, a growing season produces a 0.45-kg (1-lb) catfish from an egg in 18 months or less. Pond culture also requires large areas of land with the correct topography and soil type for economical construction and operation of the ponds. Even when all these resources are available, production will not be profitable without the proper industry infrastructure in place. Variable operating costs, such as feed costs and marketing, must be dealt with for profitable production of fish.

Pond Site Selection

Site selection is important in earthen ponds. Characteristics of an ideal site include adequate supply of good quality water, impermeable soil (i.e., clay) with no chemical residues, flat land, and a location outside of a flood plain area.

Water Supply

Water is the greatest limiting factor in aquaculture. Water quality characteristics (salinity, total alkalinity, and hardness) should be taken into account when considering water sources (See Chapter 3). The preferred source of water is a shallow well (less than 76.2 m [250 ft]), but spring water may also be used where available. Surface water from streams, rivers, lakes, and reservoirs is also a good source but may contain wild fish and their eggs and diseases. Surface water can also have high silt levels. These undesirable elements can be eliminated with the use of proper filters. Water sources can also contain pesticides and contaminants that can kill fish. Chemical analysis and bioassays can determine if the water is suitable before pond construction begins.

Intensive fish culture requires 236.5–473.0 L per min per ha (25–50 gal per min per acre), depending on the location and climate. In the southeastern United States, the ideal amount for levee ponds is 284 L per min per ha (30 gal per min per acre). This amount is needed to provide enough water to fill a pond in 2 weeks, in addition to replacing losses from seepage and evaporation. The absolute minimum amount required is 142 L per min per ha (15 gal per min per acre). In the southeast, this amount is approximately twice the daily maximum evaporation rate allowing maintenance of water during droughts with moderate seepage (Tucker and Robinson 1990). One well will usually supply four levee ponds. Tucker and Robinson (1990) calculated that if each pond is 7 ha (17 acres), the well must pump 3,785 L (1,000 gal) per min to satisfy the 142 L per min per ha (15 gal per min per acre) minimum amount. If each pond were 1.2 m (4 ft) deep, 15 d of continuous pumping would be required to fill one pond.

When using surface water, consideration should be given to the location of the intake pump. The use of a sump pump or concrete inlet may reduce the congesting of the inlet with mud or debris. Flood stage records should also be obtained to establish the minimum elevation of the pump. Water supply lines installed to each pond should be large enough to carry all the available water in the area. This is very important when rapid filling is required (Figure 2).

Pond Size and Shape

Production ponds are usually rectangular with the length of the pond being determined by the land. In hilly country, ponds are built parallel to the contour. Warmwater aquaculture, especially for channel catfish, uses production ponds up to 8 ha (20 acres) or more. Larger ponds are less expensive to build, but are more difficult to manage. Catfish fingerling ponds are usually 0.8–2.0 ha (2–5 acres) in size and brood ponds can range from 0.8 to 8.0 ha (2–20 acres). While food fish ponds may be any size, most are now 4–8 ha (10–20 acres).

FIGURE 2. Water supply to a pond.

Accessibility and Drainage

Topography of the site is an important consideration. Pond construction on land with only moderate relief will be more economical than moving soil on rolling or hilly land. Pond size will depend on species, production type, and cost. Warmwater fish culture (i.e., channel catfish) generally uses large rectangular shallow ponds, whereas cool and coldwater culture (i.e., salmonid culture) use smaller ponds or raceways that require large volumes of water. For fingerling ponds, 0.8–2.0 ha (2–5 acres) are recommended; 2–8 ha (5–20 acres) are needed for grow out ponds. Flat bottom ponds are usually built with levees around the ponds. Levees should be at least 5–6 m (16–20 ft) wide to prevent erosion. The steepness of the levee will depend on the soil type. For most soil types that are properly compacted, a 3:1 inside slope is satisfactory. For ponds greater than 3 ha (8 acres), a 4:1 slope is needed since erosion tends to be greater due to wind action across larger ponds.

Ponds should be drained by gravitational flow (Figure 3). There are several types of drainage systems, but, regardless of the one used, the drainpipe must allow the pond to completely empty. Two of the most common types of drains used are inside swivel drains and outside valve drains (Tucker and Robinson 1990). Of the two, outside drains are more convenient to use and maintain. Drains should have at least 1% slope

FIGURE 3. Pond draining by gravitational flow.

and be at least 30.5 cm (12 in) in diameter for ponds larger than 4 ha (10 acres).

Levee Ponds

Correct levee construction is important for good pond management. Using proper planning, ponds can be constructed next to each other, requiring fewer levees and saving space and construction costs. Ponds constructed at right angles to prevailing winds will reduce the damage of wave action on levees. Levees not only contain water in ponds, but also permit vehicular traffic around and between ponds (Figure 4). Levees can be constructed by using a bulldozer, pan (dirt scoop), or drag line. The preferred methods are the bulldozer and pan. Earth is pushed or carried to the pond sides to form the levees. In these methods, the earth is compacted by the machinery and the pond bottom is smooth. The dragline method requires earth to be taken from either side of the levee creating narrow ditches and rough-sided levees that are not as desirable as the other two methods. The dragline method is mainly used in wet and boggy areas where heavy equipment cannot be used.

Preparation of the base is important in levee construction. The levee core should penetrate to the impervious layer below the topsoil, be of good quality clay, and be free of plant material and roots. Obstructions

FIGURE 4. Pond levees with roads.

in the base can lead to failure of the levees due to seepage. The tops of the levees should be flat, at least 3.0–3.7 m (10–12 ft) across to allow vehicles to pass and at least 0.6 m (2 ft) above the water level. The outside slope of the levee should be 2:1 and the inside slope can range from 2.5:1–4:1, but 3:1 seems to be the norm. Steeper slopes have the advantage of using less earth in the levee and, therefore, allowing less shallow water, keeping vegetation to a minimum. The disadvantages of steep slopes are the increase in erosion rates and the difficulty mowing. Outside slopes should be seeded with grass, and water supply lines should be placed in the levees below the frost line.

Sealing Pond Bottoms

Pond bottom soil type is very important, with clay being the best. Seepage tests are important in ponds that will not have continuous water flow, such as with warmwater fish culture. Next to sandy soil, poor soil compaction while building the levees is the main cause of seepage. Dry soil does not compact well, so water must be added periodically to the levee, with a water truck, during construction. Ponds built with a pan have better levee compaction than ponds built with a dragline. There are several approaches that can be used to seal ponds when they will not hold water. A clay layer can be added to the pond bottom or bentonite, a special kind of clay, can be tilled in with the soil at a rate of 9.8–39.0 kg/m^2 (2–8 lbs/ft^2), depending on the soil type, before compaction. Bentonite will swell when exposed to water, thus sealing the pond bottom. Polyvinyl liners are a solution in very porous soil, but are quite expensive (Figure 5). The size of the liner required depends on the size and

FIGURE 5. Polyvinyl-lined pond under construction.

depth of the pond, but should extend above the maximum water level. The liner should be covered with soil and the edges secured.

Pond Draining

Harvest methods need to be taken into account during the design of pond drains. Ponds must be drainable and may contain a concrete kettle or collection area for harvesting fish (Figure 6). For harvest methods that use seining and draining of the pond, rapid draining of the pond is important, especially in hot weather. A majority of the fish will be removed by seining, but the pond must be drained to remove all the fish. Slow draining at this point can lead to low dissolved oxygen (DO) and turbid water resulting in fish kills. For all ponds, the bottom should slope toward the drainage area at a slope of at least 2.5 cm per 30.5 m (1 in per 100 ft) (15.0 cm per 30.5 m [6 in per 100 ft] is the norm). If the drainlines have turns, a clean-out arrangement should be added at the turn. Standpipes, dam boards, or valves are needed to control draining of the ponds for seining or during high water episodes. Ponds in areas subjected to heavy rains should have appropriate drainage. Watershed ponds require emergency spillways. The design will depend on the location, and recommendations should be obtained from the local extension agent or Natural Resource Conservation Service personnel.

FIGURE 6. Harvesting fish in concrete kettle.

Pesticide Residue

Soil at the pond site should be analyzed for pesticide residue before any construction begins. Local extension agents can provide assistance in contacting soil testing laboratories and showing how and where to collect samples. Several samples should be taken and analyzed for chlorinated hydrocarbons, especially toxaphene and endrin. The extension agent may also recommend testing for additional pesticides, depending on your area. Low lying areas where runoff water collects should receive special consideration.

Watershed Ponds

Watershed ponds or hill-type ponds are built into hillsides, usually following a natural contour. Many times these ponds are initially built for irrigation, watering of livestock, or recreation, but they can be constructed specifically for seine-through fish culture. Since the water supply is mainly runoff and the ponds are usually irregular in shape, these ponds are not as extensively used for intensive fish culture. However, in areas such as the Blackbelt region of east Mississippi and west Alabama, watershed ponds are normally used for catfish production due to the topography. In constructing watershed ponds, altering the contours to form a uniform basin will allow easier seining of the pond. Placement of levees at the upper end of watershed ponds can lead to control of runoff water entering the pond. Ponds can be placed in such a way as to permit water to flow by gravity from upper to lower ponds. Pumps can be used to replenish water, when necessary, from wells or other water sources.

Operation

Pond design is necessary to ensure maximum harvest and, therefore, profit. The preparation of ponds before stocking is also critical to success. Preparation includes filling and fertilization to ensure good water quality and fish survival at stocking. Harvesting is an important aspect of aquaculture that is frequently overlooked during facility design and production. Harvesting is usually the most labor-intensive part of fish-farming and damage to fish can reduce their value. Therefore, harvesting methods and techniques should be planned ahead. Harvesting fish should also not begin without a marketing plan already in place. All arrangements for harvest, transport, and sale should be made weeks before harvesting begins. Harvested fish are normally transported alive to a processing plant or for restocking into other ponds for additional growth or for recreational fishing. Advanced planning is a key element in a successful fish culture operation.

Prefilling Preparation

When possible, complete draining and drying of ponds between stockings is desirable. Many compounds, such as nitrite (NO_2^-), hydrogen sulfide (H_2S), and organics, can accumulate in ponds, but are oxidized when exposed to air. When allowed to accumulate, soil organic matter increases oxygen demand and the resulting anaerobic soils can release reduced compounds such as iron (Fe^{++}), manganese (Mn^{++}), NO_2^-, and H_2S into the water, which may be toxic to fish. Older ponds, which have more organic material, may need to have the bottom disked or turned over to promote aerobic breakdown during the drying period. Before filling, it may be necessary to recompact the bottom to prevent leaking. Aerating ponds has also been shown to reduce the accumulation of organics. Ayub et al. (1993) reported that aerated ponds did not accumulate as much organics in the bottom soils as control ponds. They also found that the addition of urea to manured ponds did not accelerate carbon loss from pond bottoms.

During the drying period, lime can be added to raise the alkalinity or buffering capacity of the water, elevate pH in the water and bottom soils, enhance phosphorus solubility, and accelerate organic matter decomposition. Ponds should be limed during late fall or winter, especially if a fertilizer will be applied in the spring. If low alkalinity ponds are fertilized before adding lime, phosphorus may be bound to the bottom mud and not available for plant production. Therefore, liming should be done before fertilization.

Three basic types of lime are used in fish ponds: agricultural limestone ($CaCO_3$), burnt lime (CaO), and hydrated (slaked) lime ($Ca(OH)_2$). Agricultural limestone is made by pulverizing limestone. It should be ground finely enough to pass through a size-10 mesh sieve. Small particles will more readily dissolve in water than larger particles. Burnt lime is made by putting limestone in a kiln to remove carbon dioxide, and hydrated lime is simply the addition of water to burnt lime.

Lime can be applied in two ways. It can be applied to the pond bottom during the drying period or applied over the water surface. When applying lime to the pond bottom, the type of lime used depends on the objective. Burnt and hydrated lime are added to bottom soil to raise the pH high enough to kill any pathogens that are present. Caution should be used when using burnt or hydrated lime. When used in high enough concentrations to increase total alkalinity and hardness, these limes can raise the water pH to more than 10, which is detrimental to fish production. Agricultural limestone is applied to acidic pond bottom soils when the objective is to increase pH and stimulate organic decomposition. Agricultural limestone will not increase soil pH high enough to kill microorganisms, therefore, it is best suited to enhancing organic decomposition and improving soil condition.

When agricultural limestone is added directly to the pond water, a small portion dissolves immediately and the remainder goes to the bottom where it slowly dissolves. Since the solubility of agricultural limestone is low, it will not increase total hardness or alkalinity above 50 mg/L. Therefore, if ponds have an alkalinity above 40–50 mg/L, there will not be a substantial benefit to adding limestone. Hydrated and burnt lime are more soluble than agricultural limestone, but, again, the user must be cautious to not raise the pH to lethal levels.

The application rate of limestone should be determined by analyzing pond mud samples. Random samples should be taken from deep and shallow areas over the entire length and width of the pond. Samples can be collected using PVC pipe to take plugs. Twenty-five to thirty-eight samples per ha (10–15 samples per acre) should be taken (Wynne 1993). The samples should be mixed together, dried, and then placed in a labeled container. Laboratories to analyze the samples can be found through the local county extension office. Lime application rates will usually range between 1,134–2,835 kg of lime per ha (1,000–2,500 lb of lime per acre). The application rate should be determined by the following equation (Boyd and Masuda 1994):

$$\text{Application rate} = \text{liming rate}/(\text{neutralizing value} \times \text{efficiency rating}/100)\,.$$

The efficiency rating and neutralizing value of limestone varies. Information on what type of lime is best for a particular situation can be provided by your county extension office.

Filling

Just before ponds are filled, herbicides can be added to prevent the emergence of rooted aquatic plants or dyes can be added to shade the bottom. Chemical and organic fertilizers (discussed below) can also be added to encourage growth of zooplankton and phytoplankton. Larval fish require live zooplankton in sufficient numbers in order to survive and grow to stockable size. Phytoplankton is used to shade the pond bottom and retard the growth of vegetation. Phytoplankton is also a food source for zooplankton.

Care must be taken when filling ponds to prevent the entrance of wild fish since many are predators; they can also carry parasites and disease. This is usually not a problem with well water or springwater, but surface water must be filtered. Two types of filters can be used. A sock-type filter can be made by forming Saran cloth into a cylinder with nylon cord drawstrings on each end. A sock filter 0.9 m (3 ft) wide by 3.7 m (12 ft) long can handle 3,785 L per min (1,000 gal per min), as long as the inlet pipe is near the water surface (Huner and Dupree 1984). The drawstring cords will allow easy cleaning and removal. A box-type filter is also used and can be positioned at the pump inlet or outlet. Saran cloth is attached to a wooden frame that is 0.6 m (2 ft) deep, 0.9 m (3 ft) wide, and 2.4 m (8 ft) long. The box-type filter can also handle 3,785 L per min (1,000 gal per min). The filter is supported on a wooden grid with 0.3- to 0.6-m (1- by 2-ft) openings. The box-type filter can be floating or fixed in the pond.

When ponds cannot be drained or if wild fish remain in potholes after draining, water can be treated with rotenone or chlorine. Rotenone is available in liquid and powder forms; the liquid is preferable. Liquid rotenone (5%) is applied to provide a concentration of 1.0 ppm. A 0.4-ha (1-acre) pond with an average depth of 0.9 m (3 ft) would require 3.8 L (1 gal) of 5% rotenone. Temperature affects rotenone toxicity (more toxic at higher temperatures) so directions should be followed carefully.

Chlorine can also be used to kill wild fish in ponds. Calcium hypochlorite (70% available chlorine) is used at a concentration of 10 ppm active ingredient. At this concentration, it would take 14 g/m^3 of calcium hypochlorite (39 lb per 1 acre-foot of water) to kill all the fish. Chlorine will rapidly breakdown in the water and is usually neutralized after a few days.

When filling ponds, the water may be low in DO. To raise oxygen levels, water should be splashed over screens or baffles. Mechanical aeration can also be used to bring the DO level up to an acceptable level. Paddlewheels (electric and tractor driven) and electrical aerators are often used for supplemental aeration (Figure 7 and 8).

Barkoh et al. (1993) found that filling largemouth bass ponds in stages, rather than all at once, allowed managers to reduce the total quantity of fertilizer applied while maintaining existing fertilization rates. This may be beneficial for fry and fingerling producers who use fertilization to ensure large zooplankton and phytoplankton blooms.

Fertilization

The fertilization of ponds can be affected by a number of biological, chemical, and physical factors and, therefore, responses are not always predictable. Biological factors to be considered include what type of fish and plants are present and what type of food is available for the fish. Chemical factors include hardness, alkalinity, pH, composition of bottom mud, and possible chemical interactions. Physical factors include

FIGURE 7. Paddlewheel aerator.

FIGURE 8. Electric aerator.

turbidity, rate of water exchange, temperature, depth of pond, and amount of shoreline. These factors need to be considered when deciding if a pond should be fertilized, since not all ponds will benefit from fertilization. Ponds should not be fertilized if they are turbid, muddy (light penetration <15 cm [6 in]), or have a high water exchange rate. Fertilization should also not be considered in ponds where the fish would not benefit from the zooplankton and phytoplankton produced. For example, food fish are fed a commercial diet and, therefore, will not benefit greatly from the zooplankton that fertilization will produce. Ponds with lower temperatures also may not produce enough zooplankton and phytoplankton to make it economically feasible to fertilize.

Fertilization involves the addition of inorganic or organic compounds, mainly in the form of phosphate and nitrogen, to encourage growth of food organisms (especially zooplankton) for pond fry and fingerlings. There are many different schools of thought on what kind and how much fertilizer should be used. Should you use organic or inorganic fertilizer or a combination of both? The answers to these questions depend upon the age and type of fish being raised, pond conditions, and availability of certain types of fertilizers. Phosphate is usually the main nutrient responsible for controlling phytoplankton blooms in ponds. The amount of fertilizer recommended for use in ponds has changed over time. Piper et al. (1982) recommend an initial application of 2,499 kg per ha (2,204 lb per acre) chicken manure, 1,729 kg per ha

(1,543 lb per acre) alfalfa meal, 494 kg per ha (441 lb per acre) of meat meal, and 124 kg per ha (111 lb per acre) of super phosphate. Anderson (1993) estimates that the fertilization rates recommended by Piper et al. (1982) yielded nitrogen and phosphate levels of 5,919 mg N/L and 2,451 mg P/L. These rates are significantly higher than those recommended by Culver et al. (1993) and Boyd (1997). Culver et al. (1993) demonstrates that the optimal inorganic fertilization rate for percid pond production was 600 mg N/L (NH_4^+ + NO_3^-) and 30 mg P/L as PO_4^{-3}. Using these inorganic fertilizer levels, in addition to filling ponds immediately before stocking and adding no organic fertilizer, Culver et al. (1993) increases percid total yield fourfold and survival tenfold. Boyd (1997) recommends that pond fertilization levels in freshwater contain 1 part nitrogen to 3 or 4 parts phosphate (N:P ratio of 1:1.3 or 1:1.7). Research at Auburn University in Alabama has demonstrated that large amounts of nitrogen in the fertilizer are not necessary. Even when using phosphorus-only fertilizer, research found that ponds still had N:P ratios that ranged from 5:1 to 10:1 (Boyd 1990).

Organic fertilizers can include composted plant material, meals, manure, slaughterhouse waste, and sewage (Table 1). Organic fertilizers should have a low carbon to nitrogen ratio to allow rapid decomposition and release of nutrients. They are recommended only for fingerling

TABLE 1. Composition of several organic fertilizers (Snow et al. 1964; Piper et al. 1982).

Fertilizer	Nitrogen	Phosphorus	Potassium	Protein	Carbohydrates
Alfalfa hay	2.37	0.24	2.05	14.8	33.5
Grass hay	1.12	0.21	1.20	—	—
Peanut vine hay	1.62	0.13	1.25	10.1	38.5
Cottonseed meal 36%	5.54	0.83	1.22	34.6	24.5
Cottonseed meal 43%	7.02	1.12	1.45	43.9	15.8
Fish meal	10.22	2.67	0.40	63.9	2.1
Peanut meal	6.96	0.54	1.15	43.5	31.3
Meat scrap	8.21	5.15	—	51.0	3.5
Soybean oil meal	7.07	0.59	1.90	44.2	29.0
Horse manure	0.49	0.26	0.48	—	—
Cow manure	0.43	0.59	0.44	—	—
Chicken manure	1.31	0.40	0.54	—	—
Sheep manure	0.77	0.39	0.59	—	—
Green Italian ryegrass	0.50	0.09	0.40	3.1	11.5
Green rye	0.42	0.10	—	2.6	12.9
Green oats	0.42	0.09	0.50	2.6	13.5
Green vetch	0.67	0.07	0.41	4.2	8.1
Green, white clover	0.82	0.09	0.38	5.1	6.6

production to enhance zooplankton population, particularly in new ponds. The advantages of using organic fertilizers include

- a shorter cycle for zooplankton production than inorganic fertilizers,
- they can be used as a supplemental source of feed,
- their decomposition liberates carbon dioxide, which plants use for growth, and
- they are an aid in clearing turbid waters (mainly silt).

Although manure (animal and plant) have long been used as fertilizers to increase pond production, inorganic fertilizers have many advantages. Inorganic fertilizers have a more consistent composition, are more widely available, they obtain a more predictable response, require smaller applications, use less labor, and are easier to store.

Inorganic fertilizer comes in many different forms. Phosphate fertilizers in the solid form should be placed on a platform and allowed to slowly dissolve. If broadcast over the pond surface, they will settle quickly to the bottom where they will be absorbed by the mud. Liquid fertilizers will dissolve quickly, but since they are heavier than water, they need to be sprayed over the surface and not poured into the pond. Tucker and Robinson (1990) recommended applying 2.3–4.5 kg per ha (2–4 lb per acre) of liquid fertilizer (10-34-0 or 13-38-0; nitrogen-phosphorus-potassium) every other day for 8–14 d or until a phytoplankton bloom appears. Kastner and Boyd (1996) found that controlled-release fertilizers (i.e., Scotts Company, Marysville, Ohio) had only 25% of phosphorus and nitrogen inputs compared with conventional fertilizers, but fish production was not significantly different between the two. Controlled-release fertilizers have a polymer coating that allows slow release of the nutrients, allowing less frequent and smaller applications compared with conventional fertilizers.

Urea and ammonia fertilizers are widely used as a nitrogen source since they are cheaper than other nitrogen fertilizers, particularly nitrate. Boyd (1997) has recommended using sodium nitrate as an alternative to urea or ammonia fertilizers. The advantages of nitrate fertilizers over ammonia:

- they are not acid forming;
- they do not cause oxygen demands;
- they are not toxic to fish or shrimp; and
- the presence of nitrate in the pond effluent is more desirable than ammonia.

Although nitrate fertilizers are more costly than urea or ammonia fertilizers, the advantages listed above may render them preferable in the long run.

Harvesting

Harvesting of ponds to remove fish can be performed by draining the ponds, seining, or a combination of both. Harvest efficiency of larger ponds by seining, in the channel catfish industry, averages 80%, therefore, the remaining 20% are either removed by draining the pond or are carried over to the next harvest. Fish not removed by seining in smaller research ponds are easily harvested from collection basins in the ponds after draining (Figure 6); draining large commercial ponds is more difficult. In a study comparing harvesting methods for channel catfish, Seok et al. (1995) found no differences in net fish production, mortality, average fish size at harvest, or feed conversion between seining with or without draining ponds over a 3-year period. Therefore, the method used depends on size and species of fish and production system.

Fish are sensitive to stress and shock during handling and harvest. To reduce stress that can cause disease and mortality, handling of the fish should be minimal. The best time to remove fish by draining the pond is in the fall, but if harvesting in the summer, the best time is at night or early morning. This will avoid the stress of increasing water temperature and direct sunlight. Well water pumped into the harvest basin of the pond will help regulate temperature. Aeration should also be supplied in ponds and transportation tanks. Feed should be withheld for, at least, 24 h prior to harvest. Fish should not be overloaded into nets or tanks as this can damage scales and skin, increasing the risk of disease and causing direct mortality from low oxygen levels. A 0.5–1.0% salt solution has been found to reduce stress and should be used in handling and transportation tanks (Mazik et al. 1991).

Harvesting Methods. Fry and fingerling are usually stocked in large numbers and are removed in a single harvest. In this method, the full pond is seined repeatedly to remove as many fish as possible. The pond is then drained and the remaining fish collected. In multiple harvests, fish are baited into a small area with food and confined with a seine. Fish are then removed with dip nets. Fall harvest is preferred due to cooler temperatures. If fish are harvested in the summer, precautions must be taken to reduce the stress from high water temperature and low DO (see above).

Larger fish, such as food size channel catfish, are usually harvested by seining. In commercial ponds, fish may be harvested periodically as they reach market size. In this continuous system, smaller fish are added to the ponds to replace the harvested fish. These ponds may be completely drained every 6–8 yr, usually only when levee repairs are needed. Generally, seines should be 1.5 times the maximum depth and at least 1.5 times the width of a pond. Typical channel catfish production ponds average 0.9–1.2 m (3–4 ft) in depth and are harvested with a seine that can run 2.4 m (8 ft) deep and 61–488 m (200–1,600 ft) in length. Seines are typically constructed from knotted nylon netting or polyethylene. Since catfish have spines that are easily caught in untreated nylon nets, these seines are coated with tar, asphalt, or a commercial product. Polyethylene nets do not need to be coated. Seines have a float line with floats (foam or plastic) every 46–61 cm (18–24 in) and a mud line along the bottom edge rather than a bare lead line. The mud line has 56.7-g (2-oz) lead weights every 30.5–46.0 cm (12–18 in), but tends to ride on the surface of the pond bottom rather than digging into the soft mud as a bare lead line would. There are three types of mud lines used on seines. Mud rollers consist of rollers every 7.6 cm (3 in) along the bottom of the seine. These rollers are usually rubber coated cylindrical weights. Regular lead weights are interspersed at 0.3 m (1 ft) intervals. Rope "many ends" mud lines consist of 10–15 pieces of 1.3–1.6-cm (1/2- to 5/8-in) nylon rope bound around the lead line to form at 7.6-cm (3-in) diameter bundle. The rolled-netting mud line consists of fine-mesh knotless nylon netting rolled bound around the lead line to form a 7.6-cm (3-in) diameter bundle (Tucker and Robinson 1990).

Long seines are usually pulled by tractors, whereas short seines can be hauled by hand. Since each pond is usually 4–8 ha (10–20 acres), tractors and hydraulic reels are used to pull the seines from one end of the pond to the other (Figure 9). Ponds are usually seined from the deep to the shallow end. As the seine is pulled the length of the pond, the float

FIGURE 9. Seining fish during harvest.

line should be closely watched. A disappearing float line indicates either a snag or that the mud line has dug into the bottom. To alleviate this problem, a boat equipped with a "seine catcher" is used to catch the float line and pull the seine forward, dumping the mud as it accumulates. One or two people should also stand on the mud line near shore to keep it from lifting and allowing fish to escape. Upon reaching the shallow end of the pond, the seine is reeled in to confine the fish. Two different methods, live cars and cutting seines, can be used to complete the harvest.

Live cars (or "socks") are attached to the seine used in the harvest of food size channel catfish (Figure 10). Live cars are mesh holding bags higher than 29.4°C (85°F) would necessitate lower holding densities (Tucker and Robinson 1990). Once the fish are contained in the live car, hydraulic boom mounted baskets holding 227–454 kg (500–1,000 lb) of fish can then be used to load the fish, eliminating the use of dip nets (Figure 11). The baskets are usually attached to a scale on the end of the boom so the fish can be weighed as they are loaded. The fish are released into the hauling tanks through a quick release door at the bottom of the basket (Figure 12).

Cutting seines can also be used to crowd fish for final harvest. Fish are seined as described above until they are confined to the desired area. The mud line is anchored to the bottom and the float line secured above

FIGURE 10. Live car or sock.

FIGURE 11. Channel catfish harvest.

FIGURE 12. Hydralic boom loading fish onto truck.

water level to prevent fish from escaping. Fish are allowed to grade for a desired time period. Aeration should be provided in the confinement area during this time. After grading, cutting seines are then used to crowd fish for loading onto the trucks. Further grading can also be done using cutting seines of different mesh size. Fish can then be loaded by dip nets or hydraulic boom mounted baskets.

Although boom baskets are the usual method of moving fish to a hauling tank, fish pumps are another option. Fish can be pumped into a sorting or grading bin before going into the hauling tank. This would allow small fish to be returned to the pond and the market size fish retained. This is advantageous for multiple harvest ponds where fish are of different sizes and in polyculture situations.

Watershed Ponds. Due to the irregular size and bottoms and greater depth of watershed ponds, the seining methods of earthen ponds cannot be used. Overall, the most effective method of harvesting is to drain the pond until it is shallow enough to seine, then seine and drain until the pond is empty. Because of this, most watershed catfish ponds are managed with a "batch" system and completely harvested only once a year.

Watershed ponds can also be effectively harvested with the trapping method. In this method, fish are baited with feed into one area of the pond. Fish can then be corralled with a seine and harvested. Another method is the drop-seine method developed at the Fish Farming Experimental Station, Stuttgart, Arkansas (Huner and Dupree 1984). This method is similar to the corral method, but sections of the seine are lifted off the pond bottom and the ends are attached to the shore. When the fish are baited into the feeding area, the net falls and the fish are trapped. These methods cannot be used more than once a week since the fish become wary of the net. To overcome this problem, harvesting can be rotated between ponds and in different areas within a pond. Other methods used include lift nets and submergible seines, panel traps, slat traps, hoop nets, and electrical stunning devices. See Huner and Dupree (1984) and Tucker and Robinson (1990) for more details on these methods.

Harvesting Scaled Species. Baitfish, bass, goldfish, and sunfish are some of the scaled fish species harvested from earthen ponds. Care must be taken to minimize damage from crowding since loss of scales can lead to infection and death. Therefore, harvesting of these species is usually done by draining the ponds and seining into a catch basin or kettle (Figure 6). Ponds are usually drained as quickly as possible to minimize stress and prevent temperature and DO problems. Seines are used to confine fish near the kettle, catch basin, or the deep end of the pond. Untreated, knotless seines need to be used to avoid abrading the skin and removing scales.

Fish should be moved in small amounts in containers full of water. Sloshing water will cause fish to be thrown against the container and each other. If dip nets are to be used, they should be small mesh, untreated, and knotless. A 1% salt solution is recommended when hauling and handling fish to reduce stress (Mazik et al. 1991).

Trapping is another method used in the harvest of scaled fish. The V-trap is the most widely used trap in warmwater aquaculture (Figure 13). The trap is usually placed in front of the outlet pipe and secured so it does not impede the draining of the pond. The traps are buoyant to allow them to fall with the water level. Fish are attracted to the V-trap by the current and can be easily harvested from the trap. The advantages of these traps include the better overall condition of the fish since they are not seined and they are less labor intensive. The disadvantage is trapping does not supply a reliable number of fish compared with seining.

Production

Production methods in earthen ponds are similar regardless of the species, and the procedures are applicable to warmwater, coolwater, and coldwater fish culture. Growth rates, carrying capacity, feeding, and inventory are just some of the details that need to be considered in fish production (See Chapter 2 for more details).

FIGURE 13. V-trap with glass box.

Raceways

Raceways are culture units in which the water flow is sufficient to induce a current to which fish respond and in which detrital material can be transported (Figure 14). The earliest raceways were earthen channels or streams fitted with screens, weirs, and bars to contain fish within a section. Water depth is typically 1 m (3.2 ft) or less for salmonids, but raceways of only 7.6 cm (3 in) in depth have been used for some turbot (Cripps and Poxton 1992). Earthen streams and channels would rapidly erode and were lined with wood and rocks to stabilize the structure. These early raceways were located in, or adjacent to, flowing water. The gravity flow of water was often dependent upon the amount of rainfall or snowmelt in the watershed. Raceways have historically been used to culture coldwater species such as salmonids and only recently have they been used to culture warmwater species such as catfish.

Raceways have advantages over other culture systems, such as ponds, including increased production of fish per unit space and ease of harvesting, feeding, grading, and disease treatment. Disadvantages include the vast volume of water required and the associated cost of pumping if recirculation is required.

FIGURE 14. Raceway production of golden trout.

Types of Raceways

A raceway is a linear fish production system with a continuous water flow from one end to the other. Raceways may be constructed from various materials such as earth, plastic, and metal, but the most common material used is concrete. Raceways are composed of individual sections. Sections are single fish grow-out units that average 3–5 m (10–16 ft) wide by 7–9 m (24–30 ft) long by 1.1 m (3.5 ft) deep. Minimum water volumes to each section range from 4,732–7,570 L (1,250–2,000 gal) per min (Mattei 1994; Tucker and Robinson 1990). Sections are arranged in a series or line called a tier. The number of tiers in a system depends upon water and land availability. A complete set of tiers followed by settling tanks and lagoons is the raceway system. Raceway sections can be arranged in single pass, parallel, and series configurations.

Single-Pass Raceways

Single-pass raceways (Figure 15) require a large volume of high quality water. In this raceway system, water flows through each section at 5,678–7,570 L per min (1,500–2,000 gal). Raceway tiers are constructed with a 0.6-m (2-ft) drop between tiers. This drop replaces approximately 50% of the oxygen consumed by fish in the previous section.

FIGURE 15. Single pass raceway.

Parallel and Series Raceways

In parallel and series raceways, water flows, usually by gravity, from a header pond into the raceways. The water exchange rate is about 6–7 exchanges per min. The water is discharged from the raceways into a series of ponds to settle solids, oxidize nitrogenous wastes, and add oxygen back to the water. Water is then pumped back into the header pond. In a parallel system (Figure 16), water flows from the header ponds to raceway sections that are parallel to each other and then exits. In series systems (Figure 17), water flows from the header ponds into raceways sections. The water flows from section to section until it exits into the ponds.

In-Pond or Floating Raceways

An innovation in raceway design, most appropriate for fry and fingerlings, is the floating raceway. This unit consists of a plastic or wooden rectangular box or a cylinder cut lengthwise and floated in a pond or other water body. Water for the raceway is pumped from the pond over a weir into the raceway (Figure 18) and passes through a fine screen before discharging back into the pond. Airlift pumps are used to move zooplankton-rich water from the pond into the raceway to provide natu-

FIGURE 16. Parallel raceways.

FIGURE 17. Series raceways.

FIGURE 18. Air-lift pumps supply water to floating raceway.

ral forage for fry and juvenile fish (Masser 1997a). Advantages offered by this system include the ability to

- maintain fish of different ages or sizes in a series of raceways in a common pond without mixing the sizes,
- optimize zooplankton production before fish are stocked,
- remove fish from any raceway without draining the pond and losing the plankton bloom,
- easily monitor and inventory fish in each raceway,
- easily protect fish from predators, and
- culture fish in open systems such as bays or lakes that are too large to manage as culture units.

Disadvantages include

- increased threat of theft or predation since fish are confined to one area, and
- mechanical failures could cause loss of fish.

Operation

Water Flow

The growth and survival of fish in raceways is greatly influenced by the water quality, flow, and hydraulics of the system. Depending upon water velocity and physical structure of the raceway, water flow may: (1) be laminar, with water moving faster in the center and more slowly along the sides and bottom, (2) move as a nonmixing plug from the influent to the effluent, or (3) be turbulent and lightly mixed with currents moving some of the water from the effluent end back upstream to the influent end. Plug-flow conditions are the most difficult to achieve, yet provide the highest quality with the minimum flow.

Raceways arranged in series are commonly placed with an elevation drop of about 0.6 m (2 ft) between units to provide aeration, and baffles may be placed in a raceway to increase velocity and remove solids (Figure 19). Baffles have been shown to significantly reduce mortality in raceways, but did not alter DO or growth of fish.

In a typical raceway, water flows down the long axis of the unit and water quality degrades (DO decreases, ammonia and carbon dioxide increase) progressively from the influent to the effluent. In a cross-flow raceway, water moves across the short axis of the unit and waste products are quickly removed. Cross-flow raceways operate similarly to a set of parallel raceways from which the common walls have been removed. The water is introduced along one long side of the raceway and withdrawn from the opposite side.

FIGURE 19. Baffles in a raceway.

The production of fish is largely dependent upon the quality of water available. In raceways, water is discharged from the tail end to remove metabolic wastes, uneaten food particles, and other contaminants. Aerated freshwater is added at the head end to supply oxygen and to replace the water discharged from the tail end of the raceway. The oxygen required in a raceway is influenced by the size and number of fish. The metabolic rate and, therefore, the production of ammonia, carbon dioxide, food material, and other waste products is greater on a per weight basis for small fish than for large fish and also increases with temperature. These relationships are represented by the equation of Piper et al. (1982) as follows:

$$F = W / (L \times I)\,, \tag{1}$$

where F = the flow index, W = weight of fish in pounds, L = length of fish in inches, and I = water inflow in gallons per min. The metric form of this equation (denoted by the subscript "m") is

$$F_m = W_m / (L_m \times I_m)\,, \tag{2}$$

where F_m = the flow index, W_m = weight of fish in kilograms, L_m = length of fish in centimeters, and I_m = water inflow in cubic meters per min.

The flow index is derived empirically by establishing the maximum weight of fish of any size that can be reared in a raceway with a constant inflow of water. Once the flow index has been established, the formula can be rewritten, as follows, to predict weight of fish produced for any given size of fish or flow of water:

$$W_m = F_m \times L_m \times I_m . \qquad (3)$$

This form of the equation is used to calculate the weight of fish that can be safely maintained as fish length increases or water flow changes. Most hatcheries of the U.S. Fish and Wildlife Service are operated with a flow index equivalent to 10–25 when F_m is calculated in metric units (1.0–2.5 when F is calculated in English units). Hatchery efficiency declines at flow indices below 10, and water quality deteriorates at indices above 25.

Engineers measure water turbulence in units defined as Reynolds numbers. Reynolds numbers less than 2,000 indicate water flow of minimum turbulence and plug-flow conditions, while Reynolds numbers greater than 2,000 indicate turbulent flow with downstream water mixed with upstream water in a mixed flow reactor (Wheaton 1977). Most raceways for salmonids have been designed for fish loads of 32–48 kg/m^3 (2–3 lbs/ft^3), but densities of 544 kg/m^3 (34 lbs/ft^3) have been achieved experimentally (Table 2). The usual swimming speed for larval and juvenile fish ranges from 1 to 10 body lengths/s at flows characteristic of Reynolds numbers of 10–10,000 (Webb and Weihs 1986). Flow rates in raceways should, therefore, normally equal about ½ body lengths/s for the fish being cultured. At 50 water exchanges per hr, velocity in the floating raceway would be 1.7 m/s (5.5 ft/s). However, if water was

TABLE 2. Comparison of calculated water exchanges, calculated water velocity, and achieved fish loads in intensive culture systems of various sizes.

System	Tank volume m^3 (ft^3)	Inflow (gpm) 1/min	Water exchanges per hr.	Velocity cm/s (ft/min)	Fish load (density) Kg/m^3 (lb/ft^3)	Kg/1pm (lb/gpm)
Hatchery	0.006	2.27	50.0	1.1	544.0	1.5
jar[1]	(0.22)	(6.0)	–	(0.22)	(34.0)	(6.0)
Gallatin	19.8	1136.0	3.75	14.2	60.0	1.0
raceways [2]	(700.0)	(300.0)	–	(2.8)	(10.0)	(22.3)
Georgia	113.0	1892.0	1.0	8.5	12.0	0.72
raceways [3]	(4000.0)	(500.0)	–	(1.67)	(0.75)	(6.0)
Floating	19.8	4953.0	15.0	50.8	288.0	1.15
raceway [4]	(700.0)	(1308.0)	–	(10.0)	(18.0)	(9.6)
Cross-flow	1.72	52.7	185.0	4.9	21.7	0.71
raceway [5]	(60.6)	(13.9)	–	(9.6)	(1.36)	(5.9)

Sources: 1, Buss et al. 1970; 2, Collins et al. 1984; 3, Hill et al. 1973; 4, Wayne Caillouet, Clarendon, Arkansas, personal communications; 5, Watten and Beck 1987.

introduced on one side and flowed across the width of the raceway without turbulence in a plug-flow pattern, the water velocity would be less than 10 cm/s (3.9 in/s). It is possible to affect a high exchange rate, yet maintain a low velocity of water flow in a small culture unit. For a successful cross-flow raceway, the water would need to enter uniformly through one side, flow across the raceway, and exit uniformly on the other side to keep the velocity at a low level.

Limiting Factors in Raceways

Production of fish in raceways is limited by several factors, such as low dissolved oxygen (DO), the accumulation of metabolic waste products, high temperature, disease, unfavorable water chemistry, feeding rate, and fish behavior. The DO frequently becomes the first and one of the most serious limiting factors. Although oxygen cycles over day–night periods in ponds and other bodies of water, it can be closely regulated and controlled in raceways. The amount of oxygen required depends on both water temperature and fish size. The DO requirements of fish typically double for every 10°C (18°F) increase in temperature. Because smaller fish have a higher metabolic rate, they consume considerably more oxygen per unit of body weight than do large fish.

Since oxygen consumption of fish is influenced by the DO concentration in water, when the DO concentration is low, fish reduce their activity, cease to feed, and consume much less oxygen than when the DO concentration is high. Well-fed fish also require more oxygen than do fish that have not been fed, and oxygen consumption peaks shortly after feeding. In heavily stocked culture units, the feeding activity of fish may rapidly deplete DO at the feeding site, and, as the DO concentration drops, the fish cease feeding. In channel catfish, oxygen declines of 1.5–2.0 mg/L (ppm) in a few minutes may stress channel catfish and cause them to cease feeding. Therefore, in raceways or other intensive culture systems, feed must be offered to fish over a large area, feeding activity must be spread over time (multiple feedings per day), or additional oxygen must be added at the time of feeding. Demand feeders—units that deliver feed only when activated by the fish—have been effectively used to overcome the problems of feeding fish in raceways (Figure 20).

Supplemental Aeration

Oxygen levels are usually not a problem in raceways unless the water is recirculated.

Because water does not wear out, but simply accumulates waste products, various systems have been used to replenish oxygen, filter waste, and recycle and reuse it (Wheaton 1977; Muir 1982; Parker 1983).

FIGURE 20. Demand feeders in raceways.

Oxygenation can be achieved by aeration or by injecting pure oxygen into the system. Aeration devices include air stones, spargers, packed columns, baffles, airlift pumps, agitators, spray nozzles, and splashboards (Figure 21).

Supplemental aeration may be needed if the water supply has high levels of nitrogen. Supersaturated levels of nitrogen and oxygen can cause gas bubble disease, similar in effect to a diver with the bends. Small nitrogen bubbles can form in the blood and block blood flow. Severe cases can cause significant mortality. Aeration, using a packed column, can reduce nitrogen levels below saturation. Packed columns are usually tubes "packed" with screen or other loose material to help aerate incoming water. Gas supersaturation can be a problem in winter or when cold water is heated, but is rarely a problem at temperatures above 20°C (68°F). Although the saturation curves are similar for oxygen and nitrogen, it is the combined total gas pressure that should be of primary concern to aquaculturists managing systems in which fish develop gas bubble trauma.

Stocking

The flow index (described above) is used to estimate the maximum loading or maximum weight of fish at a given size and water flow rate. Pro-

FIGURE 21. Kettles with catch basins and built-in raceways for sorting fish in a Texas Parks and Wildlife hatchery.

ducers can use this index to determine the best stocking density of fish for their particular system.

Feeding

Demand feeders seem to be the norm in raceway culture. Producers have found that, occasionally, fish can empty demand feeders as soon as they are filled. This burst of feeding activity can temporarily deplete oxygen in the lower raceway tiers and needs to be closely watched. Higher volumes have been fed after considering water temperature, fish size, and, most importantly, fish density in the raceway section immediately below the section being fed.

Automation and Control in Intensive Culture

A major advantage of cultivating fish in raceways is the high degree of automation and control that can be achieved, as compared with that in other systems. In a highly automated commercial trout hatchery in northern Italy, 12 parallel raceways were spanned by a rolling bridge equipped with a feeder above each unit. Feed, automatically loaded from storage bins at the end of the raceways into feeders on the bridge, was delivered to the fish throughout the raceways as the bridge rolled from one end to the other. The rolling bridge also was equipped with crowding screens, grading bars, and a series of brushes and pumps used to clean the raceways as the bridge slowly traveled over the 12 units. The entire system was electrically operated. The intensively cultured fish were protected from bird predation by netting material placed above the raceways and above the traveling bridge.

Labor costs have been greatly reduced in these culture units through the use of automation and mechanization. In areas where labor is not expensive, this degree of automation might not be economical, especially if skilled specialists are not readily available to maintain the equipment.

Harvesting

Harvesting and grading can be easily accomplished by using grading bars and crowding fish to one end of the raceway. Fish can be removed with a basket attached to a crane, backhoes with lift baskets, fish pumps, or dip nets. Catfish producers are harvesting 4,536–6,350 kg (10,000–14,000 lb) of fish from each raceway section. Based on water flow, production is 640,746–800,932 kg/m^3/s (40,000–50,000 lbs/ft^3/s) (Tucker and Robinson 1990).

Production

Producers rearing fish in intensive culture systems face risks proportionate to the degree of complexity of the system and the stocking density. As in almost any investment, the greater the risk, the greater the potential return. Producers continually push production systems to the upper limit and must continually be alert for signs of trouble to avoid disaster. In intensively stocked systems, there is little room for error, and management determines the likelihood of success or failure.

Planning and attention to detail increase the chances for success in intensive fish culture. Shepherd and Morris (1987) provide a checklist and a review of emergency procedures for culturists. Raceways in which water flow is maintained by pumps, or other electromechanical devices are used to supply oxygen or filter water, are extremely vulnerable to mechanical failure. In addition to the obvious requirements for backup systems, emergency power, and more, producers must locate a source of feed and fingerlings before they establish a new business. They also must locate the market for their fish and determine how to get the fish from the production facility to the market. Planning and attention to detail are the keys to success.

Cages

To minimize confusion about the term cages, sea cages, marine enclosures, and net-pens, in this chapter, we will refer to freshwater enclosures as cages and seawater enclosures as net-pens. Cage culture is the method of raising fish in enclosures that are either floating on the surface or are attached to the bottom of a water column (Figure 22). The advantages and disadvantages of cage culture are listed below (Masser 1997b, Tucker and Robinson 1990).

Advantages:

- cages can be placed in water bodies not suitable for seining or systems that are not easily drained such as farm ponds, lakes, or strip pits;
- cage culture is an inexpensive way to develop fish culture skills before investing in more expensive production systems;
- pond construction costs are eliminated when existing water bodies are used;
- growth and disease problems are easier to monitor; and
- harvesting is easier.

Disadvantages:

- total production is less than in ponds;

FIGURE 22. Cages in a pond.

- water quality problems due to high stocking densities can occur;
- vandalism problems and damage to cages can cause fish to escape;
- disease problems can spread quickly; and
- daily access to cages is needed.

In most situations, the disadvantages outweigh the advantages. Cages are found in freshwater ponds, lakes, and reservoirs and can be constructed of plastic, wire, netting, or wood. The type and size of cage depends on the size and species of fish and environmental conditions.

Types of Cages

Wood

Early cages were made of wood or bamboo, but due to their high rotting potential, they have generally been replaced by more impervious materials. Wooden cages were considered good starter cages due to low cost, ease of construction and availability of materials. Drawbacks include the fact that since wood rots in water, these cages have less than half the life expectancy of other types. Wood also has increased biofouling, causing slippery conditions and increased drag. Due to these problems, wood cages have been replaced by cages built of more resistant material.

Polyvinyl Chloride

Polyvinyl chloride (PVC) cages are the most popular freshwater cages. These cages have a PVC frame and nylon mesh sides and bottom. The most common material is raschel nylon or polyester netting. Raschel netting is woven like cloth and not twisted or knotted like ordinary mesh netting. Most mesh is plastic coated for durability. These cages do not have the size constraints of other types of cages, are lightweight, inexpensive, fairly durable, and easy to handle. The main problem involves damage from predators. This can be alleviated by the addition of a larger rigid cage (such as wire mesh) around the mesh cages.

Plastic

Plastic cages are medium priced and can be circular or square. They are resistant to corrosion and biofouling and are buoyant. Since they are flexible, they do well in larger water bodies with wave action. Problems include stress points at the joints. Due to their flexibility, they are not ideal work structures. Workers leaning on them to pull up a net can end up in the water if they lean too heavily on the sides.

Wire Mesh

Wire mesh, such as galvanized and plastic-coated welded wire mesh, is durable, rigid, easier to clean than flexible material, and more resistant to fouling. Wire mesh is also more expensive than netting, heavy, and more cumbersome. Flotation is needed with wire mesh cages if they are not anchored above the pond bottom.

Cage Design and Construction

Cage size can vary from 1 m^3 (35.3 ft^3) to greater than 28.3 m^3 (1,000 ft^3), although smaller cages are more common and economical. Larger cages cost less per unit volume, but production per unit volume decreases due to a lower water exchange rate. Cages can either be standing or floating, cylindrical, square, or rectangle. The most common cage sizes are: cylindrical, 1.2 × 1.2 m (4 × 4 ft) (diameter × depth); square, 1.2 × 1.2 × 1.2 m (4 × 4 × 4 ft) and 2.4 × 2.4 × 1.2 m (8 × 8 × 4 ft) (length × width × depth); and rectangle, 2.4 × 1.2 × 1.2 m (8 × 4 × 4 ft) and 3.7 × 1.8 × 1.2 m (12 × 6 × 4 ft) (length × width × depth). Standing cages are anchored by stakes driven into the bottom substrate. Floating cages require flotation devices to remain on the surface of the water. Flotation devices can be sealed PVC pipe, Styrofoam, metal, or plastic bottles.

The mesh size of the sides and bottom of the cage have a significant impact on production. The mesh size needs to be large enough for good circulation in order to remove waste and provide oxygen, but not so large that fish escape or wild fish can enter the cages. A minimum mesh size of 1.3 cm (½ in) should be used to maintain good water circulation in the cage while holding small fingerlings (10–13 cm [4–5 in]). Larger mesh sizes can be used with larger fingerlings.

Operation

Placement

Cages can be placed individually in a water body or be linked together into groups. They should be located where water movement is greatest; usually the windward side, but care should be taken since too much movement can cause problems. Stagnant water areas should be avoided. Cages should be at least 1 m (3.2 ft) off the bottom and 2.7–4.6 m (9–15 ft) apart to optimize water quality and alleviate problems from accumulated food and the resulting low DO. Covers are important to prevent fish loss from bird predation or jumping. In lieu of covers, the cage sides can be elevated 0.3–0.6 m (1–2 ft).

Larger lakes and ponds (greater than 2 ha [5 acres]) are usually better suited for cage culture than smaller systems since larger systems are more stable and are able to handle the fish waste without a negative impact. Highly eutrophic systems are an exception due to the high organic content. Smaller ponds (0.4–2 ha [1–5 acres]) can be used, but emergency aeration and water exchange are required. Ponds should not have chronic weed problems, runoff from crops, or livestock; livestock should not have direct access to the pond.

Stocking

The rate of stocking fish in cages depends on species, cage size, mesh size, pond size, aeration, and final desired fish size. Masser (1997b) suggests calculating the stocking density of cages on the kg per surface ha (lb per surface acre) of pond and kg/m^3 (lbs/ft^3) of cage that can be reared. Ponds without supplemental aeration produce from 567 to 1,701 kg per ha (500–1,500 lb per surface acre). With aeration, this number increases to 2,802–4,484 kg per ha (2,500–4,000 lb per surface acre). Catfish cage production will generally range between 80 and 128 kg/m^3 (5–8 lbs/ft^3) (Masser 1997b).

In addition to using the proper densities, other management practices that can impact fish production include variability in fish weight at stocking and periodic fish grading. Schwedler et al. (1989) found that stocking fish within a 3 cm (1 ¼ in) length group can significantly reduce variability in weight at harvest. Lazur (1996) indicates that periodic grading and removal of market size catfish during the growing season could increase cage production and lower feed conversions. Care should be taken to not unduly stress the fish during grading.

Feeding

Feeding rings are used to retain floating feed to reduce waste. Floating rings are made of small size mesh (0.3 cm [1/8 in or less]) and should occupy only a portion of the cage surface. The feeding ring should extend down in the water at least 46 cm (18 in). The size of the ring should be small enough to not impede water movement through the cage, but large enough so all fish can feed. Research has indicated that feeding catfish in cages at dusk improves feeding efficiency (Masser 1997c). This feeding cannot be done without nighttime aeration. Dusk feeding begins 1 h before dusk and continues for at least 1 h after sunset. Feeding the correct amount is very important as overfeeding wastes feed and can deteriorate water quality. Generally, most warmwater fish will feed to satiation within 10–30 min at or above 21.1°C (70°F). Since caged fish may not feed immediately upon being offered feed, a good practice is to offer half of the previous day's amount. Add additional feed at 20–30 min intervals until feed is left untouched. This practice will avoid overfeeding (Masser 1997c).

Behmer et al. (1993) evaluated a cone-bottom cage to facilitate removal of solid waste and phosphorus from cage-reared rainbow trout.

Removal of solids from the cages by pumping was effective in stagnant water, such as a pond, but not as effective in strong currents. This method may be advantageous in small ponds where water exchange is not readily available.

Fouling

Cage fouling is more of a problem in marine environments than brackish and freshwater. Freshwater cages can be cleaned by scraping the outside of the cage. Care should be taken, as this process can be stressful to the fish if performed too often. Carp or tilapia can also be added to the cages, depending on the climate, to remove biofouling from the inside of the cage, if this is a problem.

Harvesting

Fish are harvested from cages using a long-handled dip net. Fish should be taken off feed 2 d before harvesting to allow the digestive system to empty. The advantage of harvesting in cages is the ease of removing fish since smaller cages can be brought to the shore for complete harvesting. In larger cages, fish must be crowded with a screen or net and may be partially lifted out of the water to facilitate harvest.

Production

Generally, yields (kg/ha or lb/acre) are greater in pond culture than cage culture, but in waters unsuitable for open pond culture, cages are an alternative practice. Production in cages can be maximized by stocking fingerlings of similar size, grading fish, removing larger fish, and providing aeration. For channel catfish, cage production will normally range between 80 and 128 kg/m^3 (5–8 lbs/ft^3). Small cages (< 7.6 m^3 [270 ft^3]) can have maximum production of about 224 kg/m^3 (14 lbs/ft^3), and large cages (> 7.6 m^3 [270 ft^3]) can have a maximum production of about 176 kg/m^3 (11 lbs/ft^3)(Masser 1997c). See Chapter 2 for more information.

Net-Pens

Net-pens are the most economical way of rearing fish, such as salmonids, in a marine environment (Figure 23). There are four designs of net-pens: fixed, floating, submersible, and submerged. Fixed net-pens are anchored to the sea floor, floating net-pens have buoyant material to keep then on the water surface, submersible net-pens can be lowered below the water surface, and submerged net-pens are anchored below

FIGURE 23. Net-pens.

the water surface. Advantages of net-pens include containment of fish for ease in harvesting, feeding, and consistent temperature, salinity, and DO in offshore sites. Disadvantages include exposure to predators, vandalism, pollution, and damage from storms.

Types of Net-Pens

Typical net-pens have supporting structures of wood, galvanized metal, PVC, or plastic. A typical farm may consist of 50 pens arranged in a rectangle or square, each pen from 3 to 6 m (10–20 ft) wide and 6–12 m (20–39 ft) long. Walkways of 1–2 m (3.2–6.6 ft) can provide access to the pens. Circular pens can also be used and are usually clustered together. Nets are hung from railings on the inside of the walkways. Since these railings are 1 m (3.2 ft) above the walkways, the nets are high enough to keep fish from jumping out of the pen. Nets are generally 5–8 m (16.5–26.0 ft) deep with a 1–3 cm (1/2 to 1 1/4 in) mesh size.

Wood

Early net-pens were built of wood and are considered good starter pens due to low cost, basic design, and easy construction. Disadvantages include shorter life span than other materials, attraction of algal growth, and slippery surfaces when wet. Wood cages must be used in sheltered inshore sites and are smaller in size than those made of steel, plastic, or rubber.

Plastic

Plastic net-pens are probably the most popular pen in use today. Plastic pens are light and, therefore, almost neutral in buoyancy. These pens are mid-priced and come in rectangle and circular shapes. Advantages of plastic pens include the fact that they are corrosion proof, do not support algal growth, and can sustain heavy wave action well. The main disadvantage is the stress points at joints, similar to steel net-pens.

Steel

Steel net-pens tend to be used more in open-ocean fish culture than wood or plastic ones. These pens are more expensive, but tend to be easier to manage due to wide walkways and sturdy handrails. Steel net-pens need to be manufactured with a region's weather in mind. Steel net-pens in the East Coast region need to withstand heavy icing conditions and gale force winds. Since the joints are the stress points of these pens, designs need to take into consideration the force of wave action and weather patterns in a particular area in order to make a system weather-secure.

Rubber

Rubber net-pens are constructed of high-density rubber tubing clamped together to provide support in the horizontal plane. These pens hold large volumes of fish and are able to ride heavy seas with less stress than other types of pens. However, the large size of these pens (20 m [65.6 ft] octagonal pens with 30 m [98 ft] deep nets), demand a different type of management than smaller pens.

Operation

Location

Net-pens need to be located in areas of good water quality. The fish will produce ammonia and reduce DO within the pens, so sufficient currents are needed to provide a rapid water exchange. If a pen is well flushed, the only change in water quality should be within the pen itself. Since the currents will be reduced within the pens due to drag, a current velocity of at least 10 cm per s (4 in per s) should be present throughout the tidal cycle. Maximum currents should not exceed 50–100 cm per s (1.6–3.3 ft/s). Currents higher than this can lead to structural damage and stress to the fish. Pens should be at least 1–5 m (3.2–16.4 ft) above the ocean floor during low tide (Stickney 1991). This will allow dispersal of excess feed and fecal material that can result in low DO and disease.

Nearshore net-pens should be sheltered from high winds and severe storms. Open ocean pens should be designed to withstand the weather for a particular area. High winds and waves can cause structural stress on pens if not properly designed.

Stocking

Stocking densities are variable, depending on locale and, especially, tidal exchange. A typical Norwegian net-pen is usually stocked with smolts at a density of approximately 1 kg/m^3 (0.06 lbs/ft^3). After a few months, fish are redistributed into three pens. Average densities range from 15 to 20 kg/m^3 (0.9–1.2 lbs/ft^3), with some as high as 50 kg/m^3 (3 lbs/ft^3)(Stickney 1991). In net-pen productions in the Puget Sound near Washington State salmon are reared in freshwater hatcheries for 18 months and then stocked into net-pens for 7–12 months at densities of 15 kg/m^3 (0.9 lbs/ft^3). Higher densities tend to increase disease and stress-related problems. Since pens are fairly inexpensive, the risk of high densities does not outweigh any possible economic benefit.

Feeding

Feed costs in the salmon industry represent 45–60% of the operating costs of a commercial net-pen farm. Salmonids are fed by hand 4 h/d in the winter and 8 h/d in the summer. The industry feed conversion ratio (FCR = kg of supplied food/kg weight gain) is 1.5 for Atlantic salmon and 2.0 for chinook salmon (Isaksson 1991). Hand feeding is a labor-intensive method used to ensure maximum feeding with minimal waste. Fish are fed until they no longer actively feed. Other feeding methods include demand feeders and timer-controlled feeders. Demand feeders allow the fish to determine the feeding frequency, whereas timer-controlled feeders provide a set amount of feed at specific times of the day. These two methods are much less labor intensive, but mechanical failure due to humidity is a possibility. High-tech methods, including hydroacoustics and image analysis, have also been used to determine uneaten feed in an attempt to regulate feeding. When uneaten feed is detected, feeding is terminated, resulting in better feed conversions and less waste.

The amount of feed fed and frequency of feeding varies with temperature and fish size. Particle size is very important. Feed too large cannot be consumed and feed too small may not be consumed. In either case, feed is wasted, water quality may be impaired, and fish growth is reduced. Dry, wet, and semimoist feeds are all used in the industry. Dry feeds are used in Norway with about 85% of the salmon receiving dry feeds. Dry feed is more convenient than wet or semimoist feeds and gives better growth at temperatures above 5°C (41°F) (Isaksson 1991). Dry feeds are usually fed with automatic feeders. Wet and semimoist feeds are still commonly used in Atlantic salmon culture. These feeds usually consist of chopped fish mixed with binders, vitamins, and minerals. Wet and semimoist diets are usually fed by hand, although some automatic feeders have been developed to handle these feeds.

In situations where flesh color is important, as with salmon, pigments can be added to the feed to provide an acceptable color. Two commercially synthesized pigments, canthaxanthin and anthaxanthin, are routinely added at 4–6 mg per 100 g of feed. The feed must be fed at least 4–6 months before harvest.

Net Fouling

The multifilament net used in net-pens is an ideal material for fouling. The large surface area is an ideal place for biological organisms to attach, and the large nutrient load of the pens from feed and waste material allows rapid growth. Depending on the geographic area, nets must be removed for cleaning; cleaning can be required every 5–8 d in the summer. Removing nets for cleaning can damage nets, is very labor- and cost-intensive, and there is a risk of lost stock and disturbance of feeding regimes.

Antifouling agents can be used, but these products can cause environmental concerns since many are toxic, copper-based formulas. Also, these agents have a limited life span and are not effective against masses of drifting algae. Underwater scrubbing is not common due to the expense and the fact that underwater scrubbing of surfaces can select for fouling organisms that grow faster after repeated cleanings. New technology is being developed for in situ net cleaners, but cost and effectiveness is still a factor.

Harvesting

Harvesting from net-pens is an essential part of the operation. Harvest techniques can vary from dip nets for smaller inshore pens to rafts, pumps, and cranes for large offshore pens. Harvested fish, especially at offshore sites, need to be transported and processed in a timely manner. This may necessitate the use of rafts or barges to store and haul the fish to the processing facility.

Production

Production in the Atlantic Ocean is mainly concentrated in the Bay of Fundy in New Brunswick and Nova Scotia, Canada. The major limiting factor is the availability of suitable sites. Most of the coastal Atlantic is too cold and some suitable sites can have occasional lethal winter temperatures. Smolt supply is also a limiting factor that needs to be addressed.

Production in the Pacific Ocean is concentrated in British Columbia, Canada, and Washington State. The main species produced are Pacific salmon, primarily coho salmon and chinook salmon. It is generally accepted that producing market size fish is more difficult in the Pacific than the Atlantic. Pacific salmon are more prone to disease, and experience has shown, in British Columbia, that only 5–8 kg/m^3 (0.3–0.48 lbs/ft^3) of Pacific salmon can be grown in net-pens as compared with 15–20 kg/m^3 (0.9–1.2 lbs/ft^3) for Atlantic salmon. Harvest of salmonids in the Puget Sound is higher at approximately 15 kg/m^3 (0.9 lbs/ft^3). This includes pan-size fish (230 g [0.5 lb]) and larger 2- to 5-kg (4.4- to 11-lb) fish. Pan-size fish can be produced from 4 g (0.14 oz) fingerlings in approximately 280 d at 15°C (59°F) (Stickney 1991).

The use of local stocks is important in Pacific Ocean net-pen production. Atlantic stocks have been extremely susceptible to diseases that are not found in the Atlantic coast. Since conditions in British Columbia and Washington State are similar to Norway, the major producer of salmonids in net-pens, success should be achievable with planning and the use of native stocks.

References

Anderson, R. O. 1993. Effects of organic and chemical fertilizers and biological control of problem organisms on production of fingerling striped bass, *Morone saxatilis*. Journal of Applied Aquaculture 2:119–149.

Ayub, M., C. E. Boyd, and D. Teichert-Coddington. 1993. Effects of urea application, aeration, and drying on total carbon concentrations in pond bottom soils. The Progressive Fish-Culturist 55:210–213.

Barkoh, A., R. O. Anderson, and C. F. Rabeni. 1993. Effects of pond volume manipulation on production of fingerling largemouth bass, *Micropterus salmoides*. Journal of Applied Aquaculture 2:151–170.

Behmer, D. J., R. W. Greil, D. C. Greil, and B. P. Fessell. 1993. Evaluation of cone-bottom cages for removal of solid wastes and phosphorus from pen-cultured rainbow trout. The Progressive Fish-Culturist 55:255–260.

Boyd, C. E. 1990. Water quality in ponds for aquaculture. Alabama Experimental Station, Auburn University, Auburn, Alabama.

Boyd, C. E. 1997. Practical aspects of chemistry in pond aquaculture. The Progressive Fish-Culturist 59:85–93.

Boyd, C. E., and K. Masuda. 1994. Characteristics of liming materials used in aquaculture ponds. Journal of the World Aquaculture Society 25(1):76–79.

Buss, K., D. R. Graff, and E. R. Miller. 1970. Trout culture in vertical units. The Progressive Fish-Culturist 32:187–191.

Collins, C. M., R. L. Schweinforth, and G. L. Burton. 1984. Summary of channel catfish and rainbow trout production at the Gallatin Waste Heat Aquaculture Facility, 1979–1980. Tennessee Valley Authority, Gallatin Steam Plant, Gallatin, Tennessee.

Cripps, S. J., and M. G. Poxton. 1992. A review of the design and performance of tanks relevant to flatfish culture. Aquacultural Engineering 11:71–91.

Culver, D. A., S. P. Madon, and J. Qin. 1993. Percid pond production techniques: timing, enrichment, and stocking density manipulation. Journal of Applied Aquaculture 2:9–31.

Hill, T. K., J. L. Chesness, and E. E. Brown. 1973. Growing channel catfish, *Ictalurus punctatus* (Rafinesque), in raceways. Proceedings of the Southeast Association of Fisheries and Wildlife Agencies 27:488–499.

Huner, J. V., and H. K. Dupree. 1984. Third report to the fish farmers. United States Department of the Interior, Fish and Wildlife Service, Washington, D.C.

Isaksson, A. 1991. Culture of Atlantic salmon. Pages 87–129 *in* R. R. Stickney, editor. Culture of salmonid fishes. CRC Press, Boca Raton, Florida.

Kastner, R., and C. E. Boyd. 1996. Production of sunfish (*Lepomis spp.*) in ponds treated with controlled-release fertilizers. Journal of the World Aquaculture Society 27:228–234.

Lazur, A. M. 1996. The effects of periodic grading on production of channel catfish cultured in cages. Journal of Applied Aquaculture 6:17–24.

Masser, M. P. 1997a. In-pond raceways. Southern Regional Aquaculture Center Publication Number 170, Mississippi State.

Masser, M. P. 1997b. Cage culture: species suitable for cage culture. Southern Regional Aquaculture Center Publication Number 163, Mississippi State.

Masser, M. P. 1997c. Cage culture: handling and feeding caged fish. Southern Regional Aquaculture Center Publication Number 164, Mississippi State.

Mattei, E. 1994. Twenty years and holding: an assessment of catfish raceways. Aquaculture Magazine. 20(3):48–55.

Mazik, P. M., B. A. Simco, and N. C. Parker. 1991. Influence of water hardness and salts on survival and physiological characteristics of striped bass during and after transport. Transactions of the American Fisheries Society 120:121–126.

Muir, J. F. 1982. Recirculated water systems in aquaculture. Pages 357–446 *in* J. R. Muir and R. J. Roberts, editors. Recent advances in aquaculture. Room Him, London.

Parker, N. C. 1983. Air-lift pumps and other aeration techniques. Pages 24–27 *in* C. S. Tucker, editor. Water quality in channel catfish ponds. Southern Cooperative series Bulletin 290. Mississippi Agriculture and Forestry Experiment Station, Mississippi State University.

Piper, R. G., I. B. McElwain, L. E. Orme, J. P. McCraren, L. G. Fowler, and J. R. Leonard. 1982. Fish hatchery management. United States Department of the Interior, Fish and Wildlife Service, Washington, D.C.

Schwedler, T. E., J. R. Tomasso, and J. A. Collier. 1989. Production characteristics and size variability of channel catfish reared in cages and open ponds. Journal of the World Aquaculture Society 20:158–161.

Seok, K., S. Leonard, C. E. Boyd, and M. F. Schwartz. 1995. Water quality in annu-

ally drained and undrained channel catfish ponds over a three-year period. The Progressive Fish-Culturist 57:52–58.
Shepherd, B. G., and J. G. Morris. 1987. A review of practical emergency procedures for fish culturists. Aquaculture Engineering 6:155–169.
Snow, J. R., R. O. Jones, and W. A. Rogers. 1964. Training manual for warmwater fish culture. U.S. Bureau of Sport Fisheries and Wildlife, National Fish Hatchery, Marion, Alabama.
Stickney, R. R. 1991. Growout of pacific salmon in net-pens. Pages 71–83 *in* R. R. Stickney, editor. Culture of salmonid fishes. CRC Press, Boca Raton, Florida.
Tucker, C. S., and E. H. Robinson. 1990. Channel catfish farming handbook. Van Nostrand Reinhold, New York.
Watten, B. J., and L. T. Beck. 1987. Comparative hydraulics of a rectangular cross-flow rearing unit. Aquacultural Engineering 6:127–140.
Webb, P. J., and D. Weihs. 1986. Functional locomotor morphology of early life history stages of fishes. Transactions of the American Fisheries Society 115:115–127.
Wheaton, P. R. 1977. Aquacultural engineering. Wiley-Interscience, New York.
Wynne, F. 1993. Liming of ponds. Journal of the World Aquaculture Society 24(1):15.

6

Controlled Systems: Water Reuse and Recirculation

Steven Summerfelt, Julie Bebak-Williams, and Scott Tsukuda

Overview

Limited water resources, as well as more stringent water pollution control and water use permitting, have made systems that clean and reuse water (i.e., recirculating systems) more appealing (Summerfelt 1999). Increasing competition and declining farm-gate prices have also focused attention on niche marketing, technologies to produce high value species, and selecting more cost-effective production strategies for the intensive culture of fish.

Here, recirculating systems provide several advantages, given that they (1) minimize water use, allowing fish production in regions where water is scarce; (2) allow greater control of the rearing environment, especially water temperature, than is possible in conventional flow-through or pond applications; (3) are more amenable to implementation of biosecurity measures than outdoor systems; (4) increase biosecurity, or "hazard reduction through environmental manipulation," by reducing the risk of pathogen introduction into a fish farm through the water supply; (5) allow fish farms to locate in better market areas or where niche markets provide a better price for local or fresh fish; (6) concentrate waste solids, which reduces the volume of wastewater to be treated and, thus, the size and cost of wastewater treatment; and (7) significantly reduce the waste load discharged (i.e., the total maximum daily load = TMDL) in the farm effluent, allowing recirculating systems to produce fish in regions that contend with strict environmental regulations.

Recirculating systems have disadvantages due to their requirement for additional equipment to treat the water for reuse. The additional equipment requires extra capital investment, can increase operating costs, and expands the risk of catastrophic loss due to the use of a more complex operating system (Kazmierczak and Caffey 1995; Dunning et al. 1998; Losordo 1998a, 1998b; Blancheton 2000). Increased intensification and system scale are two methods that have been used to reduce the cost of fish production using recirculating systems (Wade et al. 1996; Timmons 1997; Losordo 1998a, 1998b). Most recirculating systems have relatively small and concentrated discharges (although a completely closed recirculating system would have no discharge); these recirculating system discharges usually require treatment to meet discharge limits, especially concentration limits (Chen et al. 1997; Summerfelt et al. 1999; Cripps and Bergheim 2000). However, use of recirculating systems does not always increase operating costs. For example, many salmon smolt farms have recently converted from heated flow-through systems to heated recirculating systems in order to decrease water heating and groundwater pumping costs (Wilton and Boschman 1998). Operating costs are reduced by conserving energy previously expended in heating water and sometimes by reducing pumping costs by decreasing reliance on pumped flow from deep wells.

Water reuse systems have been under development for fish culture since the 1960s. Much of the early and truly groundbreaking recirculating system design work was developed to meet the needs of state and federal public fish hatcheries (Burrows and Combs 1968; Liao and Mayo 1972, 1974; Speece 1973), which produce fish for fisheries management. Since then several large public hatcheries that reuse water have been constructed (Mayo 1981, 1991), including the Dworshak National Fish Hatchery (Ahsahka, Idaho) where three open systems recirculate over 3.1 m^3/s (50,000 gal/min) of water to produce 1-year-old steelhead trout (Owsley et al. 1988). Application of recirculating systems for commercial finfish production became more widespread between 1970 and 1980 (Chiba 1981; Fridman 1981; Parker 1981; Rosenthal 1981a; Kim and Jo 1998). However, many large commercial finfish producers that were using recirculating systems have been notable in their failure (Lee 1992; Ismond 1996).

Improvements continue in the development of commercial recirculating systems (Muir 1981, 1982; Rosenthal and Black 1993; Summerfelt 1996; Twarowska et al. 1997; Eikebrokk and Ulgenes 1998; Losordo 1998a; Muir 1998; Blancheton 2000; Losordo et al. 2000; Summerfelt et al. 2000a; Timmons 2000a, 2000b) and apparently successful commercial systems have been reported (Kim and Jo 1998; Losordo 1998b; Blancheton 2000; Smith 2000). Taking tilapia for example, by the end of the 1990s, recirculating systems accounted for more than 75% of the more than 8,000 metric tons of annual tilapia production in the United States (American Tilapia Association, Arlington, Virginia). Also, significant levels of eel and African catfish are produced within European recirculating systems.

Clearly, for recirculating systems to be economical, they must produce a valuable fish. Currently, freshwater recirculating systems are used to raise high value species or species that can be effectively niche marketed, such as salmon smolt and ornamental fish, as well as fingerling and food-sized tilapia, hybrid-striped bass, sturgeon, yellow perch, eel, rainbow trout, walleye, African catfish, channel catfish, and Arctic char (to name just a few). Additionally, saltwater reuse systems are being used to produce many species at both fingerling and food-size, including summer flounder, sea bass, turbot, and Atlantic halibut. Water reuse systems are also used to maintain many kinds of coldwater and warmwater broodfish.

Types of Water Reuse and Recirculating Systems

Recirculating systems consist of an organized set of complementary processes that allow at least a portion of the water leaving a fish culture tank to be reconditioned and then reused in the same fish culture tank or other fish culture tanks (Liao and Mayo 1972; Wheaton 1977; Muir 1982; Timmons and Losordo 1994; Lawson 1995). Dissolved oxygen (DO) supplementation is usually the first process applied to prepare water for further use, because DO is often the first water quality variable to limit production in most intensive systems (Colt 1991). However, once the availability of DO has been increased, other fish wastes can begin to accumulate to concentrations that must be reduced to maintain a healthy fish culture environment (Colt et al. 1991). At this level of intensification, especially when the water is reused, several complementary water treatment processes may be required to reduce waste accumulations in order to maintain a healthy fish culture environment (Figure 1). Water treatment processes are used to change the character of the water passed through the process. Sometimes water treatment processes can change more than one characteristic of the water. For example, water flowing through a trickling biofilter can gain DO and nitrate, while dissolved carbon dioxide and un-ionized ammonia (NH_3) are removed (Wheaton et al. 1991; Summerfelt et al. 2000d).

Recirculating systems may require water treatment processes (Tables 1 and 2) to control dissolved gases (oxygen, carbon dioxide, and nitrogen), pH, pathogens, and fish metabolites such as solids (both dissolved

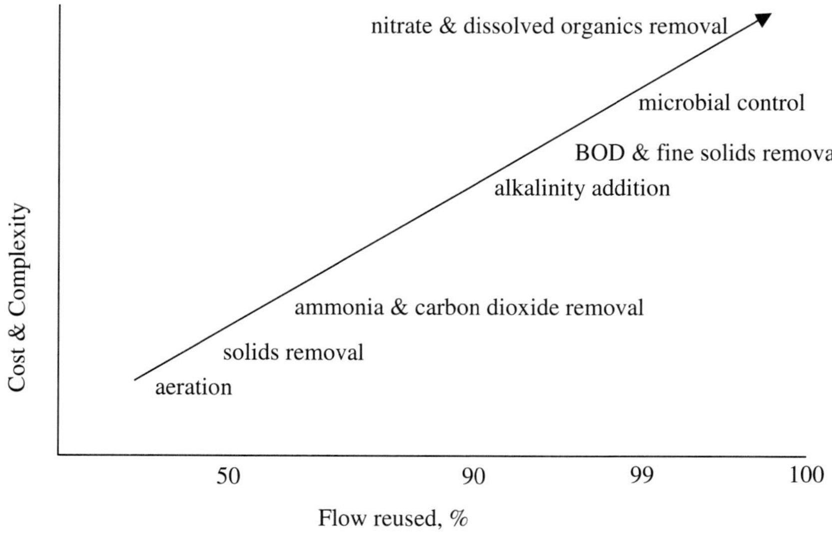

FIGURE 1. Increasing the percentage of flow reused will increase the cost and complexity of water quality control processes requirements (after Muir, 1981).

and particulate) and dissolved nitrogen compounds (ammonia and nitrite). Water reuse systems generally require at least one or more of the following treatment processes, depending upon their water-use intensity (Figure 1) and species-specific water quality requirements:

- sedimentation units, granular filters, or mechanical filters to reduce particulate solids,
- biological filters to abate dissolved wastes,
- strippers/aerators to add DO and decrease dissolved carbon dioxide or nitrogen gas to levels closer to atmospheric saturation,
- oxygenation units to increase DO concentrations above atmospheric saturation levels,
- advanced oxidation units (i.e., ultraviolet [UV] filters or units to add ozone) to disinfect, oxidize organic wastes and nitrite, or supplement the effectiveness of other water treatment units,
- pH controllers to add alkaline chemicals for maintaining water buffering or reducing dissolved carbon dioxide levels, and
- heaters or chillers to bring the water temperature to a desired level.

These individual treatment processes and their placement within the different types of water reuse systems will be discussed in this chapter. First, an overview of the three basic types of tank-based water reuse systems, that is, serial reuse, partial reuse, and fully recirculating, will be presented.

TABLE 1. Unit processes used in intensive aquaculture systems and their effects on water quality and fish health (from Noble and Summerfelt 1996).

Component: *Purpose.*	Significant conditions within the unit and their action on water quality	Importance to fish health
Culture unit: *To contain fish during grow-out, allow fish to feed, and flush feces.*	• Fish respiration reduces levels of dissolved oxygen and increases levels of ammonia and carbon dioxide, which reduces pH. • Acid-base equilibrium (i.e., pH) controls the fraction of unionized ammonia and carbon dioxide present. • Feed introduces solids into the system. • Excretion produces particulate and dissolved solids. • High fish densities. • Fish may or may not carry pathogens. • Tank cleaning and fish handling or grading may occur that involve physical interaction with the fish.	• High fish densities and the deterioration in water quality (low dissolved oxygen and elevated carbon dioxide, ammonia, nitrite, and solids levels) may stress fish. • Physical interactions with the fish may produce fish stress or physical damage. • Horizontal pathogen transmission.
Clarification: *To remove solids via settling, sieving, flotation, or filtration.*	• Most conventional clarifiers do not remove fine solids (<20 μm). • Some clarifiers store organic solids, which can produce anoxic or anaerobic conditions, exert an oxygen demand, support microbes, and leach nutrients.	• Fish pathogens may be associated with solids. • Elevated levels of solids and ammonia may induce gill pathology.
Biofiltration: *To provide surface area where microorganisms can establish; when the reused flow passes across these surfaces, the microbes remove a portion of the dissolved wastes.*	• Microorganism metabolism lowers oxygen levels and produces carbon dioxide, which reduces pH. • Oxygen must be present for bacteria to oxidize ammonia and nitrite. • Increased levels of nitrite leaving the biofilter can occur if the biofilter is overloaded. • Biofilters may take 3-6 weeks to develop the population of bacteria to convert ammonia to nitrite and an additional 2-4 weeks to develop the population of bacteria required to convert the nitrite to nitrate. • Microbial growth produces solids within the biofilter. • Biofilm and biosolids produced in the biofilter may slough and be carried out with the recirculating flow, which contributes to the microbial counts and the concentration of fine solids in the water.	• Biofilms may support certain fish pathogens, which may be passed to fish in the culture tank with sloughed biofilm carried in the reused flow. • Certain anaerobic by-products (e.g., sulfides) are toxic to fish. • Improved water quality reduces fish stress, which makes fish more resistance to disease.

TABLE 1. (Continued).

Component: *Purpose.*	Significant conditions within the unit and their action on water quality	Importance to Fish
Stripping/aeration: *To contact water with air at near atmospheric pressures.*	• Shifts concentrations of dissolved carbon dioxide, nitrogen, oxygen, and ozone towards equilibrium (i.e., adds oxygen, removes carbon dioxide, nitrogen, ozone, and gas supersaturations). • Strips little ammonia at typical pH levels (pH < 9.0).	• Elevated carbon dioxide reduces the capacity of blood to transport oxygen and may induce nephrocalcinosis.
Oxygenation: *To contact water with purified oxygen at pressures greater than or equal to atmospheric.*	• Generally used to create supersaturations of oxygen. • High gas pressures can be produced when off-gas not vented. • Little carbon dioxide is removed due to insufficient gas exchange with respect to the volume of water treated.	• Gas supersaturations can produce gas bubble disease. • Increased oxygen levels can support much higher fish loadings in the culture tank, which adds stress.
Ozonation: *To oxidize constituents in the water.*	• Oxidation can reduce levels of nitrite, organic matter, microbes, water color, odor, or off-flavor compounds. • Organic matter and nitrite react with ozone, which makes sustaining an ozone residual difficult.	• Disinfection reduces risk from infectious diseases. • Improves water quality. • Ozone and certain of its by-products are toxic.

Serial Reuse Systems

Traditionally, commercial and public coldwater finfish culture requires relatively large water resources to produce fish in flow-through systems (Figure 2). Serial reuse systems are flow-through systems that reuse water as it flows by gravity through a sequence of culture units, where water discharged from one or more tanks is input into one or more tanks located downstream (Figures 2 and 3). Raceway culture situations are often suited for serial reuse systems. Reusing water through a series of raceways requires a large water supply at the correct culture temperature and sufficient water pressure to allow the water to flow by gravity from one raceway unit to the next. As a rule of thumb, traditional flowing water systems can produce approximately 6 kg of fish annually for every 1 L/min of water flow (i.e., 50 lb of fish annually for every 1 gpm of water flow). Although large water resources are used, water is not consumed and 100% of the water is returned to the environment.

TABLE 2. Unit processes typically used to handle wastes (from Summerfelt 1996).

Unit process	Example types
Solids removal	• microscreen filters (drum, Triangel™, and disk) • settling basins • tube/plate settlers • roughing filters (packed with random rock or plastic, and with structured plastic) • swirl separators • pressurized filters (sand, activated carbon, and plastic bead) • gravity filters (high rate sand and slow sand) • flotation/foam fractionation
Biofiltration	• fluidized-media reactors (sand and plastic bead) • rotating biological contactors • trickling filters • submerged large media reactors • pressurized bead filters
Stripping/aeration	• mechanical-surface mixers • diffusers • columns (open to atmosphere or enclosed with forced ventilation) a. packed or tray b. spray • shallow air-lifts • corrugated inclined plane • stair-type drops
Ozone contractors	• U-tubes • columns (atmospheric pressure and pressurized) a. multistaged (e.g., low head oxygenators) b. packed or tray c. spray • oxygenation cones • oxygen aspirators • diffusers • enclosed mechanical-surface mixers

Today, almost 32 million kilograms of trout (mostly rainbow trout) are grown in the United States, primarily for the food fish market (U.S. Trout Farmers Association 2000). More rainbow trout are produced within land-based, flowing water systems than any other coldwater finfish in the United States. The vast majority of U.S. trout are grown in Idaho, which accounts for an estimated 73% of U.S. trout production. In 1992, Idaho had 120 commercial trout rearing facilities in operation (MacMillan 1992) and, in 1995, 19.5 million kilograms of trout were produced in Idaho. Clear Springs Trout Company (Buhl, Idaho), established in 1966, pro-

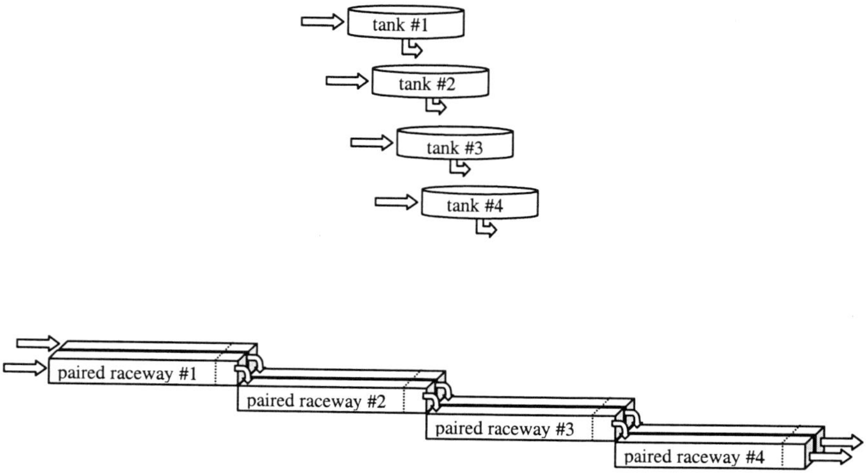

FIGURE 2. Single-pass tank systems (top) and serial-reuse raceway systems (bottom) have been traditional choices for producing coldwater finfish.

FIGURE 3. Water discharged from raceway units is often serially reused after settleable solids are captured in quiescent zones at the end of each unit and the water is reaerated as it cascades into a lower raceway, as shown at Leo Ray's fish farm near Buhl, Idaho (from Summerfelt et al., 2000d).

duced 8.2 million kilograms of trout in 1990 and is the world's single largest producer of rainbow trout for human consumption (MacMillan 1992). According to MacMillan (1992), Clear Springs uses 22.6 m^3/s (360,000 gal/min) of gravity-fed springwater flow for up to five serial-reuses through concrete raceways. When water is reused serially, the water exchange is required to transport DO to the fish and to flush waste metabolites from the system (Phillips et al. 1991). Typically, NH_3 is controlled simply by dilution with fresh makeup water. However, to capture settleable solids, replenish DO, and strip dissolved carbon dioxide, many serial reuse raceway systems use quiescent zones and 30–120 cm hydraulic drops between raceway units (Figures 2 and 3) (MacMillan 1992; IDEQ 1998). The hydraulic drops allow for cascade aeration or oxygenation and the quiescent zones[1] are used to capture settleable solids. In some instances, water can be reused through a series of more than 10 raceways before being discharged, especially in regions with waters of relatively low alkalinity and pH values less than 7.0.

Partial-Reuse Systems

Partial-reuse systems will generally settle or filter solid matter from the flow leaving the culture tanks before pumping a large portion of the flow, 50–85% of the total system flow, back to the head of the system for further treatment and reuse. Compared to serial-reuse raceway systems, partial-reuse systems use much smaller flows to achieve production goals and can significantly increase waste capture by concentrating wastes into much smaller discharges. Partial-reuse systems, such as shown in Figure 4, can produce as much as 48 kg of fish annually for every 1 L/min of makeup water flow (i.e., 400 lb of fish annually for every gallon per minute of water flow) (Summerfelt et al. 2000a).

Partial-reuse systems use relatively large makeup water flows (compared to fully recirculating systems) in order to dilute the NH_3 produced, preventing it from accumulating to toxic levels without the use of biofilters (Summerfelt et al. 2000a). The high volume of makeup water flow maintains the water temperature in the partial-reuse system at levels similar to the temperature of the makeup water, unless large volumes of system water are stored in outdoor ponds or extreme differences in ambient air and water temperatures are encountered. Water treatment usually consists of processes for solids removal, aeration, and oxygenation (Figure 4). However, as the fraction of NH_3 declines with declining pH levels, removing less dissolved carbon dioxide with aera-

[1]Quiescent zones are regions within a raceway where fish are excluded to promote capture of settleable solids.

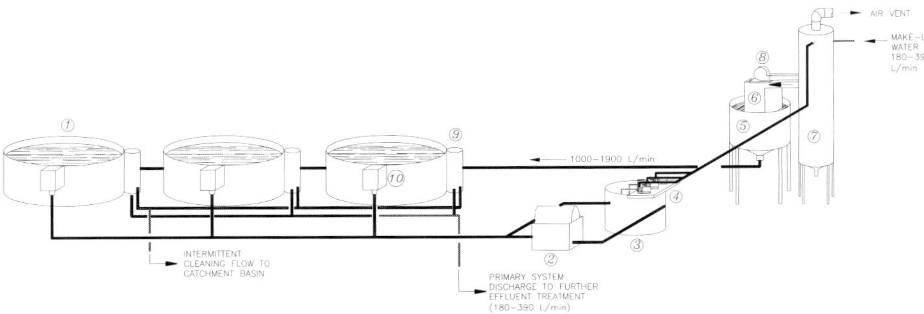

FIGURE 4. A partial-reuse system at the Freshwater Institute (from Summerfelt et al., 2000a). The units are defined according to the following: ① 3.7 m × 1.1 m culture tank; ② drum filter; ③ 1.8 m × 1.2 m pump sump; ④ three 1.5-HP reuse pumps; ⑤ header tank (with cone bottom to improve cleaning); ⑥ LHO; ⑦ carbon dioxide stipping column (with cone bottom to improve cleaning); ⑧ fan; ⑨ triple standpipe sump (to direct bottom flow & observation of waste feed); ⑩ "Cornell-type" sidewall drain. The illustration is courtesy of PRAqua Technologies, Nanaimo, British Columbia, Canada.

tion can help to keep the pH down and, thus, the NH_3 at lower levels in a partial-reuse system (Summerfelt et al. 2000a). Notwithstanding, sufficient aeration must be provided to prevent dissolved carbon dioxide from accumulating to toxic levels.

Fully Recirculating Systems

Fully recirculating systems reuse the water to a high degree (generally >95–99% of flow is reused), and, therefore, require water treatment processes to control the accumulation of wastes. At a minimum, fully recirculating systems will require clarifiers or filters to reduce particulate solids, biological filters to abate dissolved wastes (primarily ammonia), and aerators to add DO and decrease dissolved carbon dioxide to safe levels (Figure 1).

Considerable research has been done on unit processes development and their integration into functional water-reuse systems (Timmons and Losordo 1994). The ultimate goal of this research has been to make finfish production more cost competitive within recirculating systems. Consequently, a variety of commercially viable recirculating system designs and technologies exist, and the selection of water treatment units is largely dependent upon the water quality required and the reliability and cost-effectiveness of the technologies. However, technology selection also depends upon their cost and the size of the application (Wade et al. 1996; Timmons 1997; Losordo 1998a,b), because at larger scales many water treatment units are not available, cannot be built, or do not function as effectively as smaller units.

There have been efforts to develop "turnkey" recirculating systems by carefully integrating unit processes in a manner that could be easily replicated and which, theoretically, would be ideally suited to produce a certain type of fish under most conditions common to a given region. If successful, turnkey systems would certainly help standardize aquaculture production. Unfortunately, a number of turnkey systems that have been marketed have not proven successful. Reasons for these failures can range from technology problems to inadequacies in management and marketing. Also, it appears that many of these systems were simply not large enough to pay off their debt and support the required profit margin. Additionally, the turnkey system approach may not provide a good fit for the culture requirements or management ability specific to all situations, particularly, because turnkey systems are often defined by and limited to a fixed package of hardware. For these reasons, modular and scaleable recirculating system designs are often tailored to fit each specific application. This approach is finding wider acceptance as it provides the flexibility to select and adapt technologies to fit the scale and requirements of each application. The information set down in this chapter is intended to help the reader better understand, identify, and apply technological options for use within recirculating systems.

Centralized Water Treatment Versus Distributed Treatment (at Rearing Units)

There are many ways to arrange individual water treatment units within recirculating systems (Rosenthal and Black 1993). The water treatment units can be positioned to receive water flow in series, in parallel to each other, or to the culture units. Recirculating systems can have their treatment components grouped in a central location, that is, centralized treatment (Figure 4 and 5), or distribute the treatment components to every culture tank, creating a large number of independent systems, that is, distributed treatment (Figure 6). Other types of recirculating systems can be hybrids of the two, where certain treatment components are distributed at every culture tank (e.g., oxygenation units or swirl separators), but other components are provided in a central water treatment system (e.g., microscreen filters, biofilters, and aeration units).

Recirculating systems that have centralized processes (Figure 5) for treating and returning water flow to a row of culture units can have several advantages. Primarily, combining the water discharged from several culture tanks into one large flow can produce economies of scale that can reduce fixed and variable costs compared to the costs of treat-

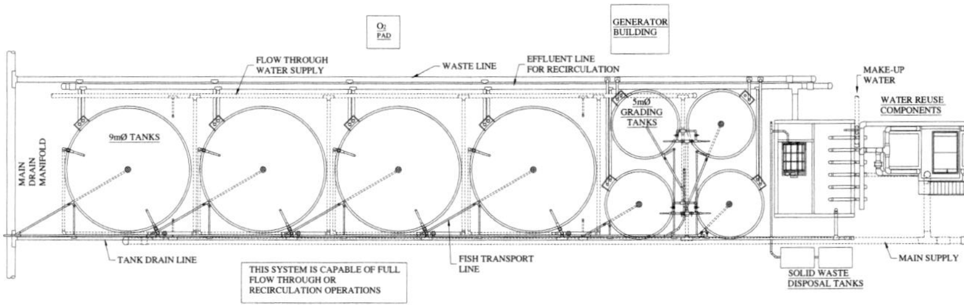

FIGURE 5. The 10,000 L/min salmon smolt production unit at Target Marine Hatcheries in British Columbia is a good example of a relatively compact fully recirculating system with centralized water treatment (from Summerfelt et al., 2000d). Drawing courtesy of PRAqua Technologies, Ltd., Nanaimo, British Columbia, Canada.

ing the same flow if it had been kept in many small distributed systems. Typically, large economies of scale can be obtained for biological filters, solids filters, aeration units, and oxygenation units. Also, a centralized treatment system provides the same water quality to all culture tanks, which reduces the number of inlet water quality samples that must be

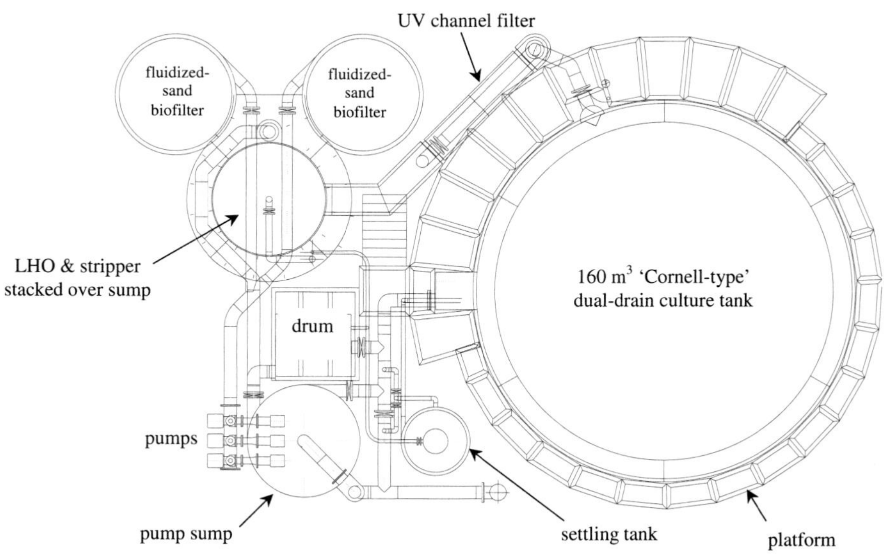

FIGURE 6. The 4,800 L/min recirculating system at the Freshwater Institute is an example of distributed treatment, where a culture facility would contain at least several of these modular and independent systems. Drawing courtesy of Marine Biotech, Inc., Beverly, Massachusetts.

collected, analyzed, and tracked. Using centralized water treatment processes will also reduce the number of probes and controllers (e.g., DO, pH, oxygen-reduction potential, dissolved ozone, gas or water flow, etc.) that must be purchased and maintained, along with the associated pumps, valves, and pipes required to monitor and control water quality during treatment.

Combining tank flows and transporting them to a central location for treatment (Figure 5) is especially appealing when considering processes such as pH control, ozone addition and control, UV disinfection, biofiltration, microscreen filtration, and heating/chilling, because these processes are relatively expensive and would have to be replicated at each distributed water treatment system. Incorporating a centralized fish grading, counting, and redistribution station can also provide labor savings. However, recirculating systems with centralized treatment processes have two main disadvantages: increased risk and decreased flexibility. Combining the water, centrally treating it, and then redistributing it, places the fish in the distributed culture tanks at a greater risk from infectious disease outbreaks or catastrophic failure at one of the treatment processes. Also, supplying the same water quality to all culture tanks does not provide as much flexibility to vary fish densities or specific water quality variables that are sometimes necessary for different life stages.

Distributed systems supply all of the water treatment units necessary to maintain water quality to every culture unit, which creates a large number of completely independent water recirculating systems (Figure 6). Advantages of distributed systems include: more control over water quality variables within individual culture tanks, lower risk of pathogen transfer from tank to tank, and lower risk from a catastrophic failure in any one pump or treatment process. Disadvantages are primarily due to the higher cost of replicating and managing large numbers of distributed recirculating systems.

Hybrid recirculating systems contain both centralized and distributed water treatment processes in order to maintain flexibility while decreasing the relative cost of some of the more expensive water treatment technologies available. In a hybrid system, a portion of the treatment processes (such as oxygenators and particle traps) are located at every culture tank in order to provide greater control of the fish culture environment or enhance waste removal. However, a portion of flow from all the culture tanks is collected and sent to a central system where the remaining water treatment requirements are provided (e.g., biofiltration, aeration, heating or chilling, pH control, disinfection) at some economies of scale. These hybrid systems are often relatively large in order to take advantage of economies of scale. However, hybrid systems are still at risk from pathogen transfer or a catastrophic pump or treatment process failure.

Water Quality Issues in Recirculation Systems

In aquaculture systems that reuse water, water quality variables should be maintained at levels sufficient for supporting healthy and fast growing fish (Losordo 1991; Noble and Summerfelt 1996). Operating a fish farm under limiting water quality conditions can reduce the profitability of fish production (Kazmierczak and Caffey 1995), because the water quality problems can be lethal or lead to stress, and the resulting deterioration of fish health will reduce growth and increase the risk of infectious disease outbreaks and catastrophic loss of fish (Wedemeyer 1996). These problems can be created by high or low water temperature, low DO levels, elevated waste metabolite concentrations, gas supersaturation, measurable dissolved ozone levels, and the presence of certain cleaning chemicals or chemotherapeutants in water (Piper et al. 1982; Klontz 1993; Wedemeyer 1996). Waste metabolites of concern include NH_3, nitrite, nitrate (to a lesser extent), dissolved carbon dioxide, suspended solids, and nonbiodegradable organic matter. Of these waste metabolites, fish produce roughly 1.0–1.4 mg/L total ammonia nitrogen (TAN), 13–14 mg/L dissolved carbon dioxide, and 10–20 mg/L total suspended solids (TSS) for every 10 mg/L of DO that they consume.

Nitrite and nitrate are produced when ammonia is oxidized by nitrifying bacteria concentrated within a biological filter, but these bacteria are also found throughout the water column and on surfaces within the recirculating system (Hagopian and Riley 1998). Consequently, fully recirculating systems convert practically all of the TAN produced into nitrate-nitrogen. Denitrifying bacteria (which convert nitrate to nitrogen gas under anoxic conditions), algae, and plants can remove some of the nitrate produced, depending upon the type of recirculating system in operation. Nonbiodegradable dissolved organic matter can also accumulate in the recirculating system water if it is degraded too slowly by the heterotrophic microorganisms in the biological filter.

The water's pH is also important as dissolved carbon dioxide production and nitrification continuously depresses pH (Kaiser and Wheaton 1983; Bisogni and Timmons 1991; Loyless and Malone 1997). The water's pH controls acid-base equilibrium, which proportions the TAN and inorganic carbon that exist into the more toxic forms of NH_3 and carbon dioxide, respectively. Unfortunately, exact quantification of the water quality requirements for a given fish species and life stage in intensive systems can be difficult because disease outbreaks or reduced growth can be caused by stress due to nonlethal levels of contaminants (Wedemeyer 1996). Different fish husbandry procedures and biosecurity protocols used at different fish farms can also confound quantification of water quality limits.

Poor water quality within a water reuse system can result from improper design or management of any of the treatment units (Tables 1 and 2). Therefore, water quality should be monitored closely in a recirculating system so those problems with the water treatment units can be detected early and corrected. Water quality is also of concern if the effluent characteristics (e.g., biochemical oxygen demand, suspended solids, phosphorus, or nitrogenous compounds) of the culture facility must be controlled to meet water pollution requirements (Bergheim and Åsgård 1996; Cripps and Kelly 1996; Chen et al. 1997; Summerfelt 1999).

Temperature

Temperature control can be used for many purposes in aquaculture. It is often used to eliminate the natural temperature variation that would occur in surface waters. Relatively constant warm or cool temperatures can be maintained to promote or limit egg development or fish growth. For the specific temperature requirements of various species, see Table 17 in Piper et al. (1982). More dynamic water temperature regimes can also be used to stimulate the reproductive cycle in broodfish. However, rapid changes in temperature can prove fatal to aquatic organisms. A rule of thumb is to limit temperature changes to a maximum of 1°C per day, although the tolerance limits can vary considerably for different species. Temperature changes also change the solubility of oxygen and nitrogen in water (Colt and Orwicz 1991). Heating water decreases the concentration of oxygen and nitrogen that are soluble in the water, which temporarily creates supersaturated gas pressures until the water has had an opportunity to degas and come back to equilibrium with the partial pressures of oxygen and nitrogen in the surrounding atmosphere.

Oxygen

Depending upon the species, for optimum growth fish require a minimum DO concentration of approximately 5 mg/L (warmwater species) to 7 mg/L (coldwater species) (Piper et al. 1982; Wedemeyer 1996). The minimum DO concentration that is safe for fish also depends on the concentration of dissolved carbon dioxide present. The minimum safe DO level should be increased by 3–4 mg/L if carbon dioxide concentrations are high (Colt et al. 1991), for example, if dissolved carbon dioxide exceeds 30 mg/L for salmonids or exceeds 40–50 mg/L for certain warmwater species. For these reasons, DO is usually the first water quality variable to limit culture tank carrying capacity.

Salmonids consume an average of 0.2–0.3 kg of DO per kilogram of wet feed consumed, but oxygen consumption rates can be higher for warmer temperatures (Wedemeyer 1996). In commercial high-density warmwater fish culture, Colt and Watten (1988) reported that mean oxygen consumption rates of 1.0 kg per kilogram of wet feed consumed are assumed. However, peak oxygen demands can be 1.4–1.5 times higher than the average oxygen demand, depending upon activity and feeding schedule (Westers 1981). For design purposes, the daily oxygen consumption rate in a culture tank, $r_{DO,fish}$ (kg DO per day), can be estimated from this average feed-specific oxygen consumption constant, a_{DO} (kg DO per kg feed), along with the culture tank's volume, V_{tank} (m^3), fish density, r_{fish} (kg fish per m^3 culture volume), and feeding rate, r_{feed} (kg feed per kg fish per day), using the following equation:

$$r_{DO,\,fish} = V_{tank} \times r_{fish} \times r_{feed} \times a_{DO}\,. \tag{1}$$

Without surface aeration or significant photosynthesis, the mass of DO available to the fish is entirely dependent upon how much oxygen the water flow carries into the tank. Here, the daily mass of available DO, $r_{DO,avail}$ (kg per day) can be calculated by multiplying the water flow rate, Q (L/min), by its available DO concentration, which is the inlet DO concentration, DO_{in} (mg/L), minus the minimum allowable DO concentration at the tank outlet, DO_{out} (mg/L):

$$r_{DO,\,avail} = Q \times \left(DO_{in} - DO_{out}\right) \times \frac{1{,}440\,\text{min}}{\text{day}} \times \frac{\text{kg}}{10^6\,\text{mg}}\,. \tag{2}$$

For the fish to have enough DO, the rate that available oxygen is delivered to the culture tank must be greater than or equal to the rate that fish consume oxygen in the tank; that is,

$$r_{DO,avail} \geq r_{DO,fish.} \tag{3}$$

Example. The DO concentration in the water supply entering a 100 m^3 (V_{tank}) culture tank supporting 250 g hybrid striped bass at a density of 70 kg/m^3 (r_{fish}) can be estimated by equating and rearranging equations (1) and (2):

$$DO_{in} = DO_{out} + \left(\frac{V_{tank} \times r_{fish} \times r_{feed} \times a_{DO}}{Q} \times \frac{\text{day}}{1{,}440\,\text{min}} \times \frac{10^6\,\text{mg}}{\text{kg}} \right). \tag{4}$$

A flow of 6,670 L/min (Q) will be supplied to the culture tank in order to exchange the culture tank volume four times every hour, which is the general practice for raceway systems. The fish are fed 1.5% body weight per day (r_{feed}) and consume 0.25 kg DO per kilogram feed (a_{DO}). The DO concentration exiting the culture tank is limited to 7 mg/L. Substituting these values into equation (4) provides an estimate of the mean DO concentration that must be supplied to the culture tank:

$$DO_{in} = 7.0\,mg/L + \left[100\ m^3 \times \frac{min}{6{,}670\,L} \times \frac{70\,kg\ fish}{m^3} \times \frac{1.5\,kg\ feed}{100\,kg\ fish \times day} \times \frac{0.25\,kg\ DO}{1\,kg\ feed} \times \frac{day}{1{,}440\,min} \times \frac{10^6\,mg}{kg}\right] = 9.7\,mg/L$$

If the flow rate to the culture tank were cut by 50% to 3,335 L/min, then the mean DO concentration that must be supplied to the culture tank would have to rise to 12.5 mg/L.

According to equation (2), the culture tank's carrying capacity can be doubled by doubling the water flow (*Q*) through the tank because this doubles the rate that DO is supplied to the tank ($r_{DO,avail}$). Unfortunately, increasing the water flow through a culture tank may not always be possible and can increase the farm's fixed and variable costs because it may require larger pipes, increased pumping, or higher water pressure. Alternatively, using pure oxygen transfer technology to increase DO concentrations to supersaturated levels is an effective, and often cost-effective, method to increase the mass of oxygen supplied to a culture tank (Colt and Watten 1988). Using the previous example, the mass of available oxygen supplied to the culture tank (and, thus, its carrying capacity) can be doubled simply by using pure oxygen transfer technology to increase the DO concentration entering (DO_{in}) the culture tank from an initial 9.7 mg/L (without oxygenation) to a final 12.4 mg/L (with oxygenation). Colt and Watten (1988) and Boyd and Watten (1989) have provided additional information on estimating oxygen demand and methods for meeting this demand with oxygenation technologies.

Other fish metabolites such as dissolved carbon dioxide, ammonia, and suspended solids can limit culture tank carrying capacity when DO is no longer limiting. Colt et al. (1991) showed that intensive fish farms can use clean (with low carbon dioxide levels) flowing water, without worry of ammonia or carbon dioxide limitations (assuming no biofiltration to remove ammonia or air-stripping to remove dissolved carbon dioxide), up to a cumulative DO consumption of approximately 10–22 mg/L (Figure 7), depending upon pH, alkalinity, temperature, fish species, and life stage. After reaching this cumulative oxygen demand, the water flow must be treated to remove waste metabolites before further use.

Ammonia, Nitrite, and Nitrate

Ammonia and nitrite are toxic and their accumulation in water reuse systems can result in noninfectious problems. Ammonia is the primary nitrogenous metabolite excreted by fish. Ammonia is converted into nitrite and then nitrite is converted into nitrate by autotrophic bacteria

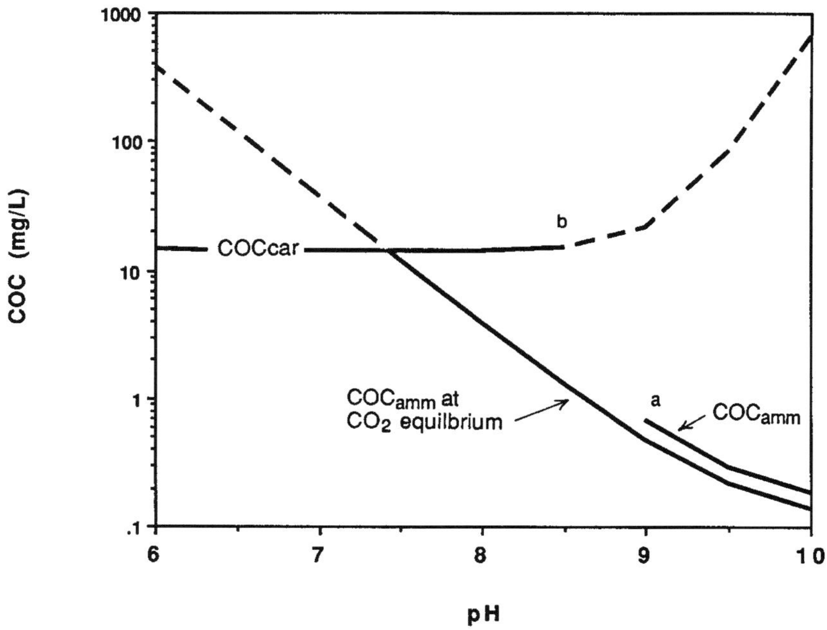

FIGURE 7. Colt et al. (1991) showed that the fish's maximum cumulative oxygen consumption (COC) within an initially clean culture water would be limited by accumulations of either un-ionized ammonia (COC_{amm}) or carbon dioxide (COC_{car}), depending upon pH. To develop this graph, Colt et al. (1991) assumed that the un-ionized ammonia and dissolved carbon dioxide became limiting at 0.016 mg/L and 20 mg/L, respectively.

within biological filters (Hagopian and Riley 1998). Therefore, nitrate is the primary form of dissolved nitrogen discharged from fully recirculating systems, while ammonia is the dominant form of dissolved nitrogen discharged from serial-reuse systems and partial-reuse systems. Elevated levels of ammonia and nitrite typically occur only during biofilter start-up and during periods of biofilter malfunction or overload (Wheaton et al. 1991). Therefore, the sum of nitrite-nitrogen and TAN concentrations in these reuse systems and their effluents are usually less than 2–10 mg/L, depending upon type of reuse system, the pH of its water, and the species being raised. Fortunately, nitrate is much less toxic to fish than either ammonia or nitrite (Russo and Thurston 1991; Wedemeyer 1996), and accumulations of 100–1,000 mg/L of nitrate-nitrogen are not uncommon within recirculating systems. Eggs and certain fish species may be more sensitive to nitrate, especially marine species, and require much lower nitrate limits (Russo and Thurston 1991).

In general, fish produce 0.25–0.35 kg of ammonia per kilogram of feed consumed. However, ammonia production can depend upon factors including the fish species and life stage, feed formulation, and even the feeding frequency (Wedemeyer 1996). In recirculating systems, ammonia is also produced by biological deamination of organic compounds (waste feed and feces), endogenous respiration, and cell lysis.

Ammonia exists in two forms, either as un-ionizied ammonia (NH_3) or as the ammonium ion (NH_4^+), and these two forms are in acid-base equilibrium with each other. Therefore, the concentration of either form is dependent upon the TAN present and the water's pH (see Tables 3 and 4, and Figures 8 and 9). Un-ionized ammonia is much more toxic than ammonium (Russo and Thurston 1991). Unfortunately, NH_3 can kill rainbow trout and catfish at concentrations of 0.32 mg/L and 3.10 mg/L, respectively. For salmonids, long term exposure to lower concentrations (0.05–0.2 mg/L) of NH_3 can significantly reduce growth and optimal conditions require an NH_3 concentration less than 0.01–0.03 mg/L (Wedemeyer 1996).

Nitrite oxidizes the iron in hemoglobin, which reduces how much oxygen can be transported by blood and can cause asphyxiation even when DO levels are not limiting (Russo and Thurston 1991). Nitrite toxicity is often referred to as brown blood disease because fish exposed to toxic nitrite levels will have gills that are an unusual chocolate-brown color. Nitrite toxicity depends upon fish species and life stage, as well as on the pH and presence of other anions in the water, as discussed by Russo et al. (1981), Russo and Thurston (1991), and Wedemeyer (1996). Nitrite is much more toxic for salmonids than for most warmwater fish, e.g., it can kill rainbow trout and channel catfish at concentrations above 0.20–0.40 mg/L and about 13 mg/L, respectively (Wedemeyer 1996). Nitrite toxicity is suppressed by other anions such as chloride. Sodium chloride is used to control nitrite toxicity in fish culture systems, as the chloride ions compete with nitrite for transport across the gill. For example, water containing at least 50 mg/L of dissolved calcium or chloride ions can increase the tolerance of fish to nitrite by a factor of 50 (Wedemeyer 1996). Dissolved nitrite is in acid-base equilibrium with nitrous acid (Tables 3 and 4), which is more toxic to aquatic organisms than nitrite. A shift to lower pH values increases the nitrite toxicity as the acid-base equilibrium shifts towards more nitrous acid. At normal culture pH, however, equilibrium favors essentially only the nitrite ion.

Toxic ammonia and nitrite accumulations can occur in water reuse systems during a 1–2 month initial lag period that results as the microbiological populations in a biofilter first develop (Wheaton et al. 1991). The biofilter start-up period takes longer at colder water temperatures than in warmer water. Ammonia spikes can occur as the bacterial population responsible for converting ammonia into nitrite (i.e., *Nitrosomonas*) becomes established during the first 2–4 weeks of operation. Nitrite ac-

TABLE 3. Acid-base equilibrium relationships for estimating un-ionized ammonia, carbon dioxide, and nitrite concentrations from measurements of pH and total ammonia nitrogen (TAN), alkalinity (Alk), and total nitrite nitrogen (TNN), respectively.

Parameter		
Ammonia		
	Equilibrium	$NH_4^+ \leftrightarrow NH_3 + H^+$
	Equilibrium constant	$K_{NH_3} = \frac{[NH_3][H^+]}{[NH_4^+]}$
	Equilibrium concentration	$NH_3(\text{mg/L as N}) = \frac{\text{TAN (mg/L)}}{\left\{1+10^{(pK_{NH_3}-pH)}\right\}}$
Carbon dioxide		
	Equilibrium	$CO_2 + H_2O \leftrightarrow H_2CO_3$ $H_2CO_3 \leftrightarrow HCO_3^- + H^+$ $HCO_3^- \leftrightarrow CO_3^= + H^+$
	Equilibrium constant	$K_0 = \frac{[H_2CO_3]}{[CO_2]}$ $K_1 = \frac{[HCO_3^-][H^+]}{[H_2CO_3]}$ $K_2 = \frac{[CO_3^=][H^+]}{[HCO_3^-]}$
	Equilibrium concentration	$CO_2(\text{mg/L}) = 44{,}000\left\{\frac{\text{Alk(mg/L, as,}CaCO_3)}{50{,}000} - 10^{(pH-pK_W)} - 10^{(-pH)}\right\} \times \left\{\frac{1}{10^{(pH-p(K_0K_1))} + 2 \times 10^{(2 \times pH - p(K_0K_1)-pK_2)}}\right\}$
Nitrite		
	Equilibrium	$HNO_2 \leftrightarrow NO_2^- + H^+$
	Equilibrium constant	$K_{NO_2^-} = \frac{[NO_2^-][H^+]}{[HNO_2]}$

TABLE 3. (CONTINUED).

Equilibrium concentration	$NO_2^-(\text{mg/L as N}) = \frac{\text{TNN(mg/L)}}{\left\{1 + 10^{\left(pK_{NO_2^-} - pH\right)}\right\}}$
Water	
Equilibrium	$H_2O \leftrightarrow H^+ + OH^-$
Equilibrium constant	$K_W = [H^+][OH^-]$
Equilibrium concentration	$pH = -\log[H^+]$

cumulation can occur in a second 2–4 week lag period before the bacterial population responsible for converting nitrite into nitrate (*Nitrobacter*) become established. To avoid exposing fish to potentially toxic accumulations of ammonia and nitrite, populations of both *Nitrosomonas* and *Nitrobacter* should be established within the recirculating system's biofilters well before fish are stocked into the system. Ideally, biofilters should be started up before fish are placed in the fully recirculating system. Biofilters can be started up using a supply of wastewater from an operating fish production unit or by mixing urea or ammonia salt and organic carbon solutions and recirculating these solutions (with frequent replenishment) through the biofilter for around 4–8 weeks, depending on water temperature and other conditions. Alternatively, bacterial populations are developed in biofilters while fish are present in the recirculating systems. In such cases, NH_3 and nitrite toxicity problems are countered by a combination of increased dilution with makeup water, salt addition, or pH control. Controlling pH in the range of 6–7 will reduce

TABLE 4. Temperature dependence of equilibrium constants defined in Table 3.

	Temperature (°C)			
Equilibrium constant	15	20	25	30
pK_{NH_3}	9.564	9.403	9.246	9.095
$p(K_0K_1)$	6.419	6.381	6.352	6.327
pK_2	10.43	10.38	10.33	10.29
$pK_{NO_2^-}$	–	–	3.3	–
pK_W	14.35	14.17	14.00	13.84

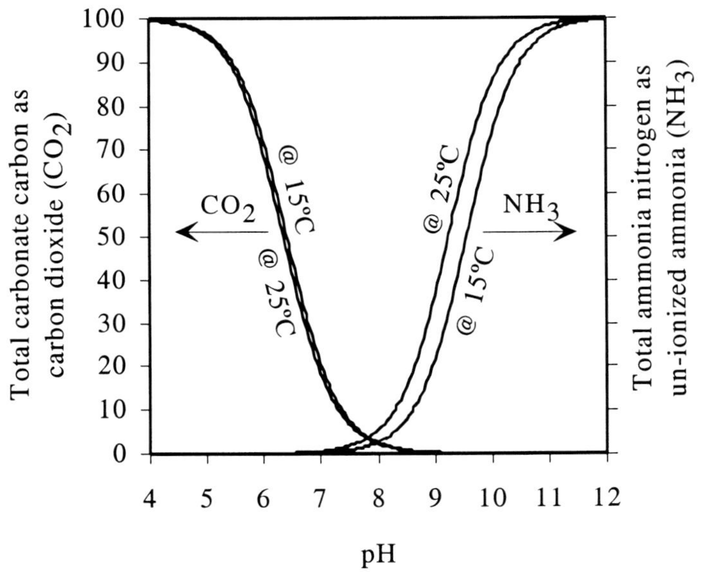

FIGURE 8. The percentage of dissolved carbon dioxide and un-ionized ammonia is a function of pH (after Summerfelt, 1996).

the toxicity of accumulating TAN concentrations because the lower pH levels reduce the fraction of NH_3 (Figures 8 and 9).

Ammonia and nitrite can accumulate to toxic levels in recirculating systems when a biofilter is exposed to certain chemotherapeutants used for disinfection or for controlling epizootic outbreaks (Noble and Summerfelt, 1996). Ammonia and nitrite accumulations are also caused by an inadequate oxygen supply to the biofilter and by overloading the biofilter with particulate matter, dissolved organic matter, ammonia, or nitrite. High nitrite levels can occur if ozone addition to a fully recirculating system is abruptly stopped. Ozone stoichiometrically oxidizes nitrite to nitrate; therefore, addition of ozone to recirculating systems is beneficial because it reduces nitrite levels. The fact that ozone decreases the levels of nitrite is a substantial benefit on those occasions when bacterial conversion of nitrite to nitrate in the biofilter is lost. However, because ozone reduces the nitrite concentration going to the biofilter, it also reduces the quantity of bacteria converting nitrite to nitrate and thus reduces the total acclimated nitrite removal capacity of the biofilter. Nitrite concentrations can rapidly accumulate within fully recirculating systems when ozone addition is interrupted, because ozone can be responsible for removing a fairly large fraction of the total nitrite produced.

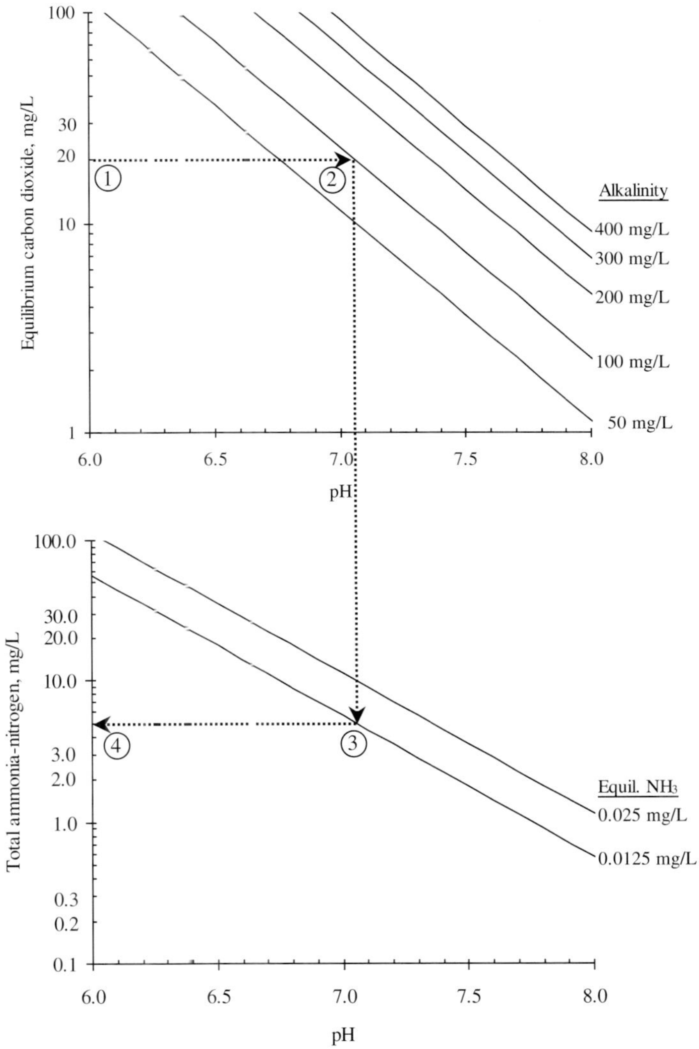

FIGURE 9. These two nomographs illustrate the dependence of dissolved carbon dioxide concentrations on pH and alkalinity (top graph), and the dependence of un-ionized ammonia concentrations on pH and total ammonia-nitrogen (bottom graph), according to acid-base equilibrium at 15°C (from Summerfelt et al., 2000a). The heavy-dashed lines indicate how setting a maximum limit of 20 mg/L on the concentrations of dissolved carbon dioxide (1) and of 0.0125 mg/L on un-ionized ammonia (3), at an alkalinity of 100 mg/L (2), will set the minimum operating pH (about 7.06) and the maximum allowable TAN (4) (about 5.0 mg/L).

Carbon Dioxide

Carbon dioxide (CO_2) is a gas that is very soluble in water—over 40 times more soluble than oxygen. Fish ventilate dissolved carbon dioxide (a by-product of metabolism) through their gills, producing about 0.3–0.4 kg of carbon dioxide for every kilogram of feed consumed. Therefore, shifts in dissolved carbon dioxide concentrations can be quite rapid, corresponding to periods of peak oxygen consumption. Elevated and toxic levels of carbon dioxide can occur in intensive fish culture systems using oxygenation to obtain high fish densities if these systems have inadequate water exchange and aeration. Low intensity aquaculture systems do not use oxygenation to support high fish densities and generally have sufficient aeration and water exchange to keep dissolved carbon dioxide from accumulating to toxic levels.

Dissolved carbon dioxide is in acid-base equilibrium with the total carbonate carbon system, which means that its dissolved concentration is dependent upon the pH and the aqueous concentration of total carbonate carbon present (Tables 3 and 4, Figures 8 and 9). Adding a base such as sodium hydroxide, sodium bicarbonate, or calcium oxide will produce a pH rise and a shift in equilibrium to reduce concentration of carbon dioxide (Bisogni and Timmons 1991; Loyless and Malone 1997). Dissolved carbon dioxide can also reach equilibrium with the partial pressure of carbon dioxide in the surrounding atmosphere according to Henry's law. During aeration, if there is enough air-water contact, a significant fraction of the dissolved carbon dioxide can be removed from water and transferred to the air (Colt and Orwicz 1991). Ventilated cascade columns are often used to strip dissolved carbon dioxide because they can be designed to contact the 5–10 volumes of air required per unit volume of water flow (Grace and Piedrahita 1993; Summerfelt et al. 2000d). Therefore, aeration columns and alkaline chemical addition are both viable options for increasing pH and controlling carbon dioxide accumulation in recirculating systems (Summerfelt 2000). For more information, see the Dissolved carbon dioxide removal section below.

Safe levels of dissolved carbon dioxide depend upon the fish species, developmental stage of the fish, and other water quality variables that include alkalinity, pH, and DO levels (Wedemeyer 1996). For example, dissolved carbon dioxide begins to effect salmonids at concentrations greater than 20 mg/L, but tilapia, hybrid striped bass, sturgeon, catfish, and many other warmwater species will tolerate considerably higher dissolved carbon dioxide levels in their environment. Even the 20 mg/L recommended safe level for salmonid culture may be conservative if DO concentrations in the water are at or above saturation levels. Sufficiently elevated carbon dioxide levels can decrease the ability of hemoglobin to transport oxygen (the Bohr effect), decrease the maximum oxygen binding capacity of blood (the Root effect), and increase blood acidity (i.e., hypercapnia) (Wedemeyer 1996). As dissolved carbon dioxide concentrations approach 30–40 mg/L, the blood oxygen carrying capacity of salmonids will be depressed to the point where even high environmental DO concentrations may be insufficient to prevent decreased blood oxygen levels. For instance, if carbon dioxide levels are high (e.g., ≥30 mg/L for salmonids), then the concentration of DO that is considered safe should be increased by 3–4 mg/L to 11–12 mg/L (Colt et al. 1991). Fish may also be able to tolerate higher concentrations of dissolved carbon dioxide when reared in water with high alkalinity, pH, or both, because these conditions strengthen the fish's blood buffering capacity (Wedemeyer 1996). Yet, to be safe, dissolved carbon dioxide accumulation should be minimized whenever possible.

Clinical signs of a carbon dioxide problem in rainbow trout include: moribund fish, gaping mouths, flared operculums, and extra-bright maraschino-red gill lamella (Noble and Summerfelt 1996). Sustained exposure to elevated levels of carbon dioxide may also produce nephrocalcinosis (Wedemeyer 1996), that is, the formation of calcareous deposits in the kidneys, which cannot be detected from external observations.

Water pH

Water pH plays an important role in aquaculture, because it controls the equilibrium of the ammonia nitrogen and carbonate carbon systems (Tables 3 and 4, Figures 7, 8, and 9) and the toxicity of metals such as aluminum. Extremes in pH, outside of the range of 5–9, can cause death; more moderate acid and alkaline conditions can reduce fish performance. Fish physiological problems can occur in more acidic water (pH < 7) where sodium and hydrogen osmoregulation at the gill surface can be inhibited, causing sodium loss from the fish (Wedemeyer 1996). More alkaline water (pH > 7) can result in ammonia accumulation within the fish. General safe guidelines for freshwater pH values range from 6.5 to 9.0.

Controlling water pH is especially important in water reuse systems (Bisogni and Timmons 1991; Loyless and Malone 1997; Summerfelt et al. 2000a). If left unabated, the pH of the water in recirculating systems continuously declines due to (1) the production of dissolved carbon dioxide (a by-product of respiration), which is nearly indistinguishable from carbonic acid (a strong acid); and (2) the production of H+ acid during nitrification, which consumes alkalinity and reduces the buffering capacity of the water. Nitrification can be severely inhibited as pH declines below 6–7 and if the alkalinity in the water drops below 50–100 mg/L as calcium carbonate (Wheaton et al. 1991; Loyless and Malone 1997). Because recirculating systems use relatively little makeup water, the pH of the water in these systems can be buffered and increased by adding a base such as a hydroxide or bicarbonate salt (lime, caustic soda, soda ash, or sodium bicarbonate; Bisogni and Timmons 1991; Loyless and Malone 1997). However, methods used to increase pH and reduce dissolved carbon dioxide levels will also increase the portion of TAN existing as NH_3 (Figures 8 and 9), which is the more toxic form to aquatic life. Alternatively, maintaining a more acidic pH will minimize the fraction of TAN existing as NH_3, shifting operation of the recirculating system to allow higher TAN concentrations, but also producing higher dissolved carbon dioxide concentrations (Figures 7, 8, and 9). Therefore, pH control can really only be applied over a limited pH range in intensive recirculating systems—from 6.5 to 8.5 for most fish—depending upon TAN and total carbonate carbon levels (Figures 7, 8, and 9), and the relative toxicity of the fish to dissolved carbon dioxide and NH_3. Eels appear to be an exception, with culture environment pH of 6.0–6.5.

Ozone

Ozone is a powerful oxidizing agent that, when dissolved in water, has seen wide use in aquaculture applications for disinfection and improving water quality (Rosenthal 1981b; Wedemeyer 1996; Summerfelt and Hochheimer 1997). Although ozone has a rapid reaction rate and few harmful reaction products, it is toxic to aquatic life at low levels (Wedemeyer et al. 1979; Langlais et al. 1991). The maximum safe level of chronic ozone exposure for salmonids is 0.002 mg/L (Wedemeyer et al. 1979). A compilation of results from several studies indicates that most fish exposed to ozone concentrations greater than 0.008–0.06 mg/L will develop severe gill damage that can result in serum osmolality imbalances and can kill fish immediately or leave them more susceptible to microbial infections (Bullock et al. 1997). Ozone gas is also toxic to humans. Therefore, care must be used when transferring an ozone containing air or purified-oxygen gas mixture into water, when providing time for reaction of dissolved ozone with the targeted constituents in the water, and when considering the removal or monitoring of dissolved ozone before the water enters the culture tanks (Summerfelt and Hochheimer 1997). Only limited and expensive technology exists to continuously monitor dissolved ozone at concentrations low enough to be safe for fish. Oxidation-reduction potential (ORP) probes have also been used, with varying degrees of success, as an indirect means to monitor and control dissolved ozone levels (Bullock et al. 1997).

Solids

Solids control is one of the most critical processes that must be managed in recirculating systems, because solids decomposition can degrade water quality and, thus, directly and indirectly affect fish health and the performance of other unit processes within recirculating systems (Chen et al. 1993b; Summerfelt 1999). Suspended solids can harbor opportunistic pathogens, are associated with environmentally induced disease problems, and have been reported to cause sublethal effects such as fin rot and direct gill damage (Klontz 1993; Noble and Summerfelt 1996; Wedemeyer 1996). Suspended and settleable solids may also affect reproductive behavior, gonad development, and the survival of the egg, embryo, and larval stages of fishes (Wedemeyer 1996).

Uneaten feed, feed fines, fish fecal matter, algae, and sloughed microbiological cell masses are all sources of solids production within recirculating systems (Chen et al. 1993b). Uneaten feed and feed fines, when present, can significantly contribute to total solids production (Bergheim and Åsgård 1996). However, feed waste can be controlled with better feeding techniques (Hankins et al. 1995; Summerfelt et al. 1995), and feed fines can be reduced with extruded diets, careful handling, and separation techniques (Mayer and McLean 1995). If waste feed is minimal, the total mass of solids produced can add up to 30–50% of the total mass of feed input daily (Liao and Mayo 1974; Piper et al. 1982).

Soon after being deposited, shear forces produced by water turbulence, fish motion, and pumps can break apart fecal matter, uneaten feed, and feed fines, causing the larger particles to disintegrate into much finer and more soluble particles. For example, if solids are filtered and stored in a pressurized-bead filter (a type of granular media filtration unit) between 24-hr backwash cycles, as much as 40% of the TSS generated in the recirculating system may decay (Chen et al. 1993a). The suspended organic solids common to recirculating aquaculture systems can exert a strong oxygen demand as they degrade into smaller particulate matter and leach ammonia, phosphate, and dissolved organic matter (Cripps and Kelly 1996; Garcia-Ruiz and Hall 1996). The fine particles and dissolved compounds produced are considerably harder to remove when broken apart and dissolved than when they were contained within the original fecal or feed pellet (Chen et al. 1993b; Cripps and Bergheim 2000). This dissolution process increases the water's oxygen demand as it deteriorates the water quality within the recirculating system and in the discharged effluent.

Dissolved Organic Matter

Fully recirculating system waters can accumulate dissolved organic matter, because daily replacement of water is usually low and because a portion of the dissolved organic compounds is not readily biodegradable due to their size or chemical nature. Dissolved organic matter promotes the growth of heterotrophic microorganisms, which can foul surfaces, screens, and orifices; cause an increase in the total oxygen demand of the system; and inhibit nitrification by competing for space and oxygen in a biofilter. Some heterotrophic (saprotrophic) microorganisms in the system may consist of facultative pathogens (i.e., *aeromonas*), which can cause disease outbreaks when the fish are stressed by handling or sublethal water quality conditions (Noble and Summerfelt 1996).

Gas Supersaturation

Gas bubble trauma can result from high total dissolved gas supersaturations (Wedemeyer 1996). However, the supersaturated nitrogen and argon components of water are much more likely to produce gas bubble trauma than supersaturated oxygen and carbon dioxide (Boyd and Watten 1989). Nitrogen saturation levels as low as 102–103% can affect juvenile fish (Wedemeyer 1996). Dissolved carbon dioxide contributes little to total gas pressure problems because it is much more soluble than nitrogen and argon (Colt and Orwicz 1991). And, in the absence of dissolved nitrogen supersaturation, DO concentrations would have to approach 300% of air saturation concentrations to create gas bubble trauma (Boyd and Watten 1989). Fortunately, properly designed recirculating systems will not usually have problems with high dissolved gas pressures. Gas supersaturation problems can occur when water is heated (and the saturation concentration of gas decreases) or when gas comes into contact with water at pressures greater than atmospheric. Sometimes, gas supersaturation problems are encountered when air is aspirated into water flowing at a high velocity through improperly sealed pipes. Denitrification produces dissolved nitrogen gas and the supersaturation that results can cause gas bubble trauma. Gas supersaturation problems are also produced when air is carried into deep water (>5 m, depending on species sensitivity). Such would be the case in a U-tube oxygenator (discussed later in this chapter) that, through a design flaw, could capture air-bubbles in the inlet flow and carry them through the depths of the unit.

Nutrients

Phosphorus and nitrate (discussed above) do not usually become growth limiting in freshwater recirculating systems, unless very low water exchange rates are required, in which case, the nitrate concentration can become limiting. Phosphorus and nitrate are primarily a concern if they are contained within water being discharged, because their accumulation in receiving waters can lead to eutrophication (Bergheim and Âsgård 1996; Cripps and Kelly 1996). Phosphorus and nitrogen can be either soluble or bound with particulate matter. Most of the effluent nitrogen released (75–80%) is in the form of dissolved ammonia or nitrate, depending upon the effective use of feed and how much nitrification has occurred (Heinen et al. 1996). Fecal solids contain approximately 1–4% phosphorus (on a dry weight basis) and approximately 2–5% nitrogen (on a dry weight basis). The filterable or settleable solids contain most (50–85%) of the effluent phosphorus, but relatively little (about 15%) of total effluent nitrogen (Heinen et al. 1996). The large variability in the phosphorus fractionation between dissolved and particulate matter is largely due to variability in feed formulation, which, in turn, influences the phosphorus solubility and availability to the fish. If present in a soluble form, phosphorus can leach rapidly from feed and fecal matter (Cripps and Kelly 1996; Garcia-Ruiz and Hall 1996). Therefore, the best nutrient management plan first focuses on reducing the waste production and then focuses on removing the waste particles from the water as rapidly and as gently as possible (Summerfelt 1999; Cripps and Bergheim 2000).

Recirculation Percentage and Makeup Water Requirements

Recirculating systems contain a certain water volume (V_{system}, m^3) within their fish tanks[2], treatment processes, pump sumps, and connecting piping. Even in extremely closed recirculating systems, water must be replaced as it is lost to evaporation, to random splashing, or to the drain while washing concentrated waste solids from settling basins or filters. Relatively small volumes of water are also added to provide a continuous overflow from certain recirculating systems, to provide heating or cooling, supply alkalinity, or flush away nitrate and dissolved or fine

[2]Note that most people do not precisely estimate the culture tank water volume by subtracting out the volume that is actually occupied by fish. This correction is probably not relevant unless extremely high fish densities (>50–100 kg/m^3) are used and would only be relevant if the culture tank water volume was to be used for calculating chemotherapeutant dose requirements.

particulate organic matter. Therefore, water can be added to recirculating systems continuously or on an intermittent basis. Either way, it is important to use the correct terminology when referring to the makeup water requirements of a specific recirculating system (Rosenthal and Black 1993; Timmons and Losordo 1994). The makeup water requirements should be expressed by dividing the average volume of makeup water added daily (r_{water}, m^3/d) by the best estimate of the total recirculating system volume (V_{system}, m^3), which is then multiplied by 100 to report the value as a percentage of water volume replaced per day:

$$\text{water volume replacement rate} = 100 \times \frac{r_{water}}{V_{system}}, \ \% \text{ per day.} \tag{5}$$

The water is circulated within the system at a design flow rate that is usually selected to supply adequate DO to the fish culture tanks, while at the same time transporting waste metabolites out of the culture tanks and to the water treatment units. Some recirculating systems continuously add a considerable makeup water flow. In these situations, the percentage of water flow that is reused (R) is sometimes reported. The reverse, the percentage of flow replaced, can be estimated by dividing the makeup water flow rate (r_{makeup}, L/min) by the overall recirculating water flow rate (Q, L/min) and then multiplying by 100:

$$\text{percent of flow replaced} = 100 \times \frac{r_{water}}{Q} = 100 \times (1 - R), \ \%. \tag{6}$$

Note that as R increases, so does the cost and complexity of waste treatment (Figure 1).

Therefore, to understand what is implied by a reported recirculating percentage, the units of water replacement must be specified explicitly, whether water volume is being exchanged daily or the percent of makeup flow exchanged continuously. Although both definitions provided are useful, many prefer using the system's daily water volume replacement, as defined in equation (5).

By their very nature, reuse systems demand far less water resources and produce a much more concentrated and treatable discharge than a system of similar production capacity that only uses water once. To maintain safe and healthy water quality, however, water reuse systems must rely either on dilution with makeup water or treatment processes to control accumulations of limiting waste metabolites, that is, dissolved carbon dioxide, NH_3, and suspended solids. Water exchange in serial reuse systems is considerably different than water exchange in partial- and fully recirculating systems, therefore, the two areas will be addressed separately.

Serial Reuse in Raceway Systems

When water is reused in raceway systems, the water exchange is used to transport DO to the fish and to flush waste metabolites from the system. Serial reuse raceway systems use dilution with fresh makeup water to control NH_3 accumulations, quiescent zones to capture settleable solids, and 30–80 cm hydraulic drops between raceway units to provide some aeration (for both carbon dioxide stripping and oxygen transfer) or oxygenation (Figures 2 and 3) (MacMillan 1992; IDEQ 1998).

Under steady-state conditions and with no waste removed at the culture unit, the mean concentration of waste discharged from the culture unit, that is, waste_{out} (mg/L), can be estimated from the sum of the concentration of waste entering the raceway cell, that is, waste_{in} (mg/L), plus the mean concentration of waste produced by the feeding fish, r_{waste}/Q (mg/L):

$$\text{waste}_{out} = \{\text{waste}_{in}\} + \left\{\frac{r_{waste}}{Q} \times \left(\frac{10^6 \text{ mg}}{1 \text{ kg}}\right) \times \left(\frac{1 \text{ day}}{1{,}440 \text{ min}}\right)\right\}, \quad (7)$$

where Q is the water flow rate through the culture tank (L/min). For practical purposes, the daily waste production rate, r_{waste} (kg waste per day), can be estimated from the mean feed-specific waste production constant, a_{waste} (kg waste per kg feed), the feeding rate, r_{feed} (kg feed per day per kg fish), and the culture tank's volume, V_{tank} (m^3), and fish density, ρ_{fish} (kg fish per m^3 culture volume), as shown in the following equation:

$$r_{waste} = V_{tank} \times \rho_{fish} \times r_{feed} \times a_{waste} . \quad (8)$$

These relations are illustrated graphically in Figure 10.

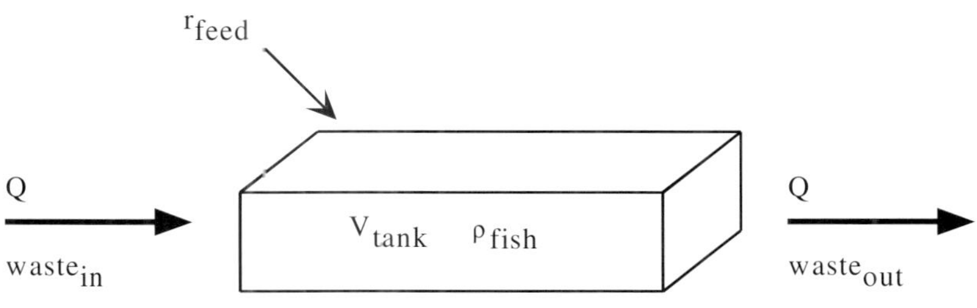

FIGURE 10. Waste production in a tank or raceway is dependent upon the feed rate (r_{feed}), tank volume (V_{tank}), and fish density (ρ_{fish}). Assuming no waste accumulation in the tank or raceway, the concentration of waste discharged from the vessel also depends on the water flow (Q) and concentration of waste entering the tank (waste_{in}).

Example: The TSS, TAN, and the dissolved carbon dioxide (DC) production and mean concentration leaving the culture tank can be calculated for a given unit. In this example, the culture tank has a volume of 100 m^3 (V_{tank}) and contains 250 g hybrid striped bass at a density of 70 kg/m^3 (ρ_{fish}). The fish are fed 1.5% body weight per day (r_{feed}) and subsequently produce about 0.27 kg TSS per kilogram feed (a_{TSS}), 0.031 kg TAN per kilogram feed (a_{TAN}), and 0.35 kg dissolved carbon dioxide per kg feed (a_{DC}). Therefore, daily production of waste TSS, TAN, and DC can be estimated in this example using equation (8), as follows:

$$r_{TSS} = 100\ m^3 \times \frac{70\ kg\ fish}{m^3} \times \frac{1.5\ kg\ feed}{100\ kg\ fish \times day} \times \frac{0.27\ kg\ TSS}{1.0\ kg\ feed} = 28.4\ kg\ TSS/day$$

$$r_{TAN} = 100\ m^3 \times \frac{70\ kg\ fish}{m^3} \times \frac{1.5\ kg\ feed}{100\ kg\ fish \times day} \times \frac{0.031\ kg\ TAN}{1.0\ kg\ feed} = 3.26\ kg\ TAN/day$$

$$r_{DC} = 100\ m^3 \times \frac{70\ kg\ fish}{m^3} \times \frac{1.5\ kg\ feed}{100\ kg\ fish \times day} \times \frac{0.35\ kg\ DC}{1.0\ kg\ feed} = 36.8\ kg\ DC/day$$

If water flow to the raceway is 6,670 L/min (i.e., 4 water volume exchanges every hour) and if the water entering the raceway contains no TSS or TAN, but contains 5 mg/L DC, then the mean concentration of waste metabolites discharged from this tank can be estimated using equation (7), as follows:

$$TSS_{out} = \{0\ mg/L\} + \left\{\frac{28.4\ kg\ TSS}{day} \times \frac{min}{6{,}670\ L} \times \frac{10^6\ mg}{1\ kg} \times \frac{1\ day}{1{,}440\ min}\right\} = 3.0\ mg/L\ TSS;$$

$$TAN_{out} = \{0\ \ mg/L\} + \left\{\frac{3.26\ kg\ TAN}{day} \times \frac{min}{6{,}670\ L} \times \frac{10^6\ mg}{1\ kg} \times \frac{1\ day}{1{,}440\ min}\right\} = 0.34\ mg/L\ TAN;$$

$$DC_{out} = \{5\ mg/L\} + \left\{\frac{36.8\ kg\ DC}{day} \times \frac{min}{6{,}670\ L} \times \frac{10^6\ mg}{1\ kg} \times \frac{1\ day}{1{,}440\ min}\right\} = 8.8\ mg/L\ DC.$$

These estimates ignore the equilibrium relationship between dissolved carbon dioxide and the total carbonate system; they also assume steady state conditions and that no TSS is stored in the raceway.

If water is reused through a series of raceway units, the water exiting the upstream raceway units is used as the water supply for the downstream raceway units. In this type of arrangement, when waste treatment processes are not used, the concentrations of waste metabolites increase from raceway to raceway in an *additive* manner directly proportional to the feed loading and, thus, the cumulative DO consumption. For every 10 mg/L of DO that fish consume, they will produce roughly 1.0–1.4 mg/L TAN, 13–14 mg/L DC, and 10–20 mg/L TSS.

It is informative to use a mass balance approach to examine the "6 kg of annual fish production for every 1 L/min of water flow" rule of thumb. Assuming steady-state DO consumption, a feed conversion of 1.2, and a DO consumption rate of 0.30 kg oxygen per kilogram of feed, we can estimate that it would only require about 4 mg/L of cumulative oxygen consumption to produce "6 kg of fish annually on 1 L/min of water flow;" that is,

$$\text{Oxygen demand} = \frac{6\,\text{kg fish}}{\text{yr}} \times \frac{\text{min}}{1\ \text{L}} \times \frac{1.2\,\text{kg feed}}{1\,\text{kg fish}} \times \frac{0.3\,\text{kg oxygen}}{\text{kg feed}} \times \frac{\text{day}}{1{,}440\,\text{min}} \times \frac{\text{yr}}{365\,\text{days}} \times \frac{10^6\ \text{mg}}{\text{kg}}$$

$$= 4.1\,\text{mg/L of cummulative oxygen demand.}$$

Furthermore, the concentration of TSS and TAN generated would be only 4–8 mg/L and 0.4–0.6 mg/L, respectively. For this reason, serial-reuse systems typically generate dilute wastewater flows. If the water used in the raceways has a relatively low pH, then the fraction of NH_3 will be relatively low (Figures 8 and 9). If the alkalinity in the water is relatively low, then the dissolved carbon dioxide produced will further acidify the water and further reduce the fraction of TAN existing as NH_3. Therefore, a slightly acidic water with a relatively low buffering capacity can sometimes maintain safe water quality levels in serial reuse systems without biofiltration (and partial-recirculating systems) even when TAN concentrations exceed 5–10 mg/L (Figure 9). This assumes that settling within quiescent zones can control the suspended solids levels and that some of the dissolved carbon dioxide accumulation can be stripped as the water cascades through 30–120 cm hydraulic drops placed between raceway units.

Example: Water with a relatively low buffering capacity (e.g., alkalinity of 50 mg/L as calcium carbonate) is used intensively within a serial reuse system so that dissolved carbon dioxide production, in conjunction with an initially acidic pH, has driven the pH down to 6.4 before the water enters the last raceway unit. At a water temperature of 25°C, the equilibrium relationships (Refer to online spreadsheets at *http://www.fisheries.org/publications/epubs/fm2e/appendices.shtml* for an ammonia calculator.) reported in Tables 3 and 4 (or Figure 9) can be used to estimate that only 0.14% of TAN would exist as toxic NH_3, but that the dissolved carbon dioxide would have reached nearly 40 mg/L. If the water contained even as much as 10 mg/L TAN, it would still not create toxic NH_3 concentrations. However, further fish culture may be limited because the 40 mg/L dissolved carbon dioxide concentration may already have reached or surpassed limiting levels for certain warmwater fish. After this intensive water reuse, the suspended solids levels in the water may have also reached deleterious levels.

Partial- and Fully Recirculating Systems

In a parial- or fully recirculating system operating at steady state, waste accumulation in the system is controlled by the rate at which the waste is produced, that is, r_{waste} (kg waste per day), by the efficiency that water treatment units remove the waste, that is, the waste fraction removed, (f_{rem}), and by the fraction of water flow that is reused, R (Figure 11). If the makeup water flowing into the recirculating system contains a measurable concentration of waste, waste_{new} (mg/L), the makeup flow will not be as effective at diluting the wastes produced in the recirculating system. However, the concentration of waste discharged from a culture tank (waste_{out}, mg/L) can still be estimated using the following equation, first developed by Liao and Mayo (1972):

$$\text{waste}_{out} = \left\{ \frac{1}{1 - R + (R \times f_{rem})} \right\} \times \left\{ \left(\frac{r_{waste}}{Q} \times \frac{10^6 \text{ mg}}{1 \text{kg}} \times \frac{1 \text{day}}{1{,}440 \text{min}} \right) + (1 - R) \times (\text{waste}_{new}) \right\}. \quad (9)$$

This equation was developed assuming that no waste accumulation can occur in a culture tank and that the recirculating system is operating at steady-state conditions, that is, water flow rates, waste production rates, and unit process treatment efficiencies are relatively constant. However, the equation can be used to estimate most day-to-day operating water quality conditions, except during rapid changes in water quality (such as DO) that occurs with feeding events or when something startles the fish.

Most warmwater recirculating aquaculture systems operating in temperate climates reuse high fractions of their flow (to conserve heated water) and generally operate with only 5–40% of the total system water volume exchanged daily—equivalent to a fraction of water flow reused R greater than or equal to 0.99. In such warmwater recirculating systems, waste accumulation depends mainly upon the f_{rem} across the water treatment units, as illustrated in Figure 12. Even in recirculating systems that use significantly more makeup water (R = 0.90 – 0.99) to maintain cool- or cold-water temperatures, the accumulation of wastes is largely controlled by the f_{rem} across the water treatment units (Liao and Mayo 1972) (Figure 12). Note, also, that when the removal efficiency

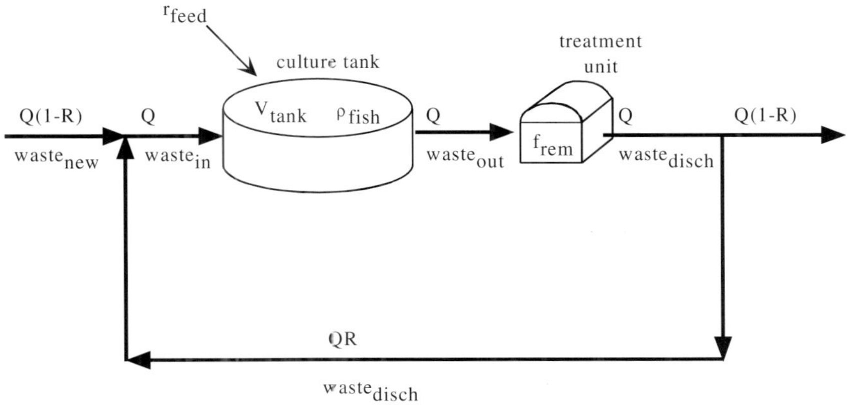

FIGURE 11. The waste concentration in the water exiting a culture tank ($waste_{out}$, mg/L) in a recirculating system is dependent upon the feed rate (r_{feed}), the water treatment unit efficiency (f_{rem}), the fraction of water flow (Q) that is reused (R), and the concentration of waste in the make-up water ($waste_{new}$) (after Liao and Mayo, 1972).

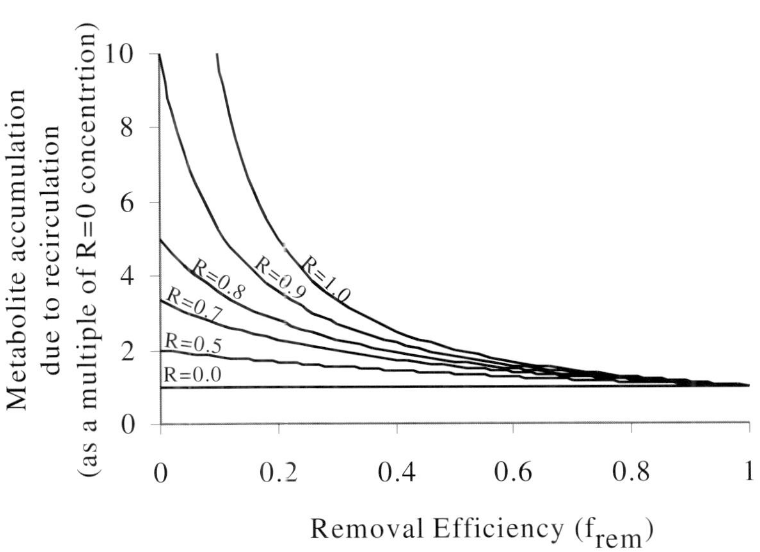

FIGURE 12. The magnitude of metabolite accumulation within recirculating systems can be expressed as a multiple of the concentration that would occur in a single-pass system; this accumulation is dependent on the ratio of water flow reused (R) and the fraction of metabolites removed (f_{rem}) in the water treatment processes (after Liao and Mayo, 1972).

of the treatment process (f_{rem}) is greater than about 60%, then the fraction of water flow reused (*R*) plays a much smaller role controlling metabolite accumulation (Figure 12). Conversely, when the treatment process does not remove a large portion of the passing waste metabolites (i.e., f_{rem} < about 40%), then the recirculating system would accumulate large concentrations of waste with respect to the concentration that would accumulate within a single-pass system (Figure 12). Therefore, maintaining the best water quality in reuse systems requires selection of water treatment processes that operate at high removal efficiencies (Liao and Mayo 1972), especially when considering toxic variables such as NH_3, nitrite, and dissolved carbon dioxide. Otherwise, these waste metabolites can accumulate in the water to multiples much greater than the concentration of waste that would be generated in just a single-pass through the culture tank, that is, at *R* equal to 0 (Liao and Mayo 1972) (Figure 12).

Nearly all of the TAN produced in fully recirculating systems is eventually converted to nitrate by bacterial nitrification. Therefore, nitrate concentrations can easily reach levels of several hundred ppm in recirculating systems that exchange only a small fraction of their water volume daily. In recirculating systems that have no appreciable nitrate removal by denitrification or by algae or plant uptake, the concentration of nitrate accumulating is a measure of water use. However, if the culture species is sensitive to nitrate concentrations of only several hundred mg/L, then the fraction of water flow reused within the recirculating system would have to be decreased or a bacterial denitrification unit added. If simple dilution is used to control nitrate accumulation, then equation (9) can be rearranged (assuming $waste_{new} = 0$ mg/L) to estimate the maximum water reuse fraction (*R*) necessary to control nitrate at a predetermined safe level. The estimate for *R* can be used to estimate the requirements for makeup water flow rate [$Q(1 - R)$].

Due to their low water exchange, partial-reuse systems and fully recirculating systems are much more effective at capturing and concentrating wastes than single-pass systems and serial-reuse raceway systems. The Freshwater Institute (Shepherdstown, West Virginia) has designed and evaluated a partial-recirculating system that incorporates "Cornell-type" dual-drain circular culture tanks to provide uniformly healthy water quality, optimized water velocities for fish health and condition, and solids removal that is rapid, gentle, and much more efficient (Summerfelt et al. 2000a). The relatively small (<15% of the total system flow) and concentrated discharge from the partial-reuse system was treated with a microscreen drum filter, which resulted in an overall suspended solids capture efficiency of 80%. However, in a fully recirculating systems that only discharges water to flush solids, 100% of the total solids, biochemical oxygen demand, and phosphorus, and nitrogen (disregarding denitrification) waste are captured and discharged with the solids wash water. Even in a fairly open cold-water recirculating system, where the makeup water continually replaces 6% of the reused flow (which amounts to a system volume exchange every 12 h), over 97% of the TSS produced daily is discharged with the low-volume microscreen filter wash water (Heinen et al. 1996). In comparison, Mudrak (1981) reported that the overall suspended solids capture efficiency was only 25–51% at three serial-reuse raceway operations that used quiescent zone settling and off-line settling ponds.

Recirculating System Components and Design

As mentioned earlier, water reuse systems must use processes and equipment to supply DO and to control the accumulation of toxic wastes such as ammonia and carbon dioxide. The water treatment units should be designed and managed so water quality does not degrade to the point where it stresses the fish, reduces growth rate, or causes mortality. Water reuse systems should also be designed to minimize waste generation or should be operated in a manner that keeps the waste in a form that can be readily removed. Fecal matter, uneaten feed, and feed fines can be rapidly broken into much finer and more soluble particles by water turbulence, fish motion, scouring along a tank/pipe bottom, and pumping. Dissolved and fine particulate matter are much more difficult to remove from water than are larger particles. Therefore, to maintain water quality within recirculating systems and, also, to help the fish farm effluent meet discharge limits, it is extremely important that solids are removed rapidly and gently to reduce particulate dissolution and leaching. To achieve this, the culture tanks, piping systems, sumps, pump placement, and treatment processes should be selected, designed, located, and operated to allow solids to be removed rapidly and with the least turbulence, mechanical shear, or opportunity for microbiological degradation.

This section reviews the application of water treatment technologies to minimize water use and concentrate wastes.

Rearing Units

The type of culture tanks used and their management have a significant influence on fish farm profitability. Culture tanks, their water distribution and collection components, support equipment, and floor space requirement constitute a large portion of a farm's capital. As well, the management of the culture tanks and the control of their waste discharge can constitute a large portion of farm operating costs. Therefore, a large financial incentive exists to select the best culture tank type, scale, and operating strategy to optimize fish farm profitability (Summerfelt et al. 2000c).

Circular Culture Tanks

Circular tanks have several desirable features: they can provide a uniform culture environment; they can be operated under a wide range of rotational velocities to optimize fish health and condition; they can be used to rapidly concentrate and remove settleable solids; they allow for good feed and fish distribution; and, they can permit designs that allow for visual or automated observation of waste feed to enable satiation feeding (Timmons et al. 1998).

The water volume in the culture tank is exchanged with fresh water at a rate sufficient to maintain water quality throughout the tank, even when operating up to maximum carrying capacity. The water inlet structure design and orientation plays a large role in determining the homogeneity of water quality and water rotational velocities throughout a circular culture tank, as well as the ability to rapidly flush solids from these tanks (Timmons et al. 1998). The tank becomes self-cleaning when a rotational flow is produced that is strong enough to create a secondary radial flow along the tank bottom that can move settleable fish feed and fecal matter to the center drain. The strength of the radial flow is dependent upon the water's rpm. Typical circular tank geometries require water rotational velocities greater than 15–30 cm/s to effectively flush solids (Timmons et al. 1998). However, fish generally require water velocities on the order of 0.5–2.0 times fish body length per second to maintain fish health, muscle tone, and respiration. Ideally, a rotational velocity can be selected that is both beneficial for the fish and can rapidly flush settleable solids from the culture tank.

The self-cleaning attribute of the circular tank is also related to the overall rate of flow leaving the bottom-center drain, the strength of the bottom radial flow towards the center drain, and the swimming motion of fish resuspending the settled materials (Timmons et al. 1998). However, only about 5–20% of the total flow typically passed through a circular tank is required to flush settleable solids through the tank's bottom-center drain. Because of their ability to concentrate settleable solids at their bottom and center, circular culture tanks can be managed as "swirl settlers." In this situation, a relatively small portion of the flow (5–20% of the total) is withdrawn through the bottom-center drain with the bulk of settleable solids, while the majority of flow is withdrawn free from settleable solids at an elevated drain (Mäkinen et al. 1988; Eikebrokk and Ulgenes 1993; Twarowska et al. 1997; Summerfelt et al. 2000a,b).

Some designs place the elevated drain in the center of the tank (Figure 13), which helps to strengthen the radial flow along the tank's bottom

FIGURE 13. A dual-drain design by AquaOptima AS (Trondheim, Norway), uses a solid plate fixed just above the bottom center drain to accelerate particles into the bottom drain; it then withdraws the majority of flow through the screened area above the approach plate.

toward the center (Timmons et al. 1998). The Cornell-type dual-drain tank, is a circular culture tank with a center drain on the tank bottom and an elevated drain partway up the tank sidewall (Figures 4, 5, and 14). Separating the two drains so that one is partway up the tank sidewall and the other is in the tank center makes the Cornell-type dual drain

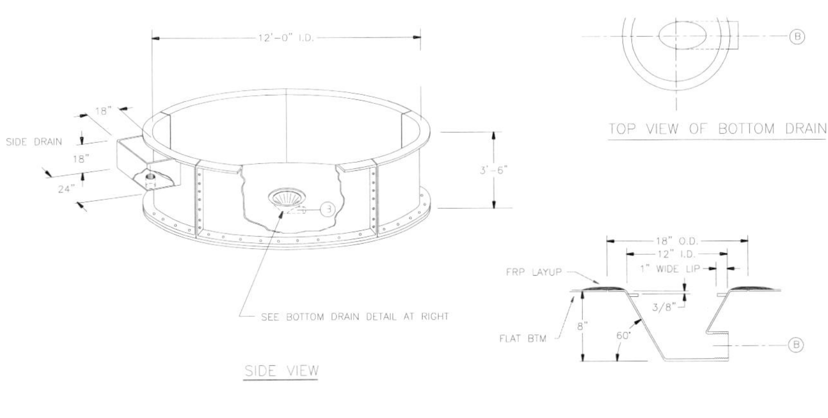

FIGURE 14. Details of a circular culture tank design used at the Freshwater Institute (Shepherdstown, West Virginia) to concentrate settleable solids at its bottom-center drain while discharging the majority of water flow, i.e., 80 to 95% of the total flow, through its "Cornell-type" side-wall drain (from Summerfelt et al., 2000d). Drawing courtesy of Red Ewald, Inc., Karnes City, Texas.

tank easy to install, even as a retrofit on existing circular culture tanks (Summerfelt et al. 2000c). Also, because the elevated drain in the Cornell-type design is partway up the tank sidewall, it does not capture many of the solids that sometimes "plume" up in the center of circular culture tanks that are operated with strong radial flows (Summerfelt et al. 2000a, 2000b). Therefore, the Cornell-type dual drain design can be effective at rapidly fractionating settleable solids to the tank's bottom-center drain (Summerfelt et al. 2000a, 2000b).

A study on use of the Cornell-type dual-drain tank within a partial-reuse system (Figure 4) at the Freshwater Institute (Summerfelt et al. 2000a) found that 78% (8.91 ± 0.75 kg/day) of the TSS produced daily were rapidly and gently flushed from the reuse system through the culture tank's bottom-center drain (Table 5). This discharge leaving the system amounted to only 12–15% of the system's total water flow, but this flow flushed the majority of particles from the system within only 1–5 min of their deposition into the culture tank. The other source of solids discharge from the partial-reuse system, the system's drum filter backwash (Figure 4), only contained 22% (2.51 ± 0.32 kg/day) of the daily solids produced (Table 5). Solids fractionation within the culture tank can be extremely effective—the TSS concentration discharged from the partial-reuse system through the three culture tanks' bottom-center drains

TABLE 5. The average concentration and mass of total suspended solids (± standard error) at key points within the Freshwater Institute's partial-reuse system (Figure 4; from Summerfelt et al. 2000a).

Sampling sites	Concentration of total suspended solids, mg/L	Mass of total suspended solids, kg/day
Bottom drain effluent		
Tank 1	25.26 ± 2.64	2.69 ± 0.27
Tank 2	27.23 ± 4.31	3.07 ± 0.47
Tank 3	26.57 ± 4.11	3.20 ± 0.55
Overall	Mean = 26.18 ± 2.07	Total = 8.91 ± 0.75
Side drain effluent		
Tank 1	2.34 ± 0.25	2.13 ± 0.25
Tank 2	2.92 ± 0.39	2.28 ± 0.31
Tank 3	2.44 ± 0.28	2.19 ± 0.28
Overall	Mean = 2.50 ± 0.17	Total = 6.60 ± 0.48
Drum filter effluent	1.20 ± 0.09	2.51 ± 0.32
Culture tank influent	1.30 ± 0.11	3.28 ± 0.34

(26.2 ± 2.1 mg/L) was 10 times greater than the TSS concentration exiting the three culture tanks' sidewall drains (2.5 ± 0.2 mg/L) (Table 5). Further treatment of the flow discharged from the tank's bottom-center drain using a commercial microscreen filter then captured 82% (±4%) of the TSS in the bottom discharge, so that only an average of 3.5 ± 0.8 mg/L TSS would be discharged from the facility.

Note that dual-drain culture tanks are not always as effective at concentrating waste solids in all culture situations, because this depends on the settling rate of waste feed and fecal matter, as well as the culture tank hydraulics. Problems can arise if finer or less dense solids do not concentrate effectively at the bottom center of dual-drain tanks.

Under the correct circumstances, concentrating solids in the relatively small underflow leaving circular tanks with dual-drains can have large economic implications; it can reduce the capital cost, space requirement, and headloss requirement of the downstream solids removal units (Timmons et al. 1998).

Raceway Culture Tanks

Raceways are found at many commercial fish farms, as well as at state and federal fish hatcheries, especially at those used to produce salmonids where aquaculture has tapped into huge water resources. Such is the case in Idaho's Snake River valley, where some of the largest rainbow trout producers in the world use large raceways that are typically 24–46 m long, 3–5.5 m wide, and 0.8–1.1 m deep (IDEQ 1998). Raceways are very convenient culture tanks for managing fish because they are usually shallow enough to work in while wearing chest waders. Also, their constant width and long aspect ratio (i.e., length:width), often 10:1, simplifies fish crowding and grading (IDEQ 1998). However, the long aspect ratio and water flow characteristics can also make distributing feed throughout raceways more difficult than it is within circular tanks (Summerfelt et al. 2000c).

Water flows through a raceway with minimal back mixing. In this type of plug-flow situation, the metabolism of fish in the raceway produces a gradient in the concentration of DO, waste solids, ammonia, and carbon dioxide. The best water quality exists where the water enters the raceway; it then deteriorates as the flow moves through the fish toward the raceway outlet. This may cause fish to congregate at the head of the raceways.

Raceways also use long aspect ratios to minimize their cross-sectional area and, thus, maximize the water velocity flushing out the tank. However, raceway water flow requirements are often based on the oxygen demand of the fish, which only require a water velocity of 2–4 cm/s to support typical raceway fish production densities (Westers 1991; IDEQ

1998). These water velocities are often large enough to exchange the water volume in one raceway unit every 10–15 min (Westers 1991), producing a hydraulic exchange rate roughly 2–6 times greater than what a typical circular culture tank uses. Therefore, to get the most effective use of the water flow, groups of raceways are often operated in series, as discussed in the Serial reuse section (Westers 1991; MacMillan 1992; IDEQ 1998). However, the water velocities in raceways are typically lower than the optimum water velocities (0.5–2.0 body length per second) recommended for fish health, muscle tone, and respiration.

Also, the water velocities are much lower than the 15–30 cm/s required to rapidly transport waste solids along a raceway's concrete bottom. So, if not for the swimming activity of larger fish that helps to resuspend the settleable solids, a substantial amount of solids would settle in the raceway's rearing area (Westers 1991). In some instances, mounting a series of baffles perpendicular to the water flow, across the width of a raceway, and spaced at intervals equaling the raceway width, increases the rate that solids are flushed from the rearing area (Westers 1991; IDEQ 1998). The baffles create a gap beneath their bottom edge and the raceway bottom, which produces a region of high water velocity (20–30 cm/s) to help sweep solids out of the raceway. However, baffles are not widely used because they must be moved to effectively manage the fish (IDEQ 1998).

Some of the solids swept from a raceway's fish rearing area can be captured within quiescent zones where fish are excluded and settled solids are undisturbed (Westers 1991; IDEQ 1998). Quiescent zones should be designed to produce a surface overflow rate less than 0.01 m^3/s per square meter of surface area (IDEQ 1998). Captured solids are removed as often as daily or as infrequently as twice per month, usually by vacuuming or flushing down a central drain. Operating labor for solids removal has been reported to exceed 25% of the total farm labor (IDEQ 1998).

A raceway's inability to rapidly flush solids allows solids to degrade into smaller or dissolved components that are much harder to remove and, thus, could readily accumulate if the raceway were used within a fully recirculating system. Therefore, this author does not consider raceways as ideal a culture environment as circular tanks for use in partial- or fully recirculating systems used to raise a finfish that is relatively sensitive to elevated levels of suspended solids.

Particulate Removal Systems

Removing particulate matter from water reuse systems is critical for maintaining fish health and the efficiency of all the other water treatment processes. Waste solids can be settled and concentrated within the rearing tank (discussed earlier) or they can be removed from the bulk flow leaving the culture tank using a treatment unit such as a settling basin, microscreen filter, or granular media filter (Summerfelt 1999; Cripps and Bergheim 2000). Ozone and foam fractionation are water treatment processes usually used to remove dissolved organic matter, so they work well to compliment solids removal via settling or filtration (Rosenthal and Black 1993; Summerfelt and Hochheimer 1997). Ozone and foam fractionation are described in the dissolved organic matter and protein removal section found later in this chapter.

Settling Basins

Settling basins are designed to capture sinking particles when they intercept the bottom of the basin, while relatively clarified water flows out of the basin. Waste solids, when freshly produced at the fish culture tanks, usually settle at a constant rate as discrete particles, because they do not change in size, shape, or density (Chesness et al. 1975). For example, fresh fecal matter from rainbow trout is generally intact and can settle relatively fast at 1.7–4.3 cm/s (Warren-Hansen 1982; Wong and Piedrahita 2000), depending upon particle size and specific gravity (typically 1.005–1.20, depending upon conditions). Waste feed pellets that are formulated to sink can settle even faster than fecal matter, at 14–18 cm/s velocities (Juell 1991), depending upon many factors associated with the specific pelleted feed. However, if the waste solids have been allowed to break apart or if the waste solids are detached biofilm material, then these solids might settle much slower at about or less than 0.05–0.1 cm/s (IDEQ 1998). Therefore, designing settling basins based on a characteristic settling velocity is made more difficult due to the many types of particulate matter that must be removed from water reuse systems, because each has its own characteristic settling velocity.

If the characteristic settling velocity of a targeted waste solid fraction is known, then the settling basin can be sized by selecting its overflow rate (v_o, in m^3/s flow per m^2 plan area) to be less than the assumed characteristic settling velocity (IDEQ 1998). The overflow rate (also termed hydraulic loading rate) of the basin is simply the water flow rate through the settling basin (Q, in m^3/s) divided by the horizontal plane area (A_h = length × width, or = π × radius^2, in square meters) of the settling basin. The surface area for the settling basin can be estimated using the following equation:

$$v_o = Q/A_h. \qquad (10)$$

Because nonideal conditions produced from wind and inlet/outlet zone turbulence can occur, the overflow rate used for settling basins design is usually at least one-half the characteristic settling velocity. This acts to oversize the basin and help prevent nonideal conditions from decreasing settling basin performance (IDEQ 1998). The Idaho Division of Environmental Quality has published waste management guidelines for Idaho aquaculture operations (IDEQ 1998), which recommends overflow rates for settling basins designed to treat

- the full flow from a fish farm ($v_o \leq 0.0040$ m^3/s flow per m^2 surface area),
- the flow leaving a raceway unit within a quiescent zone ($v_o \leq 0.0095$ m^3/s flow per m^2 surface area), and
- the backwash or quiescent zone cleaning flows in an off-line settling basin ($v_o \leq 0.00046$ m^3/s flow per m^2 surface area).

Other studies recommend v_o less than or equal to 0.0014–0.00047 m^3/s flow per m^2 surface area (Summerfelt 1999). Typically, settling pond water depths are 1–2 m and length to width ratios are 4:1–8:1 (IDEQ 1998). Settling basins are also designed to promote laminar-type plug flow and, thus, reduce turbulence and short circuiting. For this reason, settling basins may incorporate full-width weirs at the settling basin inlet and outlet to more evenly distribute and collect flow across the width of the settling basin. Settling basins are usually installed in pairs, so that one settling basin can be placed off-line for cleaning while the other treats the entire process flow.

Theoretically, settling ponds can capture more than 90% of their solids loading if resuspension of settled solids was not a factor (Henderson and Bromage 1988). However, nonideal conditions that occur in many settling ponds keep them from effectively removing solids when TSS inlet concentrations are less than 10 mg/L, which makes it difficult to achieve effluent TSS concentrations of less than 6 mg/L (Henderson and Bromage 1988). Waste removal efficiency measured under typical operating conditions may be reduced by dissolution of nutrients, resuspension of settled solids, and the production of algae within the settling basin water column (Figure 15). Particulate matter may also be resuspended by the action of fish that escape into the settling area and the microbial production of gasses that float waste solids.

Although settling basins have been the most prevalent unit process used to remove solids from aquaculture effluents, this treatment process has several problems that keep it from being an ideal method for removing suspended solids from large flows, especially within recirculating aquaculture systems. As a result of the settling characteristics of aquaculture solids (i.e., a large proportion of the particles are only slightly more dense than water and are < 100 mm), settling basins are often considered too large or not effective at removing solids from recirculating systems. The relatively high nitrate concentrations that

FIGURE 15. Solids flotation, algae production, and nutrient leaching can degrade water quality discharged from settling ponds, as shown in this radial-flow, off-line settling tank used to capture solids flushed from the quiescent zones at the Pennsylvania Fish and Boat Commission's Big Spring Fish Culture Station, Newville, Pennsylvania.

occur in fully recirculating systems can diffuse into the solids stored in settling basins and become denitrified within the solid matter. In a matter of only a few hours, the resulting nitrogen gas produced can float the solids collected within the settling basin. Modern recirculating systems do not often use settling basins to treat the full recirculating flow, because solids control is so critical and space is especially valuable if it must be housed in a building. However, use of relatively small solids traps have been reported to effectively capture settleable solids that are already concentrated within a relatively small flow (5–20% of the total flow) discharged out the bottom drain within dual-drain circular culture tanks (Eikebrokk and Ulgenes 1993; Eikebrokk and Ulgenes 1998; Losordo et al. 2000). These solids traps are often distributed one to each circular culture tank, so that the solids can be captured rapidly and with little opportunity for degradation.

Swirl settlers and tube or plate settlers are two variations on the principle of sedimentation that have been used in some recirculating systems in an effort to minimize the high settling basin surface area requirement and to increase the efficiency of solids removal.

Swirl Settlers. Swirl settlers (separators), also called tea-cup settlers and hydrocyclones, operate by injecting the water at the outer radius of a conical tank such that the water spins around the tank's center axis

(Figure 16). The primary water flow spinning inside the tank creates a secondary radial flow along the bottom toward the tank center. This appreciable radial flow carries settleable solids to the center drain and maintains the self-cleaning aspect of the swirl separator. In aquaculture, swirl separators are used in low head applications to concentrate solids in a smaller underflow, approximately 5–10% of the total flow (Eikebrokk and Ulgenes 1993; Twarowska et al. 1997; Eikebrokk and Ulgenes 1998; Losordo et al. 2000). Traditionally, swirl separators have been used to treat wastewater flows that contain particles of high specific gravity, for example, sand and grit. Because aquaculture solids have specific gravities only slightly greater than water, concentrating these solids with swirl separators is not always guaranteed and performance is strongly dependent on maintaining proper hydraulics in the swirl separator. However, with an upper and a lower drain, swirl separators do not store the solids in the flow, which is a definite advantage over a settling basin.

Tube and Plate Settlers. Tube settlers or plate settlers pass the water flow through an array of inclined tubes or plates (Figures 17 and 18). Use of tube or plate settlers increases the effective settling area per unit volume and reduces the depth a particle must settle to contact a surface, which can increase solids capture efficiencies and reduce the required settling area of the unit. Surface loading rates for sizing tube and plate settlers

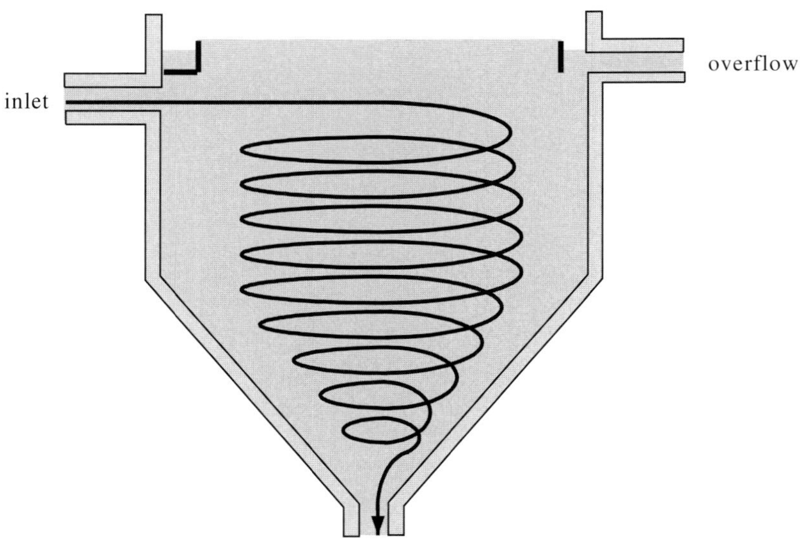

FIGURE 16. A swirl separator used to concentrate settleable solids in the relatively small (5–10% of total flow) underflow (from Summerfelt, 1999).

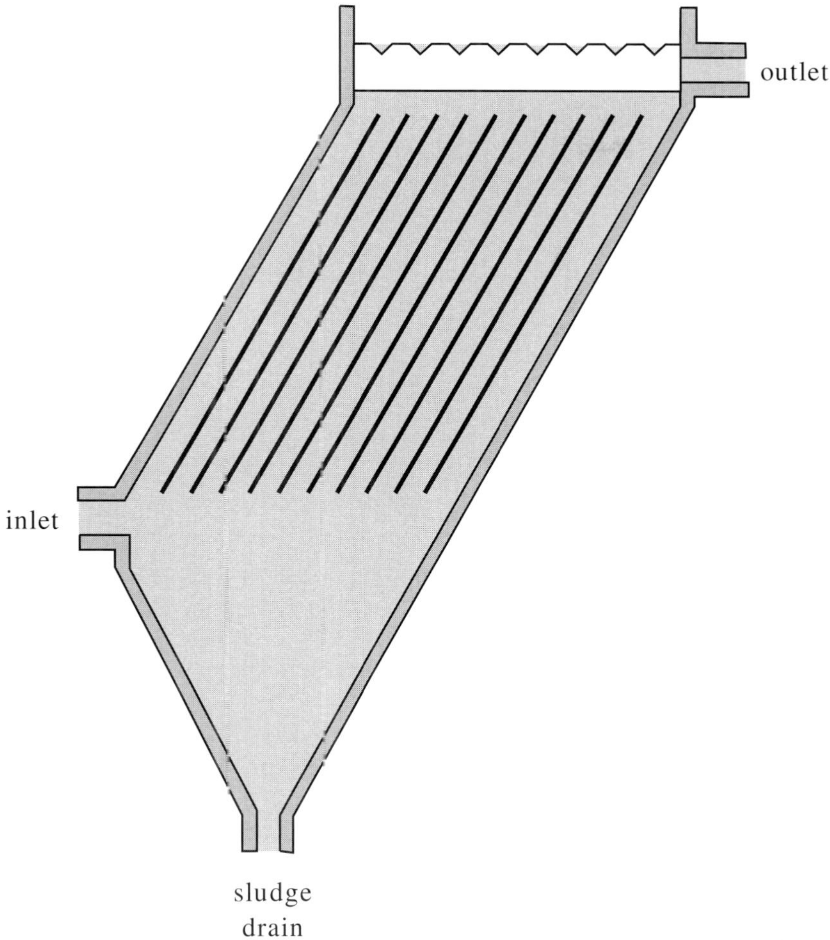

FIGURE 17. Tube or plates settlers have a large effective settling area per unit volume, which reduces the depth a particle must settle before it contacts a surface (from Summerfelt, 1999).

have ranged from 1.5 m^3/h (McLaughlin 1981), 7.4 m^3/h (Libey 1993), and 67 m^3/h (Parker 1981) flow per m^2 surface area, depending upon how conservative or aggressive the design is applied to an aquaculture application.

Tube or plate settlers have been used to remove solids in water reuse systems (Figure 18), particularly when space has been limiting and before the mid-1990s, when microscreen filter use became more widespread. However, tube or plate settlers are relatively expensive and biological growth on the tubes or plates generally requires periodic draining followed by washing down the tubes or plates to remove the captured

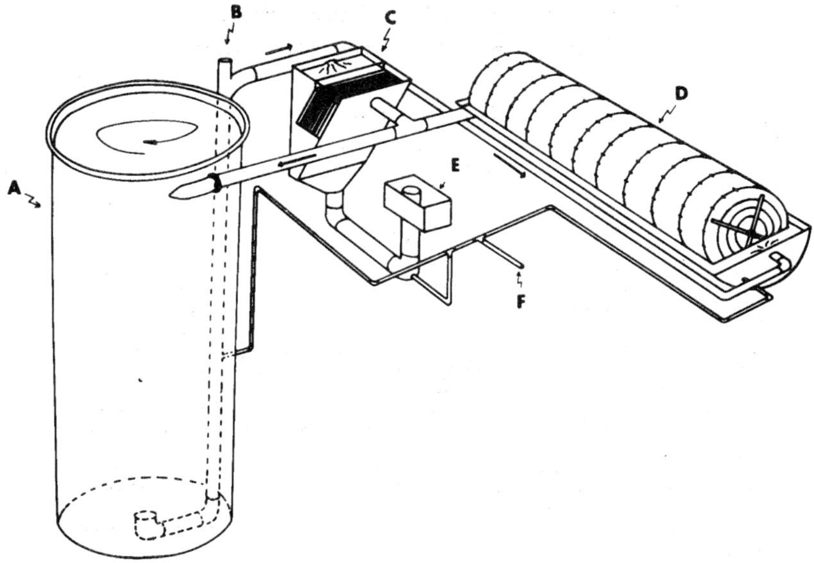

FIGURE 18. Several low-head water reuse systems have incorporated plate settlers (C) and rotating biological contactors (D), which allows them to be operated by air-lift pumps (B) as illustrated by Parker (1981). This system also contains a subsurface silo (A), sludge hopper (E), and an air input (F).

solids. Cleaning these units can be labor intensive. Also, the water and the heat it contains are both lost during the cleaning cycle, which can be expensive and amount to a considerable discharge for certain warmwater recirculating systems. Based upon operational difficulties, Timmons (2000a) does not recommend using tube or plate settlers in recirculating systems.

Microscreen Filters

Microscreen filters are sieves that strain water-bound particles larger than the filter screen openings from the process flow (Mäkinen et al. 1988; Bergheim et al. 1993; Cripps 1995; Cripps and Kelly 1996; Kelly et al. 1997; Cripps and Bergheim 2000). A cleaning mechanism removes solids from the filter continuously, periodically, or on demand. Cleaning can be by hydraulic flushing, pneumatic suction, mechanical vibration, or raking. The drum (Figure 19), disk, and incline belt microscreen filters are the three main variations used.

Individual microscreen filter units are available that can treat flows from 400 to 50,000 L/min. Economies of scale realized when purchasing large microscreen filters make them especially attractive for removing solids from flows greater than 2,000–4,000 L/min, where there are few viable solids removal technologies available. Therefore, microscreen fil-

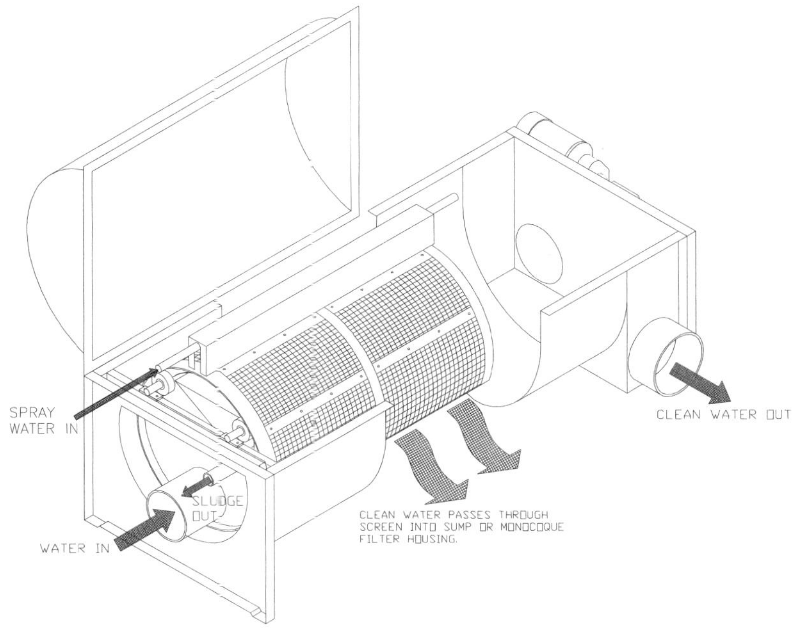

FIGURE 19. A process flow diagram for a microscreen drum filter. Drawing courtesy of PRA Manufacturing Ltd., Nanaimo, British Columbia, Canada.

ter units are generally more affordable when one large unit is centrally located to remove solids from a combined water flow that has been discharged from a row of culture tanks. Using one relatively small drum filter to treat the flow discharged from every single culture tank could be relatively cost prohibitive.

Microscreen filters are also advantageous because they require little space, have a high hydraulic capacity, have an acceptable pressure drop, and are relatively easy to install. However, one of the most important reasons that microscreen filters are selected for use within recirculating systems is that they are one of the few devices that do not store the solids captured within the flow (Summerfelt 1999). Microscreen filter units rapidly remove the captured solids from the process flow within the reuse system because the solids are washed from the screens several hundred to several thousand times daily. Required frequent backwash cycles come with an operating cost, that is, a pressure wash system (400–700 kPa) that requires maintenance (Summerfelt 1999) and that produces a sludge water volume of approximately 1% of the total passing flow for continuous washed units and as low as 0.1–0.3% of the total passing flow with intermittent flushed drum filters. In warmwater recirculating aquaculture systems, the cost of heating makeup water can be relatively high, and replacing water lost from the system during filter backwash would add to operating costs.

The performance of microscreen filters is largely dependent on the size of the microscreen panel openings, which influences the unit's hydraulic capacity, the fraction of particles removed, the sludge water production rate and concentration, and the filter wash frequency (Cripps 1995; Kelly et al. 1997). Finer screen openings remove more solids; however, selection of the minimum microscreen opening size is a compromise between treatment requirements and operational factors, such as hydraulic loading and backwash requirements. The best microscreen opening size may be situation specific, but microscreen panel openings of 60–100 μm are typically used. Reports from Norway indicate that decreasing the sieve panel opening from 60 μm to 30 μm does not improve solids capture (Cripps and Bergheim 2000). Average suspended solids removal across microscreen filter units have been reported to range from 22 to 70% within recycle systems and from 68 to 80% in flow-through systems (Summerfelt 1999). However, solids removal efficiency across a microscreen filter is largely dependent upon the concentration of solids entering the filter in the process flow. Solids capture data from four drum filters in use at the Freshwater Institute was graphed (Figure 20) and fit with a logarithmic regression model to provide an equation to estimate solids capture efficiency across a microscreen filter as a function of the filter's inlet TSS concentration:

$$\text{TSS capture percentage} = [16.2 \times \text{LN}(\text{inlet TSS concentration})] + 32.6. \quad (11)$$

Microscreen filter sizing is based on the maximum flow rate that must be treated, the concentration and characteristics of the waste solids to be removed, and the desired microscreen opening size. Unfortunately, a precise method for sizing microscreen filters has not been published and selection of filter units must be based largely on discussions with commercial distributors or those with prior relevant microscreen filter experience. However, care must be used when determining the maximum flow rate that will be loaded on the microscreen filter, because these filters will often receive all tank cleaning flows in addition to the design flow that the units were originally sized to treat. Too frequently, recirculating systems have been designed where the increases in flow volume and solids concentration during cleaning events were overlooked. In these instances, cleaning flows can overwhelm the microscreen filter and water containing high solids levels can be pumped back to the fish. Ideally, cleaning flows are best flushed completely out of recirculating systems to ensure that the fish are never exposed to the high levels of particulate matter (even after filtration) that pipe, tank, or channel

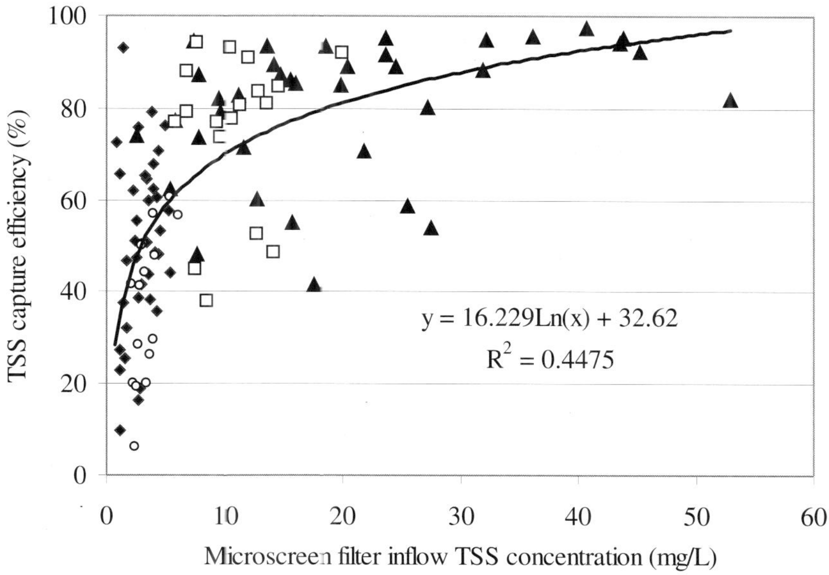

FIGURE 20. Microscreen filter removal efficiencies are dependent on the concentration of Total Suspended Solids (TSS) entering the filter. Data shown is from four microscreen filters used at the Freshwater Institute (Shepherdstown, West Virginia): one microscreen drum filter (◆) treated the flow discharged from the elevated drains of dual-drain culture tanks in a water reuse system; another microscreen Triangle™ filter (▲) sieved the bottom flow from a dual-drain culture tank in a single-pass system; another microscreen drum filter (□) treated the combined effluent from a reuse system and a single-pass system; and the other microscreen drum filter (O) sieved the effluent from a single-pass system.

cleaning can produce. However, many water reuse systems must rely on treating and reusing the cleaning flow. To handle the relatively large increase in flow volume and solids concentrations that occurs during cleaning events, the microscreen filter may need to be sized 50–200% larger than would be required to treat the steady-state recirculating flow. Alternatively, cleaning flows may be bypassed around the microscreen filter to an off-line settling basin.

Granular Media Filters

Granular media filters include sand filters, plastic-bead filters, and diatomaceous earth filters. Solids are captured in granular media beds at the pore openings at the granular media surface or at the void spaces occurring within the granular media bed, due to straining, sedimentation, absorption, or chemical bridging mechanisms. Periodically, the granular media bed must be backwashed to flush entrapped particles.

Plastic-bead filters are sometimes used in recirculating aquaculture systems to remove ammonia and solids from relatively small flows, usually less than 1,000–2,000 L/min (discussed in more detail in the Biofilter section). Plastic-bead filters can be relatively expensive per unit of flow treated, but are sometimes used to provide both solids control and biofiltration within one unit, which can simplify plumbing, save on floor space, and halve the number of unit processes required to achieve solids removal and biofiltration. However, the majority of plastic-bead filters produces a significant pressure drop, which requires pumping the solids-laden water into the units and through a media retaining screen located at the entry of the bead chamber (Figure 21). The pumping and screening exposes the fresh fecal pellets to a high level of shear before the solids even reach the bead bed. Particles captured by the filters are then stored within the bead filter until the unit is backwashed. Chen et al. (1993a) estimated that as much as 30–40% of the TSS generated in a recirculating system will decay if the solids are filtered and stored in a plastic-bead filter between 24-h backwash cycles. There is also a complex relationship between plastic-bead filter backwash intensity and frequency, solids degradation, and nitrification efficiency (Golz et al. 1999).

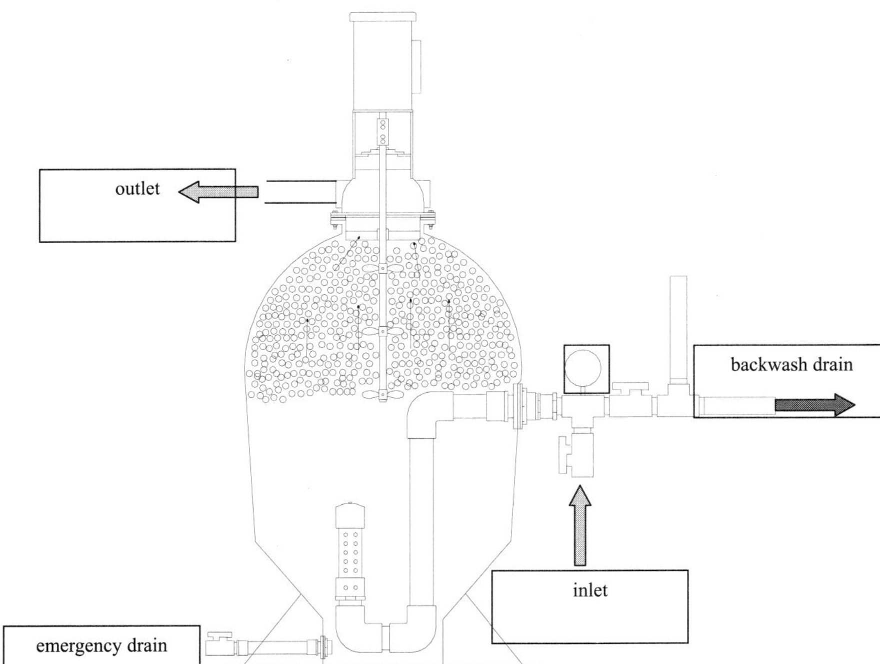

FIGURE 21. An illustration of a propeller washed bead filter (courtesy of Aquaculture Systems Technologies, New Orleans, Louisiana).

Probably for these reasons, more recent models of the plastic-bead filter are being produced to reduce the pressure drop requirement across these filters and also to allow for increased backwash frequency.

Sand filters and diatomaceous earth filters are rarely used in recirculating aquaculture systems because, although they are exceptional at removing fine particulate matter, these filters can also require complex backwashing mechanisms and have relatively large capital and operating costs per unit of flow treated. Biofouling of filter sand can also become problematic in recirculating systems and, combined with other operational variables, can cause sand filters to require prohibitively large backwash flows. For these reasons, sand filters and diatomaceous earth filters are used in only a relative few hatchery and research recirculating systems where water flows are relatively small and exceptional water quality is expected.

Temperature Control

Water temperature control has many purposes in aquaculture. Relatively constant warm or cool temperatures can be maintained to maximize or hold back fish growth during fingerling or food fish production. More dynamic water temperature regimes can also be used to stimulate the reproductive cycle in broodfish. When warmwater supplies are not available, water reuse systems can still be used to provide the optimum water temperatures for producing warmwater species such as tilapia, eel, African catfish, and hybrid-striped bass in relatively temperate climates. Whatever the circumstances, water reuse systems housed in well-insulated buildings that exchange small water volumes (5–20% daily), require relatively little energy for heating water (Timmons 1997).

The energy (kW) required to heat water at a given flow rate, Q (L/min), is directly proportional to the increase in water temperature desired, that is, ΔT (°C):

$$\text{Energy} = \frac{0.072 \times Q \times \Delta T}{e}. \tag{11}$$

For example, it would take 43 kW of energy to raise or lower by 10°C the temperature of a 60 L/min water flow, assuming 100% energy transfer (e). The power requirement (in kWh) can be calculated by multiplying the energy requirement (in kW) by the time (in hours) that the energy is applied. The cost of the power is then calculated by multiplying the energy requirement (in kWh) by the energy cost (in $/kWh).

Units used to heat water include gas or oil boilers, electric heaters, heat pumps, and "waste" cooling water from power plants or other industrial applications. Heat pumps extract heat energy from a water source and add it to the energy from the electric heating process to give off more heat than they get from electricity. Heat exchangers are also used to transfer heat from one water source to another while keeping the two flows separate. Heat exchangers can transfer heat to the aquaculture system process water from sources of heat that include geothermal waters, waste heat from industrial effluents, warm ocean waters, and from the wastewater discharged from the aquaculture system. Sometimes, air-to-air heat exchangers are also used in northern latitudes to recapture heat when ventilating humid and carbon dioxide laden air from buildings containing recirculating systems (Summerfelt et al. 2000d).

The water retained and reused in a fully recirculating system is also heated from the waste heat produced from pumping the water through the system, and from heat transfer through the pipes and tank walls when the reuse system is enclosed in a heated building. As stated above, this temperature gain can be advantageous when warmwater species are cultured in fully recirculating systems. However, a large gain in water temperature may have serious fish health consequences, including decreased growth rates and increased incidence of stress and disease, when trout, salmon smolt, or other coldwater species are produced in fully recirculating systems. For these reasons, high water exchange rates (100–200% of culture volume per day) are sometimes used in coldwater recirculating systems to keep water temperatures from exceeding 15–18°C, depending upon location and season. To reduce heat gain in warm climates and heat loss in cold climates, recirculating systems are usually housed in relatively low cost greenhouses or agricultural buildings. Insulated buildings cost significantly more than greenhouses, but longer life spans and savings in fuel (to heat) or water (to cool) can sometimes justify their selection.

Dissolved Gas Control Systems

Oxygenation and aeration are two processes used in intensive water reuse systems to add oxygen to water, increasing the mass of fish that can be supported in a given flow (i.e., the oxygen increases the carrying capacity of the flow) and making fish production more economical (Colt and Watten 1988; Boyd and Watten 1989; Colt and Orwicz al. 1991). In the aeration process, air is contacted with water to bring the concentration of DO, nitrogen, and carbon dioxide closer to equilibrium levels. In the oxygenation process, pure oxygen gas is used instead of air to increase DO concentrations to above atmospheric saturation levels.

Dissolved Oxygen Supplementation

Many of the new, intensive recirculating aquaculture systems have incorporated pure oxygen transfer technology to increase carrying capacity by increasing the concentration of oxygen delivered to the fish culture tank (Summerfelt et al. 2000d). Since the late 1970s, oxygen addition has been recognized as an economical means of boosting fish production in certain situations (Colt and Watten 1988; Boyd and Watten 1989).

Example. The cost of adding oxygen can be estimated at a commercial tank-based fish farm using the following farm and species-specific variables:

- oxygen consumption rate per unit feed consumed (e.g., 0.25 kg oxygen per kg feed),
- feed conversion rate (e.g., 1.3 kg of feed is required to produce 1.0 kg of fish),
- oxygen absorption efficiency into water (e.g., 1.0 kg of oxygen feedgas must be supplied to dissolve 0.6 kg oxygen),
- density of oxygen gas (e.g., 0.75 cubic meter oxygen gas [at 20°C and 1 atm pressure per kg oxygen]), and
- cost of oxygen gas (e.g., $0.15 per cubic meter oxygen gas [at 20°C and 1 atm pressure]).

Estimates for each of the variables listed above are used below to make a simple prediction of the cost of oxygenation assuming all DO requirements are provided by the purified oxygen feed gas:

$$\begin{aligned}\text{oxygen cost} &= \frac{1.3\,\text{kg feed}}{1.0\,\text{kg fish}} \times \frac{0.25\,\text{kg DO}_2}{\text{kg feed}} \times \frac{1.0\,\text{kg O}_2\text{ supplied}}{0.6\,\text{kg DO}_2} \\ &\times \frac{0.75\,\text{m}^3\text{ O}_2\text{ gas}}{\text{kg O}_2\text{ supplied}} \times \frac{\$0.15}{\text{m}^3\text{ O}_2\text{ gas}} \\ &= \$0.061/\text{kg fish.}\end{aligned}$$

In this example, the cost of the oxygen feed gas required to produce 1.0 kg fish would amount to around $0.061. At this same feed conversion rate, fish feed alone might cost $1.00 for every kilogram of fish produced. The cost of the oxygen would be relatively low, much lower than the cost of the fish feed requirement. Oxygen addition is still not an insignificant cost and whether or not its application is warranted depends upon conditions specific to each application.

Increasing fish production with supplemental oxygen, however, increases the loading of the by-products of fish metabolism, e.g., ammonia, carbon dioxide, and dissolved and particulate organic matter (Colt et al. 1991), which must be removed within a recirculating system by the water treatment process or these metabolic by-products will adversely affect the fish or the biofilter. Providing DO at supersaturated levels also eliminates the need for frequent drops between raceways (assuming dissolved carbon dioxide accumulations are accounted for) and provides emergency control of DO levels in the culture environment during periods of reduced water flow.

Oxygen Source. Oxygen can be purchased from commercial sources as a bulk liquid or gas, or can be produced on-site using pressure swing adsorption (PSA) equipment (Colt and Watten 1988; Boyd and Watten 1989). The purity of oxygen gas is a consideration when selecting an oxygen source; liquid oxygen is 98–99% pure, whereas, PSA systems generally produce an 85–95% oxygen-enriched gas. The cost of oxygen is another criteria considered when selecting an oxygen source. Generating oxygen on-site with PSA units requires a source of dry air at pressures of 90–150 psi; the capital costs of the compressor, dryer, PSA unit, and standby electrical generator amount to a considerable investment. The energy requirement of generating oxygen using PSA units is about 1.1 kWh of electricity per kg of oxygen produced (Boyd and Watten 1989). The cost of liquid oxygen varies with location, depending upon the grade of oxygen and the amount of liquid oxygen used in the same region. At the Freshwater Institute, the operating and capital costs of using liquid oxygen versus on-site generation were compared using their net present value over a 10-year period. Liquid oxygen costs about three times more annually than would oxygen generated on-site in this case study (E. Wade, Freshwater Institute, personal communication). However, liquid oxygen is considered the more reliable oxygen source, because of the many mechanical aspects of PSA oxygen generation that have the potential to fail. Therefore, even though its operating costs may be higher, liquid oxygen is often used because the capital investment and risk of system failure are lower than with PSA generation technologies.

Oxygen Transfer Equipment. Oxygen is a significant production cost, even though oxygen costs may amount to only a fraction of feed costs. Therefore, the transfer of oxygen into the fish culture water must be efficient. Fortunately, oxygen transfer principles, techniques, and equipment used in aquaculture are relatively well established (Colt and Watten 1988; Boyd and Watten 1989; Colt and Orwicz 1991).

Oxygen transfer equipment is designed to take advantage of the factors governing the rate of mass transfer (Fick's law): the area of the gas-liquid interface; the thickness and rate of surface film renewal (both dependent upon the energy transfer in the gas-liquid contactor); and the gradient between the saturation and existing concentrations (mole fractions) of oxygen in water. Packing, spray nozzles, fine bubble diffusers, and other mechanisms are used to increase the area of the gas-liquid interface, decrease the thickness of the surface film, and increase the rate of surface film renewal. The gradient between the saturation and existing concentrations of DO (or other gases) is the driving force for mass transfer. The saturation concentration of oxygen (or other gases) dissolved in water is temperature dependent and directly proportional to the product of the mole fraction of oxygen in the surrounding atmosphere times the absolute pressure[3] (Henry's law). Therefore, the amount and rate of oxygen transfer can be increased by using purified oxygen gas rather than air, or by increasing the total pressure during oxygenation. Transferring pure oxygen rather than air into water will increase the mole fraction of gaseous oxygen from 0.2095 (air) to 1.00 (pure oxygen). Consequently, use of pure oxygen gas at a pressure of 1 atmosphere will increase the solubility of DO to 48.1 mg/L (at 15°C), nearly fivefold over the oxygen solubility obtained when air is used (10.1 mg/L at 15°C). Also, increasing the pressure from 1 to 2 atmospheres during the transfer of pure oxygen doubles oxygen's solubility to 97.0 mg/L (at 15°C). The total pressure in which the transfer occurs is typically increased with either a pump or hydrostatic head, but this may also increase the operating costs.

The DO concentration that can be produced in flowing water is dependent upon the purity of the oxygen-feed gas, the oxygen absorption efficiency, and the volume of the oxygen-feed gas relative to the volume of water flow, that is, the volumetric gas to liquid loading rate (G:L). An oxygenation unit can potentially add 13.2 mg/L of DO to the water flow for every G:L = 0.01:1, assuming 100% oxygen adsorption efficiency.

The total mole fraction of the individual gases dissolved in water must equal 1 at equilibrium. Therefore, exposing water to a purified oxygen gas will displace dissolved nitrogen and argon gas as the water absorbs oxygen (Watten et al. 1991). Unlike air contact units (i.e., aeration units), oxygen contact units that vent the displaced nitrogen and argon gas can reduce dissolved nitrogen and argon concentrations to levels below saturation at the same time that oxygen supersaturation levels are being created, which reduces the risk of gas bubble trauma (Colt and Watten 1988; Boyd and Watten 1989; Watten et al. 1991). Venting nitrogen and argon gas is important, both to reduce the total gas pressure of the water and to increase the efficiency of oxygen transfer. The type of oxygenation equipment selected will effect nitrogen venting and the total gas pressure produced, depending on whether the oxygenation process is open and unpressurized, versus closed and pressurized. Unfortunately, the G:L loading rates used in oxygenation units (e.g., 0.005:1–0.02:1) are much too low to achieve significant carbon dioxide stripping efficiencies (Watten et al. 1991), as will be discussed in the Carbon dioxide control section.

[3]Absolute pressure is the barometric pressure plus gauge pressure.

Selection of the best oxygen transfer unit to use within a recirculating system may be controlled as much by the selection and arrangement of the culture tank, biological filter, and other water treatment processes as by the oxygen absorption and transfer efficiencies within the oxygenation unit. Mixing characteristics within the culture tank influence the levels of DO and total gas pressure found in different portions of the culture tank, which should be considered when selecting an oxygen transfer unit (Colt and Watten 1988). Circular tanks provide environments where the dissolved constituents are completely mixed, therefore, high total gas pressures and oxygen concentrations within the flow entering the tank are rapidly diluted to relatively constant concentrations throughout the culture tank, which is advantageous for the fish. Raceways, on the other hand, provide plug-flow conditions that produce high total gas pressures and oxygen concentrations at the head end of the raceway, which decrease to lower concentrations and pressures down the axis of the raceway. The hydraulics near the inlet of raceways can potentially expose fish to harmfully high total gas pressures or oxygen saturations. Therefore, oxygenation units that vent nitrogen gas and do not produce relatively high total gas pressures are safer to use with plug-flow raceways systems.

Many recirculating systems are designed to pump the water just once each pass. Therefore, the choice of the biofilter type affects the choice and placement of the oxygenation unit (and the aeration/stripping unit). Gravity-driven oxygenation units (e.g., open-type units and sometimes large U-tubes) are often used if a 1–3m elevation drop is provided between the biofilter outlet and the water elevation in the culture tank (Summerfelt et al. 2000d). When gravity-driven oxygenation units are used, an open-type cascade aeration/stripping tower is sometimes located between the biofilter and the oxygenation unit. Placing the cascade aeration/stripping towers between the biofilter and the oxygenation units serves several purposes:

- The aeration raises the DO levels in the flow up to near saturation levels just before the flow enters the oxygenation unit; therefore, nearly all of the purified oxygen added in the oxygenation unit is used for supersaturating the water flow;
- The aeration strips the dissolved carbon dioxide at the point in the recirculating system where carbon dioxide has reached its highest concentration, that is, after the biofilter, but before oxygen saturations are produced; and
- The oxygenation unit is placed to supersaturate the water flow with DO just before the flow enters the fish culture tank.

Of note, traditional fine-bubble diffused oxygenation systems can provide such poor oxygen absorption efficiencies (e.g., 3–7%; Colt and Watten 1988) that this technique is seldom used under standard operating conditions. However, some ultra-fine bubble diffusers have been recently introduced and are claimed to significantly improve oxygen absorption efficiency. Either way, oxygen diffusers are still used to oxygenate the culture tank in the case of an emergency, when water flow is lost or fish activity produces temporarily low DO concentrations. Oxygen diffuser systems can serve as a valuable backup system for emergency situations, because no power is required for diffuser operation when a pressurized source of oxygen gas is available.

U-tubes, oxygenation cones, and multistage low head oxygenation units (LHO; Figure 22) are commonly used to add pure oxygen within large recirculating systems (Summerfelt et al. 2000d). They are widely used, probably because they are readily scaled-up, easy to control, and resist fouling from organic solids. The U-tube and LHO also require only a modest hydraulic head to achieve large oxygen adsorption efficiencies. Performance and design criteria for each are given below.

Multistage Low Head Oxygenators. A multistage LHO unit maximizes oxygen transfer efficiency by reusing the oxygen feed gas through a series of contact chambers (Watten and Boyd 1990). Water flow is distributed across the LHO unit using head-controlling orifices spaced proportional to the chamber cross-sectional area throughout the distribution

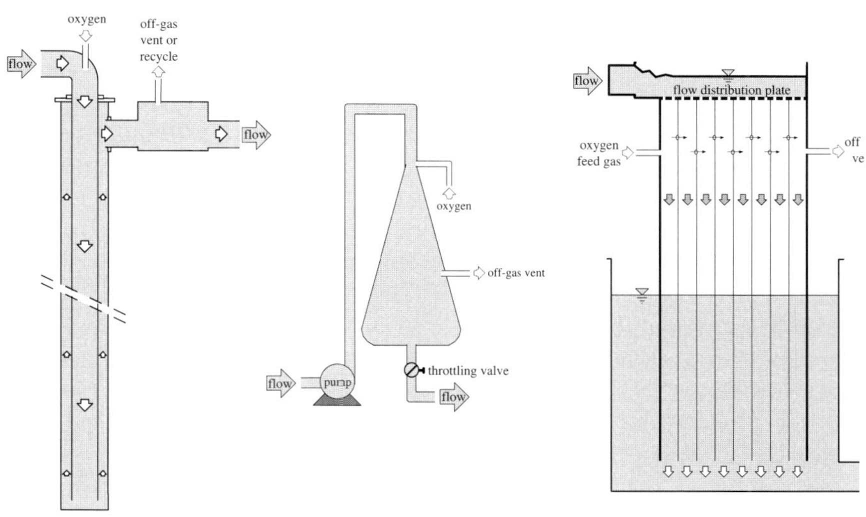

FIGURE 22. U-tubes (left), oxygenation cones (middle), and multi-stage low head oxygenators (right) are examples of technologies used to add pure oxygen in large recirculating systems (from Summerfelt et al., 2000c).

plate. Flow through the orifices typically cascades 0.3–1.2 m through the contact chambers before plunging into a pool of water at the base of the LHO. The oxygen-feed gas is held within the chambers by the water pooled above the distribution plate at the top of the LHO unit and by the water pooled inside the chambers at the base of the unit. The chambers typically do not contain packing, but can be packed with plastic media to improve mass transfer.

Low head oxygenation units are sometimes used within serial reuse raceways systems to oxygenate the water flow when it plunges down the 0.3–0.8 m hydraulic steps between raceway units. When an LHO unit is used in conjunction with one or more circular tanks, the LHO units are often installed within a small sump (Figures 4, 22, and 23) just before the culture tank. The sump provides the water seal at the bottom of the LHO chambers that prevents oxygen gas from escaping and also gathers the flow into a point source discharge that can be directed into the circular culture tanks to create the desired rotating flow. However, the LHO sump provides a quiescent zone where settleable solids can accumulate (Figures 22 and 23); these solids have to be removed periodically to maximize water quality and reduce the potential reservoir of fish pathogens that can harbor in the system. Low head oxygenation units can be installed above cone bottom tanks to allow for easy removal of accumulated solids (Figures 4 and 23). For similar reasons, the aeration/stripping column can also be designed within cone bottom tanks to provide simple and rapid flushing of settled solids (Figures 4 and 23) (Summerfelt et al. 2000d).

Low head oxygenation units have been evaluated primarily in cold water (Summerfelt 1996), for example, 12 and 17°C, where the saturation of oxygen (10.7–9.7 mg/L at 12 and 17°C, respectively) is higher than the levels that would occur at the temperatures used to raise cool water or warmwater fish. Excellent (60–90%) oxygen absorption efficiencies have been reported when LHO units are operated at gas to liquid ratios of less than 1 volume of oxygen feed gas for every 100 volumes of water flow (G:L = 0.01:1). However, if the absorption efficiency is only 60%, a G:L less than 0.01:1 would limit the increase in DO concentrations to 8 mg/L or less above saturation. Producing DO concentrations near 15 mg/L above saturation usually requires increasing the G:L to more than 0.02:1, because the oxygen transfer efficiency can drop to 40–60% at the increased G:L, even when using plastic packing within the LHO chambers. Oxygen absorption efficiencies are sacrificed when trying to double (or more) the saturation concentration of DO within the flow leaving an LHO unit (Summerfelt 1996). The LHO unit does, however, effectively displace and vent nitrogen gas when its adds oxygen, which helps prevent a large total gas supersaturation even when a large oxygen supersaturation is obtained.

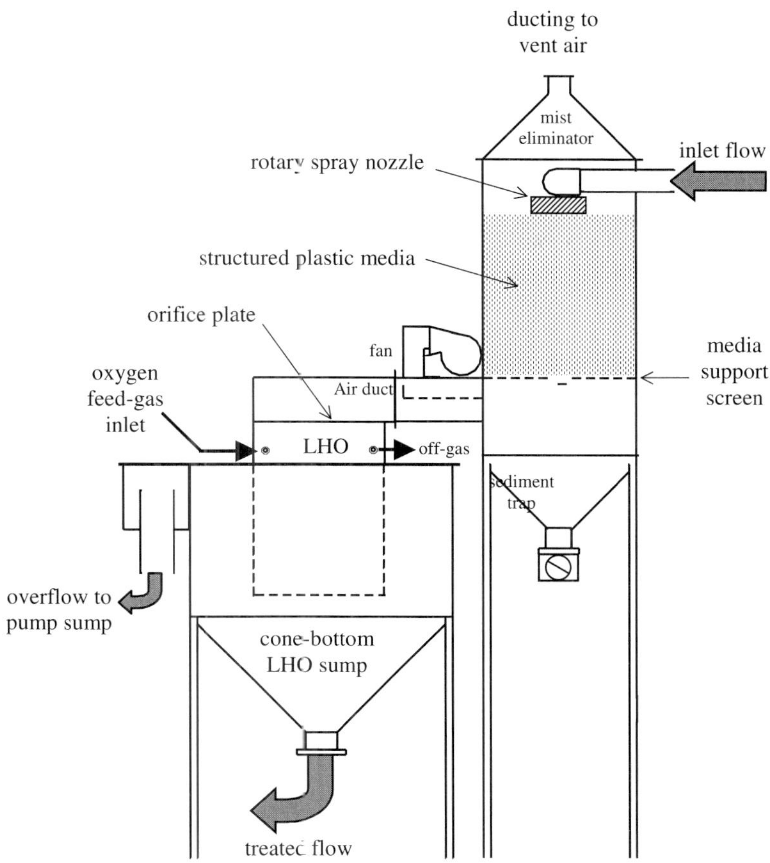

FIGURE 23. The design and placement of a Low head oxygenation unit and a stripping column within the Freshwater Institute's partial-reuse system (Figure 2) illustrates how cone bottom sumps can be incorporated to simplify sediment removal.

U-Tube Oxygenation Unit. A U-tube maximizes oxygen transfer by using 25–60 m of hydrostatic pressure to temporarily increase the saturation concentration of oxygen as the flow passes through the depth of a vertical, U-shaped conduit, or a pair of annular pipes (Boyd and Watten 1989). Pure oxygen is typically bubbled into the water entering the top of the inner of the two concentric pipes. The flow then carries the oxygen bubbles down to the bottom of the U-tube, where the rate of gas absorption is maximized due to the large hydrostatic pressure. At the bottom of the U-tube, the flow moves from the inner pipe to the annular space between the outer and inner pipes. The flow then travels up through the annular space to the top of the U-tube and exits the unit. Boyd and Watten (1989) report that the buoyant velocity of the bubbles is about 0.3 m/s, thus, the U-tube is designed to provide a cross-sectional area within the down-flowing pipe to produce a nominal velocity of 1.8 m/s. They also report that the cost of oxygenation with U-tubes is minimized at water velocities of 1.8–3.0 m/s, depths of 25–60 m, and a G:L ratio of 0.025:1.

The DO concentration in the effluent of a U-tube depends on the conduit depth, water temperature, influent DO saturation concentration, and volumetric gas to liquid ratio (Boyd and Watten 1989). The high pressure at the bottom of the U-tube can produce a relatively high oxygen supersaturation within the water leaving the U-tube, as much as 20–40 mg/L DO. A U-tube without off-gas recycling provides oxygen transfer efficiencies of only 30–50%, but off-gas recycling can increase oxygen transfer efficiencies to 55–80% (Boyd and Watten 1989). U-tubes generally require from 1 to 3 m of water pressure to operate, depending upon pipe diameter, water velocity, and U-tube depth. Pressure requirements can be estimated by adding the head losses resulting from pipe friction and two-phase flow (Boyd and Watten 1989). At a constant pipe velocity, large diameter U-tubes used to oxygenate high volume flows have less total head loss than do small U-tubes that oxygenate correspondingly lower volume flows.

U-tube advantages include ease of control, high DO outlet concentrations, relatively low pressure requirements (gravity flow is possible in some instances), and the ability to oxygenate water containing high levels of suspended solids. However, dissolved nitrogen is not stripped and vented from U-tube units as readily as within LHO units. The majority of gas transfer occurs in the depths of the conduit where nitrogen venting is not possible. Therefore, a U-tube typically produces higher total gas pressures per unit of oxygen transferred than an LHO unit, which can cause gas bubble trauma under extreme conditions. Total gas pressures large enough to harm or kill fish can also be produced if a design flaw allows air bubbles to enter U-tubes. Also, water flow through U-tubes can become stalled and gas-locked when the G:L loading exceeds a level that depends upon the operating water velocity through the down-flow pipe, U-tube depth, and hydrostatic pressure available.

Oxygenation Cones. The oxygenation cone or downflow bubble contactor is a cone-shaped cylinder that is wider at its base than at its top. Water flow and the oxygen feed-gas both enter the top of the cone within a high turbulence zone that is used to increase bubble shear and mixing. The inlet water velocity is usually designed for 1.8 m/s, but, as the flow of water and bubbles travel down into the cone, this water velocity will decrease in inverse proportion to the square of the cone's expanding diameter (Boyd and Watten 1989). The bubbles are captured within the lower half of the cone when the water velocity is reduced to less than the buoyant velocity of the bubbles, which is about 0.3 m/s. The water velocity exiting the cone is kept less than or equal to 0.15 m/s to prevent any of the captured bubbles from washing out of the cone with the oxygenated water flow (Boyd and Watten 1989).

A vent can be located in the lower half of the cone to remove some of the displaced nitrogen from the bubble zone. Bubble capture is efficient and oxygen absorption efficiencies of 30–100% (80–90% under the right conditions) can be achieved, depending upon cone geometry, operating pressures, off-gas ventilation rate, and water flow rates (Boyd and Watten 1989). Because oxygenation cones provide good oxygen absorption efficiencies (especially when pressurized) and are not susceptible to plugging with solids, oxygenation cones are used widely within recirculating systems, sometimes in conjunction with trickling filters (Figure 24), as applied in many European eel farms (Nijhof 1995) and some tilapia farms (Twarowska et al. 1997). However, pressurized oxygenation cones

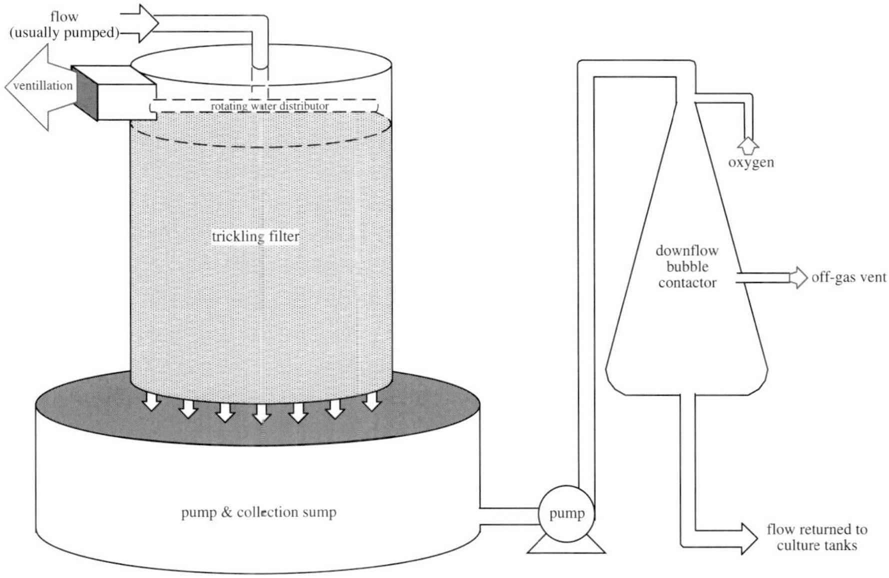

FIGURE 24. Water flow collected at the base of a trickling filter is often pumped through a downflow bubble contactor (e.g., oxygenation cone) before returning to the culture tank(s) in some recirculating systems (Nijhof, 1995; Twarowska et al., 1996). This combination has been found to be effective at both dissolved gas control and nitrification.

do not vent nitrogen gas effectively and gas bubble trauma could become an issue at high total gas pressures.

Dissolved Carbon Dioxide Removal

Historically, before the introduction of pure oxygen absorption technologies into water reuse systems, the relatively low intensity systems that were employed generally had sufficient water exchange rates and aeration rates to keep carbon dioxide from accumulating above safe levels. Dissolved oxygen, suspended solids, and NH_3 were the primary water quality concerns. However, using oxygenation technologies for the intensification of fish culture within water reuse systems has often produced accumulations of dissolved carbon dioxide in the culture tank water that are elevated enough to limit production (Summerfelt et al. 2000d). Intensive water reuse systems using oxygenation technologies are especially susceptible to dissolved carbon dioxide problems because

- pure oxygen injection technologies typically do not allow for the removal of substantial quantities of carbon dioxide;
- many intensively oxygenated recirculating systems use relatively low water exchange rates relative to the mass of feeding fish in the system; and
- many intensively oxygenated recirculating systems simply did not include an adequate aeration step, which is required to strip dissolved carbon dioxide.

Dissolved carbon dioxide concentrations can be controlled in water reuse systems by aeration to strip dissolved carbon dioxide into air (Grace and Piedrahita 1993; Summerfelt et al. 2000d); addition of alkaline chemicals to increase the water's pH, thereby, reducing the fraction of total carbonate carbon existing as carbon dioxide (see Figures 8 and 9) (Bisogni and Timmons 1991; Loyless and Malone 1997); or a combination of both aeration and alkaline addition.

Carbon dioxide is extremely soluble in water compared to oxygen and nitrogen, which makes stripping dissolved carbon dioxide from water more difficult than adding oxygen or stripping nitrogen (Watten et al. 1991). In an atmosphere of pure carbon dioxide gas, the concentration of carbon dioxide dissolved in water would be 1,993 mg/L at 15°C (Colt and Orwicz 1991). The concentration of dissolved nitrogen, oxygen, and argon in water when exposed to their respective pure gases would only be 20.95, 48.14, and 65.94 mg/L, respectively, at 15°C (Colt and Orwicz 1991). However, the mole fraction of carbon dioxide in air, that is, 0.00035, is very low when compared to the fractions of nitrogen, oxygen, and argon present—0.78084, 0.20946, and 0.00934, respectively (Colt and Watten 1988). Therefore, as Henry's law would predict, the solubility of carbon dioxide in air at 1 atmosphere would be only 0.69 mg/L at 15°C, whereas the solubility of nitrogen, oxygen, and argon would be about 16.36, 10.08, and 0.62 mg/L at 15°C, respectively.

Dissolved carbon dioxide concentrations can reach 10–30 mg/L in recirculating systems growing salmonids, and 20–50 mg/l in systems growing tilapia or other warmwater species. These levels are 15–70 times the normal saturation concentration (0.69 mg/L at 15°C), which would be found in well-aerated water that has come to equilibrium with atmospheric carbon dioxide levels. Stripping high concentrations of dissolved carbon dioxide from water can increase the partial pressure of carbon dioxide in the passing air flow to the point that the amount of dissolved carbon dioxide that can be stripped is limited (Summerfelt et al. 2000d). Under these conditions, maintaining relatively low carbon dioxide partial pressures in the passing air flow will help to maintain efficient carbon dioxide degassing. Therefore, stripping dissolved carbon dioxide requires contacting large volumes of air for every volume of water treated, generally 5 to 10 air flow volumes for every 1 water flow volume, that is, G:L of 5:1–10:1 (Figure 25). These gas-to-liquid ratios (G:L) are several hundred times larger than the ratios of oxygen gas flow volume to water flow volume used in oxygen transfer units (i.e., G:L of 0.005:1–0.03).

Oxygen and carbon dioxide can be transferred into and out of water, respectively, with any open aeration system. However, aeration units that move water droplets through air, for example, surface aerators and cascade columns, are often more effective than aeration units that move air bubbles through water, for example, diffused aeration systems and subsurface aerators (Colt and Orwicz 1991). Cascade columns are frequently used to degas dissolved carbon dioxide from water reuse systems, because they can incorporate relatively inexpensive and low power ventilation fans to move large volumes of air counter-current to the water flowing through the column (Summerfelt et al. 2000d). The pressure (and thus power) requirement to force air through the column is low, generally less than 1–10 cm of water back pressure for every 1 m of column height. In contrast, diffused aeration systems can be less efficient than cascade columns as bubbles readily saturate with carbon dioxide and the enormous amount of diffused air requires expensive blowers to work against 1–2 m of water pressure.

Carbon dioxide stripping efficiency is also directly effected by cascade column height. However, the depth of the cascade is generally limited to 1.0–1.5 m, because only diminishing improvements in dissolved carbon dioxide stripping are achieved at greater depths (Figure 25) and there are practical limitations on the use of taller columns, for example, taller ceilings and increased pump pressure requirements (Summerfelt et al. 2000c). The cross-sectional area of these cascade columns is deter-

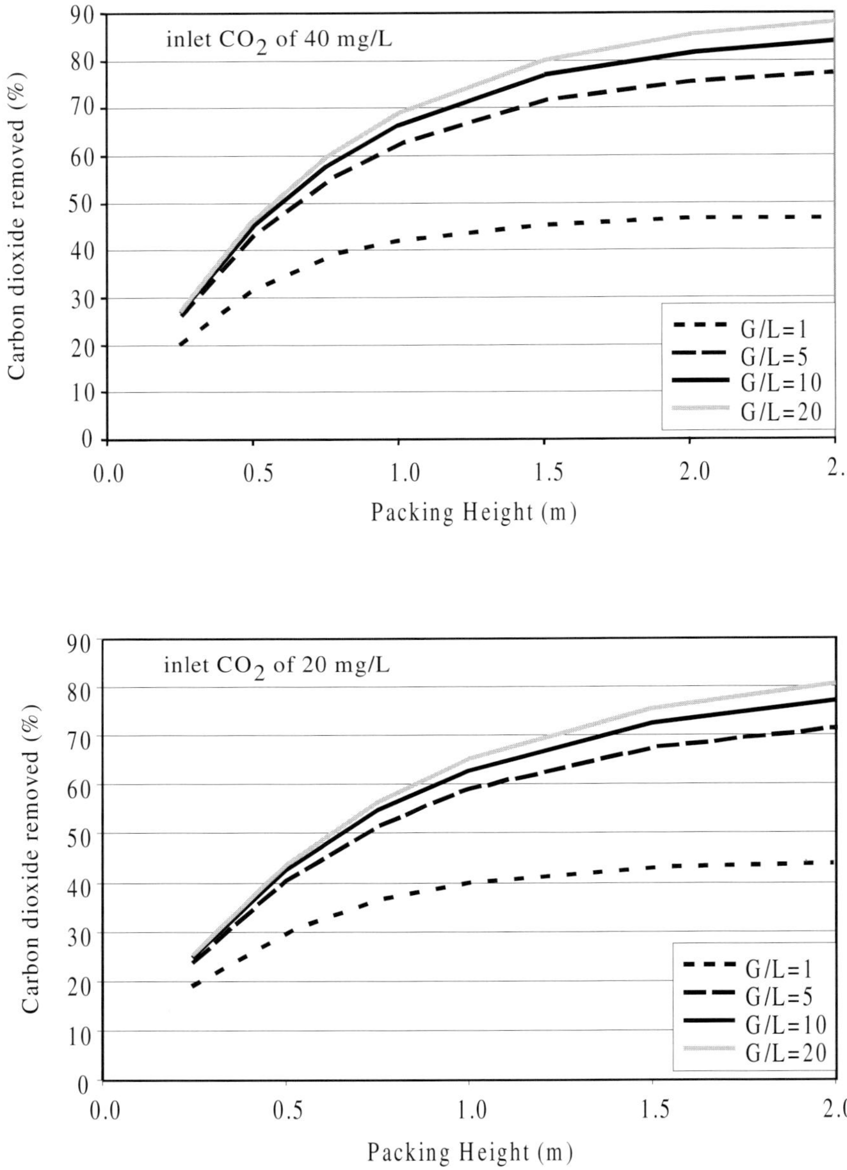

FIGURE 25. Performance of a given packing material is a function of packing depth, air to water ratio (G:L), and the inlet dissolved carbon dioxide concentration. The top and bottom graphs estimate treatment at inlet concentrations of 40 mg/L and 20 mg/L, respectively (Summerfelt et al. 2000c).

mined based on a hydraulic loading rate of 60–84 $m^3/h/m^2$ to 100–250 $m^3/h/m^2$. The lower rates are more conservative and will reduce the back pressure on the air handling fans, especially if there is potential for even a small amount (3–10 cm) of water holdup (i.e., flooding of the media voids) that could occur if biosolids accumulate in the media void spaces or between the media and the media support grid.

Operationally, carbon dioxide degassing column design requires uniform water distribution across the top of the column. Water distribution can be effective, with relatively low-pressure loss (e.g., 0.1–1.0 m water head), using either spray nozzles (Figure 26), orifice plates, or rotating distributor arms (Figure 24). Carbon dioxide degassing can also be improved by breaking up the water cascade with a plastic media or with stacked screens. However, problems with plugged media have been reported in warmwater systems (in part due to biosolids production) and in cold-water systems that contain higher concentrations of suspended solids. To avoid fouling problems, a stripping column containing splash screens rather than packing may be easier to maintain. However, elimination of packing can reduce the dissolved carbon dioxide removal by as much as 40% (Summerfelt et al. 2000d). Alternatively, use of a structured plastic media that incorporates large vertically-continuous void spaces may lessen the likelihood of plugging, while providing the media surface area to improve dissolved carbon dioxide stripping (Figures 26 and 27).

FIGURE 26. Two spray nozzles are shown. On the lefthand side is a picture of a rotary spray nozzle (Tower Tech, Inc., Oklahoma City, Oklahoma) above 5 cm NORPAC tubular media; the nozzle can distribute up to 570–1,330 L/min of water flow at a pressure drop of only 0.1 bar. On the righthand side is a picture of a crown-nozzle from LS Enterprises (Ft. Myers, Florida) next to a penny for perspective; the nozzles surface mounts through a distribution plate and each nozzle can distribute up to 40 L/min of water flow with only 15 cm of water head pooled above the plate.

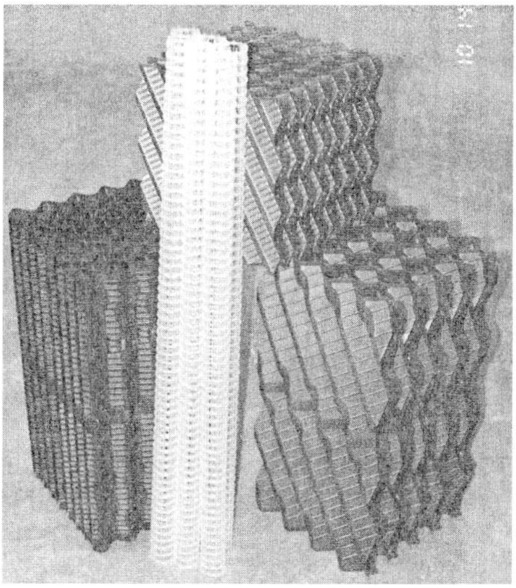

FIGURE 27. Two structure plastic media, each with large (4–5 cm) vertically-continuous void space, are being evaluated within stripping columns contained in a fully recirculating system at the Freshwater Institute (Shepherdstown, WV). The media are a 1-m tall cluster of 2-in tubular NORPAC media (NSW, Roanoke, Virginia) and a 0.9-m tall stack of structured media blocks from L.S. Enterprises, Ft. Myers, Florida.

Air discharged from stripping columns should be ventilated from buildings to maintain safe carbon dioxide levels in the work environment, that is, less than 5,000 ppm carbon dioxide is the maximum allowable time-weighted average exposure over an 8-h workday and 50,000 ppm is immediately dangerous to life and health (Vinci et al. 1998). However, a considerable amount of heat can be lost when buildings are ventilated in cold climates, resulting in higher operating costs—especially at the temperatures used to raise warmwater species. A portion of this heat can be recovered by venting the air through an air-air heat exchanger when heat loss is a serious issue.

Stripping dissolved carbon dioxide does not change the alkalinity of the water, but it does increase the pH of the water while it decreases the total carbonate carbon. Manipulation of the acid-base equilibrium between dissolved carbon dioxide and the carbonate carbon system (Tables 3 and 4, Figures 8 and 9) is another method used to control dissolved carbon dioxide levels in water reuse systems. Alkaline chemicals are added to water to shift the equilibrium of the carbonate carbon system away from carbon dioxide and, thus, reduce the concentration of dissolved carbon dioxide in the water. Addition of alkaline chemicals is already required within fully recirculating systems to counter the loss of alkalinity that occurs as a result of nitrification (Bisogni and Timmons

1991; Loyless and Malone 1997). Adding lime, caustic soda, or soda ash to the water provides a source of hydroxide ion that can directly react with carbon dioxide to produce bicarbonate alkalinity and increase pH. Adding sodium bicarbonate is simply a source of bicarbonate alkalinity and a means to increase pH, thereby reducing dissolved carbon dioxide levels. A common practice is to combine both carbon dioxide stripping and chemical addition to maintain safe levels of dissolved carbon dioxide, alkalinity, and pH.

Ammonia Removal

Serial- and partial-reuse systems flush away ammonia by dilution with makeup water. However, fully recirculating systems use such low system volume water exchange rates that dilution does not provide adequate ammonia removal. Bacterial nitrification is the primary method for removing ammonia in tank-based recirculating aquaculture systems. Although there are several species of nitrifying bacteria, two types of nitrifying bacteria are generally considered to be responsible for oxidizing ammonia to nitrate (Wheaton et al. 1991). *Nitrosomonas* converts ammonia to nitrite, while *Nitrobacter* converts nitrite into nitrate. Establishment of nitrifier populations takes 4–10 weeks; this was discussed previously, in the Nitrite section of this chapter.

Nitrifying bacteria are primarily obligate autotrophs, consuming carbon dioxide as their primary carbon source, and obligate aerobes, requiring oxygen to grow (Hagopian and Riley 1998). These nitrifying bacteria usually coexist with heterotrophic microorganisms such as certain bacteria, protozoa, and micrometazoa, which metabolize biologically degradable organic compounds. Heterotrophic microorganisms typically grow much faster than nitrifying bacteria. Therefore, when concentrations of dissolved or particulate organic matter are relatively high, heterotrophic microorganisms will have the food necessary to outcompete the nitrifying bacteria for space and oxygen in the biological treatment processes, especially at the head end of these treatment processes if plug-flow hydraulics exist. The larger heterotrophic microorganisms will also prey upon the population of nitrifying bacteria and the carbonaceous oxygen demand exerted by heterotrophic microorganisms can limit the amount of oxygen available to nitrifiers.

Ignoring cell synthesis, the stoichiometry for these energy conversion reactions are (EPA 1975)

$$NH_4^+ + 15.2\,O_2 + 2\,HCO_3^- \rightarrow O_2^- + H_2O + H_2CO_3 \qquad (12)$$

$$NO_2^- + 0.5\,O_2 \rightarrow NO_3^- \qquad (13)$$

Taking nitrifier synthesis together with nitrification, the overall relationship between ammonium, bicarbonate, oxygen consumed, cell mass, nitrate, water, and carbonic acid can be written (EPA 1975)

$$NH_4^+ + 1.86\ O_2 + 1.98\ HCO_3^- \rightarrow$$
$$0.0206\ C_5H_7NO_2 + 0.980\ NO_3^- + 1.041\ H_2O + 1.88\ H_2CO_3 \qquad (14)$$

According to the above stoichiometry, nitrification and nitrifier synthesis can consume between 6.0 and 7.4 mg of alkalinity (as $CaCO_3$) and about 4 mg of oxygen for every 1.0 mg of ammonia nitrogen that is converted to nitrate-nitrogen (EPA 1975). Therefore, supporting nitrification in fully recirculating systems can require around 0.2 kg alkalinity (as calcium carbonate) for every 1.0 kg of feed consumed within the system. This alkalinity loss may be reduced if the makeup water flow contains a significant concentration of alkalinity or if denitrification is occurring within the recirculating system, because denitrification releases alkalinity. Chemicals containing the hydroxide, carbonate, or bicarbonate are added to recirculating systems to maintain an alkalinity of 50–100 mg/L (as calcium carbonate). Sodium bicarbonate (baking soda) is usually considered the safest and easiest chemical to add, because it does not produce drastic pH increases, it is relatively nonharmful, and it dissolves rapidly and completely in water (Bisogni and Timmons 1991; Loyless and Malone 1997). Sodium hydroxide is usually the lowest cost base, but this depends on location. However, sodium hydroxide must be used with care because adding too much hydroxide can rapidly increase the pH to levels of 9–10.

Biological treatment processes use microorganisms that grow attached (in a fixed film) on wetted or submerged surfaces or that simply grow suspended in the water column (Wheaton et al. 1991; Hagopian and Riley 1998). Most recirculating systems use fixed-film biological treatment processes, although some suspended growth treatment processes (similar to activated sludge) have also been used. A physical unit containing the fixed-film biological treatment process is termed a biological filter (biofilter, for short). A wide range of biofilter types are used to treat dissolved wastes in aquaculture, but the most common include: submerged filters, trickling filters, rotating biological contactors, pressurized-bead filters, and fluidized-sand filters. These units are discussed later in this section.

Selecting a biofilter design to use within a recirculating aquaculture system affects the system's water quality, the reliability of water treatment, and the capital and operating costs. An ideal biofilter would remove 100% of the inlet ammonia concentration, produce no nitrite, require a relatively small footprint (by supporting super-dense microbial growth), use inexpensive support material, require no water pressure or maintenance to operate, and would not capture solids; solids can plug a biofilter and degrade rapidly, thereby releasing nutrients and exerting a large oxygen demand in direct competition with nitrification. Unfortunately, there is no biofilter type that can meet all of these ideals—each biofilter type has its own strengths and weaknesses.

Media. Biological filters have a capacity to remove dissolved ammonia, which is largely dependent upon the total surface area that is available for attached biological growth. In many instances, increasing the biofilter surface area should result in a corresponding increase in ammonia removal capacity. A review of studies on many different biofilters under a wide range of conditions (Table 6) indicates that 1 m^2 of surface area in a biofilter can remove between 0.03 and 0.78 g TAN per day (Eikebrokk and Piedrahita 1997). Note the tremendous variation in areal nitrification rates that have been reported in Table 6, even for a given biofilter type.

The tremendous variation in the capacity of 1 m^2 surface area to remove ammonia is a product of the sensitivity of nitrification to different conditions, which include: water temperature, pH, alkalinity, availability of DO, concentration and daily loading of TAN and organic matter, accessibility of the media surface, rate of ammonia and oxygen transfer into the biofilm, growth phase of the biofilm (lag, log, stationary, and death phases), and competition with heterotrophic microbes for space and oxygen. The variability in nitrification rates that are encountered under different conditions makes the selection of an accurate areal nitrification rate difficult, unless one has considerable experience with how a specific biofilter type performs under specific conditions that can be replicated fairly consistently. Therefore, selecting a conservative areal nitrification rate will help to ensure that the biofilter can meet its ammonia treatment goals. However, this will increase the cost and size of the biofilter.

Biofilter function can be impaired if there is too little oxygen or if the available surface area is overloaded from receiving too much ammonia or dissolved or particulate organic matter. If oxygen or area become lim-

TABLE 6. Areal nitrification rates reported for various biofilter types (from Eikebrokk and Piedrahita 1997).

Biofilter type	Areal nitrification rate (g TAN/m^2/day)
Biodrum	0.10
Fluidized bed (sand)	0.08–0.28
Packed bed (plastic bead)	0.08–0.51
Rotating biological contactor	0.19–0.78
Trickling filter	0.03–0.55
Upflow packed bed (sand)	0.06–0.15
Submerged filter (plastic)	0.36–0.42

iting when the biofilter is overloaded, *Nitrobacter* are generally the first to be displaced, as they are the last in the line of microbial consumers, but *Nitrosomonas* can also be displaced or suffer reduced removal rates from limiting conditions. Conditions that cause a reduction in the relative amount of *Nitrobacter* compared to *Nitrosomonas* will cause the nitrite concentration across the biofilter to increase. Two approaches that can be applied (sometimes simultaneously) to counter the problem of insufficient ammonia and nitrite removal in large recirculating systems: (1) size biofilters with more available surface area for microbial treatment and with consideration of oxygen requirements, so that the biofilter has both the area and oxygen supply to handle higher organic and ammonia loading rates, and (2) add ozone to the system to oxidize excess nitrite to nitrate.

Media used in biofilters must have a relatively high surface area per unit of bed volume (i.e., specific surface area in m^2/m^3) and an appreciable voids ratio. The specific surface area is important as it controls the amount of bacterial growth that can be supported in a unit volume (Wheaton et al. 1991). The voids ratio is critical for adequate hydraulic performance of the system. The voids ratio characterizes how much space is provided for the water to pass through the media in close contact with the biofilm. In some cases (e.g., trickling filters and rotating biological contactors), the voids must also provide space for air ventilation so that oxygen can be transferred to the flow within the bed. Biofilter media must also be inert, noncompressible, and nonbiologically degradable (Summerfelt 1999). Biofilters such as submerged filters, trickling filters, and rotating biological contactors often contain relatively large void spaces within media consisting of structured blocks or tubes (Figure 27). Randomly packed media can be sand, crushed rock, river gravel, some form of plastic pellet or plastic or ceramic material shaped as larger spheres, rings, or saddles. Packed or fluidized beds of plastic beads or finely graded sand use the finest media, with average equivalent diameters ranging from 1.0–3.0 mm or 0.1–1.5 mm, respectively (Summerfelt and Cleasby 1996).

Submerged Filters. The submerged biofilter consists of a bed of media (such as shown in Figure 27) through which the wastewater passes in an upward (Figure 28) or downward direction (Summerfelt 1999). The hydraulic retention time can be controlled in either the upward flowing or downward flowing mode by adjusting the water flow passing through a given submerged biofilter volume.

The surfaces within the submerged filter will capture solids that can eventually block the void spaces. Solids accumulate in static filter beds based on both physical and biological mechanisms. Settling, sieving, and surface interception can all lead to solids accumulations that fill the

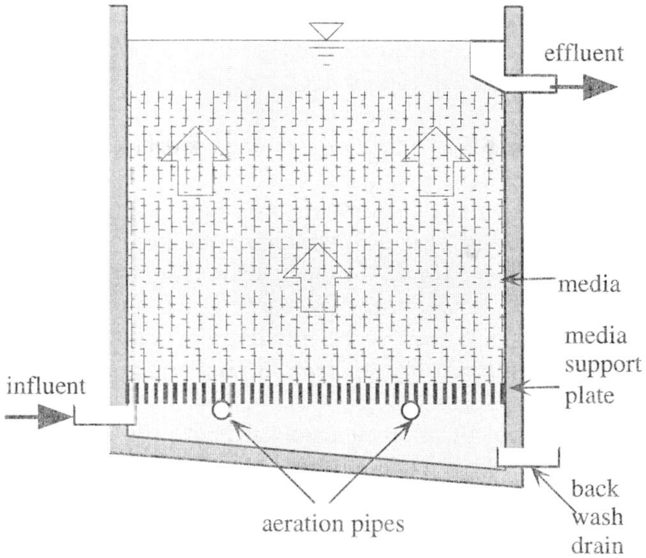

FIGURE 28. An upflow submerged biofilter with backwash capabilities (from Summerfelt, 1999).

biofilter void spaces. Cell mass is also produced on the surface of the media and in the voids between the media by the growth of microorganisms removing organic matter and ammonia passing through the biofilter. Cell mass resulting from heterotrophic organisms, though required to some extent, increases plugging problems in static media filters.

Providing a mechanism to flush solids from the biofilter is critical. Good solids flushing requires a drainable filter, a properly designed underdrain, a high velocity of a reversed flow of water, and a means for air scouring (Summerfelt 1999). Underdrains must also support the media and direct the water flow uniformly throughout the bed during regular operation and during flushing.

The characteristics of the media in a submerged biofilter ultimately influences the removal of suspended, settleable, and dissolved organics, as well as the head loss generated and the success of flushing the filter (Summerfelt 1999). Media with sufficient voids for use in submerged biofilters can be random media—such as uniform crushed rock over 5 cm diameter, or plastic media over 2.5 cm diameter—or structured media, such as plates or tubes containing large voids (Figure 27). However, 5.0-cm diameter crushed rock would only have a specific surface area of 75 m^2/m^3 and a voids fraction of only 40–50%. Random plastic media that is roughly 5.0 cm in size can have a specific surface area of 100–120 m^2/m^3 and a voids fraction greater than 95%.

Aerated submerged filters have been developed to reduce plugging problems and problems resulting from low levels of DO. Aeration serves to both raise the level of available DO for nitrification and to break up clogging to a variable extent, because vigorous aeration strips off thick biofilm growths and helps to prevent biofilm growth from clogging the pores in the filter media.

A new, relatively small (7 mm long and 10 mm diameter) and slightly buoyant (0.96 specific gravity) polyethylene tubular media has seen some recent use in heavily-aerated, submerged biofilters. The technology is termed the "moving-bed biofilm reactor," because the heavy aeration slowly but continuously moves the bed of small plastic media (Rusten et al. 1998). The combination of aeration and bed movement helps to prevent inadequate DO or plugging problems. The biofilters are generally only filled to 70% of operating volume with the plastic media, but this produces an effective surface area for biofilm growth of around 350 m^2 per cubic meter of biofilter operating volume (Rusten et al. 1998). This type of submerged biofilter requires little hydraulic head to pass the water flow, however, the heavy aeration required can amount to a significant amount of power.

Trickling Filters. The physical characteristics of a trickling filter (Figures 24 and 29) appear similar to those of a downflow submerged biofilter (Figure 28). However, the media in a trickling filter is not submerged during normal operation, because the water level is always maintained below the level of the media (Figures 24 and 29). The media is kept wet by the trickling water flow through the media (Wheaton et al. 1991). Because it is not submerged, air that is ventilated through the wet media carries oxygen to the bacteria and provides opportunity for carbon dioxide degassing (Nijhof 1995). In this case, ventilation requires little power, as there is minimal resistance to airflow in a properly designed trickling biofilter.

Trickling filters have been widely used in fully recirculating aquaculture systems. The most important reason for their popularity is ease of operation, self-aerating action, and moderate capital cost (Summerfelt 1999). Modern trickling filters use plastic media because it is light, has high specific surface areas (100–300 m^2/m^3), and contains void ratios greater than or equal to 90%. A range of trickling filter design criteria have been reported; typical values follow (Summerfelt 1999): hydraulic loading rate of 100–250 m^3/d per m^2 of horizontal plan area; media depth of 1–5 m; media specific surface area of 100–300 m^2/m^3; and TAN removal rate of 0.1–0.9 $g/d/m^2$ surface area.

High specific surface area trickling filters can sometimes have more problems than filters of lower specific surface areas due to regional plug-

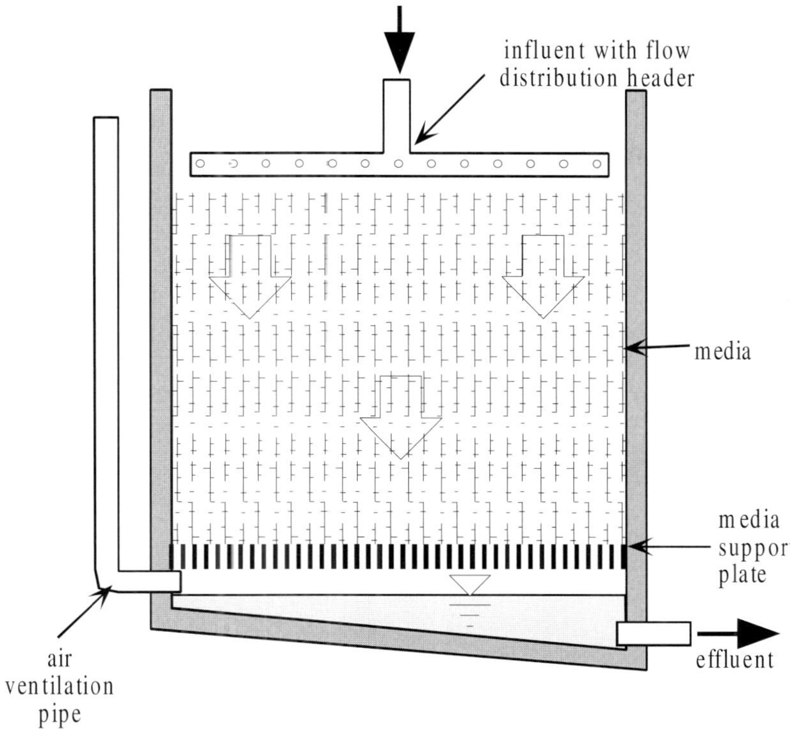

FIGURE 29. A trickling filter with a rotary spray arm to distribute the water flow above a bed of media (from Summerfelt, 1999).

ging, channeling, or hydraulic loading that is not sufficient to moisten all the area of the densely packed media (Summerfelt 1999).

Rotating Biological Contactors. Rotating biological contactors (RBC) function by rotating their partially submerged plastic media volume (which has been attached to a shaft) through the water flow containing the dissolved wastes to be removed (Figures 18 and 30). Media is typically submerged 40% of the drum diameter and is often rotated at speeds of 1.5–2.0 rpm (Tchobanoglous and Schroeder 1985). Rotation of the media through the atmosphere provides aeration of the water flow and the microbial film, helping to maintain aerobic conditions for nitrification.

Rotating biological contactors can be used to treat small flows, but they also scale to lengths of 8 m and total surface areas of approximately 8,400 m^2 to treat large flows (Tchobanoglous and Schroeder 1985). Rotating biological contactors have seen some use in fully recirculating systems because they require little hydraulic head, have low operating costs,

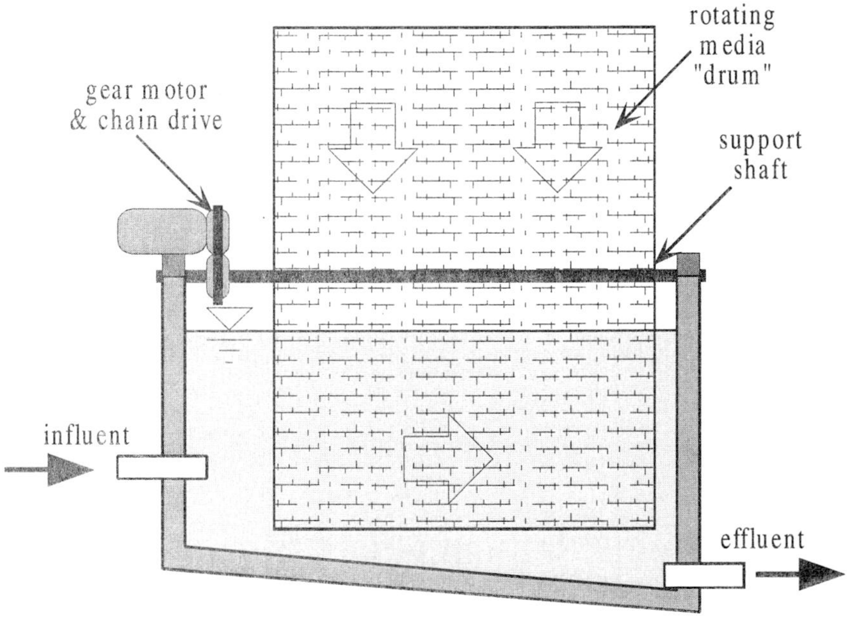

FIGURE 30. A rotating biological contactor (from Summerfelt, 1999).

and can maintain a consistently aerobic treatment environment. However, heavy biosolids growth caused catastrophic failures of either the shaft or the media in some of the early units (Tchobanoglous and Schroeder 1985; Timmons et al. 2000). Most of these problems were mechanical, and properly designed RBCs can be relatively reliable. However, the capital cost of RBCs are relatively high (Table 7), depending upon the given situation (Timmons et al. 2000).

Pressurized-Bead Filters. The pressurized-bead filter passes water up through a bed of floating plastic beads, retaining the beads with a screen

TABLE 7. Capital costs estimates associated with biofilter choices for a tilapia farm producing 1,000,000 lb (454 metric tons) annually. Cost is listed as $ per pound per year (PPY) (from Timmons et al. 2000).

Biofilter type	Farm cost	Cost, $ per PPY
Rotating biological contactor	$668,000	$0.68
Trickling biofilter	$620,000	$0.62
Bead filter	$296,000	$0.30
Conventional fluidized-sand biofilter	$124,000	$0.12
Cyclo biofilter	$76,000	$0.08

located at the top of the vessel (Figure 21). Operating pressures are typically 35–100 kPa.

The pressurized-bead filter is sometimes used to treat small or moderate flows, usually less than or equal to 1,000–2,000 L/min. Pressurized-bead filters can have considerable advantages due to their modular and compact design, generally straight forward operation, and effective ammonia removal. The pressurized-bead filter also effectively removes solids from a flow (Chen et al. 1993a) and is sometimes used only for solids removal or as a hybrid filter that combines both nitrification and solids removal (Golz et al. 1999). The bead filter is typically either bubble-washed or propeller-washed during its backwash procedure, which helps to separate trapped solids from the floating beads.

Both nitrifying bacteria and heterotrophic microorganisms will grow on the floating plastic beads and within the void spaces that surround the beads within the filter. Large populations of heterotrophic microorganisms can grow in the bead bed, because the bead filter removes solids so effectively. To maintain nitrification, solids must be periodically removed (Golz et al. 1999). Back-flushing too frequently or too vigorously, however, can remove too much of the nitrifying bacteria (Golz et al. 1999). Therefore, when solids loading are relatively high, maintaining nitrification can require proper adjustment of the frequency and intensity of back-flushing cycles.

Fluidized-Sand Biofilters. Fluidized-sand biofilters (Figures 5, 6, and 31) are seeing wide use in North America as they approach the "ideal biofilter" in several important aspects. Filter sand is usually inexpensive (\$40 to \$70/m^3) and its high specific surface area (4,000–45,000 m^2/m^3) reduces the cost of surface area (\$0.02 to \$0.001/m^2 surface area), which allows additional sand to be included at little extra cost and makes fluidized-sand biofilters exceptionally compact (Summerfelt and Wade 1998; Summerfelt 1999). Fluidized-sand biofilters can also be designed to treat large flows, with existing units treating 11,000 L/min of water flow. For these reasons, large fluidized-sand biofilters can be exceptionally cost competitive compared to other large-scale biofilters (Table 7) (Summerfelt and Wade 1998; Timmons et al. 2000), even when constructed with excess nitrification capacity. Fluidized-sand biofilters are also efficient at removing ammonia, typically removing 50–90% of the ammonia each pass in cold- and cool-water aquaculture systems. Nitrite removal is also usually excellent, probably due to the biosolids mixing in the bed, and the high available surface area. In well-engineered fully recirculating systems for cool- and cold-water culture, fluidized-sand biofilter outflows will often only contain 0.1–0.5 mg/L TAN and less than 0.1–0.2 mg/L nitrite-nitrogen.

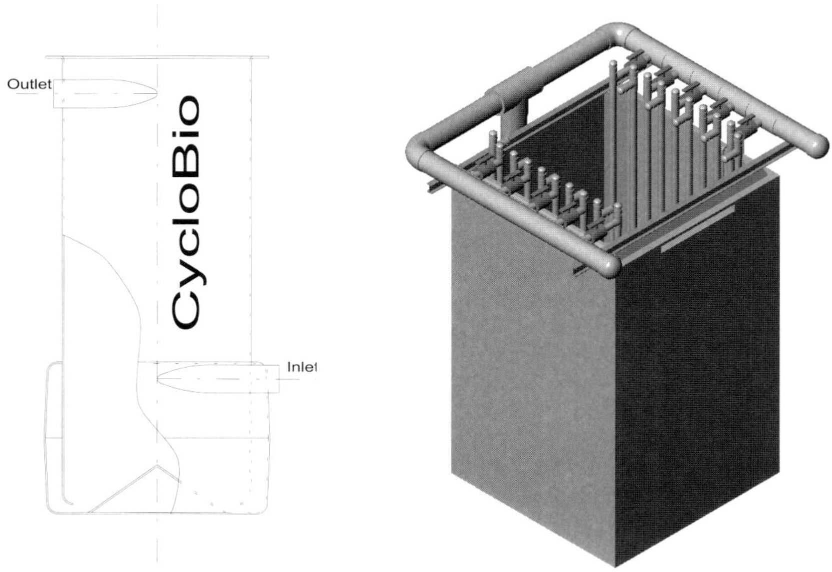

FIGURE 31. A fluidized-sand biofilter of the new 'Cyclo-Bio™' type (courtesy of Marine Biotech, Inc., Beverly, Massachussetts) on the left and a more standard pipe-manifold type fluidized-sand biofilter (courtesy of PRAqua Technologies, Ltd., Nanaimo, British Columbia) on the right are both effective at injecting water flow beneath the sand bed.

On the down-side, fluidized-sand biofilters require water pressures of 5.5–10.5 m of water head to pump water from the sump through the biofilter to its overflow (assuming a water lift of 3–4 m), but this can be a relatively inexpensive part of the total system operating costs if power costs are reasonable. Also, fluidized-sand biofilters do not aerate, as do trickling filters. Fortunately, the likelihood of oxygen limitation in fluidized-sand biofilters is reduced due to their relatively high water velocities and short hydraulic retention times. Nonetheless, fluidized-sand biofilters should always be designed with a cascade column placed immediately downstream to strip dissolved carbon dioxide and bring the DO up to near 90% of saturation.

In operation, water is injected under the sand bed with a force that overcomes the sand's apparent weight, buoying up the individual grains (Summerfelt and Cleasby 1996). The bed becomes fluidized and expands, allowing water at high velocity to surround and tumble each particle. Microorganisms that metabolize ammonia and other dissolved wastes grow as a biofilm attached to the sand surfaces; this prevents them from being flushed out, essential for good performance (Timmons and Summerfelt 1998; Nam et al. 2000). As the biofilms grow, the sand grains become less dense, and the bed expands further. Pieces of the growing biofilm are continuously abraded and are flushed to the top of the bed. With fine sands, water velocity through the bed is insufficient to lift the larger biofilm particles clear and the bed grows deeper with time. Excess biosolids produced in the bed should be siphoned from the top of the filter when necessary, as often as several times per month, and the sand lost should be replaced at least twice annually. The cost of sand replacement is negligible.

Fluidized-sand biofilters in aquaculture typically use extremely hard, whole grain, finely graded crystalline silica sand. Sands have a specific gravity 2.65 times that of water; therefore, the pressure drop across an expanded sand bed is approximately 0.9–1.0 m of water head for every 1.0 m of initially static sand depth, independent of sand size (Summerfelt and Cleasby 1996). The pressure drop across the bed remains constant at all bed expansions. During fluidization, the relative amount of bed expansion is dependent upon the media's shape (i.e., sphericity) and diameter, and the superficial velocity and temperature of the water (Summerfelt and Cleasby 1996). Fluidized-sand biofilters are often operated at expansions of 40–100% (based on clean sand).

Size grades of filter sand are available commercially, and are usually specified by a sand's "effective size," that is, D_{10} is the sieve opening size that will pass only the smallest 10%, by weight, of the granular sample and "uniformity coefficient (UC)," that is, UC is a quantitative measure of the variation in particle size of a given media, defined as the ratio of D_{60} to D_{10} (Summerfelt and Cleasby 1996). Sand used in fluidized biofilters for recirculating systems usually has a mean effective diameter in the range of 0.2–1.0 mm and a uniformity coefficient between 1.3 and 1.8. The D_{10}, an estimate of the smallest sand in the sample, is the sand that will expand the most at a given superficial velocity. The D_{90}, an estimate of the sieve size that will pass only the largest 10% of sand in the sample, is the sand fraction that will expand the least at a given superficial velocity. The D_{90} can be estimated from the effective size and the uniformity coefficient (Summerfelt and Cleasby 1996):

$$D_{90} = D_{10} \times \left(10^{1.67 \times \log(\mathrm{UC})}\right). \tag{15}$$

Because sands are not perfectly uniform, larger sands move to the bottom of the fluidized beds where they expand less than the smaller sands that have migrated to the top of the bed. The average expansion of a bed at a given velocity depends upon the size gradation within the bed (i.e., uniformity coefficient), which makes it important to predict expansion of both the largest and smallest fractions of sand. In particular, the largest fraction of sand must expand at the velocity selected. A new fluidized-sand biofilter design termed the "Cyclo-Bio™" (Figure 31) incorporates an inverted cone into the bottom of the vessel, which increases the water velocity in the same region where the larger sands accumulate. This increased velocity gradient in the region around the cone also increases the likelihood that even the largest sand collected at the bottom of the bed will be expanded at the design flow rate.

The design of the flow distribution mechanism is absolutely critical for reliable operation of fluidized-sand biofilters (Summerfelt and Cleasby 1996; Summerfelt et al. 1996). The flow distribution system must deliver an equal amount of flow across the base of the bed, prevent loss of media, operate without detrimental fouling (or have a fouling prevention system) and, in some cases, support the bed. There are a wide variety of distribution mechanisms used to inject water into the bottom of large fluidized-sand biofilters (Summerfelt and Cleasby 1996). Each mechanism differs in how flow is transported and distributed. Most mechanisms used in recirculating aquaculture systems transport the flow through a manifold, starting at the top of the biofilter, that runs down the inside of the reactor to the base of the sand.

Recently, however, the Cyclo-Bio™ introduces the water flow into the bottom of the vessel through an outside annulus (integral with the vessel wall) that supplies water into the bottom of the vessel (i.e., at the base of the sand bed) through a continuous slot that circumscribes the vessel wall (Figure 31; Timmons et al. 2000). This design is analogous to a slotted inlet used in animal housing to introduce air uniformly to the entire building. The Cyclo-Bio™ design saves energy as a result of eliminating the conventional "pipe-lateral with distribution orifices" type of distribution system (shown in Figure 31) The Cyclo-Bio™ incurs very little pressure drop due to the much lower water velocities produced across the slotted inlet. The Cyclo-Bio™ also decreases the cost associated with fluidized sand filters both in terms of equipment components and the time required to install and setup a new biofilter, because the Cyclo-Bio™ does not require assembly of a typical header and pipe-lateral system, which has been described by Summerfelt and Cleasby (1996) and Summerfelt et al. (1996) and is shown in Figure 31.

Nitrification rates and biofilter bed maintenance requirements are both dependent upon the sand size, water temperature, ammonia, and organic loading rates. Nitrification rates reported in cold-water systems range from 0.2 to 0.4 kg TAN removal per day per cubic meter of expanded bed volume (Timmons and Summerfelt 1998). In warmwater systems, TAN removal rates range from 0.6–1.0 kg per day per cubic meter expanded bed volume (Timmons and Summerfelt 1998). Fine sands with an effective sand size (D_{10}) of about 0.2–0.25 mm have been shown to be especially effective in biofilters intended for cold-water systems at 10–15°C (Tsukuda et al. 1997). However, the best effective sand size (D_{10}) for use in warmwater systems (i.e., at 25–30°C) appears to be closer to 0.5–0.7 mm, e.g., ranging from a 20–40-mesh sand to a 16–30-mesh sand (Timmons and Summerfelt 1998).

Disinfection Systems—Ozone and UV

Ozone has been used for disinfecting water in recirculating systems, but it has been used much more widely to achieve water quality improvements by microflocculating fine particulate matter (making particles that are easier to settle or filter) and oxidizing nonbiodegradable organic molecules (creating smaller and more biodegradable molecules), nitrite, and refractory organic molecules (reducing water color) (Summerfelt and Hochheimer 1997; Summerfelt et al. 1997). Ozone has seen wide use because it has a rapid reaction rate, produces few harmful reaction byproducts in freshwater, and oxygen is produced as a reaction end-product. Application of ozone to aquaculture requires ozone generation, ozone transfer into solution, contact time for ozone to react, and possibly ozone destruction to ensure that no ozone residual makes it into the culture tanks (Summerfelt and Hochheimer 1997). A system that provides ozone contacting and residual removal is illustrated in Figure 32.

Ozonation of fully recirculating systems can reduce fish disease simply by improving water quality, which reduces or eliminates environmental sources of stress, and by disinfecting the water (Wedemeyer 1996; Bullock et al. 1997). Ozone oxidation can kill microorganisms, but disinfecting the water requires maintaining a certain dissolved ozone concentration for a given contact time. Thus, disinfecting efficiency depends on the product of the ozone residual concentration times its contact time. An ozone contact vessel provides the time necessary for the ozone residual to react with and inactivate the microorganisms. Disinfecting water can require maintaining a residual ozone concentration of 0.1–2.0 mg/L in a plug-flow type contact vessel for periods of 1–30 min, depending upon the target microorganism (Wedemeyer 1996).

Unfortunately, recirculating systems contain enough organic matter and nitrite to create a large ozone demand. Therefore, achieving large microorganism reductions in recirculating systems requires much greater ozone dosages than are required for water quality control alone (Bullock et al. 1997) and requires much higher ozone dosages than are typically required for disinfecting single-pass inflows. If enough ozone has been

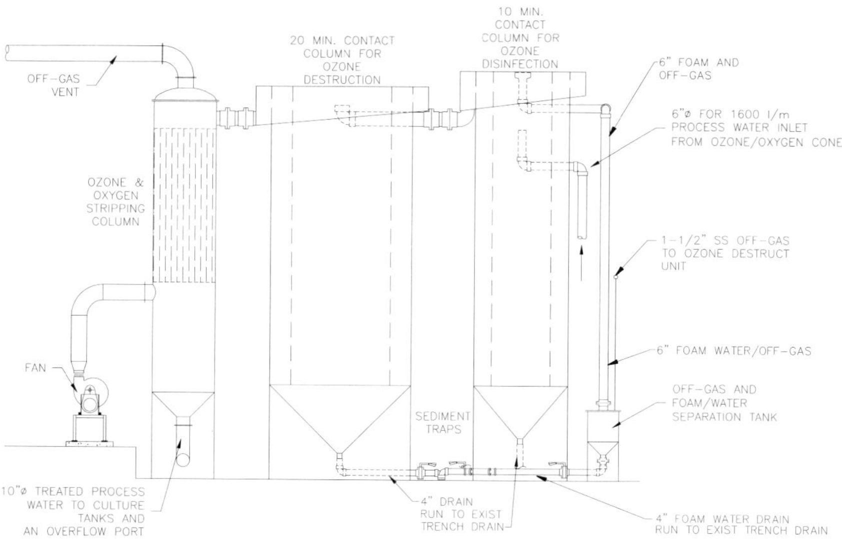

FIGURE 32. This illustration shows an ozone treatment system for disinfecting 1,600 L/min of surface water at the U.S. Fish and Wildlife Service's Lamar National Fish Hatchery (Lamar, Pennsylvania). The primary ozone treatment components (shown here) are located on the process pipeline immediately after ozone gas has been transferred into the water. First, the righthand vessel provides 10 minutes of plug flow contacting to achieve disinfection. Next, the middle vessel provides 20 minutes of plug flow contacting to achieve further disinfection and ozone residual destruction. Finally, the lefthand vessel (a counter-current, forced-ventilation column) strips any minor ozone residuals and elevated dissolved oxygen levels immediately before the flow is piped to the fish culture systems. Drawing courtesy of Oak Point Associates, Biddeford, Maine.

added to create a disinfecting ozone residual at the end of the contact chamber, then that ozone residual will need to be removed before the water reaches aquatic organisms. The ozone residual can be lethal to fish at ozone concentrations as low as 0.01 mg/L, but the actual concentration depends upon species and life stage (Table 8). Dissolved ozone can be removed by providing extended contact times, aeration and degassing, intense UV light doses, or reaction with hydrogen peroxide.

Ultraviolet light can be used to destroy ozone residuals (catalyzing the conversion of O_3 to O_2) and to denature the DNA of microorganisms, causing the microorganisms to die or lose their function. Achieving ozone destruction and microorganism inactivation with UV electromagnetic radiation depends on the wavelength of the UV light source and the

TABLE 8. Toxicity of dissolved ozone to fish, where LC50 is the concentration lethal to 50% of sample fish (from Summerfelt and Hochheimer 1997).

Species	Ozone concentration (mg/L)	Effect
Rainbow trout	0.0093	96-h LC50
Rainbow trout	0.01–0.06	lethal
Bluegill	0.01	60% mortality after 4 weeks
Fathead minnow	0.2–0.3	lethal
White perch	0.38	24-h LC50
Bluegill	0.06	24-h LC50
Striped bass (larvae)	0.08	96-h LC50

quantity of energy transmitted (Rodriguez and Gagnon 1991; Hunter et al. 1998). Ozone residuals are destroyed at UV light wavelengths ranging from 250 to 260 nm, while microorganism inactivation can be achieved at UV light wavelengths ranging from 100 to 400 nm, although a wavelength of 254 nm is most common. Ironically, UV wavelengths of 185 nm can be used to generate ozone. As in ozone disinfection, achieving UV disinfection requires maintaining a minimum UV dose that is the product of the UV light intensity times the exposure time to this constant intensity. Ultraviolet light intensity is described in terms of milliwatts per square centimeter (mW/cm^2) and UV dose in terms of $mW\text{-}s/cm^2$.

Ultraviolet doses of 60–75 $mW\text{-}s/cm^2$ have been reported to completely destroy ozone residuals as high as 0.5 mg/L (Hunter et al. 1998). Ultraviolet doses required to inactivate microorganisms can vary tremendously, from only 2 $mW\text{-}s/cm^2$ to more than 230 $mW\text{-}s/cm^2$ (at 254 nm), depending upon the target organism and the required kill rate (Wedemeyer 1996). The intensity of UV light required to inactivate microorganisms also depends upon the presence and size of particulate matter. For example, prefiltration through 50-μm mesh screens can improve bacterial inactivation by as much as 3.0 $\log_{10}$ units (Liltved and Cripps 1999). However, before the UV dose can even reach the target organism, it must be able to transmit through the water. Therefore, the lowest expected UV transmittance of the process water should be established and used to predict how much UV intensity must be generated to transmit the desired UV dose through the water between the target organism and the light source. The UV filter unit should also be sized to account for the 40% decline in bulb intensity that occurs over the typical 12 month expected lamp life.

Ultraviolet filters can be built as nonpressurized open channel units (Figure 33) or as pressurized tube-and-shell units. The UV bulbs are usually contained within quartz sleeves to allow submergence in the pro-

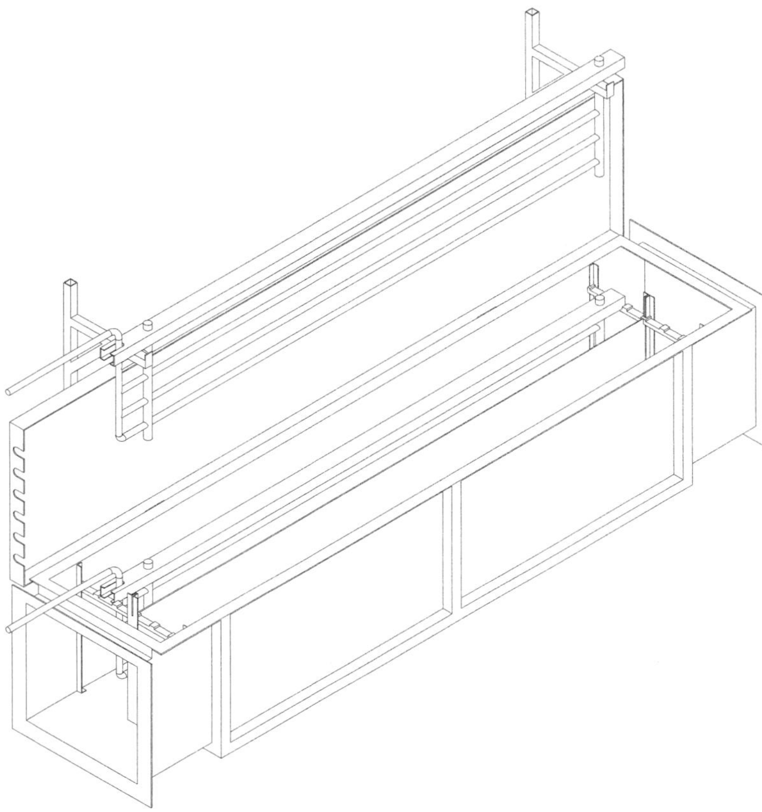

FIGURE 33. The horizontal UV channel filter shown here (with one of its lamp-racks removed for service) is being used to evaluate the influence of full-scale UV filtration on recirculating system performance at the Freshwater Institute in Shepherdstown, West Virginia. Drawing courtesy of PRAqua Technologies, Ltd., Nanaimo, British Columbia.

cess flow. The quartz sleeves must be kept clean to maintain transmittance. Using UV light does not produce toxic residuals or form byproducts that pose a risk to aquatic organisms.

Dissolved Organic Matter and Protein Removal

Dissolved organic matter and protein are not effectively removed using standard solids sedimentation and filtration technologies. Biofilters do not effectively remove nonbiodegradable organic matter. Therefore, dissolved organic matter and protein can accumulate in fully recirculating aquaculture systems; some researchers suspect that these compounds can reach levels that limit fish production. Foam fractionation and ozonation are water treatment processes that are both used, sometimes simultaneously, to help reduce accumulations of dissolved organic matter (Rosenthal and Black 1993).

Dissolved Air Flotation and Foam Fractionation

Flotation and foam fractionation are two options for removing relatively fine solids and some dissolved organic matter that are not commonly removed by settling basins, microscreen filters, or granular media filters (Timmons and Losordo 1994). In the flotation process, the solids attach to air bubbles, float to the surface, become concentrated, and are removed. The fine bubbles, when attached to a solid particle, increase the difference between the specific gravity of the particle and water by changing the particle's specific gravity from just over 1.0 to considerably less than 1.0. Flotation can be especially effective for removing hydrocarbons, protein, fibers, and algae.

There are three basic mechanisms used to produce bubbles for flotation: dispersed air, dissolved gas, and electrolytic flotation. The primary difference in the three flotation mechanisms are their cost and the size of the bubbles produced. Systems using dispersed air (often called foam fractionation) are the least expensive and generate bubbles larger than 100 μm. Systems using either supersaturations of dissolved gas or electrolytic cleavage of water are called dissolved air flotation because they can precipitate bubbles in the 10–100 μm size range. Solids capture is highly dependent upon the bubble diameter, solids concentration, air to water ratio, surface chemistry of the solids, and surfactant concentration in the water. Chemical addition, like a strong oxidant such as ozone, or a salt, or polymers can aid in the formation of a cohesive froth to improve the quality of the treated water.

Foam fractionation of fine particulate matter and dissolved organic matter has been demonstrated in recirculating aquaculture systems using dispersed air and with a dispersed ozonated-air mix (Rosenthal and Black 1993; Timmons and Losordo 1994). However, because foam fractionation does not remove many of the particles greater than 30 μm, it is usually used in parallel with a settling tank or filter.

Ozonation

Dissolved organic matter, including refractory compounds and fine particulate organic matter that accumulate in recirculating systems, can be removed by increasing the daily water exchange rates or by introducing an oxidizing agent such as ozone (Summerfelt et al. 1997). Because ozone and its reaction by-products are capable of oxidizing a great many organic substances, ozone is added to some recirculating systems in order to improve water quality without resorting to high daily water exchange rates. Ozone breaks refractory and relatively nonbiodegradable organic compounds into smaller, more biodegradable compounds, and is thought to precipitate dissolved organic molecules and microflocculate fine particulate matter (Summerfelt and Hochhelmes 1997). Ozone's action on fine particulate matter can significantly improve its removal by settling, filtration, or flotation (Summerfelt et al. 1997).

Water Movement and Transmission

Nearly all raceway and tank-based aquaculture systems must move water to carry oxygen to the fish, to receive the waste produced, and disperse or carry these wastes away from the culture environment. Water can be moved by gravity if a hydraulic gradient exists to move the water from point a to point b. Gravity water flow, an ideal situation that does not require additional energy, is used in many serial-reuse systems, but gravity flow aquaculture systems are restricted by limited water resources of this type. Pumps are often used to transfer mechanical energy to water, increasing the water pressure, and providing water the motive force necessary to move. Many aquaculture applications require pumps with low to moderate pressure requirements. Pumps employed in aquaculture include: airlift pumps, centrifugal pumps, axial flow pumps, and reciprocating pumps. Specifics on these pumps are available elsewhere (Karassik et al. 1986).

Pipes, channels, raceways, culture tanks, and other vessels contain the water while transmitting it from point a to point b. Pipes and channels are sized so that they provide the minimum acceptable conduit area that will accommodate the design flow rate while maintaining the pressure drop and water velocity within reasonable limits. With sufficient elevation drop available, both piping and channel systems can be designed to flow based on the elevation drop provided. However, most tank-based recirculating systems must rely on pumping water in at least one location to provide the pressure head required to maintain design flow rates through conduits, culture tanks, and treatment units. After pumping the water, many recirculating systems will let the water gravity flow through all of the treatment components and fish culture tanks until the water returns to the pump sump, where it is pumped again for further reuse (Figure 34).

Reasonable water velocities must be selected so that the pipe or channel size can be optimized. Selection of the best pipe or channel size is based on an economic trade-off between the installed capital cost of the piping system, which is greater for larger diameter conduits, and the

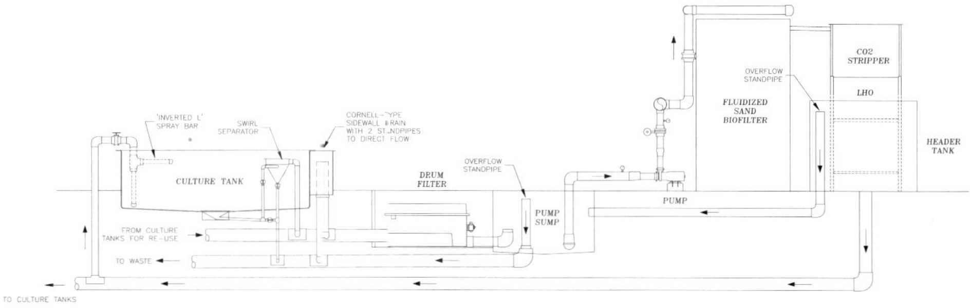

FIGURE 34. This profile view (courtesy of PRAqua Technologies, Ltd., Nanaimo, British Columbia) shows the path water takes through the major components within one type of fully recirculating system. After pumping the water through the biofilter, the water flows by gravity through all of the treatment components and fish culture tanks until the water returns to the pump sump, where it is pumped again for further reuse. Make-up water is added in the pump sump through a float valve (not shown) and excess water overflows the system from the bottom discharge of each swirl separator and/or from the pump sump.

sum of the capital plus lifetime operating costs of the pumping system, which increases with smaller diameter conduits (Nayyar 1992). A detailed piping or channel system design requires consideration of a water flow rate, viscosity, density, and conduit wall frictional drag, which are beyond the scope of this chapter, but can be found in hydraulics books such as Nayyar (1992).

In simplistic terms, the water velocity that can be created through the pipe or channel depends on the water depth, slope of the conduit, or backpressure behind the water flow. Many hydraulics books and, sometimes even pipe supplier catalogs, will supply tables containing water velocities and pipe friction loss estimates for a wide range of nominal pipe sizes and water flow rates. These are extremely useful tables for estimating pipe size when the design flow rate and acceptable water velocity are known. The trick is to select a water velocity through the pipe that is sufficient to maintain a self-cleaning pipe and keep the pipe size relatively small and less expensive, while at the same time limiting the pipe friction loss to acceptable levels. Most settled solids deposits in recirculating systems (excluding biological films) can be reduced by sizing all pipes, channels, and standpipes to produce a water velocity of at least 40–100 cm/s, depending upon the situation. Somewhat higher velocities can be used when the water is pumped, when conduits have relatively large cross-sectional areas, or when larger hydraulic drops are available.

Meticulous attention to detail is required when it comes to selecting the size, slope, and pipe clean-out points for a recirculating system, because these details not only influence the water transmission, but they also influence how "free" from solids the system can be maintained. In water reuse systems, contingencies should always be made to flush deposits of settled solids and thick biofilm growth from the pipes, channels, and basins that carry flow from the culture tank and throughout the system. When possible, pipe runs in recirculating systems should also be designed with clean-out points to allow the pipes to be frequently (daily or weekly) flushed and sometimes brushed to further reduce solids deposits and biofilm mat growth within pipes. The water flushed from pipes during clean-out periods should be discharged outside of the recycle system so that these routine flushing operations do not degrade water quality in the culture tank. For example, the sudden surge in the water flow created by pulling a stand-pipe can carry high concentrations of organic matter as the settled solids and biofilm mats are flushed from the pipe. Ideally, cleaning flow should be discharged outside of the recycle system or to a treatment unit that will remove the excess particulate matter before returning the water volume back to the culture systems.

System Operation

Operating recirculating aquaculture systems can be complex, yet capital and labor costs must be kept under control. Therefore, strategies for process monitoring, alarms, and automation should be considered along with emergency and contingency operation during failures. Further, operating practices designed to prevent the degradation of water quality and introduction and dissemination of fish pathogens are critical.

Process Monitoring, Alarms, and Automation

Process monitoring can be as simple as having a reed switch that closes a circuit when a float rises or falls. When the circuit closes, a bell, light, pump, or other device can be activated. This type of monitoring forms the basis for more complicated monitoring processes, is important to understand, and may be appropriate for many smaller systems. Details on the construction of these types of simple monitoring and control systems are available in the literature. However, the application of only simple control mechanisms may not be adequate when managing large production-scale systems. Recirculating systems may have a large number of tanks and cover a large physical area; therefore, providing sufficient skilled labor to adequately monitor every piece of equipment and respond to alarms manually can be a daunting task. Added on that are the responsibilities of making adjustments to the tank environment (e.g., temperature, water flow, oxygen, pH), properly managing feeding frequency and distribution, observing fish behavior, and responding to emergencies. While human employees remain essential to the operation, the implementation of a well thought out, automated monitoring and control system can provide more continuous coverage while reducing both manpower requirements and the likelihood of human error.

Supervisory control and data acquisition (SCADA) is the term used in industrial automation that refers to automated data collection by remote units with display of data often centrally located on a personal computer (PC). The human machine interface (HMI) is the interface by which the operator interacts with the remote device(s). Some of the functions of the HMI are to allow the operator to adjust set points, configure recipes, or turn alarms on or off. The SCADA equipment often resides in the electrical room or on the floor of the production area, and the PC(s) is placed in a practical location such as the manager's office or is centrally located in the production area.

Early applications of SCADA used a programmable logic controller (PLC) to collect data, make logic decisions, and send information back to a display connected to a PC. Although early PLCs took some effort to program, they have typically been rugged and robust units, with no moving parts. Programming typically required someone fluent in the DOS operating system, ladder logic, BASIC, or other special command language. This type of programming often required the hiring of an expensive specialist, on contract, who was also required for later reconfiguration. Therefore, the operation was at the mercy of the specialist even for rudimentary changes. In addition, the software and hardware could not always be selected independently, which meant that old equipment could not always be upgraded and that the best hardware could not be combined with the best software. Finally, it was relatively difficult to combine data from different systems or to integrate the data produced with standard office software such as spreadsheet packages or word processing software to automate reporting.

Fortunately, SCADA and HMI have been evolving and many of their weaknesses have been resolved. Some of the developments include software packages that allow the end users to configure a control system without learning complex programming languages, and using the PC for control functions over multiple dummy input/output (I/O) ports to reduce the need for multiple PLCs. In addition, similar to the merging that office software has undergone (where spreadsheet, database, word processor, and graphics programs have melded into an integrated package), much of the HMI software is also following suit by merging data acquisition, display, alarming, and trending features into one piece of software. The availability of data has been significantly improved with the development of better communication protocols between software such as Dynamic Data Exchange (DDE) and between computers such as TCP/IP. The end result is that these developments have greatly reduced the cost and enhanced the features available to the end user.

Recirculating systems can benefit from the application of SCADA for the following reasons:

- Recirculating systems are composed of many unit processes. At the simplest level, a system can be broken down into its individual units and monitored on the basis of simple values. For example, in the culture tank commonly measured variables may include DO, pH, and water level. Regardless of what is going on outside of the tank, measurements made within the tank that exceed safe ranges should generate an alarm;
- Recirculating systems operate 24-h per day. Although some employees prefer alternative work hours, more typically it can be difficult to schedule employees to work outside of the normal 8–5 workday. Using process monitoring, alarming, and perhaps some control features, a smaller number of employees can be used to complete repetitive tasks and handle emergencies;
- Recirculating systems have variables that are monitored under controlled conditions. Recirculating systems are inherently designed to work under specifically designed conditions that are generally independent of the local weather. Knowing these design conditions makes it easy to set limits on acceptable operating variables. When these design variables are exceeded, an alarm can trigger or emergency equipment can be activated to deal with the problem automatically. Some examples of relatively simple controls would be adding bicarbonate for pH adjustment or turning on a heater to control temperature.

A SCADA/HMI system can help to prevent recirculating system failure and buy extra time to respond to a crisis when the critical window or response is only minutes and the system is not fully manned. For example, at night, an automated control system can be used to monitor for conditions that lead to failure such as a drop in DO or to compensate for failure modes such as turning on a backup pump on noting failure of a primary pump.

Process Monitoring

Automated process monitoring is one step that can be used to reduce labor requirements and can alert to problems quicker than a human can diagnose. Recording data while monitoring the process can be used to analyze the operation of the system in real time, as well as to generate data for future analysis by either the production manager or an independent third party. Patterns in the data may suggest solutions to recurring or one-time problems. As long as probes are maintained properly, the computer will be able to provide accurate data 24-h per day.

Oftentimes, managers may have had poor experiences with first generation or homegrown monitoring systems that required the specialized skills of an expensive computer programmer for configuring and tweaking the monitoring equipment. Bad experiences with reliability, poor user-interface, and high costs are valid concerns that may make a manager reluctant to implement any sort of automation. The good news is that monitoring has improved greatly within the past few years, making it worthwhile for system managers to investigate their options.

Redundancy

For critical parts that are in continuous operation, implementing redundancy is one of the best actions to prevent failure. One approach for achieving redundancy in the design phase would be to design one relatively large system and provide two of everything (e.g., power and a backup generator, two pumps, automated and manual monitoring). Alternatively, multiple smaller systems could be designed and only some of the most critical components would be provided with a redundant backup, or else the most critical components in each of the systems could have common replacement parts in stock for quick replacement. Advantages of a single large system include

- one set of redundant parts,
- one point of focus with fewer locations to measure system variables such as water quality or fish behavior,
- only one system to monitor and control, and
- fewer sensors to purchase, calibrate, or replace.

Advantages of a several smaller systems include

- replacing a smaller piece of equipment that might fail will cost less than replacing a larger unit,
- components in each of the smaller systems can be substituted,
- repair requiring shutdown of an individual unit would not impact monitoring in the other units, and
- a smaller-scale system may have smaller equipment that would be easier to access or replace.

Perhaps the biggest disadvantage of a large number of small systems, rather than one large system, is that the large number of smaller systems will take more time to monitor and service. However, computer monitoring, automation, and alarming can reduce labor required to manage many small systems. The critical question to ask is whether the monitoring computer is as reliable or more reliable than a human.

Input/Output

A data acquisition and control system has a variety of Input/Output (I/O) ports that provide the connection between the sensor and the computers (Table 9). Generally, I/O ports fall into four categories: digital input (DI), digital output (DO), analog input (AI), analog output (AO). Inputs send signals from the sensor to the computer and are used for monitoring purposes. Outputs send signals from the computer to a device such as a pump for control purposes. Digital I/O, also known as discrete I/O receives DI or sends DO digital signals—for example, on/off, true/false, yes/no. Analog I/O receives AI or sends AO analog signals—for example, a continuous value such as DO concentration or quantity of oxygen supply gas to send to a oxygenator. Determining the type of I/O required by the sensor is the first step for matching the sensor and the I/O. There are many other issues to consider when matching a sensor to an I/O. Manufacturers often outline the steps in the selection process in great detail.

Sensors

Sensors connected to a data acquisition system should provide continuous monitoring and sometimes control of system variables. Sensor features to consider include drift, recommended use, and rated life span. Lower quality sensors require more frequent calibration, may not stand up to the aquaculture environment, and have a short life span. Temperature compensation is another feature that may or may not be required on temperature sensitive measurements such as pH and DO.

A sensor must be periodically calibrated to function correctly. When

TABLE 9. Some examples of I/O ports that connect computers to sensors or devices.

I/O Type	Applicable sensors or control equipment
Analog input	Dissolved oxygen sensor, temperature sensor, pH sensor, flow sensor
Digital output	Level switch, valve status (opened/closed), counters
Analog input	Activate proportional control for precise dosing of oxygen, treatment chemicals
Digital output	Activate relay to turn devices ON/OFF such as a pump, a light, an alarm, feeder

evaluating whether or not to purchase a sensor, it is important to evaluate its life expectancy, as well as its initial cost. For example, quality DO probes are expensive, but can last nearly indefinitely with regular calibration and membrane and electrolyte replacement. In contrast, many of the sealed pH probes available on the market have a design lifetime of only 1–4 years.

Not all sensors will need to be used in all locations. Dissolved oxygen may be critical and dynamic enough to justify monitoring in every tank, if the tanks are relatively large. In contrast, it may only be necessary to measure pH in one or more specific locations within a recirculating system.

Data Acquisition Hardware and Software

Data acquisition hardware and software are critical components of a monitoring system.

Input/Output Hardware. Often sensors interface directly with analog or digital inputs on the I/O modules. This simplifies wiring. In contrast, some sensors pass their signal through monitoring devices, where the signal is displayed locally, before passing the signal to the I/O at the computer. Sensors may also hook to amplifiers (to strengthen the signal) or transmitters (to convert the signal to a format appropriate for the I/O device).

Software. Choosing the proper HMI software is critical, and, like SCADA hardware, is a field containing many manufacturers. Without a computer background, it may be difficult to choose the best software for your application. It is often simplest to purchase a software/hardware package specifically designed for recirculating system aquaculture or have one specifically tailored to your system. One problem with purchasing generic software that has been tailored specifically for aquaculture is that it may be suited for only one area of aquaculture, such as environmental monitoring or feeding. Because aquaculture system designs are not yet standardized, a custom system is often required.

Alternatively, a development package can be used that has higher level programs designed to ease configuration, which greatly simplifies programming a monitoring system from the ground up. A development package has many useful features such as alarming, trending, and mathematical calculations built into the program. There are a variety of features to look for in a good development package: support, feature set, clarity of documentation, ease of use, expandability, communications features (e.g., network compatible, wireless), alarming features, graphics and trending capabilities, configurable security levels, wide hardware compatibility, steep learning curve, consistent interface, training cost, hidden costs (training, consulting fees, add-on features), scripting language for special customization, reliability, and replacement/upgrade costs.

Alarming

Alarming is a critical function of aquaculture system monitoring and can be integrated with a facility-wide plan to encompass fire, security, employee safety, and fish system status. Available software and hardware allow an infinite number of events to trigger alarms (e.g., low flow, pump failure, low DO, intrusion detection). There are a number of ways to be alerted to any function; it is important to understand the process before implementing alarming functions. It is not cost effective to monitor everything one can think of as a failure point, and it may increase the likelihood of false alarms.

Staff notification of alarms can be done many ways. Currently, most notification systems include pagers, monitoring services manned by live operators, and automated dial-out equipment. With the widespread use of communications technology, such as the Internet and cellular phones, it is possible to implement a more complex alarm notification system. For example, it is possible to send an e-mail with a description of the alarm out to a cellular phone upon the activation of an alarm. The employee can then read the e-mail and decide whether or not the alarm requires immediate response. The employee could also monitor the system remotely through the Internet or through direct dial-in.

Automation

Automation can be applied to routine aquaculture operations and to emergency responses to an alarm. Automation on a routine basis could include automatic addition of alkaline chemicals and feed, as well as automated sampling and management of environmental variables. When the process depends on a response within minutes, implementing reliable automated responses to alarm conditions can be a valuable tool. If a comfort level for controlling critical alarm variables does not exist, a manual on/off/auto switch can be used to manually bypass or activate equipment during an alarm condition. One danger of having a physical on/off/auto switch is that the setting on the switch may be forgotten. Fortunately, the system can be designed to alert how the switch is set.

Emergency Operation and Contingency Operation during Failures

Operating a profitable recirculating production system requires operating the system at close to its design loading. A single point of failure of critical equipment such as a pump, blower, oxygen generator, or water treatment unit could result in the loss of an entire crop in only a matter of minutes. Redundancy may solve the problem or buy extra time to respond and correct the problem. For example, if the power goes out and the emergency generator turns on, there is only a fixed amount of time before the generator runs out of fuel. Therefore, the line power must be restored within that time frame.

Ideally, contingency planning is considered during the design process. Questions to ask during the design process include: How reliable is my power and what power outage duration should be expected? How reliable is the water supply and how long can the system operate without makeup water? Will a water storage tower be necessary to meet large short-term demands in water? Can the facility be accessed during severe weather? Can the facility survive a 100-year flood?

In an aquaculture system, it will be easiest to prioritize responses and actions if you have a firm grasp of how the system can fail, which actions are critical to avoid failure, and the order the actions must be implemented to avoid failure. For example, in the event of a pump failure, diffused oxygen can be added to the culture tank(s) to keep the fish alive while the pump is replaced. Ideally, the system will be designed to fail in a systematic way so that operation can be easily regained.

Making rational judgment calls at or near the point of failure is critical to decrease the severity of a catastrophic event. However, contingency planning should be proactive, with several possible response scenarios prepared in advance for catastrophic events. The first step to take when planning for a catastrophe is to assign decision-making responsibilities to an individual or group of individuals before a crisis occurs. Also, simulated emergencies can be used for training purposes.

Ideally, the alarm system will be programmed to differentiate between the severity of alarms, especially when multiple alarms go off. The HMI software may be able to call specific people depending on the nature of the alarm; it can also report whether a person is responding.

Features such as automated dial-out are still important when staff are on-site, which allows the on-site staff time to assess or attempt to fix the problem rather than trying to reach somebody on the phone. As well, digital cellular phones can now receive e-mail messages that could clarify the alarm event to a person who is off-site.

Identifying Resources

Three important resources to identify before an emergency include money, manpower, and equipment. The need for money in an emergency may be critical to pay for services rendered, express delivery of equipment, or extra personnel. Having a "all hands on deck" policy may be required under certain circumstances. It is also critical to identify the availability of personnel under emergency situations. Not only should the manpower be available, but general responsibilities should be assigned, in advance, to minimize confusion and duplication of effort. External resources, such as an electrician or a supplier who can get you parts rapidly, should also be identified beforehand.

Safety

A recirculating aquaculture environment has many hazardous aspects. While the daily design and management should incorporate safe procedures and policies, these may be stretched or ignored during the panic of an emergency. The combination of high voltages and water may present a hazardous situation. Monitoring devices should be designed with low voltage and current in mind. The use of relays, waterproof conduit, and intrinsically safe equipment should be considered both during the design process and when performing contingency planning.

Other aspects of safety require locating, in advance, all emergency shutoffs for electric, gas, and pumps, and preparing a list of emergency phone numbers for gas, electric, phone service, and emergency services (if not 911). Emergency shutoffs should be located where they are readily accessible. Post signs to indicate that a breaker is off for a reason.

Ideally, hazardous operations should be approached in two-person teams, so that one person can call out in an emergency. If that is not possible, a ready alarm or communication by phone can ensure that help will arrive quickly, if needed.

Costs of Operations

According to Losordo (1998a), a recirculating system may actually have lower and more even power costs over an entire year than the power costs incurred during catfish production in ponds, because pond production can require heavy aeration during the summer months. Additionally, although labor costs for pond production may be lower than in a more complicated recirculating system, the feed costs in a recirculating system can actually be less than for pond or net-pen based operations, because feed conversion can be that much better in the tank systems. As well, labor costs in recirculating systems have the potential to drop significantly as larger recirculating systems, using more automated fish handling practices, are refined.

The reason many recirculating systems have not been successful competing with traditional fish production methods may be largely due to the reality that recirculating systems tend to have larger capital costs per mass of fish produced (Timmons 1997; Dunning et al. 1998; Losordo 1998b). Commercial recirculating system producers that have had success are generally those that have been able to achieve high production levels with relatively lower capital equipment costs and, thus, lower debt. Large economies in both capital and operating costs are now being achieved within recirculating systems by scaling up the system size (which is not as simple as it sounds) and by increasing production efficiencies through the use of continuous culture techniques and more automated fish handling practices (Wade et al. 1996; Timmons 1997; Losordo 1998a, 1998b). Note, however, that the design and management of larger recirculating systems that operate effectively under continuous high-density culture strategies are still developing processes.

Biosecurity for Recirculating Systems

Biosecurity, or "hazard reduction through environmental manipulation" (Plumb 1992), involves design features and management practices that reduce the risk of introducing and disseminating pathogens throughout an aquaculture facility. Whether a recirculating culture system or a more traditional technology such as serial reuse culture, three elements are critical to successful biosecurity: (1) that fish introduced to culture are free of pathogens, (2) that the culture system is free of pathogens, and (3) that management practices prevent introduction and dissemination of pathogens. If any of these elements is not addressed, fish health management becomes an exercise in mitigation and, hopeful, elimination of disease problems rather than a proactive program based on prevention of pathogen introduction and dissemination.

Recirculating systems have higher production costs than more traditional culture technologies To be competitive, these systems use continuous production strategies and fish loading densities that can be much greater than those used in typical serial reuse systems. In this environment, a system will be more prone to production health issues. Recirculating system culture requires consistent attention to biosecurity because the prolonged water retention that results from recirculation provides an extended residence time that will allow for the growth of microorganisms, including opportunistic and overtly virulent pathogens. As the pathogen concentration increases and is recirculated, the risk of infection, disease and epidemic loss increases. The point at which a disease outbreak will occur depends on the interaction of many variables, including environmental conditions, fish density and immunocompetence, and pathogen concentration and virulence.

Biosecurity recommendations developed for food animal industries such as poultry, swine, dairy, and beef provide a conceptual framework for developing biosecurity practices for aquaculture (Shane et al. 1995; Carlson 1998). Veterinarians play an integral role as consultants to focus attention on the implementation and conduct of critical biosecurity management practices and by proposing design features throughout the planning, construction, and operation of an aquaculture facility.

Biosecure recycle system husbandry requires that a system be designed so that it can be cleaned completely, easily, and frequently. Any surface can serve as a substrate for microorganisms. All components of a recycle system including biofilters, low head oxygenators, CO_2 strippers, pipes, and tanks should be constructed of nonporous materials and arranged to be easily accessible for cleaning and disinfection. Clean-outs should be installed to access in any part of the system for flushing of accumulated biosolids. Because wood cannot be easily and thoroughly disinfected, it should be considered only for fabrication of disposable, temporary structures. Equipment and supplies should never be transferred from other locations to the facility.

Disinfectants are used routinely as an important part of the biosecurity plan, but unfortunately, every disinfectant used in aquaculture has serious disadvantages (Noga 1996). The three disinfectants most commonly used are chlorine (sodium hypochlorite and calcium hypochlorite), quaternary ammonium compounds (QAC), and potentiated iodine compounds. Chlorine is a better overall disinfectant than quaternary ammonium compounds or potentiated iodine, but has several disadvantages. Chlorine will cause rapid deterioration of netting, corrode metal, and damage some plastics. Because chlorine is highly toxic to fish in trace concentrations, chlorine disinfected equipment must be thoroughly rinsed before use, and chlorine-laden water must be neutralized with sodium thiosulfate before release. QACs and potentiated iodine compounds are commonly used as disinfectants in foot baths and for nets, but have a more limited spectrum of disinfecting action than chlorine. Potentiated iodine is approved for disinfection of fish eggs (Amend and Conte 1982; Federal Joint Subcommittee on Aquaculture 1994).

In the tables that follow (Tables 10–14), biosecurity concerns related to various aspects of management practice are divided into several topics. These tables can serve as a guideline for producers working with or without consultation from a veterinarian on all aspects of biosecurity, including quarantine design and protocols (Citino 1988; Harms 1993), monitoring protocols, and disinfectant use. The number of "yes" answers should be maximized for each section.

Table 10. **Water and Fish.** Pathogens can enter a facility through the water supply, through the introduction of fish to a facility, and can be transmitted from tank to tank via air-borne water droplets.

Yes	No	Water and fish
___	___	Does the facility use a water supply that is free of fish, other aquatic animals, and aquatic invertebrates?
___	___	If fish and/or invertebrates are present, is the water supply disinfected using ozone or ultraviolet disinfection designed for target fish pathogens?
___	___	Is ozonation or ultraviolet radiation used to disinfect water within the recycle loop?
___	___	Does the facility design prevent the transmission of pathogens via air-borne water droplets?
___	___	Does the facility restock with eggs that are certified free of target pathogens or that were produced from broodfish that are certified free of target pathogens?
___	___	Are eggs disinfected on arrival at the facility?
___	___	If eggs are not used, is the facility restocked with fish that are certified free of target pathogens? Certified fry and fingerlings of some species are difficult to obtain if fry are reared in pond culture.
___	___	If eggs are unavailable and fish are purchased, are the fish quarantined when they arrive at the facility?
___	___	Are fish transferred to tanks in the facility without the introduction of water from the delivery truck or shipping bags?

TABLE 11. **Quarantine.** Quarantine helps protect resident production fish from infectious diseases introduced by new fish.

Yes	No	Quarantine
___	___	Was a quarantine system planned early in the design of the facility?
___	___	Is the quarantine system in a separate building, room, or area where it is physically isolated from production?
___	___	Does quarantine use an independent, isolated culture system that uses a water supply and water discharge separate from the production system?
___	___	Is a flow-through water system or a system with a high proportion of make-up water used to dilute and flush pathogens from the system?
___	___	Does the duration of quarantine account for specific pathogen incubation and development times at the water temperatures for the rearing system?
___	___	When fish are added to ongoing quarantine is the clock reset to day zero?
___	___	Is the water temperature maintained at the upper end of the fish species optimum range to speed up pathogen life cycles?
___	___	On arrival, are new fish acclimated for 4 to 7 days to observe for abnormalities in appearance and behaviour, and to confirm food consumption?
___	___	When abnormal fish are observed, are they sampled and examined?
___	___	If nothing abnormal is observed, are normal fish sampled and examined for target pathogens (parasites, bacteria, and viruses)?
___	___	As quarantine proceeds, are subsamples of moribund and healthy fish taken periodically?
___	___	Are fish transferred to a new tank within the quarantine system if dealing with the possibility of stages of organisms that can be left behind when the fish are moved from the tank?
___	___	After the initial acclimation period, are the fish held at the culture densities they will encounter in the production system?
___	___	Do personnel wash hands and arms before going between the quarantine and production areas? Disinfect footware? Change clothing?
___	___	Is work in the quarantine area saved as the last element of the work day?
___	___	Is quarantine equipment used only in the quarantine area?
___	___	Are fish acclimated to production system water by introducing it to quarantine before transfer out of quarantine?

Table 12. **Fish health.**

Yes	No	Fish health
___	___	Has a fish health monitoring plan been incorporated into the management of the facility?
___	___	Is water quality monitored and managed to maintain conditions appropriate for the species?
___	___	Is a premium quality feed fed to maintain optimal nutritional status?
___	___	Are fish handled gently to prevent injury?
___	___	Are eggs, fry and fingerlings purchased from optimum year class broodstock?
___	___	Is vaccination used as a disease prevention management tool?
___	___	Are abnormal, sick and dead fish culled to reduce the spread of pathogens?

TABLE 13. **Husbandry.** Appropriate, well-planned fish management can help prevent the spread of pathogens around the facility.

Yes	No	Husbandry
___	___	Are foot baths installed at the entry to culture areas and are they cleaned and changed frequently?
___	___	Do personnel wash hands and arms before entering the fish culture area?
___	___	Is a disinfectant and rinse area available for treating buckets, nets, dissolved oxygen meters, thermometers, and other equipment?
___	___	Is cleaned equipment stored in an equally clean area?
___	___	Are meticulous husbandry procedures followed for all life stages to remove feces, uneaten feed, algae, aquatic plants, dead fish, and decomposing debris from tank surfaces, inflow and outflow pipes, aerators, spray bars, and any other equipment inside the tanks?
___	___	Are hands washed with antibacterial soap between working on separate cohorts of fish?
___	___	Is work scheduled so that the youngest, most vulnerable stages are tended first?
___	___	For tanks that are on the same recycle loop, is each tank treated as a discrete rearing unit and the potential for cross-contamination minimized?
___	___	Are tanks and equipment disinfected before use with a different group of fish?
___	___	Is the floor regarded as "contaminated" and managed accordingly?
___	___	Are all parts of the system inspected and cleaned, if necessary, at least once per month?
___	___	Are floors cleaned on a regular basis?
___	___	Are pets, rodents, birds, other vertebrates, and insects excluded from the fish culture area?

TABLE 14. **Personnel management.**

Yes	No	Personnel management
___	___	Is frequent hand and arm washing with an antibacterial soap standard practice?
___	___	Are culture activities strategically scheduled to minimize the number of different personnel working with a particular group of fish?
___	___	Are vehicles disinfected before driving into the facility?
___	___	Has a visitor parking area been established at the periphery of the facility grounds?
___	___	Is facility access restricted to a minimum number of people?
___	___	Are the number of tours minimized and limited to small, easily managed groups of people?
___	___	Is a log book maintained?
___	___	Are visitors that have not been at an aquaculture facility within the past 48 hours required to put coveralls over clothing, put on disinfected boots and wash hands for at least 30 seconds with antibacterial soap?
___	___	Are visitors that have been at an aquaculture facility within the past 48 hours required to remove their clothing, put on clean coveralls and disinfected boots and wash hands and arms for 5 minutes with antibacterial soap?
___	___	Are visitors instructed not to touch anything in the culture room?
___	___	After every tour, are footbaths changed and the floor disinfected?

References

Amend, D.F., and F.S. Conte. 1982. Disinfection: Necessary preventive maintenance for healthy fish. Aquaculture Magazine Nov–Dec:25–29.

Bergheim, A., and T. Åsgård. 1996. Waste production from aquaculture. Pages 50–80 *in* D. Baird, M. C. M. Beveridge, L. A. Kelly, and J. F. Muir, editors. Aquaculture and water resource management. Blackwell Scientific Publications, Osney Mead, Oxford, UK.

Bergheim, A., S. Sanni, G. Indrevik, and P. Holland. 1993. Sludge removal from salmonid tank effluent using rotating microsieves. Aquacultural Engineering 12:97–109.

Bisogni, J. J., Jr., and M. B. Timmons. 1991. Control of pH in closed cycle aquaculture systems. Pages 333–348 *in* Engineering aspects of intensive aquaculture: proceedings from the aquaculture symposium. Northeast Regional Agricultural Engineering Service (NRAES) 49, NRAES, Ithaca, New York.

Blancheton, J. P. 2000. Developments in recirculation systems for Mediterranean fish species. Aquacultural Engineering 22:17–31.

Boyd, C. E., and B. J. Watten. 1989. Aeration systems in aquaculture. Reviews in Aquatic Sciences 1:425–473

Bullock, G. L., and five coauthors. 1997. Ozonation of a recirculating rainbow trout culture system: I. Effects on bacterial gill disease and heterotrophic bacteria. Aquaculture 158:43–55.

Burrows, R. E., and B. D. Combs. 1968. Controlled environments for salmon propagation. Progressive Fish-Culturist 30:123–136.

Carlson, K. R. 1998. Biosecurity: Profit for the taking. Best management practices for controlling infectious diseases. Agri-Education, Inc., Stratford, Iowa.

Chen, S., D. E. Coffin, and R. F. Malone. 1993a. Production, characteristics, and modeling of aquaculture sludge from a recirculating aquaculture system using a granular media biofilter. Pages 16–25 *in* J.-K. Wang, editor. Techniques for modern aquaculture. American Society of Agricultural Engineers, Saint Joseph, Michigan.

Chen, S., D. E. Coffin, and R. F. Malone. 1997. Sludge production and management for recirculating aquaculture systems. Journal World Aquaculture Society 28:303–315.

Chen, S., M. B. Timmons, D. J. Aneshansley, and J. J. Bisogni. 1993b. Suspended solids characteristics from recirculating aquaculture systems and design implications. Aquaculture 112:143–155.

Chesness, J. L., W. H. Poole, and T. K. Hill. 1975. Settling basin design for raceway fish production systems. Transactions of the ASAE (American Society of Agricultural Engineers) 18(5):159–162.

Chiba, K. 1981. Present status of flow-through and recirculating systems and their limitations in Japan. Pages 343–355 *in* K. Tiews, editor. Aquaculture in Heated Effluents and Recirculation Systems, volume II. Heenemann Verlagsgesellschaft mbH, Berlin.

Citino, S.B. 1988. Providing veterinary care to the commercial fish farmer. Veterinary Clinics of North America, Small Animal Practice 18:449-469.

Colt, J. E. 1991. Modeling production capacity of aquatic culture systems under freshwater conditions. Aquacultural Engineering 10:1–29.

Colt, J. E., and K. Orwicz. 1991. Aeration in intensive culture. Pages 198–271 *in* D. E. Brune and J. R. Tomasso, editors. Aquaculture and water quality. World Aquaculture Society, Louisiana State University, Baton Rouge.

Colt, J. E., K. Orwicz, and G. Bouck. 1991. Water quality considerations and criteria for high-density fish culture with supplemental oxygen. Pages 372–385 *in* J. Colt and R. J. White, editors. Fisheries bioengineering symposium. American Fisheries Society, Symposium 10, Bethesda, Maryland.

Colt, J. E., and B. J. Watten. 1988. Applications of pure oxygen in fish culture. Aquacultural Engineering 7:397–441.

Cripps, S. J. 1995. Serial particle size fractionation and characterization of an aquaculture effluent. Aquaculture 133:323–339.

Cripps, S. J., and A. Bergheim. 2000. Solids management, and removal for intensive land-based aquaculture production systems. Aquacultural Engineering 22:33–56.

Cripps, S. J., and L. A. Kelly. 1996. Reductions in wastes from aquaculture. Pages 166–201 *in* D. Baird, M. C. M. Beveridge, L. A. Kelly, and J. F. Muir, editors. Aquaculture and water resource management. Blackwell Scientific Publications, Osney Mead, Oxford, UK.

Dunning, R. D., T. M. Losordo, and A. O. Hobbs. 1998. The economics of recirculating tank systems. Southern Regional Aquaculture Center, Publication 456, Stoneville, Mississippi.

Eikebrokk, B., and R. H. Piedrahita. 1997. Nitrification performance and ammonia excretion from salmonids. Journal of Applied Aquaculture 7(4):15–32.

Eikebrokk, B., and Y. Ulgenes. 1993. Characterization of treated effluents from landbased fish farms Pages 361–366 *in* H. Keinertsen, L. A. Dahle, L. Jorgensen, and K. Tvinnereim, editors. Fish farming technology. Balkema, Rotterdam.

Eikebrokk, B., and Y. Ulgenes. 1998. Recirculating technologies in Norwegian aquaculture. Pages 129–137 *in* G. S. Libey and M. B. Timmons, editors. The Second International Conference on Recirculating Aquaculture. Virginia Polytechnic Institute and State University, Blacksburg.

EPA (U.S. Environmental Protection Agency). 1975. Process design manual for nitrogen control. U.S. Environmental Protection Agency, Office of Technology Transfer, Washington, D.C.

Federal Joint Subcommittee on Aquaculture. 1994. Guide to drug, vaccine and pesticide use in aquaculture. Texas A&M University System, College Station, Texas.

Fridman, A. L. 1981. Recirculating systems in eastern Europe. Pages 317–327 *in* K. Tiews, editor. Aquaculture in heated effluents and recirculation systems, volume II. Heenemann Verlagsgesellschaft mbH, Berlin, Germany.

Garcia-Ruiz, R., and G. H Hall. 1996. Phosphorus fractionation and mobility in the food and faeces of hatchery reared rainbow trout (*Oncorhynchus mykiss*). Aquaculture 145:183–193.

Golz, W. J., K. A. Rusch, R. F. Malone. 1999. Modeling the major limitations on nitrification in floating-bead filters. Aquacultural Engineering 20:43–61.

Grace, G. R., and R. H. Piedrahita. 1993. Carbon dioxide control with a packed column aerator. Pages 496–505 *in* J.-K. Wang, editor. Techniques for modern aquaculture. American Society of Agricultural Engineers, Saint Joseph, Michigan.

Hagopian, D. S., and J. G Riley. 1998. A closer look at the bacteriology of nitrification. Aquacultural Engineering 18:223–224.

Hankins, J. A., S. T. Summerfelt, and M. D. Durant. 1995. Impacts of feeding and stock management strategies upon fish production within water recycle systems. Pages 70–86 *in* M. B. Timmons, editor. Aquacultural engineering and waste management. Northeast Regional Agricultural Engineering Service, NRAES 90, Ithaca, New York.

Harms, C.A. 1993. Quarantine procedures for fish aquaria. Pages 70–74 in Annual Proceedings of the American Association of Zoo Veterinarians, Oct. 10-13, St. Louis, Missouri.

Heinen, J. M., J. A. Hankins, and P. R. Adler. 1996. Water quality and waste production in a recirculating trout culture system with feeding of a higher-energy or a lower-energy diet. Aquaculture Research 27:699–710.

Henderson, J. P., and N. R. Bromage. 1988. Optimizing the removal of suspended solids from aquaculture effluents in settlement lakes. Aquacultural Engineering 7:167–188.

Hunter, G. L., W. J. O'Brien, R. A. Hulsey, K. E. Carns, and R. Ehrhard. 1998. Emerging disinfection technologies: medium-pressure ultraviolet lamps and other systems are considered for wastewater applications. Water Environment and Technology 10(6):40–44.

IDEQ (Idaho Division of Environmental Quality). 1998. Idaho waste management guidelines for aquaculture operations. Idaho Department of Health and Welfare, Division of Environmental Quality, Twin Falls.

Ismond, A. 1996. Memories of simplot, visions of the future. Pages 3–16 *in* G. Libey and M. Timmons, editors. Successes and failures in commercial recirculating aquaculture. Northeast Regional Agricultural Engineering Service, NRAES-98, Ithaca, New York.

Juell, J. E. 1991. Hydroacoustic detection of food waste-A method to estimate maximum food intake of fish populations in sea cages. Aquacultural Engineering 10:207–217.

Kaiser, G. E., and F. W. Wheaton. 1983. Nitrification filters for aquatic culture systems: state of the art. Journal of the World Mariculture Society 14:302–324.

Karassik, I. J., W. C. Krutzsch, W. H. Fraser, and J. P. Messina. 1986. Pump handbook, 2nd edition. McGraw-Hill, New York.

Kazmierczak, R. F., Jr., and R. H. Caffey. 1995. Management ability and the economics of recirculating aquaculture production systems. Marine Resource Economics 10:187–209.

Kelly, L. A., A. Bergheim, and J. Stellwagen. 1997. Particle size distribution of wastes from freshwater fish farms. Aquaculture International 5:65–87.

Kim, I.-B., and J.-Y. Jo. 1998. Recirculating aquaculture systems in Korea—development of an environmentally friendly aquaculture system. Pages 139–146 *in* G. S. Libey and M. B. Timmons, editors. The second international conference on recirculating aquaculture. Virginia Polytechnic Institute and State University, Blacksburg.

Klontz, G. W. 1993. Environmental requirements and environmental diseases of salmonids. Pages 333–342 *in* M. K. Stoskopf, editor. Fish medicine. Saunders, Philadelphia.

Langlais, B., D. A. Reckhow, and D. R. Brink. 1991. Ozone in water treatment: application and engineering. American Water Works Association Research Foundation, Denver, Colorado.

Lawson, T. B. 1995. Fundamentals of aquacultural engineering. Chapman and Hall, New York.

Lee, W. 1992. Risky business: is aquaculture still a good investment? Seafood Leader, January/February:84–92.

Liao, P. B., and R. D. Mayo. 1972. Salmonid hatchery water reuse systems. Aquaculture 1:317–335.

Liao, P. B., and R. D. Mayo. 1974. Intensified fish culture combining water reconditioning with pollution abatement. Aquaculture 3:61.

Libey, G. S. 1993. Evaluation of a drum filter for removal of solids from a recirculating aquaculture system. Pages 519–532 *in* J. K. Wang, editor. Techniques for modern aquaculture. American Society of Agricultural Engineers, Saint Joseph, Michigan.

Liltved, H. and S. J. Cripps. 1999. Removal of particle-associated bacteria by prefiltration and ultraviolet irradiation. Aquaculture Research 30: 445–450.
Losordo, T. M. 1991. Engineering considerations in closed recirculating systems. Pages 58–69 *in* P. Giovannini, compiler. Aquaculture systems engineering. American Society of Agricultural Engineers, Saint Joseph, Michigan.
Losordo, T. M. 1998a. Recirculating aquaculture production systems: the status and future. Aquaculture Magazine 24(1):38–45.
Losordo, T. M. 1998b. Recirculating aquaculture production systems: the status and future, part II. Aquaculture Magazine 24(2):45–53.
Losordo, T. M., A. O. Hobbs, and D. P. DeLong. 2000. The design and operational characteristics of the CP&L/EPRI fish barn: a demonstration of recirculating aquaculture technology. Aquacultural Engineering 22:3–16.
Loyless, J. C., and R. F. Malone. 1997. A sodium bicarbonate dosing methodology for pH management in freshwater-recirculating aquaculture systems. Progressive Fish-Culturist 59:198–205.
MacMillan, R. 1992. Economic implications of water quality management for a commercial trout farm. Pages 185–190 *in* J. Blake, J. Donald, and W. Magette. National livestock, poultry, and aquaculture waste management. American Society of Agricultural Engineers, St. Joseph, Michigan.
Mäkinen, T., S. Lindgren, and P. Eskelinen. 1988. Sieving as an effluent treatment method for aquaculture. Aquacultural Engineering 7:367–377.
Mayer, I., and E. McLean. 1995. Bioengineering and biotechnology strategies for reduced waste aquaculture. Water Science and Technology 31:85–102.
Mayo, R. D. 1981. Recirculating systems in North America. Pages 330–342 *in* K. Tiews, editor. Aquaculture in heated effluents and recirculation systems, volume II. Heenemann Verlagsgesellschaft mbH, Berlin.
Mayo, R. D. 1991. Review of water reuse in hatcheries. Pages 180–197 *in* D. E. Brune and J. R. Tomasso, editors. Aquaculture and water quality. World Aquaculture Society, Baton Rouge, Louisiana.
McLaughlin, T. W. 1981. Hatchery effluent treatment. Bio-engineering Symposium for Fish Culture, (FCS Publication 1):167–173, American Fisheries Society, Bethesda, Maryland.
Mudrak, V. A. 1981. Guidelines for economical commercial fish hatchery wastewater treatment systems. Pages 174–182 *in* L. J. Allen and E. C. Kinney, editors. Proceedings of the bio-engineering symposium for fish culture. American Fisheries Society, Fish Culture Section, Bethesda, Maryland.
Muir, J. F. 1981. Management and cost implications in recirculating water systems. Pages 116–127 *in* L. Allen and E. Kinney, editors. Proceedings of the bio-engineering symposium for fish culture. American Fisheries Society, Fish Culture Section, Bethesda, Maryland.
Muir, J. F. 1982. Recirculating systems in aquaculture. Pages 358–446 *in* J. F. Muir and R. J. Roberts, editors. Recent advances in aquaculture, volume 1. Croom Helm and Westview Press, London.
Muir, J. F. 1998. The European experience with production intensification. Pages 210–227 *in* G. S. Libey and M. B. Timmons, editors. The second international conference on recirculating aquaculture. Virginia Polytechnic Institute and State University, Blacksburg.

Nam, T. K., M. B. Timmons, C. D. Montemagno, and S. M. Tsukuda. 2000. Biofilm characteristics as affected by sand size, and location in fluidized bed vessels. Aquacultural Engineering 22:213–224.

Nayyar, M. L. 1992. Piping handbook, 6th edition. McGraw-Hill, New York.

Nijhof, M. 1995. Bacterial stratification and hydraulic loading effects in a plug-flow model for nitrifying trickling filters applied in recirculating fish culture systems. Aquaculture 134:49–64.

Noble, A. C., and S. T. Summerfelt. 1996. Diseases encountered in rainbow trout cultured in recirculating systems. Annual Review of Fish Diseases 6:65–92.

Noga, E. J. 1996. Fish disease. Diagnosis and treatment. Mosby-Year Book, Inc., New York.

Owsley, D. E., J. S. Jeris, and R. Owens. 1988. Ammonia removal allows effluent reuse at fish-hatchery using fluidized bed reactors. Pages 449–457 *in* Proceedings of the 43rd Purdue industrial waste conference. Lewis Publishers, Chelsea, Michigan.

Parker, N. C. 1981. An air-operated fish culture system with water-reuse and subsurface silos. Bio-engineering symposium for fish culture (FCS Publ. 1):131–137. American Fisheries Society, Bethesda, Maryland.

Phillips, M. J., M. C. M. Beveridge, and R. M. Clarke. 1991. Impact of aquaculture on water resources. Pages 568–591 *in* D. E. Brune and J. R. Tomasso, editors. Aquaculture and water quality. Advances in world aquaculture, volume 3. The World Aquaculture Society, Baton Rouge, Louisiana.

Piper, R. G., and five coauthors. 1982. Fish hatchery management. U.S. Fish and Wildlife Service, Washington, D.C.

Plumb, J. A. 1992. Disease control in aquaculture. Pages 3–17 *in* M. Shariff, R. P. Subasinghe, and J. R. Arthur, editors. Diseases in Asian aquaculture I. Fish Health Section, Asian Fisheries Society, Manila, Philippines.

Rodriguez, J., and S. Gagnon. 1991. Disinfection: liquid purification by UV radiation, and its many applications. Ultrapure Water 8(6):26–31.

Rosenthal, H. 1981a. Recirculating systems in western Europe. Pages 305–315 *in* K. Tiews, editor. Aquaculture in heated effluents and recirculation systems, volume II. Heenemann Verlagsgesellschaft mbH, Berlin.

Rosenthal, H. 1981b. Ozonation and sterilization. Pages 219–274 *in* K. Tiews, editor. Aquaculture in heated effluents and recirculation systems, volume I. Heenemann Verlagsgesellschaft mbH, Berlin.

Rosenthal, H., and E. A. Black. 1993. Recirculation systems in aquaculture. Pages 284–294 *in* J.-K. Wang, editor. Techniques for modern aquaculture. American Society of Agricultural Engineers, Saint Joseph, Michigan.

Russo, R. C., and V. R. Thurston. 1991. Toxicity of ammonia, nitrite, and nitrate to fishes. Pages 58–89 *in* D. E. Brune and J. R. Tomasso, editors. Aquaculture and water quality. World Aquaculture Society, Baton Rouge, Louisiana.

Russo, R. C., V. R. Thurston, and K. Emerson. 1981. Acute toxicity of nitrite to rainbow trout (*Salmo gairdneri*): effects of pH, nitrite species, and anion species. Canadian Journal of Fisheries and Aquatic Sciences 38:387–393.

Rusten, B., M. McCoy, R. Proctor, and J. Siljudalen. 1998. The innovative moving bed biofilm reactor/solids contact reaeration process for secondary treatment of municipal wastewater. Water Environment Research 70:1083–1089.

Shane, S. M., D. Halvorson, D. Hill, P. Villegas, and D. Wages, editors. 1995. Biosecurity in the poultry industry. American Association of Avian Pathologists. Kennett Square, Pennsylvania.

Smith, D. 2000. Twenty-one years of commercial recirculating aquaculture with trout, and perch. Pages 42–44 *in* G. S. Libey, and M. B. Timmons, editor. The third international conference on recirculating aquaculture. Virginia Polytechnic Institute and State University, Blacksburg.

Speece, R. E. 1973. Trout metabolism characteristics and the rational design of nitrification facilities for water reuse in hatcheries. Transactions of the American Fisheries Society 102:323–334.

Summerfelt, S. T. 1996. Engineering design of a water reuse system. Pages 277–309 *in* R. C. Summerfelt, editor. Walleye culture manual. NCRAC Culture Series 101. North Central Regional Aquaculture Center Publication Center, Iowa State University, Ames.

Summerfelt, S. T. 1999. Waste handling systems. Pages 309–350 *in* E. H. Bartali and F. Wheaton, editor. CIGR handbook of agricultural engineering, volume II: animal production and aquacultural engineering. American Society of Agricultural Engineers, Saint Joseph, Michigan.

Summerfelt, S. T. 2000. Carbon dioxide. Pages 141–147 *in* R. R. Stickney, editor. Encyclopedia of aquaculture. Wiley, New York.

Summerfelt, S. T., P. R. Adler, D. M. Glenn, and R. N. Kretschmann. 1999. Aquaculture sludge removal and stabilization within created wetlands. Aquacultural Engineering 19(2):81–92.

Summerfelt, S. T., and J. L. Cleasby. 1996. A review of hydraulics in fluidized-bed biological filters. Transactions of American Society of Agricultural Engineers 39:1161–1173.

Summerfelt, S. T., J. Davidson, T. Waldrop, and S. Tsukuda. 2000a. A partial-reuse system for coldwater aquaculture. Pages 167–175 *in* G. S. Libey and M. B. Timmons, editor. The third international conference on recirculating aquaculture. Virginia Polytechnic Institute and State University, Blacksburg.

Summerfelt, S. T., J. Davidson, and M. B. Timmons. 2000b. Hydrodynamics in the 'Cornell-type' dual-drain tank. Pages 160–166 *in* G. S. Libey, and M. B. Timmons, editor. The third international conference on recirculating aquaculture. Virginia Polytechnic Institute and State University, Blacksburg.

Summerfelt, S. T., J. A. Hankins, M. D. Durant, and J. N. Goldman. 1996. Removing obstructions: modified pipe-lateral flow distribution mechanism reduces backflow in fluidized-sand biofilters. Water Environment & Technology 8(11):39–43.

Summerfelt, S. T., J. A. Hankins, A. Weber, and M. D. Durant. 1997. Ozonation of a recirculating rainbow trout culture system: II. Effects on microscreen filtration and water quality. Aquaculture 158:57–67.

Summerfelt, S. T., and J. N. Hochheimer. 1997. Review of ozone processes and applications as an oxidizing agent in aquaculture. The Progressive Fish-Culturist 59:94–105.

Summerfelt, S. T., K. H. Holland, J. A. Hankins, and M. D. Durant. 1995. A hydroacoustic waste-feed controller for tank systems. Water Science and Technology 31(10):123–129.

Summerfelt, S. T., M. B. Timmons, and B. J. Watten. 2000c. Tank and raceway culture. Pages 921–928 *in* R. R. Stickney, editor. Encyclopedia of aquaculture. Wiley, New York.

Summerfelt, S. T., B. J. Vinci, and R. H. Piedrahita. 2000d. Oxygenation and carbon dioxide control in water reuse systems. Aquacultural Engineering 22:87–108.

Summerfelt, S. T., and E. M. Wade. 1998. Fluidized-sand biofilters installed at two farms. Recirc Today 1(1):18–21.

Tchobanoglous, G., and E. D. Schroeder. 1985. Water quality. Addison-Wesley Publishing Co., Reading, Massachusetts.

Timmons, M. B. 1997. Economic engineering analysis of aquaculture production systems and broiler production. Pages 219–234 *in* M. B. Timmons and T. M. Losordo, editors. Advances in aquacultural engineering. Northeast Regional Agricultural Engineering Service, NRAES 105, Ithaca, New York.

Timmons, M. B. 2000a. Aquaculture engineering–experiences with recirculating aquaculture system technology, part 1. Aquaculture Magazine 26(1):52–56.

Timmons, M. B. 2000b. Aquaculture engineering–experiences with recirculating aquaculture system technology, part 2. Aquaculture Magazine 26(2): 56–60.

Timmons, M. B., N. Helwig, and S. T. Summerfelt. 2000. The Cyclone sand biofilter: a new design concept, and field evaluation. The third international conference on recirculating aquaculture. Virginia Polytechnic Institute and State University, Blacksburg.

Timmons, M. B., and T. M. Losordo. 1994. Aquaculture water reuse systems: engineering design and management. Elsevier, New York.

Timmons, M. B., and S. T. Summerfelt. 1998. Application of fluidized-sand biofilters. Pages 342–354 *in* G. S. Libey and M. B. Timmons, editors. The second international conference on recirculating aquaculture. Virginia Polytechnic Institute and State University, Blacksburg.

Timmons, M. B., S. T. Summerfelt, and B. J. Vinci. 1998. Review of circular tank technology and management. Aquacultural Engineering 18(1):51–69.

Tsukuda, S. M., and five coauthors. 1997. Effects of sand size on fluidized-bed biofilter performance in cold water systems. Pages 368–380 *in* M. B. Timmons and T. Losordo, editors. Advances in aquacultural engineering. Northeast Regional Agricultural Engineering Service, NRAES 105, Ithaca, New York.

Twarowska, J. G., P. W. Westerman, and T. M. Losordo. 1997. Water treatment and waste characterization evaluation of an intensive recirculating fish production system. Aquacultural Engineering 16:133–147.

U. S. Trout Farmers Association. 2000. www.ustfa.com.

Vinci, B. J., M. B. Timmons, S. T. Summerfelt, and B. J. Watten. 1998. Carbon dioxide control in intensive aquaculture: version 2.1: user's guide and source code. The Conservation Fund's Freshwater Institute, Route 1 Box 265, Turner Road, Shepherdstown, West Virginia.

Wade, E. M., S. T. Summerfelt, and J. A. Hankins. 1996. Economies of scale in recycle systems. Pages 575–588 *in* G. Libey and M. Timmons, editors. Successes and failures in commercial recirculating aquaculture. Northeast Regional Agricultural Engineering Service, NRAES-98, Ithaca, New York.

Warren-Hansen, I. 1982. Methods of treatment of waste water from trout farming. Pages 113–121 *in* J. Alabaster, editor. Report of the EIFAC workshop on fish-farm effluents. European Inland Fisheries Advisory Commission Technical Paper 41, Food and Agriculture Organization of the United Nations, Rome.

Watten, B. J., and C. E. Boyd. 1990. Gas transfer within a multi-stage packed column oxygen absorber: model development and application. Aquacultural Engineering 9:33–59

Watten, B. J., J. E. Colt, and C. E. Boyd. 1991. Modeling the effect of dissolved nitrogen and carbon dioxide on the performance of pure oxygen absorption systems. Pages 474–481 *in* J. Colt and R. J. White, editors. Fisheries bioengineering symposium. American Fisheries Society, Symposium 10, Bethesda, Maryland.

Wedemeyer, G. A. 1996. Physiology of fish in intensive culture. International Thompson Publishing, New York.

Wedemeyer, G. A., N. C. Nelson, and W. T. Yasutake. 1979. Physiological and biochemical aspects of ozone toxicity to rainbow trout (*Salmo gairdneri*). Journal of the Fisheries Research Board of Canada 36:605–614.

Westers, H. 1981. Fish culture manual for the state of Michigan. Michigan Department of Natural Resources, Lansing.

Westers, H. 1991. Operational waste management in aquacultural effluents. Pages 231–238 *in* C. B. Cowey and C. Y. Cho, editors. Nutritional strategies and aquaculture waste. proceedings of the first international symposium on nutritional strategies in management of aquaculture waste. University of Guelph, Ontario.

Wheaton, F., J. Hochheimer, and G. E. Kaiser. 1991. Fixed-film nitrification filters for aquaculture. Pages 272–303 *in* D. E. Brune and J. R. Tomasso, editors. Aquaculture and water quality. World Aquaculture Society, Baton Rouge, Louisiana.

Wheaton, F. W. 1977. Aquacultural engineering. Robert E. Krieger Publishing Co., Malabar, Florida.

Wilton, S., and R. Boschman. 1998. The Pacific Northwest experience with production intensification through recirculation. Pages 201–209 *in* G. S. Libey and M. B. Timmons, editors. The second international conference on recirculating aquaculture. Virginia Polytechnic Institute and State University, Blacksburg.

Wong, K. B., and R. H. Piedrahita. 2000. Settling velocity characterization of aquacultural solids. Aquacultural Engineering 21:233–246.

7

Broodstock Management for Imperiled and Other Fishes

J. Holt Williamson

Introduction

This chapter presents a practical conceptual and operational framework within which broodstock managers can acquire the knowledge and skills necessary to apply the language, theory, concepts, principles, and practices of their profession. These assets are the mental tools and techniques they rely on to make the decisions that result in successful broodstock management programs. This chapter is intended to educate the experienced and develop new and aspiring broodstock managers to accomplish the goal of building effective broodstock management programs that can be improved and sustained through time in a complex or total management environment (TME).

Successful broodstock managers remain students all their lives. Preparation for their work is never finished. They must keep abreast of new information and emerging technologies to improve their facilities and their operational efficiency through application of knowledge and skills to create successful broodstock management programs and to develop useful broodstocks. Above all, they view their role in natural resources management as vital and their responsibilities as an intellectual challenge. They are aware of their role, and their responsibility is complex and dynamic. As complete broodstock managers, they understand that their successful performance depends on far more than simple application of a standard set of technical practices, procedures, and protocols (PPP; Meffe 1992). This chapter is presented from the point of view of a captive propagation and broodstock manager who designs and employs strategies and tactics from a genetics conservation perspective to assist in the restoration or recovery of imperiled fish species. However, the principles discussed apply to other fish populations as well.

The Complete Broodstock Manager

This chapter on broodstock management has a central theme—management. The broodstock manager strives to be an ethical and efficient manager. Her success in that regard depends on her knowledge, skills, abilities, education, experience, and desire to manage people, fish, the environment, materials, and ideas to get a job done. A broodstock manager who excels in exhibition of these characteristics in day-to-day operations while consistently delivering a superior product is the complete broodstock manager.

Hatchery Controversy

A major change since the original *Fish Hatchery Management* (Piper et al.) was published in 1982 is the public's perception of fisheries management, especially hatcheries and their programs (Hilborn 1992, 1999; Wiley 1999; Williamson 2000). Public concern has culminated in the recent and ongoing "hatchery controversy." In essence, the hatchery controversy reflects differing opinions, public and professional, with respect to the value of hatcheries in solving complex fisheries management problems. Actually, the hatchery controversy has been going on for years within the ranks of fisheries professionals. Only recently has the controversy bled over into the broader public domain where the debate has become increasingly acrimonious. The interest and emotion are due, in part, to the public's increasing awareness, education, understanding, concern, and activism regarding environmental issues. The stakeholding public wants to be informed of, and even participate in, managing fisheries resources.

The value of hatchery programs, principally controlled fish propagation and reintroduction programs, is receiving broad critical review. From the perspective of the profession's practitioners, it is appropriate to view public critique as healthy and a challenge. The only choice for managers is to meet the challenge with improved performance. Failure to satisfactorily meet the challenge will result in further erosion of public trust. Over time, the result of lost trust is forfeiture of resource management responsibility.

Stock Concept

In 1980, the Great Lakes Fisheries Commission sponsored the Stock Concept International Symposium (STOCS). Although the purpose of the symposium was to draw attention to fisheries management problems on the Great Lakes, the effects were revolutionary and far-reaching, affecting all aspects of fisheries conservation and management. The symposium introduced the concept of the fish stock as representing the natural or biological unit of management. The biological basis of the stock concept is essentially genetic and virtually synonymous with the population as defined biologically (genetically, phenotypically, demographically) and ecologically—a local group of conspecific organisms sharing a common gene pool. A management unit, population, or stock was no longer arbitrarily or subjectively defined by the manager. The fish stock or population defined itself. The manager's perspective changed, and the scope broadened. This event coincided with and promoted genetic conservation and management principles as major considerations in defining management problems and in solving them. The definition of management success changed, and with it the criteria by which it was measured changed radically. The STOCS five objectives were to (1) develop a refined, up-to-date concept based on sound ecological and genetic principles, (2) examine the prospects and strategies for the preservation of gene pools, (3) examine the potential use of the stock concept in improving fish culture, (4) seek applications of the stock concept in rehabilitation of freshwater fish resources, and (5) develop use of the stock concept in regulating fisheries. The implications for broodstock managers were significant. A new way of doing business was called for; a new way to understand and define fisheries management problems, needs, strategies, and activities emerged. However, recognition and acceptance of the concept have not translated so easily into understanding and practice at the operational level, including the domain of the broodstock manager.

Appropriate Use and Effects Concept

> The efficient operation of a fish rearing facility requires a sufficient quantity of parent broodfish of good quality. The quantity of broodfish needed is determined by the number of eggs needed to produce the fry required, with normal losses taken into account. Quality is a relative term best defined by considering the use of the product. Persons producing fish for a restaurant or supermarket use different measurements of quality than a hatchery manager rearing fish for use in research or stocking. Most

work defining fish quality has focused on performance in the hatchery, broodfish reproduction, and progeny growth and survival under hatchery conditions. In the future more emphasis will be placed on the ability of hatchery fish to survive after release and their contribution to a particular fishery program.

The above quote is from the chapter "Broodstock, Spawning, and Egg Handling" in *Fish Hatchery Management* (Piper et al. 1982). As an observation or prediction, it was visionary in its time, setting the stage for today's broodstock manager. For the last 20 years, every fisheries management practitioner, from field biologist to hatchery technician, has been more or less aware of this management concept. The appropriate-use concept presents a legitimate rationale and strong force defining fisheries management opportunities that rely on captive or controlled propagation strategies as employed by hatcheries, from broodstock development to fish production for stocking. But, alone, the appropriate use concept is insufficient to address complex fisheries management and ecosystem management problems.

In March 1994, the International Symposium and Workshop on the Uses and Effects of Cultured Fishes in Aquatic Ecosystems was held in Albuquerque, New Mexico. Radonski and Loftus (1995) introduced the symposium:

> The theme of this symposium is fish husbandry: the culture of fishes. Usually, such a symposium would focus on husbandry techniques and the near miraculous advances in technology would be chronicled. But instead we are gathered here to discuss the uses and effects of cultured fishes in aquatic ecosystems, that is, the ethical use (heading toward ethical production) of the products of fish husbandry. Fishery managers and biologists have debated this subject for years among themselves for decades. Now the discussion has moved to the broader societal forum, the environmental community. We are all well aware that the controversy over the role of cultured fishes in natural environments has increased in intensity over the years. All aspects of fish husbandry, in the broader term, aquaculture, are in question, including facility siting, effluents, genetics, animal rights, and a host of social and economic impacts. It is a quantum leap from the early days of fish culture in the United States, described by Bowen (1976) as the product of fisheries destroyed by America's early eighteenth century industries. . . .

Although many criticisms have been leveled at hatchery programs, hatchery operations, and their products, the concerns involve appropriate use and potentially negative, often unintended, consequences. The use concept has been around, in some form or fashion, for some time. But, the effects concept is relatively new, effects meaning results or management success. Did the fish do what they were supposed to do? A more recent development in the effects concept is influenced by increasing public and professional awareness and concern about unintended consequences or "collateral damage" associated with stocking fish (Pearsons and Hopley 1999; Ham and Pearsons 2001). Appropriate-use and effects are no longer defined by strictly biological or resource management considerations, but by a larger and more involved public representing, along with ecological concerns, political, sociocultural, and economic considerations.

Total Management Environment

The management terrain is composed of four major informational domains: ecological, economic, political, and sociocultural. The unique environment in which these domains intersect and overlap form the total management environment or TME.

The ecological domain is the environment within which the broodstock manager is traditionally grounded in terms of professional interest, education, and experience. At the same time, the ecological domain is the managed habitat described by biological, physical, and chemical components and their complex multitude of interactions. The ecological domain is the spatial and temporal location wherein the managed organisms dwell, completing their life cycles after contributing to new generations. Whether watershed, lake, pond, or aquarium, ecosystem understanding of the biotic and abiotic, the microscopic and macroscopic components, their structure, organization, and attendant processes, are critical to the broodstock manager's success. Ecosystem knowledge helps the broodstock manager explain, tweak, manipulate, or otherwise influence ecosystem function and productivity. The ecological domain includes genetic and environmental components interacting with each other in a seemingly infinite array of phenotypic expressions. Genetically determined and environmentally mediated morphs and behaviors within and among groups influence their adaptive responses and evolutionary trajectories. Clearly, ecosystems represent extreme complexity (Wilson 1998).

The economic domain exerts market and nonmarket forces that affect the dollar value of an aquatic ecosystem or fishery resource. Traditionally, economic influences have infiltrated the TME through user groups, particularly recreational, commercial, and tribal fishers. The economic domain includes agency and program budgets that must be understood in terms of source, timing, and allocation, in order to set priorities, plan management activities, evaluate cost:benefit ratios, and make proper decisions. The economic domain in today's TME allows the cost of recovering an endangered fish in the Colorado River to be understood in terms of the cost of a cool glass of water to a citizen of Los Angeles. This, of course, leads broodstock manager into another domain, the political domain.

The political domain, which includes significant institutional, legal, and bureaucratic aspects, is that segment of the TME exerting influence on the management program and process through established laws and policies that formally express social contracts and agency intent, expectations, or values through the governing process. The political domain is where conflict arising unresolved from the other domains is resolved. As resource management issues mount in number and complexity, and the will to reach consensus and agreement is weak among the principles, solutions will be extracted from the political domain. In more controversial and complex natural resources management situations, particularly those involving a clash of development or private property interests involving threatened and endangered fishes and critical habitat, the Endangered Species Act of 1973 (as amended; ESA) may be invoked. In those cases, the ESA captures and directs the management process toward recovery. If a simple political "win-win" solution is not achieved through compromise and negotiation, then a solution will be handed down by the courts, a situation that has come to be known as the "lose-lose situation." Indeed, court-mandated solutions are not real solutions. They are very expensive and time consuming, and the outcomes are unpredictable and unsatisfactory for all. In today's TME, the broodstock manager is aware of economic, political, and legal realities that affect her program, and she strives to develop an appropriately empathetic and cooperative attitude to support her management decisions, strategies, priorities, and actions.

Another aspect of the political domain is the bureaucratic subdomain. The bureaucratic subdomain involves the agency's or comprehensive management program's organizational structure and the administrative, budgetary, and operational machinery driving the program and process, influencing the efficiency of decisions within the TME all the way down through the various levels to the broodstock management program. The bureaucratic subdomain in comprehensive recovery programs for high priority species can be very complex, hierarchically structured, and involve the various state and federal agencies, divisions, committees, subcommittees, standing and ad hoc workgroups, teams, coordinators, stakeholders, interested parties, consultants, experts, consumers, and lawyers that emerge from the cooperative and collaborative process within the

TME. Bureaucracies must be managed like any other complex system in order to extract a greater probability of management success from its complexity.

The sociocultural domain can be viewed as the public domain, both exterior (institutional structure and organization) and interior aspects (value systems). In modern western societies, the public domain is the dominant domain affecting, even creating, the TME. The sociocultural domain is far more compelling, ultimately, in terms of management decisions and program success than in the past, perhaps more compelling than the ecological domain. From the public domain derives the public's interest, concern, and influence alluded to throughout this chapter. The public creates, directly or indirectly, intentionally or unintentionally, the need, authority, and approval of the broodstock manager. Public opinion, often expressed through congressional action, is the ultimate arbiter and judge of management. Managers are functionaries of the public and come and go at the public's discretion. Thus, it is fair to say, as do Krueger and Decker (1993), that the sociocultural domain is the primary influence, motivator, or stimulus eliciting the manager's attention and response.

The CHIMP^{10} System

To expedite and improve the efficiency and quality of broodstock management within the TME, the broodstock manager must operate within a comprehensive management system that holistically integrates ten principle components: (1) problem, (2) purpose, (3) perspective, (4) principles, (5) program, (6) plan, (7) process, (8) performance, (9) product, and (10) progress. A management system so defined is called a comprehensive, holistic, integrated, management program and process system or CHIMP^{10} system.

There are several reasons why the broodstock management program should be considered a subcomponent of the CHIMP^{10} system. First, the broodstock is the gene pool and, as such, is both the conceptual and biological center (genetic and phenotypic components) of the integrated captive propagation program, which includes artificial genetic refuge, production broodstocks, and fish production (for genetic conservation, fisheries management, and research purposes) and evaluation components. With respect to genetic conservation, the broodstock and their offspring are both the repository for DNA and the vehicle for DNA transfer. The essential hereditary material, DNA, carries the code for fish survival, adaptation, sustained production, and evolution. Therefore, the broodstock management program is the backbone of fisheries management programs that rely on captive propagation and reintroduction. The broodstock provides the identity and directs the focus of the captive propagation program; without broodstock, there is no captive propagation and stocking program. Errors made in broodstock management are compounded and magnified as consequences of poor broodstock management decisions radiate throughout the overall recovery or management effort in all directions. Consequences of poor broodstock management decisions may be irreversible and actually expedite and ensure program failure, perhaps catastrophic failure resulting in extinction.

Second, although the broodstock management program is the essential program element within the captive propagation and stocking program, broodstock management remains just a piece of the captive propagation puzzle, just as captive propagation and stocking are pieces of the overall strategic management or recovery puzzle defined by the TME and the CHIMP^{10} system. In complex fisheries management programs, captive propagation is one of two major production approaches with several alternative or supporting strategies that may have application in addressing a number of unique fisheries management or threatened and endangered fishes recovery opportunities, depending on the special set of circumstances.

Third, adopting the CHIMP^{10} system empowers the broodstock manager to be more effective in navigating, orienting, and locating herself within the TME.

Genetic Hazard and Risk Management System

A genetic hazard and risk management system (GHRMS) is a management instrument developed to aid broodstock managers in reconciling apparently disparate strategic goals and in facilitating preferred management decisions and actions toward accomplishment of those goals, while minimizing the risk of undesirable genetic consequences. It is a device to promote identification, analysis, and description to better communicate genetic hazards and assess their respective risks within a captive propagation context. A GHRMS provides a conceptual framework to support systematic, objective decision making on a case-by-case or population-by-population basis to promote program success while reducing conflict among managers, experts, and stakeholders resulting from confusion arising from insufficient knowledge, inadequate vocabulary, and perceptual differences about hazards and risk.

There is a general perception within the discipline that fish propagated in hatcheries and broodstock facilities negatively affect the genetic character of both captive and wild fish populations. It follows, therefore, that these genetic effects may have adverse consequences for the performance and long-term sustainability of fish populations. Indeed, much of the controversy, with respect to appropriate uses and effects of hatcheries and their products, concerns the potential loss of genetic diversity and fitness in naturally spawning populations that are exposed to hatchery fish for the purpose of restoring sustainable fish populations or boosting fish production for exploitation. This is an important management concern because, as Currens and Busack (1995) point out,

> Genetic variation is the raw material for natural selection (Fisher 1930). Consequently, variation within and among populations of fish determines their capacity to persist in changing environments and to meet human wants and needs. Long-term production and usefulness of populations or fisheries depend on conserving genetic variation among and within populations.

This central assumption, however, is clouded by uncertainty, leaving it open to various and conflicting interpretations based, in large measure, on opinion and philosophical perspective. Nevertheless, genetic conservation has become a guiding principle, generally accepted as true by most fisheries scientists and managers. Campton (1995) provides a thoughtful overview of many of the assumptions that have emerged in the discipline with respect to genetic hazards, sources, and effects on fish performance. Campton concludes that there is a fundamental lack of well-designed empirical studies testing these assumptions, leaving a great void in "hard" data necessary to support this guiding principle.

As with most management interventions, the potential danger captive propagation and reintroduction programs pose to fisheries resources can be identified and characterized as hazards with associated risks. In natural resources management, risk of management failure cannot be eliminated from the management equation because it is directly, if inversely, related to the probability of management success. As in all management endeavors, there is no benefit available without risk of loss. The key is striking a thoughtful balance, one in which the benefit and cost are commensurate. To take unnecessary risk for little gain is foolish. When the situation is extreme, as is the case with imperiled fishes in altered ecosystems, the need is extreme to master a hazard and risk management system. In that context, many management decisions, including the decision not to intervene, come with considerable risk. Captive propagation and reintroduction programs, in that respect, are no different than any other management intervention.

Although the management problems associated with degraded ecosystems and imperiled fishes are extremely complex, managers have narrowed the focus of the debate on captive propagation and reintroduction programs. Attention on the broodstock manager and her program is particularly and properly intense, and there is little doubt that the role of the broodstock manager is critical in fish supplementation programs. In management programs depending on a captive propagation strategy, broodstock development and management is where reproduction lost in nature is regained, if temporarily. Broodstock development and management is where genes are saved or lost, where the evolutionary potential of populations and species is protected or squandered. Aware of the visibility and vulnerability of her role in the TME, the complete broodstock manager must be even more aware of her great responsibility in putting critical program and genetics management principles, concepts, theories, and terms into practice within a GHRMS to achieve successful results. GHRMS provides a rationale that integrates basic components with a working vocabulary to promote understanding, communication, and application of a genetic conservation strategy in the broodstock management program, plan, and process.

The particular GHRMS presented here is modified slightly from one developed by Currens and Busack (1995), based on a genetic vulnerability model. Their model provides a conceptual framework and systematic process for assessing genetic vulnerability in order to manage risk to reduce the potential for genetic losses in a captive propagation program designed to supplement Pacific salmon populations in the Columbia River. Within the management environment, their genetic vulnerability assessment model links five naturally occurring processes as sources (management decisions and actions) of four types of genetic hazards, each with its corresponding consequences for genetic identity, structure, and function. More importantly, the model introduces management safeguards (management decisions and actions) to reduce the risk of hazards at the management source while increasing the potential for recovering from their occurrence and subsequent genetic losses.

Basic Genetic Terms

Although a genetics discussion can occur at various levels of complexity (molecular, cellular, individual, population, species, and community), our discussion in this chapter focuses principally on the population level. A detailed discussion of the full spectrum of genetic principles, concepts, theories, and terms is beyond the scope of this chapter. Nevertheless, the terms defined here will facilitate meaningful interpretations by the broodstock manager. The definitions presented are generally accurate, but, as with all generalizations of complex issues and ideas, they are necessarily limited.

A *species* is a group of naturally interbreeding populations that is reproductively isolated from other such groups (Mayr 1977).

Genetic diversity or *variation* is the genetic structural complexity that characterizes the species, reflecting all of the genetic differences that exist within and among individuals, within and among populations, and within a species. Genetic differences are expressed as differences in the quantity and quality of alleles, genes, chromosomes, and gene arrangements on the chromosomes that are present within and among constituent populations. A species' genetic diversity is organized hierarchically with actual genetic structure determined by natural processes (i.e., mating behavior or geographic barriers; Meffe and Vrijenhoek 1988; Smith and Rhodes 1992).

A *population* is a group of interbreeding individuals. Ecologists refer to a population as a local population or a deme. In genetics conservation and population genetics, management biologists refer to a population as the basic or simplest biological unit of management, specifically as a genetic conservation unit (GCU). Populations differ in one or all of the following dimensions: genetic, phenotypic, demographic, ecologic, and geographic. In a captive breeding and broodstock management program, the population of greatest interest is the broodstock, frequently captive. In a gene pool maintenance program, the donor population and its derivative, the founder or broodstock population, should correspond in identity and character; that is, the two are expected to be one, or biological equivalents. The captive broodstock should reflect accurately the complex genotypic and phenotypic diversity of the natural parent population.

Deoxyribonucleic acid (DNA) is the fundamental genetic material providing the operational instructions for replicating and perpetuating life. In most cases, DNA is a double-stranded, helical molecule found in the nucleus of every cell of the body. Mitochondrial DNA (mtDNA) is an exception, being a single-stranded molecule found outside the nucleus in a cytoplasmic organelle called the mitochondria. DNA is heritable information that defines, within limits, an individual's performance potential and a population's adaptive and evolutionary potential. Ultimately, DNA variation is what the broodstock manager manages, indirectly, by managing fish, people, technology, information, and other available resources. Discrete segments of DNA are called genes.

A *gene* is the most basic, indivisible unit of inheritance; it is the smallest functionally and physically discrete piece or bit of heritable genetic information. A gene is an operationally complete segment of DNA. A simplistic interpretation of a gene is "one gene, one function," meaning a particular gene has a unique function. Genes are linearly arrayed along the length of the DNA molecule like beads on a string. A particular gene is found at a particular site or location, a locus, on the DNA molecule. Genes and their constituent DNA are found on chromosomes (except for mtDNA). A particular gene is found on a particular chromosome.

Genetic information contained in the DNA controls the function and structure of an organism. Genetic function is what a gene does. Genes are usually classified as structural or regulatory. Structural genes code for the synthesis of proteins, while regulatory genes govern the functioning of structural genes. Collectively, structural and regulatory genes, interacting with the environment, determine individual, population, and species identity.

A *chromosome* is a discrete package of DNA. It is an organelle, a physical structure found in the nucleus of the eukaryotic cell. A single chromosome can be understood as a gene complex or linkage group. Fish, like most complex organisms, generally have more than one chromosome, a "set" of chromosomes, within which each chromosome is different. In somatic cells, an individual's set of chromosomes is called a diploid set and is assigned the symbol 2N. In contrast, the set of chromosomes carried by the organism's gametes or sex cells is called the haploid set, which is half the number of chromosomes (N) found in a somatic cell (2N). In diploid organisms like fish, chromosomes are associated as homologous pairs or mates. One represents the paternal donor and the other the maternal donor; for example, the X and Y sex chromosomes in humans form a homologous pair.

A *trait* is the phenotypic expression of a gene; the gene itself is the genotype. When a trait is governed by a single gene or locus (plural of locus is loci), it is called a simple or qualitative trait. A trait controlled by several genes is often called a complex or quantitative trait.

An *allele* is the alternate form of a particular gene. In sexually reproducing organisms, such as fish, one gene copy is on the maternal DNA strand donated by the mother and the other gene copy is at the same physical location (locus) on the alternate, complementary, paternal DNA strand donated by the father. Although an individual can have no more than two alleles for a particular gene, a population may have many alleles at a particular locus in its gene pool.

An individual is *homozygous* ("same state") at a particular locus or gene when both copies of the gene are identical in form or base sequence. A locus with identical copies of the same gene is a monomorphic locus; there is no genetic variation at that site. If there is no variation at that site among all the individuals within the population, the gene is said to be "fixed," monomorphic or invariant.

An individual is *heterozygous* ("different state") for a particular gene when the alternate copies or alleles are different in form or base sequence at that locus. A locus with different alleles is a variable or polymorphic locus. Polymorphic loci are interesting, informative, and diagnostic to geneticists, especially if geneticists can detect and measure them. Polymorphic loci are also of interest to the broodstock manager, as they help us to understand differences seen among individual organisms within and among populations. Polymorphic genes are the basis of genetic diversity and contribute to phenotypic diversity.

The *genotype* is genetic information, generally of a locus or an individual. It also applies to a species or a population as opposed to the physical or phenotypic expression of the same. In the latter two instances, the term genome, rather than genotype, is most often used.

The *phenotype* is what the critter looks like, the set of observable properties of a locus, individual, population, or species expressed or reflected by the corresponding genotype in a particular environment. These phenotypic properties are behavioral, physiological, chemical, meristic, and morphometric features usually expressed at the individual or population level of organization. Individuals may express the same or very similar phenotypes and yet have different genotypes. In those cases, the genetic diversity is said to be "cryptic." Or the same genotype may result in differential phenotypic expression among individuals or within an individual in a different environment, called "genetic by environmental" interaction. That is so because the interaction of genes and their environment is ambiguous, meaning no particular phenotype is expressed all the time under all conditions. Instead, a range of phenotypic expressions is possible.

An individual's genotype, or genetic identity, does not change over its lifetime (except through mutation), but many of its phenotypic expressions will change through time. The structure of a specific protein, its primary phenotype, is constant within an individual and unaffected by environment over time. On the other hand, both a population's genotype and phenotype can change with time.

Genetic structure, in the broad sense, is the organization of genetic information or genetic diversity, usually hierarchically expressed within and among populations, as measured and analyzed at the locus, chromosomal, individual, population, and species levels.

A *genetic marker* is a gene with a known location or clear (unambiguous) phenotypic expression that is detected by analytical methods and is used as a reference point when mapping genes or the movement of genes within and among populations, or in identifying and characterizing individuals or populations.

Evolution is the cumulative change in characteristics of populations related by descent that occurs over generations due to natural selection of favorable genotypes in a particular environment. The result of evolution from natural selection is fitness or the relative ability of the population to survive and transmit its genes from one generation to the next. A fit population, one expressing superior performance in this manner, is adapted to its environment.

Identifying, Characterizing, and Measuring Genetic Diversity

Measuring genetic diversity in fish populations is key to our interpretation, understanding, appreciation, and effective management of them. Historically, genetic diversity has been measured indirectly and inferentially through controlled breeding and performance studies or by classical taxonomic or systematic analysis of phenotypic traits (i.e., morphometric and meristic characters), often with ambiguous results. The relationship between genes and their phenotypic expression is complicated and often confounded by environmental influences.

Ambiguity in the relationship between an organism's genotype and its phenotype underscores a dilemma that presents a fundamental challenge for fisheries managers. Direct, unambiguous measurement of genes, an organism's genotype, is a recent technological phenomenon. That capability resulted from the discovery and accurate description of DNA followed by methods developed in the later twentieth century to identify, characterize, measure, and analyze genes as simple gene products. Mathematical modeling and statistical analysis of genetic information have become extremely powerful tools aiding our understanding of evolution, systematics, taxonomy, phylogeny, population dynamics, gene expression, and developmental biology. Biochemical or electrophoretic analysis of proteins or allozymes was the dominant method employed from the 1960s through the 1980s and revolutionized our ability to identify, characterize, and measure genetic diversity. Although still very useful in many situations, allozyme analysis has yielded, in the 1990s, to more advanced methods permitting direct measurement and analysis of differences in DNA fragment lengths or differences in the linear sequence of bases that form the DNA molecule. The genetic data revealed by these new technologies is used, in part, to identify and characterize populations; to understand complex population genetic identity, structure, organization, and processes; and to provide genetic markers to measure differences and changes in amounts and types of genetic variation among individuals within populations and differences and changes among populations (Hillis and Moritz 1990). Methods to determine individual kinship and parental relationships are now possible. These methods provide the measuring, monitoring, and evaluation tools to manage genetic diversity.

A Fundamental Management Challenge

A fundamental challenge presented to the broodstock manager is what to make of this wealth of genetic information provided by the new and developing technologies. How can the broodstock manager avoid complexity overload and manage this information, in order to perform in a superior manner? The challenge emerges because the relationship or nexus between genotypes, as expressed by the molecular genetic markers revealed by the new technologies and the phenotype as expressed by an individual's or a population's performance, is uncertain. A genetic marker, at best, represents one or two genes, the corresponding function or phenotypic expression of which, if any, is usually unknown. Even genotypic and phenotypic patterns of inheritance and their response to selection is frequently unknown or uncertain. On the other hand, many phenotypic characters are performance characters that influence fitness or adaptive potential. These complex performance characters are generally determined by many loci interacting in unknown or unclear ways. Differences in these performance traits are influenced not only by the individual's genotype but also by the interaction of other genes (g) and gene combinations with the environment (g*e):

$$V_p = V_g + V_e + V_{g^*e} + V_{\mathrm{error}}, \tag{1}$$

where V is the variance associated with the phenotype (p), genotype (g), environment (e), interaction (g*e), and measurement error. For details, see Falconer (1981), Becker (1984), and Tave (1986).

The Hardy–Weinberg Population

A simple definition of evolution is a change in a population's allele frequencies over time (Hartle 1981). Evolutionary biologists and population geneticists use this definition when studying the simplest population genetics model to test their hypotheses and to develop and refine their operational principles, concepts, and theorems. That model is the Hardy–Weinberg population, also referred to as the ideal population (Allendorf and Ryman 1987). As we will see, in managing a gene pool maintenance program, the broodstock manager's goal is to avoid genetic changes in the captive population. The overall strategy applied to accomplish this goal is to strive for the ideal population, which has several unique characteristics of particular interest to the complete broodstock manager, especially one managing a gene pool maintenance program from a genetics conservation perspective. Theoretically, the ideal population is evolutionarily static; that is, it experiences no net genetic change from one generation to the next. Gene and genotype frequencies are stable. For that to be possible, the following characteristics must apply:

1. The ideal population is infinitely large;
2. Mutation does not occur within the population;
3. Selection does not occur;
4. There is no gene flow from immigration;
5. Matings are random, that is, each individual has the same opportunity to breed with any other individual of the opposite sex;
6. The sex ratio is one male to one female;
7. Generations are nonoverlapping;
8. Individuals are diploid; and
9. Reproduction is sexual (Hartle 1981).

The effective population size (N_e) of the ideal population is the census size (N); that is, the effective population size equals the actual number of all reproductive individuals. Inheritance of characters or alleles in an ideal population follows a Mendelian pattern whereby the transmission of one allele to the next generation is not favored over another; selection is neutral or inconsequential. Consequently, allele frequencies in an ideal population do not fluctuate or change over generations. Although evolution is effectively arrested in the ideal population, the population is in a state of dynamic equilibrium, known as the Hardy–Weinberg equilibrium. Of course, in reality, all populations are evolving—on and off the hatchery. The critical point for the broodstock manager is understanding the processes involved in genetic change and how to manage them to accomplish genetic conservation goals.

The Natural Population

As a result of natural microevolutionary processes, natural populations deviate from those characteristics that define the ideal population and also deviate from each other. These natural processes or phenomena cause alterations in the population's allele frequencies—increases, decreases, or even complete loss of alleles or changes in genotypic frequencies. There are five basic microevolutionary processes of most concern to the broodstock manager: (1) mutation, (2) drift, (3) gene flow, (4) selection, and (5) mating system. One or more of these processes may be influencing genetic changes, perhaps concurrently, often in complex interactions, within a particular population in a particular environment at any one time or among interacting populations. In a management program having a major goal of conserving genetic diversity, the processes causing genetic changes, particularly changes resulting in a loss of genetic diversity and potential losses in fitness, must be managed. With respect to the goals and objectives of a gene pool maintenance program, genetic changes in the broodstock are unwanted. Thus, the broodstock manager strives to manage, avoid, limit, or control those forces that will result in genetic change. Therein lies the rationale for the broodstock management program, plan, and process. The foundation, framework, direction, tools, and language for developing the required gene pool maintenance program on a broodstock facility is provided by a GHRMS.

A natural fish population can be described as a complex, dynamic entity (also a biological system composed of interacting individuals), performing in concert with other entities in space, time, and circumstance. The population can be described and characterized quantitatively and qualitatively, evolutionarily, temporally, spatially, genetically, phenotypically, systematically, taxonomically, geographically, ecologically, and demographically. Population size is an extremely important demographic parameter related to these various descriptors, playing an enormous role in defining a population and determining its destiny. Obviously, population size is very important to the broodstock manager as well, particularly if she is managing a gene pool maintenance program. The census population size, N, is the number of individuals within a population with potential to mate, essentially all adult or mature fish. But in natural populations (in contrast to the ideal population), not all adults with mating potential or capability of mating do, in fact, mate. Nor do those that mate leave the same number of offspring. Recognizing the discrepancies between the expectations of the ideal population and actual characteristics of the natural population, geneticists developed a concept that corrected for the differences resulting from the effects of microevolutionary processes at work within and among natural populations. That concept is the effective population size (N_e). The effective population size is the size of an ideal population that would produce the same level of inbreeding and the opportunity for drift as the real or target population under consideration and having a specific census size, geographic area, and characteristics known to affect kinship within the population. The formula for calculating N_e is

$$N_e = 4(N_f)(N_m)/(N_f + N_m)\ . \qquad (2)$$

The effective population size is a key population parameter in conservation biology and gene pool maintenance programs, in part because the rate of inbreeding (ΔF; the rate of loss of genetic variation measured as heterozygosity) is the inverse of N_e and can be written as

$$\Delta F = \frac{1}{2N_e}. \quad (3)$$

In a natural population, factors affecting the magnitude of N_e, besides the abundance of potential breeders, are dispersal distance, sex ratio of breeding individuals, variance in reproductive success, overlapping generations, breeding strategy, and variance or fluctuation in census size over time. These factors, to the extent that they deviate from the ideal state, result in N_e being less than N in most natural populations. Variation in family size (inequality in the number of offspring per individual or variance in reproductive success) is considered to be the greatest factor reducing N_e with respect to N (Falconer 1981). Fluctuation in population size is also an important determinant of N_e because N_e calculated over several generations is the harmonic mean of N_e averaged over generations. That simply means overall N_e will be closer to the lowest N_e value of any particular generation at any point in time. That is why genetic bottlenecks are so hazardous where conserving genetic diversity is important.

A clear understanding of the N_e concept provides the broodstock manager with a practical framework for discussion, a starting point and direction with focus, as she communicates, plans, implements, monitors, and evaluates her program and product. Since N_e determines the rate of genetic change (loss in genetic structure or frequency of heterozygotes) experienced by a population, most management decisions made by the broodstock manager concentrate on maximizing the broodstock's N_e. If N_e is maximized, then loss of population characteristics or genetic diversity related to a reduction in N_e is minimized. Maximizing N_e reduces the chance of genetic changes resulting from drift and inbreeding and may preserve fitness of succeeding generations.

The efficiency of managing captive populations of imperiled fish species depends, in large measure, on the broodstock manager's skill in managing small populations to maintain vital population characteristics and in limiting deviations between the donor and founding population. For example, one condition that can cause N_e to be greater than N (up to $2N - 2$) is the reduction of the variance in family size below that expected to occur at random (Falconer 1981; Allendorf and Ryman 1987). We have used this knowledge in developing reliable management safeguards to reduce the risk of specific genetic hazards in a number of captive broodstocks for several threatened and endangered fishes in the intermountain West. Variance in family size is managed by conducting one-on-one matings or pedigree matings using individually marked adults to create different identifiable families and rearing offspring as family lots separately to maintain genetic identity and integrity. In addition, we breed as many unrelated individuals as possible to further increase N_e. The critical point here is that the broodstock manager can manage N_e to reduce genetic loss by managing broodstock numbers and breeding strategies to provide an equal number of progeny from each family that are destined to become broodfish at maturity. This particular management response is undoubtedly more difficult to implement in hatchery fish production programs because the natural variation among families means the smallest family lots can determine how many fish per family lot will be managed on the facility. Are there ways to manage around this situation?

In the family equalization process, many offspring from very large families may become excess or surplus to the defined needs of the broodstock management program. The numbers may be excess even to all strategically defined needs, including stocking and research. From a genetics conservation point of view, fish produced in excess of need must be disposed of in a responsible manner. To hold fish unnecessarily poses another hazard with associated risk. In extreme cases, surplus fish are buried. Destroying fish, even when ecologically and economically appropriate, is politically unpopular. The loss of potential productivity represented by not stocking the number of fish critical for recovery may impose higher demographic risks, limiting the rate of population recovery to the degree that not stocking hatchery fish is insupportable. However, if family variance in the hatchery can be managed such that the broodstock's N_e in the hatchery is significantly increased relative to that occurring in the wild, then management strategies will be developed to accommodate this very powerful management tool. The experienced and knowledgeable broodstock manager will ensure that the broodstock and production programs are well planned, integrated, and managed to reduce waste and increase efficiency of production. For example, planning ahead reduces the expense of carrying additional fish unnecessarily. The savvy broodstock manager learns quickly that not all problems are related to technological or ecological limitations but to economic, logistical, sociocultural, and political limitations as well.

Five Microevolutionary Processes

The five natural forces influencing loss of genetic variation in fish populations are (1) mutation, (2) random genetic drift, (3) gene flow, (4) mating system, and (5) selection. Generation interval can also have substantial impact on a broodstock management facility in that genetic diversity is most susceptible to loss when it is transmitted from one generation to the next. As we will later see, spawning is a source of potential genetic loss.

1. A mutation is an alteration in the primary structure of the DNA molecule resulting from a random error in replication. Unlike the other natural processes, mutation is the ultimate source of novel genetic variation or new alleles. Mutations can be favorable, deleterious, or neutral, sometimes depending on the environment in which they are expressed. Over time, in larger, isolated populations, mutations are the only source of new genetic material. However, in small populations and time frames associated with most captive breeding pro grams, mutations are expected to occur so infrequently as to be of no management significance.
2. Random genetic drift, or simply drift, results in undirected and stochastic genetic change within a population. That is to say, drift is a random event occurring from one generation to the next in populations whose effective population size is continuously small or frequently reduced. Indeed, drift is a function of small population size, a sampling artifact wherein genetic change is due to the sampling error inherent in small sample sizes or small populations. Drift occurs when parents produce more gametes than zygotes. Drift, therefore, is due to the fact that parents produce more gametes than offspring, which means that each generation is a sample, an "imperfect" representation, a statistical function of total parental genetic information.

   Drift is the most powerful of microevolutionary processes affecting small populations (Lacy 1987). Genetic changes resulting from drift usually occur very quickly causing fluctuations in gene frequencies and loss of alleles. Loss of alleles is accompanied by increased homozygosity through random fixation of one allele and the extinction of alternate alleles, regardless of their adaptive value. The potential effects of drift in a population are affected by the number of breeding individuals or the effective population's size, the relationship among breeders, the selective value of alleles, mutation pressure, and gene flow. Before a genotype can be attributed to drift, the population must be small enough, isolated, and the genes involved subject to little selection by the environment. The founder effect is a microevolutionary and stochastic or chance event that determines the development of a new population based on a single or very few individuals or founders. The result is a large genetic drift effect. Drift is not directly related either to selection or inbreeding, although the three forces can occur concurrently within a particular population. The effect of drift is loss of genetic or allelic diversity within the population and, perhaps, in total species diversity.

3. Gene flow is generally associated with movement of breeding indi viduals from one or more different populations into the target population. The genetic change results when new migrants interbreed with original members of the target population, the population being managed or under consideration. The change reflects increased genetic diversity within the target population in the form of new alleles introduced from an outside source. At the same time, however, there is homogenization and a coincidental reduction in diversity among the affected populations as they become more similar. In that sense, there is a loss of genetic uniqueness or identity among the affected populations.
4. In multicellular organisms, mating is the pair-wise union of unisexual parents for the purpose of sexual reproduction resulting in the formation of zygotes or new individuals and the transmission of genetic material from one generation to the next. Mating systems or mating behaviors in fish can be very complex and, in large measure, can determine the complexity of the population's genetic structure, the organization of genes into genotypes, and their expression as phenotypes, which ultimately influences among-population divergence, creating a population's unique genetic identity and evolutionary potential. Alone, mating systems do not change allele frequencies. In sexually reproducing organisms, matings are either random or nonrandom.

   A random mating system or panmixia is one wherein an individual within the population has an equal probability of mating with any other individual of the opposite sex. In a large random mating popuation, genotypic frequencies are determined by chance, and the influence of microevolutionary forces that affect small populations is inconsequential. Random mating may occur in natural populations having greater than 500 reproductive individuals, particularly where the sex ratio is close to equal (Kapuscinski and Jacobson 1987; Busack and Currens 1995; Currens and Busack 1995). Random mating is a feature of the Hardy–Weinberg population.

   Nonrandom or assortative mating is sexual reproduction in which pairing between parents is determined by some trait or character. Genotypic assortative mating is based on a genotypic character, and phenotypic assortative mating is based on the expression of some phenotypic trait. If the parents are more alike than expected by chance, then assortative mating is positive, and, if more different than expected by chance, then assortative mating is negative. Negative assortative mating is rare in nature. If the genotypes of the mating individuals are more alike than expected by chance, then positive genotypic assortative mating or inbreeding occurs. If genotypes of the male and female are more different than expected by chance, then negative genotypic assortative mating or outbreeding occurs. Phenotypic assortative mating can also be distinguished as negative and positive. To the extent that phenotype accurately reflects genotype, inbreeding or outbreeding may occur.

4a. Inbreeding or positive genotypic assortative mating increases the frequency of homozygotes at the expense of heterozygotes in the population. Homozygosity is the fixation of alleles at a locus, not the loss of alleles within the population. Inbreeding alone does not affect allele frequencies, only their genotypic distribution. Even though allele frequencies do not change as a result of inbreeding, phenotypic variation within the population may increase due to the redistribution of genotypic frequencies (Hartle 1981). Relatives are more alike genetically than would be expected by chance. In its simplest sense, then, inbreeding is the mating of relatives that share greater common ancestry than if they had been drawn by chance from a large randomly breeding population ($N > 500$, Hartle 1981; or $N > 1{,}000$, Shields 1993). Inbreeding results in increased homozygosity at all loci as opposed to phenotypic assortative mating, which may affect only those loci responsible for the particular phenotypic trait.

Inbreeding mating systems differ in the rate at which heterozygosity is lost. The rate of inbreeding within any particular population is a function of the relationship among the breeders. For example, self-fertilization is the most extreme case of inbreeding with complete inbreeding occurring within two or three generations. In sequence of inbreeding intensity are backcrossing to an inbred strain, full-sibling mating, half-sibling mating, and repeated backcrossing of a single individual from a random-bred strain (Hartle 1981).

The effects of inbreeding can be quantified in terms of the population's reduction in heterozygosity. The amount of inbreeding in a population is measured by comparing the actual or observed proportion of heterozygotes in the target population with the proportion that would be expected in a randomly mating or Hardy–Weinberg population. This measurement or value is called the inbreeding coefficient, F. The inbreeding coefficient measures the fractional reduction in heterozygosity due to inbreeding with respect to a random-mating population having the same allele frequencies. The equation for calculating the inbreeding coefficient *(F)* of a population is

$$F = (H_0 - H)/H_0 , \quad (4)$$

where F is the inbreeding coefficient, H_0 is the expected frequency of heterozygous genotypes in the random-mating population (2pq), and H is the actual or observed frequency of heterozygous genotypes in the inbred population.

The inbreeding coefficient can also be defined as the probability that the two alleles at a particular locus in an individual are identical by descent, or autozygous. If the alleles at the locus are not identical by descent from a common ancestor, they are called allozygous. In this sense, the inbreeding coefficient is truly a relative concept. It measures the amount of autozygosity relative to some theoretical ancestral population. The inbreeding coefficient of an individual in the current or target donor population is the probability that the two identical alleles at the locus arose by replication of a single allele more recently than the time at which the ancestral population existed (Hartle 1981). The broodstock manager arbitrarily, or by convention, assumes alleles in the wild or in a natural donor population are not identical by descent. That is, unless otherwise known, the broodstock manager assumes that all alleles in the natural donor population are allozygous. As a first approximation, the donor population is assumed to be noninbred, ($F = 0$) and the individuals no more closely related than would be expected in a large, randomly mating population. The inbreeding coefficient in the initial founder or first generation broodstock population is also assumed to be zero. The inbreeding coefficient can apply to relationships between individuals or within and among populations. The complete broodstock manager is well versed in animal breeding practices, for example, calculating the inbreeding of individuals produced by pedigree mating (Hartle 1981).

It is generally accepted that normally outbreeding populations can suffer from prolonged or extreme inbreeding due to genetic load. That is because the increase in homozygosity results in fixation in the population of rare deleterious or lethal recessive alleles. When these alleles occur in a population in the heterozygous state, their effects are masked or hidden by the expression of a dominant alternate allele. However, in the homozygous state (resulting from inbreeding), the deleterious recessive alleles are expressed phenotypically, perhaps with a reduction in performance or fitness. Reduction in fitness or viability associated with inbreeding is called inbreeding depression. Conversely, in normally inbreeding populations, few deleterious recessive alleles exist to contribute to inbreeding depression, presumably because they have been purged from the population over time by natural selection. Normally, inbreeding populations may suffer no ill consequences whatsoever from inbreeding but, in fact, are adapted and perform very well in the environment in which they evolved. Inbreeding is their "lifestyle of choice." The complete broodstock manager understands this phenomenon, and, therefore, one of the first questions she asks from the experts is if the broodstock she is expected to develop are derived from a normally inbreeding or outbreeding species. The answer to that question will be useful in making broodstock management decisions. The proportion of polymorphic loci between outbred and inbred populations is approximately the same. It is only the genotypic frequencies that differ. Again, inbreeding alone does not alter allele frequencies. Inbreeding simply results in a reorganization of the population's genetic structure, a shift in the distribution of genotypes from heterozygotes to homozygotes. The loss in performance an outbreeding or randomly breeding population suffers from inbreeding depression can be reduced by one generation of outcrossing to a different population and eliminated, theoretically, by one generation of outcrossing with a different inbred population.

4b. Outbreeding, or negative genotypic assortative mating, is the mating of individuals with genotypes that differ from the average pair chosen at random from the population. In domestic breeding programs, outbreeding different inbred lines may result in a dramatic increase in performance of the first generation hybrid offspring for fitness characters—growth, survival, and, perhaps, fertility or fecundity.

This phenomenon of superior hybrid performance is called hybrid vigor or heterosis. Heterosis is a phenomenon exploited frequently by plant and animal breeders (e.g., hybrid corn) that have developed inbred genetic lines from normally outbred populations with the express purpose of crossbreeding them to exploit the potential for hybrid vigor in the first generation offspring. Hybrid vigor is thought to result from the maximum restoration of heterozygosity and the subsequent allelic interactions at all polymorphic loci, the masking of deleterious rare recessive alleles in particular or, perhaps, overdominance, the superior performance of the heterozygotes over homozygotes. Hybrid vigor is an ephemeral phenomenon expressed in the first generation of hybrids but disappearing in subsequent generations as genotypic frequencies are established, according to the expectations of the Hardy–Weinberg equilibrium. The role, if any, of heterosis in natural populations is little studied and poorly known. It is not usually considered a legitimate goal of any genetic conservation program for imperiled fishes, perhaps due to concerns of loss of genetic identity and outbreeding depression. Outbreeding depression is a phenomenon that is considered to be the opposite of inbreeding depression. It results from outbreeding between two different noninbred populations with negative effects on the fitness of the offspring, often called maladaptation. Outbreeding depression is thought to occur when adaptive gene complexes or "local adaptations" within each different parental population are broken up during production of the outcrossed progeny.

There are several explanations for outbreeding depression. One is that when populations interbreed, there is a relative decrease in favorable alleles at the expense of alien alleles, alleles that may be less favorable in that particular environment. Another explanation for outbreeding depression is that there may be a disruption of genetic organization or reduction of coadapted gene complexes called, generically, local adaptations (Templeton 1986). There is little empirical data reported in the literature as evidence of outbreeding depression except between very distantly related populations (Gharrett and Smoker 1993). This discussion of mating schemes is not at all esoteric to animal and plant breeders. Unfortunately, the field of animal breeding has become a long neglected discipline in fisheries management. In point of fact, a broodstock manager, for lack of a more elegant term, is, among other things, an animal breeder. The problem arises when broodstock mangers breed fish while ignorant of genetics management as it relates to basic animal breeding principles and practices.

5. Selection is a microevolutionary process resulting in genetic change. Selection occurs from the interaction of a gene pool or population genotype with its environment resulting in adaptation and genetic loss. Selection is the differential survival among genotypes due to the selective action of the environment on different heritable phenotypes. Selection is a systematic or deterministic process resulting in genetic change that is consistent and predictable in quantity and quality. Selection, the environment acting on the phenotypes, results in the nonrandom reproduction and survival of different genotypes within a population. In the absence of selection and mutation, these genotypes or gene frequencies remain constant or unchanged in a large, randomly interbreeding population experiencing no gene flow. Generally, selection results in a relative increase in frequency of those genotypes influencing fitness, such as mating choice, fecundity, vi-

ability, longevity, and emigration. Selection of some genotypes is at the expense of less favorable genotypes with the net result being a fitter individual or population within its environment. Selection, however, does not occur without cost to the population. The gain in fitness from selection is accompanied by a shedding of genetic diversity within the population and, perhaps, total species diversity.

Primary Management Challenge

The primary challenge faced by the broodstock manager is how to put these principles into practice. The broodstock manager is responsible for the unification of genetic conservation, natural resources, broodstock management, and fish production goals through a comprehensive, holistic, integrated management system that puts into effective practice the decisions and actions required for their mutual accomplishment. Given the strategic management goals for captive propagation and broodstock management within a gene pool maintenance program, any management decision or activity, planned or unplanned, intentional or unintentional, has the potential of invoking genetic changes that are undesirable. Indeed, the physical environment itself is an inherent source of genetic change resulting in genetic loss through domestication of broodstock and their progeny. In theory, domesticated offspring, though adapted to the hatchery environment, may have lost the genetic diversity to respond to the rigors of alternative environments into which they are stocked. Unable to adapt, they are unable to provide the additional resources necessary to restore and sustain the target population. Management events that decrease or alter the desired genetic diversity also alter the unique genetic identity and character of the broodstock population, its genetic structure and organization, and the quantity and type of alleles. Management events having those effects are counterproductive and diametrically opposed to the program's strategic goals and undermine program success.

Genetic Hazard and Risk Management Language

When incorporating the vocabulary and definitions from the domain of hazard and risk assessment theory with the domains of fisheries management and genetics conservation, the lack of a shared language presents a serious impediment for all participants. Words have meanings, sometimes profound meanings. The complete and conversant broodstock manager has a working risk management vocabulary that includes the following terms:

Hazard is a potentially adverse consequence, effect, or loss resulting from the occurrence of an event, either a decision or an action. A hazard may be viewed as a loss resulting from an event and is evaluated within a specific type of program or management context.

Risk is the probability of a hazard occurring.

Failure is the loss sustained by the occurrence of a hazardous event within the management system, the program, and the process.

Genetic hazard is the loss of genetic diversity described as structure or function. There are four types of genetic hazard: (1) extinction, (2) loss of within-population genetic diversity, (3) loss of among-population diversity, and (4) loss of fitness of hatchery fish through selection in the captive propagation and broodstock development and management process.

Sources of genetic hazards are the events or series of events, either management actions or decisions, from which management benefit and loss might occur. In natural resources management, the sources of genetic hazards are due to human intervention in natural processes (i.e., the process of fisheries management).

Natural processes are microevolutionary forces causing genetic changes within and among populations. Those natural processes constitute the inherent potential within the management program and process to cause genetic changes in the managed population.

Endpoints are the critical levels of biological organization, structure, or complexity (individual, population, species, community) where the magnitude and duration of management failure is measured, described, and evaluated.

Performance consequences are the effects of a genetic hazard on phenotypic performance, especially as it relates to fitness or adaptive potential.

Unresolved issues generally pertain to uncertainty or ambiguity in the relationship between genotype and phenotype and fitness.

Reliability describes the potential success of proximate safeguards consciously designed and integrated into the broodstock management program and process as protective management measures against genetic hazards.

Resilience describes the potential rate of recovery from failure, the occurrence of a genetic hazard.

Risk assessment is a strategy for identifying and characterizing hazards and thresholds; for reducing the vulnerability or risk of hazards; and for resolving societal conflict from differences in perception of relative risk, cost, and benefit of management decisions and actions. A narrow definition of risk assessment is the process of qualitative and quantitative measurement and evaluation of risk.

Genetic risk assessment is ecological risk assessment that emphasizes the systematic identification and characterization of vulnerability of losses of genetic diversity.

Currens and Busack's Genetic Vulnerability Model

An important aspect of GHRMS is risk assessment. An important aspect of risk assessment is defining critical thresholds beyond which a failure has occurred or will occur, by definition. As discussed previously, fisheries scientists or managers face considerable difficulty in knowing or defining, in quantitative or even qualitative terms, the critical biological or genetic thresholds for the losses caused by genetic hazards and their performance consequences in captively propagated fish. As this requirement is a routine expectation for other types of risk assessment, a serious limitation is imposed on the broodstock manager. To circumvent this impediment, Currens and Busack (1995) suggest that managing genetic vulnerability may be a more realistic and useful approach than a traditional risk assessment approach based on objective and quantitative thresholds or values. The following equation represents their concept of vulnerability. Although fisheries scientists and managers may be a long way from being able to quantify the variables in the equation, at least the idea is significant and very useful in its conceptual application to operations: vulnerability (V) is defined as the product of risk (R) and loss (L) associated with a particular hazard, or

$$V = R \times L \,. \tag{5}$$

Currens and Busack (1995) explain that managing genetic vulnerability in natural resources management programs differs from other vulnerability models because the desired outcome or goal is not containment to confine, isolate, or separate the object of management (the broodstock in our case) from genetic hazards. Instead, the goal is to reduce risk and limit losses that occur despite the reliability of proximate safeguards through resilience provided by ultimate safeguards. Natural, but managed, resource systems are not artificially constructed systems designed to be controlled or managed in a vacuum, isolated from danger. Their existence and future is defined continuously and instantaneously by circumstances, generally anthropogenic interactions, generally beyond the control or even knowledge of the managers and experts. Indeed, managed fish populations are exposed purposefully to sources of hazards, management decisions, and actions. Thus, a primary component of a GHRMS based on a genetic vulnerability model is evaluation of the program, plan, process, and product at the point of operations.

The genetic vulnerability model (GVM) presented by Currens and Busack (1995) is a type of GHRMS designed to integrate operationally hazard and risk management theory with genetic conservation principles and captive propagation programs, plans, processes, and practices. The components of the GVM are (1) types of genetic hazards, (2) sources of genetic hazards, (3) endpoints, (4) proximate safeguards, (5) ultimate safeguards, and (6) failure.

Types of Genetic Hazards in a Broodstock Management Program

Busack and Currens (1995) identify four genetic hazard types associated with gene pool maintenance programs employed on genetic conservation hatchery and broodstock facilities. These four genetic hazard types connect the genetic effects of microevolutionary processes, discussed above, with the broodstock and with the management source, those management decisions and actions that occur operationally, purposefully, or accidentally on hatchery and broodstock facilities, specifically, with respect to Pacific salmon and steelhead *Oncorhynchus mykiss*. As a genetics conservationist, Campton (1995) generally recognizes these four genetic hazard types, making a distinction, however, between direct and indirect genetic effects or hazards. Direct genetic effects result from biological causes intrinsic to the physical hatchery environment, presumably beyond the control of management. While indirect genetic effects result from extrinsic management practices, those decisions or activities associated with routine management events occur in the broodstock and fish propagation facility. Campton contends that most genetic effects or hazards, at least in propagation of Pacific salmon and steelhead, result indirectly from causes initiated by the manager's decisions or actions and, therefore, are correctable, to some degree. And that degree is increased through superior management. Along with Currens and Busack (1995), Campton (1995) believes superior management is based, in part, on application of genetic management guidelines for conserving genetic resources. Genetic management guidelines should be applied during all phases of the broodstock management process: in orienting, problem definition and goal development, prioritizing, planning, implementing, monitoring and evaluating, improving management decisions and actions, and, finally, correcting hatchery practices—the sources of genetic hazards.

Using a captive propagation strategy based on genetic conservation principles that guide a gene pool maintenance program, the most basic genetic hazard encountered is the loss of genetic diversity of the target population and the captive broodstock. Loss of genetic diversity equates to a confounding of natural genetic complexity with reduction in performance potential. The GHRMS model presented by Busack (1990) and adopted with minor modification by A. R. Kapuscinski, C. R. Steward, M. L. Goodman, C. C. Krueger, J. H. Williamson, E. Bowles, and R. Carmichael (unpublished data, hereafter Kapuscinski et al.), Williamson and Wydoski (1994), Williamson et al. (1994), Busack and Currens (1995), and Currens and Busack (1995) is based on our understanding of the structure and organization of genetic diversity within and among populations (Falconer 1981; Hartle 1981; Barker 1989).

Type I Genetic Hazard: Extinction. Extinction is total loss of a gene pool (a population genome). When extinction is the result of negligence or mismanagement, it is the ultimate management failure, especially in a gene pool maintenance program. Extinction is the worst case scenario or genetic hazard in a fish conservation facility. Even more extreme is the case where the "victim" population is the last remnant of the species.

The natural processes associated with extinction on a captive propagation facility are mainly stochastic and nongenetic. For example, fluctuations in population size are often random and demographic in nature and result in reduction in effective population size, perhaps culminating in extinction. An on-site catastrophe is the most common nongenetic cause of extinction and may be due to a stochastic event or, if the result of chronically poor management, extinction may be seen as the consequence of a systemic anthropogenic process. Then, there are genetic or microevolutionary processes associated with extinction as well: drift, extreme inbreeding depression, and interspecific hybridization (whereby two or more populations or even species can be lost at essentially the same time). Accompanying the genetic loss is the potential loss of fitness of hatchery fish, which, over time, results in a gradual decline leading to the extinction vortex that can culminate in extinction (Gilpin 1987).

The management sources of a type I genetic hazard on a captive propagation or broodstock facility are generally associated with mistakes or errors in the management process and are not usually of intrinsic biological origin. For example, a well goes down; a pond is drained; one species is placed in a pond with another species, consuming it, competing with it, or interbreeding with it; a disease outbreak; improper disease treatment; the list of sources goes on. The complete broodstock manager is aware of and identifies the potential sources of a type I genetic hazard within her domain and develops, implements, evaluates, and adapts management responses—guidelines, plans, PPPs—to reduce the risk of extinction due to inferior management decision and operational error. In truth, in the larger context, the imperiled population in captivity was most likely "marked" for extinction by earlier mismanagement. By the time many populations are recognized as vulnerable and listed as endangered, they are well within the extinction vortex, their fate determined by powerful anthropogenic forces arising from ignorance and neglect, long before they arrived at the captive propagation facility. If such is the case, the responsible broodstock manager cannot be complacent; she is compelled to perform in a superior manner. In critical situations, her strategy amounts to triage.

Obviously, the consequences of extinction in terms of fish performance are profound, certainly for the victim, the extinguished population. However, since conservation hatcheries are a relatively new concept in captive propagation of imperiled fishes, there are a number of unresolved management issues. What are the performance consequences for the species as a whole when a population is lost? If it is the last remnant of the species, the answer is obvious. But, what are the performance consequences for an ecosystem when an integral component and its related processes are lost? Like ripples spreading across the pond, the effects of extinction cascade through the various levels of ecosystem complexity with unpredictable consequences.

Type II Genetic Hazard: Loss of Within-Population Diversity. Loss of within-population diversity is the reduction in quantity, variety, and combinations of alleles within a population. With respect to the various dimensions of genetic diversity within a population, quantity is the proportion or frequency of an allele, variety is the number of different kinds or types of alleles; and combinations are the different cooperating genotypes and gene complexes and their organization or distribution.

The natural processes associated with a type II genetic hazard are typically drift, including a founder event, inbreeding, and selection. The more infrequent or rare an allele in the parental pool, the less likely, by chance, it will be transmitted to the next generation (i.e., the broodstock). And, of course, the smaller the sample, the more reduced is the probability of a rare allele's transmission. Finally, the sampling of gametes is generally not random. Another process associated with a reduction in N_e that results in a type II genetic hazard is an inbreeding-mating scheme that, as we have seen, results in a reduction in heterozygotes and an increase in homozygotes. Selection also results in loss of alleles, but that will be discussed later in more detail under type IV genetic hazard: loss of fitness. The concurrent impact of drift, inbreeding, and selection can compound genetic loss within a population.

The management sources resulting in a type II genetic hazard are those decisions or practices resulting in a decrease in N_e. For example, a small founding number of broodfish, unequal sex ratios, individuals breeding more than once, overlapping generations, differential survival of offspring among individuals, and population fluctuations over generations are commonly experienced phenomena that reduce the effective population size below the census size and, thus, are potential causes of lost genetic diversity within a population. A type II genetic hazard can also occur quite easily, if thoughtlessly, on a broodstock facility from another process or management source—biased sampling. Captive propagation programs that found their broodstock on a small sample from the donor population and continue to use only a few fish for broodfish from subsequent generations will suffer from a reduction in effective population size. Deviations from an equal sex ratio when breeding or spawning fish, pooling gametes, inappropriate pairing (i.e., pairing parents of different year-classes or breeding siblings), and failure to equalize variation among family lots when developing a broodstock—all of these practices may reduce N_e and, thus, are sources of potential loss of genetic diversity within the population. The potential performance consequences of type II genetic hazard are reductions in fitness, survival, and adaptation.

The most immediate and effective management response in the broodstock manager's arsenal is mastering the concept of N_e. Development and implementation of planning and operational guidelines and PPPs to identify and control sources of genetic loss as a result of diminished effective population size is a good protective strategy. For example, some appropriate management approaches involve maximizing broodfish numbers used in spawning and managing fish identity, sex ratio, age structure, and variance in progeny number among families. To reduce the risk of hazardous consequences from a small N_e, as well as from inbreeding, pedigree mating using a large number of uniquely tagged adults of known genetic relationships is a start. Pedigree mating is followed by progeny-rearing regimes designed to maintain family identity and integrity while enhancing their survival to reduce effects of hatchery selection. Further benefits can be had when these two basic management responses are guided by a genetic management plan and supported by genetic management guidelines. First, systematic acquisition of detailed genetics and performance records from parents and family lots, then relying on those data to evaluate and to improve the program, plan, and process are essential components in a standard propagation strategy for gene pool maintenance.

An effective propagation strategy is geared principally to maximize N_e while avoiding artificial selection and reducing effects of domestication. Major features of this strategy are a large random sample of founders from the appropriate donor population; equal numbers of uniquely marked and unrelated males and females paired for the diallel or factorial mating schemes necessary to support a pedigree breeding program; every fish capable of spawning contributes to the next generation; families and year classes are maintained separately (at least until individuals can be marked by family); family variance is equalized by random culling (doubles the effective population or N_e beyond $2N-1$); and propagation methods are developed to reduce effects of natural selection on the hatchery or domestication. The methods emphasize improvements in sampling randomly, incubating and rearing small family lots, marking and tagging fish as individuals and by family, and through culture methods to increase survival and growth rate. The broodstock is supplemented periodically with additional wild donors.

Type III Genetic Hazard: Loss of Among-Population Diversity. Loss of among-population diversity, or loss of genetic identity, is a reduction in the differences in quantity, variety, and combinations of alleles among populations. The principle microevolutionary process responsible for occurrence of a type III genetic hazard is migration or, to be specific, immigration, the movement of genetic material into a population from an outside source. Loss of among-population genetic diversity results from excessive or unnatural gene flow, gene flow beyond that level experienced by normally outbreeding species that depend on some degree of gene flow to maintain an adaptive level of diversity. The outside source is from a different population within the species or, occasionally, from a population of different species. Commonly, excessive intraspecific gene flow among populations is referred to by fisheries managers as introgressive hybridization.

The management source of a type III genetic hazard on a broodstock facility is a mixing of populations that results in interbreeding among populations or within or among species, intentionally or unintentionally. Type III genetic hazards in a facility are almost always of management origin. Thus, the sources of introgressive hybridization are not inherent to hatcheries but, following Campton (1995), are extrinsic causes with indirect genetic effects. The sources frequently originate in the manager's intentional desire to improve fish performance. If the broodstock manager operates a conservation facility without understanding underlying principles or their appropriate application and introduces new blood to enhance performance of a population that is not suffering from inbreeding depression, then a type III genetic hazard most likely has occurred. Occasionally, a type III genetic hazard occurs accidentally when managers dump one conspecific population on top of another in a pond or some other rearing unit. Different populations mixed in error may eventually lead to a mixing of their gametes and production of hybrid or introgressed offspring with mixed ancestry. If severe enough, a type III genetic hazard can become a type I genetic hazard, extinction of a population, and in the most extreme case, extinction of a species.

Busack and Currens (1995) explain the potential consequences of a type III genetic hazard from two perspectives. First, at the species complex or multipopulation level, reduction in genetic differences among populations may reduce overall fitness of the species rendering it less capable of adapting to environmental changes. Loss of fitness in this case may be related to decomposition or disintegration of locally adapted gene complexes. Loss of among-population genetic diversity results in "genetic homogenization" wherein the genetic identity and character of the target population, as well as the species as a whole, is reduced and its native complexity is altered. Species-level genetic diversity is reduced and simplified. Secondly, at the broodstock level, loss of among-population diversity due to intrusion of alien genes alters the genetic uniqueness of the broodstock and may alter, concurrently, its performance profile as well. The consequences for a single population or a broodstock is loss of genetic uniqueness, deviation in identity and character from the donor population with a potential reduction in broodstock performance, perhaps a reduction in response to diseases and pathogens.

The potentially negative performance consequences of type III genetic hazard for the species and its constituent populations is an overall reduction in the species adaptive response capability to a wider range of environmental differences. Negative consequences are speculated to result from outbreeding depression. If it occurs due to breakdown of coadapted gene complex, a reduction in performance will be observed in the second generation following the genetic invasion (Gharret and Smoker 1993).

Reducing the likelihood of a type III genetic hazard requires long range planning and rigorous management and monitoring of fish performance. Plans, practices, protocols, and procedures must be developed to address accurate genetic identification and characterization of all populations on the facility (genetic and performance profiles and databases); effective fish marking and tagging to permit identification of individuals and families; effective incubation and accurate rearing unit labeling; genetic family segregation free from contamination (separate and covered tanks); and complete and effective record keeping, data management, and genetic and performance monitoring; in short, development and application of a fish, tissue, gamete, and database security system ensuring the integrity and accuracy of fish provenance, identity records, and breeding and performance histories.

Type IV Genetic Hazard: Loss of Fitness from Hatchery Selection. Selection in a hatchery, regardless of the process (natural or artificial) or the intent, results in changes in the quantity, quality, or combination of alleles within the captive population and, ultimately, between the captive population and its donor source. Selection results from differential survival among genotypes. Genetic diversity within the captive population is reduced and the genetic diversity between the captive population and the donor population is increased. Potential consequences of hatchery selection include improved fish performance in the hatchery, at the expense of performance of hatchery fish released into the wild or at the expense of the wild population if interbreeding occurs and outbreeding depression results. Selection in the hatchery results in a loss of within-population diversity, but unlike losses associated with a type II genetic hazard, selection is neither random nor related to N_e, but instead, is related to specific phenotypic traits with a heritable component and the selection intensity imposed by hatchery methods and the physical environment upon these traits. There are two basic types of hatchery selection: domestication and artificial selection.

Domestication. Domestication is change in the quantity, variety, or combination of alleles within a captive population or between the founder captive population, the derivative broodstock, and the source or donor population. The basic process behind domestication is natural selection. Domestication is the direct genetic effect resulting from the inherent selective influences of the physical hatchery environment on the broodstock population, presumably beyond the effective or practical control of the manager. As a possible consequence, the founder or broodstock population may change genetically or evolve in ways different from its parental donor population due to the differential selective forces existing between the two different environments. Although the genetic effects of domestication are similar to a type II genetic hazard, there are two important distinctions (1) with respect to a type II genetic hazard, genetic changes or loss due to drift are based on an entirely different microevolutionary process or mechanism. Losses from drift are random and stochastic in nature, whereas population genetic diversity lost to domestication is directly related to environmental selection for specific phenotypic or performance characters, and (2) diversity lost to drift is at a rate proportional to the reduction in effective population size, whereas with domestication, genetic diversity is lost at a rate dependent on the genetic nature or inheritability of the traits under selection and the intensity of selection imposed by the hatchery environment (Doyle 1983).

Selection can be measured, theoretically, in terms of viability, the probability of an individual surviving from fertilization to reproductive age. Relative viabilities are called the relative fitness of the genotype compared with other less fit alternative genotypes. Selective advantage is the advantage in competition between favored and less favored genotypes. In natural selection, selective advantage of a favored genotype is the collective property of all its phenotypic characters and may have different causes, that is, production of larger progeny and better resistance of progeny to environmental factors. In an evolutionary sense, a successful or favored genotype is one favored by natural selection within a particular environment. In a gene pool maintenance program, where the management goal is genetic conservation, specifically to avoid or prevent genetic loss within the founding population, domestication results in loss of within population genetic variation.

Although diversity lost within the population in a hatchery due to domestication is based on the same basic process—selection—its origin is of an intrinsic nature. Genetic diversity is lost from phenotypic exposure to the physical hatchery environment resulting in unintentional but direct genetic effects. Campton (1995) emphasizes this distinction from artificial selection for a very important reason. If the source of the hazard originates with management decisions and actions, then steps can be taken to reduce the risk of the hazard.

As is common with all of these various types of genetic hazards, there are unresolved issues resulting in unavoidable ambiguity, even for the most learned and experienced geneticist or broodstock manager. There simply is little empirical evidence supporting the theoretical hypothesis that domestication presents a genetic hazard with serious genetic consequences resulting in impaired performance of hatchery fish stocked in nature. The definitive work necessary to test the hypothesis with unambiguous results has not been done (Campton 1995). What studies have been conducted are merely suggestive. In addition, developing performance criteria to evaluate the effects of genetic losses from domestication in the hatchery is counterintuitive. Conclusive studies would have to be conducted at the same time in two environments, in captivity and in nature, the latter uncontrolled with identical experimental fish. These studies are expensive, time consuming, and require complex design, control, and coordination.

Many experts consider potential losses associated with domestication the cost of doing business where a captive propagation strategy is part of the management equation. Hatcheries are chosen as an alternative production strategy for imperiled fishes, in part, by necessity. The natural environment no longer sustains production, so an alternative environment is sought, typically one wherein a controlled, managed environment predictably results in increased production. To eliminate the effects of domestication on a captive propagation facility presumes that our knowledge of the pristine natural environment, which may no longer exist, is complete and that we have mastered technological ways to reproduce it to match its selective forces on the hatchery. Beyond some obvious modifications, that approach appears quite foolish given that if we were able to replicate the natural environment to sustain production, we wouldn't be in the environmental predicament we are now. The solution is somewhere between willful ignorance of genetic principles and fanciful attempts to replicate the natural environment. Busack and Currens (1995) suggest that a more realistic approach is to learn enough about the types and magnitude of genotypic and performance changes due to domestication in a captive propagation facility in order to better understand short- and long-term costs and benefits. Short of replicating the natural environment in the hatchery, the broodstock manager may be able to reduce the effects of domestication by developing culture environments and employing culture methods that reduce mortality. These methods, when employed in conjunction with random sampling techniques, particularly where fish culling or reduction in numbers is required, theoretically, at least, helps reduce domestication effects. Also, reducing the number of sequential generations demanded from a broodstock with occasional replacement of additional wild fish from the donor source reduces opportunity for domestication.

Artificial Selection. In general, the process underlying artificial selection is differential survival of genotypes based on selection pressure imposed by management decisions or actions to favor or disfavor specific phenotypic, usually quantitative, characters. The indirect genetic effects from artificial selection are a reduction or loss of within-population diversity. (The result is the same, but the process is different than a type II genetic hazard where loss is primarily from drift). The immediate gain, from the perspective of the misguided broodstock manager unconscious of the cost to genetic diversity, is improved performance in the hatchery as the population responds to selection and adapts to specific management practices. The long-term consequences for the stocked offspring may be their maladaptation to the new environment. If the goal of the broodstock manager directing a gene pool maintenance program is to conserve genetic resources by avoiding loss of genetic material, then engineering an artificial selection program, condoning correctable hazardous management practices, and rearing fish in unhealthy environments is mismanagement.

In contrast to domestication (natural selection due to exposure to a hatchery environment), artificial selection is caused by management decision or action. Selection is intentionally applied by the manager to improve performance, by selecting broodfish or offspring because of certain heritable phenotypic traits or characters. The manager's goal is to intentionally change the population's identity and character in desirable ways through artificial selection. A logical consequence of artificial selection in the hatchery is the genetic departure of the broodstock from the donor source resulting in two different populations. Artificial selection does not have to be planned or result from a conscious or deliberate decision or action by the manager. Artificial selection can occur within a gene pool maintenance program quite unintentionally through any decision or action by management with inadvertent, unexpected, or unintended results. In artificial selection, selective advantage is conferred by the manager or staff, intentionally or not, on some members of the population in preference to others because they or their relatives possess a trait or phenotype deemed desirable. A potential consequence of offspring from a selected broodstock interbreeding with fish from the natural population following reintroduction may reduce the overall fitness of the donor population through a type III genetic hazard, outbreeding depression.

Examples of artificial selection include selection of early-run timing, size (sex or growth rate), condition (sex or food conversion, or aggressiveness), and grading (cannibalism). Management response or safeguards include the manager's knowledge of animal breeding concepts and applying them to determine inheritabilities and coefficients of variation of important performance traits (Tave 1986; Kapuscinski and Jacobson 1987).

The broad management point of artificial selection is to improve fish performance on the hatchery. The problem associated with artificial selection within a gene pool maintenance program is that the goals and results are incompatible at conception. Success in improving productivity through selection results in failure in genetic conservation. Achieving productivity goals within a gene pool maintenance program depends on an improvement in both culture methods and environment that results in maximizing survival during all life history stages and monitoring and evaluation of mean performance values in contrast to those of the natural population.

Another management source of artificial selection is a management practice that results in nonrandom or biased sampling. Biased sampling can occur during any stage of broodstock management and operations from the beginning, with a nonrandom sampling of founder fish from the donor population throughout the population's entire life cycle in captivity and from one generation to the next. Genetic variability lost from within the population because of biased sampling is due more to sampling error than selection for any particular phenotype. But, if the sampling error associated with biased sampling occurs frequently, systematically, and involves specific quantitative characters with high inheritabilities, the loss to biased sampling is compounded by artificial selection. The effect is simply more of the same—genetic variation within the population will be lost—associated with artificial selection instead of domestication because the loss results from a management action. The simple fact is, representative sampling during captive propagation is very difficult to achieve. If biased sampling occurs randomly and does not involve heritable traits, then genetic or fitness losses may not result. However, if the sampling is systematically associated with traits related to fitness, then selection may occur with related losses in genetic diversity. Obviously, the management response to avoid or reduce the impact of biased sampling is to implement protocols that promote truly random sampling.

Sources of Genetic Hazards

Currens and Busack (1995) have identified seven broad hazard sources, coinciding with major captive propagation program steps or activities, that exist in the life cycle of captively propagated fish produced for release into the wild, specifically Pacific salmon: (1) broodstock selection, (2) broodstock collection, (3) broodstock mating, (4) broodstock and progeny rearing, (5) progeny release, (6) juvenile migration, and (7) adult migration. Sources 1 through 4 will be discussed later in this chapter. Within each of the seven source categories is a series of management decisions and actions with the potential of causing genetic change. Specific actions and decisions may depend on the species involved. The forces of nature driving genetic changes within and among fish populations in the broodstock management domain are the natural microevolutionary processes discussed earlier: mutation, drift; migration, nonrandom mating systems, and selection.

Endpoints

Endpoints are critical management impact or control sites and decision nodes that coincide with major levels of ecological and biological organization. Endpoints are the levels of ecological and biological complexity (individual, population, species, community, etc.) where the magnitude and duration of management failure is measured, described, and evaluated. "Because impacts of management hazards in ecosystems cascade through different levels of ecological and biological organization, determining a single endpoint for risk assessment is not satisfactory" (Currens and Busack 1995). Identifying critical endpoints is an essential objective in a GHRMS and genetic vulnerability assessment. Currens and Busack (1995) present an example wherein wild steelhead, chinook salmon, and cutthroat trout spawn within several adjacent Pacific coastal rivers and where supplementation programs exist for steelhead and chinook salmon in one stream (target area). For each of the captively propagated species, the authors identify at least five endpoints. For example, with respect to steelhead, the five endpoints are (1) target population (the steelhead), (2) nontarget populations of the same species, (3) nontarget populations of nontarget species, (4) nontarget populations of other target species, and (5) nontarget populations of target species outside the basin.

Proximate Safeguards

Proximate safeguards are immediate or tactical broodstock management responses that are designed and employed as mechanisms, decisions, and actions providing reliability to the broodstock management process by ensuring that sources of genetic hazards, the major operational components or steps, are productive while limiting vulnerability of the broodstock and progeny to genetic losses. Currens and Busack (1995) have provided extensive information on four major management responses or proximate safeguards for Pacific salmon supplementation programs: (1) management guidelines, (2) descriptions of biological variability, (3) technical training for team members, and (4) logistical capability.

Ultimate Safeguards

Ultimate safeguards provide the resilience in natural resources management, broodstock management in our case, to recover from the failure of proximate safeguards to avert or reduce the risk of genetic losses. Currens and Busack (1995) identify two sources of resilience. The first source, genetic reserves or natural genetic refugia (NGR) might seem to lie outside the immediate domain of the broodstock manager. In some cases, however, NGRs may have a direct bearing on the broodstock management program. Natural genetic refugia are potential pools of genetic material available to the broodstock manager to meet her own program goals. The relationship between the NGR and management is crucial for another reason. The NGR provides a biological buffer, preserving fresh opportunities for management action following experiments or failures to rebound or respond adaptively. Adaptive management is the second source of resilience provided through the ultimate safeguard system. Adaptive management is a comprehensive approach integrated into the management processes at all levels. It is a strategy that promotes acquisition and synthesis of new information while learning from management errors or failures with the stated purpose of correcting mistakes and improving the process and the outcomes of management.

Failure

Failure within GHRMS based on the genetic vulnerability model simply means the occurrence or realization of a genetic hazard. Genetic assessment strategies, plans, and tools to monitor and evaluate the broodstock management program are essential to quickly detect or diagnose failures at the various endpoints in the broodstock management process.

To identify vulnerability of her program and broodstock to genetic hazards, the broodstock manager needs two types of data. The first data set consists of systematically acquired and analyzed genetic and demographic information that she can relate to common sources of hazards, situations that create the conditions of undesirable genetic change, as well as those critical endpoints within her management domain. These data are classified and categorized by type, duration, and amount. The data are essentially the same she extracted from the TME and the CHIMP[10] system to measure, identify, and characterize GCUs and to determine their ecological and biological status, trends, and threats to their existence and their sustainability. Genetics data derived through various biochemical and molecular methods describe genetic information from which inferences are made of genetic structure and, perhaps, function or performance. Demographic data describe distributional ranges, trends in abundance, effective and census population sizes, migration rates, spawning behaviors, mating systems, sex ratios, straying patterns, age structure, fecundity, age to maturity, and longevity. Unfortunately, data of this nature are scarce or sketchy for most imperiled fish species.

The second data set required to identify genetic vulnerability is based on the presence or absence of proximate and ultimate safeguards and the baseline data necessary to assess their effectiveness. Acquiring these data is accomplished by employing the genetic vulnerability system described herein (Currens and Busack 1995). Proximate and ultimate safeguards and their component parts are identified within the broodstock management program, plan, and process. Subsequently, they are characterized qualitatively, and perhaps even quantitatively, by their presence or absence, a narrative description, and rated based on some predetermined set of criteria or results from empirical studies.

Boodstock Management Process

The broodstock management process is logically derived from, and is a functional product of, the TME and the CHIMP[10] system. It is one of many strategic subelements of a greater, more expansive management or recovery effort. It shares, at a lower level of complexity, all essential systemic components of its parent: problem, purpose, perspective, principles, program, plan, process, performance, product, and progress. If each component of the CHIMP[10] system were logically stepped-down from the strategic level to the tactical or operational level, the broodstock management program would emerge. At the same time, each of the seven process steps of broodstock management has its logical source in the corresponding process steps of the CHIMP[10] system. This means that the systemic components and process steps of the broodstock management program are consistent with and compatible with those systemic components and process steps of the parent strategic management system and process.

The broodstock management process can also be viewed as a detour in the natural life cycle of the fish. The detour is necessary because many imperiled fish populations have lost, even if temporarily, their ability to reproduce and be replaced by young recruits. To ameliorate that condition, for the time necessary to have their natural reproductive ability restored, their path is redirected through the hatchery program and process where their genetic legacy and evolutionary potential are protected by managed reproduction or captive propagation. In that sense, the managed reproduction program for threatened and endangered fishes can be viewed as a temporary prosthetic device. Not totally desirable perhaps, but when it is the preferred production strategy and managed well, it is a device that can be effective. A major component of a managed reproduction program is the resident broodstock (captive broodstock). Managed reproduction in the hatchery revolves around the captive broodstock. The broodstock management process logically follows a circular path, usually connected to the natural life cycle of the target species. It is supported by genetics management guidelines, assumptions, and principles that are continuously and adaptively applied within a GHRMS framework through a process that enacts the broodstock management program in ways that reflect the natural character of the target species. The broodstock management process follows the seven steps associated with the CHIMP[10] system: (1) orienting and scoping, (2) problem analysis and goal development, (3) prioritizing, (4) planning, (5) implementing, (6) monitoring and evaluating, and (7) adapting.

Orienting and Scoping

Orienting and scooping is an important preliminary task in the broodstock management process. It means (for those who enjoy target shooting) "getting on the paper." The broodstock manager acquires information from the information pool, generally strategic planning and supporting documents; preliminarily identifies the fundamental management problem and its relationship to program's purpose, type (i.e., recreation, restoration, recovery, mitigation, etc.), and direction; then determines the significance of the problem, purpose, and direction to her own broodstock management program, in the broad sense. Understanding the program's primary management perspective and its guiding principles educates her about program philosophy and mission—against which she can compare her professional and ethical perspective.

Problem Analysis and Goal Development

The second step in the broodstock management process is to subject the problem to a more detailed analysis. A problem can be described as a discrepancy or a disequilibrium between two states of affair. First, the existing condition that is not desired and second, an anticipated or ideal state. Solving the problem means seeking to reduce or eliminate the existing disequilibrium in favor of the desired condition. Problem solution means the discrepancy between the two conditions has been removed in favor of the desired condition. In this second step of the process, there are five subtasks that require the broodstock manager to access, interpret, synthesize, analyze, and integrate the information gained from step one.

Problem Analysis and Goal Verification

The major purpose of this task is to better define and understand the basic management problem, particularly as it has been translated into management goals. This task requires the broodstock manager to examine the information, particularly historical, ecological, and biological trends; identify broad strategic management goals (and criteria); and verify their derivation, appropriateness, and consistency with the supporting data. Ultimately, the broodstock manager breaks problems down analytically, reducing them into pertinent problem elements that can then be translated into tangible broodstock management goals and discrete measurable objectives that will form the broodstock management program's foundation and guide the remaining steps in the process—prioritization, planning, and implementation. The broodstock manager's focus in this task, from a genetics management perspective, is the GCU, the basic unit of concern and management.

For example, the best scientific data available suggest the Gila trout is in danger of extinction. The broad recovery goals for Gila trout are to downlist, to recover, and to delist the species. Broad biological goals for Gila trout require avoidance of extinction and provisions for long-term species sustainability. Broad genetic management goals call for avoidance of Gila trout extinction and protection of natural amounts, types, and patterns of genetic diversity within and among wild and captive populations of Gila trout. Specific recovery goals from the Gila Trout Recovery Plan (USFWS 1992) are to improve and to secure the biological status of Gila trout to the point that survival of all indigenous lineages is secured and maintained. The problem is authentic and the various recovery goals are appropriate, consistent, compatible, and doable.

Species Identification and Characterization

The stock concept (STOCS) presents the hypothesis, supported by robust theory and abundant empirical evidence, that fish populations reflect a natural structural complexity. That is, a complex fish species is organized and subdivided into, more-or-less, distinct or discrete populations or stocks that reflect their fundamental genetic nature and evolutionary history (Berst and Simon 1981). The stock or population is the fundamental unit of many legally mandated conservation efforts. However, the fish stock or population concept is sometimes poorly understood, ill defined, and difficult to apply in many management situations (Loftus 1981). It is important that a stock or conservation unit mean the same thing ecologically, biologically, and legally in order for sensible conservation and management planning and implementation to fall in line with regulatory requirements. This requirement is necessary for making natural resource decisions to develop management and recovery goals and to assign priorities for funding and action.

Individuals of natural animal populations are rarely distributed randomly across and throughout the limits of their range. Their actual distributions can range from continuous to highly clumped (Barker 1989). Specific populations may contain unique genetic attributes that allow the fish to occupy and interact with their habitat, survive, reproduce, recruit, adapt, and evolve. The program, problem, purpose, philosophy, principles, mission, and goals regarding genetic management of threatened and endangered fishes are based on the assumption that the local fish population and its immediate environment are considered the natural unit of recovery and management and are the operational target of the recovery and management program, plan, process, and activities. The condition of the population and its habitat are the standards against which program performance and progress are evaluated.

As we have seen, a fish stock or population is defined as a randomly breeding group of individuals with spatial, temporal, genetic, demographic, and behavioral integrity that serves to separate it from other randomly breeding groups of the same species. This definition of a fish population applies to any one among multiple populations within a species, each reflecting its own unique identity. This holds true in most cases with respect to rivers, whether the various populations occur within an individual river basin or as a single population from among several rivers. In addition, the population maintains the Hardy–Weinberg equilibrium when that population is genetically stable with respect to potential population perturbations (i.e., gene flow from immigration; Booke 1981). The population's biological uniqueness, resulting from effective reproductive isolation, may coincide with geographic isolation. Reproductive isolation results in genetic differentiation and is an evolutionary response to unique environmental realities. Assuming this is the case, such populations have been called genetic conservation units. Genetic conservation units have also been called "evolutionary significant units" (ESU; Dizon et al. 1992).

The detailed definition of a GCU will improve our knowledge and understanding of its genetic, phenotypic, demographic, ecological, and biogeographic dimensions. Each new analytical method or descriptive technique may reveal a new aspect of its identity and provide criteria useful for more accurate and meaningful identification and characterization. An important key is the extent to which the GCU is an evolutionarily important or significant component relative to the overall identity and character of the species and its relative value in sustaining ecosystem complexity. The hierarchically structured identity and character of a fish species enhances our understanding of this issue (Hartle 1981; Chakraborty and Leimar 1987; Barker 1989; Smith and Rhodes 1992).

All available data from the TME's information pool relative to the fish's nature, identity, and character are used to identify and characterize the species and its various constituent GCUs (Utter 1981; Dizon et al. 1992; Williamson and Wydoski 1994). The data include geographic (historic and current range, and distribution), demographic (estimates of important population characteristics associated with viability or sustainability, such as total and relative abundance, age structure, birth and death rates, immigration rates, fecundity, sex ratio, and N_e), ecological (life history and life style strategies and requirements, environmental or habitat requirements, and interspecific or community interactions), and phenotypic (morphological, meristic, physiological, and behavioral information, such as spawning migration and fidelity to spawning sites) information, ontological and evolutionary history; phylogenetic history; and genetic organization and structure.

Results from genetic studies may also be available to the broodstock manager from the information pool. If such studies have not been conducted, they should be initiated as soon as possible. The recommended tasks would include

1. Determining phylogenetic, cladistic, systematic, and taxonomic interspecific and intraspecific relationships among similar taxa,
2. Determining species boundaries,
3. Interpreting and defining species genetic variation in terms of hierarchical organization and structure, diversity within populations and among populations, and overall or total genetic diversity,
4. Determining intraspecific variation to define local populations, stocks, demes, lineages, GCUs,

5. Determining heterozygosity values and other statistical estimates of genetic variation and stability,
6. Examining for introgression and hybridization and other evidence of gene flow and directionality or genetic contamination,
7. Considering inbreeding and outbreeding events and potential consequences,
8. Optimizing analytical methods to improve resolution capability or the ability to distinguish among important groups,
9. Developing nonlethal tissue sampling methods,
10. Developing genetic profiles or fingerprints for all potential broodfish,
11. Developing methods to mark or identify individuals or discrete ge netic groups, and
12. Developing and maintaining integrative genetic databases for monitoring and evaluating management activities.

Identify Genetic Conservation Units

Identify GCUs, the basic units of management, to provide evaluation targets around which measurable management goals, objectives, and other benchmarks of management progress, success, and failure can be judged. Genetic conservation units are defined based on data generated from the information pool and studies summarized above: distributional, population response or demographic, phenotypic, and genetic data (Dizon et al. 1992). To repeat, a GCU is considered synonymous with a local population, stock, lineage, deme, or any other term signifying or implying a naturally interbreeding unit or entity. In accordance with principles established earlier in the CHIMP[10] system (Williamson and Wydoski 1994; Williamson et al. 1994) there is a general precautionary practice among broodstock managers to treat questionable groups as putative GCUs even when data are less than conclusive. Reluctance to "lump" is an ultimate safeguard, and it is a common management choice to err on the conservative side by providing presumptive or provisional identity to a group in order to protect potential genetic diversity until such time as information becomes available to make an alternative decision. "Lumping" choices lack resilience and are irreversible once translated into management action. Reassessment and decisions relative to GCUs can be accommodated during the routine monitoring, evaluation, and adaptive management process steps.

Assess Status and Trends

Status and trend assessments are determined by species and by each GCU within each species, particularly with respect to historical conditions in relation to current conditions. Ecological and biological (genetic, phenotypic, and demographic) status and trend surveys and data sets are used to define historical and current health, condition, sustainability, and viability of a species, in general. The same criteria can be used to determine and assign an operational status and trend designation for each GCU. This step is important in defining factors that limit the population's long-term sustainability and the variables determining its future. This is a vital step in the management process to aid the broodstock manager in evaluating management goals and decisions of others in other program elements and in making informed management decisions herself. The purpose of relative status and trend assessments and designations is to provide a systematic, quantitative or qualitative standardized criteria set and evaluation process by which to establish management goals and objectives, set priorities among GCUs, plan solution strategies, identify corrective actions, and to determine overall health or condition of a population with respect to program success, failure, or progress.

Status and trend designations and assessments are arrived at only after preliminary efforts have been made at defining GCUs using all the criteria heretofore mentioned. Next, many of the same characters used in identifying and characterizing a GCU are examined and evaluated by some arbitrarily weighted, but biologically meaningful, set of values. Geographic distribution data, ecological variables and limiting factors, demographic characteristics and dynamics (relative abundance, sex ratio, birth and death rates, effective population size, reproduction, and recruitment), physiological characteristics, physical anomalies, behavioral characters, and genetic concerns that might impact fitness, sustained survival, and evolutionary potential are considered and weighed in the process.

Status and trend assessment and designation provide a diagnostic snapshot of the relative health of an individual GCU and collectively of the entire species. Status and trend data are vital in identifying management problems, evaluating them against previous situations, and suggesting the most appropriate course of action to plan and direct toward a remedy and away from a hazard or threat. Status and trend assessment and designation define the current condition of a GCU relative to some other time, perhaps historical, and relative to other GCUs or with respect to some desired condition.

Status and trend categories can be applied simply as

Status Category (Demographic)
I. Stable
II. Unstable
III. Critical

IV. Extirpated
V. Unknown

or

Status Category (Ecological)
I. High risk of extinction
II. Low risk of extinction
III. Extirpated
IV. Unknown

Review Recovery Goals and Criteria for Each Genetic Conservation Unit

At this point in the process, program managers usually have developed management goals and criteria or, at least preliminary or provisional qualitative goals and criteria. Now, they reexamine them for completeness, consistency, compatibility, credibility and feasibility. Are specific management goals defined for each GCU with measurable values? Ranges of values or mean values with acceptable deviation or confidence intervals are usually more sensible than absolute values or "hard numbers."

The manager assesses costs and benefits and ecological and genetic hazards and risks associated with particular decisions and actions, including the potential effects and probability of extinction, genetic drift, inbreeding, outbreeding, contamination, introgression, hybridization, artificial selection, and domestication on the hatchery at all critical endpoints.

Prioritizing

Prioritizing is a complex, difficult, and controversial step in the management process because it is the step in the management process when folks get down to business (Sparrowe and Wight 1975). Prioritizing sets the stage for dividing up the financial resources that must precede the first on-the-ground management action. Generally, no money is targeted nor allocated until GCUs, personnel, programs, and activities are identified and prioritized. Every aspect of the program and process can be broken out for prioritization.

Programmatically, priorities are set among and within program elements and within and among management strategies and among specific management actions. These priorities are often based on cost:benefit or feasibility analyses (the action's relative importance to recovery, probability of success or likelihood of failure, demand on resources in terms of money, time, personnel, data availability, expertise, technology, and infrastructure), and hazard and risk assessments.

Ideally, in fisheries conservation and management programs, prioritization should be based on information emerging from the ecological domain. Prioritization occurs within the ecological domain, among ecosystems and river basins, among tributaries within river basins, among river reaches within a river, and across these levels of natural organization. Contributing information can include consideration of the probability of imminent extinction should a single catastrophic event occur, or chronic population decline as old individuals die without replacement by recruitment. Prioritization criteria can include such considerations as urgency, contribution to overall program success, probability of success, consequences of failure, cost:benefit, potential hazards and risks, and feasibility with respect to time, money, personnel, expertise, infrastructure, and knowledge and technology base. On the other hand, prioritization can be very subjective. For example, how valuable is that particular GCU in that particular river reach to program success? And, is the answer to that question determined by ecological threats of extinction, abundance, reproduction, and recruitment? Or, is the answer based on someone's political or economic agenda, perhaps associated with the acquisition of water rights needed to retain future water and power development options for a large city or other population center. There are some things the broodstock manager can control or influence and others that are beyond her reach.

Planning

Planning is a critical process step enabling the broodstock manager to harness and manage the complexity of her domain. The more complex the program, the more organizational structure or organizational levels it exhibits, and the more complex or structured the necessary planning becomes. To be an effective planner, the broodstock manager must master certain basic management concepts (Beveridge 1950; Phenicie and Lyons 1973; Krueger and Decker 1993; Meade 1989).

Scenario Planning

Scenario planning lends itself well to the broodstock manager's planning process. In orienting to better understand the situation and her role in it, she examines data from the information pool and artfully and logically creates a management scenario. A management scenario is the most likely of all possible scenarios based on the information available to her and integrated with her own knowledge and experience. In a way, a management scenario is like a working hypothesis. The management scenario provides sufficient direction and information for her to develop an operational plan for broodstock management. For example, imagine that information identifies, as part of the resource management problem, declining or failed fisheries productivity. In some of the cases, natural productivity can be relied on to recover with only minimal habitat improvements. On the other hand, some GCUs are in extreme danger of extirpation; biological and ecological indicators suggest that what natural productivity remains must be supported artificially through captive propagation and reintroduction. In the cases calling for captive propagation and reintroduction, several management opportunities exist, each of which may coincide with a particular type of program: (1) supplementation, either restoration or augmentation, (2) artificial genetic refuge population (AGR), or (3) experimental stocking. Using the management scenario, the broodstock manager identifies the most pertinent management opportunity from which she can identify the most applicable or preferred production strategy or approach. Then, the broodstock manager reduces the strategic goals into operational objectives, the achievement of which is outlined in the stepped-down operational plan of action. Of course, the plan is developed within a GHRMS and follows genetic management guidelines.

Implementation

The fifth step or phase in the broodstock management process is implementation, "where the rubber meets the road," so to speak, in the management process. Goals are accomplished through implementation of planned actions. Implementation may range from implementing the program's strategic plan or, at a lower lever of organization, to implementing the broodstock management (action) plan, stepped-down from the strategic plan.

Monitoring and Evaluation

The broodstock manager assesses her performance during the monitoring and evaluation phase of the process. Performance is the result of actions implemented over time to achieve objectives. Successful performance suggests that tactical objectives were achieved. Performance progress is defined by successful implementation over time of the various tasks defined in the action plan. Performance progress is monitored systematically, periodically through the course of the action (Are we getting there?) and evaluated at its conclusion.

During the implementation, monitoring, and evaluation phases of the broodstock management process, the broodstock manager documents and reports her progress in status communiqués to other system participants. Communication is critical to the success of fully integrated management programs where success or failure in one program element affects success or failure in other program elements and, ultimately, the success of the entire program.

Correction and Adaptation

Based on the results and what she has learned or was unable to learn from her decisions and actions, the broodstock manager may decide to revise or terminate this aspect of her broodstock management program. This self-correcting phase of the process, wherein the broodstock manager learns from mistakes made in the past and, integrating new information from the ever-evolving information pool, improves her performance adaptively for emerging needs. Collectively, information gained during these phases rapidly and efficiently enhances, broadens and deepens, the TME information pool, the programs institutional knowledge base.

Genetic Management Guidelines for Captive Broodstocks

A broodstock management action plan for gene pool maintenance is based on guiding principles of genetics conservation (Williamson and Wydoski 1994; Williamson et al. 1994). An action plan for broodstock management designed within a gene pool maintenance program is more likely to be successfully executed if it is grounded firmly in evolutionary theory, advanced animal breeding practices, and new technologies. Further, both genetic conservation and fish production goals and objectives are more likely to be achieved when management decisions are made and propagation activities and tasks are conducted within a GHRMS framework and applied from a genetics management perspective.

First Genetic Hazard Source: Choice of Donor Population

Choice of the donor population is part of the fisheries management process. Participants, including the broodstock manager, create a management scenario that best fits the data gleaned from the TME's information pool. The management scenario identifies a preferred production strategy (either natural or artificial propagation) to address the particular management opportunity arising from the management scenario (Riggs 1986, 1990). Choosing the appropriate donor population from which to found an artificial genetic refuge (AGR) or a production broodstock is an essential part of the scenario planning process.

Traditionally, the broodstock manager is rarely consulted regarding the choice of donor, the GCU from which to develop broodstock. Even more rarely does she make the decision. However, when dealing with populations on the verge of extinction, the earlier the coordination among principals and the more preemptive the intervention, the better. It is becoming more acceptable, even necessary, for administration, upper level managers, and field fisheries biologists to involve the hatcheries and broodstock managers as early in the management process as possible to make decisions together. Choice of the donor population should be based on its similarity to the population targeted for management, and they may, in fact, be one and the same. Kapuscinski et al. (unpublished data); and Kapuscinski and Miller (1993) present these three similarity criteria to aid in the choice of the donor source: (1) genetic similarity (in genetic lineage or phylogeny; genetic complexity diversity as amounts, types, organization, or structure), (2) phenotypic and demographic similarity (patterns in demography and life history, behavior, physiology, morphology), and (3) environmental similarity (as ecological equivalents). As we have seen, the purpose, management opportunity, or type may determine the priorities with which the similarity criteria are applied and the decision made with respect to the donor GCU.

The equation

$$V_p = V_g + V_e + V_{g*e} \tag{6}$$

represents a very instructive model explaining the various influences that determine the phenotypic variation (V_p) or the variation in performance of a fish population. The more alike two populations are in the three components contributing to their identity (genetics, V_g; environment, V_e; and the interaction of the two, V_{g*e}), the more similar they are one to the other.

> Taken together, adherence to these three similarity criteria will increase the probability that the hatchery offspring will have an evolutionary history as similar as possible to that of the extirpated or hatchery stock. In turn, this increases the probability that hatchery offspring will have high fitness, thus high potential for adaptation, in the target stock's environment. Adherence to these criteria will also reduce the risk of unintentionally reducing fitness of populations genetically interacting with released fish (Kapuscinski et al., unpublished data; Kapuscinski and Miller 1993).

Selecting the donor population to found a new captive broodstock is a complex and vital decision because it represents an initial and critical decision node in the broodstock management process. A critical decision node is a point where an error in decision making can result in the occurrence of genetic hazards with enormous consequences jeopardizing the success of the program.

The two most common genetic hazard types with which the broodstock manager must be concerned when choosing the donor source are type I, extinction, and type II, loss of within population variation. Type III, loss of among population variation or loss of stock identity is also possible if a dissimilar donor is chosen to supplement an existing captive broodstock. If the donor population is over-sampled to found a broodstock, the wild population could go extinct. Likewise, loss of within-population diversity could occur if the donor population is over-sampled to found the broodstock or the broodstock is founded from a very small, biased sample of fish from the donor population. The donor population could lose genetic identity if fish from another source are stocked into it and they interbreed or the broodstock could lose its genetic identity if additional fish from another source are used to supplement it in the hatchery.

The selection of the donor population must be purposeful, suitable, and feasible to accomplish genetic conservation and fish production goals defined in the strategic recovery and captive propagation plans while reducing risks of peripheral and unforeseen consequences associated with sources of hazards in the process. For example, the plans may call for protecting and conserving genetic resources and producing fish in sufficient quantity and quality for reintroduction to augment a declining target population over a specified period of time. In that case, genetic information includes the local, or targeted, population's unique genotype or identity, as well as its genetic relationships to other adjacent conspecific populations, which includes assessment of the total amount of diversity and structure or its partitioning among the various populations sampled. Plans may also consider the ecological and behavioral relationships with other species in the aquatic community.

If protection and preservation of a species' total genetic diversity is the principal genetics conservation goal (a gene bank or AGR population) and only one captive population can be developed and maintained in the program, it is important for the broodstock manager to conduct a comprehensive examination of the genetics data to facilitate her decision or choice of "which one." These data may reveal that most, or a biologically significant portion, of the species' genetic diversity is found within local populations. If such is the case, then the difference among populations is not as important. Then, it may be reasonable, particularly where resources and facilities are limited, to found the representative broodstock from a single donor population or GCU. On the other hand, if genetic diversity is highly structured within the species as a whole—as it appears to be among Pacific salmon—and a substantial portion of the diversity is found among populations within species, then the single broodstock might be developed by systematically combining representative individuals from the major populations or genetic groups comprising the species (Krueger et al. 1981; Vrijenhoek et al. 1985; Kincaid 1986; Allendorf and Ryman 1987). The benefit of the latter decision is more economical conservation of overall or total species diversity. The cost is the loss of the genetic identity of each population with, perhaps, a loss of their unique local adaptations. Additionally, there is the loss of genetic structure or diversity among groups (occurrence of type III genetic hazard). If there is substantial evidence that inbreeding is occurring within a high priority population of an outbreeding species, infusion of genetic material from other sources may present an alternative founding strategy. Clearly, the value of genetics information is crucial in making informed decisions where the issues are complex and there is little resiliency from the consequences of management failure. Without genetic information, the broodstock manager must acquire as much alternative information about the particular species as possible from other disciplines—ecology, biology, zoogeography, systematic, taxonomy, physiology, behavior, and life history, and make the best scientific decision possible.

Management responses or safeguards associated with broodstock choice depend on the purpose of the captive propagation program. The donor population should be from the population properly identified, characterized, and targeted for conservation or management. If the target population is not available, that is, the target population is extirpated or functionally extinct (below an abundance or demographic threshold), then the broodstock manager chooses another population, the closest ecological and biological correlate, often called the "nearest neighbor," based on the three similarity criteria described above.

Collective consideration of these similarity criteria when making decisions on the choice of a donor source increases the probability that the founding broodstock will have an evolutionary history, identity, and character as similar to the target population, extirpated or depleted, as possible. A superior decision regarding this hazard source increases the likelihood that broodstock and its offspring will possess high fitness and greater opportunity for survival and adaptation in the target population's environment. At the same time and for the same reasons, a proper decision at this juncture in the program reduces the risk associated with a type III genetic hazard if hatchery fish may interact genetically with the target population or any other population with which they might interbreed.

Restoration

Restoration of an extirpated population in a historical habitat is a major recovery goal. Restoration is accomplished through captive propagation of fish from the appropriate donor source followed by stocking into suitable, historic habitat. Since the original population is extirpated, literally or functionally, the donor source for captive propagation must come from somewhere else unless an AGR exists for the extirpated population. Kapuscinski et al. (unpublished data) suggest four possibilities for choosing a donor population in priority order based on similarity to the extirpated population (Kapuscinski et al., unpublished data; Kapuscinski and Miller 1993):

Priority 1. Select as the donor source or population the nearest neighbor. Similarity criteria for this determination are (1) genetic similarity based on results from genetic surveys, (2) similarity in environmental and ecological correlates (including geographic proximity or watershed complexity), and (3) similarity in demographic and life history patterns, as well as similarities in behavior, physiology, and morphology.

Modified Priority 1. If the nearest neighboring population is at risk, that is, not sustaining or extirpated then several neighboring populations may need to be considered as multiple sources for developing a production broodstock to be the foundation of the restoration effort, particularly if fish are to be reintroduced into substantially altered habitat. This immediately presents a type III genetic hazard (genetic contamination of breeding groups) to the target population and all potentially interbreeding populations with which this synthetic population may come into contact. The results of implementing such an alternative are unpredictable. This option should be viewed as experimental and the effects should be monitored and evaluated carefully.

One way to minimize a type III genetic hazard in the hatchery is to maintain the identity of the various populations by marking and pregenotyping each individual broodfish, by population, at least. Population integrity is maintained in this fashion. In contrast to multiple stock maintenance is production and use of a single hatchery stock developed by combining offspring from fish of several sources as described by Krueger et al. (1981). Once a mixed lineage broodstock is produced in this manner, partitioning out genetic differences among different groups is impossible (Lacy 1987). The difficulty and cost of maintaining separate broodstocks in a hatchery, each representing a unique genetic resource, has been greatly reduced with the ability to mark individual broodfish (e.g., using passive identification transponder [PIT] tags). This capability is further enhanced by rotating wild fish through the production system. Once a wild fish contributes to the genetic foundation of the broodstock, it is PIT-tagged, returned to the wild, and replaced by an untagged wild adult. Hard et al. (1992) explain that this strategy has dual benefits to the recovery program of listed species. Avoidance of using adults that have already been used in founding AGRs and production broodstocks will prevent unnecessary reductions in N_e, while limiting to a single generation the exposure of any natural fish to captive conditions. Research is in the planning stage in the upper and lower Colorado River basins to cryopreserve gametes from wild threatened and endangered fish before release.

Priority 2. If suitable neighboring populations or GCUs are unavailable, then use of a nonneighboring wild or naturalized population is recommended. The population selected should meet, as closely as possible, the four similarity criteria. All decisions should consider the threat or risk of extinction in the wild to any donor source by removing adults to facilities.

Priority 3. Use a hatchery stock, one derived from the extirpated population, if possible. If not, use the hatchery stock most closely related based on genetic data, dispersal patterns, and stocking history.

Priority 4. The last choice is the use of an unrelated hatchery broodstock that has a stocking history successful in similar environments or in a diversity of environments (Kapuscinski et al., unpublished data).

Augmentation

Maintaining or enhancing declining populations in historical habitat is a recovery goal. Captive propagation and stocking, in conjunction with habitat protection and restoration, may aid in accomplishing this goal by reversing declining population trends and improving population status. Augmentation is different from restoration because the target population still exists functionally, albeit in decline or not sustaining. This means genetics, demographic, life history, and ecological data can and should be obtained to assist in making propagation-related decisions.

It is important to emphasize that failure to detect genetic differences among groups of fishes through a particular analytical method does not necessarily imply "no difference," "identity," or "equality" of the groups, particularly if other types of data indicate potential differences. Characterization data can be based on differences in life history patterns, behavior, physiology, demography, and environmental and ecological variations. In some cases, differences among presumptive genetic taxa for important survival traits can be demonstrated under controlled experimental conditions (Williamson 1983, 1986). Ignorance of such data can be counterproductive where management actions increase chances of type I and type III genetic hazards (Kapuscinski et al., unpublished data; Kapuscinski and Miller 1993). Developing multiple data sets derived from different analytical methods and sources to define populations is highly recommended.

Priority 1. If the target population is self-sustaining and removing adults does not adversely affect it (types I and type II genetic hazards), then the target population is the donor source (Krueger et al. 1981; Kapuscinski et al., unpublished data; Kapuscinski and Miller 1993). This minimizes the chance of a type III genetic hazard. Although broodstock development may be a suitable activity at this point in the process, depending on priorities, stocking on top of sustaining populations is not recommended. Nothing is to be gained in recovery by doing so since a wild, naturally sustaining population has the highest, most valued status in the recovery program. Securing and protecting habitat might be a better, more productive choice for a sustaining a population if progress toward recovery is an issue.

Priority 2. Exercise the nearest neighbor option (See Restoration, priority 1).

Priority 3. If the nearest neighbor option is not feasible, then the other options presented under Restoration should be considered. Those options best meeting the similarity criteria should be considered first.

Artificial Genetic Refuges

Artificial genetic refuges are developed to protect a particular, high priority, wild population from imminent extinction and conserve its unique genotype while retaining future management opportunities (i.e., fish produced for reintroduction). That particular population is the donor population and provides the fish from which to found the AGR.

Artificial genetic refuges can be a legitimate end in and of themselves, particularly if the target population is in immediate danger of extinction. That decision is driven by population status, priority, and availability of funds, technology, personnel, and facilities. But, the real value of reserves in "rescue and return to nature" endangered species programs is a place to hold fish until suitable habitat can be obtained and progeny produced in captivity can be reintroduced.

If the fish production option is exercised to meet other program needs at a later date, then those options may be developed in terms of objectives and tasks when AGRs are formed. For example, if a naturally sustaining population is required to meet recovery goals and reproduction and recruitment are nonexistent or severely reduced in the target population, then fish production required to implement the stocking option may be selected as a alternative management opportunity. Or, if additional high priority information can be obtained through small, strategic reintroductions with minimal risk, then the experimental stocking option may be suitable. Regardless, if wild fish are going to be used in the AGR or fish production option, genetic conservation theory supporting these two propagation activities is the same. The initial tactical objectives and tasks are the same as well. It is also true that populations with a high priority for the AGR management opportunity are generally the same populations that are most in need of supplemental reintroductions to aid in restoring or augmenting natural production.

Although development and implementation of the captive propagation strategy targets supplementation, the same processes apply to AGR development. For simplicity, development and implementation of the captive propagation option will be conducted as if any AGR population is appropriate, at some point, for hatchery propagation and stocking. Consequently, selection criteria for an appropriate donor population are the same for AGR development as for production broodstock development.

Experimental Stocking

The history of captive breeding programs in recovery efforts is brief, particularly stocking endangered fishes. Reports of successful stocking efforts are even briefer, particularly regarding Colorado River fishes (Brooks 1986a, 1986b). Consequently, little is known about the use of stocking in recovering endangered fishes. Clearly, there are genetic hazards associated with captive breeding and stocking. However, expected results from stocking and assessments of the risks are made difficult by the lack of basic information on genetics, life history patterns, limiting factors, controlling variables, and habitat suitability. This information must be obtained before stocking can be expected to play a positive recovery role (Burdick 1992).

The situation in the upper Colorado River basin presents a justification for stocking that has no direct or immediate relationship to hatchery supplementation. Rather than stocking to restore extirpated populations or to augment declining populations, the goal of experimental stocking is to introduce a limited number of marked fish into the system to obtain critical core information essential in increasing the likelihood of success of other recovery activities, including habitat improvement and restoration and augmentation stocking. Caution in using this tool is compulsory. Its success depends on experiments designed well enough to produce the information required while minimizing the risks of genetic hazards associated with it. One of the first steps in reducing risks is in selecting a donor population from which to found the production broodstock (USFWS 1992).

Priority 1. If fish are going to be stocked back into an existing population, then they must be derived from that population. This would correspond to priority 1 under Augmentation. Capturing wild fish, developing and spawning broodstock, rearing their offspring, and marking and stocking the offspring must be carried out in a manner consistent with genetic conservation principles and recovery goals. This includes assessing and reducing risks of genetic hazards associated with captive propagation followed by monitoring and evaluating the results.

Priority 2. If fish are going to be stocked into habitat no longer containing the target population, then the donor population would be the nearest neighbor population based on its conformance to the similarity criteria. This would correspond to priority 1 under Restoration. Again, taking wild fish, developing and breeding broodstock, rearing, and stocking marked offspring would be in conformance with protocols designed to protect genetic resources and reduce risks of genetic hazards associated with captive propagation.

Priority 3. Other options are less desirable but would conform to the priorities as listed under Restoration and Augmentation.

There are three rules of thumb applied when evaluating, setting, and changing priorities for choosing a donor population. If wild or naturalized populations are donor sources to be used in captive propagation, then (1) their genetic and demographic characteristics, as well as their life history patterns and productivity must be protected, (2) the donor sources must meet the similarity criteria and are least susceptible to genetic hazards, and (3) N_e of the founder population should be maximized in each generation of propagation but without reducing that of the donor population to hazardous levels (Kapuscinski et al., unpublished data; Kapuscinski and Miller 1993).

It should be noted that when dealing with endangered species, there may be populations whose status is so precarious that the risk of extinction is imminent. In those cases, at least some of the remaining adults should be taken into captivity to develop propagation methods and secure artificial refuge populations. These adults should be marked, pregenotyped, and used in breeding programs to produce offspring with which to form backup AGR populations and production broodstocks. This is an iterative process. As wild adults contribute offspring to the captive populations, they are returned to the wild and replaced in the captive breeding cycle with new, wild recruits taken into captivity. This creates a dynamic captive breeding program based on wild populations and that tracks natural evolutionary events within and among these wild populations through time. This strategy should buffer the donor and founder populations against undesirable genetic changes inside and outside the hatchery.

Second Genetic Hazard Source: Collecting Founders

Assuming that the broodstock manager made the best possible choice in selecting a donor population, then collecting fish from it as founders is the next most critical decision and, in sequence, is the second most hazardous source in executing the broodstock management action plan. "No mating, rearing, or release schemes can be devised to prevent unnatural genetic alterations if too few or a biased (i.e., nonrandom) sample of fish has been collected from the spawning run." (Kapuscinski et al., unpublished data; Kapuscinski and Miller 1993). Just as it is important that the choice of the donor population conform to the similarity criteria, so is it important for the broodfish to be collected in such a way that there is no departure from the donor population's identity (as determined by the similarity criteria). Just as in choosing a donor source, when collecting broodfish or their gametes, it is important to know if the donor population has a history and demonstrates biological evidence of previous stockings and interbreeding. If so, were released fish, including wild adults returned following hatchery spawning, marked or tagged? To avoid genetic hazards (type II and type IV) associated with repeated use of the same wild adults or their hatchery offspring in developing a broodstock, information on prior releases is important. The number of fish collected and the manner or method of collection is determined by the necessity to minimize changes in allele frequencies and departures in life history patterns between the donor and founder population. By managing the number of founders collected and their method of collection, the effects of sampling bias in reducing the risk of genetic losses can be controlled and minimized. The number of fish collected affects random processes that change gene frequencies, and collection protocols may impose selective forces that alter gene frequencies.

Number of Founders to Collect

The rule of thumb, regardless of the actual number collected, is to do it in a manner that maximizes N_e of founders—the more the better, up to a point. Artificial genetic refuge populations, backup refuge populations, and production broodstocks should be founded with a sufficient number of individuals to reflect the genetic composition of the donor population from which they were derived and to reduce drift and the rate of inbreeding within the captive population (Allendorf and Ryman 1987). Genetic hazards have tremendous effects on small sample sizes that reduce genetic diversity within populations and alter genetic variation and organization among populations within the species. The number of adults from a donor population required to found a captive population should have some relationship to the donor population's estimated N_e. Further, in most cases, a minimum number should be collected that allows retention of genetic character in both the donor population and founder populations. This number requirement is most likely to be met when broodstock are being collected from a large population. However, where the donor populations are small, effort should be made to collect at least a "minimum" number of founders to reduce risk of genetic hazards associated with it but not so many as to jeopardize the donor population.

Kapuscinski et al. (unpublished data) and Kapuscinski and Miller (1993) recommend two guidelines to arrive at a "minimum" number of founders to collect in order to develop a broodstock or AGR. First, as example, they present a table that provides the number of founders to collect for a given population size that assures 90% confidence that an allele with a frequency of 0.2 in the donor population will not be outside the frequency range of 0.16 and 0.24 (a 20% fluctuation) in the founding broodstock. The authors believe that these criteria will be adequate to retain the presence in both the donor and captive population of the more common alleles, particularly if additional wild fish are periodically used and hatchery progeny are stocked back into the donor population. The authors also present a formula for calculating sample size requirements and lists sample sizes for several sets of criteria. A number of other experts (Allendorf and Ryman 1987; Ryman and Ståhl 1980) recommend that when N_e equals N, rare alleles ($P \leq 0.05$) will be retained in the founder's offspring with 95% probability with 30 spawners, 15 males and 15 females. Kapuscinski et al. (unpublished data) and Kapuscinski and Miller (1993) suggest 50 spawners, equal numbers of males and females, as a minimum to provide more confidence that the likelihood of loss of a rare allele will not exceed 5% in the broodstock's first generation offspring.

As far as a guideline for the maximum number to remove from the donor population, the understood limit is that number above which undesirable loss of genetic variation in the donor population will result. Kapuscinski and Miller (1993) suggest that an effective population size of 50 must remain in the donor population for the same reason that they suggest 50 as the minimum population size for founders. The authors, at least for Pacific salmon, suggest that the maximum number of founders to collect is no more than 50% of the donor population.

Size of Remaining Donor Population

The broodstock manager must exercise caution in removing wild adults as founders from a donor population whose sustainable status is precarious or uncertain. Reducing N_e of the donor population below some critical threshold will have the effect of reducing genetic variability (type II genetic hazard) and altering critical life history patterns, which could drive the population to extinction (type I genetic hazard). Estimates of census population size and N_e of the donor population would permit a rudimentary assessment of risk of genetic hazards to donor populations associated with removing adults from the wild. Unfortunately, effective population size estimates of the target populations of most imperiled fishes in many environments range from unknown to uncertain. To avoid removing a proportion of fish that might adversely impact the donor population, a limited number of fish are removed during any one spawning season. These fish are spawned in captivity and their offspring reared to develop first generation (F_1) captive production broodstocks and backup AGR populations. Once spawned and their genetic contribution obtained, the wild fish, each uniquely marked, are returned to the wild. They are replaced in the captive breeding cycle during the following breeding season with additional unused recruits identified as wild by lack of a mark or tag. The cycle is repeated as often as is feasible. In this way, a broodstock is developed over time that is representative of the donor population while exposing the donor population to minimal risk to types I and type II genetic hazards. At the same time, mark–recapture data of adult fish may be useful in estimating adult population size and N_e, vital demographic characteristics useful in developing operational plans, defining recovery goals, and evaluating success.

Founder Collection Protocols, Procedures, and Practices

The manner of collecting founders is as important as collecting the appropriate number. Founders are collected randomly as an unbiased sample of the naturally spawning donor population, if possible. Important demographic and life history patterns are factored into the sampling design and methodology. Among those performance features are reproductive behavior, mating strategy, breeding season or timing, sex ratio, age structure, fecundity, longevity, and other traits that may be related to adaptation and long-term survival. This knowledge is critical if selection, intentional or not, is to be avoided during founder collection. For example, if founders are to be sampled from the donor population in an unbiased manner, then founders should be collected randomly throughout the spawning period. For the broodstock manager to accomplish this, she must have accurate data on the timing of spawning and the distribution of spawning adults throughout that period and spawning area. These data may be commonly available for many species, particularly commercial species like Pacific salmon, but are usually nonexistent for many imperiled native fishes.

Stratified random sampling, discussed in more detail by Kapuscinski and Miller (1993), as well as in any basic statistics text, is a sampling method or design to aid the broodstock manager in collecting founders in a manner that more closely approximates true random sampling, a feat that is far more difficult to accomplish in nature, even with much more information than that which typically exists for most imperiled native fishes. For example, if the spawning distribution is estimated reasonably well, in terms of time, abundance, and location, then the broodstock manager can plan sampling trips for founders in proportions that roughly mimic characteristics of the natural spawning run, in time, number, and location. For reasons discussed earlier, it is important for the broodstock manager to obtain, during the overall broodstock development process, as many males as females and a cross section of age-classes. This does not mean that equal sex ratios must be collected every time founders are collected, only that over the lifetime of the program, an equal sex ratio is achieved, and that may be over a period of many years. Protocols, procedures, and practices should be developed that reduce the stress and mortality imposed on fish captured, transported, and held for spawning (Carmichael 1984; Carmichael et al. 1984a, 1984b). No doubt, these efforts represent a substantial increase in critical management knowledge, intensity, and expense; indeed, the complexity of management increases. In addition, such efforts may impose considerable stress on fish captured and handled in the process.

Kapuscinski et al. (unpublished data) and Kapuscinski and Miller (1993) emphasize the importance of properly sampling wild adults or their gametes when collecting founders for a captive broodstock. For reasons mentioned earlier, failure to do so represents type II and type IV genetic hazards. Repeated use of the same broodfish or the same families in the breeding and stocking process results in overrepresentation of those particular individuals in subsequent generations. The result is a loss of within-population diversity. The potential for artificial selection and domestication selection also increases with incorporation of hatchery fish, offspring from the captive broodstock, as "new" founders. How can the risk of these genetic hazards be reduced and managed? Obviously, some mark or tag must be applied to fish, which are returned or released into target habitat, that allows differentiation among naturally produced fish and hatchery fish. In particular, wild adults that have been used previously as broodfish must be distinguished from wild fish that have not. Fish produced in captivity must be marked prior to release with marks that permit their identification or separation from fish produced in nature. In so far as identifying offspring produced in nature by stocked fish, that requires a natural, evolutionarily neutral, heritable mark easily and economically detectable by laboratory analysis. This technology is being developed.

Kapuscinski et al. (unpublished data) and Kapuscinski and Miller (1993) suggest that the continued use of wild fish in the captive breeding and broodstock program is justified only as long as wild fish are available and as long as their progeny survive following stocking to do what it is that management intended. In most cases, management intends for stocked fish to survive and enter the reproductive population. If stocked fish survive and reproduce in nature at the desired rate, then managers assume they are fit and are a successful match with their environment. If, on the other hand, stocked fish are not surviving to reproduce, a complete evaluation of the entire program and its strategy is necessary. Certainly, questions of habitat suitability or the effects of rearing and handling practices, particularly the stress associated with transportation and delayed mortality, must be asked and answered satisfactorily before stocking is continued (Carmichael 1984; Carmichael et al. 1984a, 1984b). If program adjustments are made that indicate continuation of stocking, then there may come a time when some balance between using wild fish and hatchery returns as broodstock is struck. A balance is the risk between the genetic losses associated with the use of hatchery fish versus the losses in population productivity, resulting from the complete depletion of wild fish—systemic losses in production that could lead to population extinction. Regardless of the final management decision, two critical elements must exist in any captive breeding and stocking program with imperiled fishes—an adequate marking or tagging system and a rigorous system to monitor and evaluate performance of stocked fish.

There are theoretical, practical, and logistical reasons for deciding between collecting wild adults, their gametes, or immature individuals, to develop a captive broodstock. If the program's need to forestall extinction is immediate production, then wild adults must be used to produce offspring for stocking (some sturgeon species reach maturity at 20 years of age). At the same time, their offspring should be used to develop AGRs and captive production broodstock. If the need for production is not urgent or is poorly defined, or removal of wild adults is not possible, then gametes taken from wild fish can be used to establish a captive broodstock with the added potential of using fish surplus to that need for other predefined reintroduction and recovery purposes. The same guidelines for collecting adults are applied to the collection of their gametes with the added technical complication of predictably extracting gametes in the field and, if immediate fertilization is called for, the difficulty in applying complex mating strategies more easily applied in controlled settings. For populations on the brink of extinction, the need for immediate increases in productivity in nature may outweigh some genetic concerns associated with the fear of genetic and demographic losses associated with collecting adults, transport, and captive propagation.

Kapuscinski et al. (unpublished data) and Kapuscinski and Miller (1993) suggest the development of a captive broodstock program using immature individuals when the following appertain: (1) the number of adults in the donor population is very small and the stocking goal is to give the population a temporary boost; (2) the young individuals can be collected so that they accurately represent the donor population in all essential characteristics; and (3) they are used as broodstock for only one generation. The principle concern in collecting immature fish to use as broodstock is the uncertainty of their familial relationships. Depending on how they were collected, such individuals could represent a biased sample of the donor population, thereby constituting a type II genetic hazard, loss of within-population diversity.

Minimum Acceptable Value for the Effective Population Size

The guiding principle in determining the minimum acceptable value for N_e is that a founder population should be large enough to accurately reflect the genetic character of the natural population from which it was derived. The concern of a limited N_e is loss of rare alleles, which is particularly likely with small founder populations. At least 30 founders are required to be 95% certain that an allele with a frequency of 0.05 will be present in founders (Allendorf and Ryman 1987). Ryman and Ståhl (1980) suggest a minimum founder population of 25 males and 25 females. Lande and Barrowclough (1987) suggest an N_e of 500 is necessary to conserve evolutionarily important variation. Lacy (1987) presents an interesting approach to conserving genetic variation, principally retaining high collective heterozygosity levels through population subdivision and maintenance, perhaps using multiple facilities.

With most species and populations targeted in recovery programs, population sizes are small. Employing management techniques already discussed will be the best way to approach an acceptable effective population size for broodstock—use as many adults as possible in paired matings, equalize sex ratios, reduce variance in offspring numbers among family lots, continuously introduce new wild adults into captive populations over time, and increase the generation interval. It may be advisable to maintain no more than 20 adults (10 males and 10 females) per population in a facility at any one time. If these are replaced periodically every two to three years, continuously, then a larger N_e may be approached. When access to new wild recruits is limited or not possible, then, increase the generation interval or productive lifetime of captive wild adults and their F_1 offspring reared as parental replacements; in other words, the broodstock manager reduces broodstock losses from attrition in the hatchery. Development of cryopreservation technology may greatly advance this effort.

There are too many uncertainties about a "universal" lower limit or minimal N_e to recommend to broodstock managers in every situation. Some recommendations remain firm. The broodstock manager must understand the theory of population dynamics and animal breeding and apply the principles and practices in operations. She must understand the concept of N_e and its relationship to genetic diversity. She must understand the fish, the natural patterns of diversity in genotype and phenotype, and their relationship one to the other, including life history traits or performance characters. These patterns of variation have a direct relationship with N_e and awareness of this relationship and effective management of diversity is the most practical way to limit the effects of genetic hazards associated with reductions in N_e in the facility.

Third Genetic Hazard Source: Mating Schemes and Number of Broodfish to Breed

Once a donor source has been chosen and founders collected and transported as marked individuals to the broodstock facility, the third step within the domain of the broodstock manager is the act of fish reproduction, exchange of gametes among fish to form the individuals of a new generation. Reproduction is a major source of genetic hazard and another critical decision node, the point at which type II, type IV, and even type III genetic hazards are likely to occur. Thus, reproduction must be under the control of the broodstock manager. Knowledge of the natural processes governing these genetic hazards, particularly the knowledge of effective population size, is essential here because N_e is an indicator of the rate at which genetic diversity within a population is lost, the most common genetic hazard (type II) for small populations. Genetic guidelines to reduce the risk of this genetic hazard focus on breeding schemes that grant the broodstock manager control over N_e. In a gene pool maintenance program, one point the broodstock manager drives home to the fish breeding team is to maximize N_e over the life of the program. Beyond that, however, is the knowledge that the broodstock should be a duplicate of the donor population.

Maintain Existing Life History Patterns

Fish bred in a hatchery should be a representative sample of the donor population. They should reflect accurately genetic, life history, and population characteristics (means, ranges, and variances of important performance traits) existing in the population targeted for stocking, extirpated or not. Performance data should be collected as soon as possible (and periodically thereafter) from the field and literature. If there is no life history information for the extirpated population, data derived from a nearest neighboring population meeting similarity criteria may be used as estimates. Examples of vital population characteristics are migrational patterns of young and adults, spawning time and distribution, spawning behavior, adult census, age structure, age to maturity, longevity, habitat preference, fecundity, birth and death rates, growth, survival, condition, and morphology. These very same data should also be obtained in the hatchery.

Maximize the Effective Population Size

Recall that N_e is defined as the size of an ideal population that would lose genetic variation at the same rate as the population under consideration (i.e., the donor population or the broodstock) (Gall 1987). An ideal population has a 1:1 sex ratio and all breeding individuals have an equal probability of being the parent of any progeny individual (Allendorf and Ryman 1987). A 1:1 sex ratio simply means that each male is bred to the same number of females. As would be expected in real fish populations, there are departures from a 1:1 sex ratio, random age distribution, and zero variance in fecundity and fertility among breeding adults. This means N_e in real populations is almost always less than the number of breeding adults. Again, the formula required to estimate effective population size, N_e, is

$$N_e = 4(N_f)(N_m)/(N_f + N_m)\,, \tag{7}$$

where N_f is the number of females used in breeding and N_m is the number of males used in breeding. For example, in a broodstock of 99 females and 1 male, the N_e is 4, while the number of breeding adults or census population is 100. This means that drift would occur in this population as if it were composed of 2 males and 2 females (Allendorf and Ryman 1987).

Effective population size of the broodstock should be maximized in every hatchery generation to prevent genetic erosion and deviation from the donor source due to a type II genetic hazard, loss of within-population diversity. Obviously, maximizing N_e at the outset in collecting and breeding the first founders is critical. The following is a list of critical decision points where a broodstock manager can employ safeguards that maximize N_e, reducing the risk of type II, type III, and type IV genetic hazards (Kapuscinski et al., unpublished data; Kapuscinski and Miller 1993):

1. Breed as many unrelated parents as possible, including all of the least numerous sex; individually mark or tag each fish prior to spawning.
2. At a minimum, breed one male per female.
3. Whenever possible, use factorial matings to increase the genetic contribution of the most numerous sex available, particularly the diallel cross scheme for extremely depleted imperiled populations.
4. Minimize or reduce variation in family size among all matings; monitor and collect data on natural patterns of variation in performance characters among families, particularly family size on the hatchery.

Except under very controlled breeding practices (pedigree or paired mating with uniquely marked individuals) and segregated rearing conditions, it is difficult to know or calculate N_e with much accuracy. The best estimate, particularly in wild populations, may be the census size (Ballou and Cooper 1992). A rule of thumb is to maximize N_e during the "turning" or creation of every new generation. It is most important to start the broodstock with as great an N_e as is feasible.

Prespawning Considerations

If possible, all fish brought into the hatchery for captive propagation should be PIT-tagged or in some way marked for individual identification, at least by the time they are to be mated or spawned. Marking fish with unique tags permits development and implementation of customized breeding strategies or mating schemes necessary to conserve genetic variation in captive breeding programs. Secondary marks or tags may also be employed for extra security and the convenience of the managers. If feasible, each broodfish should also be genetically characterized to provide its own genetic profile. Fin, muscle, gametes, and blood tissue can be sampled nonlethally for genetic analysis prior to breeding. This may be done at the same time the initial genetic surveys are conducted or the founders are collected and PIT tagged. A genetic profile or fingerprint is part of the database developed and maintained for each broodfish. The individual's database includes a performance profile as well, including origin, sex, age, length, weight, condition, fecundity, genetic contribution, and unusual characteristics and identifying marks. Both genotype and phenotype profiles correspond to the individual's unique identifying mark, perhaps its PIT tag number.

Controlled Spawning

It may be important for the broodstock manager to use or develop methods to promote, induce, and control reproduction supporting paired matings and family integrity. There are, for a number of species, the PPPs to induce or manage spawning to accommodate controlled breeding; generally, greater control requires more labor-intensive efforts and advanced technologies. Advanced technologies and methods focus on diet development and nutrition management to improve fish condition and gonad development coupled with techniques inducing gametogenesis and enhancing fertilization opportunities (short-term gamete extension or long-term gamete storage, almost exclusively sperm cryopreservation; Tiersch et al. 1997, 1998, 2000; Gorman 2000).

Mating Schemes

Designing mating strategies to maximize N_e is most useful if the broodstock manager can control breeding to some degree, usually through application of PIT tags, induced gametogenesis, manual spawning, controlled fertilization, and segregated incubation and larval rearing. Gamete extension and cryopreservation are technologies that also permit greater control over breeding. The following mating schemes rely on the broodstock manager's control, to some extent, of the above PPPs and availability of mature fish and gamete quality.

Random Mating. Some experts are justifiably concerned about nonrandom mating in captive breeding programs (Kapuscinski et al., unpublished data; Kapuscinski and Miller 1993). Underlying their concerns is the theory that departures from random mating are undesirable if genetic loss is to be averted. Nonrandom mating in hatcheries is generally undesirable, not because fish breed randomly in nature but, because nonrandom mating in a facility is due to influences most likely different from those occurring in nature. Consequently, genetic diversity of hatchery fish may be lost with a potential for reduced fitness when stocked in a different environment. Since, in most cases, scientists do not know the details of reproductive behavior in nature, the broodstock manger attempting to mimic the unknown is likely to be off the mark, thus, creating high risk of genetic hazards occurring, perhaps resulting in disruption of naturally evolved breeding behaviors and reduced fitness. Better then, according to some, that the broodstock manager resort to methods, though unnatural (i.e., random mating), that are known to aid in maximizing N_e for a given number of fish. Imperiled fish are so reduced in number "due to harvest, habitat degradation, and passage problems, this same mating pattern (naturally evolved breeding behaviors) may cause a reduction in N_e below a critical value for the population. In this situation, a 'natural' pattern may not be the best in the face of new exploitation and environmental pressures." (Kapuscinski and Miller 1993).

The success of a random mating strategy in meeting both production and genetics goals in a captive broodstock program hinges on two major requirements. First, a very large population of mature broodfish, perhaps on the order of 2,000 adults (Allendorf and Ryman 1987), is necessary. Second, the broodstock manager must be able to sample individuals in a random, unbiased manner when picking out broodfish to spawn. The difficulty in meeting these requirements in most captive broodstock facilities, particularly of imperiled native fishes, is another reason leading many broodstock managers, not to mimicry, but to more complex, labor-intensive, controlled breeding methods that require fewer fish to achieve the same N_e. Pooling gametes, sperm or eggs, presents problems in captive broodstock development, greatly reducing control of N_e. Thus, mating schemes to develop captive broodstock based on pooled gametes will not be discussed in detail in this chapter. Systematic pooling of gametes may have greater use in producing offspring for reintroduction for a number of reasons. Various strategies based on gamete pooling are described by Doyle et al. (1995), Kapuscinski et al. (unpublished data), Kapuscinski and Miller (1993), Withler (1988), and Gharrett and Shirley (1985). The following mating schemes depend on substantial control of reproductive behavior to maximize N_e.

Simple Paired Mating. Simple paired mating results from crossing one male with one female, preferably with no individual used more than once. This scheme, producing unique families, is considered to be the minimal standard for developing broodfish. If fish are used that can be marked and identified individually and their young identified by family, then pedigree matings and programs can be developed, an even higher standard. Simple paired mating is most useful when a relatively large number of unrelated mature fish are available to spawn, sperm quality is high across males, and egg supply among females is not limiting (Kapuscinski et al., unpublished data; Kapuscinski and Miller 1993; Tave 1986). Where numbers of fish, sexes, or gametes are limiting, more complex mating schemes are necessary.

Factorial Mating. Factorial mating also involves mating one male with one female but may require spawning an individual of the least numerous sex with multiple partners forming half-sibling families. This scheme is most useful when the number and sex of broodfish is limiting because it results in the use of all available broodfish and increases the diversity of alleles passed to the next generation. Factorial mating is a more complex, labor-intensive, controlled breeding strategy than the simple one-on-one, and its execution is enhanced by gamete extension and subdivision methods. For example, when females are the limiting sex, then an egg lot from a female is partitioned or subdivided into aliquots, each of which is fertilized by a different male until all males are used.

Diallel Mating. Diallel mating is a variation of the factorial mating scheme and is most useful when developing captive broodstock from an imperiled donor population whose numbers have dropped drastically low. In that case, all fish are used to make all possible combinations of matings to maximize genotypic combinations according to a matrix format. Each male is bred with each female. Although not increasing N_e above that obtained in simple factorial matings, it does maximize genotypic diversity in a population that has lost some allelic diversity and likely to lose more shortly. This has the effect of producing new genotypes, some of which may confer high fitness in the progeny (Williamson and Wydoski 1994; Williamson et al. 1994; Modde et al. 1995; Kapuscinski et al., unpublished data; Kapuscinski and Miller 1993).

Development of gamete extender solutions or media and storage techniques have greatly facilitated execution of complex factorial mating designs for imperiled fishes in the desert Southwest, including some of the native salmonids. Factorial mating schemes, and in extreme cases, diallel matings, are recommended for breeding some endangered species and are currently being employed in various captive propagation programs to aid recovery of threatened and endangered fishes in the Colorado River Basin. These include razorback sucker, Colorado pikeminnow (Williamson 1992; Kapuscinski et al., unpublished data; Kapuscinski and Miller 1993; Williamson and Wydoski 1994; Williamson et al.1994; Modde et al. 1995), bonytail, and a few other imperiled fishes of the southwestern United States such as the Chihuahua chub. The downside to the diallel or any factorial breeding strategy is that, if not managed properly, especially when used to develop captive broodstock, inbreeding in the second and subsequent generations can follow through mating of full-sibling or half-sibling families. However, the longevity of the four target species in the aforementioned recovery programs is such that wild fish and F_1 generation offspring may suffice as broodstock for as long as 20 years. In addition, wild adults are continuously introduced from the donor populations (as long as they last or permits are let to collect them) to the broodstocks to reduce the risks of inbreeding and selection (type II and type IV genetic hazards).

Variance in Family Size. With respect to its influence on N_e, family size is simply the number of offspring produced by a mating pair, one male and one female, that survive to reproduce themselves. Variance in family size (V_f) among matings reduces N_e in both natural and captive populations and results in erosion of genetic variation. Family variance may result from demographic forces, natural differences in fecundity, or through selection, a type IV genetic hazard, which, theoretically, could lead to a type I genetic hazard, extinction. Variation in family size in natural populations is assumed to be the result of natural variations in fecundity and natural selection. Selection can also occur in hatcheries leading to variance in family sizes, the number of offspring among different families, and serious reductions in N_e. The result is evolution of hatchery populations quite different from the donor population. Fortunately, in some cases, variance in family size can be controlled to some degree by captive propagation, particularly where controlled matings among marked individuals are possible. As soon as the offspring have been reared through the heavy mortality stages of captive propagation, then the numbers among family lots can be equalized. This will require promotion of survival in the less productive families and rational culling among the most productive families (Nelson and Soulé 1987). The procedure can actually increase N_e beyond the total number of parents used because the ideal population is expected to have random differences in offspring among individuals. Equalizing the number of offspring among matings (reducing family variance to zero) can almost double the N_e beyond the total number of breeding adults (as much as $N_e = 2N - 2$). Both Allendorf and Ryman (1987) and Kapuscinski and Miller (1993) have good discussions on the effect of family variance on N_e.

With respect to pooling gametes, it is a common practice in many hatcheries to mix milt from a number of males to fertilize the eggs from one or more females. The problem with mixing gametes is that it prevents the manager from knowing what N_e is and controlling it. The N_e would likely be less than the number of fish used in the matings because of differential viability of gametes among individuals and even sperm competition within and among individuals. Gamete pooling is a management decision and action that constitutes type II and type IV genetic hazards. To maintain a large N_e and minimize fertilization-induced family variance, hatchery managers should mate one male with one female (Tave 1986; Allendorf and Ryman 1987; Withler 1988; Allendorf 1993; Williamson and Wydoski 1994; Williamson et al.1994; Modde et al. 1995). In situations where N_e can be increased through factorial mating and females are limiting, the milt of a different male can be used to fertilize each separate subset of eggs from an individual (Femscinski et al., unpublished data).

Natural Spawning. Some broodstock managers permit captive broodstock to spawn in open ponds or tanks with very little, if any, controls or assistance. This is often referred to as natural spawning, an obvious and misleading misnomer. Generally, there are three reasons why a broodstock manager conducts or permits this type of mating strategy with imperiled fishes. The first is legitimate, and the other two are not. In the first case, resources (time, money, labor, facilities, and expertise) frequently limit the broodstock manager's ability to develop or execute more controlled breeding schemes. The second reason, ignorance, is not legitimate because it is not management. The third reason appears benign. It is the case of the broodstock manager believing that the less intervention or control exerted over the reproductive process in her facility, the more natural it is and the better it will be for the resource. That is mysterious or magical thinking based on an archaic romantic philosophy that has been scientifically discredited and a luxury that both managers and fish can no longer afford (Wilber 1996). As philosophies or guiding principles, laissez-faire management and applied retro-romanticism are major sources of genetic hazard in a captive broodstock facility for imperiled fishes. Studies have shown that fish spawning in ponds is nonrandom; far fewer fish actually spawn than are represented by the census size, *N*. Those that do spawn may do so repeatedly and finally, family variance is very high. These natural phenomena result in considerable reduction in N_e and, since mortality among offspring is usually high, surviving progeny are most likely the product of strong selection. Evolution is not arrested in ponds, an undesirable condition for a gene pool maintenance program.

Fourth Genetic Hazard Source: Rearing and Maintenance Environment, Strategies, and Practices

The fourth source of genetic hazard in a broodstock management program is rearing the progeny of the founders as future captive broodstock. In gene pool maintenance programs, genetic management guidelines applied within a GHRMS will reduce the risk of those genetic hazards associated with management decisions and actions during progeny rearing and maintenance (Hynes et al. 1981; Kincaid 1995).

Fish populations exposed to a hatchery environment, particularly over many generations, may undergo genetic changes resulting in a domesticated population, one well adapted to hatchery conditions. A domesticated, hatchery broodstock is typically one that has been selected over the generations of its propagation for superior performance in the hatchery. Exposure to the combined selective forces of management method and environment are followed by increased fitness, adaptation, and genetic change in subsequent generations of hatchery fish. In a gene pool maintenance program, genetic losses associated with selection and adaptation on the facility constitute a type IV genetic hazard. As we have seen, this practice or process is undesirable in a conservation hatchery engaged for imperiled fishes. In addition to the appropriate selection of the donor population, collection of founders, and breeding strategies, another recommendation is to avoid hatchery selection in future broodstock by maximizing survival of progeny (or reducing undesirable attrition of broodstock) through all life history stages and reducing the number of generations a hatchery broodstock evolves in isolation. Developing recommendations for safeguards of this type can be explained through understanding the components of phenotypic variation and the forces that mold these components.

Variation in phenotypic expression of a quantitative or polygenic character, such as growth rate, is due to genetic and environmental influences, as well as the interaction among the two. Again, this relationship can be expressed by the equation (see equation 6)

$$V_p = V_g + V_e + V_{g*e}\,. \tag{8}$$

Simply put, this formula means that the phenotypic variation observed for a particular population trait is due to the combined but variable effects of genetics, environment, and their interaction. The manager will benefit in knowing in which direction and how much each component governs the fish population's appearance and performance for all vital performance characters under the conditions prevailing at the facility. The broodstock manager can roughly estimate how much performance is affected by genetic changes and to what extent genetic safeguards will influence changes in performances.

The genetic contribution, V_g, to the expression of a particular population character is determined by any or all of the genetic processes associated with the population genotype mentioned earlier: drift, mating behavior, inbreeding, selection, and mixed gene pools. These natural processes are inherent in the manner in which fish are traditionally spawned, reared, and maintained in captivity. Breeding methods and rearing techniques must be developed to avoid loss of genetic diversity. Genetic diversity in this case means the type, amount, and organization of the diversity. The less genetic variation a population has, the less the genetic contribution to the expression of fitness characters, the less the opportunities for the genotype to resonate with the environment to produce a fit organism and population.

The formula reveals that hatchery conditions, V_e, most likely influence to some unknown degree the environmental component of such important performance traits as behavior, survival, and growth. (Hynes et al. 1981; Kincaid 1995). Hatchery fish may express phenotypes that are not entirely genetic in nature. For example, hatchery fish may feed more readily on a prepared diet than wild fish brought to the hatchery. But the genetic contribution to that trait may be minimal. If such is the case, then the difference in phenotype and feeding behavior is not due to selection, but instead to learning. In that case, feeding behavior on pelleted feed is not a heritable trait and could be altered simply by altering the physical environment.

The specific operating mechanism influencing performance among hatchery fish in a hatchery, whether genetic or nongenetic, may not be known for some time, if ever. The fact is, both processes are operating simultaneously on the phenotype and genotype to determine the population's phenotypic expression.

Some managers believe that with this uncertainty, the best approach is applying the same general set of safeguards that may address both sources of genetic and nongenetic changes in progeny performance. Some managers suggest the key to resolving this problem of unacceptable phenotypic expression in hatchery fish is reducing unnatural culture conditions or practices (see environmental simulation discussion below). The only reservation the broodstock manager may have is what is natural and what isn't. The astute broodstock manager questions the typical answer to what appears, at first glance, to be a simple problem with a simple and obvious solution, hence, an often untested solution. The most sensible approach to managing genetic vulnerability or risk of genetic hazards is by institution of more effective safeguards. For example, the broodstock manager manages to maximize fish survival during all life history stages through improved culture PPPs in a healthful rearing environment. She may also provide for variation in the culture environment while attempting to maintain phenotypic variation of progeny during rearing by maximizing survival and minimizing unintentional and unplanned mortalities. Increased survival from improved methods and environments must be coupled with periodic, rational random culling. Behavioral modification can be attempted during later stages of culture shortly before stocking to increase the rate and success of poststocking acclimation. Selection more properly occurs in the natural environment and not in the hatchery.

To that end, performance characteristics of interest to the broodstock manager are time-to-hatch; percent hatch; numbers and types of abnormalities among fry at various stages; percent survival at various stages (hatch, swim-up and first-feeding, yolk-sac absorption); and physical and physiological development at various stages under various environmental conditions. Etiology of disease during egg incubation and hatch and early larval development should be investigated. Imprinting eggs or larval fish may be necessary at this point. The type, size, and availability of food at first feeding may be important in successful larval fish rearing. An important consideration in egg incubation and fry development is the need to rear each lot (eggs and fry of a paired mating) of eggs and fry separately until they can be marked, at least by family lot. It is important to identify a fish by lot in order to help reduce variation in family size during terminal production stages. As we have seen, reducing variation in family size contributes greatly to maximization of broodstock N_e and maintenance of genetic diversity. An effective method to mark fish less than 100 mm total length in batches is vital to development of advanced captive breeding programs.

Simulate the Natural Incubation and Early Life History Environment

An alternative approach to rearing and maintaining captive broodstock and their progeny calls for simulating the natural environment and is based on the following theory and recommendations:

> Essentially, hatchery design and rearing practices must minimize 'unnatural' conditions that might impair overall expression of important post-release performance traits . . . along with a willingness on the part of hatchery managers to deviate from the traditional aquaculture practices. Although measures for improving the performance of hatchery fish in the wild may compromise the short-term, economic efficiency of hatcheries, long-term net benefits to wild and naturalized populations and, ultimately, to society will be obtained (Kapuscinski and Miller 1993).

Kapuscinski et al. (unpublished data) suggest that to avoid unnatural alterations of the environmental and genetic components of early life history characteristics, eggs should be incubated under substrate, light, temperature, and water quality conditions that might be encountered in nature. Caution is urged in designing facilities and developing strategies and methods to simulate the natural environment. In pursuit of this strategy, anticipating and recognizing natural environmental sources of, and responses to, selection can be just as unpredictable in a hatchery as in the wild, unless the programs are designed to do so based on properly controlled studies. Designing elaborate and costly hatchery production and maintenance facilities and systems to simulate something as complex and fraught with uncertainty as the natural environment may be impractical.

Simulate Natural Rearing Conditions of Subadults

Performance of fish under captive propagation is determined by genotype, rearing methods and environment, and the interaction between the components (Williamson 1986; Kapuscinski and Jacobson 1987). Knowledge of these components and their relationships adds a dimension of control to the manager not typically incorporated into fish culture programs. If the fish are expected to perform as desired in the wild following stocking, then some experts urge consideration of simulating the natural environment during rearing of subadults. To the extent that consideration is possible and practical, we agree it should be considered. Controlled studies should be designed and conducted to generate useful performance and genetics data to test the validity of this consideration and evaluate its relative cost to more traditional methods of culture. Mimicking natural conditions on the hatchery is intended to avoid unnatural selection and nonadaptive physical and behavioral characteristics that reduce fitness of stocked fish in the wild. This concern must be weighed when evaluating economic factors relating to production, particularly traditional production methods. The measure of success may not be the total number of fish produced for stocking or the cost per fish produced, but instead, (1) the quality (genetically, physiologically, and behaviorally) of fish produced and (2) the proportion of fish stocked that live to spawn. Another measure of success may be the cost per spawner as verified by long-term stocking evaluations. Important performance characteristics to evaluate are survival, growth, maturation, physical and nutritional condition, and behavior. These characteristics can be affected by water quality and movement patterns (i.e., velocity), rearing container, photoperiod, loading rates, loading densities, diet, feeding regime, feeding methods, amount of food, predators, competitors, disease, parasites, handling, and other stressful factors.

Alternatively, avoiding unnatural selection in hatcheries requires maximizing survival during all life history stages by controlling and reducing mortality, particularly the type of mortality associated with intensive culture and catastrophic events. Most hatchery-related mortality is due to stress-related events such as poor water quality, handling practices, and diet. Translated into practical terms, this means inappropriate or inadequate hatchery design, technology, and management. Managers must be aware of the most stressful aspects of the production program and modify them. Novel considerations and recommendations to modify hatchery design and production methods in order to meet quality production goals are presented in Hynes et al. (1981); Kapuscinski et al. (unpublished data); and Steward and Bjornn (1990).

Facilities, technology, personnel, and funds are often limiting in captive propagation programs. Limitations imposed by scarcity of these resources often mean that excessive numbers of fish cannot be maintained in the facility. Numbers of fish must be periodically reduced. When reduction in numbers and biomass is necessary, it must be done in accordance with recovery goals, approved fish disposition guidelines (Williamson 1992), and genetic conservation principles. That translates into random reduction of numbers within family lots. Reduction in fish numbers should not be based on phenotype. Final equilibration of numbers among families, however, is not essential until fish are mature or are stocked. The whole notion of culling and grading, both common fish culture practices, must be reevaluated. They are clearly phenotypically driven and may constitute a type IV genetic hazard.

Summary: Broodstock Development and Management Recommendations

One of the cornerstones of a genetic conservation policy should be the development of acceptable genetics management guidelines for establishing and maintaining hatchery broodstocks. Genetic management plans may vary depending on programmatic goals and objectives, species characteristics, number of founding fish, spawning and rearing technology, available facilities, experienced personnel, and funding. However, genetic management guidelines and practices are fundamental. The fundamental basic rule of gene pool maintenance programs is that the number of spawners used to propagate hatchery fish should be adequate to ensure that most of the genetic diversity present in the founding population is passed down from generation to hatchery generation. In the simplest sense, broodstock development for captive propagation of endangered species is based on a gene pool maintenance approach.

Broodstock should be derived from a locally adapted source population, if possible, by using as many wild adults as possible without adversely impacting the donor source. This may have to be accomplished over a number of years, particularly if facilities are limiting and source populations are vulnerable. This has been a useful approach for propagating endangered Colorado River fishes because (1) wild adults are rare or difficult to catch in most instances; (2) fish do not mature for several years; (3) the species appear to be very long lived; (4) adults can be uniquely marked and manually spawned with high survival of parents and offspring; (5) marked adults can be returned to the wild following contribution of offspring; (6) new recruits to the broodstock development program can replace those returned to the wild.

Principles, practices, and protocols necessary to maximize N_e of the founding population have been mentioned several times in previous sections: (1) breeding as many fish as possible using marked fish and specific mating protocols, (2) equalization of sex ratios among breeding adults, (3) equalization of family variance among numbers of offspring, (4) infusion of new genetic material from the wild into the broodstock development cycle, and (5) extending generation interval. Due to the fecundity and longevity of the target species, much genetic erosion associated with turning generations in captivity can be reduced in captive breeding programs, at least with the Colorado River endangered fishes. It is possible that first generation progeny of these iteroparous species (repeat spawners) will form the broodstocks necessary for production over many years before they have to be replaced. The Dexter National Fish Hatchery and Technology Center, New Mexico, continues to use Colorado pikeminnow one generation removed from the wild, spawned in 1974, as captive broodstock. We are currently involved in capturing additional wild adults to genetically enhance this population.

The following are some general genetics management guidelines. They apply to all aspects of captive propagation and broodstock development in a gene pool maintenance program for imperiled fishes. Detailed specific guidelines associated with particular program phases can be found in Steward and Bjorn (1990):

1. Regardless of the final choice of a donor population, found the broodstock on a representative, unbiased sample that is large enough to accurately reflect the character and identity of the donor population.
2. Maximize N_e; collect large random samples of founders from a donor source; maintain equal sex ratios of parents and conduct pedigree matings; equalize the number of offspring among family lots used as new broodstock generation.
3. Manage or control selection to reduce the effects of hatchery practices and domestication imposed by the broodstock facility or hatchery environment.
4. Maximize hatchery survival of broodstock and their offspring so that the number of unrelated hatchery broodfish successfully contributing to the next broodstock generation or to fish for stocking is as close to the original number of unrelated founders as possible.
5. Culling to reduce numbers to manageable sizes must be done randomly, and, in addition, when reduction in fish numbers is necessary in one-on-one mating schemes and pedigree breeding programs,

numbers are reduced, not just randomly, but guided by the additional objective of maintaining or restoring equality of family size among unique families.
6. Introduce genetic material periodically from additional wild fish appropriately sampled from the donor population.
7. Increase generation intervals between old and new broodstocks; or, optimize the productive lifetime of a broodstock.
8. Maintain broodstocks as separate year-classes and avoid breeding among year-classes.
9. Genetically monitor and evaluate all stages of the broodstock development and management process within and across years to determine management success and make changes when necessary.
10. Genetically monitor donor or target populations and compare with data from the broodstock management program.

Conclusion

Trends toward greater complexity in natural resources management reflect our greater understanding of nature's underlying complexity. Our greater understanding of the complexity of the total management environment coupled with alarming trends in fishery resource health and abundance only amplifies the pressing need for informed, even enlightened, management. Advances in broodstock management, often nontraditional advances, are crucial to the success of comprehensive, holistic, integrated management programs including subordinate fisheries management program components. Derivative fisheries programs of this type, particularly those that rely on controlled propagation to protect and preserve unique and imperiled fisheries resources, scientifically integrate captive broodstocks in the program. Managers depend on captive broodstocks to faithfully transmit important genotypes from one generation to the next, conserve them, and produce from them fish for research and reintroduction to restore, recover, and maintain sustainable fish populations.

The complete broodstock manager is a powerful ally of imperiled fishes. It has been the aim of this chapter to identify and develop the concept of the complete broodstock manager, a manager who is aware of her proper role and responsibilities within the total management environment. The broodstock manager is a goal-oriented problem solver, scientifically managing information she acquires, analyzing the new data to develop practical knowledge based on the advanced principles, concepts, and theories necessary to ask the critical questions and formulate effective responses and decisions. Tolerant of other perspectives and aware of her public service responsibility, she embraces a resource conservation philosophy and expresses it professionally in a manner to meet public expectation. The complete broodstock manager relies on advanced management principles and planning concepts as the rational basis for her broodstock management program, plan, and process. She manages purposefully through teamwork, relying on a knowledgeable staff skilled in the appropriate technologies required to employ those practices, protocols, and procedures necessary to enact the decisions designed to accomplish both production and genetics conservation goals. Her program and performance is creative, dynamic, and expansive requiring greater investment of resources, and, through more efficient use, result in superior product that translates into predictable and sustainable broodstock management progress.

Acknowledgments

Team Dexter and the rest of the fisheries personnel in the U.S. Fish and Wildlife Service's Southwest Region; Fred Allendorf, Tony Echelle, Dave Philipp, Richard Wydoski, and Larry Vischer; A. R. Kapuscinski, C. R. Steward, M. L. Goodman, C. C. Krueger, E. Bowles, and R. Carmichael; Gary Wedemeyer and Chris Horsch.

References

Allendorf, F. W. 1993. Delay of adaptation to captive breeding by equalizing family size. Conservation Biology 7(2):416–419.

Allendorf, F. W., and N. Ryman 1987. Genetic management of hatchery stocks. Pages 141–159 *in* N. Ryman and F. Utter, editors. Population genetics and fisheries management. Washington Sea Grant Program, University of Washington Press, Seattle.

Ballou, J. D., and K. A. Cooper. 1992. Genetic management strategies for endangered captive populations: the role of genetic and reproductive technology. Symposium of the Zoological Society of London 64:183–206.

Barker, J. S. F. 1989. Population structure. Pages 75–80 *in* W. G. Hill and F. C. Mackay, editors. Evolution and animal breeding. C.A.B. International, Wallingford, UK.

Becker, W. A. 1984. Manual on quantitative genetics. Academic Enterprises, Pullman, Washington.

Berst, A. H., and R. C. Simon. 1981. Introduction to the proceedings of the 1980 Stock Concept International Symposium (STOCS). Canadian Journal of Fisheries and Aquatic Sciences 38(12):1457–1458.

Beveridge, W. I. B. 1950. The art of scientific research. Vintage Books, New York.

Booke, H. E. 1981. The conundrum of the stock concept: are nature and nurture definable in fishery science? Canadian Journal of Fisheries and Aquatic Sciences 38(12):1479–1486.

Bowen, J. T. 1976. A history of fish culture as related to the development of fishery programs. Pages 71–94 *in* N. G. Benson, editor. A century of fisheries in North America. American Fisheries Society, Special Publication 7, Bethesda, Maryland.

Brooks, J. E. 1986a. Annual reintroduction and monitoring report for the razorback sucker *Xyrauchen texanus* in the Gila River basin, Arizona, 1985. Arizona Game and Fish Department, Phoenix.

Brooks, J. E. 1986b. Reintroduction and monitoring of Colorado squawfish *Ptychocheilus lucius* in Arizona, 1985. Arizona Game and Fish Department, Phoenix.

Burdick, B. D. 1992. A plan to evaluate stocking to augment or restore razorback sucker in the Upper Colorado River. Final report. U.S. Department of Interior, Fish and Wildlife Service, Region 6, Colorado River Fishery Project, 529 25 1/2 Road, Suite B-110, Grand Junction, Colorado.

Busack, C. A. 1990. Yakima/Klickitat production project genetic risk assessment. Yakima/Klickitat Production Project Preliminary Design Report, Appendix A. Washington Department of Fisheries and Bonneville Power Administration. Division of Fish and Wildlife, P.O. Box 3621, Portland, Oregon 97208.

Busack C. A., and K. P. Currens. 1995. Genetic risks and hazards in hatchery operations: fundamental concepts and issues. Pages 71–80 *in* H. L. Schramm and R. G. Piper, editors. Uses and effects of cultured fishes in aquatic ecosystems. American Fisheries Society, Symposium 15, Bethesda, Maryland.

Campton, D. E. 1995. What do we really know? Pages 337–353 *in* H. L. Schramm and R. G. Piper, editors. Uses and effects of cultured fishes in aquatic ecosystems. American Fisheries Society, Symposium 15, Bethesda, Maryland.

Carmichael, G. J. 1984. Long distance truck transport of intensively reared largemouth bass. Progressive Fish-Culturist 46:111–115.

Carmichael, G. J., J. R. Tomasso, B. A. Simco, and K. B. Davis. 1984a. Confinement and water quality induced stress in largemouth bass. Transactions of the American Fisheries Society 113:767–777.

Carmichael, G. J., J. R. Tomasso, B. A. Simco, and K. B. Davis. 1984b. Characterization and alleviation of stress associated with hauling largemouth bass. Transactions of the American Fisheries Society 113:778–785.

Chakraborty, R., and O. Leimar. 1987. Genetic variation within a subdivided population. Pages 89–120 *in* N. Ryman and F. Utter, editors. Population genetics and fishery management. Washington Sea Grant, University of Washington, Seattle, Washington.

Currens, K. P., and C. A. Busack. 1995. A framework for assessing genetic vulnerability. Fisheries 12:24–31.

Dizon, A. E., C. Lockyer, W. F. Perrin, D. P. Demaster, and J. Sisson. 1992. Rethinking the stock concept: a phylogeographic approach. Conservation Biology 6(1): 24–36.

Doyle, R. W. 1983. An approach to the quantitative analysis of domestication selection in aquaculture. Aquaculture 33:167–185.

Doyle, R. W., C. Herbinger, C. T. Taggart, and S. Lochmann. 1995. Use of DNA microsatellite polymorphism to analyze genetic correlations between hatchery and natural fitness. Pages 205–211 *in* H. L. Schramm and R. G. Piper, editors. Uses and effects of cultured fishes in aquatic ecosystems. American Fisheries Society, Symposium 15, Bethesda, Maryland.

Falconer, D. 1981. Introduction to quantitative genetics. Longman, New York.

Fisher, R. A. 1930. The genetical theory of natural selection. Clarendon Press, Oxford, England.

Gall, G. A. E. 1987. Inbreeding. Pages 47–87 *in* N. Ryman and F. Utter, editors. Population genetics and fishery management. University of Washington Press, Seattle.

Gharrett, A. J., and S. M. Shirley. 1985. A genetic examination of spawning methodology in a salmon hatchery. Aquaculture 47:245–256.

Gharrett, A. J., and W. W. Smoker. 1993. A perspective on the adaptive value of hybrids between even- and odd-year pink salmon *(Oncorhynchus gorbuscha)*: a test for out-breeding depression? Canadian Journal of Fisheries and Aquatic Sciences 48:1744–1749.

Gilpin, M. E. 1987. Spatial structure and population vulnerability. Pages 87–124 *in* M. E. Soulé, editor. Viable populations for conservation. Cambridge University Press, New York.

Gorman, O. T. 2000. Ecological, and genetic considerations for collection of gametes from wild fishes. Pages 319–322 *in* T. R. Tiersh, and P. M. Mazik, editors. Cryopreservation of aquatic species. World Aquaculture Society, Baton Rouge, Louisiana.

Ham, K. D., and T. N. Pearsons. 2001. A practical approach for containing ecological risks associated with fish stocking programs. Fisheries 26(4):15–23.

Hard, J. J., R. P. Jones Jr., M. L. Delarm, and R. S. Waples. 1992. Pacific salmon and artificial propagation under the Endangered Species Act. National Technical Information Service, United States Department of Commerce, 5285 Port Royal Road, Springfield, Virginia 22161.

Hartle, D. L. 1981. Principles of population genetics. Sinauer Associates, Sunderland, Massachusetts.

Hilborn, R. 1992. Hatcheries and the future of salmon in the Northwest. Fisheries 17(1):5–8.

Hilborn, R. 1999. Confessions of a reformed hatchery basher. Fisheries 24(5):30–31.

Hillis, D. M., and C. Moritz. 1990. An overview of applications of molecular systematics. Pages 502–515 *in* D. M. Hillis and C. Moritz, editors. Molecular systematics. Sinauer Associates, Sunderland, Massachusetts.

Hynes, J. D., E. H. Brown Jr., J. H Helle, N. Ryman, and D. A. Webster. 1981. Guidelines for the culture of fish stocks for resource management. Canadian Journal of Fisheries and Aquatic Sciences 38:1867–1876.

Kapuscinski, A. R., and L. D. Jacobson. 1987. Genetic guidelines for fisheries management. Minnesota Sea Grant. University of Minnesota, St. Paul.

Kapuscinski, A. R., and L. M. Miller. 1993. Genetic hatchery guidelines for the Yakima/Klikitat fisheries project (public review draft) Co-Aqua. 2369 Bourne Avenue. St. Paul, Minnesota.

Kincaid, H. L. 1986. Fish breeding manual. United States Department of Interior, Fish and Wildlife Service, National Fisheries Center, Route 3, Box 41, Kearneysville, West Virginia.

Kincaid, H. L. 1995. An evaluation of inbreeding and effective population size in salmonid broodstocks in federal and state hatcheries. Pages 193–204 *in* H. L. Schramm and R. G. Piper, editors. Uses and effects of cultured fishes in aquatic ecosystems. American Fisheries Society, Symposium 15, Bethesda, Maryland.

Krueger, C. C., and D. J. Decker. 1993. The process of fisheries management. Pages 33–53 in C. C. Kohler and W. A. Hubert, editors. Inland fisheries management in North America. American Fisheries Society, Bethesda, Maryland.

Krueger, C. C., A. J. Gharrett, T. R. Dehring, and F. W. Allendorf. 1981. Genetic aspects of fisheries rehabilitation programs. Canadian Journal of Fisheries and Aquatic Sciences 38(12):1877–1881.

Lacy, R. C. 1987. Loss of genetic diversity from managed populations: interacting effects of drift, mutation, immigration, selection, and population subdivision. Conservation Biology 1:143–158.

Lande, R., and G. F. Barrowclough. 1987. Effective population size, genetic variation, and their use in population management. Pages 87–123 *in* M. E. Soulé, editor. Viable populations for conservation. Cambridge University Press, England.

Loftus, K. H. 1981. Where are we now? Canadian Journal of Fisheries and Aquatic Sciences 38:1464–1465.

Mayr, E. 1977. Population, species, and evolution. Belknap Press, Cambridge, Massachusetts.

Meade, J. W. 1989. Aquaculture management. Van Nostrand Reinhold, New York.

Meffe, G. K. 1992. Techno-arrogance and half-way technologies: salmon hatcheries on the Pacific coast of North America. Conservation Biology 6(3):350–354.

Meffe, G. K., and R. C. Vrijenhoek. 1988. Conservation genetics in the management of desert fishes. Conservation Biology 2:157–169.

Modde, T., A. T. Scholz. J. H. Williamson, G. B. Haines, B. D. Burdick, and F. K. Pfeifer. 1995. An augmentation plan for razorback sucker in the upper Colorado River basin. Pages102–112 *in* H. L. Schramm and R. G. Piper, editors. Uses and effects of cultured fishes in aquatic ecosystems. American Fisheries Society, Symposium 15, Bethesda, Maryland.

Nelson, K., and M. Soulé. 1987. Genetical conservation of exploited fishes. Pages 345–368 *in* N. Ryman and F. Utter, editors. Population genetics and fishery management. University of Washington Press, Seattle.

Pearsons, T. N., and C. W. Hopley. 1999. A practical approach for assessing ecological risks associated with fish stocking programs. Fisheries 24(9):16–23.

Phenicie, C. K., and J. R. Lyons. 1973. Tactical planning in fish and wildlife management. Resource Publication 123. U.S. Department of Interior, Fish and Wildlife Service Bureau of Sport Fishing and Wildlife, Washington, D.C.

Piper, G. G., I. B. McElwain, L. E. Orme, J. P. McCraren, L. G. Fowler, and J. R. Leonard. 1982. Fish hatchery management. U.S. Department of the Interior, Fish and Wildlife Service. Washington, D.C.

Radonski, G., and A. J. Loftus. 1995. Fish genetics, fish hatcheries, wild fish, and other fables. Pages 1–4 *in* H. L. Schramm and R. G. Piper, editors. Uses and effects of cultured fishes in aquatic ecosystems. American Fisheries Society, Symposium 15, Bethesda, Maryland.

Riggs, L. A. 1986. Genetic considerations in salmon and steelhead planning. Northwest Power Planning Council, Technical Discussion Paper. Final Report. Genrec, Berkeley, California.

Riggs, L. A. 1990. Principles for genetic conservation and production quality. Results of a scientific and technical clarification and revision. Northwest Power Planning Council, Contract No. C90-005. Portland, Oregon.

Ryman, N., and G. Ståhl. 1980. Genetic changes in hatchery stocks of brown trout *(Salmo trutta)*. Canadian Journal of Fisheries and Aquatic Sciences 37:82–87.

Shields, W. M. 1993. The natural and unnatural history of inbreeding and outbreeding. Pages143–169 *in* N. W. Thornhill, editor. The natural history of inbreeding and outbreeding. University of Chicago Press, Illinois.

Smith, M. H., and O. E Rhodes Jr. 1992. Genetics and biodiversity in wildlife management. Pages 243–251 *in* F. L. Knopf and M. H. Smith, editors. Special Session 5. Biological diversity in wildlife management. Transactions of the 57th North American and Natural Resources Conference. Wildlife Management Institute, Washington D.C.

Sparrowe, R. D., and H. M. Wight. 1975. Setting priorities for the endangered species program. Fortieth North American Natural Resources Conference, March 17, 1975, Pittsburgh, Pennsylvania.

Steward, C. R., and T. C. Bjornn. 1990. Supplementation of salmon and steelhead stocks with hatchery fish: a synthesis of published literature. Part 2. Pages 101–126 *in* W. H. Miller, editor. Analysis of salmon and steelhead supplementation, Parts 1–3. Technical Report 90-1, Bonneville Power Authority, U.S. Department of Energy, Portland, Oregon.

Tave, D. 1986. Genetics for fish hatchery managers. AVI, Westport, Connecticut.

Templeton, A. R. 1986 Coadaptation and outbreeding depression. Pages 105–116 *in* M. R. Soulé editor. Conservation biology: the science of scarcity and diversity. Sinauer Associates, Sunderland, Massachusetts.

Tiersch T. R., C. R. Figiel, W. R. Wayman, J. H. Williamson, G. J. Carmichael, and O. T. Gorman. 1998. Cryopreservation of sperm of the endangered razorback sucker. Transactions of the American Fisheries Society 127:95–104.

Tiersch T. R., C. R. Figiel, W. R. Wayman, J. H. Williamson, G. J. Carmichael, and O. T. Gorman. 2000. Cryopreservation of sperm of the endangered razorback sucker. Pages117–122 *in* T. R. Tiersh and P. M. Mazik, editors. Cryopreservation of Aquatic Species. World Aquaculture Society, Baton Rouge, Louisiana.

Tiersch T. R., W. R. Wayman, C. R. Figiel, O. T. Gorman, J. H. Williamson, G. J. Carmichael, and O. T. Gorman. 1997. Field collection, handling, and storage of sperm of the endangered razorback sucker. North American Journal of Fisheries Management 17:167–173.

USFWS (U.S. Fish and Wildlife Service). 1992. Recovery plan for Gila trout. United States Department of Interior, Fish and Wildlife Service, Region 2, Albuquerque, New Mexico.

Utter, F. M. 1981. Biological criteria for definition of species and distinct intraspecific populations of anadromous salmonids under U.S. Endangered Species Act of 1973. Canadian Journal of Fisheries and Aquatic Sciences 38:1626–1635.

Vrijenhoek, R. C., M. E. Douglas, and G. K. Meffe. 1985. Conservation genetics of endangered fish populations in Arizona. Science 229:400–402.

Wilber, K. 1996. A brief history of everything. Shambala, Boston, Massachusetts.

Wiley, R. W. 1999. Fish hatcheries are a powerful tool of fisheries management. Fisheries 24(9):24–26.

Williamson, J. H. 1983. Comparing training success of two strains of largemouth bass. Progressive Fish-Culturist 45(1):3–7.

Williamson, J. H. 1986. An evaluation of Florida, Northern, and hybrid largemouth bass *Micropterus salmoides* under intensive culture conditions. Ph.D thesis. Texas A&M University, College Station.

Williamson, J. H. 1992. Disposition of surplus artificially propagated endangered fishes. Third draft, July 17, 1992. Recovery implementation program for endangered fish species in the Upper Colorado River Basin. United States Department of Interior, Fish and Wildlife Service, Denver.

Williamson, J. H. 2000. Response of an unabashed hatchery reformer. Fisheries 25(2):28–30.

Williamson, J. H., and R. S. Wydoski. 1994. Genetics management guidelines. Recovery implementation program for endangered fishes in the Upper Colorado River Basin. Fish and Wildlife Service, Denver, Colorado.

Williamson, J. H., R. S. Wydoski, and L. Visscher. 1994. Genetics management guidelines for listed and candidate fish species. Fish and Wildlife Service, Denver, Colorado.

Wilson, E. O. 1998. Consilience, the unity of knowledge. Vintage Books, New York.

Withler, R. E. 1988. Genetic consequences of fertilizing chinook salmon *(Oncorhynchus tshawytscha)* eggs with pooled milt. Aquaculture 68:15–25.

8

Nutrition and Feeding

Frederic T. Barrows and Ronald W. Hardy

Introduction

Nutrition encompasses the ingestion, digestion, and absorption of food. The rearing of large numbers of animals, whether they are terrestrial or aquatic, in relatively restricted areas requires a detailed knowledge of their nutritional requirements so that they can be provided with feed that supports rapid growth and optimum health. Although fish farming for conservation, fisheries enhancement, and food production has been practiced for over a century in North America, there has not been the emphasis on fish nutrition research or feed development that there has been for other livestock. In addition, the quantity of fish feed required by hatcheries and commercial fish farms has not been sufficient to justify the investment by feed companies or others of more than a minimal amount for fish nutrition research. At present, global fish feed production is less than 3% of the total amount of pelleted animal feeds. Until the 1970s, commercial fish production was a minor activity, at least in the United States. As a result, information on the nutritional requirements of fish feed formulation and on production of water-stable feeds has not kept pace with advances in animal nutrition. Nevertheless, significant progress has been made in fish nutrition and feed technology, spurred by the needs of the commercial sector and of state and federal agencies producing both traditionally cultured species (e.g., trout, salmon, and catfish) and emerging species (e.g., striped bass, tilapia, and perch). Research has led to numerous developments that have greatly advanced the field of fish nutrition, especially in the last decade.

The problem of feeding cultured fish was first studied by investigating the natural diet of wild fish, mainly brook trout. The natural diet of trout was 46% protein and 16% fat; feeds for trout were formulated to be similar to this diet until recently. Early fish rearing relied on natural prey items to supply most or all of the nutrients needed for fish to grow. As fish production increased and more fish culture stations were established, it became uneconomical and impractical to use natural foods. Because of the limited supply and uncertain nature of artificially cultured natural food organisms, fish culturists turned to more readily available and reliable food supplies. Organs of slaughtered animals were among the first ingredients used to supplement or replace natural feeds.

Hatchery operators began feeding dry meals (e.g., wheat by-products, dried skim milk, brewer's yeast, or cottonseed meal) combined with meat products to provide greater quantities of finished feed. One of the early problems with this practice was how to bind the mixtures so that they would hold together when placed in the water. In the early days of fish culture, a large proportion of these feeds leached into the water. This resulted in slow growth, poor feed conversion ratios (FCR), increased mortality, water pollution, and increased labor in cleaning ponds and raceways. The use of dry meals to reduce feed costs compounded the water-stability problem associated with these mixtures. The use of spleen and liver mixed with salt resulted in rubber-like mixtures, called meat meal feeds, that were suitable for trough and pond feeding. These were mixed in cement or bread mixers and extruded through meat grinders. This type of feed produced more efficient feed utilization, faster growth, and lower feed loss. However, considerable labor was involved in the preparation of the meat meal feeds. In addition, the use of fresh meat in the diet required their frequent shipments or cold storage facilities at each hatchery. In Pacific salmon hatcheries, fry and fingerling feeds were often made using frozen salmon carcasses collected from adult spawners. Transmission of disease, specifically fish tuberculosis, from adult carcass to fry via the feed, made this practice untenable. The ideal hatchery feed was one that would combine the advantages of the meat meal feed, but would eliminate the labor involved in preparation, reduce the expense of cold storage facilities, and reduce the possibility of feed being a disease vector.

In 1959, the Oregon State Game and Fish Commission began to use a pelleted meat meal fish feed called Oregon Moist Pellet (OMP). These pellets were developed to eliminate the transmission of disease via the feed and to produce a moist pellet that was accepted by Pacific salmon fry. Use of this feed in production was preceded by 6 years of research. The formula was composed of wet fish products (pasteurized to kill pathogenic bacteria and viruses) and dry ingredients. It had a moist, soft consistency and had to remain frozen until shortly before feeding. Many salmon hatcheries used the OMP as a standard production feed because it provided satisfactory feed conversion ratios, rapid growth, high survival, and eliminated the necessity of each hatchery having to make its own feed. The disadvantage of the OMP was that it had to remain frozen during storage until use. When thawed, OMP pellets deteriorated rapidly, losing certain nutrients, becoming moldy, and spoiling due to microbial action.

By the mid-1950s, development and refinement of vitamin fortifications made possible the "complete" dry (<11% moisture to prevent microbial growth) pelleted feeds as we know them today. Fish feeds manufactured in the form of dry pellets solved many of the problems of hatchery operations in terms of feed preparation, storage, and feeding. There are several additional advantages to pellet feeding. Pellets require no preparation at the hatchery before they are fed. They can be stored for 90–100 d in a cool, dry place without refrigeration. When a fish swallows a pellet, it receives the ingredients in proportions that were formulated in the diet. There is evidence that fish that are fed dry pellets are more similar in size within a group than those fed meat meal. The physical characteristics of the pellets provide for more complete consumption of the feed. Less feed wastage results in far less pollution of the water during feeding and a comparable reduction in the effort required to clean ponds and raceways. Pelleted feeds can be used in automatic or demand feeders. All of these qualities make dry feed the obvious choice for hatchery production of most species of fish.

Many feedstuffs have been tested as ingredients in pelleted feeds. Some have been eliminated from consideration because they result in pellets that were too hard or too soft; others do not provide the essential nutrients required by the fish or reduced the palatability of feeds when they are included in formulations. Commercial fish feeds were pelleted and marketed before the development of "open-formula" feeds. A few early commercial feeds failed to produce good economical growth and to maintain the health of the fish but, by and large, most were satisfactory. Commercial feeds are "closed-formula" feeds, meaning that the formulation is the proprietary property of the feed producer, in contrast to open-formula feeds—produced using a known formulation and defined production conditions with specifications for formulation and production generally developed by state or federal agencies.

Several items must be considered in developing a feeding program for fish. These include the nutrient requirements for different fish species, fish sizes, environmental conditions, stress factors, types of feed, and production objectives. Feeding methods are an important aspect of hatchery or farm production and will be discussed extensively in the last part of this chapter. Conservation hatcheries often have different production objectives than commercial fish farms, namely the production of fish at a specific size and at an optimum time for release. This influences the choice of feed and feeding practices for conservation hatcheries, which is discussed further in this chapter.

Hatchery fish rearing programs succeed or fail due to many aspects of feeds and feeding. No single factor is more important than another; it is a combination of many factors that results in a successful fish culture program. Application of the available knowledge of fish nutrition and feeding will result in healthy, fast-growing fish and low production costs. A fish culturist must be able to recognize the factors affecting feed utilization and develop an appropriate feeding program to meet the production objectives of the farm or hatchery.

Factors that Affect Nutritional Requirements

Determination of Requirements

The physiological functions of fish (metabolism, respiration, growth, activity, and reproduction) govern their need for essential nutrients and energy. The physiological need for nutrients is generally defined as the lowest dietary level resulting in a satisfactory physiological response, which can be growth, health, survival, absence of deficiency signs, tissue storage level, enzyme activity, or immune function.

Knowing the nutritional requirements of fish allows nutritionists and feed manufacturers to formulate diets that are complete and balanced in all essential nutrients. Excesses of nutrients result in a more expensive feed than is necessary and can increase the nutrient load in the effluent leaving the hatchery because fish excrete that portion of many nutrients present in diets in excess of their needs. In addition, excesses of some nutrients can interfere with the absorption or utilization of other nutrients through antagonistic interactions. An example of this is the reduction in zinc absorption in the gut caused by high levels of calcium and phosphorus from bones in fish meal or in rendered products, such as meat and bone meal.

Nutrient requirements are typically determined through feeding trials. These trials are designed to estimate the true metabolic requirement of a given nutrient. In these trials, various levels of a single nutrient are added to feeds that are otherwise nutritionally complete and each feed is fed to, at least, triplicate lots of fish. Experimental diets are randomly arranged within a group of tanks to eliminate the effects of tank location on fish performance, or are arranged in some other fashion to minimize or negate tank position effects (near a window, next to an aisle, under a light). Similarly, feeding should be done randomly to avoid the natural tendency of workers to spend more time feeding the first few tanks than the last few tanks. Response variables, such as weight gain, length gain, feed conversion ratio, or biochemical measurement, are measured from each tank of fish and that data is used to determine the minimum level of a nutrient that results in a maximum response.

Ideally, response variables increase up to the dietary requirement and do not increase further at higher dietary levels. However, responses are rarely clear-cut, and dietary levels chosen in feeding trials are nearly always below and above the point of maximum response, rather than right at it. Thus, a variety of statistical methods is used to estimate the dietary nutrient level resulting in an optimum or maximum response. Statistical methods, such as regression analysis, broken-line analysis, or curve fitting, are commonly used to precisely identify the dietary level yielding the best response. However, the statistical method that is used with a given data set can affect the estimated requirement and result in an over or under estimation of the requirement for a given situation. There is no general agreement among nutritionists on which type of test for determining requirements is best, nor should there probably be agreement. Different types of studies (and different response variables) require different types of statistical evaluation. The important point to remember is that there are many ways to statistically analyze data generated in feeding trials, and these analyses yield an estimation of the true requirement, which is usually good enough for practical diet formulation.

The most common and easily measured response variable in nutrient requirement studies is growth or weight gain. Growth is the ultimate goal of most feeding programs, but, in nutritional requirement studies, it is not a very sensitive measurement. For example, two groups of fish fed different experimental diets could gain the same amount of weight, but one group might gain more fat and less protein than the other. The percentage of body fat may be an important consideration in a particular feeding program that may be overlooked if total weight gain is the only response variable used to determine the dietary requirement.

Another limitation with using weight gain as a sole response variable is the amount of time required to observe differences among groups of fish fed different dietary nutrient levels. Direct measurements of tissue stores of a nutrient (e.g., vitamins and minerals) are made both to increase sensitivity of the measurement and decrease the amount of time needed to reach a conclusion. The liver is sometimes assayed for nutrient levels since it is a very active organ and sensitive to dietary changes.

However, liver nutrient levels commonly reflect recent dietary intake, so other tissues are often assayed for nutrient content. Whole body nutrient levels are sometimes measured, but specific tissues are usually more sensitive to dietary intake than whole body measurements because nutrients are not evenly distributed throughout the body. For example, phosphorus is stored in hard tissue (bones and fins) and skin of fish, so sampling these tissues will give a more accurate estimate of phosphorus status of fish than will whole body analysis. Vitamins and minerals have many functions in the body, and some are essential cofactors for metabolic enzymes. The activity of certain enzymes in fish tissue reflects the status of fish for certain vitamins and minerals more accurately than do tissue levels. Thus, activity of metabolic enzymes is sometimes used as a response variable to varying dietary intake for the purpose of estimating requirements.

Another approach to determining nutrient requirements is commonly used in human and animal nutrition and is beginning to be used in fish nutrition. This involves measurement of nutrient excretion, for example, levels in urine of fish receiving different dietary intake levels. Recently, the level of phosphorus in the urine of fish has been used to estimate the phosphorus requirement of growing fish. At dietary phosphorus intakes below the dietary requirement, virtually no phosphorus is excreted in the urine. When the required dietary intake of phosphorus is exceeded, there is an immediate and linear increase in urinary phosphorus concentration. A similar approach has been used to determine amino acid requirements by measuring plasma amino acid levels, but this approach is complicated by the transitory nature of plasma levels in the hours following a meal. Thus, most amino acid requirement estimates for fish are based on weight gain or protein gain over the course of a feeding trial.

Each response variable used in nutrient requirement studies has inherent advantages and disadvantages. The safest approach when estimating the requirement is to measure as many variables as possible and to consider a number of response variables when estimating a dietary nutrient requirement. Different dietary nutrient requirement estimates from different researchers often are the result of researchers basing their estimates on different response variables.

A key point to remember is that fish, like other animals, can store essential nutrients for periods when food is scarce. During periods when deficient diets are being fed, fish typically grow normally, using up their tissue reserves until they are gone, at which point deficiency signs often appear very rapidly. Another point to remember is that dietary requirements found in the scientific literature are not absolute—they are estimates. Precisely defining dietary requirements is of academic interest, but in practical fish feed formulation allowances are made to account for the fact that actual dietary requirements are not based on exact estimates. Thus, fish feeds are always formulated to contain a slight excess of essential nutrients, even after taking into consideration any losses that might occur during pelleting and storage.

The Effect of Life Stage and Species

Fish have what is called an indeterminate growth pattern, they never stop growing, but growth rates slow as fish become larger. The proportional increase in size is greatest in young fish and is reflected in their dietary needs. In general, the energy requirement per unit weight is higher for smaller fish than it is for larger fish. Different fish species have different metabolic rates associated with their life history and habits. Species that are less active have lower metabolic rates and generally use less energy for voluntary activity than active species. Water temperature also affects metabolic rate—metabolic rates increase with increasing water temperature. At a given water temperature, coldwater fish species, such as trout, consume more oxygen than warmwater fish species, such as channel catfish, indicating that coldwater fish have a higher metabolic rate than do warmwater fish.

Spawning, seasonal changes, and physiological changes also affect metabolic rate. Growth rates slow with the onset of sexual maturity. During maturation, nutrients are used to produce eggs and sperm, and somatic (muscle) growth essentially stops. At spawning, fish lose as much as 10–20% of their body weight. Fish also have high metabolic rates during the spawning season, associated with courtship and spawning activities. Conversely, during winter, resting fish have low metabolic rates associated with winter dormancy. This is a natural characteristic of the life cycle of many fish species and corresponds to periods of low water temperatures, shortening day lengths, and food scarcity. During this period, protein synthesis rates in fish tissues are very low, so overfeeding results in fat accumulation in the fish, no matter what the protein or fat level of the feed happens to be. The metabolic rate of starving fish is approximately 20% lower than that of actively feeding fish. Excitement, stress, and increased activity elevate metabolic rates. Pathogen loads may also influence dietary nutrient utilization, as the demands of the immune system may reduce the proportion of dietary protein and energy available to support growth. All of these factors affect the amount of energy that must be supplied by the feed.

Environmental Factors that Affect Nutrient Requirements

Apart from the feed, water temperature is the single most important factor affecting fish growth. Because fish are coldblooded, their body temperatures fluctuate with environmental water temperatures. Growth essentially stops in trout when the temperature decreases to 3.3°C (38°F). The lower limit for catfish growth is about 10°C (50°F). As water temperatures rise, growth rate, measured as gain in wet body weight or gain in length, increases to a maximum and then decreases as temperatures approach the upper lethal limit for a given species (Figure 1). The best temperature for rapid, efficient growth is that at which appetite is high and maintenance requirements (or the energy cost of living) are low. Energy needs increase with rising water temperatures, resulting in an increase in the fish's appetite. The fish culturist must adjust feeding rates (energy intake) or energy content of the feed to provide appropriate energy levels at various water temperatures. Failure to make this adjustment will result in less than optimal growth and feed waste.

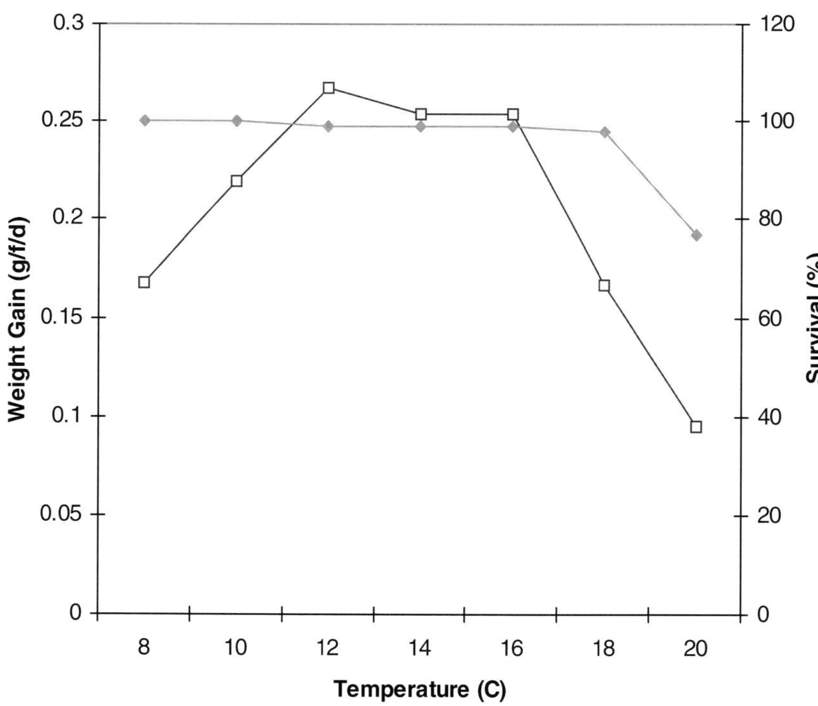

FIGURE 1. Schematic representation of the relationship between water temperature, survival (◆) and growth rate (□) in fish.

Oxygen needs associated with metabolism increase with increasing water temperature. Oxygen saturation levels in water decrease with increasing temperature. Thus, when water temperatures increase, oxygen demands for metabolism increase, but the oxygen carrying capacity of water decreases. As a result, the metabolic rate of fish typically increases until the critical oxygen level is reached, at which point oxygen demands exceed oxygen uptake from water. Above this point, fish performance falls and mortality can ensue.

Because water temperature affects fish growth and metabolic rates, it can become a complicating factor in studies designed to establish the nutrient requirements of fish. To deal with this, the National Research Council has defined standard environmental temperatures (SET) for conducting nutritional requirement studies with various species of fish. Suggested SET are 10°C (50°F) for salmon, 15°C (59°F) for trout, and 30°C (85°F) for channel catfish.

Water flow rates, water chemistry, and pollution can be stressful to fish and result in increased metabolic rates in proportion to the severity of the stress. Water chemistry; oxygen content; and the amount of other gases, toxins, and minerals in the water all affect metabolic rates. For many fish species, darkness decreases activity and energy requirements. These fish grow faster if they have "rest periods" of darkness than they do in constant light. Crowding, disease, and cultural practices also can affect the metabolic rate and well-being of fish.

Digestion and Absorption of Nutrients

Fish have digestive systems (Figure 2) that are generally similar to other simple-stomached (monogastric) animals. After feed is swallowed, it

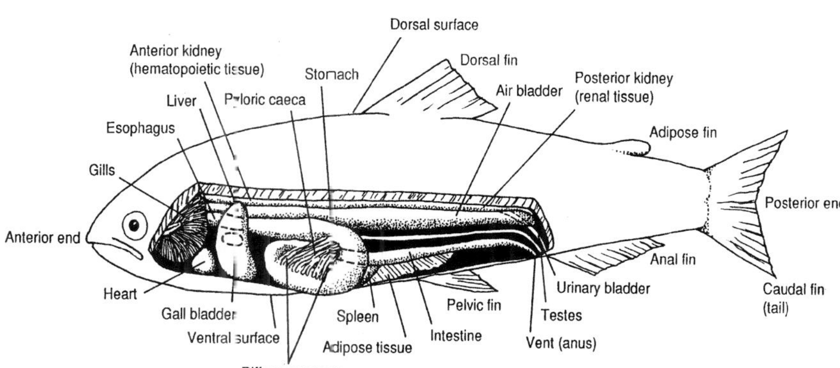

FIGURE 2. Schematic representation of the digestion system of fish.

enters the stomach where digestion begins with the addition of enzymes and hydrochloric acid to the food. Very little nutrient absorption occurs in the stomach. Food leaves the stomach and enters the pyloric ceca and intestine, where it is enzymatically broken down into basic components that are absorbed and utilized. This process is called digestion. Nutrients in feeds cannot be utilized until they are absorbed into the body and transported to cells where metabolism takes place. Some foods such as simple sugars and purified amino acids can be absorbed directly without enzymatic breakdown, but most foods must be reduced to simple components to be absorbed.

Fish can obtain some essential mineral elements directly from water by transfer across the gills or from water swallowed with food. Because the mineral content of water varies from water source to water source, fish feeds are normally formulated to supply most essential mineral elements in the diet.

The mouth is used to capture feed. Most fish do not chew their food, but gulp it down intact. Fish can bite off (tear) pieces of food from their prey, and, when cultured fish are fed pellets that are too large for them to swallow, they will bite feed into smaller pieces. Some fish species have pharyngeal teeth, which are located in the throat; these are used to grind food.

Preliminary digestion of protein occurs in the stomach, where feed is acidified by gastric acid secretion (hydrochloric acid), denatured by the acid in the case of proteins, and partially digested by enzymes (mainly pepsin), which are secreted from cells lining the stomach. Even though very little nutrient absorption occurs in the stomach, there is some evidence that free amino acids can be absorbed. Not all fish have acid-producing stomachs—goldfish (carp) are an example of an agastric or stomach-less fish.

The finger-like pyloric ceca at the junction of the stomach and small intestine are a primary source of digestive enzymes (e.g., tryopsin, chymotrypsin, carboxypeptidases, lipases). These enzymes are produced by pancreatic cells and by cells lining the pyloric ceca and break down complicated chemical structures of food into its basic components, for example, amino acids (proteins), fatty acids (lipids), and simple sugar (starch and other polysaccharides). The proximal (head end) intestine is the major site of digestion and receives digestive enzymes secreted by the liver, pancreas, pyloric ceca, and cells lining the walls of the intestine. Bile, secreted by the liver, also neutralizes the partially digested food, raising the pH to about 7–8. Most nutrient absorption occurs in the pyloric ceca and proximal intestine. Some water and mineral reabsorption occurs in the distal (tail end) intestine, but its primary function is to hold undigested materials until they are expelled as feces.

Environmental Factors (Oxygen and Water)

Oxygen and water normally are not considered nutrients to fish, but they are essential components in the processes supporting life. All vital processes require energy, which is obtained from the oxidation of various nutrients in the body. The utilization of oxygen and the resulting production of carbon dioxide and water by the tissues is the principal mechanism for the liberation of energy from feed. Oxygen consumption by fish varies with size, feed, stress, water temperature, and activity. The oxygen requirement per unit of weight decreases as fish size increases. High-nutrient feeds, density, stress, elevated water temperatures, and increased activity increase oxygen requirements of fish. As a consequence, adequate oxygen must be supplied to assure efficient utilization of the feed and optimal growth. Fish exposed to suboptimal oxygen levels in water reduce their feed intake proportionally.

Water is involved in many metabolic reactions either as a reactant or end product. Approximately 75% of weight gained during fish growth is water. Water not contained in the feed itself must be obtained from the environment. Because water always diffuses from the area of weakest ionic concentration to the strongest ionic concentration, it readily diffuses through the gills and digestive tract into the tissues of freshwater fish. As a result, fish in freshwater constantly excrete dilute urine to dispose of excess water. In marine fish, the blood ion concentration is weaker than that of seawater, so that fish lose water to the environment. Marine fish must drink seawater and excrete minerals in order to keep from becoming dehydrated.

Natural Foods and Formulated Feeds

As the name implies, natural foods are those available in natural environments. Small fish feed upon algae and zooplankton. As carnivorous fish grow, they devour progressively larger prey—insects, worms, mollusks, crustaceans, small fish, tadpoles, and frogs. Many fish remain omnivorous throughout their lives, consuming both animal and plant material.

Pond culturists take advantage of the natural foods growing in still waters. The composition of insects, worms, and forage fish that constitute natural food is mostly water (75–80%). The remaining components are protein (12–15%), fat (3–7%), ash (1–4%), and a small amount of carbohydrate (<1%), although there is considerable range in the nutritional composition of prey items, especially insects. During warm weather when insects hatch and bottom organisms are abundant, a pond can provide a considerable amount of feed for fish. This production can be increased by pond fertilization. Because the production of live food tends to be highly variable from pond to pond and from year to year, natural methods of providing food are inefficient in fish culture unless the producer is using large bodies of water. However, natural food organisms can contribute significantly to the nutritional needs of fish reared in ponds, providing that fish density is low to moderate.

Living organisms are also used in tank culture for the larval-rearing phase of some species. *Artemia* spp. (brine shrimp) can be easily hatched and fed to first-feeding larvae. Most fish readily consume this actively moving organism. Some species of fish require food items smaller than *Artemia*. For these species, one solution is to harvest rotifers or zooplankton from natural waters; rotifers can also be cultured in the hatchery. Collection or production of natural food sources has several disadvantages compared to feeding a formulated feed in tank culture. First, the supply of the natural food organisms may be variable and may not be available when needed. Second, predators and pathogens may be mixed in with the wild-harvested organisms. Once pathogens or predators are established in a larval-rearing tank, they are nearly impossible to remove without affecting the larvae one is attempting to rear. Third, once some fish species have been habituated to live food, switching to a formulated feed for the grow-out period is difficult and can result in high mortality.

Supplements have been developed to feed to *Artemia* to improve their nutritional value for larval fish. *Artemia* enrichment consists of a fatty acid supplement since *Artemia* from some sources do not contain sufficient levels of omega-3 fatty acids for larval fish. Another viewpoint is that if the broodstock fish from which the larvae were obtained were not supplied with adequate nutrition, poor quality larvae would be produced. This viewpoint holds that if broodstock feeds were improved then enrichment of *Artemia* would not be needed. Specific examples can be found to support or to refute either viewpoint, but the important point is that *Artemia* enrichment generally improves growth and survival of larvae.

The advantage of feeding live organisms to larval fish in tank culture is that, for some species, this is a more effective method to produce fry than feeding formulated feeds. The feed manufacturing technology required to produce an effective larval diet is still being developed. Although rapid advances are being made, larval diets remain inferior to live food for some larval fish species being reared in production hatcheries.

Formulated feeds are a mixture of ingredients processed into particles, pellets, capsules, granules, or other forms. They are formulated either to meet all of the nutritional needs of the fish as required in tank or raceway culture (complete feeds), or a portion of the nutritional needs of the fish (supplemental feeds). Supplemental feeds are used in ponds where natural food supplies a portion of the nutrients required by the fish. Supplemental feeds are formulated to contain adequate protein and energy, but may be deficient in vitamins and minerals that the fish are able to obtain from natural foods.

Complete feeds are formulated to contain all the essential dietary nutrients including micronutrients (vitamins and trace minerals). In high density fish rearing, whether in a pond or tank, a complete feed must be provided, as natural food is limited or absent. Formulated feeds must be delivered to fish in a form that supplies all essential nutrients. The methods of manufacturing the ingredients used in feeds and the methods used to produce feed particles are chosen to ensure the retention of essential nutrients in pellets after they are immersed in water. Simply mixing a nutrient into a feed blend does not ensure that the nutrient is present or available when the fish consumes the feed. This subject is discussed in greater detail in the section in this chapter on feed manufacturing.

Different feed particle sizes are required for different sizes of fish because fish do not normally chew their food. The texture and palatability of formulated feeds differs from natural foods, so that the optimum size of a formulated feed particle does not always correspond to the size of wild prey organisms suitable for fish of a given size.

When compared chemically and nutritionally, formulated feed and natural food clearly differ. As mentioned earlier, natural foods contain approximately 75–80% water. Dry pelleted feeds contain less than 11% water. Moist feeds usually do not contain more than 35% water. Thus, formulated feeds, on a unit weight basis, contain less water and more lipid, protein, vitamins, and minerals than do natural foods. In feeding studies with walleye, fish consumed a far greater weight of minnows than formulated feeds. Growth rates were higher for the fish fed minnows, but a feed conversion of 4.2 g of minnows per gram of gain was observed compared to 1.4 g of feed per gram of gain for the feed-fed fish. In other words, 3–4 times more minnows (ca. 70% water) are needed compared to the amount of pelleted feed required. When feed conversion ratios are adjusted to an equal moisture basis to facilitate comparison between minnows and pelleted feed, feed conversions are nearly identical. Despite this advantage with minnows, feeding pellets are preferred because the cost of minnow feeding is very high and diseases can be easily transferred to walleye from minnows.

The reasons that minnow-fed fish grow faster than do pellet-fed fish is a topic of discussion among fish culturists and nutritionists. Whether the observed difference in growth was due to lower consumption of the formulated feed, a nutritional difference, or the lower total dietary moisture content of formulated feeds compared to minnows has not been determined. In a study with coho salmon fry fed either adult *Artemia* or pelleted feed, those fed *Artemia* were 35% heavier after four weeks of feeding. Nearly all of this difference in weight was attributable to higher feed intake in fry fed *Artemia* when feed intake was expressed on a dry weight basis. The growth rate of predatory fish fed minnows may be the theoretical optimal growth rate for fish fed formulated feeds, assuming that fish would consume more feed and that the feed was nutritionally complete. With advancements in our knowledge of the nutrient requirements for particular species, increases in feed palatability through formulation and processing technology, and domestication of wild species, we can expect higher growth rates of many species fed formulated feeds.

There are several advantages of formulated feed over natural foods, when feeds are properly formulated, manufactured, and delivered to fit the fish culture situation. First, formulated feed is readily available; many feeds are shelf items of feed manufacturers. The variety of formulations that are readily available is increasing as aquaculture expands. Second, performance of fish that are fed formulated feeds is more predictable (based on past records) since abundance and quality of natural food organisms in ponds can be affected by changes in the weather and other unpredictable factors. Third, formulated feeds can be used to deliver medications or nutraceuticals, and thus treat and prevent disease.

If fish culturists do not pay close attention, problems can occur when formulated feeds are fed to excess or if a feed contains an excessive amount of undersized particles (fines). Excess feed in a tank or excess fines in a feed can lower water quality. Bacterial gill disease is the usual result of overfeeding or of a high percentage of fines in a feed.

Nutrient Requirements

The classes of nutrients to be discussed in this chapter are (1) protein (amino acids), (2) carbohydrates (sugars and starches), (3) lipids (fats), (4) vitamins, and (5) minerals. Also, carotenoid pigments, which have recently been shown to be essential for the viability of salmon fry, will be covered. A nutritionally balanced feed must contain all required nutrients in the proper proportion for the species of fish being fed. If a single essential nutrient is deficient, it will affect the utilization of the other nutrients used in tissue synthesis. In severe cases of nutritional deficiency, clinical signs of deficiency develop. Because all essential nutrients are required to maintain the health of fish, there is no logic to ranking them in terms of importance. However, deficiencies of certain nutrients have more severe consequences than do deficiencies of others. For example, a low level of protein in the feed results in reduced growth, whereas the lack of any one of the essential vitamins causes well-described deficiency signs and, if the deficiency becomes severe enough, death of fish. Essential nutrients such as protein and vitamins should be present in feeds at levels to meet or slightly exceed optimum dietary requirements, but not in excessive levels.

Protein Requirements

The primary objective of fish husbandry is to produce fish, the edible portion of which is over 50% protein on a dry weight basis. Fish digest the protein in natural food and in commercial feeds into amino acids, which are then absorbed in the gut, transported in the blood, and used by cells to synthesize tissue protein. The first priority for use of amino acids is to synthesize functional body proteins (hormones, enzymes, and products of respiration), assuming that dietary intake of energy is sufficient. They are used next for tissue repair and synthesis. Amino acids in excess of metabolic and anabolic requirements are broken down (catabolized) to supply metabolic energy or are converted to fat and stored.

There are 20 principle amino acids found in tissues, categorized as nonessential (dispensable) amino acids and essential (indispensable) amino acids. The nonessential amino acids can be synthesized from metabolic intermediates of the Kreb's (tricarboxylic acid) cycle, which operates in all cells. The essential amino acids must be supplied in the diet because fish cannot synthesize them. Fish require the same 10 essential amino acids as do higher animals: arginine, histidine, isoleucine, leucine, lysine, methionine, phenylalanine, threonine, tryptophan, and valine. Fish fed feeds lacking even a single dietary essential amino acid soon become inactive and lose both appetite and weight. When the missing essential amino acid is replaced in the diet, appetite is restored and growth resumes.

Fats and carbohydrates are the primary dietary energy sources for fish, but protein is also used for energy in the event dietary fat or carbohydrate intake is inadequate to meet metabolic energy demands. Fish are more efficient at using protein for energy than are terrestrial animals because they are highly efficient at eliminating nitrogenous wastes through the gill tissues directly into the water and using the de-aminated amino acids for metabolic energy. Nutritionists try to balance the protein and energy components of the feed to meet the requirements of the fish and to maximize the utilization of dietary protein for tissue protein synthesis. Protein is the most expensive component of fish feeds and only the amount needed to support maximum growth should be included in fish feeds; less expensive fats and carbohydrates can supply energy and spare the protein for growth.

Several factors influence the dietary protein requirement of fish, including fish size, species, feeding rate, and energy content of the diet. Larger fish have a lower protein requirement for maximum growth than do young fish. Species vary considerably in their requirements; for example, catfish need less dietary protein than do salmonids. As mentioned above, for optimal growth and feed efficiency, the protein and energy content of the feed should be balanced. Protein requirements for fish are commonly expressed as a percentage of the diet, but, in fact, fish actually require the amino acids in protein, not protein per se. Assuming that the diet contains a balanced protein (the correct proportion of amino acids), fish and other animals actually have a quantitative requirement for protein. Fish require a specific quantity of protein (amino acids) per kilogram of body weight for optimal growth—the actual amount varying with fish size, species, and so on. Fish nutrition has not yet progressed to stating requirements on a quantitative basis, but the information required to reach that goal is being developed. What the fish culturist needs to remember is that the required amount of protein being supplied to fish depends on the percentage of protein in the feed and the feeding rate. Just because a particular feed has a given percentage of protein does not assure adequate protein consumption if feeding rate is low. Adjusting feeding rate is the most practical way for a fish culturist to change the protein intake of fish.

The quality of a protein, as reflected in the amino acid content, is the most important factor determining the optimum utilization of dietary proteins. If a feed is deficient in any one of the 10 essential amino acids, poor growth and increased feed conversion ratios will result, even if the total protein level in the feed is adequate. The dietary protein source that most closely approximates the amino acid requirements of the fish has the highest protein quality value. Animal protein sources (including fish protein sources) are generally higher quality for fish than plant sources, but animal protein ingredients cost more than many plant protein ingredients. Plant proteins are typically deficient in one or more of the essential amino acids needed by fish. Synthetic amino acids (AA) can be added to feed to make up for low levels in feed ingredients, but there is some question as to how well fish utilize them. Thus, amino acid balance at reasonable cost is best achieved by blending animal protein ingredients, particularly fish meal, with plant proteins.

Fish meal is the main protein source used in feeds for carnivorous fish, such as salmon and trout, and seems to be an essential feed component for many other species. Most of the ingredients of standard trout feed formulas can be replaced with other ingredients, but when fish meal is replaced completely, reduced growth and feed efficiency generally result. Progress is being made at developing feed formulations for salmon and trout with low levels of fish meal or even lacking fish meal entirely. Feeds for catfish and tilapia in which fish meal is completely replaced with other protein sources have been available for years.

Unlike some terrestrial farm animals, fish cannot utilize nonprotein nitrogen sources, such as urea and diammonium citrate. These nonprotein nitrogen sources are used to supply the nitrogenous portion of nonessential AA during synthesis in the body. They are used in feeds for some livestock, but appear to have little value to fish and can be toxic at high dietary levels.

The chemical composition of fish tissue can be altered significantly by the levels and components of ingredients in feeds. Within narrow limits, there is a general increase in carcass percent protein relative to the amount in the feed. Further, there is an inverse relationship between the percentage of fat and that of water in the fish body, as is the case with other animals. Care must be taken when making comparisons between groups of fish of the percentage of carcass water, protein, and fat. Expressing the percent protein on a dry weight basis can lead to erroneous conclusions associated with the moisture-lipid relationship. Expressing the protein and fat percentages on a wet weight basis will prevent mistaken conclusions concerning the effect of diet or feeding level on carcass protein content

The protein and amino acid requirements for Pacific salmon and rainbow trout are similar, but not identical. The protein requirement is highest in first feeding fry and decreases as fish grow. To grow at the maximum rate, fry must have a feed that contains 45–50% available protein; at 6–8 weeks, the requirement decreases to 40–45% of the feed and to about 35–40% of the feed for yearling salmonids. The total amount of dietary protein at each stage depends on the digestibility of protein in the feed and on the dietary energy level.

Recommended protein levels in trout feeds as a percent of the diet are

• Starter feed (fry)	45–50%
• Grower feed (fingerlings)	42–48%
• Production feed (older fish)	35–40%

The natural foods of catfish are rich in protein; catfish likely metabolize some dietary protein from natural food for energy. Protein utilization is affected by the protein source and water temperature. Channel catfish utilize fish meal better than they do the best plant protein source, soybean meal, but soybean and cottonseed meals supply the bulk of dietary protein in commercial catfish grow-out feeds. A combination of protein sources yields better feed conversion ratios than any single source and catfish feeds generally contain meat and bone meal (or fish meal) in combination with soybean and cottonseed meals. In commercial catfish feeds, less than 5% of the dietary protein requirement is supplied by fish meal, and, as mentioned, commercial formulations in which all protein is supplied by plant protein sources are available.

The protein requirement of catfish is also size related. The recommended total percentage protein in feed for different-sized catfish in 23.9°C (75°F) water is

- Fry to fingerlings 35–40%
- Fingerlings to subadults 25–35%
- Adults and broodfish 26–32%

When catfish are fed as much as they will eat and the feed is balanced in amino acids and energy, lower protein levels are adequate. When the feeding rate is restricted, higher protein levels have proved beneficial. Fish fed higher protein feeds require less energy per pound of gain.

Catfish use carbohydrates as dietary energy sources quite readily, in contrast to salmon and trout. Practical catfish feeds contain a high level of carbohydrate relative to salmonids feeds. In addition, most catfish feeds are processed using cooking extrusion, which increases the digestibility of the carbohydrate to the fish.

Catfish require the same 10 essential amino acids that salmonids require. These requirements have been well established for catfish in both laboratory and pond production trials. The amino acid amount requirements for catfish are slightly lower than those of salmonids.

Low stocking rates and low standing crops of fish in ponds result in more natural food (and protein) being available to each fish. The above factors and others, such as season, fish size, feeding rate, water temperature, oxygen levels, and disease status, influence the dietary protein levels required for maximum growth. Consequently, no single protein level in feeds will meet all fish culture conditions. It remains for the fish culturist to choose a feed with a protein level appropriate to specific production situations.

Feeding trials with coolwater fish have indicated a need for similar or slightly higher protein level in the feed compared to that needed for trout. Walleye fingerlings averaging 8 g require approximately 48% protein; fish with an average weight of 50 g require 42% protein. For both walleye and muskellunge fingerlings that are consuming formulated feed well, feeding an open-formula diet with 45% protein results in good growth and survival. This feed and others have been used to raise fish to over 500 g.

Many species of fish now being cultured for both food production and for stocking purposes have only recently been produced in hatcheries. Upon initial investigation, it may appear that the previously uncultured species have a higher protein requirement than some of the domesticated species. However, as previously mentioned, the effect of feeding rate or feed consumption may affect this observation. An undomesticated species may not consume enough of a formulated feed to meet its protein requirement. Three approaches can be used to increase weight gain of these fish. First, a higher protein diet can be fed. This approach may be effective but can be expensive, require special arrangements for feed manufacture, and result in increased levels of ammonia in the rearing water if feed consumption increases. Second, diets can be formulated to include palatability enhancing substances. Palatability enhancing substances are additives, such as krill meal or betaine, that impart a better flavor to the feed and increase feed intake. Palatability can sometimes be improved by altering a feed's texture by using a different processing technology (see the section on manufacturing technology). The enhanced diets may not be needed for extended periods of time. Once fish are actively consuming a formulated feed, they can be slowly switched to a more cost-effective feed. Third, changes in culture techniques may also increase feed consumption. Undomesticated fish sometimes feed better when feed is presented at frequent intervals and when there is no human disturbance. Many types of automatic feeders are available to provide feed, more or less, continuously, without disturbing the fish. Increasing feed consumption allows the fish to consume the protein and also other essential nutrients, such as vitamins, that they require.

Carbohydrates

Carbohydrates are a major source of energy to man, domestic animals, and many omnivorous fish, such as catfish and tilapia. All of the necessary enzymes for digestion and utilization of carbohydrates have been found in fish, yet the role of dietary carbohydrates and the contribution of glucose to the total energy requirement of many fish species remain unclear. Salmon, for example, have elevated blood glucose levels for extended periods when they are fed diets containing relatively small amounts of glucose or cooked (gelatinized) starch.

Very little carbohydrate (usually <1.0% of the wet weight) is found in the bodies of fish. After being absorbed, carbohydrates are either burned for energy, stored temporarily as glycogen, used to synthesize several nonessential amino acids, or converted to fat. No carbohydrate requirements have been established for fish, therefore, carbohydrates are not considered essential nutrients.

The energy requirements of fish may be supplied by dietary fat, protein, or carbohydrate. If sufficient fat and carbohydrates are not available in the feed, protein will be used to supply energy at the expense of tissue growth and repair. The use of dietary carbohydrate for energy to save protein for other purposes is known as the "protein-sparing effect" of carbohydrate.

Carbohydrates consumed in excess of the immediate energy needs of fish and animals are converted into fat and deposited in various tissues as reserve energy for periods of food scarcity. Excessive carbohydrate intake in salmonids can lead to an elevated deposition of glycogen in the liver and eventually death. Carbohydrates are an inexpensive energy source, and there is a temptation to increase their levels in fish feeds. However, some fish (e.g., Pacific salmon and trout) seem to be incapable of tolerating high dietary levels of carbohydrates, and glycogen levels in the liver become elevated. Higher animals excrete glucose in the urine, but such excretion does not occur in trout even when blood glucose levels are greatly elevated. In trout, blood glucose levels observed after fish consume a diet containing high levels of available carbohydrates, follows the same pattern as that in diabetic humans and results in reduced feed intake, lower growth rates, and abnormal livers.

High glycogen liver is a condition resulting primarily from excess levels of carbohydrates in the feed. Fat-infiltrated livers and kidneys in salmonids are a result of fat deposition within the organ, resulting in reduced organ function from tissue destruction (Figure 3). This condition, known as fatty liver syndrome, is less common than it was 20 years ago. Fatty liver syndrome has multiple causes including excess dietary

FIGURE 3. Rainbow trout with liver lipoid degeneration (ceroidosis) of increasing severity from top to bottom. (Courtesy P. Ghittino, Fish Disease Laboratory, Tonino, Italy.)

energy from carbohydrates and other nutrients and deficiencies of vitamin involved in carbohydrate metabolism.

Carbohydrates have important dietary roles, especially as metabolic precursors for synthesis of nonessential amino acids, which are necessary for growth. Thus, in the absence of adequate dietary carbohydrates or fats, fish may use dietary protein to meet their energy and other metabolic needs. In addition to serving as an inexpensive source of energy and a source of metabolic intermediates for synthesis of nonessential amino acids, starches improve the pelleting quality of fish feeds for steampelleted and extruded feeds.

Most species of fish cannot utilize dietary fiber. Levels over 10% in salmonid feeds and over 20% in catfish feeds reduce nutrient intake and impair the digestibility of practical feeds. In addition, they increase fecal output, lowering water quality in hatcheries and in hatchery effluents. Dietary fiber in fish feeds originates from feed ingredients of plant origin and also from chitin, a component of the shells of crustaceans. Chitin is a polymer of glucosamine, a glucose molecule with a nitrogen (amine) attached.

Trout nutritionists have placed maximum digestible carbohydrate values for feeds at 12–20%. Digestible carbohydrate values are determined by multiplying the total amount of carbohydrate in the feed by the digestibility of the carbohydrates. Digestibility values of various carbohydrates for rainbow trout are: simple sugars, 100%; complex sugars, 90%; cooked starch, 60%; raw starch, 30%; and fiber, 0%.

Levels of carbohydrates up to 20% can be tolerated in trout feeds in 12–18°C (55–65°F) water, but if they are fed to trout in water below 10°C (50°F), excessive storage of glycogen in the liver and possibly death of the fish will occur. Carbohydrates should be limited to less than 15% in trout feeds. However, there are beneficial effects from the carbohydrate portion of the feed. It can supply up to 20% of the available calories in a feed, thus sparing the protein. The energy available to mammals from carbohydrates is 4 kcal/g, whereas the value for trout averages only 1.6 kcal/g, a 40% relative efficiency.

The major sources of carbohydrate in trout feeds are the starches in plant foodstuffs, such as soybean meal, grains, and grain by-products. Most animal concentrates, such as meat meal, fish meal, feather meal, and blood meal, are low in carbohydrate (<1.0%). The high percentages of milk sugar (lactose) in dried skim milk, dried buttermilk, and dried whey may cause an increase in blood sugar and an accumulation of glycogen in the liver if fed at levels greater than 10% of the feed.

Dietary carbohydrates are relatively well utilized by catfish, and catfish feeds typically contain about 50% carbohydrates. Channel catfish utilize cooked starch more readily than simple sugars. In feeds containing adequate protein, fish weight increases with the level of starch, but remains essentially the same regardless of the amount of sugar in the feed. Liver abnormalities, poor growth, and high mortality observed in salmonids fed diets containing high levels of dietary carbohydrates are not observed in catfish. No carbohydrate requirements have been established for catfish; however, carbohydrates can spare protein in catfish feeds. In the absence of adequate dietary carbohydrate or fats, catfish make inefficient use of protein to meet their energy and other metabolic needs.

Fiber is not a necessary component for optimum growth or nutrient digestibility in channel catfish production rations. Fiber levels as high as 21% reduce nutrient intake and impair digestibility in feeds for channel catfish. However, fiber in concentrations of less than 8% add structural integrity to pelleted feeds, although larger amounts often impair pellet quality. Most of the fiber in the feed ultimately becomes a pollutant in the water when it passes through the gastrointestinal tract.

Lipids

Lipids comprise a group of organic substances that include fats, oils, waxes, and related compounds. Fats are lipids that are solid at room temperature and include tallow, lard, and butter. Oils such as corn oil, soybean oil, and fish oil are liquids at room temperature. Lipids are the most concentrated energy source of the food groups, containing 2.25 times more energy per unit weight than either protein or carbohydrates. In addition to supplying energy, lipids serve several other functions such as facilitating intestinal absorption of fat-soluble vitamins and as components of cellular membranes and nerve sheaths. They provide essential fatty acids and the substrates for synthesis of hormones and are essential for gonadal development.

There are several classes of lipids in animals and plants. The storage form of lipid in tissues is triglyceride, composed of three fatty acids attached to the carbon atoms of glycerol (a 3-carbon molecule). Phospholipids are essential components of cell membranes and are composed of two fatty acids attached to glycerol and the third component (choline, inositol, or phosphatidylethanolamine) attached to the third carbon of glycerol via a phosphate bond. This gives phospholipids certain functional properties that are essential for membrane functions. Other classes of lipids include sphinglomyelin, which insulates brain cells, and waxes, a specialized form of storage lipids in some deepwater marine fish species. Of the classes of lipids, triglycerides are the most variable in fatty acid content.

Although various fish species have unique fatty acid profiles, tissue fatty acid profiles are altered by diet, up to a point. In general, the fatty acid profiles of body tissues are readily altered by dietary fatty acid intake. Tissues of fish consuming natural foods contain a high proportion of polyunsaturated fatty acids similar to those in the food. Polyunsaturated fatty acids have 3 or more double bonds and 18–22 carbons. Marine fish oils are the main source of polyunsaturated fatty acids, but they are also found in small proportions in plant oils. Plant oils are rich in 18-carbon fatty acids with 1 and 2 double bonds, except for linseed oil, which contains over 50% linolenic acid, an 18-carbon fatty acid with 3 double bonds. All fatty acids have a methyl group (CH_3) at one end of the carbon chain and a carboxyl group (COOH) at the other end. The presence of the carboxyl group makes them weak acids.

Animal fats are rich sources of saturated fatty acids (no double bonds), having 14, 16, and 18 carbons. The melting point of fatty acids increases with carbon chain length and the number of double bonds along a carbon chain. Not surprisingly, water temperature influences the fatty acid composition of the body fat of fish. When water temperature decreases, the percentage of polyunsaturated fatty acids increases, especially in membrane fatty acids. This maintains membrane fluidity at low water temperatures.

There is no single optimal level of dietary lipid in fish feeds. Rather, the species being fed, its life history stage, dietary protein level, lipid source, and goal of production, all influence the choice of dietary lipid level. Lipids are a primary source of energy for fish as previously discussed. High levels in feeds can be beneficial. However, high levels in feed mixtures reduce pellet formation; therefore, a portion of the lipids are often added by coating the pellets, a process called "top-dressing."

The quality of lipids used in fish feeds is of primary concern to feed manufacturers. Lipid oxidation causes rancidity, an ever-present problem in fish feeds. Oxidation is the process by which double bonds in fatty acids are converted to single bonds, a process called saturation. Fatty acids with three or more double bonds oxidize more readily than those with fewer double bonds. Fish feeds typically contain marine fish oils, which have a high proportion of polyunsaturated fatty acids; thus, fish feeds have a higher probability of oxidizing than do other animal feeds. Rancid lipids have a disagreeable paint-like odor and can be toxic to fish. The toxic effects are due to products of oxidation of the lipid itself and to the destructive effects that the oxidation process has on vitamins C and E. Because rancid lipids in the feed are detrimental to fish, every effort should be made to use only stable oils protected with antioxidants. Feeds should be stored in a cool, dry area to minimize lipid oxidation. Dry feeds should not be frozen to extend their storage life. Freezing dry pellets actually accelerates oxidation by concentrating pro-oxidant minerals (iron and copper) and by making it easier for oxygen to penetrate pellets. Aside from storage conditions, there is little that a fish culturist can do to prevent lipid oxidation in feeds; the burden to prevent oxidation rests with the feed manufacturer and his ingredient suppliers. Fish culturists should be alert for signs of lipid oxidation in feeds, such as an odd smell and heat production from feeds (the oxidation process releases heat).

Fatty Acids

As mentioned, fatty acids are composed of carbon atom chains with a methyl (CH_3) group at one end and a carboxyl (COOH) group at the other. Fatty acids of interest in fish and animal nutrition contain 12–24 carbon chains. If the first double bond between carbon atoms occurs at carbon number 3, counting from the methyl end, the fatty acid is called an omega-3 fatty acid. If the first double bond is found at carbon number 6, counting from the methyl end, the fatty acid is called an omega-6 fatty acid. Fish, humans, and animals are able to add double bonds to fatty acids at some positions but not at carbon number 3. Omega-3 fatty acids must, therefore, be supplied in the diet. Marine fish oils are the main source of omega-3 fatty acids in fish feeds. Plant oils, such as soybean oil, contain small amounts of omega-3 fatty acids, but mainly contain omega-6 fatty acids. The only exception is linseed oil from flaxseed, which contains 52 to 55% linolenic acid, which has 18 carbons and 3 double bonds, the first being at carbon number 3.

Fish can utilize both saturated and unsaturated fatty acids, but the digestibility coefficients of saturated fatty acids are lower than those of unsaturated fatty acids, at least for coldwater fish species such as salmon. Dietary fatty acids can be modified via metabolic processes within cells by elongation (adding carbon atoms) and desaturation (making double bonds). However, fish and other higher animals cannot transform fatty acids via these processes to form a group called omega-3 fatty acids. Omega-3 fatty acids have their first double bond at the third carbon from the methyl end and this gives them special properties. Because omega-3 fatty acids cannot be synthesized in the body, they are essential nutrients in the diet. Factors to be considered in evaluating dietary lipid sources for use in fish feeds include digestibility, content of fatty acids (omega-3 fatty acids) essential for the fish, presence of toxic substances, and the quality of the lipid (absence of oxidation, low water content, low protein content, and low free fatty acid content).

Salmon and trout require about 1% essential fatty acids of the omega-3 family in their diet. As long as at least 1% of the diet is omega-3 fatty acids, the source of the remaining dietary fat and its fatty acid composition has little effect on growth or egg hatchability. Fish tissue and eggs will reflect the dietary fatty acid profile, making it possible to alter the muscle fats to be high in omega-3 fatty acids or high in less saturated fatty acids, which increases frozen storage product quality.

Trout in the wild consume a variety of prey containing a range of fatty acid profiles. Insect lipids are 60 to 70% unsaturated, with 30 to 40% being polyunsaturated fatty acids. About 8 to 12% of these fatty acids are omega-3 fatty acids. Freshwater crustaceans (cladocerons, daphnia, copepods) contain 20 to 40% omega-3 fatty acids, as do small fish that trout consume. As mentioned earlier, the average lipid content of trout prey is 16–17% Thus, between 1.5 and 3% of the wild trout diet is omega-3 fatty acids. Insects and freshwater crustacean also contain arachidonic acid, which is an omega-6 fatty acid with 20 carbons and four double bonds, as 2–4% of their total lipids, 2–3 times more than marine prey items. Some researchers believe that the dietary fatty acid intake of Pacific salmon fingerlings in hatcheries should be similar to the fatty acid intake of wild salmon in freshwater.

As mentioned, the fatty acid profile of salmon and trout reflects that of their diet. A key point is that the body lipid profile reflects dietary intake, but, it is not equivalent in absolute levels of fatty acids. Fish cannot accumulate saturated fatty acids in their body beyond a certain level, nor can they decrease their body level of polyunsaturated fatty acids below a certain level. The main difference in body fatty acid composition associated with dietary intake is in oleic and linolenic acids (18 carbons, with 1 or 2 double bonds, respectively). This difference reflects the main differences in fatty acid composition among the three main sources of dietary lipid in fish feeds, namely fish oils, plant oils, and animal fats.

Most fatty acids are associated with triglycerides, the storage form of fat, or with phospholipids, which are biologically active lipids found in all cell membranes. Other minor classes of lipids are associated with brain and nerve tissue, waxes, and wax esters. The triglyceride content of fish tissues increases and decreases with dietary energy intake and caloric expenditures associated with activity, migration, and maturation. The phospholipid content of fish tissues is relatively constant.

The trend in recent years has been toward higher levels of both protein and fat in trout feeds, while in salmon feeds, the trend has been toward lower protein and higher fat levels (Figures 4 and 5). In both cases, feeds have become more nutrient dense, containing higher quality protein and a higher caloric content, resulting in improved feed conversion ratios and less feed handling at the hatchery or farm. Trout fed high-fat (>25%) feeds have higher visceral fat levels than trout fed diets containing 15% or less fat and this higher body fat level decreases the yield at processing by up to 4% of body weight.

High-energy feeds are desirable for many reasons, including a lower amount of feed that must be handled and lower fecal pollution, but, in

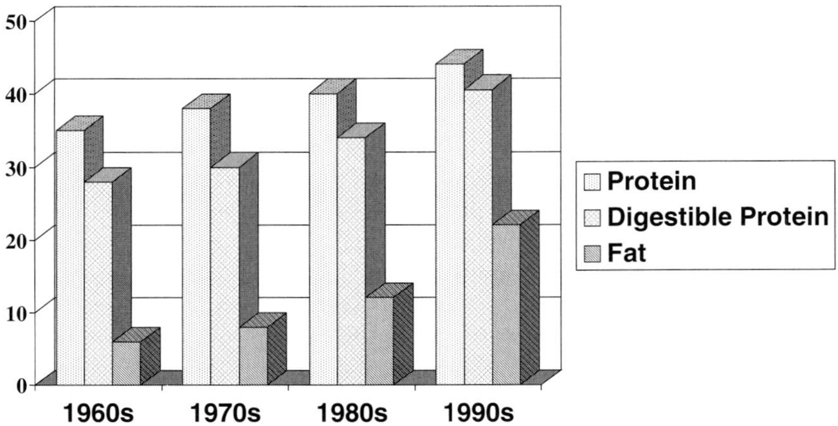

FIGURE 4. Changes in total protein, digestible protein, and lipid levels in trout feeds over the past 40 years.

salmon and steelhead trout hatcheries, these feeds can lead to high body fat levels in smolts at release. During fall and winter, salmon and steelhead trout enter a phase of growth dormancy, during which protein synthesis, controlled by hormones (growth hormone and IGF-1), essentially stops. This stage ends in late winter. During the dormancy stage, dietary protein and carbohydrate are converted to stored fat. Reducing feed intake during this stage slows growth and reduces body fat levels but can increase size variability within groups of fish. Hatchery studies are show-

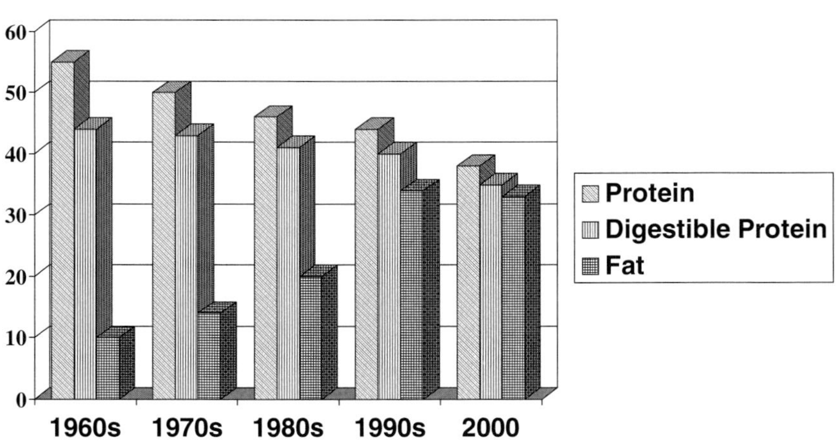

FIGURE 5. Changes in total protein, digestible protein, and lipid levels in salmon feeds over the past 40 years.

ing that these problems can be reduced by withholding feed from the fish for short (1–2 week) periods, a practice called intermittent feeding. Initial trials at the Hagerman National Fish Hatchery, Hagerman, Idaho, with steelhead trout have lowered body fat concentration by 2–3% over continuously fed fish without increasing size variability or lowering smolt quality. Research to determine the best ways to employ the intermittent feeding strategy and its effects on body composition and posthatchery release survival are continuing.

In contrast to salmon and trout, catfish are not fed high lipid feeds. This is partially due to the fact that carbohydrates can supply a large proportion of the dietary energy needs of catfish and also out of concern for product quality. Catfish require both omega-3 and omega-6 fatty acids in their feed, but smaller amounts are needed compared to salmon and trout; about 0.5% of each category is sufficient. Residual oil in soybean meal, cottonseed meal, fish meal, and ground corn supply these essential fatty acids along with added oil from fish or plant sources.

Lipids have the highest effect on sensory attributes and storage quality of fish products after harvest. Tests with animal and vegetable fats show that high levels of fish oil in feeds have a significant adverse effect on the flavor of fresh and frozen catfish fillets. Beef tallow also influenced the flavor, but does not induce the "fishy" flavor produced by fish oil. Fish fed diets containing safflower oil or corn oil have a flavor that is preferred in taste tests to fish fed diets containing beef tallow or fish oil. Catfish producers use animal fats and fish oils in fingerling feeds to obtain rapid growth and efficient deposition of protein, then change to a finishing diet containing vegetable oils to improve the flavor as the fish approach market size. Increasing the vitamin E content of catfish feeds fed near the end of production, before harvest, increases the a-tocopherol content of muscle tissue, which improves frozen storage stability and resistance to lipid oxidation.

Energy

Energy is defined in physics as the capacity to do work. Work can be mechanical (muscular activity), chemical (tissue repair and synthesis, maintenance of biological salt balance), or heat to maintain body temperature. Fish require energy for growth, activity, reproduction, and osmotic balance, all of which constitute the cost of living. Energy requirements of fish species differ, as do their growth rates and level of activity. Other factors that alter the energy requirements are water temperature, size, age, metabolic rate, composition of the diet, day length, and environmental stresses.

Food energy is expressed as calories or joules. A calorie is the amount of heat required to raise a gram of water by 1°C; a kcal (or Cal) is one thousand calories. A joule (J) is another term used to describe electrical energy and is equivalent to 0.24 calories. Thus, a calorie is equal to 4.167 J, and a kcal is equivalent to 4167 J (or 4.167 kJ). The total energy value of a food or feed is the amount that is liberated when the material is completely combusted or burned. Total energy (in nutrition called gross energy) is measured by burning a sample of known weight in a bomb calorimeter and measuring the amount of heat that results. Digestible energy of a food or feed is the amount that can be digested and absorbed in the gastrointestinal tract (Figure 6). Digestible energy is calculated by measuring the gross energy of a sample and subtracting the energy content of the feces. The difference between gross and fecal energy is digestible energy, the percentage that was absorbed by the fish or animal. In practical fish feeds, digestible energy is between 75 and 85% of gross energy. As mentioned earlier, some protein is metabolized for energy, with the amine portion of amino acids being excreted in the urine or by the gills. This excreted portion contains energy that is lost to the fish. When this energy loss is subtracted from digestible energy, the result is called metabolizable energy. Measuring metabolizable energy requires quantitative collection of urine and gill excretions in a metabolic chamber and is a time-consuming process. Even though it can result in more accurate information about feed ingredients and feeds, metabolizable energy values are not used in practical fish feed formulation; rather, digestible energy values are the standard values used to formulate feeds. Further loss of energy occurs due to the chemical production of heat in nutrient metabolism with the remainder of the energy called net energy. Net energy is further subdivided into net energy for maintenance and net energy for production. The

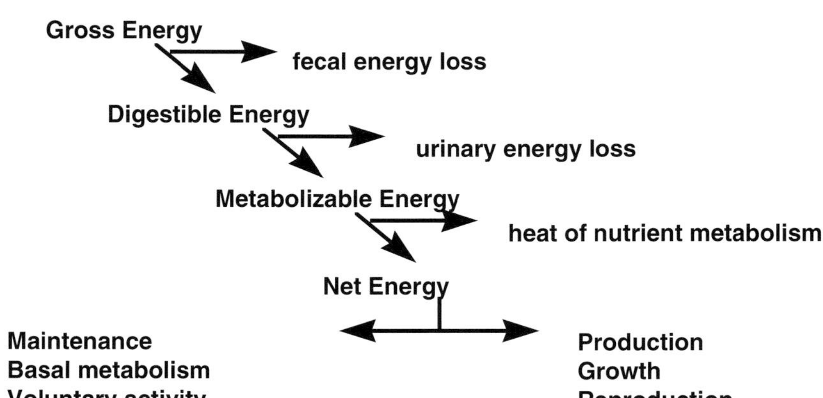

FIGURE 6. Schematic description of energy value of feed and its fate after ingestion.

net energy system is used extensively in terrestrial animal production, but has not been developed for aquatic species.

Energy is released in two forms in metabolism—heat energy and free energy. Heat energy maintains body temperature in warm-blooded animals, but this energy is lost to fish. Usually, the body temperature of a resting fish will be at or near the environmental water temperature. Free energy is available for biological activity and growth. It is used for immediate energy and for formation of body tissue. If more free energy is obtained from food than is needed, it is stored as fat (triglycerides) or, to a lesser extent, as glycogen. Fat is the long-term storage form of energy, while glycogen is used for short-term energy storage. Utilization of glycogen for energy is faster and less metabolically expensive than utilization of energy from fat. Body fat reserves contain much larger amounts of energy than do body glycogen stores.

Fish have been said to eat to satisfy their energy needs and, therefore, to adjust their feed intake according to dietary energy level. While this appears to be true to some extent, data obtained with rainbow trout fed high-energy feeds to apparent satiation show that the relationship between dietary energy level and feed intake is not clear-cut. An excessively high energy level in a feed (>25%) may result in a slight (~5%) reduction in feed intake; in well-balanced feeds, feed intake is not exactly proportional to dietary energy level. At the extremes of dietary energy level, fish gain weight at a rate comparable to those fed high-energy feeds, but they eat more feed to do so. However, within a normal range of dietary energy levels, feed intake (using satiation feeding) is more or less equivalent, and fish gain slightly more weight when fed high-energy feeds. Earlier studies, which indicated that fish eat to satisfy their energy needs, may have been complicated by the fact that feed intake was restricted. The important point to consider is that it is difficult to recommend a specific dietary energy or protein level for any fish species without considering the relative level of one to the other. Absolute figures on optimum energy requirements are difficult to state in fish nutrition because fish can be maintained with little growth on a low-energy intake or be forced to produce more weight (and more tissue fat) by feeding them to excess. To maintain optimum growth and the efficiency of a feeding program, feeding levels should be adjusted if energy levels of the feeds vary significantly. Feeding levels should be increased for low-energy feeds and decreased for high-energy feeds if the aim of production is efficient feed utilization. Energy needs for maintenance increase with rising water temperatures and decrease when temperatures are reduced, thus necessitating changes in feeding rates.

Fish normally use about 70% of dietary energy for maintenance of their biological systems and for activity, leaving about 30% available for growth. Energy requirements for vital functions must be met before energy is available for growth. Maintenance feeding programs are designed to supply the minimum energy and other essential nutrients for the vital functions and activity, with no allowance for growth. Feed conversion ratios are useful indicators of adequate dietary energy intake. In trout that are growing normally, FCR of 1.2 lb of feed to 1.0 lb of gain or less indicates that dietary energy requirements are being met. With high-energy feeds, FCRs of 1.0 or less are routinely observed in salmon and trout farming. Remember that FCRs are routinely expressed as a ratio of dry feed to wet weight gain, so values less than 1.0 can be calculated. If FCR values are expressed on a dry-weight basis for both feed and fish weight gain, values of less than 1.0 will not be found.

Brook, brown, rainbow, and lake trout have similar dietary energy requirements. Between 3,740–3,960 digestible (also sometimes called available) dietary kcal/kg of feed (1,700–1,800 kcals/lb of feed) are required to produce a pound of trout, depending upon the feed being fed and conditions under which the fish are reared. In practical terms, this corresponds to a feed containing about 4,000 kcal/kg of diet, with a dietary protein level of 42% and a dietary fat level of about 20%. The amount of available calories from fish feeds depends upon the digestibility of nutrients by the fish.

Feed category	Gross energy (kcal/g)	Digestibility (percent)	Digestible energy (kcal/g)
Protein	5.6	89	5.0
Fat	9.4	90	8.5
Carbohydrate	4.1	40	1.6

The values above show that salmonids obtain more calories of energy from fat than from protein or carbohydrates. There is some evidence that trout must use a small amount of protein for energy. In trout feeds, about 50–55% of the total digestible dietary calories are from protein.

The available calories in 100 g of a salmon or trout production feed can be calculated as follows:

Nutrient	Percent of feed	Digestible energy (kcal/g)	Energy content
Protein	45% x	5.0 =	225 kcal
Fat	22% x	8.5 =	187 kcal
Moisture	10% x	0.0 =	0.0 kcal
Ash	2% x	0.0 =	0.0 kcal
Carbohydrates	25% x	1.6 =	40.0 kcal

Total = 452 kcal/100 g (4520 kcal/kg), or 2,055 kcal/pound

A theoretical FCR can be calculated for salmonids by using the energy required to produce a pound of fish and the available calories in the feed:

$$\frac{\text{kcal to rear a pound of trout (1,700)}}{\text{Available kcal per pound of feed (2,055)}} = 0.83 \text{ feed conversion ratio}$$

In production settings, FCRs are always slightly higher than this theoretical value because of the presence of fines in most feeds, some uneaten feed, fish mortalities, and other production scale characteristics.

The number of digestible kcals required to produce a pound of catfish vary from 881 to 1,075 kcal, depending on the feed and size of fish. Growth and FCRs demonstrate that larger catfish require lower levels of protein and higher levels of energy than smaller catfish. Nutrient digestibility and energy values for catfish are

Nutrient	Gross energy (kcal/g)	Digestibility (percent)	Digestible energy (kcal/g)
Protein	5.6	80	4.5
Fat	9.4	90	8.5
Carbohydrate	4.1	70	2.9

The available calories in catfish feeds and estimated FCRs can be calculated by the same procedures as shown above for salmonid feeds, substituting appropriate values for catfish. Bear in mind that the digestibility of protein sources varies widely, both among different ingredients and also among different sources of the same ingredient. Thus, the calculations shown above are for an average feed, but these values might be different for another feed having a different formulation or for a feed produced with the same ingredients, but from different sources. An accurate value can be estimated if one knows the actual digestibility coefficients for the protein, fat, and carbohydrate sources used in a feed. Digestibility data on all feed ingredients used in fish feeds is incomplete, but for some species, especially channel catfish and rainbow trout, digestibility data is rapidly being obtained.

Vitamins

Vitamins are nutrients that are required in small quantities by all forms of plant and animal life. They were named as such because the vitamins that were first identified were amines and also vital for life, therefore, they were called "vital amines" or vitamins. Most vitamins are catalytic in nature and function as part of metabolic enzyme systems found in all cells.

Historically, vitamins were classified as fat-soluble or water-soluble vitamins, based upon how they were first extracted from foods. The fat-soluble vitamins found in lipid fractions of foods are vitamins A, D, E, and K. The water-soluble vitamins are vitamin C (ascorbic acid) and those of the B complex: thiamine (B_1), riboflavin (B_2), biotin, folic acid, cyanocobalamin (B_{12}), niacin, pyridoxine (B_6), pantothenic acid, plus inositol and choline, which mainly function as components of phospholipids.

Vitamins are present in most feed ingredients used in fish feeds. Some, such as yeast, contain high levels of B vitamins, while fish and plant oils contain high levels of vitamins A, D, and E. The vitamins supplied by feed ingredients usually have been ignored in feed formulations, with vitamin premixes formulated to supply the total vitamin needs of the fish. Recently, estimations have been made of the contribution of vitamins in feed ingredients for catfish; vitamin premixes for catfish feeds have been adjusted appropriately, thus lowering the cost of feeds.

Vitamins used in fish and animal feeds are produced commercially by chemical synthesis, extraction from natural products, or microbial synthesis. These products are chemically and biologically similar to naturally occurring vitamins. Synthetic vitamins can be added to feeds with great precision as a mixture (referred to as a premix) to complement the natural vitamins and balance the vitamin content of the finished feed.

The levels of vitamin premix added to feeds should provide for an excess for several reasons: (1) the efficiency with which fish use the vitamins that are naturally present in ingredients is unknown except for catfish to some extent; (2) some vitamins in fish feeds can be destroyed during feed manufacture and storage; (3) lipid oxidation in the feed can destroy some vitamins (C and E); and (4) some vitamins can react with other compounds in the feed and become biologically inactive. Even though vitamins represent a considerable portion of the cost of the feed, slight over-supplementation is considered a safety precaution (Table 1). If just one of the vitamins is somehow made unavailable, the whole feed will be of little value.

Because of the cost and instability of certain vitamins, much research has focused on developing protected forms of these vitamins for animal and fish feeds. In the past, loss of vitamins, especially vitamin C, was a major problem for fish feed producers. Today, protected forms of vitamin C are used, and, while losses still occur during feed manufacture and storage, these losses are much less than in the past when unprotected forms of vitamin C were used. However, feeds are still normally supplemented with extra vitamin C to ensure adequate levels remain in feeds by the time they are fed to fish.

Lipid oxidation causes the levels of naturally occurring vitamin E (α-tocopherol) to decrease continuously from the time a feed is manu-

TABLE 1. Dietary NRC vitamin recommendations for salmonids and warmwater fishes, expressed as mg or IU/pound (kg) dry diet.

Vitamin	Trout & salmon feeds	Warmwater fishes complete feed
A (IU)	1136 (2500)	1000 (2200)
D_3(IU)	1090 (2400)	500 (1100)
E (IU)	22.7 (50)	30 (66)
K	36.3 (80)	2 (4.4)
Thiamin	0.45 (1)	2.5 (5.5)
Riboflavin	7 (7)	6.1 (13.2)
Pyridoxine	2.7 (6)	5 (11)
Pantothenic acid	9 (20)	16 (35)
Biotin	0.07 (0.15)	none*
Choline	450 (1000)	125 (275)
Vitamin B12	0.005 (0.01)	0.005 (0.01)
Niacin	4.5 (10)	10 (22)
Ascorbic acid	22.7 (50)	22.7–45.4 (50–100)
Folic acid	0.9 (2)	1.0 (2.2)
Inositol	136 (300)	none*

*feed ingredients and/or synthesis by intestinal microorganisms supply these vitamins.

factured until it is fed; vitamin E serves as a sacrificial antioxidant. Protected vitamin E (a-tocopherylacetate), the form added to feeds, is not lost during feed storage, but it does not protect the feed from oxidation during storage. It only becomes biologically and chemically active in the gut after the acetate molecule is enzymatically removed from the tocopherol molecule.

Because the metabolic processes and functions of biological systems of fish are similar to those of other animals, it is safe to assume that the same vitamins are required by all species, with the exception of vitamin C, which is required by primates, guinea pigs, fruit bats, and fish, but not by other mammals or birds. However, the recommended dietary intake of various vitamins differs among fish species. The required levels of each vitamin must be present in the diet on a continuous basis to prevent deficiencies (Figures 7 and 8). Vitamin deficiencies signs in fish are well described.

Theoretically, vitamin levels should be increased in high-energy feeds compared to low-energy feeds because the nutrient density of these feeds is higher. In other words, less feed per unit weight gain is needed compared to a low-energy feed. Vitamins are usually supplemented to excess in fish feeds, but culturists should observe fish carefully, noting any loss of appetite (the first and most common sign of vitamin deficiency) or other deficiency signs. Vitamin deficiencies are still found in fish cul-

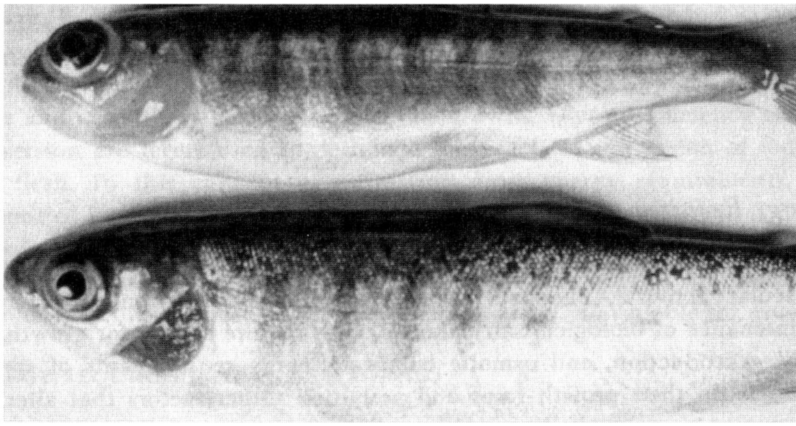

FIGURE 7. Folic acid-deficient (top) and control (bottom) coho salmon. Note the extremely pale gill, demonstrating anemia, and exophthalmia in folic acid-deficient fish. (Courtesy C. E. Smith, U.S. Fish and Wildlife Service, Bozeman, Montana.)

ture, despite the fact that the dietary needs of vitamins are well known, so fish culturists should be aware of common deficiency signs in fish.

In the past, complete catfish feeds were supplemented with levels of all of the essential vitamins in amounts required by the fish and were designed to support rapid growth of fish irrespective of the contribution of natural foods or vitamins present in feed ingredients. As mentioned, recent research has documented the contribution of certain vitamins in feed ingredients toward the requirements of the fish, and now commercial catfish feeds are formulated accordingly. Supplemental feeds contain the vitamins supplied by the feed ingredients plus a lower level of vitamin premix, as it is expected that the fish will obtain a portion of their vitamin requirements from natural foods in the pond.

Minerals

Minerals are difficult nutrients to study in fish. Nearly all essential minerals can be absorbed directly from water by the gills and skin, often in quantities sufficient to meet their metabolic needs. Phosphorus is a notable exception to this, however, probably because phosphorus is rarely found in an unbound state in the aquatic environment. Absorption of minerals in the digestive system from swallowed water also supplies essential minerals to fish. The specific qualitative and quantitative dietary mineral needs, therefore, depend upon the environment in which the fish is reared and on the type of diet being fed. Dietary requirements for some minerals have been established for fish; fish likely require the

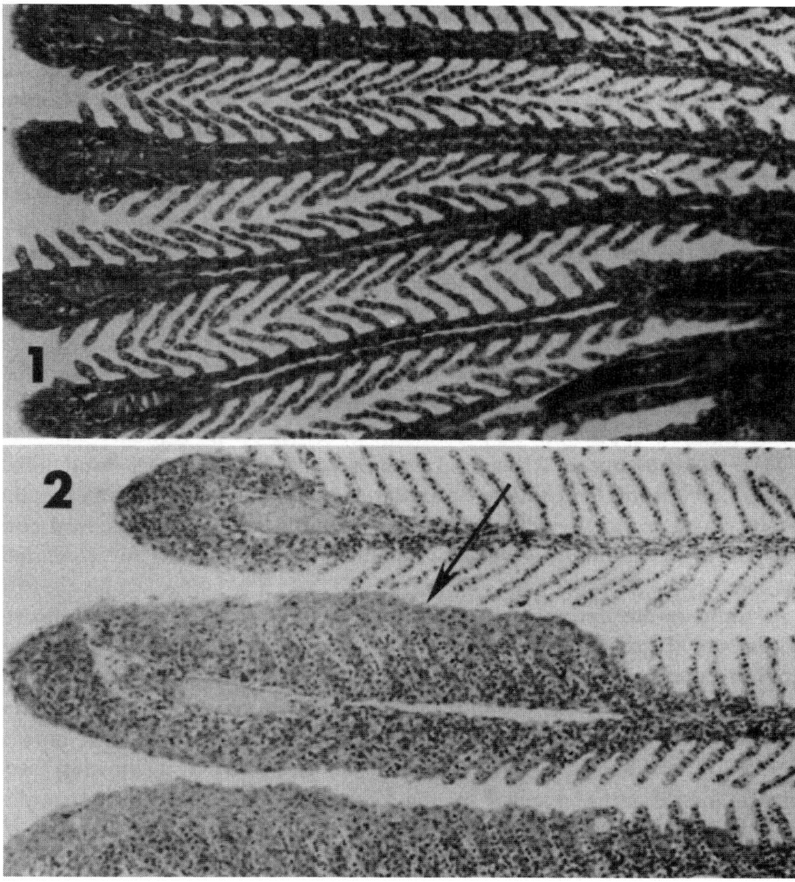

FIGURE 8. Gill lamellae from (1) a normal and (2) a pantothenic acid-deficient rainbow trout. Hyperplasia of the epithelium has resulted in fusion of most lamellae (arrow) on two filaments of the pantothenic acid-deficient trout. (Courtesy C. E. Smith, U.S. Fish and Wildlife Service, Bozeman, Montana.)

same minerals as other animals for growth and various metabolic processes. As mentioned, fish also use mineral salts and ions (e.g., sodium, potassium, and chloride) to maintain osmotic balance between body fluids and their environment.

Minerals are classified as macro or micro (trace), depending on the amount required in the diet. Seven macrominerals are required in large diet amounts (g/kg) and constitute 60–80% of the total mineral content in the body. The seven macrominerals are calcium, phosphorus, sulfur, sodium, chlorine, potassium, and magnesium.

The nine essential trace minerals are iron, copper, iodine, manganese, cobalt, zinc, molybdenum, selenium, and fluorine; all are required in mg/kg diet amounts. These essential nutrients are only required in very small or trace amounts because they function primarily as catalysts or cofactors in enzymatic reactions, rather than as major structural components of tissues. Mineral elements, both macro and trace, are interrelated and balance each other with respect to their nutritional and physiological effects. The minerals that form the hard and supporting structures of a fish's body (bone and teeth) are principally calcium and phosphorus. Small amounts of fluorine and magnesium also are essential for the formation of bones and teeth. Iron, copper, and cobalt are required for red blood cell production; deficiencies of these trace elements cause anemia. Sodium, chlorine, and potassium mainly regulate body processes and osmotic pressure. Minerals also are required for reproduction. They are transferred from female body tissues to the ovaries during maturation and must be replenished after spawning.

Under normal conditions, chloride ions are exchanged very rapidly from both food and water. Calcium and cobalt are absorbed efficiently from the water, but are utilized relatively poorly from feeds. The level of calcium in the water influences the uptake of the calcium from the food and vice versa. Feeds are the major source of phosphorus and sulfur. Inorganic phosphorus is absorbed efficiently from the intestinal tract of trout. Regulation of phosphorus levels in the body of fish is mainly accomplished by excretion of excess phosphorus in the urine in freshwater, although there is some evidence of intestinal regulation of phosphorus absorption when dietary intake is sufficiently high. The skin (including the scales) of trout is a significant storehouse for calcium and phosphorus; skin is an excellent tissue to assess phosphorus status, at least in fish having scales. As one would expect, scale-less fish have a lower dietary phosphorus requirement than do scaled fish.

Mineral deficiencies are most often caused by antagonistic interactions among feed ingredients that lower the bioavailability of certain minerals. Zinc deficiency, which is manifest as bilateral cataracts, is the classic example of this type of problem. A particular combination of dietary ingredients used in the early 1980s in Pacific salmon fingerling feeds (high calcium and phosphorus level from fish bones plus phytic acid from cottonseed meal) lowered dietary zinc availability of the feeds. This resulted in the appearance of bilateral cataracts in an estimated 40 million smolts being raised in state and federal salmon hatcheries for release as smolts to enhance salmon populations. The cataract condition was irreversible in affected fish, but preventable once the causes were understood. At the time, salmon feeds contained about 65 ppm zinc; this should have satisfied the dietary requirement of the fish, which was found to be 15–30 ppm in semipurified diets. Fortifying the diets with an additional 75 ppm (75 mg/kg diet) zinc as zinc sulfate and setting upper limits on the ash level in fish meals used in salmon fingerling feeds has prevented the condition in juvenile Pacific salmon hatcheries since then. Cataracts have remained a problem in Atlantic salmon farming, especially as diet formulations have changed to include higher levels of soybean meal, which contains the compound phytic acid, an antagonist of zinc.

In recent years, efforts to lower phosphorus levels in feeds to comply with stringent water quality guidelines have resulted in apparent phosphorus deficiencies in rapidly growing Atlantic salmon fed high-fat diets. It appears that a combination of poorly available dietary phosphorus and high-energy levels in feeds created a deficiency. This illustrates the relationship between dietary energy level and requirements expressed on a mg/kg basis. Dietary requirement would be better expressed on the basis of the dietary energy content, as is done in poultry, but fish nutrition today is not yet well enough understood to permit this.

Nonnutritive Factors

Although nonnutritive factors do not contribute directly to the maintenance, growth, or reproduction of fish, they must be considered in feed formulation as they can affect feed efficiency, feed quality, and the quality of the final marketable product, both positively and negatively. Three nonnutritive categories of feeds—carotenoid pigments, fiber, and antioxidants—warrant discussion concerning fish nutrition.

Carotenoid Pigments

Producers of salmon and trout must use feeds containing carotenoid pigments to color the flesh and skin of the fish. Salmon and trout cannot synthesize these carotenoid pigments, but rather obtain them in nature (and in captivity) from their diet. Astaxanthin and, to a lesser extent, canthaxanthin are the natural pigments that color salmon and trout. They can be added to the feed as natural products, such as shrimp, krill, specific yeast, and algae species, or as synthesized products. Xanthophylls from corn gluten meal and dried egg products will also pigment fish, giving a yellow color to trout skin and flesh. Generally, salmon and trout feeds are supplemented with pigments to achieve 45–50 ppm (μg/g diet) pigment in the diet. The rate of pigment deposition varies with fish species, as does the final concentration in the flesh. In commercial production, feeds containing pigments are fed for 6–12 months before harvest.

Higher dietary levels will pigment fish more rapidly, but also less efficiently. Rainbow trout can actually be pigmented to acceptable levels within 30 d, if necessary. Dietary carotenoid supplementation of feeds is being used increasingly in conservation and fishery enhancement hatcheries to produce salmon and trout that resemble wild fish and are better camouflaged after hatchery release, thus potentially lowering their susceptibility to predators.

Muskellunge fed fish meal based feeds have a very silver, unnatural appearance. When these fish are fed minnows, they are heavily pigmented and look much different than their pellet-fed counterparts. Studies have indicated that postrelease survival of muskellunge fed minnows for 30 d before release is higher than that of pellet-fed fish. Other studies have indicated that the difference in appearance leads to greater losses due to predation in groups fed fish meal based diets. Feeding diets high in corn gluten meal (yellow pigments) supplemented with the algae *Spirulina* (green pigments) to muskellunge results in fish with a natural appearance. A three-year stocking study conducted by the Iowa Department of Natural Resources is currently evaluating the postrelease survival of fish fed feeds with and without supplemental carotenoid pigments. If the results of this study demonstrate positive effects of feeding pigmented feeds, similar efforts may also be undertaken to evaluate the effect of adding carotenoid pigments to feeds of fish being raised for stocking.

Fiber

The digestibility of fiber by most fish species is extremely low, less than 10%, which is similar to that of poultry, swine, rats, and man. Very little microbial breakdown of fiber has been noted. Herbivorous fish (e.g., grass carp) can tolerate higher amounts of fiber than carnivorous fish. Crude fiber should not exceed 10% in fish feeds and preferably not be higher than 5 or 6%. Some fiber is useful, however, because it supplies bulk and facilitates the passage of food through the intestinal tract of fish. Since fiber is not digested, it passes through the fish; excessive dietary levels increase fecal output, requiring more effort to clean ponds and raceways and contributing to solids in hatchery effluents.

Antioxidants

Fish feeds contain high levels of polyunsaturated oils, which are susceptible to oxidation (rancidity). This can be controlled by adding antioxidants such as butylhydroxytoluene (BHT), butylhydroxyanisole (BHA), ethoxyquin, and vitamin E to fish oil and feed. The levels of BHT, BHA, and ethoxyquin allowed in feeds are not always adequate to control lipid oxidation. Therefore, feed formulators should add antioxidants to permitted levels to protect the oils in fish feeds and should also supplement feeds with vitamin E if additional protection against lipid oxidation is needed. Ethoxyquin and vitamin E function in both fish tissues and feed to control oxidation. The level of vitamin E in fish feeds must be adequate to prevent oxidation of oils and still meet the nutritional requirement of the fish. As mentioned earlier, oxidation of fish oils in feeds is a constant problem, especially in high-fat fish feeds. Culturists should be alert to the smell of oxidizing fish oil (a drying paint smell) and especially alert to feed bags that feel hot to the touch. Feed suspected of being oxidized should not be fed to fish and the feed manufacturer notified that oxidative rancidity is suspected in his feed. Oxidation in feeds is relatively easy and inexpensive to detect with chemical tests, such as peroxide value (POV) and thiobarbituric acid (TBA). Both of these chemical tests measure intermediate products of oxidation, so their levels increase during oxidation and then decrease as oxidation reactions proceed to completion. Care must be taken to avoid over-interpretation of POV and TBA values and to appreciate the fact that low POV and TBA values do not necessarily indicate the absence of lipid oxidation. Low values may indicate that oxidation has already occurred and that various oxidation reactions have gone to completion. If oxidation is suspected in feeds, samples can be sent to feed or nutrition research laboratories at universities or to commercial food testing laboratories for analysis.

Toxins and Toxicants

Numerous naturally occurring and synthetic compounds are toxic to fish. Tannic acid, aflatoxin (produced by mold), and cyclopropenoid fatty acids (found in cottonseed oil) all induce liver cancer in fish. Gossypol, a toxin present in untreated cottonseed meal, causes anorexia and ceroid accumulation in the liver. Trypsin inhibitors in soybean meal, which reduce the activity of digestive enzymes (trypsin) in the GI tract, can be destroyed by proper heating of the meal during processing. Phytic acid, which is the storage form of phosphorus in seeds, reduces the availability of zinc and other minerals. Chlorinated hydrocarbons occur as contaminants of fish meal and fish oil and can cause mortality when present in fry feeds. Broodfish transfer these compounds from the feed to their eggs, resulting in low hatchability and high mortality of swim-up fry. Toxaphene affects the utilization of vitamin C in catfish and can cause the "broken back syndrome." The environment and feed should be free of toxicants to maintain the health and efficient production of fish.

Factors Affecting Fish Quality and Off-Flavor

Substances such as algal blooms, muskgrass, chemicals, and organic compounds can produce undesirable flavors in fish. When the water temperature is high, as it is in late summer, there is a greater chance of off-flavors to occur in fish. This is principally a problem in pond-reared fish, not in those reared in spring-fed, flow-through systems. Sensory attributes of fish, for example, color, flavor, and odor, are all subject to modification by diet, as discussed earlier. Texture does not seem to be affected by diet, except when very high-fat diets are fed or when certain lipid sources are used in feeds. Fat level and fatty acid profile are quality attributes that can be easily manipulated by diet.

Formulation Goals and Feed Selection Criteria

Nutritionists formulate feeds based upon the nutrient requirements of a species, its life history stage, and practical feed manufacturing considerations. It is the fish culturist who must select the best feed to meet specific fish culture goals, such as fish growth rates and water quality criteria. Economical fish production is always important, but in some fish culture situations, economical fish production (fish survival) depends on the fish eating the feed. An example of this is during a disease outbreak, when medicated feed must be used. Some medications are bitter tasting and reduce feed intake dramatically when they are included in feeds. Adding palatability enhancers to feeds, such as krill or liver meal, can maintain feed intake by fish and thus ensure that fish ingest the necessary medication. Palatability-enhancing feed ingredients are expensive and should be used only when they are needed, such as in special-use feeds (e.g., starter diets or medicated feeds). The important point to remember is that feed formulation and selection take into account a large number of considerations, as described below.

Formulating and selecting feeds for a particular physiological state of fish is termed "phase feeding." The use of a phase feeding system allows the culturist to minimize input (feed costs) and maximize output (fish performance). Specific feeds can be formulated to maximize survival, fish growth, or fish health (immunocompetence), to optimize fish quality and body composition, or to minimize the amount of nutrients in hatchery effluent (pollution reduction). No single feed, however, can optimize each of these goals. The culturist needs to consider which goal or goals are the most important for the particular species, its life stage, the season, and the culture situation.

Another important characteristic of a feed is the type of processing used to form the pellets or particles. There are several types of feed processing technologies in use today and each has advantages and disadvantages for a given situation. The goal of feed processing is not to just form the ingredients into a pellet, but to deliver all the nutrients to the fish and, perhaps, even increase digestibility of the nutrients in the feed. In addition, pellets must retain nutrients during storage and even for a period in water until they are consumed. Water stability is not much of a problem with actively feeding species such as rainbow trout; fish eat pellets within seconds. However, first-feeding fish or slow-feeding fish, such as sturgeon, present a challenge for feed producers, the challenge being how to produce pellets that are relatively water stable without lowering nutrient availability. The effect of feed manufacturing technology on feed quality and how various manufacturing processes are used to produce various types of feeds is discussed in a later section in this chapter.

Feed formulations are logically categorized by the intended function of the feed. Naming feeds by their intended function is a good way for the feed manufacturer and the fish culturist to communicate concerning the intended use of each kind of feed.

Larval and Starter Feeds

Both larval and starter feeds are formulated with the knowledge that they will be the first feed consumed by fish after hatching (exogenous feeding). Larval feeds are generally smaller (less than 400 mm) and require special processing compared to starter feeds (400–1,200 mm). First-feeding trout, salmon, catfish, and tilapia perform well when fed starter feeds. Walleye, striped bass, sea bream, and other undomesticated species must be fed larval feeds. The cost of larval feeds is higher than starter feeds, but well worth the investment for many fish species.

The most important considerations in the formulation and manufacture of starter feeds are (1) the feed must be readily consumed by the fish, (2) the feed must be highly digestible to accommodate a developing and often not fully functional digestive tract, and (3) the feed must be water-stable to water quality and, thereby, limit bacterial contamination of the culture system. Formulation affects consumption by influencing the color, texture, taste, and smell of the feed. Manufacturing affects consumption by influencing feed particle size, shape, texture, density, and buoyancy. Feed formulation and pellet manufacturing work together to reach the goal of the feed or, if done incorrectly, work against each other. Expensive, highly palatable ingredients are of little use if the texture of the feed reduces feed consumption by fish larvae.

Maximum survival and fish health are the most important considerations during this phase of fish culture. Cost should not be a major consideration by fish culturists when selecting starter diets because only a small quantity of feed will be fed. Producing a high percentage of healthy fish from first-feeding fry in the starter phase will pay economic dividends in the long run, despite the high cost of the starter feed.

Conversion Feeds

Conversion diets are feeds used for fish that are first reared in ponds or tanks and that consume live, natural foods before being transferred to a hatchery and fed formulated feeds. These feeds are typically used for species that have not been domesticated, such as newly emerging species for aquaculture or for species being cultured for reintroduction or restoration by conservation hatcheries. The basic assumption behind the use of conversion feeds is that the stress of habituation is elevated by a radical change in taste and texture of feed and that a feed intermediate in properties between live and manufactured diets will increase feed consumption and, thus, fish survival. This assumption is supported by studies with walleye and striped bass. Feed has been shown to be a major factor affecting the survival of pond-reared fingerlings as they are introduced to formulated feed, but the importance of strong, healthy fingerlings at the beginning of feed-type transition must be considered. If the fish are in poor condition as a result of food scarcity in a pond, survival will be low no matter what feed is offered after transfer to a hatchery.

Similar to a starter feed, the primary considerations in selection of a conversion feed are (1) high palatability, (2) low levels of water pollution (water stability), and (3) high energy content to compensate for interrupted feeding. In contrast to a starter feed, conversion feeds may contain feed ingredients with normal digestibility coefficients due to the mature stage of development of the gastrointestinal tract in fingerling fish. Feed cost is more important due to the larger quantities of feed required.

A modified form of the conversion diet, referred to as the "transition diet," is used with sensitive, hard-to-rear species when they are switched from starter feeds to grower feeds. The development of transition feeds was prompted by repeated observations that a significant portion (up to 40%) of walleye fry that had been consuming very palatable starter feeds would stop eating when switched to a grower diet and die of starvation. This mortality could be avoided by offering feeds intermediate in composition between the starter and grower diets during the transition period. However, there seem to be other factors besides the type of feed used that affect fish survival during transition. Losses can occur even when water temperature is near optimal, fish are not stressed, fish have been consuming the starter diet well, are in good condition, and pathogens are not prevalent. Further research is continuing to determine conditions for which survival can be improved through use of transition feeds.

Grower Feeds

Grower feeds are the category that is used in the greatest quantity during production cycles. In fact, more than 90% of the feed consumed by fish in hatcheries or farms during a production cycle is grower or grow-out feed. Grower feeds are intended to increase the size of fish that are past early development, have a fully developed digestive system, and readily consume manufactured diets. Because of the proportionally high quantities of grower feed used in production cycles, feed cost, fish growth rate, and feed conversion ratios become the primary considerations during formulation and feed selection. Diets designed to reduce nutrient enrichment in hatchery effluent focus on this phase of production and are discussed below under low-pollution feeds.

Broodstock Diets

Research on the nutritional requirements of broodstock is both expensive and time consuming; hence, relatively little research has been conducted. The knowledge base of the specific dietary nutrient requirements for maturing fish is limited. Salmonid broodstock perform relatively well when fed slightly modified grower diets. This is in contrast to other species, such as sturgeon and a variety of endangered species, that have been studied less extensively. There are several laboratories conducting long-term studies on the nutritional requirements of maturing Pacific salmon in connection with restoration efforts of threatened and endangered stocks. The information being compiled in these studies may be applicable to similar species, such as trout or char. At present, salmon are fed grower diets supplemented with extra vitamins, carotenoid pigments, and selected trace minerals. Maturing captive salmon fed grower diets tend to accumulate excessive amounts of fat, likely in anticipation of a spawning migration that for captive fish never happens. Thus, dietary caloric intake should be reduced at selected periods during the year to avoid this problem.

Low Pollution Feeds

In recent years, salmon and trout aquaculture has been subject to increasing scrutiny by regulatory agencies for the level of nutrients leaving hatcheries in effluent water. Since hatchery effluents are a point source of pollution that is easy to sample compared to nonpoint sources, it is relatively easy to document and regulate nutrient levels in hatchery effluents. Thus, hatcheries and farms are easy targets for regulation. At present, a national regulatory plan for aquaculture is being developed. The primary nutrient and material subject to regulation, so far, are phosphorus and fecal solids. Nitrogen may be the next nutrient considered by regulatory agencies.

For years, phosphorus levels in salmonid feeds were unintentionally high for economic reasons—fish meal was the main constituent of salmon and trout feeds. Regular fish meal is produced from whole fish and, as a result, contains high levels of phosphorus in the form of fish bones. Fish meal was a high proportion of salmon and trout feeds because it was relatively inexpensive, very palatable, and supported high growth rates because of its favorable profile of essential amino acids. Reducing phosphorus levels in salmonid feeds required feed formulators to reduce the proportion of regular fish meal and this has proved to be difficult without increasing feed costs.

Several approaches have been taken to reduce phosphorus levels in effluents. The first involved more precise estimations of phosphorus requirements for all life stages of salmon and trout. It was found that very low levels of phosphorus could be fed during the final month or two of the grow-out stage in trout farming without decreasing growth rates. The fish utilized stored phosphorus reserves from their skeleton and other hard tissues to supply their needs for growth. Additional research showed that small trout have a higher dietary phosphorus requirement than do large trout; trout feeds are now formulated accordingly. These types of diets are termed finishing diets and are fed to fish reared for human consumption, rather than to fish destine for release.

Studies with rainbow trout have shown that suboptimal levels of phosphorus in broodstock diets decrease growth rate, but do not affect fertility, hatchability, or overall egg viability. Fecundity, expressed as number of eggs per kilogram of broodstock, increased when the fish were fed very low phosphorus feeds. The same effect has been observed when broodstock have been fed low-protein feeds.

A second approach to phosphorus reduction has been to increase the level of available phosphorus in feeds so that a higher percentage can be utilized by the fish. Higher phosphorus digestibility, coupled with lower total dietary phosphorus levels, results in less phosphorus being excreted in feces and urine. Alternate feed ingredients, such as wheat gluten and corn gluten meal, have been evaluated. These ingredients contain lower levels of phosphorus than does fish meal, and phosphorus digestibility is similar to that in fish meal. Modification of fish meal has also been an area of active investigation. Mechanically deboning the fish meal before drying, or employing air-classification (a process that uses air, instead of screens, to separate particles of different sizes and weights) to remove bone particles in dry meals, appears to offer two benefits. First, the total level of bone (ash fraction) and, thus, phosphorus is decreased. This allows for previously unsuitable materials, such as seafood processing waste, which are mostly heads and frames after filleting, to be processed into low-phosphorus fish meal. Second, phosphorus digestibility increases in fish meals as the bone fraction is removed. This is due to lower levels of calcium in the product. White fish meals, which have traditionally been high in ash (>20%) and, therefore, calcium and phosphorus, have been processed through equipment that removes the bones before drying. The remaining phosphorus in these products is more digestible.

Phosphorus availability can be increased using special feed additives. Phytin, also called phytic acid, is the storage form of phosphorus in seeds (e.g., corn, wheat, barley, and soybeans). Phytin phosphorus is unavailable to fish and all other simple-stomached (monogastric) animals, including humans. The shift from animal proteins to plant proteins in order to lower total feed phosphorus levels has resulted in an increase in the level of phytin phosphorus in feeds. Phytin phosphorus passes through the gut of the fish and is excreted, thus contributing to phosphorus levels in hatchery effluents or ponds. The addition of phytase, a natural enzyme in seeds that releases phosphorus from the phytin complex as seeds sprout, increases phosphorus availability in fish. Phytase is commercially produced by several companies and is added to poultry and swine feeds. Phytase activity increases with temperature; maximum activity occurs at 50–55°C. Thus, its activity in fish varies with water temperature. It is more effective in warmwater fish (e.g., catfish) than in coldwater fish. Phytase can also be added to feed ingredients to release phosphorus from phytin before their being used in feeds.

Preliminary studies have suggested that the addition of citric acid to trout feeds increases phosphorus digestibility in grow-out size rainbow trout. However, feeding diets containing citric acid to first-feeding rainbow trout lowers fish performance. This approach to increasing phosphorus digestibility in trout feeds, even though promising, needs further research before it can be implemented in practice. Citric acid increases phosphorus digestibility by lowering the antagonistic effect of calcium on phosphorus solubility in the intestine.

Another approach to reduce solids production and discharge in effluents is to increase the overall digestibility and energy content of the diet and, thus, decrease the total amount of feed fed. As mentioned, feeding very high energy diets, more than 25% total lipid, is being used successfully by Atlantic salmon and rainbow trout farmers in Europe. Fecal solids production can be further reduced by pelleting feeds using extruders or expanders that cook the carbohydrate portion of the feeds, eliminating the need to include indigestible binders in the feed. These feeds support improved feed conversion ratios compared to traditional steam pelleted feeds. Extruded and expanded feeds comprise an increasing portion of the aquaculture feed market in the United States. Very high energy feeds are not fed as widely in the United States as in Europe, perhaps due to differences in consumer acceptance of a high-lipid product produced by feeding the very high energy diets and also due to differences in size of fish (trout) at harvest.

Another area of current research is modification of feeds so that feces from fish fed such feeds remain intact long enough for removal from the effluent. Rainbow trout farms often use quiescent zones where fecal particles can settle without being resuspended by the activity of fish. Solids are removed and conveyed to off-line settling basins for further settling. However, feces tend to break up into particles that are too small to settle from the water column, and these small particles are carried out in hatchery effluents. If dietary changes that produce intact, high density fecal pellets can be identified, further reduction in total solids and, perhaps, dissolved nutrients in hatchery effluents could be achieved.

The approaches described above to reduce pollution are best applied to grow-out or broodstock fish because of the large quantities of feed used during these stages of production. The amount of starter feed fed during a production cycle is so small that even complete elimination of phosphorus excretion in fry or fingerling fish would not significantly affect total phosphorus discharges in effluents during the production cycle of the fish. The small gains in phosphorus reduction are not worth the risk of lowering fish performance in the early life stages.

Feeds Formulated to Increase Immunocompetence, Seawater Transfer

This category of feed is generally specific to individual species of fish. In salmon farming, feeds are supplemented before, during, and after fish are transferred from freshwater hatcheries to seawater farms with extra levels of several nutrients and other compounds to mitigate for the stresses of transfer. In some culture situations, disease outbreaks occur on a seasonal basis and special feeds are fed slightly before an outbreak period to help the fish fight disease. Supplements include vitamins, especially vitamin C, glucans, or other compounds that are thought to stimulate the immune system of the fish, and betaine, which is reported to assist some species of salmon make a successful transfer to seawater, especially during out-of-season transfers.

Medicated Feeds

Medicated feeds are feeds that have been supplemented with antibiotics to treat diseases in fish. Most medicated feeds are simply regular feeds to which the recommended level of antibiotic has been added, but some manufacturers adjust the formulation of medicated feeds by adding feed ingredients that stimulate feed intake to ensure that fish "swallow their medicine." Medicated feeds are highly regulated; feed companies are required to maintain accurate and up-to-date inventories of the few antibiotics approved for use in fish feeds. Medicated feeds should be stored separately from regular feeds at a hatchery or farm and fed according to guidelines of the feed company and health professionals. They should be fed for the prescribed time period (generally 10–14 d), and only to the fish for which they were prescribed. Medicated feeds should not be kept beyond their expiration date. Strict withdrawal times are mandated for fish destine for human consumption or release. Fish culturists must withdraw medicated feeds from these fish for prescribed time periods before the fish can be harvested or released.

Feeds for Recirculation Systems

Interest in using closed water recirculating systems for aquaculture has increased rapidly in the last few years. Fish farms in certain areas of the United States, such as the upper Midwest, use these systems, which employ filtration equipment to remove particulate materials and dissolved nutrients from the water. Particulates and dissolved nutrients originate from the feed and thus a feed formulation or manufacturing technique that lowers the workload of filtration systems should increase production and lower production costs. Specifications for feeds designed for recirculating systems are similar to those for pollution reduction feeds, but even more restrictive, especially with respect to dietary mineral content. Research and production of feeds formulated specifically for use in recirculation systems will be a growing sector of fish nutrition in the next decade.

Open and Closed Formula Feeds

Feeds are sometimes referred to as open or closed formulations. Open formula feeds are those for which the complete formula is disclosed and is public information. Such feeds are developed by state or federal agencies or universities. Open formula feeds have the following advantages:

- The exact ingredient and nutrient levels in the feed, including vitamin levels, are public knowledge.
- Because the same formulation and quality of ingredients are used, the feed is consistent from one production season to the next.
- Competitive bidding is possible and more specific than with closed-formula feeds.
- Feed production can be monitored through an on-site quality control program.

Open-formula feeds have the following disadvantages:

- The buyer assumes full responsibility for feed performance because the manufacturer has followed contracted instructions.
- The buyer must have precise manufacturing and formula specifications, which must be updated periodically.

For open-formula feeds to be used successfully, the buyer must be nearly as knowledgeable about dietary nutritional requirements, appropriate ingredients and levels of ingredients for a given fish species, ingredient availability and price, and feed manufacturing technology as are commercial feed producers. Open-formula feeds are commonly used for species when the requirements are not well known and for specialty diets. Specifications for common open-formula diets can be found in any of several reference texts on fish nutrition.

A closed-formula feed is one for which the feed formulation is not disclosed to the buyer. These feeds are developed and sold by private manufacturers and are also referred to as "brand name" or proprietary feeds. Feeding a closed-formula feed has the following advantages:

- The manufacturer is responsible for the formulation.
- The feed is generally a shelf item available at any time.
- The feed may be lower in cost due to large-quantity production, the option of ingredient substitution, and the skill of the feed producer at purchasing feed ingredients that constantly vary in price to reflect market supply and demand.
- The manufacturer can be liable for problems of poor production related to the diet.

Closed-formula feeds have the following disadvantages:

- The buyer has no direct control of the feed quality and the actual ingredient and nutrient content of the feed, for example, the ingredients and their proportions are known only to the feed manufacturer.

- There may be unexpected variations between batches of feed due to ingredient substitutions or formulation changes.

When a closed-formula feed is used, the buyer assumes that the feed producer will produce a high quality feed for the selling price. It is wise for fish culturists to keep records of fish growth and feed efficiency on each feed used in production. If it is feasible, it is very useful for fish culturists to compare feeds on the basis of feed cost per unit fish weight gain (economic gain). Fish culturists can contact researchers at universities or state and federal laboratories for help in designing, conducting, and analyzing the results of feeding studies. In this way, the best feed, either an open- or closed-formula, that meets the specific needs of each operation can be selected.

Feed Manufacturing Technology

Delivering nutrients to fish via manufactured feeds in aquatic environments offers problems not encountered by the terrestrial animal feed industry. In the terrestrial environment, uneaten feed remains more or less nutritionally stable. However, in the aquatic environment, feed rapidly deteriorates, loses nutritional value, and pollutes the culture system. This is an especially severe problem with starter feeds due to their small particle size and high surface area to volume ratio. The process of pelleting is used to overcome some of the difficulties of feeding aquatic animals by packaging a combination of dietary ingredients into a form that can be delivered, consumed, and effectively utilized. The interaction between the manufacturing method and the ingredients used in a feed mixture has a significant and varying effect on the final cost, stability, consistency, and acceptability of feeds. Understanding the effects of the manufacturing technique on feed performance will help the fish culturist select the feed best suited for each culture situation.

Steam Pelleting (Compressed Pellets)

Steam pelleting refers to a process by which a mixture of dry ingredients containing a binder is exposed to steam to activate the binder and add moisture to the mixture, then is forced through a circular rotating die under pressure to form a solid pellet. Binders used in steam pelleting are wood processing by-products, such as carboxymethylcellulose or lignin sulfonate, incorporated at 2 to 4% of the formula weight. Steam is added to the mixture just before pelleting, increasing the moisture content by 5 to 6% and raising the temperature of the mixture to 65.6 to 82.2°C (150 to 180°F). The combination of temperature and pressure during pelleting partially cooks the carbohydrate (starch) fraction. Moisture levels are reduced to less than 11% by blowing air over and through hot pellets immediately after pelleting. Steam pellets are hard, dense particles with a glazed appearance on the outside, caused by starch gelatinization. Steam pellets are usually not produced below 2 mm (3/32 in) diameter. Particles for feeding small fish are produced by crumbling and sifting larger pellets into various size ranges.

Most commercial salmon and trout feeds were produced by steam pelleting until the late 1980s. The primary advantage of steam pelleting is that it is the least expensive pelleting process. Disadvantages include the susceptibility of steam pellets to fracture during shipping and handling, resulting in fines (dust) and relatively rapid pellet disintegration after water exposure. To overcome the production of fines when steam-pelleted feeds are shipped, feed manufacturers have developed techniques to produce very hard pellets. These extremely hard pellets are not suitable for certain fish species. Crumbled starter feeds also have a very hard texture and sharp edges, qualities that may cause feed rejection by larval or pond-reared fish. However, steam pelleting remains a cost-effective method of feed manufacture and is preferred over other processing methods in some applications.

Cold Extrusion

Cold extrusion refers to a manufacturing process used widely throughout the food, chemical, and pharmaceutical industries in which the ingredient mixture is pelleted without added heat (steam). Advantages of cold extrusion over steam pelleting are that small particles can be formed, eliminating the need for crumbling or sifting, and that a large proportion of wet ingredients can be incorporated directly into the feed mixture. The low processing temperatures associated with cold extrusion also prevent degradation of some nutrients that occurs with pelleting processes that require heat for particle formation.

The first cold extruders used for fish feeds were meat grinders or modified pasta machines such as those used to manufacture Oregon moist pellets. These machines consist of one or two screws that move and compress a wet feed mixture down a barrel and through a flat plate, producing noodles that are cut as they emerge from the die to form pellets. Pellets as small as 1.0 mm to 1.5 mm in diameter can be produced by cold extrusion, depending upon the characteristics and particle size of the ingredients used. Water or high-moisture ingredients are added to the feed mixture before pelleting to achieve pellet formation. Pellets can be dried or left moist to produce a soft textured pellet. Feeds having 30–35% moisture can be produced, but high-moisture pellets require special precautions—for example, addition of mold inhibitors and compounds to bind water, which lowers water activity, to prevent mold formation. Otherwise, the pellets must be stored frozen. Moist pelleted feeds spoil rapidly when thawed and loss of crystalline (nonprotected) vitamin C can occur within hours after thawing, so moist pellets must be used at once after being thawed.

Cooking Extrusion (Extruded Pellets)

Cooking extrusion, also referred to as "high-temperature, short-time extrusion," has been used to manufacture breakfast cereals, snack foods, pet foods, and catfish and salmon feeds for many years. The predominant feature of cooking extruders is a chamber (extruder barrel) in which the feed mixture is subjected to direct or indirect heat (steam) under pressure high enough to change steam (a gas) into water (a liquid). A sudden release of pressure occurs as the pellet feed mixture exits the chamber through the die. Water trapped in the feed turns back into steam and vaporizes. This causes cooked starch in the feed to expand, forming a tight matrix that traps air. If enough steam and carbohydrate are incorporated into the feed, the resulting pellet will expand enough to float. Bonds are formed within the gelatinized starch to give a durable, water-stable pellet that can be made to float or sink. In pond culture, floating feeds allow the culturist to observe the feeding activity of the fish. Feeding a floating feed in tanks or raceways allows the culturist to better determine if the fish are being overfed.

The manufacture of expanded or floating pellets requires higher temperatures and pressures and more energy than the manufacture of compressed pellets. Extruded pellets must be dried using heat, rather than just air, and this adds to the cost of producing extruded pellets. The additional cost of producing floating feeds must be carefully compared to the advantages of using a floating or a cooked feed. Catfish producers prefer floating feeds because they can observe the feeding activity of the fish. This aids in pond management and reduces feed wastage due to overfeeding and loss of pellets that sink to the bottom. Uneaten feed in trout raceways is easily seen and can, therefore, be removed, and the feeding rate lowered to match feed intake.

Although the equipment used in cooking extrusion is more expensive than that used in steam pelleting or cold extrusion, cooking extrusion allows production of feeds with physical characteristics that are not possible to achieve with other pelleting methods. Cooking extrusion can produce feeds that absorb and retain added oils very effectively, as high as 40% of the diet. Fish or plant oils are sprayed onto extruded pellets after they are dried. A vacuum chamber is used to remove air from the pellets, allowing more oil to be absorbed. High-fat diets produced by cooking extrusion are used in salmonid hatcheries and net-pen farms to reduce total feed intake and, thus, lower material handling and solid waste discharge from the facility.

Cooking extruders are available with both single- and twin-screw configurations. The self-wiping design of screws used in twin-screw extruders allows for greater flexibility in manufacturing, and higher levels of lipid can be incorporated. The twin-screw extruder is the predominant design in the European aquaculture feed industry, while the single-screw design is currently predominant in the United States. This difference is probably due to the difference in total lipid levels used in the two regions, feeds in Europe generally containing higher levels of lipid.

The high temperatures and pressures employed during cooking extrusion increase digestibility of starch over that in other types of pelleting methods. This has the advantage of increasing total energy available to the fish, yet it may lead to excess accumulation of glycogen in the liver of some coldwater species, particularly in small fish. Gelatinized starch provides sufficient binding in extruded (and expanded) pellets, so the indigestible binders used in steam pelleting are not needed. The cooked carbohydrate (gelatinized starch) is very stable in water, resulting in pellets that resist disintegration. This characteristic makes extrusion the process of choice for shrimp feeds and may be the best choice for feeds for recirculating systems. Pellets produced by cooking extrusion are very hard, unlike pellets produced by cold extrusion; this lowers the proportion of fines caused by pellets breaking during shipping and storage.

The porous consistency of extruded pellets facilitates top-dressing of oil onto the surface after pelleting. Application of oil to the pellet after formation provides the following benefits: possible increase in palatability, heat-sensitive substances (enzymes, pigments, and vitamins) can be added after pelleting. The total lipid added to pellets can be very high (up to 40%). Expanded pellets (discussed below) and steam pellets can be top-dressed similarly to extruded pellets, but steam-pelleted feeds will not absorb as much oil as pellets produced by the other two methods.

Expanded Pellets

This processing technology is used to some extent in aquaculture feeds, but is more common in feeds produced for terrestrial animals and poultry, especially in developing countries. The expander itself is used in conjunction with a steam pellet mill. The expander operates similarly to a cooking extruder in that frictional stress and heat are added to the feed mixture and starch is thereby cooked (gelatinized). When the feed mixture leaves the expander, it is fed into a compression pellet mill. In this way, the pellet mill forms the pellet, and the expander provides preconditioning or cooking of the feed mixture.

Universal Pellet Cooker

A new system of feed pelleting has recently been developed that combines some aspects of compressed pelleting and cooking extrusion—the universal pellet cooker (UPC). The equipment operates very much like cooking-extrusion equipment. The UPC system involves enhanced preconditioning for 2–3 min, during which the addition of steam results in 40 to 50% starch gelatinization. Less steam and water are added during preconditioning than in cooking extrusion, resulting in a feed mixture with 16 to 18% moisture, compared to 22 to 25% in cooking extrusion. The other major difference between the UPC system and cooking extrusion is the modified action in the barrel of the pelleter, where the auger, or rotor, turns 2–3 times faster than in a conventional cooker extruder. This increases feed production levels to 18–20 metric ton per h compared to 3–5 metric ton per h for similar-sized cooker extruders. In addition, the fast turning rotor adds much more energy to the feed mixture; the frictional energy associated with the faster turning rotor and lower moisture in the feed mixture further gelatinizes the starch to 60–80%. Further, the combination of frictional energy and steam pasteurizes the feed mixture, unlike compressed pelleting, even when expanders are used. Because the moisture content of the feed mixture is lower than in cooking extrusion, there is less water (steam) entrapped in the mixture and, thus, less expansion when pellets exit the die. This results in a higher bulk density for UPC pellets than for cooking-extrusion pellets, but the UPC system can produce pellets from 400 to 600 g/liter, depending on how it is operated (Figure 9). Pellets can be dried in a cooler, rather than a dryer. This reduces the amount of equipment needed and the operating cost of pellet production.

The UPC system produces higher-density pellets than cooking extrusion, but without the glazed surface associated with compressed pelleting. Thus, high-fat levels can be added to the dried pellets by top-dressing—up to 30%. An additional benefit of the UPC pelleting system is that the dense pellets can be produced to resemble crumbles by simply turning the external cutting knife faster. This eliminates conventional crumbling by rollers and the inevitable production of fines. The UPC process is not likely to totally replace either compressed pelleting or cooking extrusion, but, for some sectors of aquaculture, it may produce a product that fills a niche between compressed and extruded pellets.

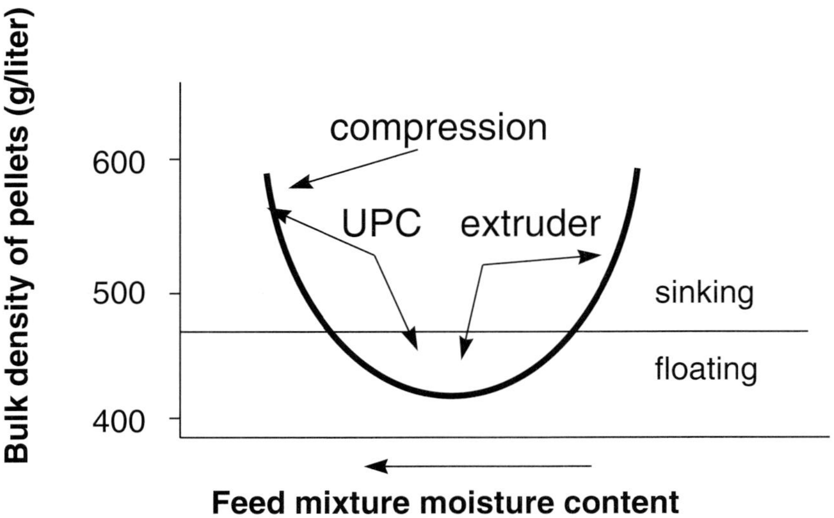

FIGURE 9. Schematic relationship between feed pelleting method and feed density (buoyancy).

Comparing the Types of Pellets

The main issues important to fish culturists concerning feed pellet type are starch digestibility, pellet density (buoyancy), water stability, durability (amount of fines), nutrient destruction during pelleting, and cost (Tables 2 and 3). Each pelleting process has advantages and disadvantages, but for each major species of farmed fish, there is generally a clear advantage to one of the types of pelleting processes (Table 4). In addition to considering the most appropriate type of pellet for a given fish culture situation, fish culturists must remember that the volume of a

TABLE 2. Physical qualities of compressed, cooker-extrusion, and UPC pellets.

Physical quality	Compressed	Expanded	Extruded	UPC
Density (g/liter)[1]	590	680	400–550	400–600
Max. temperature (C)	95	135	150	150
Time exposed to steam	< 1 min.	< 1 min.	2–5 min.	2–3 min.
Starch gelatinization (%)	<40	65–70	>80	60–80
Maximum fines (%)	2–3	<1	<1	<1
Max. fat level (%)	18	25	38	30

[1]480 g/liter is the breakpoint for pellets to float; higher density pellets sink, lower density pellets float in freshwater. Values shown above are average values; density can be altered somewhat by altering pelleting conditions.

TABLE 3. Attributes of pellet types important to fish farmers.

Attribute	Compressed	Expanded	Extruded	UPC
Starch digestibility[1]	low	high	high	medium to high
Pellet buoyancy	sinking	sinking	floating/sinking	floating/sinking
Water stability	low	low	medium/high	medium/high
Durability	low	medium	high	high
Nutrient destruction[2]	low	low	medium	medium/low
Cost of pelleting	lowest	low	highest	medium

[1]starch digestibility is a function of the degree of gelatinization, which can be measured by chemical test.
[2]nutrient destruction during pelleting is caused by high temperature, high pressure, length of time that feed mixture is exposed to high temperature, and is most important for certain vitamins and the carotenoid pigment, astaxanthin.

given weight of feed (for example, a ton) is lower for compressed pellets than for extruded pellets. To visualize this, think of the size of a bag of popcorn compared to the size of a bag containing the same weight of corn before it is popped. Thus, fewer tons of extruded feed will fit into a silo or bin than compressed (steam) pellets.

Microencapsulated and Microbound particles

Microencapsulated and microbound particles are two general classifications of particle types used for specialty starter feeds. These are feeds used for species requiring very small feed particles in the first stages of feeding, generally particles less than 400 mm in diameter. These feeds are often termed larval feeds, instead of starter feed, but larval feeds are simply starter feeds for larvae of fish species that do not have a fully-developed digestive system at first feeding. The high surface area to volume ratio of larval feeds makes producing a feed particle with the required performance a challenge for feed manufacturing technology.

TABLE 4. Current applications of various pellets for selected types of aquaculture.

Aquaculture sector	Best pellet type	Reason for selection
Catfish, pond culture	extruded	starch, floating
Salmon, marine net-pen culture	extruded	slowly sinking, high fat level
Rainbow trout, raceway culture	compressed, extruded, expanded	cost, density, durability medium fat
Shrimp, semi-intensive culture	compressed, extruded	water stability, cost
Sea bass/sea bream	expanded-extruded	slowly sinking, starch

Characteristics desired in larval feeds include very small particle size (~10–400 microns); high particle and nutrient stability; high particle yields; homogeneous mixture of water soluble and insoluble nutrients; high palatability; and good handling, shipping, and feeding characteristics. Of course, the nutrients in larval feeds must be digestible (available) to larval fish.

Microencapsulated feeds are particles with a capsule or shell around each feed particle. These capsules can be made with various materials, but lipid and protein-walled capsules are the most common for larval feeds. The microcapsules can be manufactured to be practically insoluble in water or to release the internal payload at a predetermined rate. This method of manufacturing appears to be preferred for filter-feeding fish and bivalves. The technology for producing microcapsules is progressing, but is currently relatively expensive. However, since so little larval feed is required during a production cycle, fish performance is more important than the cost of the feed.

Agglomerated or microbound particles are held together by ingredients or binders that are mixed throughout the particle. There are two subtypes of particles within this category, crumbled and on-size particles.

Crumbled microbound particles are sometimes referred to as "crumbled cake" particles. This is because these particles are formed by producing a large agglomeration in a sheet or biscuit form, followed by grinding and separating the resulting material into various sizes of particles by sifting through different-sized screens. Digestible and indigestible ingredients can be used as binders, including egg albumin, carrageenan, and zein protein from corn. This methodology is currently being used successfully on a commercial basis. Nutrient stability, particle stability, palatability, and buoyancy can vary by changing the binder, the formulation, and the conditions of production.

One-size, microbound particles can also be produced by several methods, each producing a variety of different particle characteristics. Three examples of this type of particle are micro-extrusion/marumerization (MEM), particle-assisted rotational agglomeration (PARA), and spray beadlets.

The MEM method involves cold extrusion. Cold extruders can also be designed to discharge from the side of the barrel (called "radial discharge extruders") or through a twin-domed screen (LCI Inc., Charlotte, North Carolina) to produce noodles as thin as 0.5 and 0.3 mm diameter, respectively. The MEM method was adapted from the pharmaceutical industry for the production of small diameter aquatic feeds. Thin noodles produced through cold extrusion are placed into a rotating cylinder (the marumerizer) that contains a variable-speed grooved plate on the bottom. As the plate spins, the noodles are broken into rods with a length roughly equivalent to the diameter; the particles are then rounded through impact with each other and the plate grooves. The exact shape of the final particle depends upon diet formulation and moisture content. Spheres can be produced, but an elongated egg-shape particle is more common. Centrifugal forces acting upon the feed during marumerization cause a migration of water and finer ingredients to the particle exterior, resulting in surface densification. This imparts a smooth texture and high water stability to the diet—characteristics that are beneficial in starter feeds. These particles are fairly dense and sink rapidly, which can be a disadvantage in some feeding situations.

The PARA process uses only a marumerizer, so capital and operational costs are reduced compared to the MEM method. A charge of wet feed mash is placed in the marumerizer along with a charge of inert particles (beads). As the marumerizer spins, the inert particles exert a force on the mash and form feed particles. Changes in the loading rate of inert particles, the density of the inert particles, ingredient characteristics, moisture content, marumerizer speed, groove spacing in the plate, and other variables change the characteristics of the feed particle. This is a cold, low-pressure processing method, which results in a low-density, slow-sinking feed.

Spray beadlets are produced using pressure and the binding characteristics of specific ingredients. In this method, a liquid or a suspension is sprayed through a device into a bath of another solution. The sprayed beadlets form into discrete particles in the bath. Like the MEM and PARA particles, the feed is sifted into appropriate sizes after drying.

A type of larval feed particle that can take advantage of several pelleting methods is the complex particle. Two or more types of particles described above are combined to form a complex particle. For example, impermeable microencapsulated particles containing water-soluble nutrients may be embedded inside a slow-leach microbound particle. This complex particle will deliver a known quantity of nutrient in the embedded microcapsules, and the leaching from the carrier particle increases the palatability of the feed. Other methods may be combined for different effects.

A series of laboratory tests can be used to compare manufacturing technology, diet formulation, and the interaction of these two factors in larval feeds. The tests include particle stability, nutrient stability, density, sink rate, measures of palatability, and digestibility. Feed was processed using two microbound processing technologies; the effect of the processing method on particle stability, density, and sink rate was compared (Table 5). A commercial crumbled cake feed was also evaluated. The feeds with lower density had a much slower sinking rate. Processing method also affected particle stability. The MEM method is a high-

TABLE 5. The effect of larval feed processing method on the physical characteristics of feed.

Feed	Density (g/cc)	Sink rate (cm/sec)	Particle stability, (%) 5 min[1]	24 hr[1]
MEM[2]	0.72 ± .04	1.66 ± .04	72.6	67.5
PARA[3]	0.67 ± .03	1.11 ± .04	67.4	64.7
Commercial	0.50 ± .02	1.03 ± .15	Agglomerated	

[1]Submerged in water for noted time, filtered, material dried and sifted to original 400-700 um size. Value represents weight of material recovered expressed as a percentage of original weight.
[2]Micro-Extrusion Marumerization
[3]Particle-Assisted Rotational Agglomeration

pressure agglomeration technique and produced a more dense particle with higher particle stability than feed produced with the PARA method.

Feed Handling and Storage

Formulated feeds for many species, especially salmon and trout, contain high levels of protein and unsaturated fish oils. These feeds are manufactured to remain stable until they are used, but they can rapidly deterioration if optimum handling and storage procedures are not followed. Normally, feeds are packaged in multiwalled paper bags to protect the flavor, aroma, and color. The bags also reduce exposure to air, moisture, and contamination. Plastic liners are used in bags for feeds containing oil levels over 12% to prevent oil seepage through the paper bags and to retard moisture uptake.

Many fish producers receive their feed in bulk, storing it in large bulk bins (Figure 10). Whether feed is in bags or bulk, proper handling and storage procedures must be followed to maintain the quality of the feed. Because fish feeds are fragile in comparison to feeds for other animals, up to 3% fines, by weight, can be expected from normal handling. Excess fines are the result of rough handling or poor physical characteristics of the feed.

Bagged feed should never be thrown, walked on, or stood upon. Use of a motorized belt-type bag conveyor to move bags of feed from delivery trucks to storage on a farm or hatchery causes the least damage to bagged feed. Close-spaced roller gravity conveyers work well, but the wide-spaced rollers or wheel rollers found on box conveyers are not suitable for bags and cause breakage of the granules and pellets. For handling bulk feed, a bucket elevator is preferred, followed by air sys-

FIGURE 10. Bulk storage of pelleted fish feeds. Dust and fines are screened out and collected (arrow) and can be repelleted. This type of storage is preferred to bins that require the feed to be moved by auger, which fractures pellets. (U.S. Fish and Wildlife Service photo.)

tems; screw-type augers are least satisfactory. These considerations are most important with steam-pelleted feeds. Extruded and expanded feeds are more durable than steam-pelleted feeds, but mishandling can damage even these feeds and thereby lower water quality, feed efficiency ratios, and fish health when the damaged feed is used.

If feeds are not stored properly, the likelihood of lipid oxidization can increase, as mentioned. However, other factors can cause deterioration of feed, such as physical conditions (moisture, heat, light), microorganisms (molds, bacteria, yeast), and enzymatic action.

Dry feed in bags or bulk should be stored in a cool, dry area. Low humidity must be maintained because moisture above 10% enables mold and other microorganisms to grow in feed. Molds cause feed spoilage, and many molds produce toxins. Any sign of mold on feeds is cause to dispose of the feed. Never feed moldy feed to fish! If a bag of feed has even a small area where mold is evident, discard the entire bag. The presence of visible mold is a likely sign that mold spores are growing throughout the bag, but have not yet reached the visible stage.

Oxidation of oils and deterioration of vitamins is accelerated by high temperatures. As mentioned, oxidizing fish oils can be toxic, may destroy other nutrients such as vitamins, can cause off-flavor of the feed, and can reduce product quality and shelf life. To prolong feed quality, the feed storage area should be kept clean, cool, dry, and well ventilated. Stored feed should be protected from rodents, insects, and contamination. A common misconception is that if cool storage is good then frozen storage should be better. This is a particular problem in hatcheries equipped with walk-in freezers. Dry feeds (<10% moisture) should never be stored in the freezer. Even though lipid oxidation slows when the feed is in the freezer, freezing can actually enhance lipid oxidation by making it easier for air (oxygen) to penetrate the pellets. What generally happens in hatcheries is that dry feed reaches the end of its guaranteed storage life. Rather than discard it, the hatchery staff decides to save money by freezing the feed and using it the following year, thinking that the feed will remain unchanged in the freezer. This practice is more common with starter feeds than with grower feeds. If the feed could be instantly frozen below –30°C (–22°F), this approach might be viable. However, feed doesn't freeze instantly and the process of slowly freezing concentrates pro-oxidant minerals, such as iron and copper, in tiny water pockets in the pellets. In addition, although hatchery freezers are usually set at –12°C (10°F), temperatures increase during defrost cycles. Over time, oxidation proceeds in the feed, even in the freezer, generally using up any sacrificial antioxidants that might be present. As soon as the feed is removed from the freezer, oxidation rapidly accelerates, especially in the absence of antioxidants. Oxidation produces heat. Reports of feed bags actually heating to the point of being too hot to touch within less than 24 h after removal from frozen storage are evidence of how rapidly feed can oxidize. Many hatcheries were constructed when only meat meal feeds (horsemeat, liver, etc.) and moist feeds were available, and, thus, have large freezers on site. Access to a large freezer is a temptation to the culturist to try to extend the storage time of dry feed by freezing it. Never do this. Dry feeds can be stored longer under proper conditions without being frozen than by being frozen.

Ideal conditions for storing bagged dry feed include stacking the bags not over ten high on pallets so the bags are 3–4 in off the floor. Space should be provided between the stacks for air circulation and rodent control. Low relative humidity and low air temperatures in the storage area extend the shelf life of feeds.

The recommended maximum storage time for most dry pelleted feeds is 90–100 d. If less than optimal storage conditions exist, the storage time should be shortened. Keeping and feeding a feed after the recommended storage period is a dangerous practice. The culturist risks much more than is saved by feeding unused feed. Unused feed can be composted and used for garden fertilizer or fed to other animals such as swine.

Bulk feed should be stored in clean bins free of contaminants or spoiled feed. The bins must be in good condition to protect the feed from water and weather. Bins located in shaded areas remain cooler than those lacking shade. Bins can be fitted with a screening unit on the discharge to remove dust and fines from the pellets. In many cases, the fines can be returned to the feed mill for repelleting or used to fertilize ponds or lawns.

Moist pellets should be stored in the freezer at temperatures below –20°C (–5°F) until they are needed, then thawed just before feeding. Follow the manufacturer's guidelines for the suitable length of storage while frozen in the unopened package and once the package is opened. Many frozen feeds are sealed in plastic bags that have been flushed with nitrogen. The nitrogen flush displaces oxygen and increases storage time. Once the bag is open, air (oxygen) enters and this shortens storage time. Therefore, once a bag is opened, it should be used within a few days or discarded.

Feed Evaluation

Feeds can be selected on the basis of nutrient content, life stage of the fish, or physiological stage (e.g., broodstock) as has been discussed earlier. All of the planning and choosing, however, should be followed by an evaluation of the performance of the chosen feed. Many factors can affect the performance of fish fed a particular feed, and the exact outcome of these factors is difficult to predict. It is wise to keep careful records of fish performance so that evaluations of the feeds and feeding programs are possible, even when feed from the same manufacturer has been fed for many years. Feed manufacturers change equipment, staff, and feed formulations, all of which can alter feed quality in positive and negative directions.

In production hatcheries, a variety of performance characteristics of fish can be easily measured and used to compare feeds. The most commonly observed measure of fish performance is weight gain; many other measurements can be calculated using weight gain data. Knowing the amount of feed fed to a group of fish over a period of time, the total weight gain of the group of fish, and the number of fish in the group allows the feed conversion ratio (FCR) to be calculated, as described below. In salmon or trout culture, a high ratio, such as 2.0:1 (commonly expressed as 2.0), suggests that there is a problem with the feed, the feeding program, water quality, or the fish. Low dissolved oxygen levels in the water or a disease epizootic reduces the feed consumption and increases the FCR. If water quality and fish health have been examined and are found to be satisfactory, then the feeding program and feed should be examined.

Feed conversion ratios of 0.9:1 have been observed when feeding with some of today's high-quality feeds. Remember that this ratio compares dry feed (<10% moisture) to wet fish (~75% moisture). Although culturists should strive to reduce feed conversion ratios, having the lowest possible FCR is not always the best strategy in fish culture. Rainbow trout fed 75% of the ration that results in maximum feed intake (Figure 11) and growth rates have the lowest FCR. These fish will utilize the feed more efficiently and release less nutrients into the effluent, but overall weight gain will be lower than in fish fed to satiation. Highest weight gain and lowest FCR are separate goals and are mutually exclusive, but compromises can be made based on the goals of the hatchery program.

Feed efficiency is a term sometimes used when discussing feed conversion ratios. Feed conversion is the ratio of feed fed to weight gain. Feed efficiency is also a ratio and is calculated with the same informa-

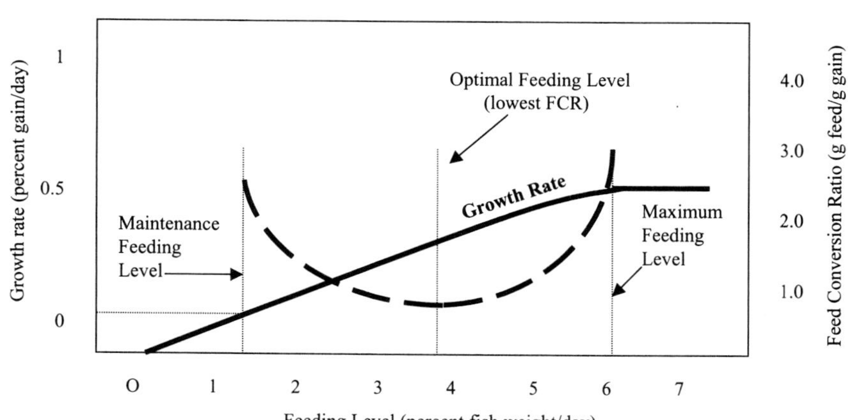

FIGURE 11. Schematic relationship between feed intake, fish growth rate, and feed conversion ratios.

tion used to calculate FCR, but the ratio is reversed. Feed efficiency ratio is the amount of gain (kg or lb) divided by the amount of feed (kg or lb) over a specific period of time. This term is often used by nutritionists and a high value indicates higher efficiency, unlike FCR where a low value indicates efficient conversion of feed into weight gain.

A commonly conducted measurement of feed quality is called proximate analysis. This analysis determines the levels (percentages) of moisture, crude protein, lipid, crude fiber, and ash in the feed. A sixth category, nitrogen-free extract, is not actually measured, but rather calculated by subtracting the total (percent) of the other five ingredients from 100. Nitrogen-free extract is mainly composed of water-soluble carbohydrates. The six proximate categories lump together nutrients and other compounds based on their physical or chemical properties (Table 6).

It is important to remember that proximate analysis is not an analysis of specific nutrients, but classes of nutrients. Proximate analysis is often conducted to determine if feeds are in compliance with specifications for agencies that purchased feed on contract; the analysis is an effective measure. Many times, a hatchery will observe slow fish growth, check the composition of the feed using proximate analysis and find the feed is 1 or 2% lower in crude protein than specified. Whether the lower protein level lowers growth rate depends on the feed formulation. Most feeds, especially contract feeds, are formulated to be slightly higher in protein than the specifications call for to give a margin of safety. Thus, being a percent or two lower than the specified level of crude protein should not be a problem. Proximate analysis is a valuable tool in feed analysis, but the results of proximate analysis (e.g., proximate composition) provide limited nutritional information and are often overinterpreted by people who do not appreciate its limitations.

TABLE 6. Nutrient and compound partitioning within proximate analysis categories.

Nutrient or component	Constituents
Moisture	Water
Crude protein	Protein, amino acids, chitin, nucleotides
Crude fat	Triglycerides, phospholipids, fat-soluble vitamins (A, D, E & K), waxes, free fatty acids, cholesterol
Crude fiber	Cellulose, hemo-cellulose, lignin, chitin
Ash	Macro-minerals, trace minerals, sand
NFE	Starch, sugars

Note that some compounds are counted twice. This is one of the reasons that this system of chemical evaluation is called proximate (approximate) analysis. It is not exact.

Chemical analysis of fish can be a useful tool to further understand the effect of feeding a particular feed. Conducting proximate analysis of a sample of fish at the beginning and at the end of a growing period provides information on the effects of the feed on fish. Two feeds may support the same weight gain and FCR, but the composition of the weight gain may be different. The main difference is usually in the proportion of fat in the weight gain. The level of visceral fat in fish can be a measure that separates the performance of two feeds. Selection of the feed that supports lean body weight gain could result in considerable economic advantage. Fish with excessively high body fat levels are more susceptible to stress, so it is just as important for culturists rearing fish in conservation hatcheries for release as it is for culturists raising food fish to be aware of the effects of diet and feeding level on body composition.

Knowing the composition of the fish at the beginning and end of a production period also allows researchers to calculate protein and energy retention values. These are more precise measures of feed quality since they reflect the percentage of dietary protein that is used to produce body protein gain in fish. The quantity of feed fed, the proximate composition of the feed, and the proximate composition and weight of the fish at the beginning and end of the feeding period are all necessary values to determine retention: protein retention = (wt. of protein gained/ wt. of protein fed) × 100. Modern, high-energy feeds yield protein retention values of 45–50% in trout and salmon culture when careful feeding practices are followed. Energy retention and fat retention is calculated in a similar way.

Retention values are particularly helpful when the amount of nutrients released into the effluent is being estimated, because the amount of nutrients released into effluents is simply the amount retained by fish subtracted from the total input (feed) into the system. This approach is used successfully in monitoring nitrogen and phosphorus discharges from hatcheries. It does not provide information of whether the nutrient is dissolved in the water or is present as insoluble fecal particles, but does provide a comparison between feeds. Calculating phosphorus retention is similar to calculating protein retention. The amount of phosphorus in the feed is measured and the amount of feed fed over a period of time is recorded, yielding the amount (g) of phosphorus fed over the feeding period. The amount (g) of phosphorus in the whole carcass of fish at the beginning and end of the feeding period is also measured in samples. From these values, the amount of phosphorus gained by the fish is calculated. Phosphorus or any other nutrient retention can be calculated on a tank, raceway, or pond basis, or on an average fish basis by dividing the total amount of feed fed and the total gain per tank, raceway, or pond by the number of fish. It is critical that the samples of fish taken at the beginning and end of a feeding period are as close to the average size of fish in the larger population as possible, to obtain a representative sample. This approach can also be used for dietary trace elements, an important consideration in recirculation aquaculture systems.

Conservation hatcheries produce fish for restoration and recovery programs. The performance of fish after hatchery release is of the utmost importance and is directly affected by the feed and feeding program used in the hatchery. Composition of fish is important since it reflects the fish's nutrient stores and ability to withstand stress, but other measurements of fish quality, such as fin integrity, are also affected by diet. Studies have indicated that even though rearing conditions affect the severity of dorsal fin erosion in salmonids, trace mineral composition and availability in the feed is a predisposing factor. These relationships are still being studied. Fin quality may be a visible indicator of other aspects of fish quality that may be correlated to fish survival after hatchery release. As many measurements of fish quality as practical should be taken from fish before release in order to compare feeds within a given production cycle.

To summarize, some measurements used to evaluate feeds at production hatcheries are

- fish growth (weight and length gain over a fixed period),
- feed conversion (or efficiency) ratios,
- nutrient retention (protein, phosphorus, etc.),
- feed cost to rear a pound (kg) of fish,
- protein and calories required to rear a pound (kg) of fish, and
- mortality, dietary deficiency symptoms, or other physical qualities such as dorsal fin height.

Broodstock feeds are evaluated somewhat differently than grow-out feeds in that fish weight gain is less important than fish fecundity and gamete quality. Criteria for evaluating broodstock feeds include (1) survival to spawning, (2) fecundity (number of eggs per fish), (3) relative fecundity (number of eggs per pound of body weight), (4) egg weight, (5) egg diameter, (6) percentage of eggs fertilized, (7) percentage of eggs to eyed stage, and (8) percentage hatch.

Feeding

Feeding was once considered a simple task and was usually assigned to the least experienced fish culturist at a farm or hatchery. The chore consisted of merely feeding all that the fish would appear to consume and then giving a little more to assure that all fish had an opportunity to feed. Even though fish were typically given more feed than they appeared to consume, fish often were underfed because some of the feed broke apart as it hit the water and was not eaten by the fish. In addition, the point at which fish are finished eating is a subjective determination. It is tempting to decide the fish are finished eating before they are actually full, especially in bad weather when feeding seems to take forever.

Nutrition is not solely a matter of feed composition. While it is true that fish cannot grow if essential nutrients are lacking in the feed, it is equally true that a feed cannot efficiently produce fish unless it is consumed. As mentioned, the conversion of feed into fish growth (FCR) is the measure that is most commonly used to judge the efficiency of a feeding program in a hatchery. If the FCR is regarded as a measure of the efficiency of feeding, what can be done to ensure good conversion ratios?

Mistakes associated with feeding fish in hatcheries are overfeeding or underfeeding. Overfeeding is wasteful in terms of unconsumed feed, but underfeeding is just as wasteful in terms of lost production. To obtain maximum production and feed efficiency during a growing season, careful attention must be given, on a daily basis, to the amount of feed the fish are being fed and are eating.

The quantity of feed fed to a group of fish (tank, raceway, pond) is conveniently expressed in terms of percentage body weight (or biomass) per day. Because the metabolic rate per unit weight of fish decreases as the fish grow, the percent of body weight to be fed per day also decreases. Feeding rates vary with fish size and with water temperature, feed formulation, and fish species. Feed manufacturers and some agencies provide feeding guides that contain suggested feeding rates for various species at different sizes and in a range of water temperatures. These feeding guides are just that—guides that must be fine-tuned by the fish culturist for specific hatcheries and farms, strains of fish, and other factors.

Time of Initial Feeding

There are considerable differences of opinion among fish culturists as to when fry should start on feed. The most common practice with salmon and trout is to offer food when the fry swim up. Swim-up occurs when the fry have absorbed enough of their yolk sac to enable them to rise from the bottom of the trough and maintain a position in the water column. A considerable amount of work has been conducted to determine when various salmonid fry first take food. Brown trout begin feeding feed approximately 31 d after hatching in 11°C (52°F) water, while feed was first found in the stomachs of rainbow trout fry 21 d after hatching in 10°C (50°F) water. The upper alimentary tract of rainbow trout fry remains closed by a tissue plug until several days before swim-up. Thus, feeding of rainbow trout fry before this plug is gone is futile.

Yolk absorption is the most useful visual guide to determine the point at which initial feeding of most species of fish should begin. Most studies (Table 7) indicate that feeding of fry at the first sign of swim-up does not provide them with any advantage over fry that are fed later, after the yolk sac has been nearly absorbed. The latter point occurs when the yolk sac is completely surrounded by skin tissue on the ventral side of the fish. Fish at this stage are said to be "buttoned-up." Many culturists start feeding when 50% of the fry in a group are swimming up because if fry are denied feed much beyond yolk-sac absorption, some will refuse to feed. Starvation will lead to a weakened fry that cannot feed even when feed is abundant.

The point of initial feeding for warmwater fishes is much more critical than for coldwater species because metabolic rates are much higher at warmwater temperatures. This leads to more rapid yolk absorption

TABLE 7. Initial feeding times for various species of fish.

Species	Initial feeding (days post-hatching)	Water temperature (°F)	Remarks
Brook trout	23–35[a]	52	Several fry had food in gut on 23rd day; all fry were feeding on the 35th day.
Brown trout	31	52	Evidence of food in stomach on 27th day; all fry feeding on 31st day.
Cutthroat trout	23	47–51	Evidence of food in stomach on 14th day; all fry feeding on 23rd day.
Rainbow trout	20–30[a]	47	Evidence of food in stomach on 21st day (16 days at 50°F).
Channel catfish	at swim-up	–	5 to 10 days after hatching, depending on temperature.
Tiger muskie	9	68	Food presented at swim-up; most of yolk sac absorbed after 8 days.
Northern pike, walleye, Muskellunge	at swim-up	50–70[a]	Food presented at swim-up, up to 12 days post-hatching.

[a]Various reports include a range of initial feeding times and water temperatures. It is important to note that, n some instances, feed was not observed in the stomach did not occur until several days after swim-up.

and a shorter window between the point when the fry will begin to consume feed and the point when the fry become weak from lack of feed.

Feeding Frequency

The frequency with which fish should be fed is governed by the size of the fish and how rapidly they consume the feed. When fish are first started on feed, it is desirable to give small amounts of small-sized particles at frequent intervals. Larval and starter feeds for some species should be provided by an automatic feeder, as frequently as every 5 min.

Several factors influence how quickly fish consume feed. The type of feed, the way it is introduced into fish-rearing tanks, and the type of trough or pond in which it is fed affect feed consumption by fish. Feeds that sink in water must be fed with more care than those that float. Once a sinking feed reaches the bottom, many species of fish will ignore it. Sinking feeds should be fed slowly and at greater frequency than floating feeds in order to avoid feed waste and potential water quality deterioration caused by disintegrating feed.

For many species of fish, it is necessary to supply feed at a rate that is many times greater than the amount that can be consumed by the fish. This is because these fish, at least at early life history stages, will only consume feed that is present in the water column, rather than feed at or near the bottom of tanks. Overfeeding is also practiced with several commercially important fish species during the fry and fingerling stages. For example, juvenile Atlantic salmon are typically fed to excess with automatic feeders in order to obtain maximum growth and to avoid a wide size range within a group of fish. The excess feed can cause water quality and disease problems if it is not quickly removed from tanks. Therefore, frequent tank cleaning is essential when feeding fish to excess. As the fish grow and feed more readily, feeding rates can be reduced to rates more closely associated with maximum consumption rates.

Trout and Pacific salmon generally are fed small amounts at hourly intervals throughout an 8–10 h day when they first start to feed. Some fish culturists feed fry at half-hour intervals, gradually reducing the number of feedings as the fish increase in length. Once salmon and trout reach the fingerling stage (3–4 weeks), the general practice is to feed three times a day until they are 5 in long (20 fish/lb, or 22.7 g/fish). Feeding is then reduced to twice daily and broodfish are usually fed once per day (Table 8). Coolwater fishes, such as northern pike and tiger muskellunge, benefit from 5-min feeding intervals for the initial feed, at 10-min intervals when fry are 2 in long, and at 15-min intervals when they are 4 in long, using automatic feeders during daylight hours.

Salmon and trout fry and fingerlings can consume more than 1% of

TABLE 8. Suggested feeding frequencies for salmonids (Source: Washington Department of Fish and Game)

	Fish size (number/pound)								
Species	1,500	1,000	750	500	250	125	75	30	10–larger
	(Times per day)								
Coho salmon	9	8	7	6	5	3	3	–	–
Fall chinook salmon	8	8	8	6	5	4	3	–	–
Trout	8	8	6	6	6	4	4	3	2

their body weight at a single feeding. A rule of thumb used by some fish culturists is to feed 1% of the body weight per feeding to ensure that all fish, not just the most aggressive fish, have an opportunity to feed. Therefore, if the fish are being fed at a rate of 10% of body weight, they would receive 10 feedings per day. If they receive 1% of body weight in feed per day, that amount would be fed in one feeding. Naturally, this guideline can be adapted to suit any group of fish or any hatchery. Channel catfish reared in raceways gain more weight when fed twice daily than when they are fed only once daily. In pond rearing, more than two feedings per day does not improve the feed consumption ratios or growth rates in catfish.

The following statements relate to feeding frequency:

- Frequently fed fingerlings utilize their feed more efficiently than those fed less frequently, resulting in better (lower) FCRs.
- Frequent feeding of fingerlings reduces the percentage of starved and stunted fish in a group. Generally, more frequent feeding results in greater uniformity in fish size, providing that a sufficient quantity of feed is fed at each feeding.
- The accumulation of waste feed on the bottom of a rearing unit due to infrequent feeding of large amounts of feed is the principal factor causing inefficient utilization of feed.
- When uneaten feed lies on the bottom of the tank, water-soluble nutrients leach out, resulting in poor utilization of the feed, if it is consumed at all, and causing pollution of the tank.
- In general, the number of feedings per day should be greater for dry feed than for soft, moist feeds.
- A rule of thumb is that 90% of the feed at each feeding should be consumed by the fish in 10 min or less, preferably 5 min or less.

Feed Sizes

The size of feed particles is a critical aspect of successful feeding of fish. If particles are too large, the fish will not be able to ingest them until they disintegrate or are bitten into smaller sizes. When this occurs, a high proportion of feed is wasted. When the particles are too small, the fish may not consume all of the pellets before they sink to the bottom or are flushed out. An appropriate range of feed sizes for fish during their various growth stages must be provided to enable the fish to grow rapidly.

Granules or crumbles are made in a range of sizes for various sizes of trout fry and fingerlings (Tables 9–11). Hard pellets are cracked into granules and the various sizes of granules are separated into size ranges using screens. Crumbles are screened into size ranges rather than to a specific size.

When fish are shifted from a small crumble to the next larger size, the shift should be gradual rather than abrupt. The change may be made either by mixing the two sizes together and feeding them at the same time, or by feeding the two sizes separately, starting with a few feedings of the larger size each day and gradually increasing the frequency until only the larger particle is fed. Any given group of fish contains a range of sizes of fish. Mixing two crumbles or small pellet sizes together for a week or so allows fry at the small end of the size distribution within a trough, tank, or raceway to feed and continue growing, rather than presenting them with feed that is a challenge for them to ingest.

Feeding Methods

In nature, most fish are opportunistic feeders. When food is available, fish eat as much as they can. Rarely is food availability not limiting in nature. In hatcheries and farms, however, feed is available on a daily basis. Feed delivery to the fish must be accomplished in a way that suits the nature of the fish, but also in a way that is efficient and economical. Feeding methods can be categorized into several categories: (1) hand feeding, (2) mechanical (automatic) feeding, and (3) demand feeding.

TABLE 9. Recommended sizes for dry pelleted feed for trout[a].

Granule or pellet size	Screen size	Number/pound (g/fish)
Starter granule	30–40	>2,000 (<0.227g)
No. 1 granule	20–30	2,000–800 (2.27–0.5g)
No. 2 granule	16–20	800–250 (0.5 g–1.8g)
No. 3 granule	10–16	250–100 (1.8g–4.5g)
No. 4 granule	6–10	100–30 (4.5g–15 g)
1/8″ pellets	–	30–10 (15g–45g)
3/16″ pellets	–	<10 (>45 g)

[a]Feed sizes: U.S. Fish and Wildlife Service Trout Feed Contract Specifications, Arlington Square Building, 4401 North Fairfax Drive, Arlington, Virginia 22203.

TABLE 10. Recommended sizes for Abernathy dry pelleted feed for Pacific salmon fry and fingerlings (Source: L. G. Fowler and R. E. Burrows 1971).

Granule or pellet size	Fish size Number/pound (g/fish)
Starter granule	>800 (<0.5g)
1/32-inch granule	800–500 (0.5g–0.9g)
3/64-inch granule	500–200 (0.9g–2.25g)
1/16-inch granule	200–100 (2.25g–4.5g)
3/32-inch granule	100–75 (4.5g–6.0g)
3/32-inch pellet	75–50 (6.0g–9.0g)
1/8-inch pellet	50–20 (9.0g–23 g)
3/16-inch pellet	<20 (>23 g)

Hand feeding simply involves the delivery of feed to the fish using hand labor. Feed is broadcast over the surface of a tank, raceway, or pond using a scoop or other device. Feeding continues until the ration is fed or until the worker determines that the fish have eaten as much as they want. The advantage of hand feeding is that it connects a human being with the fish, so that any change in feeding habits or appearance of the fish can be noticed. Also, fish do not eat the same amount every day; hand feeding allows the amount of feed delivered to a tank, raceway, or pond to be adjusted to match the feeding activity of the fish on a daily or several times per day basis. The disadvantage of hand feeding is the cost of labor involved. In high-wage countries, such as Norway, salmon farmers invest incredible amounts to eliminate hand feeding, while in low-wage countries, hand feeding is the main method of feeding.

Mechanical feeding, involves the use of machines, often called automatic feeders, to deliver and broadcast feeds to a tank, raceway, or pond. In small tanks in which fry and fingerling are being raised, mechanical feeding is very useful because feeding frequency can be nearly continuous, if necessary. Large raceways and ponds can be fed using blowers towed by tractors (Figure 12). This allows a large area to be fed with a minimum of labor. Salmon farms often use air-driven mechanical feed-

TABLE 11. Optimum feed particle sizes for small channel catfish. Crumbles or pellets should be kept to the maximum size that the fish can ingest (Source: Lovell 1998).

Crumble or pellet size	Fish size (inches)
00 crumble (starter)	Swim-up fry
No. 1 crumble	0.5–1.5
No. 3 crumble	1.5–2.5
1/8-inch pellet	2.5–6.0

FIGURE 12. Mechanical feeding of formulated pelleted feed to catfish in a large rearing pond.

ers that blow feed through pipes from a central hopper to each cage. The advantages of mechanical feeding are that feed can be delivered frequently, with little labor. The disadvantage of mechanical feeding is that feed is delivered whether or not the fish are hungry. Thus, it is relatively easy to overfeed fish with mechanical feeders, especially small fish. This can lead to water quality problems and bacterial gill disease.

Demand feeders are mechanical devices that deliver feed when fish activate a rod or trigger. Thus, the fish feed themselves. If they are hungry, they feed. If they are not, no feed is delivered. Feeding activity of rainbow trout is usually spread out over a relatively long period when demand feeders are used, presumably because fish chose to eat at different times, possibly in connection with their position in dominance hierarchies. Prolonged feeding activity suggests that fish have equal access to feed, a suggestion that is borne by observations that size ranges within tanks or raceways is lower when fish are fed using demand feeders, compared to hand or mechanical feeding. Another advantage of demand feeding is that abrupt changes in oxygen levels associated with frantic feeding activity when hand or mechanical feeding is conducted are avoided. A disadvantage of demand feeders is that they are sensitive to motion, such as wave or wind action. Thus, they cannot be used in situations where motion will trigger delivery of feed. In addition, demand feeders must be adjusted to deliver a set and relatively small amount of feed per trigger event. Otherwise, they tend to dump too much feed per event, thus wasting feed.

Economic Considerations

We have covered all of the biological and technical factors that must be considered when the appropriate feed for a given feeding situation is being chosen. These factors included nutrient requirements, life stage of the fish, specialty feeding situations, feed manufacturing, handling and storage, and feed evaluation. The reality of fish production is that budgets are always limited. This can cause the fish culturist, purchasing agent, or financial manager to ignore some of the information we have discussed and focus on total feed costs. This viewpoint is usually counterproductive. The reality is that feed costs must be considered, but combining biological evaluation with the economic consequences can allow a facility to maximize production and minimize the costs of production.

As discussed earlier, feeds can be classified by type according to their intended function. These types include larval, starter, grower, broodstock, and specialty feeds. The cost of each of these feeds is commonly expressed on a per pound (ton) or kilogram (metric ton) basis, and these feeds differ greatly in cost. Larval feeds are many times more expensive than grower feeds. As a consequence, financial personnel are tempted to recommend that the amount of larval feed used in production be reduced or even eliminated. These concerns can be put into perspective by considering the cost of each feed type based on its percentage of total feed cost, expressed on a production or annual basis. As mentioned earlier, 90% or more of the feed used in a production cycle is grower feed. When analyzed this way, reducing the cost of larval feeds by 50% saves very little of yearly feed costs since, proportionately, a very small quantity of larval feed is used. However, a change of only 5% in the cost of a grower feed has a large impact on overall feed costs. It is counter-productive to feed a less expensive larval diet when a more expensive one has been found to be best for the fish. Not only would this action have little effect on overall feed costs, but it would likely lower fish survival and quality, causing production problems later in the cycle.

It should be clear that the cost of grower feeds greatly affects overall feed cost. However, comparison of grower feeds on a cost per kilogram basis is not necessarily the only effective method of feed selection. Comparing grower feeds based upon cost per unit output (fish weight gain) is a much more useful comparison, especially in food fish production. Another basis of comparison is feed cost per unit salable product, a comparison that takes into account visceral fat content, which is lost during processing. Some feeds cost more per kilogram, but result in higher feed efficiency under production conditions. This results in more product at less cost or more product for the same cost. Either of these outcomes benefits the producer.

Broodstock feeds can also represent a major proportion of total feed costs. Again, evaluation of feed cost on product produced, in this case, viable eggs and healthy fry, is the better approach to evaluating the economic effect of feed selection.

Many state and federal hatcheries purchase fish feeds through contracts and a bidding process. This process results in the contract being awarded to the feed producer that can deliver the lowest cost feed. The bidding process was designed to reduce overall costs and can be effective in reaching that goal. Unfortunately, the bidding process can lead to other unintended consequences. When open-formula feeds are specified, feed costs are sometimes elevated by the supplier to protect from losses that might occur as a result of the changing nature of the commodity market and the inflexibility of open-formula feeds. If the contracting agency modifies formulation yearly to adjust to changing markets, these problems can be avoided. Unfortunately, this is not usually practiced.

Feed contracts can be used to reduce total feed costs, but contracts need to be flexible to allow hatchery managers to purchase specialty feeds from other sources as needed. These feeds are rarely needed beyond only a small percentage of total feed needs, so allowing hatchery managers to purchase such feeds without having to undergo competitive bidding ought to have a minor effect on the total cost of fish production. Feed contract specifications must be reviewed and adjusted annually to take advantage of recent developments in nutrition and feed technology and to account for the inevitable swings in commodity prices, especially that of fish meal.

References

Babbitt, J. K., R. W. Hardy, K. D. Reppond, and T. M. Scott. 1994. Processes for improving the quality of whitefish meal. Journal of Aquatic Food Product Technology 3:59–68.

Embody, G. C., and M. Gordon. 1924. A comparative study of natural and artificial foods of brook trout. Transactions of the American Fisheries Society 54:185–200.

Fowler, L. G., and R. E. Burrows. 1971. The Abernathy salmon diet. Progressive Fish-Culturist 33:67–75.

Halver, J. E. In press. The vitamins. In J. E. Halver and R. W. Hardy, editors. Fish nutrition, 3rd edition.

Halver, J. E., and R. W. Hardy, editors. In press. Fish nutrition, 3rd edition. Academic Press, New York.

Hardy, R. W., and F. T. Barrows. In Press. Diet formulation and manufacturing. In J. E. Halver and R. W. Hardy, editors. Fish nutrition, 3rd edition. Academic Press, New York.

Hardy, R. W., and K. D. Shearer. 1985. Effect of dietary calcium phosphate and zinc supplementation on whole body zinc concentration of rainbow trout (*Salmo gairdneri*). Canadian Journal of Fisheries and Aquatic Science 42:181–184.

Hendricks, J. D., and G. S. Bailey. 1989. Adventitious toxins. Pages 605–651 *in* J. E. Halver, editor. Fish nutrition, 2nd edition. Academic Press, New York.

Higgs, D. A., J. S. Macdonald, C. D. Levings, and B. S. Dosanjh. 1995. Nutrition and feeding habits in relation to life history. Pages 159–315 *in* C. Groot, L. Margolis, and W. C. Clark, editors. Physiological ecology of Pacific salmon. UBC Press, Vancouver.

Hublou, W. F. 1963. Oregon pellets. Progressive Fish-Culturist 25:175–180.

Ketola, H. G. 1979. Influence of dietary zinc on cataracts in rainbow trout (*Salmo gairdneri*). Journal of Nutrition 109:965–969.

Kim, J., K. C. Massee, and R. W. Hardy. 1996. Adult *Artemia* as food for first feeding coho salmon (*Oncorhynchus kisutch*). Aquaculture 144:217–226.

Leitritz, E., and R. C. Lewis. 1976. Trout and salmon culture. California Department of Fish Game Fish Bulletin 164.

Lellis, R. E., F. T. Barrows, and R. W. Hardy. In press. Low-phosphorus grower diets for rainbow trout. Aquaculture Research.

Lovell, R. T. 1998. Nutrition and feeding of fish, 2nd edition. Kluwer Academic Publishers, Boston.

National Research Council. 1993. Nutrient requirements of fish. National Academy Press, Washington, D.C.

O'Keefe, T. M., and R. L. Noble. 1978. Storage stability of channel catfish (*Ictalurus punctatus*) in relation to dietary level of a-tocopherol. Journal of the Fisheries Research Board of Canada 35:457–460.

Richardson, N. L., D. A. Higgs, R. M. Beames, and J. R. McBride. 1985. Influence of dietary calcium, phosphorus, zinc and sodium phytate level on cataract incidence, growth and histopathology in juvenile chinook salmon (*Oncorhynchus tshawytscha*). Journal of Nutrition 115:553–567.

Roberts, R. J., and R. W. Hardy. 2000. Nutritional pathology of teleosts. Pages 000–000 *in* R. J. Roberts, editor. Fish pathology. W. B. Saunders, London.

Rumsey, G. L. 1994. History of early diet development in fish culture 1000 B.C. to A.D. 1955. Progressive Fish-Culturist 56:1–6.

Smith, R. R. 1971. A method for measuring digestibility and metabolizable energy of fish feeds. Progressive Fish-Culturist 33:132–134.

Sugiura, S. H., and R. W. Hardy. 2000. Environmentally friendly feeds. Pages 299–310 *in* R. R. Stickney, editor. Encyclopedia of aquaculture. Wiley, New York.

Sugiura, S. H., F. M. Dong, C. K. Rathbone, and R. W. Hardy. 1998. Apparent protein digestibility and mineral availabilities in various feed ingredients for salmonid feeds. Aquaculture 159:177–202.

Sugiura, S. H., F. M. Dong, C. K. Rathbone, and R. W. Hardy. 1998. Effects of dietary supplements on the availability of minerals in fish meal: preliminary observations. Aquaculture 160:283–303.

Torrissen, O. J., R. W. Hardy, and K. D. Shearer. 1989. Pigmentation of salmonids- carotenoid deposition and metabolism. Reviews in Aquatic Sciences 1(2): 209–225.

Vielma, J., K. Ruohonen, and S. P. Lall. 1999. Supplemental citric acid and particle size of fish bone-meal influence the availability of minerals in rainbow trout *Oncorhynchus mykiss* (Walbaum). Aquaculture Nutrition 5:65–72.

Wood, J. W. 1979. Diseases of Pacific salmon, their prevention and treatment. Washington State Department of Fish, Olympia. Figure 12. Mechanical feeding of formulated pelleted feed to catfish in a large rearing pond.

9

Fish Health Management

James R. Winton

Introduction

Maintaining the health of hatchery fish requires a program of good management and continuous attention to detail. In its most basic sense, disease can be defined as a deviation of the body from its normal or healthy state, and disease in fish can be caused by infectious agents or by noninfectious conditions. Thus, optimal fish health is best achieved by rearing fish in a good environment, with good nutrition, with a minimum of stress, and isolated from sources of infectious agents. Whenever possible, prevention of infectious disease by avoidance of contact between the host fish and a pathogen should be a critical goal. This is best achieved with a pathogen-free water supply, the use of certified pathogen-free stocks, and strict attention to sanitation.

The principal reasons for control of infectious diseases in hatcheries are to prevent losses in production; to prevent the introduction of pathogens to new facilities when eggs, fry, or broodstock are moved; to prevent the spread of disease to wild fish via the hatchery effluent or when hatchery fish are released or stocked out; and to prevent the amplification of pathogens already endemic in a watershed. Although only a few fish pathogens cause human health problems (i.e., are zoonotic agents), certain hatchery practices can create a hazard when toxic compounds are used or if fish are contaminated with drugs or chemicals when they are released or sold as food. If a disease outbreak should occur, a serious attempt should be made to eradicate the infectious agent from the facility or to change the conditions that led to disease, rather than to rely upon frequent treatments to control future outbreaks.

Since the first edition of this book in 1982, some very significant changes have occurred in the field of fish health. These include an improved fish health infrastructure for the diagnosis of fish diseases, the discovery of important new fish pathogens, expansion of the host and geographic ranges for some of the known pathogens, increased restrictions on the use of drugs and chemicals in aquaculture, the application of molecular methods for improved diagnosis of diseases, elucidation of important epizootiological features of fish diseases, creation of new regulations to prevent the spread of important diseases, and the development of novel methods for control of some diseases, including the commercial availability of effective vaccines. These topics are included in this new version of the chapter.

Disease Characteristics

Factors Leading to Disease

Conceptually, it is useful to think of disease or health as a set of interactions among the host, the pathogen, and the environment. Host factors include the species of fish, size or age, strain or stock, immune status, and general physiological condition. For the pathogen, factors include the number of infectious organisms present, the strain of the pathogen, and its virulence. The difference between health and disease typically depends on the balance between the pathogen and the host, and that balance is greatly influenced by environmental factors such as temperature and water chemistry. For example, an adverse environmental factor such as high temperature may speed the replication of a pathogen and cause stress on the host resulting in reduced resistance.

Stress plays a major role in the susceptibility of fish to infectious disease. Wild fish are typically in equilibrium with the pathogens that are endemic within a geographic area under normal conditions, and natural outbreaks are rarely observed unless major environmental changes occur. However, fish in intensive culture are frequently affected by adverse environmental conditions (e.g., temperature or water quality) and by management practices (e.g., crowding or handling) that can impose significant stress on the defense mechanisms of most fishes. Whereas some pathogens of fish are highly virulent and cause disease even in a well-managed facility, many diseases are enhanced by stress. For example, although several species of opportunistic bacterial pathogens are present in many hatchery water supplies, disease typically will not occur unless environmental quality or the defense systems of the fish have deteriorated. Prevention of these diseases can best be done through good hatchery management. Environmental stresses and associated disease problems are minimized by high water quality, low rearing densities, and excellent nutrition.

Two other types of disease are important to fish culturists in addition to those caused by pathogenic organisms. The first is nutritional in origin and is usually caused by inadequate diet or poorly stored feeds. The second is caused by environmental factors, including bad hatchery practices and poor water quality. Both of these classes of disease can result in significant losses if severe enough or uncorrected for a prolonged period. While these diseases are covered in other chapters of this book, it is important to remember that such noninfectious diseases may act synergistically with infectious agents to produce exceptionally high losses.

Disease-Causing Organisms

Organisms that cause diseases in fish include a large number of viruses, bacteria, fungi, protozoans, and metazoans. Generally, they can be categorized as either pathogens or parasites, although the distinction is not always clear. Viruses are submicroscopic agents that are dependent upon living cells for their replication and often have highly specific requirements for a particular host and even for certain tissues within that host. Viruses are always maintained in a host and are found free in nature only in a nonreplicating state. Bacteria may also be obligate pathogens if they cannot live outside a host; however, many bacterial fish pathogens are found free-living in nature and only affect fish as opportunists. Protozoans and metazoans are considered parasites and can reside either inside or on the surface of the host. Pathogens or parasites do not always cause disease in fish, but may be present in a subclinical or carrier state.

Disease Recognition

Only a few fish diseases produce behaviors or clinical signs that are pathognomonic (specific to a given disease). Nevertheless, careful observation of signs exhibited by the fish will often allow one to restrict the cause of the disease to a particular type (or group) of causative agent(s).

Some obvious changes in behavior of fish suffering from disease include (1) loss of appetite; (2) abnormal distribution in a pond or raceway, such as swimming at the surface, along the tank sides, or in slack water, or crowding at the head or tail screens; (3) flashing, scraping on the bottom or sides of the rearing unit, darting, whirling, twisting, or loss of equilibrium; and (4) weakness, loss of vitality, and loss of ability to withstand stresses during handling, grading, seining, loading, or transportation.

In addition to changes in behavior, disease may produce physical signs such as lesions that can be seen by the unaided eye. Such gross signs may be external, internal, or both. Gross external signs of disease include darkening or discolored areas on the body; eroded areas, ulcers, or sores on the body, head, or fins; cysts, tumors, or swelling on the body or gills; exophthalmia (popeye); and hemorrhages, especially around the head and at the base of fins. Gross internal signs of disease are changes in the color of organs or tissues (pale liver or kidney or congested organs); hemorrhages in organs or tissues; changes in the texture or size of organs or tissues; accumulation of fluid in body cavities; and cysts, tumors, or lesions.

Besides changes in behavior and gross physical signs, information such as the disease history of the facility, geographic area, species and age of fish, water temperature, and time of year can assist in the formation of a presumptive diagnosis. If a serious disease problem is suspected, a fish health professional should be contacted immediately for assistance in identifying the causative agent and in helping to develop an appropriate strategy to reduce potential losses.

Defense Mechanisms of Fish

As with all living organisms, fish stay healthy only if they prevent excessive growth of microorganisms on their external surfaces and invasion of their tissues by pathogenic agents. Invasion of microorganisms is inhibited by physical barriers, and replication within the host is held in check by natural and acquired immune mechanisms.

Barriers

Physical barriers are an important first line of defense. Fish eggs are protected by the structurally tough and chemically resistant chorion. However, during oogenesis, the egg may become infected or contaminated with certain viruses or bacteria present in the female. Once hatched, the fry are then likely to have become infected. Transmission of infectious agents from adult to progeny via the egg is termed vertical transmission.

Fishes are protected from injury and invasion of disease agents by the external barriers of mucus, scales, and skin. For example, the skin protects against microbial colonization by continuously producing and sloughing mucus, which allows infectious agents only temporary residence on the host. Mucus also may contain nonspecific antimicrobial substances, such as natural antibodies, lysozyme, or other antimicrobial factors, as well as specific antibodies produced in response to prior infection or vaccination.

Gill tissue also contains mucus cells that can serve the same purpose as those in the skin. However, irritants may cause accumulation of mucus on the gill tissue and lead to asphyxiation. This is an example of a defense mechanism that can work against the host.

Immune Mechanisms

The immune system of the fish can be divided into nonspecific defense mechanisms (also termed natural immunity), which are effective against a broad range of organisms, and specific defense mechanisms, which are usually acquired upon recovery from infection with a given pathogen or following vaccination. In addition, both nonspecific and specific defense mechanisms may be innate (present continuously) or induced by infection or vaccination. One of the primary natural defense mechanisms is the inflammatory response of the vascular system, which is triggered by invasion of pathogenic agents and other irritants. Dilation of capillaries increases the supply of humoral and cellular agents at the focus of infection. The inflammatory response proceeds to dilute, localize, destroy, and remove the agent that stimulated the response.

Fish, like most animals, also have an important defense mechanism in the form of fixed and wandering phagocytes in the lymphatic and circulatory systems. Phagocytes are cells capable of ingesting bacteria, foreign particles, and infected cells. Fish also have natural killer cells, natural antibodies, and other innate humoral and cell-mediated defenses against infectious disease that are intrinsic to the species and individual. An example is the production of interferon in response to viral infection. This induced response is nonspecific and increases resistance to other virus infections, as well as to the agent that stimulated it. These types of defenses may account for some portion of the differences in the natural resistance of various species and strains of fish to certain diseases. Some of these differences have been shown to have a genetic basis as resistance to a pathogen may differ between families of the same species of fish and the trait shows significant heritability.

Fish have a relatively well-developed specific immune system and are able to produce both circulating antibodies (humoral response) and a cell-mediated response to antigenic stimuli. These humoral and cellular arms of the immune system are required to respond in differing degrees for various viral, bacterial, protozoan, and fungal pathogens. The fish's immune system shows proliferation of specific cells involved in the immune response and an immunological memory to a limited degree. These specific, adaptive immune responses are also stimulated by vaccines that are commercially available and that can be used to protect fish against certain diseases. The immune response of cold-blooded animals depends heavily upon environmental temperature. Lowering of the water temperature below a fish's optimum usually reduces or delays the onset of the immune response. Other environmental factors that stress the fish, as well as various physiological states (e.g., spawning) can reduce the magnitude or efficacy of the immune response. Age is also an important factor. The fish's specific immune system develops after hatching and, depending upon species and temperature, may not be fully functional until the fish is several months of age.

Genetic Resistance

Significant heritabilities have been shown for resistance to various fish diseases, and, under controlled environmental conditions, some increased resistance to a disease can be expected from a managed breeding program. However, selection for resistance to several pathogens can be difficult, except perhaps for closely related diseases. Intentionally or unintentionally, specific disease resistance has been increased at hatcheries by the continued use of survivors of outbreaks as broodstock. Potential exists for genetic selection and breeding to increase the survival of propagated fishes, but certain risks must be anticipated in any major breeding program that may be counter to programs designed to maintain the genetic composition of wild stocks. In addition, changes in gene frequencies resulting from selection for disease resistance may cause changes in the frequencies of other genes that are unrelated to disease resistance leading to a concomitant loss in some other desired trait (e.g., growth or fecundity). Thus, managers always run the risk of altering the fitness of their stock in unwanted ways during selective breeding programs.

In any natural population, individual fish may be found that are more or less resistant to the common diseases. In a population of wild fish, this natural genetic variation among individuals is important to ensure that not all fish in the population are equally susceptible during changes in the environment or in the levels or types of pathogens present. Natural recombination within the breeding population assures that these variations are maintained. In the absence of important needs to the contrary, any artificial propagation program should ensure that a similar genetic variability is maintained within the population as its lack may put the entire population at risk.

Diseases of Fish

The first edition of this book had photographs and relatively detailed descriptions of many of the major infectious diseases of hatchery fish that were known at the time. It also included recommendations for treatment, sometimes using drugs or chemicals that were not approved for use in food fish. Because the field of fish health was relatively new, there were few books or reference materials that could be consulted by the hatchery manager. Also, the fish health infrastructure available to assist hatchery managers was poor in many areas. In the nearly 20 years following publication of the first edition, the amount of information on fish diseases has expanded exponentially and many excellent references have become available that describe the clinical signs, diagnosis, and treatment of fish diseases. Furthermore, the list of diseases affecting cultured fish has continually grown as new species are reared or as established species are cultured in new geographic areas. Finally, the network of fish health professionals has improved greatly, and several organizations have published books containing standard methods for the diagnosis of fish diseases. For these reasons, this new edition will only briefly describe some of the most significant diseases of the fish species cultured in North America.

The material in the following sections has been developed from information contained in references listed at the end of this chapter. For further information, these references should be consulted. It is assumed that a progressive hatchery manager will have a library containing many of these references and an established relationship with a fish health professional who can provide expert advice in the diagnosis and treatment of fish diseases.

Viral Diseases of Fish

Among the infectious diseases of fish, those caused by viruses are important because they can result in explosive, uncontrollable losses and are often the subject of state, federal, and international regulations. Because viral diseases of fish are untreatable and there are few approved vaccines for their prevention, avoidance of the pathogen remains the major control option available. This is best accomplished by rearing fish in a virus-free water supply, using only certified virus-free stocks, paying strict attention to sanitation, and having sound quarantine practices. While requiring effort, it is still easier to avoid viral pathogens than to deal with their effects or to try to eradicate them from a facility once introduced. For hatcheries having open water supplies and that are located in an endemic area, viral diseases can present insurmountable problems. For example, until methods were developed for avoidance of infectious hematopoietic necrosis virus (IHNV), sockeye salmon could not be propagated successfully in Alaskan hatcheries having open water supplies because the virus was present in wild, spawning adults in nearly all populations tested.

The host specificity of fish viruses can vary widely. Some have a broad range of species in which they replicate, while other viruses have a very narrow host range that may include only certain tissues within a single host. Similarly, there may be significant genetic differences among strains of fish viruses within certain virus groups resulting in substantial differences in virulence or in reactivity with serological reagents. Many of these different strains occur within a defined geographical area making it important to restrict viruses to locations in which they are endemic.

The diagnosis of viral diseases has traditionally relied upon growth of the virus in cell culture and identification by various serological methods including neutralization, fluorescent antibody, and enzyme-linked immunosorbent assays. These methods are time-consuming, relatively labor intensive, and require specialized training and equipment. In recent years, the techniques of molecular biology have produced new diagnostic tools having great speed and precision. Among the new approaches are the use of monoclonal antibodies, DNA probes, and the polymerase chain reaction (PCR) for identification of a viral pathogen. However, many of the required certification examinations still rely upon cell culture methods because they are considered validated or benchmarked tests. In some cases, nonlethal methods have been developed that typically rely upon detection of neutralizing antibodies in a population of fish and, more recently, the use of PCR to detect viral genomes in nonlethal samples such as blood or mucus.

While a diagnosis of a viral disease is relatively straightforward, the management implications are not. Because fish viral diseases can be caused by strains of the causative agent having differences in host specificity, virulence, serological reactions, or growth characteristics, it may be necessary to examine the causative agent further before a recommendation can be made as to the disposition of the stock. For example, fish harboring a low level of IHNV should not be moved to other watersheds, even those containing fish infected with IHNV, unless it has been shown that an identical strain of the virus is present. Similarly, the emergence or introduction of a different strain of virus at a hatchery should be treated with caution.

During the time since the first edition of this book, the number of identified fish viruses has increased very substantially and only a few will be discussed below. Other viral diseases may be important locally, and the conscientious fish hatchery manager will become familiar with those diseases at risk of being found among the species cultured in the geographic area in which the facility is located. Many of the following viral diseases are listed by the Office International des Epizooties as being either "Notifiable" or as "Other Significant Diseases" of finfish, while a few others are included that are of interest to aquaculture.

Rhabdoviral Diseases

The rhabdoviruses that infect finfish generally cause viremia with resulting hemorrhage and widespread necrosis of major organ systems, which may result in very high losses. These viruses have a relatively broad host specificity and are among the oldest and best studied viral pathogens of fish. The diseases caused by these viruses typically show both a size and temperature dependence.

Viral Hemorrhagic Septicemia

Viral hemorrhagic septicemia (VHS) is best known as an infectious disease of rainbow trout, brown trout, grayling, whitefish, northern pike, and turbot. The disease is caused by viral hemorrhagic septicemia virus (VHSV, synonym: Egtved virus). Formerly thought to be confined to western Europe, in the last decade, VHSV has been isolated from an increasing number of free-living marine fish species in the North Pacific Ocean, the North Atlantic Ocean, the North Sea, and the Baltic Sea.

Typically, isolates of VHSV from marine fish are not distinguishable from normal freshwater isolates by serological means; however, isolates of VHSV from the Pacific Ocean are genetically distinguishable from the European isolates. There appear to be several distinct genetic lineages among the European isolates. In contrast to the European freshwater strain, isolates from wild marine fish species in both North America and Europe typically induce low mortality in infection trials with rainbow trout fry. Several European marine isolates are pathogenic to turbot fry, and isolates from marine species in the Pacific are highly pathogenic to Pacific herring.

Among rainbow trout infected with the European freshwater strains, infection is often lethal due to the impairment of the osmotic balance and occurs within a clinical context of edema and hemorrhages. Virus multiplication in endothelial cells of blood capillaries, leukocytes, hematopoietic tissues, and nephron cells underlies the clinical signs.

The reservoirs of VHSV are clinically infected fish and covert carriers among cultured, feral, or wild fish in the marine environment and in freshwater. Virus is shed in the feces, urine, and sexual fluids, whereas kidney, spleen, heart, liver, and digestive tract are the sites in which virus is most abundant. Several factors influence susceptibility to VHSV. Among each fish species, there is individual variability in susceptibility, and the age of the fish appears to be of some importance—the younger the fish, the higher the susceptibility. In susceptible fish stocks, however, overt infection is seen in all sizes of fish.

Water temperature is an important environmental factor. Disease generally occurs at temperatures between 4 and 14°C (39.2 and 57.2°F). Low water temperatures (1–5°C [33.8–41.0°F]) generally result in an extended course with low daily mortality, but high total mortality. At high water temperatures (15–18°C [59.0–64.4°F]), the disease generally takes a short course with an initial acute mortality, but a more modest accumulated mortality. VHS outbreaks occur during all seasons, but are most common in spring when water temperatures are rising or fluctuating.

Control methods for VHS rely upon avoidance of the virus through official health surveillance plans coupled with control measures, such as eradication. Effective use of these measures has resulted in the eradication of the disease from several parts of Europe. At present, effective vaccines are in the experimental stage as are genetic approaches using selection for disease-resistant stocks.

Infectious Hematopoietic Necrosis

Infectious hematopoietic necrosis (IHN) is an important infectious disease of rainbow or steelhead trout; Pacific salmon, including chinook, sockeye, chum, cherry (also known as amago and yamame), and coho salmon; and Atlantic salmon. Historically, the geographical range of IHN was limited to the western parts of North America, but the disease, caused by a rhabdovirus (IHNV), has spread to continental Europe and the Far East via the importation of infected fish and eggs. IHN has become a cause for concern because of its clinical and economic consequences in trout and salmon farming and its effects on wild stocks. Infection is often lethal due to the impairment of osmotic balance and occurs within a clinical context of edema and hemorrhages. Virus multiplication in endothelial cells of blood capillaries, hematopoietic tissues, and nephron cells underlies the clinical signs. High levels of virus are shed from infected juvenile fish. Older fish are increasingly resistant to infection, but adult fish at spawning may shed virus in sexual products. Survivors of IHNV infection demonstrate a strong protective immunity with the synthesis of circulating antibodies to the virus and, in certain individuals, a covert carrier state. Variations in the virulence of IHNV strains have been recorded during both natural cases of disease and in experimental infections.

The reservoirs of IHNV are clinically infected fish and covert carriers among cultured, feral, or wild fish. Virus is shed via feces, urine, sexual fluids, and external mucus, whereas kidney, spleen, encephalon, and the digestive tract are the sites in which virus is most abundant during the course of overt infection. The transmission of IHNV between fish is primarily horizontal; however, cases of vertical or "egg-associated" transmission have been recorded. Horizontal transmission is typically by direct exposure, but invertebrate vectors have been proposed to play a role in some cases. Egg-associated transmission is significantly reduced by the now common practice of surface disinfection of eggs with an iodophor solution; but egg-associated transmission is the only mechanism accounting for the appearance of IHN in new geographical locations among alevins originating from eggs that were incubated and hatched in virus-free water. Once IHNV is established in a farmed stock or in a watershed, the disease may reside among carrier fish.

Among individuals of each fish species, there is a high degree of variation in susceptibility to IHNV. The age of the fish appears to be extremely important—the younger the fish, the more susceptible to disease. As with VHSV, good overall fish health condition seems to decrease the susceptibility to overt IHN, while handling and other types of stress frequently cause subclinical infections to become overt.

The most prominent environmental factor affecting IHN is water temperature. Clinical disease occurs between 8°C and 15°C (46.4°F and 59.0°F) under natural conditions.

Control methods for IHN currently rely on avoidance of exposure to the virus through the implementation of strict control policies and sound hygiene practices. The thorough disinfection of fertilized eggs and the incubation of eggs and rearing of fry and alevins in virus-free water supplies are critical for preventing the occurrence of IHNV in a defined fish production site. At present, vaccination is at an experimental stage; however, several new vaccine preparations have shown substantial promise in both laboratory and field trials.

Spring Viremia of Carp

Spring viremia of carp (SVC) is a rhabdoviral disease of several carp and other cyprinid fish species. Overt infections have been recognized in common carp, grass carp, silver carp, bighead carp, crucian carp, goldfish, tench, and sheatfish. The common carp is the most susceptible of these species and is the principal host. The geographical range of SVC includes countries of the European continent that experience low water temperatures during winter, although the disease may be present in Asia and other regions where carp are traditionally reared or have been introduced.

As with other rhabdoviruses of fish, infection by SVC virus (SVCV) can be lethal due to the impairment of water balance that occurs in a clinical context of edema and hemorrhages. Virus multiplication, especially in endothelial cells of blood capillaries, hematopoietic tissue, and nephron cells, underlies the clinical signs.

Overcoming SVCV infection results in a strong protective immunity associated with the presence of circulating antibodies detectable by methods such as virus neutralization, immunofluorescence, or enzyme-linked immunosorbent assay. In certain individuals, this health status also results in a covert virus carrier state. Differences in virulence of virus strains have been recorded during natural cases of disease and experimental infections. A related virus, the northern pike fry rhabdovirus, shares antigens that cross-react in some serological tests, but the two agents can be distinguished by serum neutralization tests and by molecular methods.

The reservoirs of SVCV are clinically infected fish and covert virus carriers from cultured, feral, or wild fish. Virulent virus is shed in the feces, urine, gill and skin mucus, and from the exudate of skin blisters or edematous scale pockets. However, liver, kidney, spleen, gill, and encephalon are the organs in which SVCV is most abundant during the course of overt infection.

The principal mode of transmission for SVCV is horizontal from fish to fish, but vertical or egg-associated transmission cannot be ruled out. Animate vectors and fomites are also involved in transmission of SVCV. Among animate vectors, the parasitic copepod *Argulus follaceus* and the leech *Piscicola piscicola* are able to transfer SVCV from diseased to healthy fish. Once SVCV is established in pond stock or pond farm stock, it may be very difficult to eradicate without destroying all kinds of life on the production site.

There is a high variability in the degree of susceptibility to SVC among individuals of the same fish species. Apart from the physiological state of the fish, age or age-related status of innate immunity appears to be extremely important: the younger the fish, the higher the susceptibility to overt disease, but even adult broodfish can be susceptible to infection. Water temperature is a critical environmental factor. In yearling or older fish, overt infection is not often observed above 17°C (62.6°F), whereas fry can be affected at temperatures as high as 22–23°C (71.6–73.4°F). The implementation of hygiene measures and control policies are the only control methods currently feasible. Vaccination is still mostly at the experimental stage.

Herpesviral Diseases

Fish herpesviruses usually have a rather narrow host specificity. The diseases caused by herpesviruses vary from external, fluid-filled blisters, to systemic infections affecting several organ systems and the production of tumors on the body of the fish. Some of these herpesvirus diseases are described below.

Channel Catfish Virus Disease

Channel catfish virus disease (CCVD) is caused by a herpesvirus designated Ictalurid herpesvirus 1, but the commonly used name is channel catfish virus (CCV). Channel catfish virus affects channel catfish in the United States. Channel catfish virus disease is of importance because of its clinical and economic consequences in channel catfish farming where it causes high mortality rates in populations of fry and juvenile catfish in the summer months. Diseased fish demonstrate ascites, exophthalmia, and hemorrhage in fins and musculature. Histologically, the most remarkable damage occurs in the kidney with extensive necrosis of renal tubules and interstitial tissue.

In survivors, CCVD results in a strong protective immunity, the synthesis of circulating antibodies to the virus, and a covert latent carrier state. During this latent carrier state, the virus is undetectable by traditional culture or antigen detecting means, even when adults are immunosuppressed during spawning. Some variation in the virulence of CCV strains has been recorded during natural outbreaks of disease and has been demonstrated experimentally. Additionally, molecular data indicate genetic variation among isolates of the virus.

Reservoirs of CCV are clinically infected fish and covert carriers. Infectious CCV can be detected in the water from tanks of experimentally infected fish, but the route of shedding has not been determined. The sites where the virus is most abundant during the course of overt infection are posterior kidney, skin, gill, spleen, and intestine, respectively, in decreasing magnitude. The transmission of CCV is horizontal and vertical. Animate vectors and fomites could also act in CCV transmission. Vertical transmission is thought to be common, but the mechanism of vertical transmission is not known as infectious virus has not been detected on the skin or in the sexual products of spawning adults. Once CCVD occurs in a fish population, it becomes established among carrier fish.

Channel catfish and blue catfish are the only species naturally infected with CCV, and variations in susceptibility to CCV have been recorded depending on fish strain. The age of the fish is extremely important for overt infection. Although experimental data suggest that older fish are susceptible to natural outbreaks of acute CCVD, the disease occurs almost exclusively in fish that are less than 1 year of age and generally less than 4 months of age. Water temperature is the critical environmental factor. The mortality rate is high above 27°C (80.6°F), but readily decreases and ceases below 18°C (64.4°F).

Control methods currently rely on maintaining relatively low stocking densities and avoiding stressful handling of young fish during the summer months. Also, control policies and hygiene practices have been used, where practical, in catfish husbandry. The incubation of eggs and rearing of fry and juveniles in facilities separated from carrier populations are critical for preventing the occurrence of CCVD in a CCV-free fish production site. Because the virus is only detected during active outbreaks, defining CCV-free status has been done largely from historical data or by identifying populations that are seronegative to the virus.

Recent use of PCR and DNA probes to detect latent CCV genomic DNA suggests that CCV is present in many populations that have no history of the disease. Vaccination, although experimentally promising, is not in use at this time.

Salmonid Herpesviral Diseases

Two herpesviruses of salmonids are currently recognized. These are salmonid herpesvirus 1 (formerly termed *Herpesvirus salmonis*) and salmonid herpesvirus 2 (also known as *Oncorhynchus masou* virus [OMV]). Salmonid herpesvirus 1 was originally isolated from rainbow trout broodstock in Washington State. Subsequently, steelhead trout in California were also found to be carriers of the virus. However, the virus has been shown to be of relatively low pathogenicity for salmonids and is now regarded as being of limited concern.

In contrast, *Oncorhynchus masou* virus disease (OMVD) is an oncogenic and ulcerative condition among salmonid fish in Japan and probably in the coastal rivers of eastern Asia that harbor Pacific salmon. In Japan, the causative agent has been given a variety of names, but is now termed *Oncorhynchus masou* virus or salmonid herpesvirus 2. Fish species that are susceptible to OMV include kokanee salmon, cherry salmon (also known as masu salmon), chum salmon, coho salmon, and rainbow trout. On the basis of antigenic studies conducted with neutralizing polyclonal rabbit antisera, OMV can be easily distinguished from salmonid herpesvirus 1.

Clinically, the initial infection by OMV appears as a systemic and frequently lethal infection that is associated with edema and hemorrhages. Virus multiplication in endothelial cells of blood capillaries, hematopoietic tissue, and hepatocytes underlies the clinical signs. Four months after this first clinical condition, a varying number of surviving fish exhibit epitheliomas occurring mainly around the mouth (upper and lower jaw) and, to a lesser extent, on the caudal fin, opercula, and body surface. These neoplastic growths may persist for up to 1 year post-infection. In the case of coho salmon, 1-year-old infected fish, in particular, show ulcers on the skin, white spots on the liver, and tumors around the mouth parts or body surface. In rainbow trout, the diseased fish exhibit almost no external signs, although some fish manifest ulcerative lesions on the skin. Internally, intestinal hemorrhages and white spots on the liver are observed.

Following the septicemic phase, an immune response takes place that results in the synthesis of neutralizing antibodies to OMV. A carrier state frequently occurs, which leads to virus shedding via the sexual products at the time of spawning.

The reservoirs of OMV are clinically infected fish and covert carriers among groups of cultured, feral, or wild fish. Infectious virus is shed via feces, urine, sexual products, and probably skin mucus; whereas the kidney, spleen, liver, and tumors are the sites where virus is the most abundant during the course of overt infection. The transmission of OMV is horizontal and possibly egg-associated. Disinfection of the eggs just after fertilization or at the eyed stage is effective in preventing OMV infection. OMV disease has not been reported in alevins originating from disinfected eggs that had been incubated and hatched in virus-free water.

Salmonids are the only fish species susceptible to OMV. The age of the fish is critical. One-month-old alevins are the most susceptible to infection. The main environmental factor favoring OMV infection is low water temperature (below 14°C [57.2°F]).

Control methods currently rely on the implementation of avoidance and hygiene practices in the operating of salmonid husbandry. The thorough disinfection of fertilized eggs and the incubation of these eggs and rearing of fry and alevins in premises completely separated from those harboring virus carriers and free from contact with fomites are the key measures needed to decrease contamination of OMV in a defined fish production site.

Cyprinid Herpesvirus Disease

The disease caused by cyprinid herpesvirus (CHV, formerly *Herpesvirus cyprini*) has been known as carp pox or *epithelioma papulosum* and is most often seen as benign, raised, white to gray, hyperplastic papillomas on the body surface. The lesions are smooth and may eventually cover large areas in fish that are stressed. Koi (ornamental strain of common carp) and common carp are the most often affected; however, the disease has been reported from other cyprinid species. The disease probably has a worldwide distribution wherever cyprinids are native or have been introduced. It is common in Europe and Asia and has been reported from areas of the United States where cyprinids are commercially reared. In addition to monetary losses from the unsightly lesions, the disease can produce mortality in juvenile cyprinids. The causative virus is a typical herpesvirus and can be detected by isolation in carp cell lines. Serum neutralization tests and molecular probes have been used to confirm many features of the life cycle of the virus.

Cyprinid herpesvirus is probably shed from the external lesions to affect juvenile fish in the rearing water. Infection of juvenile carp can result in a distended abdomen, darkening, exophthalmia, hemorrhages, and high mortality. Fish that recover often become carriers and will show external lesions much later in life. The induction of these lesions, like cold sores in humans, is thought to be stress related. At lower temperatures (15°C [59.0°F]), the carp immune system is less effective and virus replication can lead to production of the characteristic lesions. However, at warmer temperatures when conditions are favorable for the host (20–

30°C [68.0–86.0°F]) and the fish immune system is fully operational, the lesions will regress and disappear.

There is no control method other than avoidance. Carp to be shipped into an area where the disease has not been reported should come from stocks that have been carefully screened and that have no history of the disease. The long incubation period and latency of the virus make pathogen-free certification difficult.

More recently, an important emerging viral disease of cyprinids has caused mass mortality of koi carp in Israel and the eastern United States. The principal external signs of the disease were pale and irregularly colored gills. Microscopically, the gills showed severe hyperplasia and necrosis of the epithelial cells. A new herpesvirus was isolated from diseased fish using a koi carp cell line. Experimental infections produced high mortality in koi, and the virus was reisolated from several tissues of the dead fish. This new herpesvirus of cyprinids can be differentiated from *Herpesvirus cyprini* by serological means.

Birnaviral Diseases

The aquatic birnaviruses are small, icosahedral viruses belonging to the family Birnaviridae and have a genome composed of two segments of double-stranded RNA. Birnaviruses have been isolated from a very large number of fish, crustacean, and mollusk species in both freshwater and marine environments. Serological and genetic testing has revealed that there are very substantial differences among the various isolates. Within this collection are viruses that seem to cause little mortality within the hosts that have been experimentally infected, as well as viruses that cause very high mortality in important species for aquaculture.

Infectious Pancreatic Necrosis

Infectious pancreatic necrosis (IPN) is a highly contagious viral disease of salmonids. The disease, caused by infectious pancreatic necrosis virus (IPNV) most characteristically occurs in fry of rainbow trout, brook trout, brown trout, Atlantic salmon, and several Pacific salmon species. Susceptibility generally decreases with age, except for Atlantic salmon smolts, which can suffer from the disease after transfer to seawater. The first sign of an outbreak is frequently a sudden and, usually, progressive increase in daily mortality, particularly in the faster growing individuals. Clinical signs include darkening, a pronounced distended abdomen, and a spiral swimming motion. Cumulative mortalities may vary from less than 10% to more than 90% depending on the virus strain, size and species of host, and environmental conditions.

The disease is transmitted both horizontally through the water route and vertically via the egg. Surface disinfection of eggs is not entirely effective in preventing vertical transmission.

The disease has a wide geographical distribution, occurring in most, if not all, major salmonid-farming countries of North and South America, Europe, and Asia.

Different isolates of IPNV display wide antigenic diversity and marked differences in degrees of virulence. Other birnaviruses showing serological relatedness to IPNV have been reported to cause diseases in some farmed marine fish species, such as yellowtail, turbot, and halibut; subclinical covert infections have been detected in a wide range of estuarine and freshwater fish species, such as loach, northern pike, and numerous other species in the families Anguillidae, Atherinidae, Bothidae, Carangidae, Catostomidae, Cichlidae, Clupeidae, Cobitidae, Cyprinidae, Esocidae, Percichthyidae, Percidae, Poeciliidae, Sciaenidae, and Soleidae.

Control methods currently rely on the implementation of control policies and of hygiene practices in salmonid husbandry, through the avoidance of the introduction of fertilized eggs originating from IPNV-carrier broodstock, and through the use of a protected water supply (e.g., spring or well) where the ingress of fish, particularly virus carriers, can be prevented. In outbreaks, reducing the population density may help to reduce the overall mortality. No treatment is available. A recombinant DNA vaccine is commercially available in Norway to protect Atlantic salmon, and other vaccines are in development and appear promising.

Orthomyxoviral Diseases

A good example of an emerging viral disease is found among the orthomyxoviruses. Until the 1980s, the only observation of a disease thought to be caused by a member of this group of viruses was recorded among eels. Later, increasing losses among Atlantic salmon in Norway resulted in the discovery of an important pathogen that has now been identified in salmonids from several countries.

Infectious Salmon Anemia

Infectious salmon anemia (ISA) is a disease of Atlantic salmon caused by an orthomyxovirus. It has, so far, been reported to occur in Norway, the United Kingdom, Canada, and the United States. Atlantic salmon is the only susceptible fish species known to develop the disease, but the ISA virus (ISAV) may survive and replicate in brown trout (also known as sea trout), rainbow trout, and Atlantic herring, which could act as carriers of the virus. Clinically, the initial infection appears as a systemic and lethal condition that is characterized by anemia, ascites, congestion, and darkening and enlargement of the liver and spleen, as well as peritoneal petechiae (small, inflamed red lesions). Hemorrhages in the eyes may also be seen. Hepatocellular degeneration and necrosis, tubular necrosis, and hemorrhages in the kidneys are consistent histopathological findings in typical outbreaks. The infection is mainly observed in fish held in seawater or in fish exposed to seawater, but indications of disease outbreaks in fish held in freshwater have also been reported. The infection spreads slowly and the virus is considered to be of low virulence, although high mortality can ensue in some outbreaks of disease.

The reservoirs of ISAV are not known with certainty, but spread of the disease has occurred from the movement of subclinically infected Atlantic salmon smolts from farm to farm and from the effluent of slaughterhouses where organic material (especially blood and processing water) from ISA-infected fish has been discharged directly into seawater without further treatment. The presence of natural reservoirs of ISAV among marine fish species is supported by the isolation of genetically distinct strains of the virus from Atlantic salmon cultured in sea cages in different geographic areas. Few environmental factors have been identified that can be directly linked to outbreaks of the disease. In a latent carrier population, various stress factors, such as treatment against salmon lice, cestodes, or infectious diseases may be followed by disease outbreaks some 2–3 weeks later.

The incidence of ISA may be greatly reduced by implementation of general regulatory measures regarding the movement of fish, mandatory health control, and slaughterhouse and transport regulations, as well as specific measures including restrictions on affected, suspected, and neighboring farms; epizootiological studies; enforced sanitary slaughtering; generation segregation ("all in/all out"); and disinfection of offal and wastewater from fish slaughterhouses and fish processing plants. Development of an effective vaccine will be a critical step toward the eventual control of ISA in marine aquaculture and a killed viral vaccine shows significant promise in field trials.

Iridoviral Diseases

Another good example of emerging diseases of finfish is seen in the increasing number of iridoviruses isolated from outbreaks of disease among fish and amphibians on several continents. For many years, the only iridoviral diseases of significance among finfish were lymphocystis and viral erythrocytic necrosis (VEN). These diseases caused external warty growths (lymphocystis) among many freshwater and marine species or an erythrocytic anemia (VEN) among many marine species. In recent years, severe losses have been associated with a group of iridoviruses that are related to the amphibian pathogens that are members of the *Ranavirus* genus of the family Iridoviridae.

Systemic Freshwater Ranaviruses

Epizootic hematopoietic necrosis (EHN) is a systemic iridovirus (ranavirus) infection of redfin perch (also known as Eurasian perch), rainbow trout, sheatfish, and catfish. The disease is caused by one of a group of closely related iridoviruses: epizootic hematopoietic necrosis virus (EHNV), European sheatfish virus (ESV), and European catfish virus (ECV). The geographical range of EHNV is currently restricted to Australia. The ECV and ESV agents have only been detected among fish in Europe. In North America, a virus has been isolated from largemouth bass that appears similar to these agents suggesting this group of viruses from fish and amphibians is distributed worldwide. Although closely related and inducing similar diseases in their respective hosts, these viruses can be distinguished from each other by molecular techniques.

Epizootic hematopoietic necrosis is generally lethal in redfin perch, while rainbow trout are relatively resistant and only a small proportion of individuals become infected. Infections with ESV and ECV can induce high morbidity and mortality in their catfish hosts. The largemouth bass virus (LMBV) has been associated with large natural mortalities among bass in the southern United States. The disease caused by these systemic iridoviruses is characterized by necrosis in the liver, spleen, hematopoietic tissue of the kidney, and other tissues. Experimental exposure studies have shown that rainbow trout can be infected by ECV and ESV without showing morbidity and mortality, while infections of largemouth bass with LMBV produce darkening, spiral swimming, abdominal distention, listlessness, and high mortality.

The viruses can be identified by immunofluorescence, enzyme-linked immunosorbent assay, or immunoelectron microscopy. The viruses share several antigens with each other and at least one antigen in common with the amphibian iridoviruses from North America (Frog Virus 3) and Australia (Bohle Iridovirus). Polyclonal antibodies against EHNV detect all of these viruses; however, nucleotide sequence analysis, sodium dodecyl sulfate-polyacrylamide gel electrophoresis (SDS-PAGE), Western blotting, and PCR assays can differentiate, to varying degrees, among the viruses.

The factors modulating the susceptibility of fish to systemic iridovirus infection are poorly understood. Clinical outbreaks due to EHNV are associated with poor water quality. In rainbow trout, infection occurs naturally at water temperatures from 11°C to 17°C (51.8°F to

62.6°F) and experimentally from 8°C to 21°C (46.4°F to 69.8°F). Disease in redfin perch does not occur at temperatures below 12°C (53.6°F). The following fish species were found to be susceptible to EHNV following bath exposure: redfin perch, rainbow trout, Macquarie perch, western mosquitofish, silver perch, and mountain galaxias. Both juvenile and adult redfin perch may be affected in outbreaks, but juveniles may be more susceptible to the disease. EHNV has now been detected in diseased rainbow trout ranging from hatchery fry to table-sized fish, although mortalities are most often seen in 0+ fish up to 125-mm fork length.

White Sturgeon Iridoviral Disease

The white sturgeon iridoviral disease (WSIVD) is a significant cause of mortality among farm-raised juvenile white sturgeon in North America and among Russian sturgeon in Europe. Lake sturgeon have been experimentally infected. White sturgeon iridovirus (WSIV) was originally described among hatchery-raised white sturgeon in North America; the source of WSIV was assumed to originate from adult wild white sturgeon broodstock collected from the Sacramento River in California and the Columbia River and its tributaries in Oregon, Washington, and Idaho. The virus also has been detected among wild juvenile white sturgeon collected from the lower Columbia River, and the virus is potentially endemic in wild white sturgeon populations throughout the Pacific Northwest of North America. An iridovirus similar to WSIV has been identified in Russian sturgeon from Northern Europe where it may be endemic among cultured populations of several species of sturgeon.

The WSIV infects tissues of the skin, gills, and upper alimentary tract. Infections of the oral mucosa and olfactory organ epithelium are presumed causes of the cessation of feeding that leads to a progressive emaciation or starvation of the fish—the principal external sign of the disease. Cumulative mortality of up to 95% has been reported among groups of infected fish in the hatchery and secondary infections with external protozoa or bacteria often contribute to the overall mortality. Infected fish with moderate to severe emaciation began dying 2–3 weeks following exposure to the virus at water temperatures of 17–19°C (62.6–66.2°F). Hemorrhages on the abdomen and the ventral scutes may be present, but these are not specific for WSIVD. There are no specific internal signs of infection as the virus does not invade systemically. Viral infection is evident on microscopic observation of stained tissue sections from infected fish. The integument and, particularly, the skin may show focal or diffuse hyperplasia with characteristic amphophilic to basophilic enlarged Malpighian cells. The cells are filled with virus particles as demonstrated by electron microscopy.

The modes of transmission of WSIV are not completely understood, but horizontal transmission via water has been demonstrated in the hatchery and experimentally in the laboratory. There is strong circumstantial evidence that the virus is transmitted vertically from adult broodstock, but the virus has never been isolated or observed in adult fish.

There appears to be little antigenic relationship of WSIV to the systemic iridoviral agents represented by EHNV or the red sea bream iridovirus from Asia. The larger size and inner membrane structure of the WSIV virion, host cell line specificity, type of cytopathic effect, and the location of the target host cells (epitheliotropic and not systemic) distinguish the agent from other groups of fish iridoviruses.

Control methods currently rely on avoidance of the agent where possible. Because there are currently no methods for detecting the virus in adult broodstock, quarantine and testing of juveniles suffering mortality are the principal means to detect WSIV in young fish.

Lymphocystis

Lymphocystis disease, although rarely lethal, is of interest because of its wide range of occurrence among many propagated and free-ranging fish species. Many marine and freshwater fish species are susceptible, but the disease has not been reported among salmonids. Among freshwater fishes, walleyes and most centrarchids are susceptible.

Lymphocystis is a chronic viral disease causing generally granular, wart-like or nodular tissue lesions composed of greatly enlarged host cells and their covering membrane. Cells of infected tissue may attain a size of a millimeter or more and resemble a spattering of sand-like granules or, when larger, a raspberry-like appearance.

The causative agent of the disease is an iridovirus maintained in susceptible host fishes. Healthy fish may be exposed when infected cells burst and the virus particles are released. This release of virus can occur intermittently through the duration of infection or can be massive upon death and decomposition of infected fish. Lymphocystis lesions are persistent and commonly remain for several months; some may continue for a year or more.

No method of treatment is known and avoidance remains the only control strategy. Fish with the disease should be removed from the population to control the spread of the infection.

Bacterial Diseases of Fish

Bacterial diseases of fish have been recognized for many years, although new examples are continually being discovered as new species of fish are placed into aquaculture for the first time or as traditional species are cultured in new geographic areas. Today, perhaps more than 100 bacterial species have been isolated from diseased fish and many of these have been demonstrated to be pathogens. In a few cases, these bacterial species can infect humans and are termed zoonotic agents.

The diagnosis of bacterial diseases has traditionally been accomplished by light microscopy and by culture of the bacteria on appropriate artificial media followed by identification using biochemical or serological testing. More recently, techniques of molecular biology (e.g., PCR) have provided rapid and sensitive alternatives.

While many bacterial diseases of fish are treatable with drugs or chemicals, the widespread use of these compounds has become of increasing concern (see the section on drugs and chemicals) and substantial restrictions are now being enforced in North America and Europe. These restrictions have significantly limited the numbers of compounds available and the specific conditions under which they can be legally used. In addition, withdrawal times for approved uses in food fish (or fish that may enter the human food chain) must be respected. Fortunately, several efficacious vaccines are now commercially available to prevent some of the most important bacterial diseases of fish and others are expected in the near future.

As stated earlier, the resistance of fish to infectious diseases is significantly reduced by stress; stress is often necessary for disease to occur from opportunistic bacterial pathogens that are commonly present in the aquatic environment. For these diseases, hatchery management practices are the difference between health and disease. The following are examples of some bacterial diseases of cultured fish. The list is not inclusive, and the conscientious hatchery manager will seek advice from a fish health professional to become familiar with bacterial diseases affecting the species of fish cultured in manger's geographic area and the approved methods for their prevention or treatment.

Flavobacterial Diseases

Several important diseases are caused by long, thin (0.5 × 10 mm), gram-negative, rod-shaped bacterial pathogens of the genus *Flavobacterium*. While the taxonomy of this group has undergone significant revision in the last decade, the diseases that these bacteria cause have been recognized for many years. Most flavobacterial species are common in the environment and become opportunistic pathogens when susceptible fish are present and the conditions are suitable. The nature of the disease caused by flavobacteria depends upon the location in the host in which they are growing. Some are external, causing gill diseases or external lesions, while others can result in bacteremia and systemic disease. Because these bacteria are common inhabitants in soil and water, avoidance of these diseases is difficult, even with a pathogen-free water supply. Perhaps more than any other, these diseases are best prevented by good hatchery management.

Bacterial Gill Disease

Bacterial gill disease is a common disease of cultured trout and salmon and can result in explosive losses; it also is an occasional disease of warmwater and coolwater fish reared in ponds. Sudden lack of appetite, lethargy, flared opercula, and gathering near the surface of the water are typical signs of fish infected with bacterial gill disease. Gills show proliferation of the epithelium that may result in clubbing and fusing of lamellae or even entire filaments. The disease is greatly exacerbated by stress and by adverse water conditions, especially high ammonia levels, where mortalities can rise rapidly to 50% or greater per day. Microscopic examination of affected gill tissue reveals long, thin bacteria arranged in thick patches over the epithelium. Necrotic gill tissue may be grayish-white, and many of the filaments may be completely eroded. The causative agent is typically one of a large number of the flavobacteria, especially *Flavobacterium branchiophilum*, that are common in the soil and water and can be isolated from diseased fish.

A combination of large numbers of bacteria and gill epithelial proliferation differentiates bacterial gill disease from other gill problems caused by nutritional or environmental conditions (nutritional gill disease or environmental gill disease). Crowding, abrasive silt in the water supply, and dusty starter diets are important factors that contribute to outbreaks of the disease and may be a necessary predisposing cause in some outbreaks. Water temperatures above 12°C (53.6°F) are favorable for the bacteria. Yearling and older fish are less susceptible than fry, but outbreaks can be acute in all ages of fish.

Bacterial gill disease is exacerbated by stress and seldom is a problem among warmwater fish, particularly those being reared in earthen ponds. It occasionally becomes a problem when young channel catfish, largemouth bass, bluegills, or redear sunfish are held in crowded conditions in tanks or troughs for extended periods. After the problem is under control, the fish population should be thinned or the water flow increased. Unless the management practice that precipitated the outbreak is corrected, bacterial gill disease will reappear.

Columnaris

The causative agent of columnaris is *Flavobacterium columnare*. The agent is a long, thin, gram-negative bacterium that moves in a creeping or flexing action and that has a peculiar habit of stacking up to form distinctive columns, hence the name "columnaris."

Columnaris most commonly involves external infections, but can occur as an internal systemic infection with no visible external signs. Externally, the disease starts as small, grayish lesions on the body, gills, or fins. The lesions rapidly increase in size and become irregular in shape. As the lesions get larger, the underlying musculature can be exposed. The margins of the lesions and, occasionally, the centers may have a yellowish color due to large aggregations of the bacteria. Frequently, the most lethal infections involve mouth and jaws. In Pacific salmon and warmwater fish, particularly catfish, lesions may be confined to the gills. Lesions on the gills are characterized by yellowish-brown necrotic tissue beginning at the tip of the filaments and progressing toward the base.

Columnaris disease usually is associated with some kind of stress condition such as high water temperature, low oxygen concentration, crowding, and handling. Under appropriate conditions, columnaris may take an explosive course and cause catastrophic losses in 1–2 d after the first appearance of the disease. Therefore, it is incumbent upon the fish culturist to maintain the best possible environmental conditions for the fish and to minimize any stress conditions.

Although columnaris disease attacks practically all species of freshwater fish, catfish are particularly susceptible. In warmwater fish, most outbreaks of columnaris occur when the water temperature is above 18–20°C (64.4–68.0°F), but the disease can occur at any time of the year. Columnaris disease is common in salmonids held at water temperatures above 15°C (59.0°F). Progress of the disease usually is faster at the higher temperatures.

Flavobacteria are common inhabitants of soil and water and commonly are found on the surface of fishes, particularly on the gills. The stress of crowding, handling, spawning, or holding fish at above-normal temperatures, as well as the stress of external injury, facilitates the transmission and eruption of columnaris disease. A disease resembling columnaris occurs in the marine environment and is caused by *Flavobacterium maritimus*.

Presumptive diagnosis of columnaris is accomplished best by microscopic examination of wet mounts of scrapings from lesions and detection of many long slender bacteria that move by flexing or creeping movements and form "haystacks" or "columns."

Preventative measures include maintenance of optimum water temperatures for salmonids, reduced handling during warm weather, maintenance of the best possible environmental conditions, and avoidance of overcrowding fish.

Bacterial Coldwater Disease

Bacterial coldwater disease (BCWD) or low temperature disease is one of the more common diseases of young salmonids where it produces low to moderate mortality. The disease typically manifests itself in the form of external lesions on the dorsal surface or on the caudal peduncle, hence the name "peduncle disease"; however, the bacteria frequently become systemic where they can invade the vertebral column and central nervous system leading to darkening and spinal deformities that can mimic IHN or whirling disease. As its name implies, the disease occurs at low water temperatures in the range of 4–10°C (39.2–50.0°F). Affected fish become darkened, and lesions may develop on the back near the dorsal fin, on the caudal peduncle, or on the isthmus anterior to the pectoral fins. The caudal fin may be completely destroyed. A lesion usually starts on the caudal peduncle behind the adipose fin, where it causes inflammation, swelling, and gradual erosion. The disease progresses posteriorly and the caudal fin may be eroded.

Coldwater disease is caused by *Flavobacterium psychrophilum.* The bacterium is water borne and can be transmitted from carrier fish in the water supply. The bacterium can also be transferred vertically from adult to progeny via the egg. The disease sometimes occurs without external signs and may be due to the bacterium being acquired by vertical transmission. Crowded conditions stimulate a disease outbreak, but are not necessary for the disease to appear. While all salmonids (and a few nonsalmonids) are thought to be susceptible, coho and chum salmon are the most vulnerable and, in sac fry of these species, the yolk sac may be ruptured. Typically observed in young fish during the spring when water temperatures are low, larger fish may be affected if conditions are favorable for the disease. Efforts to develop a vaccine are underway.

Gram-Negative Septicemias

Many members of the *Enterobacteriaceae* (enterics), *Vibrionaceae* (vibrios and aeromonads), and *Pseudomonadaceae* (pseudomonads) cause disease in fish. Different bacterial species within these families grow optimally at very different temperatures and typically cause disease in hosts that are cultured at those temperatures. The diseases are typically gram-negative septicemias with hemorrhagic lesions or liquifactive necrosis of organs. They can result in high losses. The first commercially licensed fish vaccines were developed as formalin-killed bacterial cultures (bacterins) to control some of these diseases.

Furunculosis

Furunculosis was traditionally known as a septicemic disease of salmonids reared in freshwater, but significant losses now occur among salmonids cultured in marine net-pens. The causative agent of furunculosis is *Aeromonas salmonicida,* and virtually all trout and salmon hatcheries with open water supplies are probably exposed to the bacterium at one time or another. Furunculosis has been reported in most salmonids of all ages. Among salmonids, brook trout, brown trout, rainbow trout, and Atlantic salmon are most often affected. While furunculosis is endemic in many hatcheries, severe outbreaks are now less common due to advances in fish culture, drug therapy, and use of vaccines.

The disease is characterized by a generalized bacteremia with focal necrotic lesions in the muscle, often seen as swellings under the skin that are not true "furuncles." The swollen skin lesions are filled with pink fluid containing blood, and necrotic tissue may have a purple or iridescent blue color. These lesions are especially apparent in chronic infections, but similar lesions may occur from other diseases caused by gram-negative bacteria. Hemorrhaged and frayed fins also are common. Internally, diseased fish may exhibit petechiae in the lining of the body cavity and especially on the visceral fat. The pericardium usually is filled with bloody fluid and is inflamed. The spleen, normally dark red in color, often will be a bright cherry-red and swollen. The lower intestine often is highly inflamed and a bloody discharge can be manually pressed from the vent.

The incidence of furunculosis generally follows the seasonal temperature pattern with most outbreaks occurring in the warmer months. When conditions are optimal for the disease, it occurs as an acute form in which death results from massive bacteremia before gross lesions can develop. In such outbreaks, only a few clinically sick fish may be seen at any one time in spite of the high death rate. Acute cases of furunculosis have incubation periods of 2–4 d with few apparent signs. Chronic cases usually occur at temperatures below 14°C (57.2°F) and may have an incubation period of one to several weeks, depending upon the water temperature. Latent cases may develop during low temperature periods and flare up with greater severity, displaying many typical signs, when water temperatures rise.

Fish exposed to furunculosis may become carriers of the disease and nonsalmonids in the water supply should be considered likely reservoirs of infection. Furunculosis may break out in virtually any freshwater fish population, including warmwater species, if conditions such as high temperature and low dissolved oxygen favor the pathogen.

Sanitation provides the most important long-range control of furunculosis. If a population of fish at a hatchery is free of furunculosis and if the water supply does not contain fish that harbor the pathogen, strict sanitation measures should be used to prevent the introduction of the disease via incoming eggs or fish. Eggs received at a hatchery should be disinfected upon arrival with iodophors used as recommended. Maintenance of favorable environmental conditions for the fish is of prime importance in preventing furunculosis outbreaks. Proper water temperatures, adequate dissolved oxygen, efficient waste removal, and avoidance of overcrowding must be observed.

A presumptive diagnosis of furunculosis takes into consideration the frequency of outbreaks in a certain area, presence of typical lesions, and the occurrence of short gram-negative rods in the lesions, kidneys, spleen, and blood. A confirmatory diagnosis can be made only after *Aeromonas salmonicida* has been identified as the predominant organism isolated.

Drugs are effective only in the treatment of outbreaks, and drug resistance is a common problem. Recurrences of furunculosis are likely as long as *A. salmonicida* is present in the hatchery system and environmental conditions are suitable. In recent years, effective vaccines have been licensed for control of furunculosis. While these vaccines are most effective when delivered by injection in an adjuvant, among Atlantic salmon farms, the vaccines provide important protection and are highly cost-effective.

In addition to salmonids, an atypical form of furunculosis has been reported in many warmwater species, mainly ornamental fish, including carp, Koi, goldfish, and eel. The disease often produces hemorrhages on the body surface that can lead to deep ulcers of the skin and musculature. This atypical form of furunculosis is caused by a different subspecies of the organism and the disease is known as erythrodermatitis of carp, carp furunculosis, ulcer disease, or atypical furunculosis.

Enteric Redmouth

Enteric redmouth disease (ERM) refers to an infection of rainbow trout and Atlantic salmon by the gram-negative enteric bacterium *Yersinia ruckeri*. *Yersinia ruckeri* produces a systemic infection that results in marked inflammation and erosion of the jaws and palate. Trout with ERM typically become sluggish and dark in color and show inflammation of the mouth, opercula, isthmus, and base of fins. Reddening occurs in body fat and in the posterior part of the intestine. The stomach may become filled with a colorless watery liquid and the intestine with a yellow fluid. This disease often produces sustained low-level mortality, but can cause large losses. Large-scale outbreaks occur if chronically infected fish are stressed during hauling, exposed to low dissolved oxygen, or exposed to other poor environmental conditions.

The disease has been reported in rainbow, steelhead, and cutthroat trout, and coho, chinook, and Atlantic salmon. Evidence suggests that the spread of the disease is associated with the movement of infected fish, and fish-to-fish contact provides transfer of the bacterium to healthy trout. Because spread of the disease can be linked with fish movements, the best control is avoidance of the pathogen. Fish and eggs should be obtained only from sources known to be free of ERM. This can be accomplished by strict sanitary procedures and avoidance of carrier fish.

While drug therapy has been used in the past to control ERM, the emergence of drug resistance strains of *Y. ruckeri* is a significant problem. Highly effective vaccines are now commercially available for the control of ERM. These are efficiently administered by immersion vaccination of fry and fingerlings providing cost-effective protection for fish in waters containing the bacterium. These bacterins produce long-term immunity and should be viewed as the recommended method for control.

Motile Aeromonas Septicemia

Motile aeromonas septicemia (MAS) is a ubiquitous disease of many freshwater fish species and is typically caused by one of several, gram-negative, motile bacteria that are members of the genus *Aeromonas*. Some of the species frequently isolated in outbreaks are *A. hydrophila*, *A. sobria*, and *A. caviae*. Occasionally, various species of pseudomonads, especially *Pseudomonas fluorescens*, can cause an indistinguishable form of the disease. A definitive diagnosis of MAS can be made only if the causative agent is isolated and identified because the signs of MAS can be easily confused with several other diseases.

When present, the most common signs of MAS are superficial circular or irregular grayish-red ulcerations with inflammation and erosion in and around the mouth as in ERM. Fish may have a distended abdomen filled with a slightly opaque or bloody fluid (dropsy) or exophthalmia if fluid accumulates behind the eyeball. Other fish, minnows in particular, may have furuncles like those in furunculosis, which may erupt to the surface, producing deep necrotic craters. Fins also may be inflamed.

In addition to the presence of fluid in the abdominal cavity, the kidney may be swollen and soft and the liver may become pale or greenish. Petechiae may be present in the peritoneum and musculature. The lower intestine and vent often are swollen and inflamed and may contain bloody contents or discharge. The intestine usually is free of food, but may be filled with yellow mucus.

Motile aeromonas septicemia occasionally takes an acute form in warmwater fish, and severe losses can occur even though fish show few, if any, clinical signs of the disease. In general, most outbreaks in warmwater fish are in the spring and summer, but the disease may occur at any time of year. Temperature is very important. Largemouth bass and channel catfish are susceptible particularly during spawning and during the summer if stressed by handling, crowding, or low oxygen concentrations. Aquarium fish can develop the disease at any time of the year. Among salmonids, rainbow trout seem to be the most susceptible and outbreaks are associated with handling stress and crowding of fish. Fish and frogs that recover from the disease usually become carriers and may contaminate water supplies if not removed. The disease has been identified throughout the world and apparently infects virtually any species of freshwater fish under conditions favoring the bacteria.

Observation of strict sanitary practices and the elimination of possible carrier fish from the water supply are extremely important to the control of MAS at trout and salmon hatcheries. For warmwater fish, everything possible should be done to avoid stressing the fish during warm weather. Because of the many types of bacteria associated with MAS, it is unlikely that a vaccine will be developed in the near future. Also, due to the common occurrence of drug resistance, therapeutic drug treatments cannot be assumed to be effective.

Enteric Septicemia of Catfish

Enteric septicemia of catfish (ESC) or edwardsiellosis is caused by the bacterium *Edwardsiella ictaluri*. It is one of the most important infectious disease problems in the commercial catfish industry in the southern United States. Most reported cases of ESC occur among channel catfish, but the bacterium has been isolated from white catfish, yellow bullhead, black bullhead, and brown bullhead. Enteric septicemia of catfish has also been reported from *Clarias batrachus* in Thailand and from several ornamental species, and the susceptibility of other species, including salmonids, has been shown experimentally. *Edwardsiella ictaluri* should not be confused with *E. tarda*, another member of the same genus that is frequently found in aquatic animals and is responsible for opportunistic infections in fish and mammals, including humans.

Acute outbreaks of ESC occur within a limited temperature range, from 18 to 28°C (64.4 to 82.4°F). Low-level mortality due to ESC can occur in carrier populations outside of this temperature range. This results in seasonal fluctuations, spring and autumn being the critical periods. Other environmental factors (water quality, organic compounds, stocking density, and stress factors) are likely to modulate the virulence of the agent In spite of these characteristics, *E. ictaluri* is generally considered to be a true obligate pathogen.

Clinical signs may differ according to the fish species. Two clinical forms of ESC have been described. The infection seems to begin most often in the olfactory sacs through the nasal route and to progress slowly upward generating granulomatous inflammation in brain tissues. This chronic meningoencephalitis can account for behavioral manifestations with alternating listlessness and chaotic swimming. In late stages of this disease manifestation, "hole in the head" is observed—an ulceration developing from the inside through the dorsocranial part of the head exposing the brain. Although general infection can result from such chronic forms, it is more often associated with acute septicemia following infection through the enteric route. It has been demonstrated that *E. ictaluri* can cross the intestinal mucosa very easily. Then, as in many other bacteremias, skin petechial hemorrhages are observed around the mouth, on the throat, and at the base of the fins. Multifocal distinct 2-mm diameter, raised, hemorragic cutaneous lesions that progress to depigmented ulcers also occur. Anemia, moderate gill inflammation and exophthalmia are common signs. Internally, hemorrhages and necrotic foci are scattered in the liver and other internal organs. Hemorrhagic enteritis, systemic edema, accumulation of ascitic fluid in the body cavity, and enlargement of the spleen are nonspecific signs. Histological examination reveals a systemic infection of all organs and skeletal muscles with diffuse granulomatosis.

Fish that recover from ESC generally remain infected, even though protective immunity and circulating antibodies can be clearly evidenced. *Edwardsiella ictaluri* has been detected in the kidney of such fish well over 4 months after exposure, suggesting that carrier fish act as the natural reservoir for the organism. It is believed that shedding with feces is the main means of dissemination into the environment. The pathogen persistence and the common practice of continual partial harvest and stocking within a production pond have contributed to the success of this pathogen and the prevalence of ESC in the industry. Moreover, the agent can survive in pond sediments for an extended period of time, and this may be another important factor in disease recurrence in given areas.

Enteric septicemia of catfish may be controlled through chemotherapy or prophylactic measures. The most common antimicrobial treatments are oral application of potentiated sulfonamide (sulfadimethoxineormethoprim) or oxytetracycline, but plasmid-mediated resistance to these antibiotics has developed recently and many producers are now focusing on alternative methods to reduce losses. This relies on management to reduce stress in fish and on vaccination, which has shown promise and remains an active area of research.

A related enteric organism, *Edwardsiella tarda*, also causes mild to severe disease in catfish and among salmonids at warmer temperatures. In catfish, the disease progresses from small cutaneous lesions to large abscesses filled with necrotic tissue, a condition termed "emphysematous putrefactive disease" which indicates the foul smell occurring in fish affected by the disease. In addition to the potential to cause disease in many fish species reared in warm water, the agent is capable of infecting humans. Control methods for *E. tarda* are similar to those used for *E. ictaluri*. Because *E. tarda* is not an obligate pathogen and is relatively common in the environment, it is difficult to eliminate the infection from open ponds.

Vibriosis

Vibriosis is a systemic disease of marine, estuarine, and (occasionally) freshwater fishes. It is known also under the names of red pest, red boil, red plague, or salt water furunculosis. Many species of vibrios have been isolated from diseased fish; however, the most important include *Vibrio anguillarum*, *V. ordalii*, *V. salmonicida*, *V. damsela*, and *V. vulnificus*. Depending upon the causative species, vibriosis can occur any time of year, even in water temperatures as low as 4°C (39.2°F; coldwater vibriosis). However, it is most prevalent in the temperate zones during the warmer summer months, and outbreaks can be expected when water temperatures reach 15°C (59.0°F).

Signs of the disease usually do not become evident until the fish have been in salt water for 2 weeks or more under crowded conditions. Diminished feeding activity is one of the first noticeable signs. Lethargic fish gather around the edges of holding units; others swim in erratic, spinning patterns. Diseased fish have hemorrhages around the bases of their pectoral and anal fins or bloody discharge from the vent. When a fish is opened for necropsy, diffuse pinpoint hemorrhages of the intestinal wall and liver may be evident. The spleen frequently is enlarged and may be two to three times its normal size.

Vibriosis is worldwide in its distribution and virtually all species of marine and estuarine fishes are susceptible. Among salmonids, pink salmon and chum salmon are the most susceptible, but serious outbreaks have occurred in coho salmon, rainbow trout, and Atlantic salmon. Stresses associated with handling, low oxygen, and elevated temperature can predispose fish to vibriosis. Vibrios are ubiquitous in marine and brackish waters and infections probably are waterborne and may be spread by contact. Salmonids usually die within a week after exposure. Fish of all ages are susceptible.

Diagnosis of vibriosis requires isolation of a slightly curved, gram-negative, motile, rod-shaped bacterium on salt-containing medium. The organism may be identified by serological or biochemical reactions. There is no reliable presumptive diagnosis of vibriosis because of its similarity to other gram-negative septicemias.

Prevention of vibriosis depends on good sanitation, no crowding, and minimal handling stress. While drug therapy has been used in the past to control vibriosis, drug resistance is a problem. Highly effective vaccines are now commercially available for the control of vibriosis and were the first commercially licensed vaccines for fish. These vaccines are efficiently administered by immersion vaccination and result in significant economic benefit. These bacterins produce long-term protection and should be viewed as the recommended method for control.

Bacterial Kidney Disease

Bacterial kidney disease (BKD) is a chronic insidious infection of salmonid fish. The disease is slow to develop, but once established, it may be difficult to control and virtually impossible to cure. The causative agent of BKD, *Renibacterium salmoninarum,* is a small, nonmotile, nonacid-fast, gram-positive diplobacillus.

The course of kidney disease is that of a chronic bacteremia. Once the pathogen enters the fish, either from vertical transmission through the egg or from contact with other infected fish in the water supply, the bacteria multiply slowly within macrophages and other cells. Foci of infection develop in the kidney and in other organs such as the liver, spleen, and heart. White cellular debris collects in blisters, and ulcers that develop in these organs are seen easily. Lesions developing in the posterior kidney are easiest to spot and may reach a centimeter or more in diameter. Some lesions extend into the musculature and result in externally visible blisters under the skin. If the disease has reached the stage in which gross lesions are apparent, therapeutic treatment has little effect.

Bacterial kidney disease has been found in many species of salmonids in North America. A tendency toward seasonal periodicity has been noted, but the incidence varies at different hatcheries. Brook trout and chinook, chum, pink, and sockeye salmon are highly susceptible. Infected or carrier fish are considered to be sources of infection. Experimentally, from 1 to 3 months will elapse before mortality begins, depending upon the route of administration. Vertical transmission via the egg has been proven and is an important source of infection. Such infections within the egg are impossible to eliminate via iodophor treatment of eggs.

Historically, diagnosis of BKD has been based on the demonstration of small, gram-positive diplobacilli in infected tissues. More recently, serological procedures such as fluorescent antibody techniques and enzyme-linked immunosorbent assays have become widely used and are considered more reliable. Molecular methods such as the PCR assay have been published and will probably become standard methods in the future.

Control methods for BKD rely upon reducing hatchery conditions that exacerbate disease and in reducing the infection levels among broodstock. Complete avoidance of *R. salmoninarum* has proven difficult because the bacterium is present at subclinical levels in many stocks of salmonids and is transmitted *intra ovum*. Thus, only the most highly sensitive detection methods have been successful at identifying and avoiding infected broodstock (broodstock segregation). Drug treatment is typically only effective in delaying mortality and not in eliminating infections. Among hatcheries rearing salmonids, low-level infections should be anticipated; however, should a serious outbreak occur, strict quarantine and eradication measures are recommended. The development of vaccines against BKD remains an active area of research.

Piscirickettsiosis

Piscirickettsiosis is a disease of salmonids caused by *Piscirickettsia salmonis*, a gram-negative, highly fastidious, intracellular bacterial pathogen of fish. Thus far, the disease has been described from Chile, Ireland, Norway, and both the west and east coasts of North America. *Piscirickettsia salmonis* has been detected in coho salmon, chinook salmon, sakura salmon, rainbow trout, pink salmon, and Atlantic salmon. Coho salmon are believed to be most susceptible.

The first evidence of disease may be the appearance of small white lesions or shallow hemorrhagic ulcers on the skin. Affected fish appear dark and lethargic, hanging at the net sides. The major gross pathological changes are gill pallor, peritonitis, ascites, an enlarged spleen, a swollen gray kidney, and a liver with large pale necrotic lesions. The mechanisms of transmission are not completely understood. The disease has been primarily reported in marine fish farms, but has also been observed in freshwater facilities. Horizontal transmission in salt water and freshwater has been demonstrated. Transmission by vectors remains a consideration, and the role of vertical transmission is presently under investigation.

The identification of *P. salmonis* is based on isolation of the causative agent in cell culture with subsequent testing for characteristics of rickettsiae. *P. salmonis* occurs in cytoplasmic vacuoles in the host cell. The identity of *P. salmonis* isolated in cell culture or observed in smears from diseased tissue may be confirmed by means of the fluorescent antibody test using type-specific antiserum. Molecular diagnostic methods using the PCR have been developed.

The implementation of hygienic measures and sound management policies are the only control methods currently available. Eggs may be disinfected as part of good hatchery practice, and elimination of infected stocks may be of some use unless the fish are continually exposed to carriers present in the marine environment. Although antibacterial treatment provides some benefit, it is not entirely effective as a means of controlling the disease, and the emergence of drug resistance has been reported to be an increasing problem. Intensive efforts are underway by various groups to develop an effective vaccine.

Mycobacteriosis

Mycobacteriosis of fish is typically a chronic, granulomatous disease that is caused by one of several freshwater or marine species of mycobacteria, including *Mycobacterium fortuitum, M. chelonae, or M. marinum.* The disease is important both from a fish health perspective and because these bacteria are zoonotic agents capable of causing disease in humans.

Mycobacteriosis is usually seen in fish reared in temperate conditions, and outbreaks have been reported in striped bass, sea bass, sea bream, cod, red drum, mackerel, tilapia, and salmonids. Clinical signs include lethargy, skin discoloration, popeye, and a distended abdomen. Internally, grayish to white nodular lesions are typically observed in the liver, spleen, or kidney. Mortality is typically chronic and unrelenting. Occasionally, the disease assumes an acute fulminating form with extensive necrosis of internal organs.

Diagnosis is by observation of acid-fast organisms in the lesions or by culture on special media. There is no effective therapy or vaccine for control of this disease. Removal of fish and complete disinfection of facilities are the only control options.

Streptococcal Infections

In the United States, *Streptococcus iniae* is responsible for significant losses in tilapia, striped bass, and other warmwater species. In freshwater species, it produces an acute disease characterized as meningitis (in trout) and meningoencephalitis (in tilapines). In the marine species red drum, infections are characterized as an acute to subacute disease with multiple necrotic lesions affecting the skin and internal organs. Bacteria have been found in both diseased and asymptomatic fish. *S. iniae* is considered a zoonotic agent capable of infecting humans, causing a localized form of cellulitis; cases have been identified in the United States and Canada.

The slow growth and similarity of *S. iniae* to other species of gram-positive cocci complicate identification by conventional methods, and PCR assays have been developed. Formalin-killed bacterins have been tested in the laboratory and in the field. Further research on vaccine development is ongoing.

Parasitic Diseases of Fish

Parasites probably cause more disease problems in fish culture than any other type of fish pathogen. It is common to find parasites of many taxonomic classes in or on wild fish, and fish reared under intensive conditions rarely are without some parasites. When present in small numbers, they usually produce no obvious damage and are of limited concern; however, under intensive rearing conditions, parasite numbers can build up rapidly, leading to problems. Here, vigilance is important. If problems can be anticipated and effective measures taken early, parasite losses may be moderated, but under adverse conditions, parasites, even normally benign ones, can result in high mortality.

Traditionally, parasites are identified or classified by microscopic observation of morphological features using fresh or fixed material. More recently, the application of molecular techniques has revealed differences among parasites with similar morphologies, as well as the identity of parasites previously thought to be entirely different organisms. These methods have also been used to develop more rapid methods for confirmation of a few parasitic diseases that formerly had to be diagnosed by more labor-intensive means, including histology.

While, literally, thousands of species within the many parasite genera affecting fish have been described, only a few examples of those appearing commonly and of major importance to fish husbandry are presented below. It should be remembered that almost all species of parasites can cause problems if allowed to reach very high levels, but, typically, significant losses due to parasites are usually a sign of inattention to good hatchery practices. Over time, hatchery managers should become familiar with the types of parasites that can present problems for the species being reared at a given location. In general, the same problems will occur routinely and can be anticipated. Because many parasites can be identified with the aid of a simple light microscope, the progressive hatchery manager will be proactive in examining fish in order to take appropriate action or to seek early assistance from a fish health professional.

Dinoflagellates

Dinoflagellates are protozoans that occur frequently in the aquatic environment. Some produce potent toxins that have caused mass mortalities among fish. While most are free living or parasites of invertebrates, several have been reported to cause disease in fish, including *Amyloodinium* and members of several related genera.

Amyloodinium is a serious parasite of fish living in warmwater marine and brackish waters. It occurs worldwide and has been associated with large losses of both cultured and wild fishes. The disease caused by *Amyloodinium* is often referred to as marine velvet disease; the parasite affects the gills and skin causing a dusty or velvety appearance. Death occurs by suffocation in heavy infestations of the gills. Most species of marine fish are susceptible.

Diplomonads, Kinetoplastids, and Amoebas

The diplomonads, kinetoplastids, and amoebas are parasitic protozoans that produce external and internal infections of many vertebrates, including fish. The best known fish pathogens of the group include species of *Cryptobia, Hexamita, Ichthyobodo, Trypanosoma,* and several amoebas.

Hexamita is the only common flagellated protozoan found in the intestine of salmonids of all ages and occasionally of ornamental fish and grass carp. Infection can occur at all temperatures and results in poor growth and elevated mortality in small fish. All species of salmonids are susceptible to infection. Because there are no well-defined signs, a diagnosis of hexamitiasis must be made by microscopic examination of gut contents from the anterior portion of the intestine and pyloric caeca. The flagellates are minute, colorless, pear-shaped organisms that dart rapidly in every direction. Gross signs of infected fish may include nervous behavior, including swimming in a cork-screw pattern and a dark emaciated condition.

Species of *Ichthyobodo* (formerly *Costia*) are very small flagellated ectoparasites easily missed during routine microscopic examinations of gills and body scrapings. These protozoans are free-swimming, move by means of long flagella, and are about 5×12 μm in size. Two species, *Ichthyobodo pyriformis* and *I. necatrix,* are commonly seen and produce "blue slime" disease of fish. The characteristic blue slime or bluish sheen is caused by increased mucus production in response to irritation. Virtually all species of fish are susceptible at all water temperatures.

An early sign of an *Ichtyobodo* infection is a drop in appetite of the fish and a general listlessness. "Flashing" may be evident if the skin is infected, but only rarely if just the gills are involved. Signs of the disease sometimes are mistaken for bacterial gill disease. Heavily infected fish often develop a bluish slime over the entire body; however, fish less than 3 or 4 months old may die before this condition develops.

Ichtyobodo can be a serious problem on all species and sizes of warmwater fish, particularly channel catfish. This flagellate can cause problems any time of year, but is most common on warmwater fish from February to April in the Northern Hemisphere. Infections with the parasite can be treated, but prevention through improved hygiene, adequate water flows, and continuous vigilance is typically a better choice.

Ciliates

Among the more that 7,000 species of ciliated protozoans are a few of the better-known fish pathogens. These include *Chilodonella, Ichthyophthirius, Trichodina,* and *Trichophyra*. Infections are typically external and have been successfully treated with a variety of chemicals. These parasites affect a wide range of fish species and typically do limited harm. When mortality occurs, it is usually a result of adverse conditions that stress the fish making them more susceptible to infection.

Ichthyophihirius multifilis, also termed white spot or "Ich," is a large ciliated protozoan exclusively parasitic on fish. It causes serious disease in catfish, but also is a common parasite of other warmwater fishes and can be a problem for ornamental fish and salmonids. Ich can be seen by the naked eye; when fully grown it may be as large as 1.0 mm in diameter and appears as gray-white grains of salt. Positive identification is based on the finding of a large, ciliated protozoan, with a horseshoe-shaped macronucleus, embedded in gills, skin, or fin tissue.

The feeding stages, or trophozoites, of Ich are found in the epithelium of the skin, fins, and gills. When mature, the adult parasites drop off the host and attach to the bottom or sides of the pond. Once encysted, they reproduce by multiple fission and, within 2 to several days, depending upon temperature, each adult may produce up to 1,000 ciliated tomites. Upon contact with the fish, the tomites penetrate the skin and begin to feed and grow into adults. At optimal temperatures of 20–23°C (68.0–73.4°F), the life cycle may take as few as 3–4 d. The cycle requires 2 weeks at 15°C (59.0°F), more than 5 weeks at 10°C (50.0°F), and months at lower temperatures.

Ich is known as "salt and pepper" and "white spot" disease by aquarists because of the gray-white specks that appear on the skin. However, in some species of warmwater fish, Ich is found almost exclusively on the gills. On rare occasions, Ich infections on catfish also may be restricted to the gills. In severe outbreaks, losses may precede the appearance of the mature parasites on the fish. Young fish exhibit considerable flashing off the bottom and often show erratic spurts of activity, jumping out of the water and thrashing about due to irritation caused by the parasites. Successful treatment of Ich depends upon the elimination of parasite stages that are free in the water and the prevention of reinfection. Tomites and adult parasites leaving the fish are, therefore, the target of therapeutic efforts.

The best control for Ich, as for any disease, is prevention. Hatchery water supplies always should be kept free of fish. If possible, any warmwater fish brought onto a hatchery should be quarantined for at least 1 week at 20°C (68.0°F) and coldwater fish for at least 2 weeks at 15°C (59.0°F) to determine if they are infested with Ich.

Ich is more difficult to treat because the encysted forms are resistant to treatment; only the free-swimming forms are vulnerable. Successful treatment usually is long and expensive. There are several pond treatments for either warmwater fish or salmonids that can be used successfully if started in time.

Species of *Chilodonella* are small, oval, colorless protozoans, 50–70 μm long, which may be found in vast numbers on the skin, fins, and gills of virtually all species of fish at all water temperatures. Under high magnification, faint bands of cilia can be seen over much of the organism. The parasite primarily affects weakened fish and is a significant problem for warmwater fish during the winter when lower temperatures suppress host resistance. Heavily infected fish are listless, produce excessive mucus, do not feed actively, and may flash. Mortality is often associated with heavy gill infestations.

Trichodinids are saucer-shaped protozoans with cilia around the margins of their body. These protozoans live on the skin, fins, and gills of fish and, when abundant, cause severe irritation and continual flashing. Salmon yearlings, if left untreated, develop a tattered appearance. Secondary bacterial infections may develop in untreated cases.

Species of Trichophrya sometimes are found on the gills of fish and can cause serious problems in catfish and occasionally in other warmwater species. They have rounded to pyramid-shaped bodies (30 × 50 mm) and are distinguished by food-catching tentacles in the adult stage. Live organisms have a characteristic yellowish-orange or yellowish-brown color that makes them very conspicuous when wet mounts of gill tissue are examined using a microscope. Affected fish gills are pale and clubbed and may be eroded. Infected fish will be listless, as if they were suffering from an oxygen deficiency.

Microsporidia

Microsopridia are intracellular protozoan parasites characterized, as their name implies, by the production of very small spores. Species within 11 genera affect a wide variety of fish worldwide; the most significant ones are *Pleistophora, Enterocytozoon, Glugea,* and *Loma.*

Several species of *Pleistophora* infect hatchery fish. *Pleistophora* spores are about the size of large bacteria, 3–6 mm long and somewhat bean shaped. Severe infections have been reported in the gills of rainbow trout and in the ovaries of golden shiners. In golden shiners, the parasites infest up to about half of the ovary and significantly reduce the fecundity of broodstock populations.

The only known control for *Pleistophora* is prevention. Rainbow trout or their eggs should not be transferred from infected to uninfected hatcheries. Broodstocks known to be infected should be phased out and the rearing facilities disinfected. Because there are few stocks of golden shiners free of *Pleistophora ovariae*, proper management is the best answer to this problem. The severity of infections increases with age, so only 1-year-old broodstock should be used and all older fish destroyed.

Myxosporea

The family Myxosporea is a large and important group of parasites, most of which occur in fish. More than 1,250 species have been described from marine, brackish, and freshwater fish species worldwide. Many species of Myxosporea have a narrow host range, but others can affect several species of fish within a family (e.g., whirling disease in salmonids). An interesting feature of the myxosporeans is the presence of alternating hosts in the life cycle of several species that have been studied extensively. In addition to the fish, an invertebrate host (e.g., a tubificid worm in the case of whirling disease) is an essential part of the life cycle producing the actinospore that infects the fish. Important genera affecting fish include *Myxidium*, *Ceratomyxa*, *Sphaerospora*, *Chloromyxum*, *Parvicapsula*, *Myxobolus*, *Henneguya*, and *Kudoa*.

Whirling Disease

Myxobolus cerebralis is the causative agent of whirling disease, a serious condition of both wild and hatchery-reared salmonids. The disease was initially endemic in central and western Europe, but has been spread by movement of infected fish to several other countries including the United States. The obvious sign of tail chasing (whirling) becomes evident about 40–60 d after infection and is caused by erosion of the cranial cartilage, particularly around the organ of equilibrium, by the trophozoite stage of the parasite. In general, only trout less than a few months old when infected go on to exhibit whirling disease because cartilage becomes ossified after a few months making the fish more resistant to neural damage. However, older fish can become infected even though they show no clinical signs. Mortality has varied greatly among outbreaks, sometimes minor, sometimes devastating. The severity of disease among wild and hatchery fish has been shown to be related to the dose received and the age of the fish at infection.

The complete life cycle of *M. cerebralis* has been established and involves an alternate host, the aquatic oligochaete worm *Tubifex tubifex*. Triactinomyxon spores released by the worm infect the fish skin where the sporoplasm leaves the spore, develops into the trophozoite, and migrates to the developing cartilage where it sporulates to form the typical myxospores observed in diagnostic examinations.

External signs alone are not adequate for diagnosis of *M. cerebralis* infections as the signs mimic those of several other conditions including BCWD and IHN. Traditionally, verification required labor-intensive identification of the myxospore stage in the cartilage of fish heads, a stage that does not appear until several months after infection. The myxospores are ovoidal (front view) or lenticular (in profile), and have two pyriform polar capsules containing filaments at the anterior end. The diagnosis of whirling disease and detection of *M. cerebralis* in both the fish and worm host has been aided greatly by the development of a PCR assay.

Because of the seriousness of whirling disease for both hatchery and wild fish, control and treatment measures must be rigorous. Infected fish should not be stocked and earthen rearing units should be converted to concrete. Hatcheries where whirling disease has occurred should be subjected to complete disinfection of facilities and equipment with high concentrations of such chemicals as sodium hypochlorite or calcium oxide. Allow the treated area to stand for 4 weeks, clean thoroughly, and repeat disinfection. New eggs or fry should be obtained from a known uncontaminated source and raised in spore-free ponds or raceways provided with a pathogen-free water supply.

Henneguya

Several species of *Henneguya* have been described from a wide variety of North American freshwater fishes. *Henneguya* infections are histozoic and localize in specific tissues. Infections may appear as white cysts within the gills, barbels, adipose fins, skin, gallbladder, connective tissue of the head, subcutaneous tissues, or sclera and muscles of the eye.

Spores of *Henneguya* grossly resemble spermatozoa; they possess two anterior polar capsules and an elongate posterior process that may or may not separate along the sutural plane. *Henneguya salminicola* has been found in cysts in the body or musculature of coho, pink, and chinook salmon. Chum salmon also are subject to infection.

In channel catfish, *Henneguya* infections are categorized with respect to the tissue parasitized and the site of spore formation. An intralamellar bronchial form develops cysts within gill lamellae. A cutaneous form causes large lesions or pustules within the subcutaneous layers and underlying musculature of the skin; a granulomatous form causes large tumor-like lesions. An integumentary form causes white cysts on the external body surface. A gallbladder form develops within that organ and may obstruct the bile duct. An adipose fin form localizes solely within the tissue of that fin.

The interlamellar form of *Henneguya* develops spores within basal cells between gill lamellae. This form, in contrast to the intralamellar form, has caused large losses among very young channel catfish. Mortalities of 95% or more among fingerlings less than 2 weeks old have been reported. Loss of respiratory function accompanies acute infections. Fish exhibit signs of anoxia, swimming at the surface of ponds with flared gill opercula. Infected fish are unable to tolerate handling. Most attempts to treat with parasiticides have resulted in additional losses.

As with other myxosporean infections, prevention is the only control measure because no chemical treatment is effective. The disease has been spread from hatchery to hatchery with shipments of infected fingerlings. Confirmation of the interlamellar form in a catfish population may warrant destruction of the infected fish and disinfection of the rearing facilities involved.

Ceratomyxa

Ceratomyxa shasta is a serious myxosporean parasite of salmonids in western North America that causes severe losses of rainbow trout, steelhead, and chinook salmon. Heavy mortalities of adult salmon have occurred just before spawning, and outbreaks have been reported among some wild salmonid populations. Natural infections also have been found in cutthroat, brook, and brown trout, and in chum, coho, pink, sockeye, and Atlantic salmon. The natural life cycle is now known to involve the freshwater polychaete worm *Manayunkia speciosa*, which releases the actinosporean stage that infects fish.

The spores of *C. shasta* are elongated and found in great numbers in the gut and in cysts in the liver, kidney, spleen, and muscle of infected fish. The disease is contracted by adult salmon upon entering infected freshwater and by outmigrating juveniles passing through an area where the infective stage is present. The first signs of infection include lack of appetite, listlessness, and movement to slack water. The fish may darken, shed fecal casts, and the abdomen swell with ascites. Exophthalmia often occurs. The first internal changes appear as small, whitish, opaque areas in the tissue of the large intestine. As the disease progresses, the entire intestine becomes swollen and hemorrhagic. Mortality generally occurs at water temperatures greater than 10°C (50°F).

There is no known treatment for *C. shasta*; however, significant differences have been shown in the susceptibility of various stocks of salmonids to infection. Water supplies known to harbor infected fish should not be used for hatchery purposes without pretreatment unless such resistant stocks are used. The diagnosis of *C. shasta* infections in salmonids relies upon the observation of the typical kidney bean shaped mature spores in wet mounts or histological sections. Recently, the development of a PCR assay has greatly facilitated the diagnosis of *C. shasta* infections, especially the pre-spore stages.

Proliferative Gill Disease

In farmed channel catfish alevins, a condition known as "hamburger gill" or proliferative gill disease causes large losses. The parasite *S. ictaluri*, once suggested to be a species of *Sphaerospora*, normally sporulates in the kidney; however, the presence of the pre-spore stages in the gill cause a strong inflammatory response, leading to necrosis of the gill cartilage and hyperplasia of the gill epithelium. When conditions favor the parasite, high mortality can occur from massive loss of gill function. Initial signs are usually loss of appetite and listlessness with increased susceptibility to low oxygen levels. As the infection proceeds, cyst-like lesions can be seen on the gills with the naked eye, giving the gill a characteristic appearance. The disease is most severe in catfish ponds at 16–20°C (60.8–68°F).

Proliferative Kidney Disease

A condition known as proliferative kidney disease (PKD) affects rainbow trout and steelhead, as well as chinook and coho salmon. Infection of fish by the pre-spore stages of the causative parasite, termed PKX, causes massive swelling of the kidney. The disease appears to be an inflammatory response to the parasite; it typically develops in the spring. Mortality is directly affected by water temperature and can be exceptionally high. Initial clinical signs are a swollen abdomen from the highly swollen kidney tissues. Infected fish appear to be more susceptible to secondary infections with bacterial or other pathogens in the water. The inflammatory response leads to a chronic granulomatous response and necrosis of kidney tubules; in severe infections, other organs may also become necrotic. Fish surviving the disease may become normal when water temperatures subside; these fish are then resistant to reinfection. The disease has now been shown to be caused by a *Tetracapsula* species having a bryozoan host. Because mature spores are rarely observed in rainbow trout, some have suggested this species is an abnormal host.

Monogenetic Trematodes

Monogenetic trematode parasites of fish can complete their life cycles on fish without involving other species of animals. Although most are too small to be seen by the naked eye, some species may reach 5 mm in length. The posterior organ of attachment, the "haptor," is used in identification of different genera and species. There often are marginal hooklets around the margin of the haptor and either zero, two, or four large anchor hooks.

Species of the family Gyrodactylidae generally are found on the body and fins of fish, rarely on the gills. These parasites move around freely. The members of this family give birth to live young similar in appearance to the adults. *Gyrodactylus* can be identified by the developing embryo inside the adult, as well as by their lack of eye spots. The haptor has 2 large anchor hooks and 16 marginal hooklets. Diagnosis is made from microscopic examination of wet mounts of fin tissue or skin scrapings. The parasites may occur in large numbers and cause skin irritation and lesions. Fish with large numbers of *Gyrodactylus* may appear listless, have frayed fins, and flash frequently. In ponds, they may gather in shallow water in dense schools.

Dactylogyrus is but one genus of several dactylogyrids commonly found on the gills of warmwater fish. These worms are particularly serious parasites of cyprinids. *Dactylogyrus*, a small gill parasite, can be identified by the presence of 4 eye spots, 1 pair of anchor hooks, and 16 marginal hooklets. No embryos will be found internally, as these worms lay eggs. These parasites feed on blood and can cause serious damage to the gills of warmwater fish when numerous. Clinical signs easily can be mistaken for those caused by an oxygen deficiency or other gill infections.

Cleidodiscus spp. is common on the gills of catfish and a variety of other warmwater fish species. Like *Dactylogyrus*, it has eye spots, but has four large anchor hooks and lays eggs, which may be seen within the adult worm. *Cleidodiscus* is found only on the gills where, when numerous, it causes respiratory problems by severely damaging the tissue. Signs of infection, therefore, are those of gill damage and may be similar to those seen when oxygen is low.

Digenetic Trematodes

Digenetic trematodes require one or more animal hosts, in addition to fish, to complete their life cycles. These parasites can be divided into two major groups: (1) those that live in fish as adults, producing eggs that leave the fish to continue the life cycle, and (2) those that penetrate the skin of the fish and live in the fish as larvae, usually encysted in the tissue, until the fish is eaten by the final host.

Blood flukes *Sanguinicola davisi* live as adults in arterioles of the gill arches of salmonids and other fish species. These tiny worms lay eggs that become trapped in the capillary beds of the gills and other organs, where they develop into miracidia, which have a characteristic dark eye spot. When fully developed, the ciliated miracidia burst from the gill to be eaten by an operculate snail, the only intermediate host in the life cycle. Cercaria emerge from the snail and penetrate fish to complete the cycle. Control of blood flukes is difficult. It depends upon either continual treatment of infected water supplies to kill the cercaria or eradication of intermediate host snails. In most cases, however, blood flukes are debilitating, but not the cause of serious losses of fish. It is conceivable that large numbers of miracidia leaving the gill at one time could cause a significant loss of blood and damage to the gills. Eggs and developing miracidia also interfere with the circulation of blood in the gill capillaries and in the capillary beds of the kidney and liver.

Copepods

Most copepods in freshwater and salt water are an important part of the normal diet of fish. Certain species, however, are parasitic on fish and the sites of their attachment may become ulcerated and provide access for secondary infections by fungi and bacteria. Crowded hatchery rearing units provide ideal conditions for infestations by copepods because of the dense fish populations and rich environmental conditions.

Argulus spp. have been given the common name of fish lice because of their ability to creep about over the surface of the fish. On first glance, they look like a scale, but on closer examination, are seen to be saucer shaped and flattened against the side of the fish. They have jointed legs and two large sucking discs for attachment, which may give them the appearance of having large eyes. Argulids have an oral stylet that pierces the skin of the host fish. They then inject a cytolytic substance and feed on blood. If these organisms become abundant, even large fish may be killed.

Lernaea spp. are most commonly found on warmwater fish. However, one species, *L. elegans,* lacks host specificity and even attacks frogs and salamanders. Heavy infestations have caused massive mortality in carp and goldfish populations. The parasite penetrates beneath scales and causes a lesion at the point of attachment. The damage caused is associated with loss of blood and exposure to secondary infections by fungi, bacteria, and possibly viruses. *Lernaea* are long (5–22 mm) slender copepods that, when attached, give the appearance of a soft sticks with two egg sacs attached at the distal ends. Actually, the head end is buried in the flesh. This end has large, horn-like appendages that aid in identification of the parasite.

Fungal Diseases of Fish

Fungi are common in the aquatic environment and are frequently encountered by all fishes at one time or another during their lives. Under cultural conditions, certain fungi can be particularly troublesome. Species of *Acklya, Aphanomyces, Branchiomyces, Leptomitus, Phoma, Pythium,* and *Saprolegnia* have been reported as pathogens. These fungi grow on many types of decaying organic matter and are widespread in nature. Fungi infesting fish or eggs generally are considered to be secondary invaders of tissues damaged by injury; but once they start growing, fungi may spread to healthy tissues and can cause death. Fungi often attack dead fish eggs and will typically spread to the adjacent live eggs, killing them.

Species of the family Saprolegniaceae affect both fish and fish eggs. In fish, *Saprolegnia parasitica* (or other species) is typically a secondary invader of tissues damaged by trauma, infectious disease, or poor environmental conditions. Conditions that suppress the fish's normal immune response (e.g., spawning in adult Pacific salmon or holding warmwater fish in cold water) also can render fish more susceptible to fungal infections. Winter saprolegniosis of catfish is a good example of the latter.

The presence of *Saprolegnia* infections on fish or fish eggs is noted by a white to off-white, cottony growth. This growth consists of a mass of hyphal filaments; these hold the sporangia containing flagellated zoospores that escape to begin infections on other fish or eggs. Unless control measures are taken, the expanding growth ultimately may cover the entire surface of the fish or every egg in the incubator.

Good sanitation and cleanliness are absolutely essential to effective control of fungi under intensive culture conditions. Mechanical and chemical methods have been used for the control of fungal infections on eggs. The mechanical method is used on both salmonid and catfish eggs and involves picking dead and infected eggs at frequent intervals during incubation. This, however, is time consuming, and some healthy eggs may be injured in the process. Chemical control of fungal infections on eggs can be achieved, and formalin is now approved for use to control *Saprolegnia* on eggs of all fish species.

Gill rot, a disease caused by fungi of the genus *Branchiomyces,* is a major threat to fish culture in Europe. Primarily a disease of cyprinids in warmer water, it has been found in trout, bass, northern pike, pumpkinseed, and guppies in the United States. Clinical signs associated with branchio-mycosis include red-brown coloration of the gills with severe necrosis, fish gasping at the surface, and losses that can exceed 50% in a few days.

A presumptive diagnosis of branchiomycosis can be made by microscopic examination of gill tissue. The red-brown, nonseptate hyphae and round spores of the fungus are seen in surface scrapings or in capillaries and tissue of the gill lamellae. There is no control for branchiomycosis except destruction of infected fish and disinfection of facilities.

An important emerging fungal disease of fish is caused by species of the genus *Aphanomyces* that typically affect fish in marine or brackish waters. In Asia, a disease termed epizootic ulcerative syndrome or EUS is rapidly spreading and causing severe losses among species cultured in brackish water. Initially thought to be caused by viral or bacterial pathogens, the disease is characterized by deep ulcerative lesions that show an intense inflammatory response. In severe cases, internal organs are affected. Similar disease conditions in Australia, termed red-spot disease, and in Japan, termed mycotic granulomatosis, have been described and attributed to *A. invadens*. More recently, an ulcerative mycosis has been associated with losses of marine species in estuaries of the eastern United States, as well as the northwestern Atlantic Ocean.

Diagnosis of *Aphanomyces* infections can be accomplished by light microscopy of scrapings from ulcerative lesions, as well as by culture and the PCR using specific primer sets.

Drugs and Chemicals

Drugs, disinfectants, pesticides, diagnostic reagents, and vaccines are needed to ensure the health and productivity of cultured fish stocks. These regulated products must be used in a manner to avoid risks to public safety and animal health or potential loss of consumer trust. It is the responsibility of everyone using, prescribing, or recommending the use of regulated products to know which products legally can be used under federal, state, and local regulations.

Much of the material in this section was taken (in some places verbatim) from the Web site of the Aquaculture Network Information Center (AquaNIC). This important resource (see electronic references) sponsored by the Joint Subcommittee on Aquaculture should be consulted frequently for current information about the status of drugs and chemicals used in aquaculture because there is a high likelihood that the list of approved products will change frequently. This is due to the conflicting pressures to reduce the use of chemicals released into the environment or of animal drugs that may give rise to antibiotic resistance in human pathogens and the substantial effort by many supporters of the aquaculture industry to increase the numbers of approved drugs and chemicals for fish.

Legal Framework

Various federal and state agencies are involved in regulating drugs, vaccines, pesticides, and other products used in aquaculture. In the United States, the Food and Drug Administration (FDA) is responsible for ensuring the safety, wholesomeness, and proper labeling of food products; ensuring the safety and effectiveness of human and animal drugs; and protecting consumers from economic fraud. The FDA's regulatory programs are intended to ensure compliance with existing laws, and enforcement activities include actions to correct and prevent violations, remove illegal products or goods from the market, and punish offenders. The testing of domestic and imported aquaculture products for drug and pesticide residues is part of these enforcement activities.

The FDA's Center for Veterinary Medicine (CVM) regulates the manufacture, distribution, and use of animal drugs. CVM is responsible for ensuring that drugs used in food-producing animals are safe and effective and that food products derived from treated animals are free from potentially harmful residues. CVM approves new animal drugs based on data provided by a sponsor (usually a drug company). To be approved, an animal drug must be effective for the claim on the label and safe when used as directed for (1) treated animals, (2) persons administering treatment, (3) the environment, including nontarget organisms, and (4) consumers. CVM establishes tolerances and withdrawal periods as needed for all drugs approved for use in food-producing animals. CVM has the authority to grant investigational new animal drug (INAD) exemptions so that data can be generated to support the approval of a new animal drug.

The Environmental Protection Agency (EPA) is responsible for registering or licensing all pesticides used in the United States under the Federal Insecticide, Fungicide, and Rodenticide Act (FIFRA). The FIFRA requires that the EPA register pesticides for specific uses, provided that the use does not pose an unreasonable risk to human health or the environment. Any pesticide sold or distributed in the United States must be registered by the EPA. Products regarded as pesticides include algicides, disinfectants, fish toxicants, and herbicides.

The Animal and Plant Health Inspection Service (APHIS) of the U.S. Department of Agriculture regulates all veterinary biologics distributed in the United States. This includes vaccines, biologics, and test kits for the diagnosis of disease. States may impose additional requirements on the use of veterinary biologics. For example, APHIS requires that state approval for distribution of products be obtained before APHIS authorizes field trials with experimental products or before a conditionally licensed product is marketed in the state.

The FDA and EPA have some areas of mutual regulatory responsibility. A memorandum of understanding provides guidance in the area of aquaculture, particularly as to which agency has jurisdiction over a particular substance for its intended use. The EPA has jurisdiction over disinfectants, sanitizers, and aquatic treatments used solely for the control of algae or bacterial slime and over any other aquatic treatments used solely for pest control that do not include claims for control of parasites or diseases of fish. The EPA or a delegated state regulatory agency also regulates the National Pollutant Discharge Elimination System (NPDES), which controls the discharge of drugs or pesticides into receiving waters. The FDA has jurisdiction over new animal drugs, including products intended to treat or prevent parasites or diseases of fish, anesthetize aquatic species, and alter the sex or regulate the reproduction of aquatic species. The FDA has taken the position that if a pesticide registered by the EPA for aquaculture or aquatic site use is being used properly (i.e., the labeled conditions in fact exist in the facility or site at the time the pesticide is used, and the compound is not misused under FIFRA), the FDA will not object to that proper use even though the pesticide may have a secondary therapeutic benefit.

State departments of agriculture or other designated state agencies may also register federally approved pesticides to permit their legal distribution and sale within a state or territory. States may have additional regulatory requirements, including the provision of further data or additional restrictions on use and licensing. Some states license or impose additional regulations on the use of certain veterinary biologics. Some states may not allow the use of specific products or may require their administration by licensed veterinarians. States also participate in the approval of field trials of veterinary biologics in their respective jurisdictions and in the experimental use of certain veterinary biologics. The use of a drug under an INAD exemption may require approval by a state agency to comply with any local, state, and regional EPA discharge regulations. Discharge approval is intended to ensure that the possible impacts of a discharge (effluent) containing a specific compound or its residues are addressed.

Legal Options for Obtaining Drugs

According to the Federal Food, Drug, and Cosmetic Act, a drug is defined as an article that is intended for use in the diagnosis, cure, mitigation, treatment, or prevention of disease in man or other animals; an article (other than food) intended to affect the structure or any function of the body of man or other animals; or an article that is recognized in official drug compendia. A new animal drug is a drug intended for animals that is not generally recognized by qualified experts as safe and effective for the uses recommended on the label. At present, no drugs used in aquaculture are considered by FDA to be generally recognized as safe (GRAS) or effective (GRAE) for their proposed uses. For a compound to be classified as GRAS or GRAE, general recognition by experts must be supported by published scientific studies that meet strict FDA standards.

There are several options for legally obtaining and using drugs and chemicals. Use of drugs in a manner other than the options discussed below is subject to regulatory action by the FDA.

1. FDA-approved new animal drugs. Only a few new animal drugs are currently approved by the FDA for use in food-producing aquatic species. Each drug is approved for specific species, for specific disease conditions, and at specific dosages. Refer to Table 1 for a listing of these approved drugs and their approved uses. This list will change occasionally and current information should be obtained from the AquaNIC Web page or other sources.
2. Investigational new animal drugs. These drugs are used under an investigational new animal drug exemption. Investigational new animal drug exemptions are administered by the FDA CVM to allow for the purchase, shipment, and use of unapproved new animal drugs for collection of effectiveness and safety data that will support a decision on drug approval. Numerous requirements must be met for the establishment and maintenance of INADs. An INAD exemption is provided with the expectation that useful data will be generated and submitted to FDA CVM. The Food and Drug Administration authorizes specific conditions of use and monitors all drugs under each INAD exemption. Adequate drug accountability, data recording, and reporting are foremost among the requirements for INAD exemptions. Additional information concerning INAD exemptions may be found on the CVM Web site. It must be stressed that INAD exemptions are granted with the expectation that meaningful data will be generated to support a new animal drug approval and must not be seen as a method of circumventing current law. For a

TABLE 1. FDA-approved new animal drugs.

Trade name	NADA number	Sponsor	Active drug	Use and species
Chorulon	140-927	Intervet, Inc.	Chorionic gonadotropin	Aid in improving spawning function in finfish.
Finquel (MS-222)	042-427	Argent Laboratories, Inc	Tricaine methanesulfonate	Temporary immobilization (anesthetic) for Ictaluridae, Salmonidae, Esocidae, and Percidae.
Formalin-F	137-687	Natchez Animal Supply Co.	Formalin	Control of external protozoa and monogenetic trematodes in trout, salmon, catfish, largemouth bass, bluegill. Control of fungi of the family Saprolegniacae on salmon, trout, and esocid eggs.
Paracide-F	140-831	Argent Laboratories, Inc.	Formalin	Control of external protozoa and monogenetic trematodes on trout, salmon, catfish, largemouth bass, and bluegill. Control of fungi of the family Saprolegniacae on salmon, trout, and esocid eggs.
Parasite-S	140-989	Western Chemical, Inc.	Formalin	Control of external protozoa and monogenetic trematodes on finfish. Control of fungi of the family Saprolegniacae on finfish eggs. Control of external protozoan parasites on cultured penaeid shrimp.
Romet 30	125-933	Alpharma, Inc.	Sulfadimethoxine and ormetoprim	Control of enteric septicemia in catfish. Control of furunculosis in salmonids.
Sulfamerazine in fish	033-950	Alpharma, Inc.	Sulfamerazine	Control of furunculosis in rainbow trout, brook trout, and brown trout. Comments: According to sponsor, this product is not presently being distributed.

TABLE 1. (Continued.)

Terramycin for fish	038-439	Pfizer, Inc.	Oxytetracycline	Control of bacterial hemorrhagic septicemia and pseudomonas disease in catfish. Control of gaffkemia in lobster. Control of ulcer disease, furunculosis, bacterial hemorrhagic septicemia, and pseudomonas disease in salmonids. Marking of skeletal tissue in Pacific salmon.
Tricaine-S	200-226	Western Chemical Inc.	Tricaine methanesulfonate	Temporary immobilization (anesthetic) for Ictaluridae, Salmonidae, Esocidae, and Percidae.

listing of current INAD applications, consult the Web page of the National Aquaculture Drug Coordinator (see electronic references).

3. Unapproved new animal drugs of low regulatory priority. Neither an approved new animal drug application (NADA) nor an INAD exemption is required for drugs in this category. Although the FDA is not aware of safety problems associated with the specific uses of these substances, their uses have not been shown to be safe and effective in well-controlled scientific studies. Regulatory action is unlikely if an appropriate grade is used, good management practices are followed, and local environmental requirements are met. Refer to Table 2 for a listing of these compounds. It is unlikely that there will be many additions to this category.
4. Extra-label use of an approved new animal drug. Under the provisions of the Animal Medicinal Drug Use Clarification Act (AMDUCA), licensed veterinarians, within a valid veterinarian-client-patient relationship, can prescribe FDA approved new animal drugs or new human drugs for an extra-label use. Extra-label use of a drug refers to the actual or intended use of an approved new animal drug in a manner not described on the label. This includes, but is not limited to, species, frequency, route of administration, dose, withdrawal time, and conditions of use. This extra-label use is limited to treat modalities when the health of an animal is threatened or when suffering or death of the animal may result if not treated. Extra-label use of FDA-approved new animal drugs or new human drugs that are not prescribed by a licensed veterinarian is a violation of law.

TABLE 2. Unapproved new animal drugs of low regulatory priority for FDA.

Common name	Permitted use
Acetic acid	Used as a dip at a concentration of 1,000–2,000 mg/L for 1–10 min as a parasiticide for fish.
Calcium chloride	Used to increase water calcium concentration to ensure proper egg hardening. Dosages used would be those necessary to raise calcium concentration to 10–20 mg/L calcium carbonate. Also used to increase water hardness up to 150 mg/L to aid in maintenance of osmotic balance in fish by preventing electrolyte loss.
Calcium oxide	Used as an external protozoacide for fingerling to adult fish at a concentration of 2,000 mg/L for 5 s.
Carbon dioxide gas	Used for anesthetic purposes in cold, cool, and warmwater fish.
Fuller's earth	Used to reduce the adhesiveness of fish eggs in order to improve hatchability.
Garlic (whole)	Used for control of helminth and sea lice infestations in marine salmonids at all life stages.
Hydrogen peroxide	Used at 250–500 mg/L to control fungi on all species and at all life stages of fish, including eggs.
Ice	Used to reduce metabolic rate of fish during transport.
Magnesium sulfate (Epsom salts)	Used to treat external monogenetic trematode infestations and external crustacean infestations in fish at all life stages. Used in freshwater species. Fish are immersed in a solution of 30,000 mg/L magnesium sulfate and 7,000 mg/L sodium chloride for 5–10 min.
Onion (whole)	Used to treat external crustacean parasites and to deter sea lice from infesting external surface of fish at all life stages
Papain	Used as a 0.2% solution in removing the gelatinous matrix of fish egg masses in order to improve hatchability and decrease the incidence of disease.
Potassium chloride	Used as an aid in osmoregulation to relieve stress and prevent shock. Dosages used would be those necessary to increase chloride ion concentration to 10–2,000 mg/L.
Povidone iodine compounds	Used as a fish egg disinfectant at rates of 50 mg/L for 30 min during water hardening and 100 mg/L solution for 10 min after water hardening.
Sodium bicarbonate (baking soda)	Used at 142–642 mg/L for 5 minutes as a means of introducing carbon dioxide into the water to anesthetize fish.
Sodium chloride (salt)	Used as a 0.5–1% solution for an indefinite period as an osmoregulatory aid for the relief of stress and prevention of shock. Used as a 3% solution for 10–30 min as a parasiticide.
Sodium sulfite	Used as a 15% solution for 5–8 min to treat eggs in order to improve hatchability.
Thiamine hydrochloride	Used to prevent or treat thiamine deficiency in salmonids. Eggs are immersed in an aqueous solution of up to 100 ppm for up to 4 h during water hardening. Sac fry are immersed in an aqueous solution of up to 1,000 ppm for up to 1 h.
Urea and tannic acid	Used to denature the adhesive component of fish eggs at concentrations of 15 g urea and 20 g NaCl/5 L of water for

Table 2. (Continued.)

	approximately 6 min, followed by a separate solution of 0.75 g tannic acid/5 L water for an additional 6 min. These amounts will treat approximately 400,000 eggs.

Note. FDA is unlikely to object to the use of these low regulatory priority substances if the following conditions are met:

1. The drugs are used for the prescribed indications, including species and life stage where specified.
2. The drugs are used at the prescribed dosages.
3. The drugs are used according to good management practices.
4. The product is of an appropriate grade for use in food animals.
5. An adverse effect on the environment is unlikely.

FDA's enforcement position on the use of these substances should be considered neither an approval nor an affirmation of their safety and effectiveness. Based on information available in the future, FDA may take a different position on their use. Classification of substances as new animal drugs of low regulatory priority does not exempt facilities from complying with other federal, state, and local environmental requirements. For example, facilities using these substances would still be required to comply with National Pollutant Discharge Elimination System requirements.

Further Restrictions on Use

New animal drugs that are added to aquaculture feed are subject to FDA regulation and must be specifically approved for use in aquaculture feed. Approved new animal drugs may be mixed in feed only for uses and at levels that are specified in FDA medicated-feed regulations. It is unlawful to add drugs to feed, even under extra-label use provisions, unless the drugs are approved for feed use. The FDA requires that certain mills producing medicated feeds be licensed by the FDA. There are more details on these regulations on the CVM Web site.

Many of the chemicals used in aquaculture are applied directly to water where the FDA or EPA restrictions are determined by the intended use of the product. Although some products may be beneficial when applied to aquaculture systems at low concentrations, they may also act as irritants or even become toxic at higher concentrations. The improper use or application of water treatments can cause severe stress, which can lead to an animal disease outbreak or even death. Some compounds can accumulate in the animal and may cause illegal chemical residues in tissues intended for human consumption. Illegal residues can also result from the improper use of products to control weeds or unwanted fish or to alter water quality. To prevent possible fish losses and illegal chemical residues from excessive treatment levels, always read and strictly follow product label directions.

Product withdrawal times must be observed to ensure that products reaching consumers are safe and wholesome. All federally approved products list any specific required withdrawal times to ensure that products used in aquaculture do not exceed legal tolerance levels in animal tissues. Withdrawal information is found on the product label, package insert, or feed tag. An exception to withdrawal requirements is made for products used in an extra-label manner. Extra-label use may require the same or a different withdrawal time from that listed on the label, depending on the species, treatment, and other conditions. Withdrawal times for the extra-label use of an approved product are not listed on the label and must be determined by the prescribing licensed veterinarian.

Quality Assurance and Record Keeping

Record keeping is a critical element of quality assurance programs. Record keeping may also be required of producers by processors purchasing fish from them to comply with hazard analysis critical control point or HACCP requirements. A good record keeping system helps producers keep track of specific treatments and their results with identifiable, known populations or stocks of aquatic animals, as well as the specific water and land areas involved. Good records provide a basis for sound, cost-effective management decisions. The treatment status of animals, ponds, and other areas is known at all times. Records are needed to determine dosage rates and certify withdrawal times. Processors may require records to demonstrate that all drugs and chemicals have been used properly. Records provide valuable evidence and protection in liability cases. Accurate record keeping also is required for any producer using an INAD exemption in clinical field trials. Quality assurance programs have a high level of participation among many of the segments of the aquaculture industry.

Using Drugs and Chemicals

Product Label

Always read and understand the product label before using any compound. Label directions and information are important for two reasons. First, they describe the conditions of use under which the product can be expected to be effective and safe. Second, labels for approved products describe uses allowed by law. The product label and package inserts provided with regulated products present information on proper storage, mixing, dosage, and administration; date of expiration; diluting or reconstituting the product; safe disposal of the unused product and product containers; and withdrawal times. Pesticide labels list precautionary statements on environmental, physical, and chemical hazards. Prescribed aquatic-use information is usually not found on the product labels of those substances determined by the FDA to be unapproved new animal drugs of low regulatory priority (LRP) because these substances typically are not marketed specifically for aquaculture use.

Economic Considerations

Drugs, pesticides, and vaccines are used to control or prevent specific diseases, water quality conditions, and pest (e.g., weed) or stress problems. These treatments should be used only when needed. Each treatment has an economic value in terms of treatment cost and expected economic benefit. The proper use of regulated products, some of which are quite costly, can be important in preventing significant economic losses. Such losses are more likely to occur if the actual problem is incorrectly diagnosed, if precautions for treatment are ignored, or if treatments are improperly applied. The use of best management practices in aquatic animal husbandry and water quality maintenance can reduce the use of regulated products. Using regulated products at less than the concentrations or dosages specified on the label can cause the treatment to be ineffective or only partially effective. Of particular concern is the emergence of antibiotic resistant strains of bacteria. Using these compounds at concentrations or in dosages greater than those specified on the label (overdosing or over treating) wastes the product and can cause unwanted side effects, including stress and toxicity problems in aquatic stocks and nontarget organisms, as well as environmental damage. Persons applying regulated products should recognize their legal responsibility for any harm to nontarget aquatic and nonaquatic species and for off-site damage.

Storage and Handling

Always follow label directions for storing, handling, mixing, diluting, reconstituting, and disposing of regulated products and their containers. This preserves the activity and quality of the product and helps prevent misuse, damaging effects on plants and animals, human injury, and environmental contamination. Disinfectants, pesticides, and most drugs should be stored in a locked cabinet in a dry, well-ventilated utility area away from children, animals, food, feed, and living areas. Some drugs and veterinary biologics require refrigerated storage; other products require storage in a freezer or at room temperature. All disinfectants, pesticides, drugs, and veterinary biologics should be stored away from bright light, because light can cause inactivation or deterioration of the product.

Regulated products should be stored in their original containers with the original label attached. Dampness in storage areas can cause paper packages to deteriorate, metal containers to rust, and metal and glass containers to lose their labels. Disinfectants, pesticides, and drugs should not be stored where flooding is possible or in sites where they might spill or leak into the environment. High-temperatures can cause excessive pressure in and bursting of sealed containers. Exposure to high temperatures can also result in product deterioration, shortened shelf life, premature inactivity, and inactivation.

Proper mixing, diluting, and reconstituting are essential for the effectiveness of products requiring such steps, as well as for reasons of safety. Improper dilution may cause the concentration or dosage administered to be too great or too small. Incomplete mixing can cause variations in the concentration or dosage applied or administered, with uneven effects ranging from ineffectiveness to overdose and toxicity. Some veterinary biologics are supplied with diluents that should be used for reconstitution and the proper concentration of materials.

Safety Considerations

The use of any pesticide (and some other regulated products) requires adequate protection from exposure. Users should always read the product label for information on recommended personal protective equipment. Common-sense precautions should be followed, such as wearing gloves, long-sleeved shirts, long pants, socks, shoes or boots, a hat, and goggles, protective glasses, or a face shield. Powders may be harmful or toxic if they are inhaled as dusts; fumes and evaporating ingredients may also be harmful or toxic. Some pesticides may require use of a respirator. Persons mixing or applying pesticide or working in an area where pesticides are being applied or have recently been applied should shower and wash their clothes after actual or possible exposure.

Users should not mix different regulated products unless this is specifically recommended on the label. The combining of products can have many—mostly undesirable—effects. One or both products can be inactivated, and chemical reactions can produce harmful gases or other reaction products and by-products, some of them toxic. Following appropriate precautions can prevent accidental poisoning from pesticide contact with bare skin or from the inhalation of fumes or dust. The pesticide label provides important product-specific information on mixing, diluting, storage, and disposal, as well as on first aid in the event of accidental poisoning. Material Safety Data Sheets, provided by product manufacturers, are a source of additional information on safety precautions. It is important that unused portions of a regulated product and empty containers be disposed of properly. The best approach is to purchase only the amount of material that will be used within a reasonable time period and to use all of the product for its intended purpose. Empty containers must be disposed of, however, and often a quantity of the product is left over. Product labels provide instructions for safe disposal; these instructions should be followed. Improper disposal can result in toxicity, environmental contamination, and liability problems.

Applicators and persons near treatment areas can be affected by various regulated products through contact, exposure to evaporated material in the air, or exposure to dusts or aerosols. Treated waters or airborne drift can carry restricted products to distant locations, where the products may affect nontarget organisms and sites. Accidental self-injection of some veterinary biologics and injectable drugs can cause local tissue reactions, allergic reactions, or infections. Use of common sense and strict compliance with product label directions can minimize undesirable effects in humans, nontarget plants and animals, and the environment. Seek professional advice when in doubt.

Disease Treatments

In general, it is important to diagnose fish disease problems as early as possible and to begin treatment as soon as practical. If routine disease problems can be recognized early by the hatchery manager, treatment can begin sooner than if a diagnosis is required from a pathology laboratory. Broad-spectrum treatments based on a poor diagnosis are ill-advised, but timely treatments based on keen observation and awareness of signs can mean the difference between losing just a few fish or losing many thousands.

There are two classes of treatments for fish disease—control and therapeutic. Control treatments are measures designed to prevent an outbreak from occurring or increasing in magnitude. Such treatments are used primarily against external parasites, bacteria or fungi. Therapeutic treatments are begun only after clinical signs appear in a significant number of animals. When frequent control treatments are needed for external parasites or bacterial gill disease, it may be an indication of poor hatchery management. The general principle should be to manage first and treat as a last resort. Fish disease treatments are a serious undertaking, and caution should be taken to avoid disastrous results. All drugs and chemicals used to control infectious organisms can be toxic to fish if concentrations are too high. All treatment calculations should be double-checked before being implemented.

Two important principles should be kept in mind regarding the treatment of fish diseases. Firstly, treatment with various medications and chemotherapeutic agents is for the purpose of tipping the balance in favor of the host, not for killing 100% of the disease organisms present. Medications may kill or retard the growth of the pathogen, but in the end, it is the fish's own immune mechanisms that must overcome the disease if the treatment is to be successful. Secondly, there is increasing concern regarding the use of drugs and chemicals in aquaculture. These concerns center around possible environmental harm and the development of antibiotic resistance in aquatic microorganisms that might be transferred to human pathogens. To help allay these concerns, the use of drugs and chemicals should be reduced as much as possible and antibiotic treatments rotated among various compounds and continued for the full treatment period to reduce the development of drug-resistant bacteria.

Treatment Methods

Before making a decision to treat a group of fish, the following questions should be asked:

1. Does the loss rate, severity, or nature of the disease warrant treatment?
2. Is the disease treatable, and what is the prognosis for successful treatment?
3. Is it feasible to treat the fish where they are, considering the cost, handling, prognosis?
4. Is it worthwhile to treat the fish or will the cost of treatment exceed their value?
5. Are the fish in good enough condition to withstand the treatment?
6. Will the treated fish be released or moved soon and is adequate withdrawal or recovery time available?

Before any treatment is started, four factors must be considered. The culturist must know and understand (1) the water supply, (2) the fish, (3) the chemical, and (4) the disease. Failure to take all these factors into consideration can result in a complete kill of all of the treated fish or a failure to control the disease with a resultant loss of many fish and wasted money.

1. *Water supply.* The volume of water in the rearing unit to be treated and the flow rate to the unit must be calculated accurately before any treatment is applied. An overestimation of the water volume or flow rate means too much drug or chemical will be used, which may

kill all the fish. An underestimation of the volume or flow means not enough of the drug or chemical will be used, thus, the disease-causing organism may not be controlled. Often overlooked is the fact that water quality factors, such as total hardness, pH, and temperature, will increase the activity of some chemicals and decrease that of others. In ponds, the amount and type of aquatic vegetation or organic matter also must be taken into consideration before any chemical is applied.

2. *Fish.* Fish of different species and ages may react very differently to the same drug or chemical. Certain species are much more sensitive to a particular chemical than others. The age of fish may also affect how they react to a specific treatment. Regardless of whether or not a particular chemical or drug has been used to treat fish at the hatchery in the past, it is always a good idea to test it first on a small number of fish before an entire pond or holding unit is treated. This can be done in tanks or in small containers such as large plastic wastebaskets.
3. *Chemical.* The toxicity of the chemical should be known for the species to be treated. The effect of water chemistry on the toxicity of the chemical also should be known. Some chemicals break down rapidly in the presence of sunlight and high temperatures and, thus, are less likely to be effective during summer months than during the cooler months of the year. Mixing chemicals may enhance or intensify the toxicity of one of them. Also, certain chemicals are toxic to plants and can cause oxygen depletion if used in ponds at the wrong time.
4. *Disease.* Although the correct diagnosis of a disease may be self-evident, confirmatory testing or a second opinion from a fish health professional is important. Most chemicals used to treat fish diseases are expensive and, generally, are effective only against certain groups of organisms. Use of the wrong chemical or drug usually means that several days to a week may pass before one realizes the treatment was not effective. During this time, large numbers of fish may be lost unnecessarily. In addition, bacterial pathogens may be resistant to one or more antibiotics and this should be determined before treatment is begun.

When it is apparent that a treatment is necessary, the following checklists may be useful:

Pretreatment checklist

1. Accurately determine the water volume, flow rate, and temperature.
2. Accurately determine the number and total weight of fish in the rear-

ing unit.

3. Confirm the identity, purity, age, and stock strength of the drug or chemical to be applied.
4. Double-check concentration figures to be delivered. Beware of cofusion from mixing metric and English units.
5. Have aeration devices ready for use if needed.
6. Make sure of the route by which chemical solutions are discharged from the holding unit.
7. Test treatment on a pilot group of fish.

Treatment checklist

1. Dilute the chemical with rearing water to the correct concentration before applying it.
2. Ensure the chemical is well-mixed in the units or ponds.
3. Observe fish closely and frequently during treatment for signs of distress.
4. Monitor temperature and dissolved oxygen levels in the rearing unit during treatment (aeration may be required).
5. Discontinue treatment and resume water flow immediately if fish become distressed.

Posttreatment checklist

1. Observe fish frequently for at least 24 h following treatment.
2. Do not stress treated fish for at least 48 h.
3. Recheck fish to determine success of treatment.

Various methods of treatment and drug application have been used in the control of fish diseases. There is no one specific method that is better than others; rather, the method of treatment should be based on the specific situation encountered. Here, experience is exceptionally valuable. A fish health professional or other knowledgeable source should be consulted if one is unfamiliar with the disease or treatment proposed.

Dip Treatments

During the dip treatments, small numbers of fish are placed in a net and dipped in a strong solution of chemical for a short time, usually 15–45 s, depending on the type of chemical, its concentration, and the species of fish being treated. Metal containers should not be used to hold the treatment solution because some chemicals can react with the metal and form toxic compounds, particularly if the water is acidic. This method of treatment is dangerous because the difference between an effective dose and a killing dose often is very small. Other disadvantages to this method include its high labor costs and stress on the fish due to handling. However, if done properly, it is very effective for treating relatively small numbers of valuable fish.

Flush Treatments

Flush treatments are simple and consist of adding a solution of the treatment chemical at the upper end of a holding unit and allowing it to flush through. It has been used at trout and salmon hatcheries, but is applicable only with raceways, tanks, troughs, or incubators for which an adequate flow of water is available, so that the chemical is completely flushed through the unit or system within a predetermined time. Highly toxic chemicals should be avoided because there is no way to assure a uniform concentration within the unit being treated.

Prolonged Bath

For prolonged-bath treatments, the inflowing water is cut off and the correct amount of chemical is added directly to the unit being treated. After a specified time, the chemical is flushed out quickly with fresh water. This treatment can be used in any unit that has an adequate supply of fresh water that can flush out the treatment within 5–10 min. Several precautions must be observed with this method to prevent serious losses: (1) Because the water flow is turned off, the oxygen concentration of the water may be reduced to the point that the fish are stressed and losses occur to anoxia. The more fish per unit volume of water, the more likely this is to occur. Aerators of some type must be available or installed in the unit being treated to ensure an adequate oxygen supply during the treatment period; (2) In areas with warm weather or intense sunlight, the temperature of the rearing unit may rise significantly putting the fish at risk; (3) Regardless of the treatment time that is recommended, the fish always should be observed throughout the treatment and, at the first sign of distress, fresh water must be added quickly; (4) The chemical must be uniformly distributed throughout the unit to prevent the occurrence of "hot spots" of the chemical. Fish being treated may be killed or severely injured by overdoses if they swim through hot spots. Conversely, fish that avoid these hot spots may not be exposed to a concentration high enough to be effective. The method used for distributing the chemical throughout the unit will depend on the kind of chemical being used, type and size of unit being treated, and equipment and labor available.

Indefinite Bath

Indefinite bath treatments usually are used to treat ponds or other large volumes having low (or no) flow rates. A low concentration of a chemical is applied and left to dissipate naturally. This generally is one of the safest methods of treatment. One major drawback, however, is that large quantities of chemicals may be required that can be expensive to the point of being prohibitive. Another drawback relates to the possible adverse effects on the pond environment. Some treatment chemicals are algicidal or herbicidal and may kill enough plants to ultimately cause an oxygen deficit. Other chemicals, such as formalin, may reduce dissolved oxygen levels as they degrade.

As in prolonged-bath treatments, it is important that the chemical be evenly distributed throughout the culture unit to prevent the occurrence of hot spots. For dry chemicals that dissolve rapidly in water, such as copper sulfate or potassium permanganate, burlap or any coarse-weave bags can be used. The required amount of chemical is put into a bag and towed behind a boat so that the chemical dissolves in the wake of the boat. Liquids and wettable powders can be applied evenly with hand or power sprayers or can be siphoned over the edge of a boat into the prop wash. As with the prolonged-bath method, there is no one correct way to apply a chemical evenly to the unit of water to be treated. Rather, the application will depend on the kind of chemical being used, the equipment available, and the type of unit to be treated. Again, experience is the best guide.

Constant-Flow Treatments

Constant-flow treatments are useful in raceways, tanks, or troughs in situations where it is impractical or impossible to shut off the inflowing water long enough to use prolonged baths. The volume of water flowing into the unit must be determined accurately, and a stock solution of the chemical metered into the inflowing water to obtain the desired concentration. Before the metering device or constant flow siphon that delivers the chemical is started, enough chemical should have been added to the water in the device to give the desired concentration. Upon completion of the desired treatment period, the inflow of chemical is stopped and the unit is flushed by allowing the water flow to continue. The method by which the chemical is metered into the inflowing water will depend on the equipment available and the type of unit to be treated. Although the constant-flow method is very efficient, it can be expensive if large volumes of water must be treated.

Feeding

Treatment of certain diseases, such as systemic bacterial infections and certain internal parasite infestations, requires that the drug be introduced into the fish's body. This usually is accomplished by incorporating the medication in the feed. Treatment by feeding is usually based on body weight; standard treatments are typically given in grams or micrograms of active drug per kilogram of fish per day. Medicated food should be purchased commercially from an approved source. Beginning treatment early is important because once an outbreak is underway, many of the fish may be clinically ill and cease feeding, making it impossible to stop the losses in a timely manner. Once feeding of medicated food is begun, it should be continued for the prescribed treatment period.

Injection

Large and valuable fish, particularly small numbers of them, sometimes can be treated best with injections of medication into the body cavity (intraperitoneal) or into the muscle tissue (intramuscular). Most drugs work rapidly when injected intraperitoneally. For both types of injections, but particularly intraperitoneal ones, caution must be exercised to ensure that internal organs are not damaged. The most convenient location for intraperitoneal injections is the base of one of the pelvic fins. The pelvic fin is partially lifted, and the needle placed at the fin base and inserted until its tip penetrates the body wall. The needle and syringe should be held on a line parallel to the long axis of the body and at about a 45 degree angle to avoid internal organs. One can tell when the body wall has been penetrated by the sudden decrease of pressure against the needle. As soon as the tip of the needle is in the body cavity, the required amount of medication should be injected rapidly and the needle withdrawn. For intramuscular injections, the best location usually is the area immediately ahead of the dorsal fin. The syringe and needle should be held on a line parallel with the long axis of the body and at about a 45 degree angle. The needle is inserted to a depth of about 0.5–1.0 cm and the medication is injected directly into the muscle of the back. The injection must be done slowly, otherwise back pressure will force the medication out of the muscle through the channel created by the needle. For adult salmon broodstock, some drugs (e.g., erythromycin) are injected directly into the dorsal sinus.

Immunization of Fish

Prevention of disease through vaccination is an important adjunct to the maintenance of fish health. In addition to the commercial availability of several new vaccines, it has become increasingly clear that reliance on the use of antimicrobial drugs and chemicals in fish culture will have to be reduced. Firstly, the list of antibacterial drugs and chemicals that can legally be used is extremely small (especially in the United States and Canada) and the effectiveness of the few available antibacterial drugs is being diminished because of the development of antibiotic resistance among bacterial fish pathogens. Secondly, the prospects for enlarging the list are limited due to the extensive testing needed to meet stringent licensing requirements and concerns that antibiotic resistance might be transmissible to microorganisms of public health importance. Finally, none of the viral and many of the bacterial, fungal, or protozoan infections in fish can be successfully treated using the antibiotics or chemicals currently available or anticipated in the near future.

Faced with the foregoing problems, fish culturists should consider other measures that might help to ensure the health of the aquatic animals under their care. One obvious approach is immunization. Advantages of immunization are several. Firstly, immunization does not generate antibiotic resistant microorganisms; secondly, it can be applied to control viral, as well as bacterial diseases; thirdly, fish may be vaccinated economically and conveniently while still very small; and fourthly, protection conferred by vaccination is more durable than that resulting from chemotherapy and persists for a considerable period. Finally, with killed vaccines, at least, the requirements for licensing the vaccines may be less stringent than those required for the registration of antimicrobial drugs.

Initially, the biggest factor working against the widespread use of fish vaccination was the lack of a safe, economical, and convenient technique for vaccinating large numbers of fish. Recent advances have remedied this obstacle and mass-delivery methods are now routinely used. In addition, improvements in adjuvant formulations and the development of multivalent vaccines have led to the widespread use of vaccines for prevention of several of the most important bacterial diseases of fish. Recent advances in the molecular biology of fish pathogens and the development of novel approaches in vaccinology have opened the way for a new generation of fish vaccines that promise further improvements in the ability to protect populations of cultured fish against significant diseases.

Killed Vaccines

The ideal fish vaccine is global in scope (i.e., composed of antigens common to all strains) and capable of stimulating specific, long-lasting, humoral and cellular immune responses that result in solid protection. Traditional fish vaccines have used killed preparations of bacteria, viruses, fungi, or parasites delivered by waterborne exposure or by injection with adjuvants and today, this strategy forms the basis for most of the fish vaccines that have become commercially successful. A new generation of killed vaccines is expected as better delivery systems and new adjuvants become available.

A significant advantage of killed vaccines is their high degree of safety and ease of development. This has also made them relatively easy to license. In fact, the main requirements for autogenous vaccines (killed cultures of a pathogen isolated from animals at an individual farm and restricted to use on that farm) is that they be produced in a licensed facility and used under the supervision of a veterinarian. The disadvantages of killed vaccines are that they are sometimes only effective when delivered by injection in the presence of an adjuvant and that protection is often less solid or of shorter duration than that resulting from natural infections. In addition, killed viral vaccines that must be produced in cell culture are probably not cost-effective for small fish.

Attenuated Vaccines

Attenuated vaccines for fish have been developed by traditional methods of serial passage in culture, by using naturally occurring mutants and cross-reacting strains or by mutagenesis of wild-type organisms. While effective in laboratory trials, acceptance has been slow due to concerns about the release of live organisms into the environment and the perceived potential of the vaccine strain to revert to virulence, to replicate in the fish in unwanted ways, or to infect nontarget species. This means that the testing required for attenuated vaccines can be extensive, making development costs relatively high. In addition, without a genetic marker to distinguish the vaccine strain from natural isolates, diagnostic examinations may be compromised and the vaccine will be difficult to protect commercially. More recently, molecular approaches have been used to circumvent some of these concerns. Following initial development, attenuated vaccines for fish can be produced inexpensively and delivered efficiently by waterborne exposure. Because the attenuated strain replicates in the fish, it will stimulate both humoral and cellular immunity that is often superior to that provided by killed vaccines.

The Next Generation of Fish Vaccines

In the last decade, the techniques of molecular biology have been used to develop a novel set of vaccines for fish. While several of these new approaches have been effective in stimulating specific immunity against challenge in the laboratory, only a few have been tested in field trials, and more work is needed to develop better delivery systems or to overcome potential regulatory concerns. For fish, the most promising of these new approaches include subunit vaccines, recombinant vectors, and genetic immunization.

Subunit vaccines consist of only a portion of a pathogen that will stimulate protective immunity. This is typically a protein that bears a protective antigen or the immunogenic region of such a protein. While subunit vaccines can be created by purification of the native antigen directly from cultures of the pathogen, development of a subunit vaccine by recombinant DNA technology usually involves inserting all or part of the gene coding for the appropriate antigen into a bacterium, yeast, or virus that can produce large amounts of the protein in vitro. Because no infectious agent is present, subunit vaccines are regarded as having a high level of safety, and the cost of producing the vaccine can be quite low after initial research and development costs. Although relatively easy to produce and license, the subunit vaccines that have been tested in fish have generally been less effective than desired. Subunit vaccines can also be made by chemical synthesis of peptides that mimic the antigenic epitope. However, like other subunit vaccines, they are more effective when delivered by injection with an adjuvant and may require coupling to a carrier molecule. Another drawback is that creation of peptide vaccines requires a detailed knowledge of the epitope structure of the protective antigens of the pathogen.

Recombinant vectors make use of DNA technology to insert copies of the genes from protective antigens into a virus or bacterium that will be able to infect the host and replicate without causing disease. While replicating, high levels of the recombinant antigen are produced by the vector that will stimulate the host immune system. One concern is that the method may be somewhat difficult to license owing to the use of a replicating recombinant organism. However, the strategy promises to be highly effective, and several human and animal vaccines are being developed with this approach. A significant advantage of recombinant vectors is that they readily lend themselves to the creation of multivalent vaccines that could stimulate protection against several pathogens simultaneously. Also, because of the replication of the vector, other aspects of host immunity may be co-stimulated.

Genetic immunization by injection of plasmid DNA coding for protective antigens represents an exciting new area of vaccine development. Demonstrated in laboratory trials to be effective for viral, bacterial, and parasitic diseases of higher animals, the major advantage of this approach is that cells transfected with the plasmid produce the antigen in its authentic form. To the immune system of the animal, these cells appear to be infected and a full immune response is produced resulting in strong, long-lasting protection. Currently, these preparations have to be delivered by injection or by "gene gun" making them either more labor intensive or somewhat less efficient than desired; better delivery methods are needed if this technology is to become widely used in fish. Finally, because this technology is new, the potential for licensing these vaccines is difficult to evaluate; however, most concerns are expected to ease as DNA vaccines are developed for human and animal diseases.

While substantial progress has been made in the molecular biology of fish pathogens and in the construction of novel vaccines that are effective in the laboratory, much remains to be done. The high cost of development and licensing of these vaccines is a major problem, especially if the market for the vaccine is limited to a small geographic area or to a minor species in aquaculture. Other significant problems that need to be overcome are the lack of optimal adjuvants and mass delivery systems for killed and subunit vaccines and an incomplete understanding of critical elements of the fish immune system.

Although it is difficult to predict the future of fish vaccination, recent experience suggests that there will be a trend toward the use of attenuated vaccines, recombinant vectors, and genetic immunization. The genetic engineering of stable attenuations with essentially no likelihood of reversion, no antibiotic resistance genes, and simple genetic markers will provide attenuated vaccines with significantly improved safety and acceptance. The low cost and high efficiency of such attenuated vaccines make them attractive candidates for commercial development. Genetic immunization, likewise, appears highly promising, especially for complex, conformation-dependent epitopes that serve as single protective immunogens; however, improved delivery systems will be required before this method can gain widespread use. Work will also continue on the development of recombinant vectors that can provide mass immunization. Finally, there will be a trend toward use of multivalent vaccines in an effort to lower costs.

Vaccination Methods

Vaccines may be delivered to fish by intraperitoneal or intramuscular injection using a needle and syringe, by oral delivery where the vaccine is incorporated into the feed, or by direct immersion of fish into the vaccination solution. These methods work well for various commercially available preparations, but further improvements are needed. Not all vaccines are equally efficacious when delivered by these methods and where many small fish must be vaccinated economics dictate that a low-cost method (oral or immersion) must be used. Recent advances in vaccine delivery promise to improve the efficacy of the methods available and to provide additional methods in the future.

Mass inoculation by intraperitoneal or intramuscular injection works well with fish that are 5–10 g or larger, and teams of operators are able to vaccinate several thousand fish per hour. The cost of this delivery method is reasonable for large or highly valuable animals, but the number of fish that can be treated is limited by the manpower available and by technical limitations such as the area available for setting up the anesthetic baths and inoculating tables. In some areas of the world, however, vaccination by injection is practiced on a large commercial scale with substantial efficacy.

Oral immunization has been shown to be effective in the laboratory, but is not yet used commercially. The method is low cost and will be the preferred method for the future. One difficulty has been that the level and duration of protection offered by oral immunization has typically been less than that offered by injection or immersion methods. Microencapsulation of the vaccine to avoid inactivation by the low pH in the stomach and other methods to improve uptake of the antigens in the lower intestine offer new approaches that will improve efficacy of preparations delivered by this method.

The immersion method allows vaccination of several thousand fish quickly and safely in a short period. Weighed volumes of fish are lightly anesthetized and placed briefly into a solution of commercially prepared vaccine diluted in the rearing water. The vaccine is taken up by exposed tissues in the skin, gill, mouth, or anus and transported to cells that process the antigen and stimulate the immune response. Most of the commercially successful vaccines are killed bacterial preparations (bacterins) that can be effectively delivered in this manner.

Today, monovalent and multivalent bacterins against several fish pathogens are licensed for commercial use in the United States (Table 3). While these are highly effective, additional vaccines are needed to prevent other important diseases in aquaculture. Following the relatively rapid development and successful application of the bacterins listed, creation of additional vaccines has proven more difficult than hoped. This has probably been due to the failure of these newer preparations to stimulate the appropriate protective responses in the fish and the need to vaccinate large numbers of small animals economically. Recent re-

TABLE 3. Bacterins and vaccines.

Trade name	Product name	USDA product code	Licensee	Species	Disease
Alphadip 1100, Alphaject 1100	Yersinia Ruckeri Bacterin	2638.01	Alpharma, Inc.	Salmonids	Yersiniosis (enteric redmouth disease)
Alphadip 1100C	Yersinia Ruckeri Bacterin	2638.00	Alpharma, Inc.	Salmonids	Yersiniosis (enteric redmouth disease)
Alphadip 2100	Vibrio Anguillarum-Ordalii Bacterin	2858.02	Alpharma, Inc.	Salmonids	Vibriosis
Aquavac-ESC	Edwardsiella Ictaluri vaccine	1531.00	Intervet, Inc.	Catfish	Enteric septicemia
Autogenous bacterin	Autogenous Bacterin	2051.00	Alpharma, Inc.	Fish	Bacterial diseases
Autogenous bacterin	Autogenous Bacterin	2051.00	Jerry Zinn, Aqua Health, Ltd. (Canada)	Fish	Bacterial diseases
Autogenous vaccine	Autogenous vaccine	1015.00	Alpharma, Inc.	Fish	Viral diseases
Autogenous vaccine	Autogenous vaccine	1015.00	MicroTechnologies, Inc.	Fish	Viral diseases
Biojec 1500	Aeromonas Salmonicida Bacterin	2035.00	Alpharma, Inc.	Salmonids	Furunculosis
Biojec 1800	Aeromonas Salmonicida-Vibrio Anguillarum Bacterin	2037.00	Alpharma, Inc.	Salmonids	Furunculosis, vibriosis
Biojec 1900	Aeromonas Salmonicida-Vibrio Anguillarum-Salmonicida Bacterin	2137.00	Alpharma, Inc.	Salmonids	Furunculosis, vibriosis
Biovax 1200	Vibrio Salmonicida Bacterin	2870.00	Alpharma, Inc.	Salmonids	Vibriosis
Biovax 1600	Vibrio Anguillarum-Salmonicida	2868.00	Alpharma, Inc.	Salmonids	Vibriosis
Ermogen	Yersinia Ruckeri Bacterin	2638.00	Jerry Zinn, Aqua Health, Ltd. (Canada)	Salmonids	Yersiniosis (enteric redmouth disease)
Escogen	Edwardsiella Ictaluri Bacterin	2637.00	Jerry Zinn, Aqua Health, Ltd. (Canada)	Catfish	Enteric septicemia
Fryvacc	Flavobacterium Columnare Bacterin	2974.00	Jerry Zinn, Aqua Health, Ltd. (Canada)	Salmonids	Columnaris disease

Table 3. (Continued.)

Furogen	Aeromonas Salmonicida Bacterin	2035.01	Jerry Zinn, Aqua Health, Ltd. (Canada)	Salmonids	Furunculosis
Furogen 2	Aeromonas Salmonicida Bacterin	2035.00	Jerry Zinn, Aqua Health, Ltd. (Canada)	Salmonids	Furunculosis
Furogen B	Aeromonas Salmonicida Bacterin	2035.02	Jerry Zinn, Aqua Health, Ltd. (Canada)	Salmonids, Koi carp	Furunculosis, ulcer disease
Lipogen forte	Aeromonas Salmonicida-Vibrio Anguilarum-Ordalii-Salmonicida Bacterin	2138.01	Jerry Zinn, Aqua Health, Ltd. (Canada)	Salmonids	Furunculosis, vibriois
Lipogen triple	Aeromonas Salmonicida-Vibrio Anguilarum-Ordalii-Salmonicida Bacterin	2138.00	Jerry Zinn, Aqua Health, Ltd. (Canada)	Salmonids	Furunculosis, vibriosis
Multivacc 4	Aeromonas Salmonicida-Vibrio Anguilarum-Salmonicida Bacterin	2137.00	MicroTechnologies, Inc.	Salmonids	Furunculosis, vibriosis
Vibrogen	Vibrio Anguillarum-Ordalii Bacterin	2858.02	Jerry Zinn, Aqua Health, Ltd. (Canada)	Salmonids	Vibriosis
Vibrogen-2	Vibrio Anguillarum-Ordalii Bacterin	2858.03	Jerry Zinn, Aqua Health, Ltd. (Canada)	Salmonids	Vibriosis

search on the fish immune system and the powerful methods of molecular biology to produce recombinant antigens and DNA vaccines offer hope that a new generation of vaccines can be created to prevent the most common diseases of fish.

Fish Disease Policies and Regulations

The last 25 years have seen significant changes in disease control with the emphasis shifting from treatment to prevention. International, national, and regional regulations have been passed to minimize the spread of certain contagious diseases of fish, and both agricultural authorities and fisheries agencies have increased their emphasis on protecting the health of aquatic animals raised by both the public and private sectors.

The Office International des Epizooties (OIE) in Paris, France, is the world animal health organization charged with developing guidelines and procedures for reducing the spread of infectious animal diseases on behalf of more than 150 member countries. In 1960, the OIE created the Fish Diseases Commission in recognition of the increasing importance of aquaculture and the potential for the international spread of important diseases of aquatic animals. The OIE publishes an international aquatic animal code and diagnostic manual for listed diseases of aquatic animals. Member countries are encouraged to use these guidelines to establish disease-free zones and farms in order to reduce the risk of spreading diseases associated with international shipment of aquatic animals. Many of the OIE materials are available via the Internet (see electronic resources at the end of this chapter).

Following creation of the European Community (EC), a series of directives was issued, beginning in 1991, to prevent the introduction and spread of aquatic animal diseases among member states. These directives (91/67, 91/492, 91/493, 92/53, 93/54) specify testing methods and control measures to be applied to fish and shellfish products to be imported into the EC. It is expected that these directives will be updated regularly to ensure that they are consistent with the guidelines of the OIE.

In the United States, the Code of Federal Regulations, Title 50, Chapter 1, Part 16, Injurious Wildlife, was amended in 1993. The revised text states that salmonids and their eggs can only be imported into the United States if accompanied by a certificate stating the shipment is free of the viral pathogens OMV, VHSV, IHNV, and IPNV. Salmon that have been eviscerated, processed by certain methods, or captured commercially in the open sea are exempt. The director of the U.S. Fish and Wildlife Service maintains a list of persons who are authorized to conduct the specified inspections and issue the required certificates. In addition, all live salmonid eggs imported into the United States must be disinfected within 24 h before shipment by immersion for 15 min in an iodophor solution containing at least 75 ppm active iodine. These regulations will also require regular updating.

In 1997, Canada amended the federal Fish Health Protection Regulations that address international and interprovincial movement of salmonids and their eggs. The Canadian regulations deal with all species and hybrids of fish in the family Salmonidae. Live fish to be imported into Canada or shipped between provinces must be tested and found free of any filterable replicating agent able to cause cytopathic effects in specified cell cultures including, but not limited to, VHSV, IHNV, and IPNV. Also listed are the bacterial pathogens *Aeromonas salmonicida* and *Yersinia ruckeri* and the protozoan pathogens, *Myxobolus cerebralis* and *Ceratomyxa shasta*. Live eggs must be tested for viral agents and surface disinfected before shipment. Dead, uneviscerated fish must be tested and found free of VHSV and *M. cerebralis*. More recently, several other countries have begun to draft regulations affecting the importation or internal movement of aquatic animals. An important goal for the future is to ensure these regulations are harmonized.

Within the United States, disease-control programs are also administered by the Colorado River Wildlife Council, the Great Lakes Fishery Commission, the Pacific Northwest Fish Health Protection Commission, the U.S. Fish and Wildlife Service, several tribes, and numerous states. Most of the state, regional, and national programs include regulations to restrict the movement of fish unless inspected and found free of certain pathogens. While these programs represent important cooperative efforts among the various management agencies, there is a lack of agreement among the various entities as to the pathogens and host species for which inspections must be conducted. It will be important to make the list of pathogens, host species, and inspection methods consistent among the various states and regions of the United States in order to develop a more uniform national program in fish health.

Fish Disease Control Programs

A fish disease control program should emphasize all aspects of good health, including infectious diseases, nutrition, physiology, and environment. The first step of any program must be the establishment of long-range goals. These goals may be broad in concept or may dictate eradication of specific pathogens. The latter is more difficult to achieve and it is possible to have disease control without pathogen eradication.

After the goals of disease control have been established, it is necessary to design a policy that is compatible with other fishery resource priorities. The backbone of the policy should be a monitoring program that will determine the range of serious fish pathogens and detect new outbreaks of disease. Control and containment of fish diseases require the periodic examination of hatchery populations, as well as fish that are free-ranging in natural waters. The health of hatchery fish extends beyond their cultural confinement to natural populations, which they affect either through direct contact or via the effluent from hatchery operations. A monitoring program should include

1. Fish health laboratories equipped to test for disease agents, nutritional deficiencies, and tissue residues;
2. A corps of competent, qualified individuals trained in inspection and laboratory procedures;
3. A training program in fish health for all persons involved in fish

husbandry;

4. Agreements between various government agencies and private groups to establish lines of communication, as well as the storage and cataloging of data derived from the monitoring program;
5. Specific guidelines for laboratory procedures to be followed and for qualifications of persons and laboratories doing the inspections and testing;
6. The development of specific steps for disease reporting and a list of reportable pathogens; and
7. Preplanned courses of action that will be used to control or eradicate a reportable disease when it occurs.

The U.S. Fish and Wildlife Service has established a national program for fish disease control administered through a series of regional fish health centers. The plan provides for determining the distribution of specific pathogens within the National Hatchery System, restricting dissemination of fish pathogens, and eradicating certain disease agents from federal fish hatcheries. The policy also provides for research and training in epidemiology, prevention, control, and diagnosis of various fish diseases. Recently, the agency has initiated a comprehensive national wild fish health survey. Many states, tribes, and regions have developed similar disease control programs for hatchery facilities under their authority.

The Fish Health Professional

Initially, those knowledgeable about matters of fish health were typically hatchery managers or workers who developed an interest in disease diagnosis and treatment in order to be more successful at rearing aquatic species in captivity. In addition, a small cadre of academics was interested in fish parasites and pathogens as objects of study. Eventually, through personal contact and training, usually provided by academic institutions or federal laboratories (e.g., the "short" and "long" courses in fish health taught at the Leetown laboratory), a corps of fish health experts developed who could assist hatchery managers in diagnosis of infectious disease and recommend appropriate treatment.

Today, the fish health professional may be a veterinarian with specialized training in aquatic animal medicine; an academic with research expertise in infectious diseases of fish; a state, federal, or tribal fish biologist with specific training in fish health from various sources; and others who have an interest in this expanding field. The Fish Health Section (FHS) of the American Fisheries Society maintains a certification program for fish pathologists and fish health inspectors who have the required training and experience to provide expert advice and to conduct fish health inspections. These professionals often work for federal, tribal, or state agencies in facilities that are specialized for work with fish diseases. Several universities and private individuals have also established laboratories that can help in disease diagnosis and provide inspection services to the private aquaculture industry.

Fish Health Inspections

During an on-site fish health inspection at a hatchery, the fish health professional will collect random samples of fish tissues for analysis. The tests to be conducted will vary according to the type of testing requested and should follow the standardized procedures of the most recent editions of the Fish Health Section of the American Fisheries Society "Bluebook" officially titled *Procedures for the Detection and Identification of Certain Fish and Shellfish Pathogens*, the *Fish Health Protection Regulations, Manual of Compliance* published by the Department of Fisheries and Oceans, Canada, or the *Diagnostic Manual for Aquatic Animal Diseases* published by the Office International des Epizooties in Paris, France.

The inspector typically takes tissues from a specified number of fish from each population at the hatchery. In most cases, each fish sampled must be killed. The minimum sample size from each population will follow a statistical plan that provides a 95% confidence for including an infected fish in the sample where the incidence of infection is equal to, or greater than, a specified level, usually 2 or 5%. The inspector will appreciate some advance preparation on the part of the hatchery manager and assistance in providing equipment needed to sample the rearing units and a place to work.

The sample population is determined on the basis of hatchery variables such as species, age, and water source. Generally, fish are sampled by "lot." A lot being defined as fish of the same species, originating from the same spawning population that have always shared the same water supply. Thus, when fish are held in different water supplies, each group has to be sampled as a separate population. All broodstock of the same species held in a single water supply can be considered one population. Broodstock are usually inspected during the period when eggs are being obtained.

All fish on hand at the time of inspection constitute the population and are sampled accordingly. Samples are collected from each tank or rearing unit. Suspect fish (moribund specimens) are collected along with healthy individuals. Fish should be alive when collected. Necropsy procedures assume that the same fish may provide tissues for the various laboratory tests (bacterial, viral, and parasitic). A modified procedure may be required for very small fish. Material to be examined for external parasites must be taken before any antiseptic or disinfectant procedures are applied. After the body has been opened aseptically, tissues for bacterial cultures and virus tests are collected. These tissues should be refrigerated or stored on ice and must not be frozen. Finally, any histological samples are taken and placed in fixative.

At least 2 weeks are required for the laboratory analyses to be completed. Upon completion of the tests, a certifying official will issue a report specifying the samples taken, the laboratory tests conducted, and the findings. The exact type of report can vary according to the governmental agency involved and the circumstances of the inspection.

Diagnostic Examinations

Correct diagnosis depends upon accurate and detailed information regarding the fish, the signs observed, the history of the facility, and the conditions under which the fish were raised. The more information that is available, the more likely that the diagnosis will be correct. If space permits, a small area of the hatchery should be devoted to fish health examinations. This space can be used by hatchery personnel to monitor levels of external bacteria or parasites and to perform standard microbiological procedures. The space will also be needed by a fish health professional conducting inspections or diagnostic examinations. In addition to being dry and well-lit, the space should have a bench or table for preparation of specimens, a compound light microscope and various slides, stains, incubator, culture media, and small equipment that may be needed to perform on-site diagnostic tests.

Even after a preliminary diagnosis by the hatchery staff, a fish health professional should be contacted who may visit the hatchery or request that specimens be sent to the laboratory for verification. Although the symptoms may seem typical, another disease may be present and while treatments may be effective for one condition, the other disease may still be uncontrolled. Hatchery personnel should furnish the fish health professional with correctly collected and handled material, including all available information, at the earliest possible date. To assist in obtaining a correct diagnosis, specimens should be collected before any treatment is given or started. Only a few fish need be collected for on-site examination or sent for examination, but these should be collected with the utmost care. Dead fish or fish that appear to be normal are nearly worthless. The most desirable fish are those that show most typically the signs of the disease in question. Moribund, but still living, fish are the best for diagnostic purposes.

Shipping Live Specimens

When it is necessary to ship live specimens for diagnostic purposes, (1) assure that everything possible is done to ensure that the specimens will be received alive, and (2) take extra precautions to ensure that other parcels will not be damaged by water leakage. Shipments should bear the notation "Live Fish—This Side Up" and should be sent by the most rapid means possible.

If air express shipments are needed, packing should allow for gas expansion that occurs at altitude. Plastic bags containing about one-fourth water and half or less air or oxygen usually provide room for expansion. A general precaution is to use a double bag system, one bag filled and sealed within another. It is best to ship a minimum number of specimens. Sick fish and coldwater species, such as trout, require greater volumes of water than healthy or warmwater fish. Twenty volumes of water for each volume of fish usually will be adequate for healthy fish, but greater volumes should be provided for sick fish. During extreme hot or cold weather, insulated containers may be required. Coldwater species usually ship better if ice is provided. The ice should be packed in double plastic bags so that it will not leak when it melts, especially if the ice is made with chlorinated city water and placed in the water containing the fish.

Shipping Preserved Specimens

Preservatives typically are corrosive and odorous. Containers should be unbreakable and absorbent material should be provided in the event leakage occurs. A good procedure is to fix the fish in a proper fixative for a day or two, then place the preserved fish, with a very small volume of fixative, in a plastic bag. The sealed bag should be placed within a second plastic bag, which also should be sealed. This durable package has minimal weight. Select representative specimens. Examine them carefully and record relevant data. Buffered formalin or Davidson's fixatives are preferred to Bouin's solution that contains picric acid that may become explosive when improperly stored.

The volume of the fixative should be at least five to ten times that of the fish or tissue. Fish and tissues should be left in the fixative for at least 24 h and then the fixing solution replaced with 70% ethyl alcohol. However, if alcohol is not available, retain the specimens in fixative. To facilitate fixation, fish should be slit down the abdomen from the anus to the gills. The air bladder should be broken to permit fixation of the kidney. The kidney of fish 15 cm or larger should be split along its entire length. The intestines and other organs should be slit if the fish are larger than fingerlings. It also is desirable to cut the skin along the back of the fish. If the fish are larger than 15 cm, the cranial cap should be opened to facilitate fixation of the brain. The importance of these incisions cannot be overemphasized. If the fish are too large to ship whole, cut pieces from individual tissues (gill, heart, liver, etc.), and especially any lesions observed. These pieces should not be larger than 1 cm square and 0.5 cm thick.

Commercial formalin also can be used for preserving specimens and should be mixed with nine parts of water to make a 10% formalin solution. Unless the lesions are very clear and obvious, always preserve several healthy specimens of the same size and age as the sick fish and send them at the same time in a separate container.

Decontamination of Equipment and Facilities

Equipment Decontamination

Equipment sometimes must be decontaminated. One of the best and cheapest disinfectants is chlorine. A solution of 200 ppm will be effective against most pathogens in less than a few minutes while one of 10 ppm may require up to several hours for sterilization. Chlorine levels are greatly reduced by organic material such as mud, slime, and plant material; therefore, it is important to thoroughly clean equipment before it is exposed to the solution. A chlorine solution also loses strength when exposed to the air or sunlight, so it may be necessary to add more chlorine or make up fresh solutions during disinfection.

Chlorine may be obtained as sodium hypochlorite in either liquid or powdered form. The latter is the more stable of the two, but it is more expensive. The amount of chlorine added to water depends on the percentage of available chlorine in the product used. As an example, powdered chlorine may contain either 15, 50, or 65% available chlorine, while liquid solutions typically contain from 5.6% (commercial laundry bleach) to 10% available chlorine. The following amounts would be needed to make a 200 ppm solution:

- 100 g (3.53 oz) of 15% available chlorine powder to 75 L (19.8 gal) of water;
- 100 g (3.53 oz) of 50% available chlorine powder to 250 L (66.04 gal) of water; and
- 100 ml (3.4 fl. oz) of 10% available chlorine solution to 50 L (13.21 gal) of water.

Chlorine is acutely toxic to all life. In addition to insuring protection to workers applying the material, residual chlorine must be completely neutralized before it is allowed to drain or to enter waters containing fish. One hundred liters (26.4 gal) of 200 ppm chlorine solution can be neutralized by 150 g (5.3 oz) of sodium thiosulfate. Neutralization can be easily determined with chlorine test paper or with a standard swimming pool test kit.

Facility Decontamination

With the increase in global trade, the risk of importation of exotic diseases has become an increasingly important threat. Also, the expansion of aquaculture has led to the emergence of new diseases for which control methods are not established. For those diseases that presently cannot be treated (e.g., most viral diseases), the only successful control is complete elimination of all infected fish from a hatchery, thorough decontamination of the facility, development of a new stock of disease-free fish, and maintenance of disease-free conditions throughout all future operations. Hatchery decontamination has been successful in removing pathogens; however, this method is practical only at those hatcheries having a controlled water supply originating in wells or springs that can be kept free of fish or a water supply treated with ozone or ultraviolet light to render it pathogen-free.

Preliminary Operations

Before chemical decontamination of the hatchery is started, a quarantine zone should be established with strict entry and exit controls. This may involve decontamination of all vehicles leaving the facility. The volumes of all incubators, troughs, raceways, or ponds must be known or measured accurately. The areas of all contaminated surfaces in the buildings are calculated, and allowance is made for disinfectant solution to be applied on walls and floors. Then, the quantity of sodium hypochlorite needed for a 200 ppm solution is computed. If the chlorine solution will enter fish-bearing waters after leaving the hatchery, it will have to be neutralized. Commercial sodium thiosulfate, used at the rate of 1.5 g for each liter of 200 ppm chlorine solution, will suffice.

All loose equipment should be brought from storage rooms, scrubbed thoroughly with soap and warm water, and left near a raceway for later decontamination. Such equipment includes nets, buckets, pans, small troughs, tubs, screens, seines, boots, and rain gear. During this operation, any worn-out, unnecessary, or difficult to disinfect equipment (e.g., wooden dam boards) should be burned or otherwise destroyed. The walls of all raceways and rearing units should be scrubbed and the bottom raked. Particular attention should be given to removing any remaining fish food, pond scum, or other organic substances having a high chlorine demand.

Elimination of Fish

Prior to chemical decontamination, all fish should be destroyed by deep burial and covered with lime. The burial grounds should be so located that leaching cannot recontaminate the hatchery water supply. All stray fish left in pipelines will be killed by chlorine, but it is important that their carcasses be retrieved and destroyed.

Decontamination

Decontamination methods should assure that the full strength (200 ppm) of the chlorine is maintained for at least 1 h and that a concentration of not less than 100 ppm is maintained for several hours. Many hatcheries are so large that total decontamination cannot be completed in one day. Treatment then must be carried out by areas or blocks and started at the upper end of the hatchery.

Before chlorine is added, all ponds, raceways, tanks, and troughs are drained. Additional dam boards, plugs, or caps are installed to hold the water to the very top of each unit. The required quantity of chlorine then is added gradually to the incoming water. The solution flows to the various rearing units, which are allowed to fill and overflow until there is about 5 cm of the chlorine solution on the floor. The incoming water then is turned off or bypassed. The chlorine solution is pumped from the floor and sprayed on the sides and bottoms of all tanks and racks, the walls and ceiling, head trough, and any other dry equipment for 1 h. The same procedure must be used in all rooms of every building. Underground pipelines must be filled and flushed several times. If the hatchery must be decontaminated in sections, the work should be planned and timed so that all buildings, springs, supply lines, and rearing units contain the maximum level of chlorine at the same time, so that no contaminated water can enter parts of the system already treated. While a maximum concentration of chlorine is being maintained, all loose equipment such as pails, tubs, trays, splashboards, and other material may be immersed in the raceways.

Throughout the course of the project, checks should be made on the approximate chlorine strength with a chlorine test kit or chlorine test paper. If any section holds a concentration below 100 ppm chlorine after 1 h, the solution should be fortified with additional chlorine.

Hatchery Maintenance

After a hatchery has been decontaminated and is pathogen free, recontamination must be prevented. The movement of live fish into the hatchery should only originate from known disease-free stocks at a facility with a strict disease inspection program. Ideally, production should be restarted with disinfected eggs from a certified pathogen-free source. The spread of disease can be prevented only by rigid cleanliness. All hatchery equipment should be double-checked to ensure it has been decontaminated thoroughly before it is used and strict internal quarantine procedures must be followed. At some facilities, this may include the decontamination of vehicles and equipment before they are allowed to enter the hatchery. Finally, the hatchery staff should be impressed with the idea that one mistake may nullify all previous efforts.

Acknowledgments

The author is deeply indebted to Drs. Ronald Hedrick and Rosalie Schnick for their thoughtful review of the manuscript and their useful suggestions.

References

Bruno, D. W., and T. T. Poppe. 1996. A colour atlas of salmonid diseases. Academic Press, New York.

Couch, J. A., and J. W. Fournie. 1993. Pathobiology of marine and estuarine organisms. CRC Press, Boca Raton, Florida.

Ellis, A. E., editor. 1988. Fish vaccination. Academic Press, New York.

Federal Joint Subcommittee on Aquaculture. 1994. Guide to drug, vaccine and pesticide use in aquaculture. Texas Agricultural Extension Service, College Station, Texas.

Ferguson, H. W. 1989. Systemic pathology of fish. Iowa State University Press, Ames.

Gudding, R., A. Lillehaug, P. J. Midtlyng, and F. Brown, editors. 1997. Fish vaccinology, volume 90. Developments in biological standardization. Karger, Basel, Switzerland.

Halver, J. E. 1989. Fish nutrition, 2nd edition. Academic Press, New York.

Herwig, N. 1979. Handbook of drugs and chemicals used in the treatment of fish diseases. Thomas, Springfield, Illinois.

Hoffman, G. L. 1999. Parasites of North American freshwater fishes, 2nd edition. Cornell University Press, Ithaca, New York.

Hoole, D., P. Bucke, P. Burgess and I. Welby. 2000. Diseases of carp and other cyprinid fish. Blackwell Science, Oxford, UK.
Iwama, G., and T. Nakanishi. 1996. The fish immune system. Academic Press, San Diego, California.
Lasee, B. A., editor. 1995. Introduction to fish health management, 2nd edition. U.S. Fish and Wildlife Service, Onalaska, Wisconsin.
Leong, J. C., and J. L. Fryer. 1993. Viral vaccines for aquaculture. Annual Review of Fish Diseases 3:225–240.
Noga, E. J. 2000. Fish diseases: diagnosis and treatment. Iowa State University Press, Ames.
Newman, S. G. 1993. Bacterial vaccines for fish. Annual Review of Fish Diseases 3:145–185.
Office International des Epizooties. 2000. Diagnostic manual for aquatic animal diseases, 3rd edition. OIE, Paris.
Plumb, J. A. 1999. Health maintenance and principal microbial diseases of cultured fishes. Iowa State University Press, Ames.
Post, G. 1990. A textbook of fish health. T.F.H. Publications, Neptune City, New Jersey.
Roberts, R. J. 1989. Fish Pathology, 2nd edition. Bailliere Tindall, London.
Roberts, R. J., and C. J. Shepherd. 1997. Handbook of trout and salmon diseases, 3rd edition. Blackwell Scientific Publications, Oxford, England.
Stoskopf, M. K. 1993. Fish medicine. Saunders, Philadelphia.
Thoesen, J. C., editor. 1994. Suggested procedures for the detection and identification of certain finfish and shellfish pathogens, 4th edition. Fish Health Section of the American Fisheries Society, Bethesda, Maryland.
Warren, J. W. 1991. Diseases of hatchery fish, 6th edition. U.S. Fish and Wildlife Service, Washington, D.C.
Wedemeyer, G. A. 1996. Physiology of fish in intensive culture systems. Chapman and Hall, New York.
Wolf, K. 1988. Fish viruses and fish viral diseases. Cornell University Press, Ithaca, New York.
Woo, P. T. K. 1995. Fish diseases and disorders, volume 1. Protozozan and metazoan infections. CAB International, New York.
Woo, P. T. K., and D. W. Bruno. 1998. Fish diseases and disorders, volume 3. Viral, bacterial and fungal infections. CAB International, New York.
Wood, J. W. 1974. Diseases of Pacific salmon: their prevention and treatment, 2nd edition. Washington Department of Fisheries, Seattle.
Yasutake, W. T., and J. H. Wales. 1983. Microscopic anatomy of salmonids: an atlas. U.S. Fish and Wildlife Service, Resource Publication 150, Washington, D.C.

Electronic Resources

American Fisheries Society	http://www.fisheries.org
American Veterinary Medical Association	http://www.avma.org
Aquaculture Health Page	http://www.geocities.com/CapeCanaveral/Lab/7490/index.html
Aquaculture Network Information Center	http://aquanic.org
Aquarium Aqualink	http://aqualink.com
Center for Veterinary Medicine	http://www.fda.gov/cvm
Fishlink	http://fishlink.com

10

Fish Transportation

Gary J. Carmichael, Joseph R. Tomasso, Jr., and Thomas E. Schwedler

Introduction

Fish transportation techniques have now been developed that allow for safer movement of several species. The impetus for the development of safe techniques came from candid observations that often entire shipments of fish would die during or soon after transport. Fish losses related to transport events were especially critical because the shipper generally had full investment in the animals at the time of harvesting and transporting. Losses associated with transport also interrupted stocking and processing schedules.

Fish transportation research has identified several actions (or lack of action) during the harvesting, transporting, and posttransport process that stress or kill fish. Mortality related to transport or the delivery of fish in poor condition may be the result of one severe stressor (a physical or chemical agent that causes pain or discomfort), several mild stressors, or infectious diseases (usually occurring 1 or 2 weeks after transport) induced by the stress associated with transport techniques.

Severe stressors associated with harvesting and transporting that may kill fish in a short period of time are usually actions associated with inadvertent mismanagement of harvesting and transport activities. One common mistake is concentrating fish (either by seining or draining the ponds or tanks) in a small volume without adequate aeration causing animals to die from hypoxia Other examples of severe stressors include allowing fish to become covered in mud while awaiting loading, moving fish from water of one temperature directly to water of a very different temperature, and crushing fish by loading dip nets or boom baskets too heavily.

Even if fish are carefully harvested and loaded into a transport tank at the correct density, a group of mild stressors may act together to cause death during or within a few days following transport. Some examples of these mild stressors include shaking as the transport vehicle moves, crowding, and less than optimal water quality (high ammonia levels, high carbon dioxide levels, low dissolved oxygen (DO) concentrations, or extreme temperatures). These stressors may act together to cause what is termed the "stress response," a complex series of physiological changes. The stress response in fish is very similar to what occurs in humans and other animals. To a fish culturist, the most important consequences of the stress response are an increase in metabolic rate or energy use and a suppression of the immune system (Pickering 1993). The increased metabolic rate has the effect of accumulating toxic metabolites such as lactic acid within the cells of the animal, as well as depleting its energy reserves. When energy reserves are depleted, the fish can no longer maintain the correct ionic (salts) balance within the blood; the animal may die from what we think is a combination of ionic imbalance and the accumulation of toxic wastes in cells (Wood et al. 1983; Carmichael et al. 1984; Weirich et al. 1992). Suppression of the immune system allows pathogens that fish normally resist to initiate the disease process that may cause mortality several days to weeks after the fish are delivered (Pickering 1993).

To avoid stress and mortality, some general guidelines for harvesting and transporting fish should be followed:

1. reduce the number of stressors
2. reduce the severity of stressors
3. minimize the duration of stressors
4. minimize plasma ion disturbances
5. minimize increases in metabolic rate

In the following sections, a general description of how a fish culturist can follow these guidelines will be presented, some equipment that will aid in following the guidelines will be described, case studies will be presented to demonstrate the success of certain techniques, and techniques and information for transporting different species and varying conditions will also be presented.

General Description of Tank Transportation Procedures

Before beginning the process of handling and transporting live fish, it is important to plan the entire procedure in a step-by-step manner. For individuals who are new to aquaculture (and for those who are not so new but have lost fish during transport), it is advisable to conduct a "walk-through" of the entire procedure. A list of equipment and events is often a helpful first step (Table 1). As with any endeavor in animal husbandry, planning should emphasize animal care and not worker convenience.

Once the procedure is planned, the initial phase may begin. This consists of either seining ponds or concentrating fish that are being held in raceways, tanks, troughs, or cages. This phase should not be initiated if fish are unhealthy or should be stopped after concentrating has begun if it is discovered that fish are unhealthy. Unhealthy fish are generally more susceptible to stressors than healthy fish. It is recommended that a few fish be captured before the scheduled harvest and examined. If seining must occur during the summer months, it should be conducted during the coolest times of the day. Seining should be planned to allow the fish to be concentrated in an area of the pond where aeration is available or where aerated water can be added to the confined fish. The aerated water may also be used to cool fish during summer harvesting. On hot days shade may be provided with a tarp. Details of harvesting are presented in Chapter 2.

After the fish are concentrated, they should be quickly placed in the transport tank. This is usually accomplished using nets or buckets. Often the nets and buckets are equipped with a trap door on the bottom for unloading and are attached to a hydraulically powered boom (Figure 1). When using nets, do not overload because fish on the bottom will be crushed. Nets are acceptable for catfish in many cases, but most nonsalmonid scaled fish require a "wet transfer" using buckets or other containers with water. When emptying the nets and buckets into the transport tanks, avoid dropping fish into the water. Instead, place the net or bucket in the transport water and then release the fish. Also, be sure to avoid temperature shock by matching the water temperatures in the pond and transport tank before loading fish. During wet transfer try to avoid adding algae and mud to the transport tank. Before loading, DO levels in the transport tank should be at saturation. The most stressful segment of the transport sequence is often the loading segment. Therefore, special care should be taken to avoid crushing, temperature shock, and stress due to low DO concentrations.

All equipment on the transport truck should be checked immediately after loading and then the truck should depart. Time en route should be kept to an absolute minimum. For long hauls, use at least two drivers and restrict stops to checking the transport system and refueling. While en route, oxygen concentrations and temperatures should be monitored constantly. This is accomplished by using a DO meter with a long-lead probe. The meter can be kept in the cab of the truck while the probe is in the transport tank. The meter should be recalibrated during fuel stops.

TABLE 1. A checklist to be verified before fish harvesting from ponds and a transport event is initiated. Culturists should add to or delete items from this list for their operation. Capture and loading items may be modified for fish cultured in raceways and tanks.

Capturing and loading

- ☐ fish not fed for at least 24 h prior to harvesting (exception for larval and small fishes)
- ☐ seines present, correct mesh size and texture
- ☐ all equipment for pulling seine present and functional
- ☐ live box or car correctly attached to seine
- ☐ cutting seines present and mesh size correct
- ☐ aeration equipment in place and working
- ☐ pond drain working
- ☐ dip nets present
- ☐ boom buckets and nets present and appropriate
- ☐ boom truck present and working
- ☐ all chemicals to be used available and premeasured
- ☐ cool time of day

Transport system

- ☐ tank properly mounted on truck or trailer
- ☐ trailer properly hitched to truck (including lights and brakes)
- ☐ spare tires available (with tire changing equipment)
- ☐ truck contains adequate fuel
- ☐ oil and other fluid levels checked
- ☐ tank filled to proper level with water
- ☐ water temperature in tank similar to pond water
- ☐ water additives in tank water
- ☐ unloading equipment on truck
- ☐ oxygen/temperature meter calibrated, in place, and operating (with a reserve)
- ☐ primary aeration system functioning
- ☐ oxygen bottles full; adequate oxygen for trip
- ☐ emergency aeration system in place and workable

Unloading

- ☐ tempering pump present and functional
- ☐ receiving tanks ready and filled
- ☐ aeration available for receiving tanks
- ☐ dip nets available and appropriate

Personnel

- ☐ pond crew trained in seining procedures
- ☐ loading crew trained in best handling procedures
- ☐ drivers know the route
- ☐ people at receiving point are expecting shipment

FIGURE 1. Net used in the channel catfish industry to lift fish from pond to transport tanks. Note the trap door on the bottom for unloading and the integrated scale above the net.

Before harvesting, the shipper should make detailed arrangements with the receiver. Arrangements should include where and when fish will be delivered, specific environmental requirements (e.g., temperature, salinity) at the receiving point, and the need for any specialized equipment at the receiving point. Arrangements should be verified before the truck leaves the production site and again while en route during long trips.

Specific requirements for the species being transported should be carefully observed. In particular, the maximum density of the fish in the transport tank is important. Density calculations can be verified with a short computer program (Fries 1992). Plasma ion disturbances may be reduced during transport by adding various salts to the transport water. Generally 5–10 g/L sodium chloride is used for most freshwater fish. Anesthetics dissolved in the transport water produce a calming effect and reduce the metabolic rate and, thus, energy use of the fish. Decreasing metabolic rate has the dual effect of reducing oxygen usage and conserving energy for use in maintaining ion balance. Caution is advised, however. Animals can be over anesthetized and die either from an overdose of the drug or from anoxia due to animals in deep anesthesia being concentrated in corners of the hauling unit with little oxygenation. Ideally, the anesthetized fish will be able to remain upright in the water column and will slowly respond to prodding or picking up by hand. Some common transport densities and water additives are presented later in this chapter. Currently, many of the traditionally used chemical additives are no longer approved by regulatory agencies for use in fish culture. Before using chemical additives, culturists should carefully check labels for approved uses or contact state extension services.

Fish Transport Equipment

The important equipment required for fish transport activities (exclusive of harvesting equipment) includes loading nets and buckets (also called boxes), dip nets, transport tanks, oxygenation and/or aeration systems, DO/temperature meters, and an unloading system. Occasionally, tanks are equipped with water-chilling units; however, most experienced fish transport specialists prefer relying on insulated tanks and cooling with ice to maintain temperatures for species that prefer cool water.

Loading buckets or boxes are usually constructed of aluminum because it is lightweight and strong. Considerations for box construction: (1) the inside of the box should be smooth, with all welds and other rough areas ground and buffed; (2) sides and bottoms of the box should be sloped toward a quick-open door on the bottom; and (3) the bucket or box should be attached to the boom cable by at least three points to prevent tipping.

Loading nets should be designed so animals will drop smoothly toward a bottom- mounted trap door. As with the loading boxes, at least a three-point attachment to the boom cable is required. Often a spring-loaded or electronic scale is mounted above the box or basket to allow weighing of the fish during loading.

Mechanical fish pumps have recently become available and are sometimes more efficient and less stressful than loading buckets or boxes (Figure 2). Pumps should be selected to be compatible with fish size and number of fish to be loaded in a given amount of water.

Nets used during the entire operation (dip nets, loading nets, seines, live cars) should be of a type that are compatible with the species being transported. Two characteristics are generally the most important—mesh

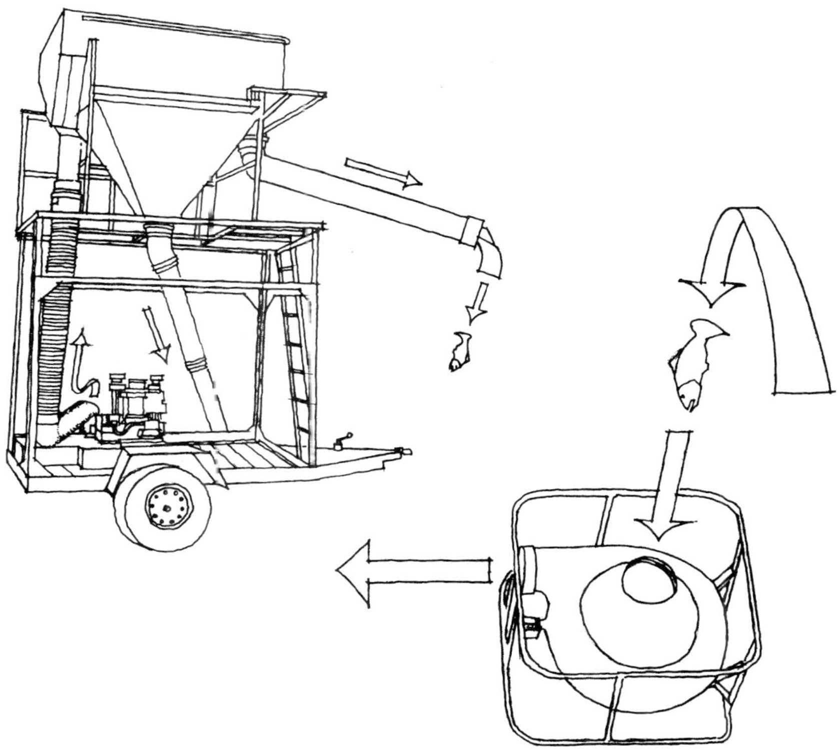

FIGURE 2. Example of a fish pump (upper left) used to load fish into transport tanks. Fish enter the pump head (bottom right), and are pumped in water to a grader box. Small fish fall between grader bars while larger fish are retained in the upper part of the box. Both groups of fish are either loaded into a transport tank or sent elsewhere in water by way of a large-diameter pipe (see arrows for direction of flow).

size and texture. Fish will slip through mesh sizes that are too large, be "gillnetted," or become entangled with part of their body, through the net. Mesh sizes that are too small will result in fins and spines becoming entangled. Ideally, the mesh size should be small enough to prevent animals from completely or partially passing through and large enough to prevent fin and spine entanglement. Texture of nets is determined by the material used, the means used to form the mesh (knots or weaving), and net coatings (Figure 3). Catfish require a coated net (usually asphalt or resin) to prevent entanglement of pectoral and dorsal spines. Knotted netting is also acceptable for catfish. Scaled fish, however, require soft, uncoated, knotless nets to limit damage due to scale loss from abrasion.

Fish transport tank designs have undergone a great deal of improvement during the past decade. Although designs are varied, we have identi-

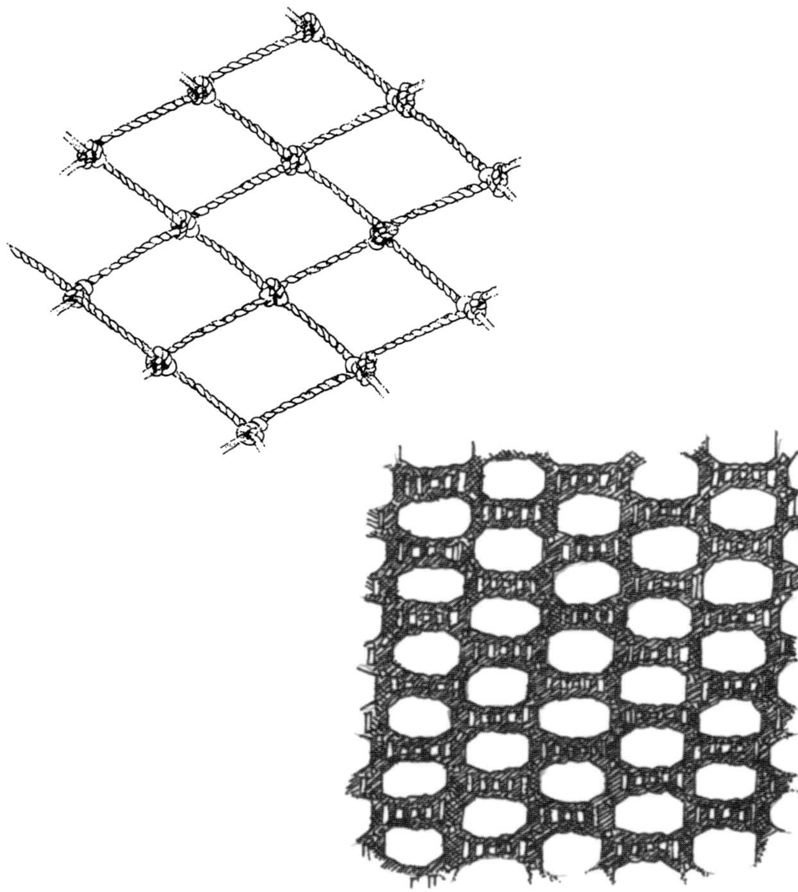

FIGURE 3. Knotted (upper left) and knitted (bottom right) net mesh.

fied the desirable characteristics of transport tanks (Carmichael and Tomasso 1988). Our following discussion considers these characteristics and presents the plans for a small transport tank (Figure 4). These plans, however, can and have been scaled to much larger capacity transport systems.

Transport tanks may be located either on the bed of a truck or on a trailer. A recent trend has been to mount tanks on trailers, especially in situations where the need for tanks is seasonal. Permanent truck-mounted tanks usually limit the use of the truck when no fish need to be transported. However, in areas where harvesting is conducted year round, such as in the catfish farming industry, truck-mounted tanks are appropriate due to their relative safety and ease of handling. Regardless of

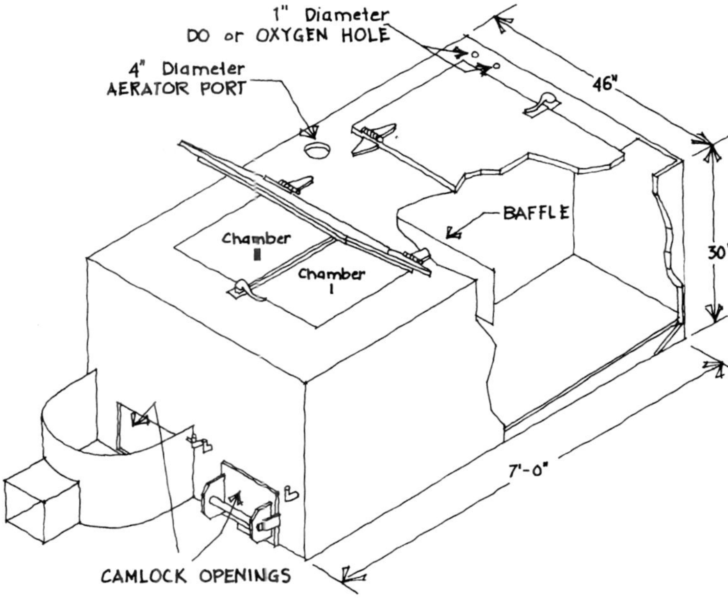

FIGURE 4. A model transport tank based on research and user surveys. See text for details.

whether the tank is truck- or trailer-mounted, one very important point should be kept in mind—water is heavy (8.3 lb/gal, 1 kg/L). It is critical that loaded tank weights be matched to the truck and/or trailer. Truck-mounted tanks, when loaded, simply need to fall within the guidelines of the truck manufacturer. The weight of a loaded trailer-mounted tank should be within the guidelines of the trailer manufacturer, and both the total weight and tongue weight of the trailer and loaded tank should be within the guidelines of the towing vehicle. Several governmental agencies have adopted the use of tanks mounted on goose-neck trailers equipped with electric brakes and towed by heavy duty tandem-wheel pickup trucks (Figure 5).

Most tanks produced today are constructed of fiberglass or a wood and fiberglass combination. These materials are light compared to some metals, provide a smooth internal surface that prevents abrasion of fish, are durable, and provide insulation against rapid temperature changes. Tanks are typically subdivided into two or more independent compartments. Most total tank volumes range from 50 to 500 gal (190–1900 L). Cubical and rectangular box shapes are most common. Urethane foams are often used to insulate the tanks. A sloped tank bottom allows the water and fish to be drained into a large diameter (depending on the size of the fish) drain line that may be extended directly into the receiv-

FIGURE 5. An example of a goose-neck trailer-mounted transport tank.

ing water. Several companies produce acceptable tanks, and, often, local fiberglass shops can produce excellent tanks.

Two aeration systems, a primary and an emergency backup, should be part of all fish transport systems. The use of bottled oxygen is the most common primary system. In cases of heavy daily use, such as in the catfish industry, liquid oxygen is more cost effective. However, bottled oxygen can be stored for long periods between transport events. In a given volume, more oxygen can be stored in the liquid form than gaseous form; thus, liquid oxygen is better for long distance transport of large quantities of fish. To efficiently use oxygen, a very fine pore diffuser must be used in the bottom of the tank. Several good diffusers are commercially available. Mechanical agitators that run off the 12 volt (direct current) truck battery are commonly used emergency systems. However, agitators are not as efficient in oxygenating water as pure oxygen systems at oxygen concentrations required for safe fish transport. Agitators can also cause foaming (especially in salt water), which interferes with carbon dioxide diffusion from the water into the atmosphere. Commercially available antifoaming compounds help alleviate the problem.

Other components of a well-constructed transport tank include internal baffles, vents, and locking hatches. Internal baffles reduce water movement in the tank. Vents on the top of the tank allow carbon dioxide to escape from the tank. If carbon dioxide builds up in the airspace above the water in the tank, carbon dioxide diffusion from the water is slowed. Loading hatches should lock securely to prevent them from opening during sudden stops or maneuvers.

Transportation of Largemouth Bass: A Case Study

Research conducted at the San Marcos (Texas) National Fish Hatchery and Technology Center in the mid-1980s (e.g., Carmichael et al. 1984) demonstrates how an integrated stress reduction program can drastically reduce transport-induced losses. A fish transport simulator (shown to give results similar to real transport situations), consisting of a transport tank equipped to rock as if it were on a moving truck, was used to investigate several different ways of loading, transporting, and unloading fish. In all of the experiments, fish weighed about 227 g, were acclimated to 22°C water in raceways, and were transported at a density of 0.18 kg/L. The simulated transport lasted 30 h (to simulate a real stocking commitment), and the fish were delivered to indoor tanks containing 22°C water. After delivery, the fish were monitored for up to 28 d. Treatments during each experiment were applied before (pretransport), during (transport), or after (posttransport) transport.

Experiment one (Table 2) consisted of no pretransport treatment other than fasting the fish for 36 h. The fish were then crowded in the raceway (flowing water maintained saturated oxygen concentrations), captured using dip nets, and loaded, at a density of 0.18 kg/L, into the simulator. Transport was then simulated for 30 h. Bottled oxygen maintained oxygen levels near saturation. The transport temperature was 20°C. After transport, the fish were placed directly into the posttransport tanks. All of the fish survived the transport event, but 83% died during the follow-up period.

Experiment two was similar to experiment one with no pre- or posttransport treatments. However, the fish in experiment two were transported at 20°C in a solution containing salts to aid in ion balance and an anesthetic to reduce energy consumption. Total mortality in this experiment was reduced to 12%.

Experiment three used pretransport, transport, and posttransport treatments. The pretransport treatment consisted of a 72 h fasting period, a 10 d prophylactic disease treatment, and preloading anesthetic treatment. The fish were then transported in a solution of salts and an anesthetic. A 10 d posttransport prophylactic disease treatment in well water followed (Table 2). Total mortality in this experiment was 5%.

Experiment four was similar to experiment three except that the water temperature in the transport tank was gradually lowered to 16°C during the early part of the transport event. Upon arrival the temperature was gradually raised and salts reduced. Total mortality in this experiment was 0%.

This study demonstrated that some simple actions could tremendously reduce losses due to transport. Note that in the above experiments, mortality was reduced from 83% to 12% simply by modifying the transport water. Pre- and posttransport treatments further reduced the losses, but if circumstances prevent pre- and posttransport treatments, modifications of the tank water alone will offer great benefits.

Specifics for Transporting Various Species

In this section, we present specifics for transporting several species. The information is taken primarily from a survey of fish culturists throughout the United States who transport fish (Carmichael and Tomasso 1988).

TABLE 2. Additives used during the pretransport, transport, and posttransport phase of the largemouth bass study.

Pretransport	
Experiment 1	none
Experiment 2	none
Experiment 3	10 mg/L copper sulfate[a], 1 h daily for 10 d; 50 mg/L tricaine methane sulfonate (MS-222) immediately before loading
Experiment 4	same as Experiment 3
Transport	
Experiment 1	well water
Experiment 2	92 mg/L calcium, 40 mg/L magnesium, 179 mg/L potassium, 2,409 mg/L sodium, 320 mg/L bicarbonate, 114 mg/L sulfate, 3,825 mg/L chloride, 116 mg/L phosphate, 15 mg/L MS-222[b]
Experiment 3	salts as in Experiment 2; 25 mg/L MS-222
Experiment 4	same as Experiment 3 except lower temperature
Posttransport	
Experiment 1	none
Experiment 2	none
Experiment 3	10 mg/L copper sulfate[a], 1 h daily for 10 d
Experiment 4	10 mg/L copper sulfate[a], 1 h daily for 10 d salts[c] as in Experiment 2, Transport

[a] Treat under static conditions in tanks or raceways, then flush system. Be sure to provide aeration. This level (10 mg/L) of copper sulfate is nontoxic to species such a largemouth bass, but may be extremely toxic to others such as salmonids. The intent of the chemical use was to rid the fish of parasites and external bacteria. Other more appropriate or species-specific chemicals may also be used.

[b] To prepare salt solution (assuming no or few ions initially in water): 5,950 mg/L sodium chloride, 252 mg/L potassium chloride, 164 mg/L potassium phosphate, 143 mg/L magnesium sulfate, 151 mg/L calcium chloride, 210 mg/L sodium bicarbonate. These salts are based on those that naturally occur in fish blood.

[c] Temper salts from system over a period of two to three days.

When using the following information, it is important to understand that fish transportation is far from being an exact science. Before using any of the presented techniques, it is imperative that individuals first try the techniques on a small number of fish to verify that the techniques will work under their specific circumstances.

An attempt has been made to provide information in a format that summarizes general tendencies. For instance, the range of hauling temperatures and densities is presented first and then followed by an indication of the more common temperatures and densities within each range. An exception is that we present only the ranges of fish sizes transported for each species. Not enough information is available for each species to allow separate analysis of transport techniques for different size classes within each species (Carmichael and Tomasso 1988). Generally, densities transported are lower for smaller fish, in warmer water, and for longer transport times. Densities reported here were determined by dividing the total weight of fish hauled by the reported volume of the tank. Other methods have been used to calculate densities. When comparing two transport methods, be sure to determine if densities are calculated in the same way.

Rainbow trout at 1,100 to 2.2 fish per kg[1] were transported in water at temperatures of 2–19°C. Temperatures of 4–7°C (22%), 7–9°C (29%), and 10–12°C (32%) were the most common. Hauling densities ranged from 0.02 to 0.30 kg/L; 23% of the densities were between 0.07 and 0.12 kg/L, and 49% of the densities were 0.13–0.24 kg/L. Transport time ranged from 0.25 to 16 h, and 71% of the trips were 5 h or less. An anesthetic (MS-222 at 8–15 mg/L) was used by 2 of the 72 respondents. Sodium chloride (1–10 g/L) was used 45% of the time. Of those specifying a salt concentration, 33% used 10 g/L. Antifoaming agents were used 43% of the time.

Channel catfish at fry to 0.9 fish per kg were transported in water at temperatures between 4 and 29°C. Temperatures of 16–21°C (41%) and 21–26°C (30%) were the most common. Hauling densities ranged from 0.01 to 0.48 kg/L; 44% of hauling densities were between 0.01 and 0.12 kg/L, and 42% were from 0.13 to 0.24 kg/L. Transport time ranged from 1 to 16 h; 54% of trips were 4 h or less, and 32% of the trips lasted from 4.1 to 8.0 h. Only 1 of 54 respondents used an anesthetic (MS-222, with no concentration specified). Sodium chloride (2–10 g/L most commonly reported) was added to hauling water 39% of the time, and antifoaming agents were used 26% of the time.

Largemouth bass at 2,200 to 4.8 fish per kg were transported in water at temperatures ranging from 10 to 29°C. Temperatures of 16–21°C (31%) and 21–26°C (49%) were the most common. Hauling densities ranged from 0.01 to 0.29 kg/L; 46% of the densities were 0.06 kg/L or less, and 36% ranged from 0.07 to 0.12 kg/L. Transport time ranged from 1 to 10 h, and 49% of the trips took 5 h or less. Only 2 of 41 respondents used an anesthetic (MS-222, with no concentration specified). Sodium chloride was used 69% of the time (1–5 g/L), and 5 g/L was the most common concentration (53% of those using salt). Antifoaming agents were used 28% of the time.

---

[1]Fish ranging in size such that 1,100 of the smallest fish would weigh 1 kg and 2.2 of the largest fish would weigh 1 kg.

Walleye at 3,520 to 8.8 fish per kg were transported in water at temperatures ranging from 10 to 26°C; temperatures of 21–26°C were the most common (68%). Hauling densities ranged from 0.01 to 0.10 kg/L, and 83% of the densities were 0.06 kg/L or less. Transport time ranged from 0.5 to 10 h, and 88% of the trips took 5 h or less. Anesthetic usage was not reported. Sodium chloride was used 68% of the time, at 1.0–2.5 g/L. Antifoaming agents were used 20% of the time.

Bluegill at 5,500 to 55 fish per kg were transported in water at temperatures of 13–29°C; temperatures of 13°C (19%) and 16–20°C (65%) were the most common. Hauling densities ranged from 0.01 to 0.18 kg/L; 46% of the reported densities were 0.06 kg/L or less, and 68% of densities ranged from 0.07 to 0.12 kg/L. Transport time ranged from 1 to 10 h, and 65% of all trips were 5 h or less. Anesthetics were not reportedly used. Sodium chloride was used 65% of the time at 3–5 g/L (most commonly reported), and antifoaming agents were used 27% of the time.

Brown trout at 88 to 1.1 fish per kg were transported in water at temperatures from 4 to 12°C; temperatures of 9–11°C were most common (84%). Hauling densities ranged from 0.05 to 0.29 kg/L, and 79% of the densities were between 0.12 and 0.22 kg/L. Transport time ranged from 1 to 10 h, and most of the trips lasted 5 h or less. No anesthetics were reportedly used, and salt, at unspecified concentrations, was used 16% of the time. Antifoaming agents were used 32% of the time.

Brook trout at 110 to 1.1 fish per kg were transported in water at 9–15°C; 9–11°C was most common (86%). Hauling densities ranged from 0.04 to 0.36 kg/L; 77% of hauling densities were between 0.12 and 0.19 kg/L. Transport time ranged from 1 to 12 h; 64% of the trips lasted 5 h or less. No anesthetics were reportedly used, and salt was used 7% of the time (no specified concentrations). Antifoaming agents were used by 43% of the respondents.

Grass carp at 4.8 to 0.88 fish per kg were transported in water at 14–27°C; temperatures of 16–20°C were reported 77% of the time. Hauling densities were 0.05–0.27 kg/L; 38% of hauling densities were from 0.05 to 0.12 kg/L, and 62% were 0.16–0.27 kg/L. Transport times ranged from 2 to 60 h; 62% of the trips lasted less than 5 h, and 23% lasted 30–60 h. Anesthetics were used by 31% of respondents, and 50% of those reported quinaldine use at unspecified concentrations (quinaldine is not approved for use on sport or food fish). All respondents reported using sodium chloride (2.5–5.0 g/L most commonly reported). Antifoaming agents were used 31% of the time.

Striped bass at fry to broodfish (6.8–13.6/kg) were transported in water at temperatures of 10–24°C. Transport temperatures of 21–24°C (55%) and 16–21°C (27%) were the most common. Hauling densities ranged from 0.02 to 0.29 kg/L, and 64% of the densities were 0.06 kg/L or less. Transport time ranged from 1 to 12 h; 67% of the trips lasted 4 h or less. One of twelve respondents reported using an anesthetic (quinaldine) at an unspecified concentration. Sodium chloride (2.5–5.0 g/L was most commonly reported) was used 82% of the time, and 18% of respondents used calcium chloride (no concentrations specified) and sodium chloride simultaneously. Antifoaming agents were used 55% of the time. Striped bass appear to be sensitive to sudden changes in light (as in opening a transport tank hatch on a sunny day). Working in shade may help limit "light shock."

Hybrid striped bass at fry to 0.66 fish/kg were transported in water at 18–27°C; temperatures of 18–21°C were the most common (50%). Hauling densities ranged from 0.04 to 0.23 kg/L, and 60% of hauling densities were 0.04–0.12 kg/L. The duration of all trips was reported to be 5 h or less. An anesthetic (25 mg/L MS- 222) was used in 1 of 11 cases reported. Sodium chloride (2–10 g/L) was used by 90% of respondents. The most commonly reported concentration of sodium chloride used was 2.0–2.5 g/L (60%). Antifoaming agents were used 30% of the time. Recent research suggests that hybrid striped bass best tolerate the stress of handling and transport when these activities are carried out in water containing 8 g/L sea salts. High (>100 mg/L) calcium concentrations also are helpful when handling fish in freshwater.

Shipping in Sealed Containers

Larvae and small fishes can be successfully shipped in sealed containers. Typically, a commercially available Styrofoam shipping box containing a heavy-duty, square-bottom plastic bag is used. The bag is filled with 5–10 cm of water, fish or larvae are placed in the water, and pure oxygen is introduced into the bag until the air is purged and the bag is inflated. The top of the bag is twisted, folded over, and sealed with a fish bag banding tool or extra heavy-duty rubber bands. The lid is placed on the Styrofoam container and the entire container is sealed with duct tape and placed in a cardboard box for shipping. The box should be clearly marked to indicate which side is up. Salts, anesthetics, buffers, cooling packs, and bacteriostats may be added to the water depending on the needs of the species being transported. The density of fish in the boxes depends on species and transport time. Generally, larvae and fish can be shipped successfully in sealed containers if the total time in the sealed container is less than 2 days. The boxes are usually accepted by air freight and overnight delivery services. Before shipping this way, it is important to check transport and delivery schedules with the shipping company and verify that someone will be available to receive the shipment.

Some reports have been offered that opening the sealed containers in bright light conditions has caused losses of very young fishes.

Commercially available shipping boxes can be used to transport fish eggs. The boxes are constructed to keep fish eggs cool and moist, but not submerged in water. This is accomplished by placing eggs on trays in the box and then placing a tray of ice above them. As the ice melts, the cold water drips on the eggs. It is critical that the boxes be marked to keep the ice side up. Successful reports using these containers vary widely. Planning of shipping and receiving is very important for high success.

Wild, Threatened, and Endangered Fish and Other Special Circumstances

More and more, traditional fish hatcheries are working with wild brood fishes or threatened and endangered species that may require specialized transport. As fisheries managers and culturists guide conservation efforts on rare or valuable fish, methods must be developed to care for the specialized needs of the various fishes. Gametes, both sperm and unfertilized eggs, are regularly transported now. Embryos are increasingly being transported. We offer, in this section, examples of how some various transport needs were successfully met. Often the cost of such specialized transport is much higher than traditional methods. Sometimes, methodology seems to drop back a century such as with transport of Gila trout from remote wilderness areas via pack mules to fish hatcheries (for parts of the trip), and sometimes very modern technologies are used, such as using helicopter transport of fish from the Grand Canyon to a fish hatchery. All specialized methods are governed by the same rules that apply to standard fish transport methods, although transport densities are often very different. Many agencies are now developing protocols for any activities that involve sensitive, rare, threatened, or endangered species. Often these protocols include recommendations for handling and transport (e.g., Shepherd and Berezay 1987; J. H. Williamson, U.S. Fish and Wildlife Service, unpublished; G. J. Carmichael, unpublished data).

Gametes, both sperm and eggs, may need to be collected, handled, stored, and transported from remote field sites to hatcheries when movement of broodfish is not feasible. Sperm transport, collection, handling, refrigerated storage, and even cryopreservation have been addressed for many species (e.g., Guest et al. 1976; Scott and Baynes 1980; Stoss 1983; Kurokura et al. 1984; Cloud et al. 1990; Tiersch et al. 1994; Tiersch et al. 1997). Sperm is generally transported in small vials or plastic bags containing sperm and a sperm extender or buffer (such as Hanks' balanced salt solution) using an ice bath to keep the specimens cool; oxygen should be injected every few hours. Success has been achieved in shipping buffered sperm for common species and also for rare species such as bonytail and Colorado pikeminnow. Short-term egg transport has been accomplished with varying results. Unfertilized eggs of razorback sucker have been successfully moved from the Colorado River to the Willow Beach National Fish Hatchery and stored for short periods before fertilization (total time, 8 h). The eggs were stripped from females into quart-sized plastic food storage bags containing a buffer solution of calcium-free Hanks' balanced salt solution, then chilled in an ice bath inside an insulated cooler and transferred by boat to Willow Beach. However, these trials demonstrated that only limited time may be available for transport and storage of unfertilized eggs in order to obtain viable embryos later for species such as razorback sucker (Tiersch et al. 1997). Eggs of other species have been more successfully transported, but success seems dependant on individual effort with little proven methodology available.

Fertilized Rio Grande cutthroat trout eggs and eggs of other species have been successfully transported from remote spawning sites to fish hatcheries. Newly fertilized Rio Grande cutthroat trout eggs, strip-spawned and fertilized at streamside from wild broodfish and water hardened for 1 h, were transported for 3 h in jars completely filled with distilled water and eggs, and sealed to prevent sloshing. The sealed jars were placed in insulated coolers containing soft foam packing and ice and then transported by pickup trucks to fish hatcheries. Upon arrival, eggs were tempered to receiving unit temperature, disinfected with 100 mg/L providone iodine, and transferred in dim light conditions. Greater than 95% hatching was obtained from eggs that were not exposed to extreme temperatures or low DO concentrations.

Razorback sucker embryos and fertilized eggs from wild and captive broodfish have been successfully transported using sealable plastic quart-size freezer bags. In experiments using razorback suckers, between 200 and 300 eggs were fertilized in plastic sealable bags. The bags were transported in ice chests or buckets from the river to a hatchery, up to 3 h en route. The bags were also transported between hatcheries in an attaché case using commercial airlines (Carmichael et al. 1996). Densities can be modified to transport or hold more or fewer eggs or embryos dependant upon time (Carmichael et al. 1996).

Bonytail broodfish and humpback chub have been transported successfully at low densities (20 mg/L) in 42–120 L insulated coolers. Medium-sized bonytail and humpback chubs (<300 g) have been transported in coolers by boat and truck and humpback chub in helicopters and golf carts. Temperatures were maintained by placing cold packs in the coolers. The cold packs were sealed in plastic food storage bags to prevent leakage. Shipping water contained 5.0 g/L NaCl, 15–25 mg/L MS-222, 0.26 mL/L Stress

Coat (Aquarium Pharmaceuticals, Chalfont, PA), and 10 mg/L Furacin (6.0–8.0 mg/L Chloramine-T has also been used). Oxygen levels were kept at saturation. In transporting bonytail, battery-operated aerators using airline tubing and small air diffusers were used. A spare aerator and batteries was carried. In transporting humpback chub, small oxygen cylinders, airline tubing, and diffusers were used for aeration. To avoid pinching it, the airline tubing was inserted through a hole drilled in the lid of the cooler. Small oxygen cylinders, tubing, and diffusers were carried for emergency backup. The water temperature always remained similar to, or slightly lower than that used for rearing the fish. For short transit time (<3 h), the hauling density did not exceed 50 grams of fish per liter of water; densities should not exceed 20–30 g/L for long transit times (4–10 h).

Gila trout broodfish and fish intended for translocation have been shipped via helicopter. Shipping in a suspended bucket was recommended for transport from remote streams to transport trucks for low densities (50 g/L) of fish. Temperature was maintained using dechlorinated ice in the bucket. The water contained 5 g/L NaCl, 0.5 mg/L Polyaqua or 1.0 mL/L Stress Coat, 6 mg/L MS-222, and 5 mg/L Chloramine-T. The oxygen level was maintained at saturation. Small oxygen cylinders, tubing, and diffusers were carried for emergency backup. The water temperature was similar, or slightly lower, to that of the stream. In one instance, Gila trout were "rescued" from a forest fire, very stressed by high water temperatures. The fish were successfully hauled, but required posttransport salt treatments for several weeks to reduce losses following the move to a hatchery. Standard practice is then to keep fish in 5 g/L salt for about one week after transport of the wild fish. Both standard flow–through and static bath treatments using salt have been accomplished. Dissolved oxygen was kept at saturation during all transport events.

Gila trout have also been shipped using pack animals. Fish were transported in insulated hard panniers lined with plastic bags (see bag shipping equipment, above) containing 15–25 L water; densities of 50 g/L were recommended for transport. Temperature was maintained by placing ice in the pannier. The water contained 5 g/L NaCl, 0.5 mg/L Polyaqua or 1.0 mL/L Stress Coat, and 6 mg/L MS-222. The oxygen level was at saturation. Small oxygen cylinders, tubing, and diffusers were carried for emergency backup. The water temperature was similar to, or slightly lower than, that of the stream.

Many small nonsport and nonfood fish have historically been transported by pail or bucket using ice to cool the water. Generally, very low densities are hauled using traditional methods and success varies widely, dependant on the species transported. The method developed to haul largemouth bass, presented in this chapter, was adapted to use during a "rescue" of fish threatened by dewatering of a stream. Rio Grande silvery minnow, an endangered cyprinid, is a small fish (<9 mm, standard length) that was transported from one site on the Rio Grande River to another. A fish hauling truck with a 1,890-L tank was used to store and haul fish during and after collection efforts. Water was treated with 5 g/L NaCl, 0.26 mL/L Stress Coat, and 10 mg/L Furacin. Several fish were exposed to the mixture of salt, antibacterial compound, and Stress Coat for 1 h before exposing the larger population. Fish were seined over a 5 h period and were placed in the transport tank to hold before transport. Oxygen was kept at saturation and temperature was kept similar to that of the stream using dechlorinated ice. A total time of 9 h in the truck during storage and transport preceded successful moving of the fish to an area of the river not threatened with dewatering.

Summary

The chances of successfully transporting fish from one point to another can be greatly enhanced by following the general guidelines listed in the introduction. We recommend that the information presented here not be "cookbooked" or blindly followed, but be used and improvised depending on particular situations, as a basis to improve techniques. Each situation will have its own set of requirements for safely collecting, harvesting, and transporting fish. An experienced fish culturist will always seek to determine what is best for the situation at hand. Eventually, with a combination of modern science, common sense, and experience that reduces the severity of stressors, minimizes the duration of stressors, reduces the number of stressors, minimizes plasma ion disturbances, and minimizes increases in metabolic rate, fish transportation may become a low-stress experience for both the fish and culturist (Figure 6).

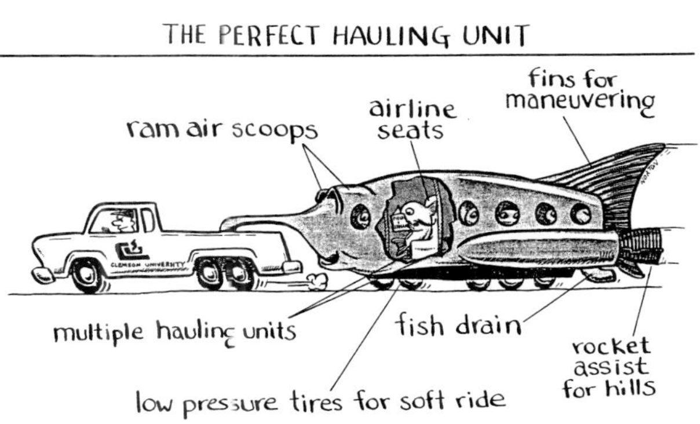

FIGURE 6. The perfect fish transport tank.

References

Carmichael, G. J., J. R. Tomasso, B. A. Simco, and K. B. Davis. 1984. Characterization and alleviation of stress associated with hauling largemouth bass. Transactions of the American Fisheries Society 113:778–785.

Carmichael, G. J., and J. R. Tomasso. 1988. Survey of fish transportation equipment and techniques. Progressive Fish-Culturist 50:155–159.

Carmichael, G. J., J. H. Williamson, O. T. Gorman, and T. R. Tiersch. 1996. Field propagation techniques for the endangered razorback sucker. North American Journal of Fisheries Management 16:963–966.

Cloud, J. G., W. H. Miller, and M. J. Levanduski. 1990. Cryopreservation of sperm as a means to store salmonid germ plasm and to transfer genes from wild fish to hatchery populations. Progressive Fish-Culturist 52:51–53.

Fries, J. N. 1992. Samcalc: a computer program for fish culturists. U.S. Fish and Wildlife Service Publication, San Marcos Information Leaflet 92–03, San Marcos, Texas.

Guest, W. C., J. W. Avault, and J. D. Roussel. 1976. Preservation of channel catfish sperm. Transactions of the American Fisheries Society 105:469–474.

Kurokura, H., R. Hirano, M. Tomita, and M. Iwahashi. 1984. Cryopreservation of carp sperm. Aquaculture 37:267–273.

Pickering, A. D. 1993. Endocrine-induced pathology in stressed salmonid fish. Fisheries Research 17:35–50.

Scott, A. P., and S. M. Baynes. 1980. A review of the biology, handling and storage of salmonid spermatozoa. Journal of Fish Biology 17:707–739.

Shepherd, B. G., and G. F. Berezay. 1987. Fish transport techniques in common use at salmonid enhancement facilities in British Columbia. Canadian Manuscript Report of Fisheries and Aquatic Sciences 1946.

Stoss, J. 1983. Fish gamete preservation and spermatozoan physiology. Pages 305–350 *in* W. S. Hoar, D. J. Randall, and E. M. Donaldson, editors. Fish physiology, volume 9, part B. Academic Press, San Diego, California.

Tiersch, T. R., C. A. Goudie, and G. J. Carmichael. 1994. Cryopreservation of channel catfish sperm: storage in cryoprotectants, fertilization trials, and growth of channel catfish produced with cryopreserved sperm. Transactions of the American Fisheries Society 123:580–586.

Tiersch, T. R., and five coauthors. 1997. Field collection, handling, and storage of sperm of the endangered razorback sucker. North American Journal of Fisheries Management 17:167–173.

Weirich, C. R., J. R. Tomasso, and T. I. J. Smith. 1992. Confinement and transport-induced stress in white bass x striped hybrids: effect of calcium and salinity. Journal of the World Aquaculture Society 23:49–57.

Wood, C. M., J. D. Turner, and M. S. Graham. 1983. Why do fish die after severe exercise? Journal of Fish Biology 22:189–201.

11
Stocking

KYLE J. HARTMAN AND BRET PRESTON

Introduction

Fish stocking is viewed by fisheries managers as one of the many tools to use to compensate for factors that limit natural production of fish and to achieve management objectives in rivers and streams (Radonski and Loftus 1995). Hatchery fishes are used to introduce species into new waters; to supply year-classes when natural reproduction is poor or has failed; to develop or maintain recreational, commercial, or tribal fisheries; or as a protein source. Yet, with the exception of an American Fisheries Society symposium on the subject (Schramm and Piper 1995), stocking is one of the least studied aspects of hatchery and fisheries management, even though it is one of the most critical aspects. Factors such as the genetics of the species being cultured relative to wild stocks have gained increasing exposure in recent years. Other aspects of stocking, such as the need for quantifiable goals and objectives for the stocking program and other biological and ecological considerations, are only recently gaining acceptance as important considerations in stocking programs.

Genetics and Stocks Concepts

Probably the most important considerations in stocking natural systems are the genetics of stocked versus wild fish and concerns over possible inbreeding, outbreeding, and domestic selection (see Chapter 12). In many ways, whether it is realized or not, the hatchery program coordinator is the lifeguard in the gene pool of fisheries. Fisheries managers must consider if the genetics of stocked fish will influence the local gene pool. Further, consideration must be given to whether, by selecting only a subset of fish for hatchery rearing, all the genetic variability in the wild population has been accounted for, or whether enhanced survival of a fraction of the genetic variability will lead to genetic drift. As used here, genetic drift occurs when the gene frequencies in a population are drifted from the "stock" by the swamping of the stock through increased survival of potentially nonrepresentative hatchery-reared fish. A simple, yet critical, question a fisheries manager must ask is: Do the genetics of the hatchery broodstock reflect those of the release site? If the genetics of the broodstock differ from wild fish, they may potentially depress locally adapted genotypes. Overall, genetic risks can be classified into four broad types: (1) extinction, (2) loss of within-population variability, (3) loss of among-population variability, and (4) domestication (Busack and Currens 1995). With respect to stocking, it is usually the loss of within- and among-population variability and domestication that is a potential conservation issue.

Simply procuring broodstock from the same system being stocked is no guarantee of minimizing among- and within-population variability risks. Without careful genetic study of a large segment of the wild population, we cannot always be sure. A population may consist of multiple stocks adapted to different areas within a system and the broodstock in question may represent only a portion of those stocks. Likewise, the stock within the population that requires restoration may not be of the same genotype as those used as broodstock. For example, in the Great Lakes, lake trout populations may consist of several stocks, among which are nearshore and deepwater stocks. If the offshore stock is depressed and stocking is prescribed to augment the population, stocking of the wrong genotype (in this instance the nearshore genotype) could easily result in a less successful, or even failed, recovery effort

It is not always possible to distinguish biological effects from biologically independent effects of stocking transfers. Many of the genetic problems attributed to hatcheries and hatchery fish may be solvable by hatchery practices that establish guidelines for conserving genetic resources. Past use and abuse of hatcheries, without care for conservation of genetic resources, may have been harmful to local populations (Campton 1995). Of course, in cases where stocking is used simply to increase angler opportunities without reproduction of stocked fish (e.g., put-and-take or put-grow-and-take), these genetic considerations are less important.

Contemporary Stocking Programs

Stocking for Recreational, Commercial, and Traditional Use Fisheries

Put-and-take. Put-and-take stocking programs are used solely for increasing angler opportunities. Typically, put-and-take stocking is used in urban lakes, ponds, impoundments, or streams where conditions do not permit establishment of growing or reproducing populations of fish. These systems are usually ones in which, at some time of the year, environmental conditions prevent continued survival of fish. For example, put-and-take fisheries may be established by stocking into streams that get too warm for trout in summer or small impoundments with either too much angling pressure or too low productivity to promote growth of fish. Put-and-take stocking is usually used for recreational fisheries and is seldom used for commercial or traditional fisheries due to the high cost to produce fish and the fact that stocking must be continued to maintain the fishery.

Put-and-take stocking is generally considered a relatively expensive means of increasing angler opportunities. This type of stocking is expensive because fish must be stocked at a "harvestable" size since growth after stocking is limited or not possible. Production of fish of harvestable size is necessarily more expensive due to feed and production costs. Although the cost per fish may be higher in stocking harvestable-sized fish, substantial angler benefits may be recognized through the creation of limited fishing opportunities, such as winter trout stockings in small warmwater impoundments. Angler satisfaction may be higher than in other stocking approaches because a majority of the stocked fish may be returned to the creel in a relatively short time.

Put-grow-and-take. Stocking of fish in a put-grow-and-take (PGT) situation is another means of increasing angler opportunities. Put-grow-and-take stocking is used where reproduction of fish is low or nonexistent, but where conditions are favorable for growth. Production costs are generally lower for fish used in PGT fisheries due to the smaller size at stocking. In PGT stocking, fish may be stocked from larval through juvenile stages, with the intention that additional growth to harvestable size will occur after stocking.

Put-grow-and-take stocking has been used in many waters within the eastern United States where various trout, catfish, walleye, and muskellunge species have been stocked into waters where reproduction is either not possible or severely limited due to poor habitat or environmental conditions. Similarly, stocking of striped bass and hybrid striped bass in freshwater impoundments and tailwaters has resulted in quality fisheries, but seldom is reproduction of these species successful, necessitating continued PGT stocking to maintain the fisheries (Figure 1).

Put-grow-and-take fisheries can support valuable sportfishing communities. New York began stocking salmonids into Lake Ontario in 1968 to control alewife abundance and to provide sportfishing opportunities. From 1968 to 1993, 85 million trout and salmon were stocked into Lake Ontario at an annual cost of $778,000 (1993). Anglers took an annual average of 885,000

FIGURE 1. In many urban regions, put-and-take or put-grow-and-take fisheries provide increased angler opportunities where traditional fisheries cannot be supported.

trips, valued at $110 million dollars, in 1988 (Lange et al. 1995). This represents an impressive cost:benefit ratio for this PGT fishery.

Although generally considered a less expensive alternative to put-and-take fishing (due to the smaller size of fish stocked into PGT and the production that occurs poststocking), other studies have pointed out that PGT stocking may be more cost-effective when stocking larger fish. Most studies that examined the survival and return to the creel of stocked fish have found that both of these are proportional to fish size at stocking. Larger fish have higher harvest rates and better survival. Costs to produce fish do go up with fish size, as does the risk of a catastrophic event at the hatchery, but these costs and risk vary from species to species. Thus, it may be worthwhile to grow certain fish to a larger size for PGT stocking. Given the high cost of producing fish, stocking programs should probably conduct more studies on the cost-effectiveness of producing different size fish for PGT stocking.

Part of these cost-effectiveness studies must include survival and return to creel of different species, strains, and sizes of fish. Each of these factors, as well as time of year in which stocking takes place, may influence the economics of stocking. In addition, the ability of the system to support this species or how much of the species in question can be supported, as well as possible interactions with existing species should be considered. Survival is a key component of the economics equation. A study of PGT stocking for channel catfish in Washington State found survival of 50–200 mm total length (TL) fish to average only 9% 13–15 months poststocking (Bonar et al. 1997). However, this is lower than reported in most studies. Survival rates vary widely based upon species, sizes, season, and many other biological and environmental factors.

Stock augmentation. Stock augmentation refers to "augmenting" or increasing an existing stock of fish. This is one of the most common reasons for stocking and is often used as a management action to help recover a population that may be at low levels or to support a population that is suffering poor recruitment (Figure 2). Augmentation may be most successful where low adult numbers, perhaps due to overfishing, have limited recruitment. In this case, increasing the adult biomass through stocking may increase the number of spawners and spawning success in future years and provide the impetus needed to rebuild the population, provided the mortality factor limiting adult survival can be alleviated or regulated. It is hoped that once adult numbers and recruitment increase through augmentation stocking, stocking may be reduced or discontinued. In situations where stocks are augmented to support a fishery where recruitment is poor, augmentation becomes a long-term management action unless the causative factor in poor recruitment can be addressed and corrected.

Genetics are of great importance in stocking initiated to augment stocks. If the purpose of stocking is to increase a reproducing stock, then care must be taken to not swamp out the locally adapted genotype through the increased survival of progeny from a few pairs of fish produced in the hatchery. In systems with low adult numbers, it is very easy to concede

FIGURE 2. In some systems, fry are stocked to help augment the natural population, which suffers from poor or highly variable recruitment. Such augmentation may help reduce year-to-year variability in catches for anglers.

that the high survival of fish from a few parents from the hatchery imparts a serious genetic bottleneck when those fish make up a high percentage of the surviving year-class in a given system.

It is especially important in stock augmentation to have a clearly defined, quantifiable objective to determine when stocking should be reduced or eliminated. Without a clear objective (e.g., to recover a naturally producing fish in Lake "Y") and clearly defined checkpoints (e.g., double adult populations, at which time stocking will be discontinued because adult stocks should be sufficient to produce viable year-classes), the augmentation stocking may be doomed to become a put-and-take or PGT fishery as users become accustomed to the increased fishing success that may accompany stocking.

Augmentation stocking may also take place when juvenile and adult survival and growth are good, but where survival of eggs and fry is highly variable. In these cases, managers may choose to stock larvae or very small juveniles each year to help minimize year-to-year variations in year-class size in support of a more stable adult fishery. In stocking these early life stages, much of the genetic drift may be alleviated because a larger number of individual adults may be used to produce the stocked offspring because fish are released well before space and water volume becomes a limiting factor at the hatchery. Stocking of larval red drum was used in Texas to help curb declines in population numbers. Results of the larval stocking were encouraging, but massive stockings were needed to affect local variations in wild populations (McEachron et al. 1995).

Stock augmentation has also taken place in systems where different strains of fish have been introduced either to provide greater genetic diversity or to extend angler opportunities to seasons not previously possible. For example, in eastern Lake Erie, the New York Department of Environmental Conservation (NYDEC) historically stocked fall-run Washington strain steelhead into tributaries of Lake Erie. This stocking provided a fall stream fishery in many stocked and even unstocked creeks as fish strayed into nearby systems. Recent studies suggest that approximately 78% of adult fish are of hatchery origin, and 22% have spawned naturally (Goehle 1999). In the 1980s, NYDEC began stocking the same systems with a spring-run strain of steelhead that is now providing stream anglers with additional angling opportunities during spring.

Special regulations. In some instances, fish may be stocked for areas with special regulations. Usually the goal is to provide a fishing opportunity to a segment of the population that either does not have easy access or cannot easily access the resource. Many states stock sections of streams or urban ponds with fish in support of "Hooked on Fishing Not on Drugs" outreach activities, or "kids only" fishing areas, or areas where physically challenged anglers can be provided convenient and safe access to the water. Usually these types of stockings have regulations that permit only the target group to fish the area and usually these stockings are of a put-and-take or PGT nature. Special regulations may also be used to reduce stocking. By segregating harvest and nonharvest anglers, overall stocking can be reduced and used primarily to support consumptive-use anglers (Van Vooren 1995).

Fish-out ponds. Stocking of fish-out ponds is usually done to provide income to the pond owners. Such ponds are referred to as "pay lakes." Fish, oftentimes catfish, are stocked at a harvestable-size and, usually, at high densities. Anglers pay for either the opportunity to fish or per each fish caught. Usually anglers must keep and pay for all fish that they catch (Figure 3). Such fish-out ponds are obviously not a natural system and residence time and fish density usually prevent significant natural reproduction. Thus, fish must be supplemented with food; the system must be continually stocked to maintain the size structure and angling opportunities desired by users.

Of special concern with fish-out ponds is health. Fish are usually stocked in extremely high densities, so stress (from hauling, handling, and density) may be high and stocks may be more susceptible to diseases and parasites brought in with the fish or existing in the receiving waters. For more information on fish health see Chapter 9.

FIGURE 3. In fish-out ponds ("pay lakes") anglers pay a fee for the opportunity to fish or may pay a fee per fish caught.

Charter boat services. Although in some cases the majority of a recreational fishery is composed of charter boat services, few fisheries agencies would specify it as such. Most would refer to it as stocking for "recreational fisheries." Very few examples exist where stocking is conducted primarily to benefit charter fisheries. The salmonid fisheries that have sprung up on the Great Lakes, particularly Lake Michigan, once salmon were stocked as part of an alewife management tool, is one example. Most of the fishery is maintained by stocking, with limited natural reproduction such that the stocking is something between an augmentation and a PGT fishery.

In the Patuxent River, Maryland, mitigation stocking of striped bass juveniles by the Potomac Electric Power Company (PEPCO) served as a stocking to support charter boat services for at least a few years following the 1990 reopening of the striped bass fishery in Chesapeake Bay. For several years after reopening the fishery, the charter fleet, located in Solomons, Maryland, experienced catches of striped bass greater than 18 in, which were composed of over 50% hatchery fish (J. Van Tassel, Maryland Department of Natural Resources, personal communication). Although this stocking was not intended to support charter boat services, but rather to help recover a stock, the PEPCO-reared fish supported this fishery for several years.

Limited stockings have taken place to benefit commercial fisheries. Mathias et al. (1992) evaluated stocking of walleye fry to enhance a commercial fishery for walleye in a large Canadian prairie lake. They found that of 7.1 million fry marked and released only 0.04% survived to age-3 (harvestable size). These fish represented a harvest of 80 kg with a cost:benefit ratio of 2.6:1.0. Given a natural production of fry in the lake of 230–249 million fish, this supplemental stocking was deemed not cost effective. However, in other situations stocking may be effectively used to benefit commercial fisheries. Such stocking should be evaluated on a case-by-case basis with careful socioeconomic analysis to ensure acceptable cost:benefit ratios.

Population Enhancement and Introductions

Restoration of depleted stocks. In many systems, the species of highest economic or recreational value have been fished down to depleted levels and require strong management actions to restore. Oftentimes, the management action chosen (often one of a suite of actions chosen) is stocking to enhance depleted stocks. This stocking may be coupled with regulations or restrictions in the harvest of these species. In the case of striped bass in Chesapeake Bay during the 1980s, all commercial and recreational fisheries were closed and stocking of striped bass juveniles was initiated in an attempt to rebuild depleted adult stocks. By 1990, the trigger (juvenile survey catch per unit effort)was achieved that permitted a limited fishery. Although the fishery closure was a large part of the recovery effort, stocking in the tributaries of the bay that historically had the most variable recruitment helped the recovery in these areas. Currently, the striped bass population in Chesapeake Bay is arguably the largest in the last 50 years.

As with most other stocking methods, stocking for restoration requires clear, definable goals and objectives with specific target criteria for termination of stocking (Figure 4). The termination criteria should include both a recovery termination criteria (stocks have now reached a particular level,

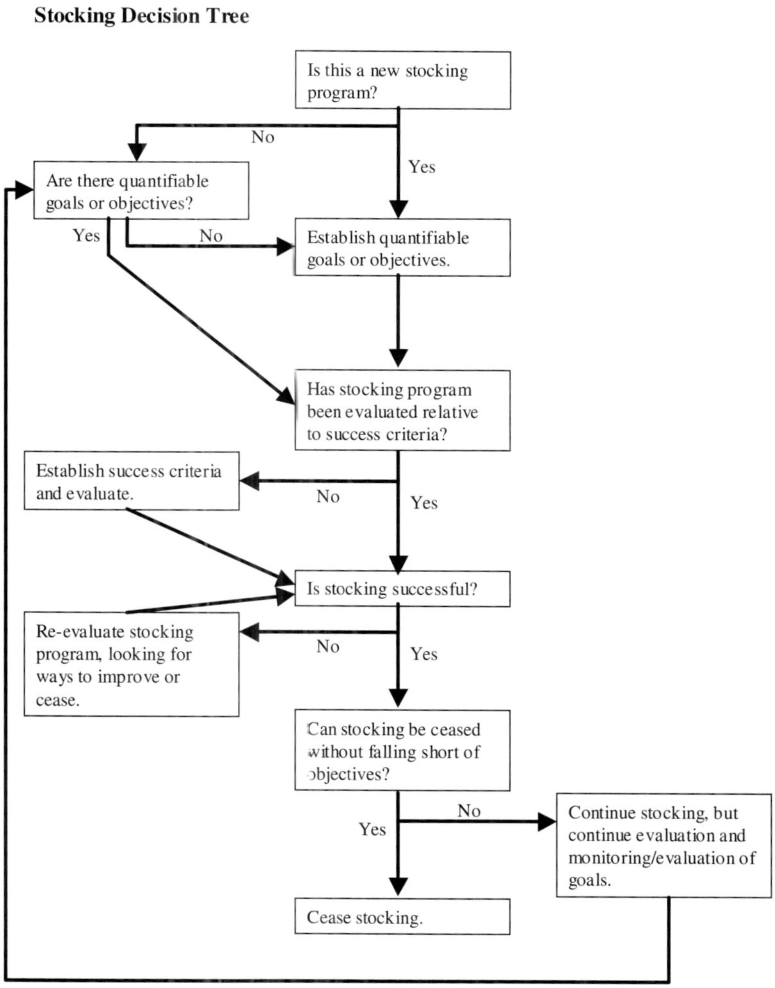

FIGURE 4. A stocking program decision tree for establishing, maintaining, and evaluating a stocking program.

therefore, we can stop stocking), as well as failure criteria (we have stocked for 5 years with no appreciable response in the stock's abundance, therefore, we will cease stocking and use those resources elsewhere). Failure to set such standards for termination criteria not only makes justification on federal aid reports difficult, but also tends to push the restoration stocking effort toward a long-term stocking project without measurable success. Resources spent on stocking without measurable success may be better spent on determining what other factors might be limiting the species abundance—spawning, rearing, overwintering habitat, overharvest, bycatch, disease, or pollution.

Recovery of threatened and endangered species. Stocking can be used in recovery efforts for rare, or threatened and endangered species. The U.S. Fish and Wildlife Service maintains 28 hatcheries involved with recovery of threatened and endangered species (Radonski and Loftus 1995). The National Marine Fisheries Service and the Idaho Department of Fish and Game have established captive brood stocks to aid in the recovery of Snake River sockeyes. Several augmentation plans also exist for the endangered razorback sucker in the Colorado River system. Genetics are extremely important in these small populations, but if they are given consideration (e.g., Snake River sockeye salmon [Flagg et al. 1995]), or if the population is deemed to be imminently threatened with extinction, then stocking may be the best alternative. Stocking of threatened and endangered species has been conducted across much of North America. In the West, numerous species of salmonids and other endemic species have been stocked in an attempt to maintain adequate adult numbers (Figure 5). In the East,

FIGURE 5. Stocking of threatened and endangered species. Rearing and stocking young of threatened or endangered species of fish represents a safer alternative to trap and transfer approaches. It is promising to take gamete material from sensitive species and use the progeny to stock species such as this Arkansas River shiner into suitable, but unoccupied habitats.

stocking of Atlantic salmon, paddlefish, sturgeon, and various alosid species has been done in an attempt to increase the number of spawning adults to viable levels.

Where stocking of threatened and endangered fish may be most appropriate is in the reestablishment of a species in areas where it is has been extirpated or in new areas that offer suitable habitat. By removing reproductive material from a few adults in one system, another suitable system can be seeded with the threatened and endangered species. Although genetics are critical here, this approach may be preferred over trapping and transferring adults and juveniles from the natal system, as it would have a less deleterious effect on the existing populations of threatened and endangered species.

Gamete banking. Although not yet widely done, egg banking may offer an important contribution to stocking. Recent advances in cryopreservation and in genetic engineering may make egg and sperm banking a common practice for storing gametic and genetic material from particular stocks or threatened and endangered species. Cryopreservation of fish spermatozoa has been conducted with some successful hatching of fertilized products. Bart et al. (1998) froze blue catfish spermatozoa for 12 months with the frozen product averaging 32% of the successful fertilization relative to fresh spermatozoa with fresh catfish eggs. Even higher relative fertilization rates were possible under higher frozen spermatozoa densities.

More common forms of gamete banking include capture of returning adults of endangered populations and hatchery rearing of all gametes. Such gamete banking has occurred for seriously imperiled populations of sockeye salmon (Flagg et al. 1995) and chinook salmon (Bugert et al. 1995) in the Snake River and is planned for the razorback sucker in the upper Colorado River system (Modde et al. 1995).

Population Assessment and Gear Efficiency Studies

Marking and tagging. In addition to their traditional use in fisheries, stocked fish commonly have been used to evaluate populations of wild fish and to determine fishing gear efficiency. Provided that some basic assumptions are met,[1] an estimate of population size for wild fish or fish already present in the system from prior stocking can be made using recapture data and the ratio of marked and unmarked fish. The Peterson method or a modification of the Peterson method is usually used in estimating population

---

[1]Assumptions of the Peterson method include: marked fish mix and disperse with the unmarked fish; the mark or tag is not lost or the loss is quantified; mortality rates are similar between marked and unmarked fish; and every fish has the same capture probability.

size. Although much more complex models and assumptions can be made, the basic Peterson equation is

$$N = S/r \cdot M,$$

where N is the number of fish present before stocking of the marked fish (M), r is the number of marked fish recaptured in subsequent sampling, and S is the total number of fish collected (marked and unmarked) in the samples. An estimate of the number of fish in the system is obtained by solving for N in the equation above.

Stocked fish also have been used to calibrate or to measure the efficiency of fishing gears. By stocking an area with a known number of marked fish and then sampling that area with fishery gear such as electroshocking, trap nets, cove rotenone samples, gill nets, bottom trawls, or weirs, the resulting catch can be used as a measure of the percentage of the population (of that species, size, and under those environmental or water quality conditions) that is effectively sampled by that gear.

For example, 50 fingerling rainbow trout are stocked into a 100 m section of stream that is block netted at either end to prevent fish movement into and out of the area. Next, the area is electrofished (using standard techniques, Reynolds 1996) in a three-pass removal technique (Hayne 1949) to estimate abundance of fish in the stream section. Given that we stocked 50 marked fish into the stream section, the ratio of the total number of marked recaptures to the number stocked is the gear efficiency. Knowledge of this gear efficiency can be used to correct other population estimates in which no marked fish were used.

Population or gear efficiency studies requires a distinct mark or tag on the stocked fish to help identify the stocked fish from others that might already be present in the system. Fish are often marked by clipping a pectoral or adipose fin, branding, internal anchor tags, coded wire tags, pit tags, subcutaneous latex dyes, chemical marks or temperature checks on calcified structures, or subcutaneous numbered tags (see Guy et al. 1996 or Parker et al. 1990 for more details about marking and tagging methods and uses). Whatever the mark or tag used, it must be consistently identifiable and persistent throughout the duration of the study period. If the mark or tag can be shed or lost, then studies of the rate of this loss must be conducted to account for this loss.

A special case of tagging or marking of stocked fish occurs with some fishing tournaments. Usually fishers pay a fee to enter the tournament and then win prizes for turning in tags. Higher prizes or cash amounts may be awarded if the fish is caught with a particular gear or lure of that prize's sponsor. Thus, there is a commercial aspect of this type of stocking.

Monitoring population dynamics. Fish have also been stocked into systems in efforts to estimate key population measurements of the wild stock.

Measurements such as mortality rate or exploitation rate can be estimated from declines in standardized abundance or catch per unit effort (CPUE) of fish given the time at large since stocking. In these cases, the slope of the regression line, with time as the independent variable and CPUE as the dependent variable, provides a measure of the total mortality rate (see Figure 6). Estimates of the number of fish creeled by anglers or the commercial fishery from other surveys can provide an estimate of the proportion of the stocked fish that were harvested. The natural mortality rate can be calculated by subtracting the fishing mortality rate from the total mortality rate (for further details on mortality and exploitation rate calculations, see Ricker 1975).

Genetic monitoring. Marking and tagging can also be used to evaluate the relative success of fish of different strains, or of fish originating from different hatcheries, or to evaluate success criteria. By identifying each strain or genetic stock, or each hatchery, through unique tags or marks, the relative contribution of each strain or hatchery can be evaluated through subsequent sampling of the populations or through creel returns. If stocks from a particular genetic strain are more suited to the conditions facing it in the stocked system, it should have better survival or growth. A

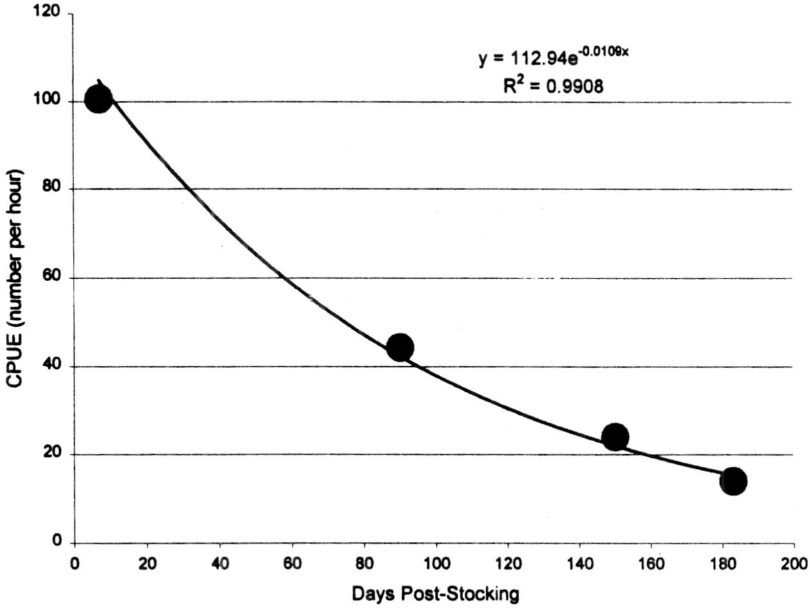

FIGURE 6. Relationship between CPUE of marked, stocked fish and total daily mortality rate. This mortality rate may be used to study populations of the wild fish in question. In this example, the daily mortality rate is –0.0109 or about 1.09% per day.

more successful strain may also represent a greater proportion of the population in terms of CPUE in sample gears or may contribute a greater proportion in angler creels. Similarly, if some aspect of rearing or stocking from one hatchery is more stressful than from another hatchery, tracking of the returns in samples or creels may point out where changes need to be made or where evaluations of stocking processes need to be made.

Stocking of fish with rare alleles can also be used to monitor success of the stocking program. Red drum with uncommon alleles were stocked annually in Texas bays and when the uncommon allele frequency rose above baseline, the stocking program was considered successful (Ward et al. 1995).

Genetics policies. Concerns regarding the impacts of stocked fish have led fisheries managers to consider more closely the implications of stocking. Increased review for certain stocking strategies is often required for compliance with state and federal laws, especially when stocking is partially funded through the Sportfish Restoration Program of the U.S. Fish and Wildlife Service.

The primary purpose of most stocking programs is to increase recreational angling opportunities. Fisheries agency policies and guidance are directed at providing these opportunities through the stocking programs previously described. Recently, greater attention is being paid to ensuring the protection of native stocks or minimizing impacts upon existing successful fisheries. Identifying discreet stocks or subpopulations has become important to fisheries managers, most notably for western U.S. salmonids and northeastern U.S. and European Atlantic salmon. Agencies may determine that certain populations are imperiled as a result of habitat loss, environmental degradation, or overfishing; or that certain stocks are valuable, either for recreational or commercial purposes.

The difficulty faced by many state agencies is the effort and cost associated with identifying specific strains or discrete subpopulations. Extensive field surveys and carefully designed protocols for fish collection are required. Following data collection and analysis, agencies may develop guidance for fisheries management that reduces potential genetic impacts. They may elect to avoid stocking cultured fish of the same species or their hybrids within the range of the fish imperiled or to be protected in order to minimize loss of genetic material. Many agencies have developed management policies based upon watersheds using documented fish distribution records. The key component in determining stocking strategies to avoid or minimize genetic impacts is to decide the level of risk stocked fish may have on certain populations, balanced with the demands for the fishery. For further information on genetics policies see, "A Conceptual Framework for Conservation Hatchery Strategies for Pacific Salmonids," *http://www.nwfsc.noaa.gov/pubs/tm/tm38/tm38.htm.*

Traditional Stocking Practices

Transport and Conditioning

Survival of stocked fish depends, at least in part, upon proper transportation and conditioning (Figure 7). Although these aspects are discussed in greater detail in Chapter 10, the major factors to consider—stress management and maintenance of water quality—will be mentioned briefly. Anything that will adversely affect water quality or elevate stress in fish should be avoided or minimized to reduce the risk of mortality.

Some general rules are:

- *Do not feed fish before handling and transport*. Feeding elevates metabolism and waste product production and may reduce water quality in the hauling tanks.
- *Try to minimize all handling and hauling times*. The longer the fish are handled or hauled, the lower the survival rate (all else being equal). This may mean mobilizing a small army or using transport devices such as helicopters and fixed wing aircraft. Studies with juvenile striped bass and walleye fry have shown survival in the absence of predation to be inversely related to hauling time (Pitman and Gutreuter 1993).
- *Haul fish in a saltwater solution* (0.5–1.0% NaCl plus, in soft water, enough $CaCl_2$ to bring hardness up to 50 mg/l). This helps reduce stress.

FIGURE 7. Transport and conditioning. State-of-the-art fish hauling trucks like this one make handling stress and transport of fish much less risky. This truck has baffles within the tanks to minimize water motion, oxygen tanks to maintain DO levels within safe limits, and insulated sides to maintain water temperatures.

- *Anesthesia.* You may wish to use some form of fish anesthetic, such as MS-222 or clove oil, to relax the fish and relieve stress. Note that MS-222 should not be used in fish that may be consumed by humans within 22 d after exposure to this chemical and that clove oil has not been FDA cleared.
- *Maintain good water quality in the hauling tanks.* Ammonia and nitrites, temperature, and dissolved oxygen (DO) are the major factors to monitor since most fish are hauled in water from the hatchery. Ammonia and nitrites should not exceed 0.01 mg/l and DO should be maintained at levels near 75% of saturation. Dissolved oxygen levels can be maintained at near 100% saturation by periodically bubbling oxygen into the tanks.
- *Temperature.* Haul during cooler times of the year or day, especially for cool water species. This helps minimize thermal changes and thermal stress in the hauling tanks.

Fish Release

There are several factors related to fish release that will help ensure the best opportunities for survival of the stocked fish. The most important factors include acclimation of fish to the release site, protection from predators and anglers, and timing the release to coincide with availability of natural foods.

Acclimation of the fish refers to the minimization of differences in hauling water and release site water (Figure 8). Properly acclimate fish to the water conditions (temperature, pH, etc.) before release by mixing hauling and release site water over a period of up to several hours. If significant differences exist between rearing and release site water conditions, a more protracted acclimation procedure may be required. Survival of hybrid striped bass fry has been associated with DO levels at stocking sites and cumulative changes in pH and conductivity exposure during stocking (Pitman and Gutreuter 1993).

Oftentimes, stocked fish may experience a high initial predator-related mortality. The predators may be resident piscivorous fish, fish eating birds, or anglers. In the Patuxent River, Maryland, older, resident striped bass fed heavily on hatchery fingerlings following release (Andreasen 1995) and in Ohio, resident largemouth bass and crappie populations often inflict casualties upon stocked esocids (Wahl et al. 1995). Whatever the ultimate source of predation, in most cases significant losses can be reduced by carefully planning the stocking.

In situations where predators may concentrate and feed on stocked fish, a change in the timing of stocking may reduce losses. If predators are a species known to feed only during daylight, fish could be stocked at dusk or at night to provide at least a short period of time to acclimate and find cover before predators are active. In other cases, stocking when alter-

FIGURE 8. Careful handling of fish including "soft release" and careful acclimation of the fish to the water conditions at the release site will help ensure better survival of stocked animals.

nate prey availability is high may be sufficient to minimize the effects of piscivores (e.g., during the fall when the abundance of most age-0 prey fishes is high).

Similarly, stocking fish to coincide with prey availability will enhance survival. As any hatchery manager will volunteer, the right type and size of food is needed for proper feeding and growth in the hatchery and the same is true with stocked fish at the release site. For example, walleye become piscivorous at about 25 mm TL. Stocking walleye too early in the spring, before other young-of-the-year fishes become available, will necessarily result in poor survival. For other piscivorous species, the best time to stock may be in late summer or early fall when age-0 clupeids are very abundant and are nearing the proper size for consumption by young predators.

Depending upon the goal of stocking, different strategies may be employed to distribute the stocked fish in a system. When stocking for a recreational fishery, stocking may either be done at a minimum of locations, if the goal is to maximize angler success; or fish may be widely distributed, if the goal is either to maximize angler opportunities or increase chances of survival and subsequent growth as in a PGT fishery. When stocking juvenile fish, some researchers have found that stocking all fish into one or two locations and saturating the predators in those locations may reduce predation mortality. In other cases, such as stocking a novel system, it may be best to stock the fish at a variety of locations to maximize the growth of individuals by reducing competition. In each situation, the personnel in charge of stocking must evaluate the situation and closely match the release design to the system and goals of the stocking.

Records

It is important to maintain good records of the stocking practices used in each release to help evaluate the methods and practices used in stocking. It is never a good idea to trust one's recollection for matters such as survival or mortality of fish and the hauling, stocking, and release conditions. Oftentimes, the success of a particular stocking is not known for several years until the stocked fish recruit to the fishery. If accurate records are not maintained, then the likely reason for a successful or an unsuccessful stocking may not be learned. Without good stocking records, we are doomed to repeat the mistakes of the past and are hard-pressed to replicate our successes.

The basic information needed to document stocking:

- Numbers, sizes, condition, and stocking density of fish.
- Water chemistry (temperature, DO, pH, conductivity, hardness, salinity, ammonia, and nitrites) during rearing, in tanks, and at the release site.
- Hauling time.
- Acclimation methods.
- Dates, times, and locations of stocking.
- Hatchery origin and genetic strain.
- Names and affiliations of all persons associated with the stocking should also be recorded in the event further information is required at a later date.

An example of a form used by the West Virginia Department of Wildlife in their fish stocking program is included in Figure 9.

Stocking Rates

Stocking rates are dependent upon a combination of human needs; biological needs; and biological, societal, and economical constraints. In an ideal system, we could produce as many fish as required to meet the human needs, while balancing the biological constraints of a system. However, as much as we strive to make it otherwise, the business of producing fish for stocking is an uncertain science. Although we may have targets for how many of each species we need for stocking within a year, often, events either prevent us from achieving targets or, in an effort to account for this event, we produce more than are needed.

Underlying the determination of stocking rates are economic considerations. The hatchery manager is required to balance human needs and

FISH DISTRIBUTION REPORT

Hatchery 01 02 03 04 05 06 07 08 09 10 11 Site 12 13 14

Water body/stream system 15 16 17 18 Stream 19 20 21 22

Tributary Of 23 24 25 26 County 27 28 29 30

Air Temp. 31 32 Date 33 34 35 36 37 38 Trip Miles 39 40 41

Time Loaded 42 43 44 45 Time Stocked 46 47 48 49

Truck Water Quality

Water Temp 50 51

pH 52 . 53 54 Cond. 55 56 57

Hardness 58 59 60

Receiving water quality

Water Temp 61 62

pH 63 . 64 65 Cond. 66 67 68

Hardness 69 70 71

Species	**Strain**	**Number**	**Number/lb.**	**Size**	**Condition**
72 73	74 75	76 77 78 79 80 81	82 83 84	85 86	87 88 89
90 91	92 93	94 95 96 97 98 99	100 101 102	103 104	105 106 107
108 109	110 111	112 113 114 115 116 117	118 119 120	121 122	123 124 125

Truck No. 126 127

Truck Driver

Superintendent

Witness

Witness

Comments ______________________

FIGURE 9. Accurate record keeping is a must in any stocking operation. This form is one used by the West Virginia Division of Natural Resources to record information needed to evaluate hauling and acclimation conditions and stocking success.

biological needs in the management of limited aquatic resources. In the event that insufficient numbers are produced, the fisheries manager, in consultation with the hatchery manager, is usually faced with allocating fewer resources among the areas to be stocked. Usually reductions are proportional across the systems to be stocked. However, that may not always be the best approach. Although societal pressures, in some cases, may dictate the need to stock all the targeted systems, biological needs may suggest otherwise. In some cases, it may be that there is either a minimum fish density that may be required for stocking to have an effect in a system, or there may be some systems that have a much greater need for stocking. In the event of hatchery shortages, these latter two points should be considered, especially in light of the great cost to produce fish and the stocking evaluation.

In the event of hatchery shortages, guidelines can be established to prioritize stocking sites and densities. Oklahoma Department of Wildlife Conservation implemented stocking guidelines for largemouth bass, smallmouth bass, hybrid striped bass, walleye, sauger, and channel catfish that use a point system for various physiological and biological criteria and fisheries characteristics for candidate reservoirs (Gilliland and Boxrucker 1995). These guidelines prioritize stocking into areas where it is most likely to have a successful impact and includes a reevaluation period whereby stocking is continued or discontinued in a particular water body.

The species, sizes, and stocking rates used by managers vary greatly from state to state in the United States. Largemouth bass, bluegill, channel catfish, and hybrid striped bass are stocked in the majority of states, but, even so, the sizes and rates vary considerably (Smith and Reeves 1986). In most cases, smaller individuals are stocked at higher densities than larger fish, but, within a species, the stocking rates often vary from 2- to 60-fold within and among states (Smith and Reeves 1986). A more complete listing of species and stocking rates used by North American fisheries managers can be found in the book *Inland Fisheries Management in North America* (Heidinger 1999).

Stocking Evaluation

In these days of diminishing availability of resources, it is especially important that no stocking program be considered complete without some form of evaluation. Without an evaluation, there are no opportunities or justifications for improvement in current stocking programs. At a minimum, stocking evaluations should include periodic checks on mortality and checks on quantifiable goals and objectives.

Instantaneous and Delayed Mortality

Following handling or any stressful event, mortality can be expected in fish. This mortality can be of two types: instantaneous mortality or delayed mortality. Instantaneous mortality can be relatively easily quantified by holding a portion of the stocked fish in an enclosure or net-pen for a short time (usually several hours to several days) and recording any mortality that occurs. However, mortality may not occur within the first few hours after stocking. Often mortality is delayed and may not occur for several weeks following the stress event. Many studies with catch-and-release fisheries have shown that the delayed mortality associated with handling can be over three times higher than the instantaneous mortality. Thus, it would be expected that delayed mortality following stocking may be at least as high as that found immediately following stocking.

Observations of dead or floating fish are not a prerequisite for mortality of stocked fish. Birds, mammals, fish, frogs, and turtles will all prey upon dead or dying fish and may remove all evidence of mortality before biologists can detect the event. For that reason, the net-pen or enclosure studies should be considered to help quantify initial and delayed mortality in stocked systems. An example of this occurred in a mortality and tag retention study involving stocked largemouth bass at Palestine State Fish Hatchery in West Virginia. There, 421 bass were tagged with a latex polymer and a standard internal anchor tag, then released into ponds at the hatchery. Fish were fed fathead minnows, which were also stocked into the ponds. During several checks, hatchery workers found 22 dead bass during the course of the 31 d study. After 31 d, the ponds were drained and the fish removed, counted, and examined. There were 257 bass recovered alive from the pond. This left 142 bass unaccounted for in the study, probably consumed by predators or eaten by scavengers. The survival rate was 61.0% for the stocked bass [(421 stocked – 22 found dead – 142 unaccounted for)/421 stocked], however, if we had relied on only observed losses from surveys around the pond we would have estimated survival to be 94.8% [(421 stocked – 22 found dead)/421 stocked]. Thus, by doing the pond study, we determined that actual mortality in the pond was about 7.5 times higher than expected by counting fish carcasses. The source of the additional mortality was unknown, but other animals consuming the bass or the bass carcasses, no doubt, obscured the apparent levels of mortality.

Monitoring and Evaluation

The single most important aspect of stocking evaluation is to set a priori goals and objectives, together with determinants for success or failure of the stocking program. Fisheries resource users are a complicated lot. Some would like to see all systems stocked and stocked heavily. Others might prefer to see stocking limited or eliminated. A good way of getting in trouble with resource users is to give the appearance of bias by determining that stocking popular species X in Lake Y was a failure without a priori setting clearly defined, quantifiable criteria for evaluating the stocking program.

In monitoring and evaluating stocking, the method or criteria that is used will depend upon the goals of the stocking program. In some cases, the criteria might be a particular target density of fish in cove rotenone samples. For other situations, some minimum CPUE or angler catch rate is the criteria. In an augmentation stocking of larval American shad, contributions of hatchery-produced larvae to the juvenile stages were used to evaluate stocking success. Juvenile shad that originated from hatchery stocking of larvae declined from 99% to 44% of all juveniles between 1985 and 1991. Here, successful reproduction of adults which had been physically "lifted" around a hydroelectric dam, shifted mitigation efforts from stocking to installation of fish ladders. High contributions by stocked larvae at the beginning of the efforts showed that good survival of young shad was possible if the upstream areas were used as nursery areas (Hendricks 1995).

In a put-and-take or PGT situation, the ultimate measure of stocking success may be the return to the creel for anglers, total angler hours, or angler catch rates. Here, an age-based assessment of angler catch can be related to numbers stocked, in order to estimate the return rate on stocked fish.[2] Similarly, in salmon ranching in western North America, a measure of success may be the number of salmon returning to weirs at their stocking site. Over time, these measures of evaluation (e.g., return rates) can be related back to stocking records to evaluate different stocking conditions such as handling and hauling, stock or strain, and sizes or density stocked. From these relations, stocking and rearing practices can be improved to maximize the effectiveness of the stocking program.

For stockings that have stock recovery as a goal, the measure of success may be the establishment of spawning populations or increasing indices of (wild) juvenile abundance as success criteria. With stocking to restore threatened and endangered species, the evaluation may be more difficult, since any type of sampling may be deleterious to the population. Underwater observations or increases in transect counts may be the criteria for evaluating the success of stocking.

---

[2]Return rate is calculated as $N_{Y\ creeled}/N_{X,\ Y\ stocked}$, where $N_{Y\ creeled}$ is the number of fish caught by anglers from the stocking in year (Y) throughout the time these fish were available to the fishery, and $N_{X,\ Y\ stocked}$ is the number of fish of that age (X) stocked in year (Y).

Throughout all of these evaluations and stocking practices, it is important to consider that these stocked fish will ultimately interact with each other and resident species. What is successful for one species may be at the expense of other species in the system. These interactions can be used to the benefit of the system or fishery, as in the case of chinook salmon (also known as king salmon) stocking to control alewife populations in Lake Michigan, but potential negative effects should be considered and the consequences weighed in considering any stocking program.

Biological and Ecological Considerations

Another major consideration for fisheries managers and biologists is that stocking deals with the way these fish may influence existing fish species and other biota. Major concerns when stocking systems in which a species already is present and reproducing are issues of genetics and genetic drift. Will the genetic integrity of a stock be maintained by stocking the system with these fish? Hopefully, the answer is yes. If the answer is no or unknown, stocking should not take place until the answer is yes. In studies with rainbow trout, hatchery-raised offspring and their progeny with wild partners were shown to be more vulnerable to predation than wild progeny (Berejikian 1995; Dellefors and Johnsson 1995; Johnsson et al. 1996). This "dulling" of survival instincts is one of the facts pointed out by resource users opposed to stocking into waters with naturally reproducing fish, and it may be considered an example of why care must be taken regarding what and when to stock. However, conservation hatchery guidelines suggest dulling of survival instincts should be counteracted with antipredator training of fish in the hatchery (Flagg and Nash 1999).

Other biological aspects deal with how the stocked fish will interact in the new system. These interactions include both direct interactions (e.g., predation and competition) and indirect interactions (e.g., disease or pathogen vectors). Many studies have shown stocked fish to displace smaller wild fish from focal feeding positions in streams or to increase density-dependent dispersal into less favorable habitats. Stocked fish may also compete with other species at some life stage, and stocking of a competitor species may lead to alterations of the fishery potential for another species. Stocked fish may also prey on other species. In the case of chinook salmon in Lake Michigan, this was by design. However, there are several cases in the western United States where walleye or largemouth bass have been stocked and later eradication efforts were made in efforts to "save" indigenous threatened or endangered species from predation. Stocked fish, or fish collected for broodstock, should also be carefully screened to minimize the threat of spreading harmful fish pathogens or other aquatic nuisances.

Spread of whirling disease in salmonids and of grubs in baitfish has been enhanced by moving fish between water bodies.

Another important aspect of stocking is the establishment of exotic species. In many cases, the game species that are stocked are not native to the area of stocking (e.g., rainbow trout and Pacific salmon in the eastern United States; brook trout in the western United States). Although opinions differ with respect to stocking nonnative species for angling opportunities, it is likely due to their popularity with fishermen that such stockings will not be eliminated any time soon. However, in many cases, nonnative species have been stocked intentionally with favorable results. For example, California has only 10 native sport fishes, but an additional 30 species of sport fish have been stocked. These stocked fish account for 42–77% of all angling effort (Lee 1995). In other cases, commercial applications of stocking have begun growing species such as Atlantic salmon in western North America streams. In 1999, Atlantic salmon were known to have escaped from pen culture and may now be establishing runs in the Amor de Cosmos Creek on northeastern Vancouver Island. The potential impact of these "accidental" stockings upon native fishes is unimaginable. Species such as grass carp are widely stocked for vegetation control. Grass carp tend to change water quality, habitat complexity, and primary and secondary production in systems. These changes may be desired by landowners, but if they are released into open waters, these changes may alter community structure and, potentially, native and other species (Cassani 1995). Escapement of these reportedly sterile, triploid grass carp, has resulted in them establishing breeding populations in the Mississippi River in Kentucky, Illinois, and Mississippi; the Missouri River; and the Ohio River.

Many stocking programs have gone on to produce successful fisheries or have restored depleted populations. However, there have been some failures and some negative results due to stocking. Therefore, given the potential for deleterious effects of introductions, the impact of stocked fish or pathogens they may carry into a system must be seriously considered before releasing fish into a body of water.

References

Andreasen, L. L. 1995. Predation and cannibalism on hatchery-reared striped bass in the Patuxent River, Maryland. Pages 553–554 *in* Schramm and Piper (1995).

Bart, A. N., D. F. Wolfe, and R. A. Dunham. 1998. Cryopreservation of blue catfish spermatozoa and subsequent fertilization of channel catfish eggs. Transactions of the American Fisheries Society 127:819–824.

Berejikian, B. A. 1995. The effects of hatchery and wild ancestry and experience on the relative ability of steelhead trout fry (*Oncorhynchus mykiss*) to avoid a benthic predator. Canadian Journal of Fisheries and Aquatic Sciences 52:2476–2482.

Bonar, S. A., J. Pahutski, B. D. Bolding, D. Fletcher, and M. Divens. 1997. Survival and growth of channel catfish stocked in Washington lakes. North American Journal of Fisheries Management 17:773–778.

Bugert, R. M., C. W. Hopley, C. A. Busack, and G. W. Mendel. 1995. Maintenance of stock integrity in Snake River fall Chinook salmon. Pages 267-276 *in* Schramm and Piper (1995).

Busack, C. A., and K. P. Currens. 1995. Genetic risks and hazards in hatchery operations: fundamental concepts and issues. Pages 71–80 *in* Schramm and Piper (1995).

Campton, D. E. 1995. Genetic effects of hatchery fish on wild populations of Pacific salmon and steelhead: what do we really know? Pages 337–353 *in* Schramm and Piper (1995).

Cassani, J. R. 1995. Problems and prospects for grass carp as a management tool. Pages 407–412 *in* Schramm and Piper (1995).

Dellefors, C., and J. I. Johnsson. 1995. Foraging under risk of predation in wild and hatchery-reared juvenile sea trout (*Salmo trutta* L.). Nordic Journal of Freshwater Research 70:31–37.

Flagg, T. A., and C. E. Nash, editors. 1999. A conceptual framework for conservation hatchery strategies for Pacific salmonids. NOAA (National Oceanic and Atmospheric Administration) Technical Memorandum NMFS (National Marine Fisheries Service)-NWFSC (Northwest Fisheries Science Center) 38.

Flagg, T. A., C. V. W. Mahnken, and K. A. Johnson. 1995. Captive broodstocks for recovery of Snake River sockeye salmon. Pages 81–90 *in* Schramm and Piper (1995).

Gilliland, E., and J. Boxrucker. 1995. Species-specific guidelines for stocking reservoirs in Oklahoma. Pages 144–151 *in* Schramm and Piper (1995).

Goehle, M. A. 1999. Assessment of natural recruitment in the mixed rainbow/steelhead fishery of Cattaraugus Creek and tributaries. Master's thesis. State University of New York, College at Buffalo, Great Lakes Center, Buffalo.

Guy, C. S., H. L. Blankenship, and L. A. Nielsen. 1996. Tagging and marking. Pages 353–384 *in* B. R. Murphy and D. W. Willis, editors. Fisheries techniques, 2nd edition. American Fisheries Society, Bethesda, Maryland.

Hayne, D. W. 1949. An examination of the strip census method for estimating animal populations. Journal of Wildlife Management 13:145–157.

Heidinger, R. C. 1999. Stocking for sport fisheries enhancement. Pages 375–398 *in* C. C. Kohler and W. A. Hubert, editors. Inland fisheries management in North America, 2nd edition. American Fisheries Society, Bethesda, Maryland.

Hendricks, M. L. 1995. The contribution of hatchery fish to the restoration of American shad in the Susquehanna River. American Fisheries Society Symposium 15:329–336.

Johnsson, J. I., E. Petersson, E. Jonsson, B. T. Bjornsson, and T. Jarvi. 1996. Domestication and growth hormone alter antipredator behaviour and growth patterns in juvenile brown trout, *Salmo trutta*. Canadian Journal of Fisheries and Aquatic Sciences 53:1546–1554.

Lange, R. E., G. C. leTendre, T. E. Eckert, and C. P. Schneider. 1995. Enhancement of sportfishing in New York waters of Lake Ontario with hatchery-reared salmonines. Pages 7–11 *in* Schramm and Piper (1995).

Lee, D. P. 1995. Contribution of nonnative fish to California's inland recreational fishery. Pages 16–20 *in* Schramm and Piper (1995).

Mathias, J. A., W. G. Franzin, J. F. Craig, J. A. Babaluk, and J. F. Flannagan. 1992. Evaluation of stocking walleye fry to enhance a commercial fishery in a large, Canadian Prairie Lake. North American Journal of Fisheries Management 12:299–306.

McEachron, L. W., C. E. McCarty, and R. R. Vega. 1995. Beneficial uses of marine fish hatcheries: enhancement of red drum in Texas coastal waters. Pages 161–166 *in* Schramm and Piper (1995).

Modde, T., and five coauthors. 1995. An augmentation plan for razorback sucker in the upper Colorado River basin. Pages 102–111 *in* Schramm and Piper (1995).

Parker, N. C., and five coeditors. 1990. Fish-marking techniques. American Fisheries Society, Symposium 7, Bethesda, Maryland.

Pitman, V. M., and S. Gutreuter. 1993. Initial poststocking survival of hatchery-reared fishes. North American Journal of Fisheries Management 13:151–159.

Radonski, G. C., and A. J. Loftus. 1995. Fish genetics, fish hatcheries, wild fish, and other fables. Pages 1–4 *in* Schramm and Piper (1995).

Reynolds, J. B. 1996. Electrofishing. Pages 221–253 *in* B. R. Murphy and D. W. Willis, editors. Fisheries techniques, 2nd edition. American Fisheries Society, Bethesda, Maryland.

Ricker, W. E. 1975. Computation and interpretation of biological statistics of fish populations. Department of the Environment Fisheries and Marine Service, Bulletin 191, Ottawa.

Schramm, H. L., Jr., and R. G. Piper, editors. 1995. Uses and effects of cultured fishes in aquatic ecosystems. American Fisheries Society, Symposium 15, Bethesda, Maryland.

Smith, B. W., and W. C. Reeves. 1986. Stocking warmwater species to restore or enhance fisheries. Pages 17–29 *in* R. H. Stroud, editor. Fish culture in fisheries management. American Fisheries Society, Fish Culture Section and Fisheries Management Section, Bethesda, Maryland.

Van Vooren, A. R. 1995. The roles of hatcheries, habitat, and regulations in wild trout management in Idaho. Pages 512–517 *in* Schramm and Piper (1995).

Wahl, D. H., R. A. Stein, and D. R. DeVries. 1995. An ecological framework for evaluating the success and effects of stocked fishes. Pages 176–189 *in* Schramm and Piper (1995).

Ward, R., T. L. King, I. R. Blandon, and L. W. McEachron. 1995. Use of gene marking to assess stocking success of red drum in Texas bays. Page 551 *in* Schramm and Piper (1995).

12

Management Considerations and Economics

James W. Meade

Introduction

This chapter is an overview of business management concepts and processes that may be useful and provide perspective for more informed decision making. Many of the discussions have been adapted from Meade (1989). Those who want to pursue detailed economic analyses should refer to more complete authoritative texts, especially *Economics of Aquaculture* by Jolly and Clonts (1993) and *Aquaculture Economic Analysis: An Introduction*, by Shang (1990).

One important matter that this chapter does not address, but that will eventually affect all of fish culture, is the technological revolution currently upon us. For example, one might search in the electronic media for more information on topics discussed in this chapter. Electronic technology can facilitate organizational effectiveness through the use of public, as well as internal, information. Technology is changing how we do business and how the public sets expectations and makes demands. Applied genetic technology is changing how we work with food organisms and especially how we manage public resources and our biological heritage. Kosinski (1999) provides an example of this new technology related to aquaculture education. Some of us quickly acknowledge that we must be ready and willing to change and adapt to new technology, and we prepare to do so. Others, perhaps with more foresight, realize that the tidal wave has arrived. Traditional ways of finding and handling processes are quickly going the way of the chisel and stone after the development of papyrus, and the manuscript copiers after Gutenberg.

Management

Never doubt that a small group of thoughtful, committed individuals can change the world, indeed it's the only thing that ever has. —Margaret Mead

Scope and Meanings

Management is the planned wise use of resources, including human, capital, equipment, and materials to accomplish goals. Management functions include planning, organizing, staffing, directing, and controlling. The manager is responsible for the results, good or bad. Productivity combines efficiency and effectiveness central to the purpose of management. Effective and highly productive managers often ask for help and input from subordinates, as well as from superiors. The scope of management includes technical aspects of culture, such as allocating water use, determining water quality, determining what to produce and how to produce it, and identifying and solving production and fish health problems. Management also includes personnel aspects (staffing, training, motivating, and controlling), financial responsibilities, and accounting. Good managers make a point of understanding the people they manage and the motivating power of people's needs. Generally, performance expectations are developed mutually by the employee and supervisor into plans built around management objectives. Plans provide constant direction, aid in control, act as a job contract, and provide a systematic method to enhance communication and reduce judgment. The following sections highlight some important, and at times overlooked, aspects of fish hatchery management.

Responsibility and Ethics

As a fish hatchery manager, clearly you make a difference. There is considerable choice, however, in the nature of the difference you make. Accepting responsibility is an immediate distinction between a laborer and a manager. Risk is involved in every decision. If outcomes were known, there would be no need for decision makers; and undesirable outcomes sometimes follow "good" decisions. The manager, not a subordinate, is ultimately responsible for the outcome, especially an undesirable one.

It can be argued that maintaining high ethical standards and behavior pays both short-term and long-term dividends. The hatchery manager is responsible for the integrity of achieving funded goals, for following ethical approaches to business, for the proper uses of environmental resources; and, of course, hatchery managers are responsible to ensure that animals in their care are treated humanely and appropriately. In the public sector, ethical standards require that one avoid any action that might result in, or create, the appearance of using public office for private gain, giving preferential treatment, impeding efficiency or economy, or making a government decision outside official channels.

Stewardship of national fisheries resources has changed, but several factors remain central: a growing awareness of the importance of fishery resources for food and recreation; a greater recognition and understanding of the limits of fishery resources and the adverse effects of overfishing and habitat damage; and the need for comprehensive, coordinated management. Opinions held by large numbers of people or by their representatives affect approaches to resource management. For instance, the U.S. Congress has directed the Department of the Interior to manage and conserve natural resources, giving government managers decision and valuation guidelines and charging managers with accomplishing work based on those guidelines. Some ethical questions for the public manager then deal with effectiveness in situations such as applying resources to exercise a mandate to protect or restore fish. If a restoration hatchery is producing little positive impact on the target population, how then can the manager justify its operation? Are resources being applied for maximum effectiveness, and if not, is the manager compromising a public trust?

Important ethical questions center on the use of resources and the associated environmental effects and may be of particular concern for international production systems. Continued development of aquaculture is expected to have considerable impact on natural environments (Tisdell 1999) and will, undoubtedly, raise questions of environmental tradeoffs or losses in behalf of market gains. Certain practices can lead to degradation or destruction of natural habitats or to the introduction of undesired species, including pathogens. In macroeconomic terms, when concepts such as social welfare are concerned, some analyses have suggested that development of natural systems into private (for profit) aquaculture operations have resulted in reduced net monetary benefits, as in the case of reallocation of mangrove areas for shrimp farms in the Philippines (Nickerson 1999). In other cases, as in salmon farming, convincing arguments indicate that benefits might far exceed the costs (Lindbergh 1999). Boyd and Schmittou (1999) address the question of whether aquaculture is sustainable in its environmental impact, as it expands to support or replace the global capture fishery already at its sustainable capacity. They argue that industry should be proactive in developing and implementing management practices to prevent or mitigate adverse environmental effects. Industry standards could then be used as a basis for future regulation in nations not yet ready or able to develop adequate regulation.

Productivity

Productivity, broadly defined, is the measure of individual or organizational performance. It's not only efficiency (ratio of outputs to inputs), but also effectiveness (the extent to which the output satisfies program objectives). An example might be found in the hatchery that produces fish for population restoration. Growing and shipping a great mass of fish represents low productivity if the hatchery-produced fish contribute little to restoration of a self-perpetuating fishery. Productivity not only involves the amount of a product produced or service delivered, and at what cost, but also quality, timeliness, and responsiveness to the need. This definition includes increasing efficiency, usefulness, effectiveness, and responsiveness to demand, as well as decreasing costs, production time, and delivery time.

Most managers have felt the demand for more product, higher quality, and more timely delivery, all for the same or lower costs. The manager does not simply oversee a process, but plays the pivotal, central role in production. A manager's strengths and weaknesses, victories and fumbles, are amplified through the lever of organization. Thus, an appropriate assessment of a manager's performance is through measurement of organizational accomplishment. Here are ways a manager might increase productivity:

1. Involve all employees and top management, seek their support, involve them in identifying processes that seem ready for improvement, keep them informed, and be sensitive to their concerns. Given a predetermined goal, your staff can help plan ways to achieve it. Be open; don't merely present your views and ask for approval.
2. Make purposes clear from the start: what you intend and why. You may not be entirely sure, but explain your general strategy, impressions, timetable, and whom you will ask for help. To make sure all understand, write down the approach and realistic goals. Subordinates must know they will have opportunity to participate.
3. Measure and analyze pertinent information.
4. Locate models in other organizations and try them on a small scale.

Accept the importance of improving productivity, as not much will happen unless you are committed, and then make it happen.

Financial Accountability and Managing for Effectiveness in the Public Sector

In the private sector, there is a constant minimum threshold and a continuing pressure for efficiency in the forms of solvency and profit. Profit, an accounting term that represents the surplus remaining after all costs are paid, is the return to management for adding value through a production process. How can the public sector hatchery manager approach financial accountability and measure effectiveness and efficiency? First, the manager must clearly establish and define objectives and goals whose accomplishments will fulfill the operational purpose of the hatchery. Here it is important to screen and eliminate objectives and goals that may sound like or deal with production, but that do not actually contribute to the bottom-line purpose.

Once this set of accountability goals is defined, the manager must determine objective ways to measure the extent to which the goals can be met. Regarding efficiency, costs and outputs can be evaluated and compared with like, past, or expected costs or returns. Finally, accomplishment of goals with a surplus, such as coming in under budget verses a shortfall of funding, is something comparable to a profit. Terminology may not be important, but measured accountability is. Instead of profits, public accountability for carrying out processes and achieving management goals plays a central role in planning, funding, and management decisions. Increasingly, public managers may need to recognize and consider the challenge of measuring outputs in meaningful economic terms.

Financial accountability and economic decision criteria are concerns not only of farm and hatchery managers, but also of societies, represented by governments that must weigh national and international concerns. Shang and Tisdell (1997) offer a perspective on the scope of micro and macro factors. However, generally, all economics issues faced by hatchery managers deal with microeconomics.

Microeconomic Principles

Things either are what they appear to be; or they neither are, nor appear to be; or they are, and do not appear to be, or they are not, and yet appear to be. —Epictetus

Fixed and variable costs and marginal costs are important concepts that can be used to analyze and improve returns to production processes. Here we discuss enterprises and the concept of enterprise mixing, then address cost tradeoffs in decision making, the production function, marginal analysis, partial budgeting, costs of ownership, and benefit–cost analyses.

The public sector management has an important economic peculiarity. As mentioned earlier, instead of profits, public accountability for carrying out processes and achieving management goals plays a central role in planning, funding, and management decisions. Decision support information based on amortized costs, depreciation, taxes, or other similar data, may not apply to public sector situations. For example, renting a piece of equipment or contracting for a service is often an alternative to purchasing a major piece of equipment. Not uncommonly, renting or contracting may be the clear, rational choice based on private sector economics.

However, ongoing costs of services are usually defined as variable costs and can be a perpetual problem and target for criticisms directed at management. Public officials can point to new construction that puts local or regional folks to work. Thus, when opportunities arise to access construction funds, while at the same time ongoing costs are seen as a concern, one must consider opportunities to shift costs from rentals and contracts to low variable cost (albeit high fixed) construction budget items. A rational choice in some situations in the public sector may be to add to the fixed costs of a construction opportunity rather than to use the profit model to show savings in the long run, a savings that may have little weight in the decision process.

On the other hand, in recent years, there is a trend in both the private and public sectors to reduce fixed costs and shift into leasing, rentals, or service contracts, and public hatchery managers have been encouraged to use mechanisms put in place for that purpose. In the public sector, the economic danger here is that the "sunk" costs tied up in buildings or equipment cannot be recovered. The sunk costs represent future paid up leasing or rental costs, and, unless anticipated maintenance or storage costs are greater than the costs of new leasing, the shift to leasing will add to, rather than reduce, costs. Similarly, in the short run, service contracts may add costs, unless they represent real savings, as in salaries. A manager's challenge is to identify and understand these subtle dilemmas to avoid added costs sometimes inherent in following a current trend.

Enterprises and the Concept of Mixing

Fish hatchery business plans (Griffin 2000), financing plans (Klinefelter 2000), and operations center around production strategies and a set of enterprises. An enterprise produces a marketable output, and a variety of enterprises are usually involved in a business. The selection and mixing of enterprises is fundamental to the efficient use of resources. The available resources limit and help define the feasible alternatives for enterprises. An important criterion for selection of enterprises is operator preference, as operators are more likely to work well in preferred enterprises than in those they dislike. Additional selection criteria are markets, temperature and quality of water, and available capital. Production strategies vary greatly to take advantage of resources, as well as to provide the requirements of the cultured organisms. Production strategies should support opportunities for complementary mixing of enterprises. Some familiarity with concepts of examining and mixing enterprises can be helpful in developing strategies and in managing hatchery production operations.

Selecting the Enterprise Mix

The manager's job in setting objectives to reach organizational goals includes that of selecting or recommending the mix of enterprises. In any given location, market opportunities and resources, such as the available water quantity, quality, and temperature, will dictate the array of feasible, profitable enterprises. Thus, the choice of enterprises often seems limited to those of fresh or salt water, and coldwater or warmwater species. There are, however, a myriad of choices.

Possession or control of suitable abundant water resources enables volume production of harvestable fish. Once reared, animals no longer require nursing and culture procedures become relatively more routine. Though quality can never be overlooked, quantity becomes the key word in production and staffing. The manager does not need highly trained biologists or experienced and perceptive hatch house specialists to drive trucks and haul seines. For a discussion of optimal harvesting models see Bjorndal (1990). However, the development and selection of broodstock requires technical expertise, and most of the challenges in fish culture have historically centered on early life history. Care of eggs and fry is labor intensive, but above that, it's an artisan's job.

A culturist who seems to empathize with the cultured organisms, who sees cleaning as an opportunity to improve the lot or life of infant fish, and who is concerned about the health of the smallest organisms is rare. An organization that identifies and develops such talent has the wherewithal to specialize in seed and juvenile production. An investigation of new or alternative enterprises and optimization of the enterprise mix does not require the use of new or additional species. A manager might consider adding or eliminating one or more of the common enterprises. For instance, most hatcheries keep and spawn brood, incubate eggs, nurse fry, rear juveniles, and grow at least one species to table size. In addition, the operator may transport live fish, harvest ponds, or do other related enterprises, such as operating a fish out (fee fishing) pond, process fish, or use part of the facility to wholesale fish into or out of the area.

The manager's challenge is to efficiently and fully use all available resources, including human resource time, throughout the period when resources are available. The larger the operation, the more options there are in enterprises and staffing, but also the greater the chances for time and labor inefficiencies. To help maximize labor efficiency, considering the usually large seasonal differences in activities, the manager should choose enterprises that are complementary. That is, in addition to the use of temporary or seasonal labor and good planning and scheduling of full time personnel, enterprises should be selected and combined in a way that allows an enterprise with high resource use to fill in during the period of low resource use of another enterprise.

Once a mix of feasible and operator-preferred enterprises is established, the manager can begin to optimize for maximum returns to the overall hatchery operation. The manager asks the question of where to put efforts among the alternatives or multiple product options. To optimize the mix of enterprises, the manager looks closely at the returns to resources from each enterprise. Enterprises that pay the highest returns should be maximized to the exclusion of those that yield the lowest returns. The equimarginal principle states that the highest total return to a scarce resource is achieved when the return per unit of distributed resource is equal in all alternative uses. First, it is necessary to narrow the options to a set of primary or best alternatives to pursue. Allocation of scarce resources, such as dollars, will depend not on added returns versus added costs, but rather on the most economical allocation to the alternatives. The manager distributes the resource among enterprises or production activities starting with the one that pays the highest net return. This principle is the basis for selecting and sizing enterprises for the optimum mix, yielding the greatest returns to the available resources (Meade 1989). See Kusumastanto et al. (1998) for a large-scale study to determine the best mix of shrimp aquaculture enterprises in Indonesia.

Enterprise Budgets

Budgets are written plans of action, with estimates of the expected results, used as analytical tools for short- and intermediate-range planning. Cash flow analyses allow development of budgets. Each potential enterprise can be scrutinized through budget development and the culture system examined as an aggregate of enterprises. With the array of alternative uses for resources, as described by the enterprise budgets, the manager is in position to review goals and resource inventories to support decision making. The first step in developing a budget is to set the intended magnitude of production for the enterprise, then estimate yield or output. Yields are usually estimated from similar operations in the local or a comparable area.

Familiarity with an array of enterprise budgets will provide perspective for managers or others interested in selecting enterprises. Examples of enterprise budgets germane to the geographic region or area and species produced are usually available from extension agents and from many major or land grant universities. Discussions and examples of enterprise budgets are available in several texts, including those of Jolly and Clonts (1993), Meade (1989), and Engle et al. (1997).

Ownership Costs

Taxes, Interest, Maintenance, and Insurance

Added to income taxes are personal property taxes on machinery and real property taxes on buildings and land. Prudent tax management can take many forms. Businesses use accrual accounting methods, but farmers may use the simpler cash method. Informal income averaging can be accomplished by selling more in low income periods and postponing sales, making repairs, and handling more expense items in high income periods. Investment tax credits can be better than depreciation, as money otherwise paid in taxes can be used to expand or improve through capital investment. Tax calculations should be prepared early to allow for credit and expenditure decisions. It is prudent to consider using a professional tax adviser, whose services often save tax dollars. Interest needed to manage opportunity costs may be computed on equity capital and on borrowed money by determining the average investment and multiplying by the current interest rate. The annual maintenance and repair budget should be about 1.5% for new and up to 5% for old equipment. Insurance costs for property, fire, weather, machinery, and cultured stock are fixed costs of up to 1% of the purchase price.

Depreciation and Total Costs of Ownership

Depreciation is an accounting procedure to allocate the used-up value of durable assets over the time they are owned or until salvaged. It represents a loss in value that has been used by the producer to generate income. To be depreciable, an item must be personal tangible property or buildings used in business; have a useful life expectation over one year; and then become useless, worn out, or obsolete. This includes pond banks, pumps, and wells, whether productive or not. The accelerated cost recovery system (ACRS) and the modified ACRS are usually used for immediate tax savings. The straight line and declining or double declining balance methods preserve more depreciation for later and may be useful for a new business with the expectation that incomes will increase. To calculate annual straight-line depreciation, subtract salvage value from the original cost (or basis) and divide by the expected useful life. Sum the costs to estimate total annual ownership cost. Costs for a new $20,000 truck might be: taxes $60, interest $1,100, maintenance $300, insurance $100, and depreciation $1,800 for a total annual ownership cost of $3,360. The manager then decides, perhaps with the aid of partial budgeting or other techniques used to compare the cost of periodically renting or leasing with those of ownership, if the costs of owning are justified.

The Production Function

The manager generally attempts to maximize returns to each enterprise, and the production function can be a useful model. We use resources, called inputs, to achieve a goal or to produce a product, the output. Inputs can be fixed or variable. A fixed input does not change, and, likewise, its amortized cost does not vary over the amortization period. Generally, fixed costs consist of and constitute the overhead and are often thought of as costs of ownership. Fixed costs are important, but they are often unrelated to output decisions, whereas variable costs change directly with output. Food costs are variable and change yearly, seasonally, and even daily. All culture systems have both fixed and variable inputs. A fixed input might be the production facility and its major components, such as the ponds, tanks, or raceways. Another fixed input, if limited, is water, although water could be a variable input if the available amount is large relative to the facility's capacity and assuming there is cost associated with pumping or increases in the use of water. But the two common and primary variable inputs in fish hatcheries are food and labor.

One important principle is the law of diminishing returns. By applying the law in a financial framework, we can determine the most effective (public) or profitable (commercial) level of production. As units of a variable input are added to one or more fixed inputs in a system, output first increases exponentially, then increases at a decreasing rate, and finally decreases absolutely. The behavior of this relationship is referred to as the production function (Figure 1).

The production function, or total product curve, is used to derive the average product (AP), as output divided by input, and the marginal

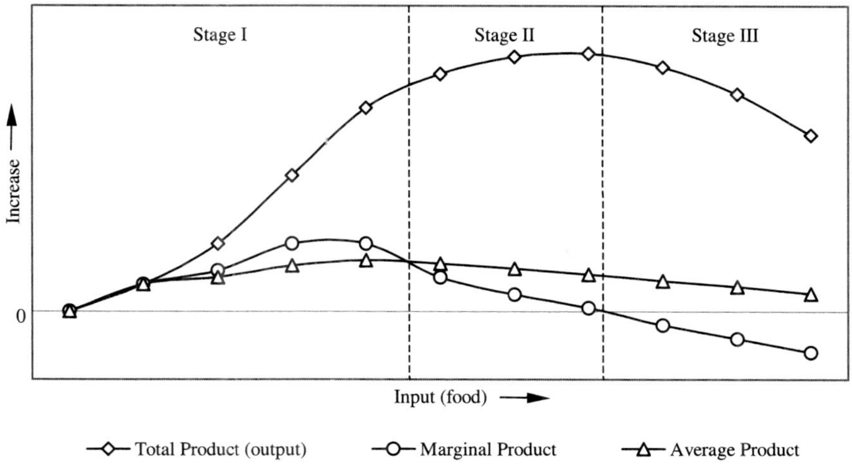

FIGURE 1. The production function.

product (MP), as the change in output divided by the change in input (Figure 1). The function can be divided into three stages. In stage one, efficiency of the variable input is increasing and MP is greater than AP. In stage two, efficiency of the variable input is declining, but marginal productivity is still positive. The most effective or profitable use of the variable input is in stage two. In stage three, total production is decreasing. The production function concept can be used to determine the most effective use of resources in an operating hatchery through marginal analysis.

Marginal Analysis and Maximizing Profit

An example will illustrate the law of diminishing returns and the use of marginal analysis. Let's assume a fish hatchery production (grow out) unit of fixed size and water flow, and we want to maximize returns to that unit using a variable food input of up to 1 ton. Hatchery managers often gear production to food conversion rates. Table 1, third column, shows food (to flesh) conversion rates the manager recorded in the past, when feeding similar production lots (in practice the manager may never feed at, or have records that reflect, levels analogous to the extreme upper and lower values in this table). Food conversion efficiency is the inverse of the food conversion rate and is also equal to AP. To achieve the greatest food conversion efficiency, the manager would use 400 lb of food. By feeding 500–900 lb, the manager could produce more fish, as shown in column 2, but to maximize output, the manager should use

TABLE 1. Marginal analysis for total production (TP) output from an aquaculture unit receiving variable input of food (100–1,000 lb).

Input (food)	Output (fish)	Food conversion rate	AP	MP	Price or MR	MC	AC	VC	Profit
100	40	2.50	0.40	0.40	0.89	1.10	0.95	38	−2.40
200	100	2.00	0.50	0.60	0.89	0.73	0.76	76	13.00
300	200	1.50	0.67	1.00	0.89	0.44	0.57	114	64.00
400	300	1.33	0.75	1.00	0.89	0.44	0.51	152	115.00
500	350	1.43	0.70	0.50	0.89	0.88	0.54	190	121.50
600	375	1.60	0.63	0.25	0.89	1.76	0.61	228	105.75
700	380	1.84	0.54	0.05	0.89	8.80	0.70	266	72.20
800	360	2.22	0.45	−0.20	0.89	−2.20	0.84	304	16.40
900	320	2.81	0.36	−0.40	0.89	−1.10	1.07	342	−57.20
1000	260	3.85	0.26	−0.60	0.89	−0.73	1.46	380	−148.60

Notes: AP (average product) = output/input; MP (marginal product) = Δoutput/Δinput; MR (marginal revenue) = \$0.89/lb; MC (marginal cost) = \$44/100 lb of food/MP; AC (average cost) = VC/output; VC (variable cost) = \$0.44 x input; profit = (price x output) – VC; fixed costs are not shown and should not affect results.

700 lb of food. Looking at output, one can see the law of diminishing returns at work and ask, "How much food should be used?"

Marginal analysis can be used to determine the level of production or amount of the variable input, food, that will maximize profits. Using the AP and MP columns in Table 1, we see that AP surpasses MP, and, thus, production function stage two begins between 400 and 500 lb of food. And MP goes negative, beginning stage three, at 800 lb of food. The point of greatest returns, then, exceeds 400 lb, but is less than 800 lb. Marginal cost (MC) is the price of food (\$0.44/lb) divided by the marginal product (MP), where MP is the change in output divided by the change in input. To maximize profit, we operate where MC = MR. In this example, MC = MR occurs between 500 and 600 lb. By interpolation, we can see it is very close to 500 lb. This can be confirmed by calculating actual profit as shown in the last column of Table 1.

Partial Budgeting

A partial budget is a modified subset of a larger budget in which expected changes in costs and revenues associated with introducing a change in the production operations are shown for the purpose of assessing profitability. Partial budgeting is used with an enterprise budget or entire hatchery budget to evaluate a contemplated change, and it is especially useful for relatively small operations. With complete enterprise budgets at hand, a partial budget is a quick and useful tool to help the manager decide whether to change an enterprise, purchase or rent machinery, grow yearlings, or sell fingerlings. For instance, the high cost of owning equipment, especially equipment that is used infrequently, must be recognized. When considering the annual ownership costs for a tractor, does usage justify the cost? Would it be cheaper to rent equipment when needed? If the money spent for the annual cost were placed in an interest bearing account, the annual amount of interest would represent the opportunity cost.

Partial budgeting indicates whether the change a manager is considering is more cost effective or less than the status quo. The analysis addresses four items for the alternative (contemplated change): cost reductions, cost additions, return reductions, and return additions. To prepare a partial budget, one inspects the enterprise (or culture system) budget and lists the amount of the expected change for each of the four categories. Then all additional returns, plus the amounts of all reductions in costs are summed. From that sum, all additional costs and the amounts of all reduced returns are subtracted. The partial budget is then complete and reflects the net increase or loss expected. See Jolly and Clonts (1993) for more explanation and examples.

A related approach, the least cost combination or substitution principle, is used for decision making when there is some fixed or required level of production and some degree of substitution among the inputs. Those interested should refer to an economics text, such as Shang (1990) for examples. Feed manufacturers use the principle to provide the nutrient mix in a diet with proximate analysis and minimum content specifications. However, the "interchangeable" aspects of certain nutrients or any set of inputs may involve incorrect assumptions and should be closely scrutinized.

Benefit–Cost Analyses

Benefit–cost analyses can be useful if well understood. There are six traditional techniques, though the present value analysis is basic and should be considered first. All analyses must include at least three steps: identify alternatives, determine cost and benefit for each alternative, and compare the costs and benefits among the alternatives. A more complete set of steps, though all are not always required, is: define a detailed, attainable objective; formulate assumptions; identify alternatives, and determine costs and benefits of each; compare costs and benefits of each; test outcome sensitivity to major uncertainties; synthesize results; and recommend an alternative. The analyses are infused with assumptions, which should be formulated only to bridge gaps due to identified uncertainties. Assumptions should be realistic, essential, and distinguishable from facts. Typical examples of assumptions used in nearly all analyses include the estimate of future resources use, economic life of the project, and the period of comparison.

Costs generally fall into three categories. Sunk costs are those already expended and are not included in an economic analysis. Nonrecurring costs include such things as construction and purchases of long-life items. Recurring costs are exemplified by the operating costs associated with labor and materials.

Once the manager has considered and addressed these preliminaries, established the objective, identified alternatives, recognized assumptions, and estimated costs and benefits of each alternative, the alternatives can be compared. The present value analysis must be used in every benefit-cost analysis when the economic life exceeds 3 years, but it should be used for all projects regardless of economic life. The analysis values all life cycle costs and benefits at present worth.

Present Value Analysis

Usually all factors to be considered in an analysis do not favor the same alternative. It is then necessary to put all alternatives on a common basis of time and cost to make a valid comparison. Present value (PV) analysis is the simplest way to compare alternatives. The PV is the estimated current worth of future benefits or costs derived by discounting the future values (Jolly and Clonts 1993). The alternative with the lowest PV cost is considered the least costly over the project life. Zucker and Anderson (1999) used PV in a model they developed to help plan production and marketing for aquaculture. The following assumptions apply in using PV analysis as the sole basis for decision making:

1. Benefits not put in monetary terms are equal for all alternatives. (When benefits are unequal, the least costly alternative may not be the best, which may be the one producing greater benefits.) For unequal benefits, the decision should not be based solely on PV.
2. The intended estimated life of the alternative is specified.
3. Service lives of alternatives are equal, or they must be placed on equal terms. (Use the common multiple approach or a comparison using the shorter life and imputing the residual value of the longer life asset. Derive monthly discount factors for costs and benefits dis-

tributed into the future, and multiply the factors by monthly cost or benefit to give PV.)

To calculate the PV for each alternative, the costs for each future month in which costs will be incurred are determined. Then a discount factor for each month, available in texts (Jolly and Clonts 1993), is multiplied by the monthly cost to determine PV costs for each month. All PV costs are summed. From the total costs, one can see in today's money which alternative would be the least expensive. For instance, if we had to pay (by establishing a trust fund) for all future costs today, we would need $X to buy, $Y to rent, or $Z to continue as is for the next 6 years.

Additional Analyses

If further evaluation is needed, five additional techniques may be used: benefit–cost ratio, uniform annual cost, savings–investment ratio, discounted payback analysis, and break-even analysis. The benefit–cost ratio is the only analysis that is useful when the benefits from alternatives are unequal. The break-even analysis is popular because the associated graphs are straightforward, easily understood, and allow the manager to visually compare the alternatives at any time. Various texts, as well as Meade (1989), offer more discussion of these analyses. Because of uncertainties in the analysis, one will want to know whether a decision would change if one or more of the inputs varied. Sensitivity analysis provides this information. Loomis and Fix (1999) offer an example of a benefit–cost comparison for catchable trout.

Considerations for Aquaculture Yet to Come

Imagination is more important than knowledge. —Albert Einstein

The Market

Engle (1999) offers marketing and economic considerations in a concise, down-to-earth summary. Hatch and Hanson (2000) provide a thorough overview of market issues. So-called globalization of national economies brings with it inherent concerns for coordination, integration, and regulation, be it self, industry, government, or multinational. Ling et al. (1999) uses a resource cost approach to show social opportunity costs, exchange rate, and national geographic location effects on market share advantages for Asian shrimp farms. Jakobsen (1999) points out that despite increasing economic globalization, local conditions still affect how firms structure business (product and marketing) strategies. In the United States and elsewhere, healthful aspects of fish- and seafoods provide a market niche and, perhaps, an advertising advantage for aquatic products. The advertising aspect of marketing should consider satisfaction of personal needs beyond food; that is, fish aren't marketed, but rather health, longevity, fitness, attractiveness, recreation, and happiness. The message to the hatchery manager is to consider the specific situation and resources in regard to your market outlook.

Waste and Beyond

Aquaculture wastes and environmental degradation are increasingly important hatchery management issues. In the recent past, a business plan and a production process needed to be technically feasible and economically sound. It seems that, increasingly, the plan and the process will also need to be environmentally acceptable and socioeconomically sound. Generally, most waste problems have been linked, directly or indirectly, to the feeding operation, and it could be argued that one of the best ways to reduce waste nutrients and solids in a production process is to prevent their introduction through improved feed formulation and optimization of the feeding processes (Sugiura and Hardy 2000). Matching and balancing feed ingredients to the needs of the production organisms appears to be an area of overlooked opportunity. Waste not only represents a discharge problem, but it indicates the level of efficiency or inefficiency of the feeds and feeding processes. Managers may benefit their future production operations by addressing this issue.

Integrity

Personal integrity is invaluable in the currency of professional interactions. Important in all business is the ability to work with people, and a key to working with people is to respect the people with whom you work and to develop and engender credibility and trust. Partnering, merging, and developing business agreements and contracts are expressions of action upon opportunity that likely can be facilitated through demonstrated integrity. Organizational integrity can reflect the personal integrity of top management, and, again, integrity is invaluable in the currency of professional interactions.

In the public sector, multiple uses of resources is a concept basic to resource management, and, in both the public and private sectors, multiple and opposing views matter. A major need exists for developing literature on public hatchery assessment and accountability systems. How does management determine what needs to be measured and analyzed? How does management develop appropriate criteria to determine production efficiency and effectiveness of a public hatchery operation?

Managing Information

The shift from paper to electronic transmission and management of information allows for new and greater empowerment of individuals. It also represents an opportunity for managers to increase productivity and delegation of responsibility, as well as to more fully use available human resource capabilities. Reliable information catalyzes the thinking process and is essential to management and decision making. The manager's challenge is to harness the capabilities of staff and associates to increase utility of information for organizational goals, that is to support business processes and decision making.

Thinking and Learning

When faced with a problem or its mirror image, opportunity, we often find solace in knowing we have a large tool kit at our disposal. But, the ability to think through a challenge is more important than just possessing tools; in fact, thought is prerequisite to selecting a tool. The idea of having a tool kit, while perhaps shoring up confidence, can help limit us to the use of only familiar tools, many of which become obsolete with the advance of new technology. It is likely that information resources, systems, and expertise will become ever more important with time. And as ever, education, comprehension, and perspective are all critically important in the application of available information. Progressive hatchery management will continue to be an iterative process that depends on learning and using new information.

Global fisheries statistics indicate that stocks are declining, that a gradual shift to landings of lower trophic level stocks is occurring, and that even this undesirable situation is unsustainable (Pauly et al. 1998). It is likely that the continued inadequate supplies and possible collapse of major portions of the world fisheries will continue to drive the growth of aquaculture. Growth potential is one aspect of an array of related issues that will present themselves to hatchery managers, as well as to resource managers, aquaculture industry leaders, and decision makers in the industry and in government. Issues will deal with markets, but also with production processes and, importantly, with personal, organizational, and industry integrity. As aquaculture managers, leaders, and decision makers are faced with the imminent problems and opportunities of the early twenty-first century, the time is at hand to access and use the pertinent information with integrity for the long term good of the industry and the society upon which it is interdependent.

Acknowledgments

Carol Engle provided thorough and insightful review, and Kristine Playfoot provided the graphics and manuscript preparation support.

References

Bjorndal, T. 1990. The economics of salmon aquaculture. Blackwell Scientific Publications, Cambridge, Massachusetts.

Boyd, C. E., and H. R. Schmittou. 1999. Achievement of sustainable aquaculture through environmental management. Aquaculture Economics and Management 3(1):59–69.

Engle, C. R. 1999. Teaching aquaculture economics and marketing. World Aquaculture 30(2):17–21.

Engle, C. R., R. Balakrishnan, T. R. Hanson, and J. J. Molnar. 1997. Economic considerations. Pages 377–395 *in* H. S. Egna and C. E. Boyd, editors. Dynamics of pond aquaculture. CRC Press, New York.

Griffin, W. L. 2000. Economics, business plans. Pages 253–261 *in* R. R. Stickney, editor. Encyclopedia of aquaculture. John Wiley and Son, New York.

Hatch, U., and T. Hanson. 2000. Market issues in the United States aquaculture industry. Pages 507–519 *in* R. R. Stickney, editor. Encyclopedia of aquaculture. John Wiley and Son, New York.

Jakobsen, S. E. 1999. Development of local capitalism in Norwegian fish farming. Marine Policy 23(2):117–130.

Jolly, C. M., and H. A. Clonts. 1993. Economics of aquaculture. Haworth Press, New York.

Klinefelter, D. 2000. Financing. Pages 367–372 *in* R. R. Stickney, editor. Encyclopedia of aquaculture. John Wiley and Son, New York.

Kosinski, R. J. 1999. Computers in aquaculture education. World Aquaculture 30(2):19–21.

Kusumastanto, T., C. M. Jolly, and C. Bailey. 1998. A multiperiod programming evaluation of brackishwater shrimp aquaculture development in Indonesia. Aquaculture 159:317–331.

Lindbergh, J. M. 1999. Salmon farming in Chile: do the benefits exceed the costs? Aquaculture Magazine 25(2):33–45.

Ling, B. H., P. S. Leung, and Y. C. Shang. 1999. Comparing Asian shrimp farming: the domestic resource cost approach. Aquaculture 175:31–48.

Loomis, J., and P. Fix. 1999. A statistical approach to estimating costs of propagating hatchery rainbow trout. North American Journal of Fisheries Management 19:110–119.

Meade, J. W. 1989. Aquaculture management. Van Nostrand Reinhold, New York.

Nickerson, D. J. 1999. Trade-offs of mangrove area development in the Philippines. Ecological Economics 28:279–298.

Pauly, D., V. Christensen, J. Dalsgaard, R. Froese, and F. Torres. 1998. Fishing down marine food webs. Science 279:860–863.

Shang, Y. C. 1990. Aquaculture economic analysis: an introduction. World Aquaculture Society, Baton Rouge, Louisiana.

Shang, Y. C., and C. A. Tisdell. 1997. Economic decision making in sustainable aquacultural development. Pages 127–148 *in* J. E. Bardach, editor. Sustainable aquaculture. John Wiley and Son, New York.

Sugiura, S. H., and R. W. Hardy. 2000. Environmentally friendly feeds. Pages 299–310 *in* R. R. Stickney, editor. Encyclopedia of aquaculture. John Wiley and Son, New York.

Tisdell, C. 1999. Overview of environmental and sustainability issues in aquaculture. Aquaculture Economics and Management 3(1):1–5.

Zucker, D. A., and J. L. Anderson. 1999. A dynamic, stochastic model of a land-based summer flounder aquaculture firm. Journal of the World Aquaculture Society 30(2):219–235.

INDEX

A

B

H

J

K

L

M

Q

S

W